Springer Geology

The book series Springer Geology comprises a broad portfolio of scientific books, aiming at researchers, students, and everyone interested in geology. The series includes peer-reviewed monographs, edited volumes, textbooks, and conference proceedings. It covers the entire research area of geology including, but not limited to, economic geology, mineral resources, historical geology, quantitative geology, structural geology, geomorphology, paleontology, and sedimentology.

Tieguan Wang

Meso-Neoproterozoic Geology and Petroleum Resources in China

科学出版社
Science Press

Springer

Tieguan Wang
China University of Petroleum (Beijing)
Beijing, China

ISSN 2197-9545 ISSN 2197-9553 (electronic)
Springer Geology
ISBN 978-981-19-5665-2 ISBN 978-981-19-5666-9 (eBook)
https://doi.org/10.1007/978-981-19-5666-9

Jointly published with Science Press
The print edition is not for sale in China (Mainland). Customers from China (Mainland) please order the print book from: Science Press.

Translation from the Chinese language edition: "zhongguo dongbu zhong-xin yuangujie dizhixue yu youqi ziyuan" by Tieguan Wang, © Science Press 2016. Published by Science Press. All Rights Reserved.

This Springer imprint is published by the registered company Springer Nature Singapore Pte Ltd.
The registered company address is: 152 Beach Road, #21-01/04 Gateway East, Singapore 189721, Singapore

This monograph is one of a series of publications dedicated to the celebration of the one hundredth anniversary of the Geological Society of the founding of the Geological Society of China.

The Geological Society of China

The Geological Society of China (GSC) *was founded in Beijing on February 3, 1922. It is* *one of the four earliest national academic societies in China. As a 4A social organization in China and a member of the International Union of Geological Sciences (IUGS), GSC is incorporated under China Science and Technology Association (CSTA). The Society is the Chinese national learned and professional society for geosciences with a country-wide fellowship of more than 50,000 members, and constituted of 60 specialized commissions, research branches and working councils, and provides professional guidance to the 31 provincial societies across the country.*

Preface

Menchikoff (1949) and Pruvost (1951) initially proposed the stratigraphic term "Infracambrian" involving the strata that underlay the lower limit of known trilobites-bearing Cambrian formations, but unconformably overlay generally metamorphosed basement, which includes Meso-Neoproterozoic plus partial Lower Cambrian strata. Until 1950s, it was generally accepted that sedimentary rocks of Infracambrian or Precambrian age located within basins, geosynclines and platforms could not contain hydrocarbon deposits. The absence of pre-Palaeozoic Era life in these Infracambrian stratigraphic column was the most often presented reason why Precambrian terranes should be ignored by the petroleum geologist (Dickes, 1986).

For recent more than half a century, along with the research advances on the Early Earth life, the Meso-Neoproterozoic biodiversity has been verified and the scientific evidence of material basis provided for the Meso-Neoproterozoic petroleum prospectivity available. Meanwhile, the global Meso-Neoproterozoic geology has made a great step forward, and the significant breakthrough discoveries were achieved in petroleum exploration. As it turns out, the Infracambrian sedimentary strata are not only indeed rich in organic matter, but also contain numerous oil and/or gas seeps, bituminous sands/sandstone and asphalt veins. Up to present totally dozens of oil-and/or gas-fields are found in some areas of the world, such as the Siberia Craton in Russia, the Chuanzhong Uplift in China, Oman and Indian Basins, etc., where the geological reserves of oil, condensate and/or gas can be individually up to a few hundred million tons to thousand million tons of oil equivalent, and the daily output of individual gas well as high as more than one million cubic meters.

Currently, the Meso-Neoproterozoic oil and gas resources have caused more attention of international geological circle, so that a series of special international conferences and related workshops on the Meso-Neoproterozoic geology and petroleum system were held, and the conference proceedings published, which have aroused more concerns of international petroleum geologists and explorationists on the Infracambrian geology and petroleum resources.

The Meso-Neoproterozoic sequences are very well developed and more completely preserved in China, and the Meso-Neoproterozoic stratigraphy was earlier studied by Chinese geologists in the world; e.g., Lee and Chao (1924) and

Kao et al. (1934) originally investigated and divided the outcropped "Sinian" stratigraphic sections, respectively, at the Yangtze Gorges in Hubei, South China, and at the Jixian in Tianjin, North China. Since then, both sections became the important reference sections for the international Precambrian stratigraphic correlation over quite a long period of time. Particularly, since 1960s on, Chinese geologists have achieved significant progresses in various geoscientific disciplines and interdisciplines, such as Meso-Neoproterozoic biostratigraphy, sequence stratigraphy, chronostratigraphy, palaeooceanography, palaeoglaciology, Cambrian biological explosion event, activistic tectono-palaeogeography as well as petroleum geology and geochemistry. As far as sedimentary stratigraphic sequences are concerned, a uniform Chinese Meso-Neoproterozoic stratigraphic framework, including seven system-level stratigraphic units, i.e., Changchengian (Pt_2^1), Jixianian (Pt_2^2), Xiamaling Formation (Pt_2^3x) (in North China), Qingbaikouan (Pt_3^1), Nanhuan (Pt_3^2) and Sinian (Pt_3^3) (in South China) in ascending order, as well as the stratigraphic correlation of equivalent formations, are established, which would be favorable to the research and assessment of oil and gas resources within the new Meso-Neoproterozoic stratigraphic framework. Moreover, the palaeooceanographic, sedimentopalaeogeographic and regional tectonic studies would also be conducive to the approach and understanding of the generation, evolution and preservation mechanisms of Meso-Neoproterozoic petroleum resources and of the development setting of source-reservoir-seal bed assemblages.

Especially, a series of Meso-Neoproterozoic oil and gas have been discovered one after another, i.e., the Weiyuan Gasfield with gas reserve of 408.61×10^8 m^3 (P1) from 1942 to 1965, the large-scale Longmenshan asphaltic veins with prognostic asphalt reserve ca. 8×10^6 m^3 during 1966–2015, the numerous Yanliao "live" oil-seeps, asphalt, bituminous sands/sandstone and associated fossil-oil-reservoirs in 1978–1979 and 2007–2010 and the Anyue Gasfield plus associated fossil-oil-reservoirs containing gas geological reserves of 8488×10^8 m^3 (P1) and *ca.* 1.2×10^{12} m^3 (3P) during 2011–2016, which have constantly encouraged Chinese petroleum geologists and explorationists to keep on the exploring of Infracambrian petroleum resources. Compared with international similar studies, however, the Chinese Meso-Neoproterozoic research would face not only the older strata and a great geological age span over 1.3 Ga, but also quite a large area of stratigraphic distribution under the complicated tectonic and/or highly thermo-evolutional settings in China, resulting in more difficulties for indigenous oil and gas preservation and exploration. Consequently, how to evaluate the spatial and temporal distribution of Chinese Meso-Neoproterozoic petroleum resources would be an immediate problem.

Therefore, the research of Chinese Meso-Neoproterozoic oil and gas resources would concurrently contain the advantages of more completely developed strata and previous profound scientific accumulation, the serious challenge of more complicated geological conditions and thus the wider opportunities of scientific innovation. In such a context, the 444th Xiangshan Science Conference was held in Beijing on November 13–15, 2012, and its proceeding, entitled *Meso-Neoproterozoic Geology and Petroleum Resources in East China* (Chinese version, Sun Shu and Wang Tieguan eds.), was published by Science Press, Beijing, in 2016. Since then, on the advises of Chinese colleagues and readers, and in order to facilitate international academic exchanges, the present editor invited some authors to update, recompile and rewrite each related chapters of a new monograph entitled *Meso-Neoproterozoic Geology and Petroleum Resources in China* (English version), and by comparison with the Chinese version (2016), the English monograph only involves three Meso- and/or Neoproterozoic sedimentary basins, i.e., the North China and the Yangtze Cratons as well as the Tarim Block in China, while the Meso-Neoproterozoic geology includes the tectonic division, chronostratigraphic framework, stratigraphic sequences and dividing history, palaeoglaciology, special Neoproterozoic palaeooceanographic environments and magmatism. In addition, the petroleum geology-geochemistry of Anyue-Weiyuan Gasfields, the Longmenshan asphaltic veins and the Yanliao oil-seeps, bituminous sands/sandstone and associated fossil-oil-reservoirs are discussed in detail.

The publication of this book coincides with the 100th anniversary of the birth of the Geological Society of China. As one of the four earliest academic organizations in China, the Geological Society was founded on the February 3, 1922. Therefore, the present authors dedicated this monograph to the 100th anniversary of the Geological Society of China. Meanwhile, the monograph is also in memory of the late academician Sun Shu (1933–2018) in Chinese Academy of Sciences, who was a celebrated geologist and the one of two co-editors for the Chinese version.

This monograph would have been very difficult to achieve without the full cooperation of numerous co-authors, and the book is far better for their contributions, thanks, in no particular order, to Mingguo Zhai, Pingan Peng, Xianhua Li, Maoyan Zhu (Chinese Academy of Sciences), Linzhi Gao, Pengjiu Liu (Chinese Academy of Geological Sciences), Shixing Zhu (Tianjin Institute of Geology and Mineral Resources), Jinhu Du, Keyou Han, Shuichang Zhang [China National Petroleum Corporation (CNPC)], Li Su (China University of Geosciences), Shunshe Luo (Yangtze University), Ningning Zhong, Daofu Son (China University of Petroleum-Beijing) and their research groups.

Beijing, China
January 2022

Tieguan Wang
Academician in Chinese Academy of Sciences; Professor in China University of Petroleum, Beijing

References

Dickes AB (1986) Precambrian as a hydrocarbon exploration target. Geosci Wisconsin 11:5–7

Kao CS, Hsiug YH, Kao P (1934) Preliminary notes on Sinian stratigraphy of North China. Bull Geolog Soc China 13:243–288.

Lee LS, Chao YT (1924) Geology of the Gorge district of the Yangtze (from Ichang to Tzeckuei) with special reference to the development of the Gorges. Bull Geolog Soc China 3(3–4):351–391

Menchikoff N (1949) Quelques traits de I'hisoire geologique du Sahara occidentail. Ann Hebert et Haug 7:303–325

Pruvost P (1951) L'Infracambrien. Bulletin de la Société Belge Geologie Palaeontologie et Hydrologie. 60:43–65

Contents

Chapter 1
Advances in Meso-Neoproterozoic Isotopic Chronostratigraphy in China

Linzhi Gao, Xiaozhong Ding, Heng Zhang, Xiufu Qiao, Chongyu Yin, Xiaoying Shi, and Chuanheng Zhang

Abstract It is a long-term objective for Chinese geologists to make the Meso-Neoproterozoic strata in North China Craton (NCC) the international reference section of the Late Precambrian chronostratigraphic stratigraphy. However, the establishment of highly precise and reliable Precambrian chronostratigraphic frame would be a top priority in current chronostratigraphic study and also a basis for Precambrian stratigraphic correlation and tectonic interpretation in each continent of the globe. This chapter will be focused on the achievements of the Chinese Meso-Neoproterozoic chronostratigraphy, especially for the new calibration of Meso-Neoproterozoic stratigraphic division in North China Craton (NCC) as well as for the new isotopic dating of Neoproterozoic chronostratigraphy in Yangtze Craton (YC in southern China) and Tarim Block (northwest China). Those new isotopic dating data would result in a complete change in the Meso-Neoproterozoic chronostratigraphic calibration, division and correlation among above three regions, and update the regional geological cognition and the metallogenic mechanism interpretation. In recent years, the Meso-Neoproterozoic petroleum geological study and exploration put forward the precise chronostratigraphic division and correlation of sedimentary cover and crystalline basement of the petroliferous basins. In view of the achievements in the chronostratigraphic isotopic technique, it is possible to obtain the complete data of precise and reliable isotopic dating for Precambrian cover and basement in NCC, YC and Tarim Block (TB), so that more complete Meso-Neoproterozoic isotopic chronostratigraphic data would be available for the vest number of geologists in this chapter.

Keywords Meso-Neoproterozoic · chronostratigraphic frame · North China craton (NCC) · Yangtze craton (YC) · Tarim block (TB)

L. Gao (✉) · X. Ding · H. Zhang · X. Qiao · C. Yin
Institute of Geology, Chinese Academy of Geological Sciences, Beijing 100037, China

X. Shi · C. Zhang
China University of Geosciences, Beijing 100083, China

T. Wang, *Meso-Neoproterozoic Geology and Petroleum Resources in China*, Springer Geology, https://doi.org/10.1007/978-981-19-5666-9_1

1.1 Introduction

The current new thought on the Precambrian chronostratigraphy draws lessons from the methodology of the geological events in supper-continental researches, which has made the Precambrian stratigraphy an integral part of geodynamic system (Lu 2002). The studies of geological events must be supported by the highly precise chronostratigraphic dating, particularly for the zircon sensitive high-mass resolution ion microprobe (SHRIMP) dating of volcanic ash layer and diabase in the Precambrian key strata. New Precambrian chronostratigraphic time scale was published by Plumb (1991) in the journal *Episodes*, which has set up some new rules for the Precambrian time scale, with a of 0.3 or 0.2 Ga for each stratigraphic unit "system", as follows in ascending order: Palaeoproterozoic sequences consist of Siderian (2.5–2.3 Ga), Rhyacian (2.3–2.0 Ga), Orosirian (2.0–1.8 Ga) and Statherian (1.8–1.6 Ga); Mesoproterozoic have three systems, i.e., Calymmian (1.6–1.4 Ga), Ectasian (1.4–1.2 Ga) and Stenian (1.2–1.0 Ga); Neoproterozoic is composed of Tonian (1,000–850 Ma), Cryogenian (850–635 Ma) and Neoproterozoic III (or Ediacaran, 635–542 Ma).

It's worth noting that the international Cryogenian working group had assigned the bottom global standard stratotype section and point (GSSP) to the diamictite at 720 Ma (Graham et al. 2016), and attributed the period of 1.8–1.6 Ga to Statherian as the top of Palaeoproterozoic in the International Chronostratigraphic Chart (International Commission on Stratigraphy 2013; Ogg et al. 2016). Whereas, based on the essential characteristics of Precambrian sedimentary cover in North China Craton (NCC), Chinese geologists referred the 1.8 Ga to the initial point of Mesoproterozoic sedimentation (National Commission for Stratigraphy of China 2001) instead of the 1.6 Ga specified in the International Chronostratigraphic Chart (International Commission on Stratigraphy 2013; Ogg et al. 2016).

On the global scale, so far the most systematic and complete Meso-Neoproterozoic stratigraphic sections are known only in three regions, i.e., the Jixian stratotype section in NCC (Kao et al. 1934), the Ural section in Russian Siberia Craton (Shatsky 1945) and the Yukon section in Canada Shield (Thorkelson et al. 2005).

The Meso-Neoproterozoic in Jixian stratotype section (1.8–0.8 Ga) is a continuous sequence with near 10 km in thickness (cf. Chap. 2). It is widely exposed without the effect of metamorphism but with simple geological structure and clear top and bottom boundaries in the NCC (Chen et al. 1980; Fig. 1.1).

This outcrop section is not only the Meso-Neoproterozoic stratotype section in northern China, but also one of rare and complete Late Precambrian sections in the world. Therefore, it would prove to be a unified reference section with systemic chronostratigraphy and completely isotopic datings for the Chinese and international Precambrian stratigraphic correlation.

Both the stratigraphic sections, i.e., the Jixian stratotype section in NCC and the Three Gorges section in Yangtze Craton (YC; Fig. 1.1), have always been regarded as the Meso-Neoproterozoic reference sections for the studies of Meso-Neoproterozoic stratigraphy and isotope chronology in China, based on which, the breakthrough points of chronostratigraphic frame optimized. With respect to the cognition on

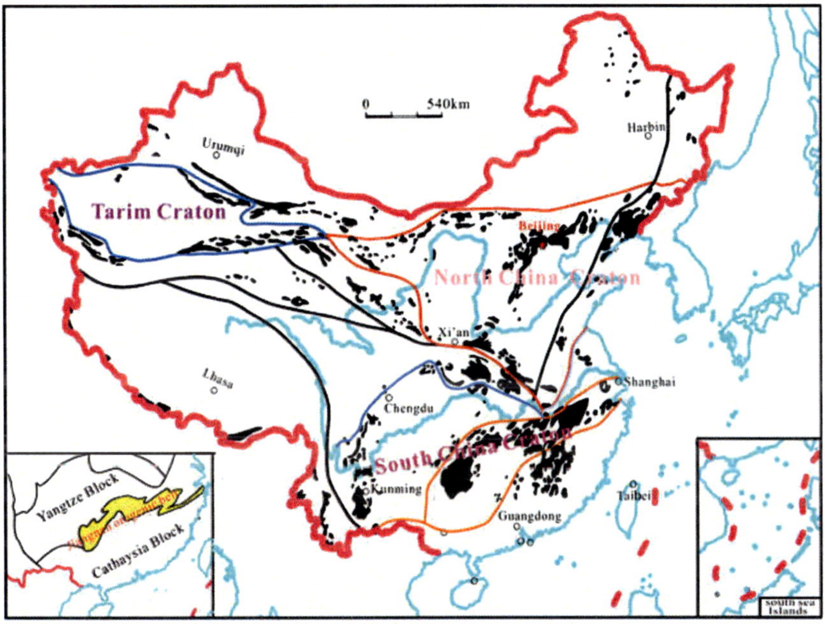

Fig. 1.1 Distribution of Meso-neoproterozoic tectonic units and stratigraphic outcrop area in China

Mesoproterozoic bottom boundary, different isotopic methods have been applied for years, and numerous isotopic data constantly added to the Meso-Neoproterozoic stratigraphic frame in NCC. And thus, there are two evident constraint factors for the Mesoproterozoic bottom age in NCC as follows. ① With isotopic age of 1.9–1.85 Ga, the alkaline basic dikes, intruding into Palaeoproterozoic metamorphic rocks, are attributed to the behavior of magmatic activity during the startup time of Yanliao Faulted-Depression Zone (YFDZ; Lu et al. 2008); ② The minimum peak age of detrital zircons for the terrigenous basal sandstone of Changchengian is 1.85 Ga (Wan et al. 2003). Consequently, the Changchengian bottom age should be latter than 1.85 Ga in NCC, by means of numerous zircon SHRIMP U–Pb dating, recently the ages of Meso-Neoproterozoic stratigraphic frame are newly corrected and the dating of Mesoproterozoic rapakivi granite pluton developed in NCC has been recalibrated. Internationally, it used to be widely accepted that the rapakivi granite would be uncorrelated with orogenies, but closely associated with assembly of supercontinent Columbia (Vigneresse 2005). While the zircon Hf isotopic data of Miyun rapakivi granite indicate a new origin of partial magmatic melting from the Neoarchean juvenile crust (Yang et al. 2005). For this reason, the granite should be caused by the magmatic emplacement under the dynamic conditions of extensional environment. Therefore, the Miyun rapakivi granite would have a startup time consistent with YFDZ, which also represent the initial timing node of a new tectonic circle, and is equivalent to the initial developmental age of the Mesoproterozoic in NCC, so that this granite can also be a startup indicator of supercontinental breakup or rifting. The zircon U–Pb age 1,685 ± 15 Ma of Miyun rapakivi granite is a significant SHRIMP dating, which could effectively limit the bottom age of Mesoproterozoic in NCC (Gao et al. 2008a, b, c, d).

As to the Qianxi Complex within the metamorphic basement of NCC, its zircon $^{207}Pb/^{206}Pb$ weighted mean age 2534 ± 9 Ma is consistent with the SHRIMP U–Pb age 2533 ± 14 Ma in the inheritable zircon of diabase intruding into the Chuanlinggou Formation (Geo et al. 2009c). While two zircon SHRIMP U–Pb ages 1626 ± 9 Ma and 1638 ± 14 Ma were repeatedly obtained from the volcanic rock in the Dahongyu Formation and from the diabase in the Chuanlinggou Formation, respectively (Gao et al. 2008c), both of which imply following two issues: ① The Mesoproterozoic bottom age in China Geological Time Scale should be older than the Mesoproterozoic age (1.6 Ga) specified in the International Geological Time Scale. ② The Mesoproterozoic bottom age in NCC is perhaps not older than 1.7 Ga. According to the above-mentioned ages and in combination with the Miyun rapakivi granite age which could be used to calibrate the Mesoproterozoic breakup age of NCC, the present authors try to approach the calibration of a series of strata in the Chinese Meso-Neoproterozoic stratigraphic frame. Currently, systemic and highly precise zircon U–Pb dating would make the evolution of tectonic events more clear and more definite in NCC, among which, the first precisely measured tuffaceous zircon SHRIMP U–Pb age 1368 ± 12 Ma within the Xiamaling Formation (Gao et al. 2007a, b), which was initially established by Yin (1920) at the Xishan section in Jingxi Depression, southwest YFDZ, has made a significant breakthrough on the Meso-Neoproterozoic chronostratigraphic frame in NCC (Gao et al. 2009c), and resulting in a chronostratigraphic transformation of the Xiamaling Formation (1.4–1.32 Ma) from the original Neoproterozoic Qingbaikouan (10–7.8 Ga) to Mesoproterozoic Era (17–10 Ga). Moreover, the second tuffaceous zircon age 1366 ± 9 Ma of the Xiamaling Formation is again acquired at the Zhaojiashan section in Xuanlong Depression, northwest YFDZ (Gao et al. 2008c). Both the ages show the existence of wide-spread and synchronal volcanic ash layers within the Xiamaling Formation and a speculation of new tectonic framework in YFDZ (Qiao et al. 2007). In November 2009, National Commission on Stratigraphy of China (NCSC) has subdivided the Meso-Neoproterozoic column of China into 6 stratigraphic units. i.e. Changchengian (Ch or $Pt_2{}^1$) is limited in 1.8–1.6 Ga; Jixianian (Jx or $Pt_2{}^2$) 1.6–1.4 Ga; Xiamaling Formation ($Pt_2{}^3x$) 1.4–1.32 Ga; Qingbaikouan (Qb as $Pt_3{}^1$) 1.0–0.78 Ga; Nanhuan (Nh or $Pt_3{}^2$) 780–635 Ma and Sinian (Z or $Pt_3{}^3$) 635–542 Ma (Table 2.1). In addition, Qiao et al., (2007) proposed to attribute the Xiamaling Formation to "Xishan System" (or "Xishanian", $Pt_2{}^3$) with the age 1.4–1.2 Ga, and Gao et al. (2015a, 2018; b) suggested 'the "Yuxi System" (or "Yuxian", $Pt_2{}^4$; "Yuxi" in western Yunnan Province) with the age 1.2–1.0 Ga.

The new Proterozoic time scale of China (Table 1.1) highlights following stratigraphic sequences: Changchengian ($Pt_2{}^1$), Jianian ($Pt_2{}^2$), Xiamaling Formation ($Pt_2{}^3x$), "Yuxian" ($Pt_2{}^4$), Qingbaikouan ($Pt_3{}^1$), Nanhuan ($Pt_3{}^2$) and Sinian ($Pt_3{}^3$) in ascending order. The new time scale would act on significant calibration for the Precambrian stratigraphic division and correlation, and causing significant changes in the interpretation on regional stratigraphical correlation, tectonic setting and strata-bound ore deposits.

Concerning tectonic interpretation, the precise dating and calibration for a series of magmatic intrusive bodies would be particularly emphasized. As above mentioned, both ages of the Miyun rapakivi granite 1685 ± 15 Ma (Gao et al. 2008a, b, c, d) and the diabase sill 1638 ± 14 Ma in Chuanlinggou Formation (Gao et al. 2008c) are obtained, which would limit the breakup relationship between the supercontinent Columbia and the YFDZ to the maximum extent. Meanwhile, as for a series of measured diabase sill dating in NCC, e.g., two ages 1345 ± 12 Ma and 1353 ± 14 Ma found in the Wumishan Formation (Zhang et al. 2009), one age 1320 ± 6 Ma obtained the limit of the Xiamaling Formation (Li et al. 2009a, b) and other SHRIMP U–Pb ages 928 ± 8 Ma (Liu et al. 2005) and 930 ± 10 Ma (Gao et al. 2009b) for the Neoproterozoic diabase sills in the Xuzhou (City)-Huai River Valley on eastern margin of NCC, could basically reveal the genetic relationship between

Table 1.1 Meso-neoproterozoic time scale in China (Gao et al. 2009c, modified)

								Ma
		Cambrian		Cambrian		Є		542
Proterozoic	Neo-proterozoic Pt_3	Ediacaran	Neo-proterozoic Pt_3	Sinain	Pt_3^3	Z_2		550, 570
						Z_1		615
		Cryogenian (720 Ma)		Nanhuan	Pt_3^2	Nh_3	Pt_3^{2c}	635, 660
						Nh_2	Pt_3^{2b}	725
						Nh_1	Pt_3^{2a}	780
		Tonian		Qingbaikouan	Pt_3^1	Qb_4	Pt_3^{1d}	820
						Qb_3	Pt_3^{1c}	870
						Qb_2	Pt_3^{1b}	930
						Qb_1	Pt_3^{1a}	1000
	Meso proterozoic Pt_2	Stenian	Meso proterozoic Pt_2	Yuxi	Pt_2^4	Yx		1200
		Ectasian		Xishan*	Pt_2^3	Xs		1400
		Calymmian		Jixian	Pt_2^2	Jx		1600
	Paleo-proterozoic Pt_1	Statherian		Changcheng	Pt_2^1	Ch		1700, 1800
		Orosirian	Paleo-proterozoic Pt_1	Hutuo	Pt_1^3	Ht		2000
		Rhyacian		Gaofan	Pt_1^2	Gf		2300
		Siderian		Unnamed	Pt_1^1			2500

the breakup of supercontinent Columbia and the assembly of supercontinent Rodinia, and thus, the basic fracture of tectonic evolution and continental dynamics could be reconfirmed in the YFDZ, which would be a big step forward the establishment of an unified and high precise chronostratigraphic frame. In addition, a series of zircon SHRIMP U–Pb dating achievements of volcanic rocks, including the potassium-rich trachyte age 1626 ± 9 Ma in the volcanic rock interval of the Dahongyu Formation (Gao et al. 2009b), the tuff age 1560 ± 6 Ma in the Gaoyuzhuang Formation (Li et al. 2010); Two zircon SHRIMP U–Pb data 1483 ± 13 Ma and 1487 ± 17 Ma had been gain from the bentonite in the Wumishan Formation (Li et al. 2014), the bentonite ages 1437 ± 61 Ma in the Tieling Formation (Su et al. 2010) and 1368 ± 12 Ma in the Xiamaling Formation (Gao et al. 2007a, b), have played a particularly important role respectively in stratigraphic calibration along whole the stratigraphic column.

In the Mesoproterozoic chronostratigraphic research of NCC, therefore, there are three significant advances on zircon chronostratigraphy of diabase veins/sills in ascending order as follows: ① the zircon U–Pb age of diabase sill is limited to 1638 ± 14 Ma in the Changchengian Chuanlinggou Formation; ② the zircon U–Pb age 1353 ± 14 Ma and baddeleyite age 1345 ± 12 Ma of the diabase sill/vein in the Jixianian Wumishan Formation; ③ the baddeleyite U–Pb age of diabase sill 1320 ± 6 Ma in the Xiamaling Formation. Those dating achievements provide a series of systemic reliable "anchor points" in chronostratigraphy, which are favorable for precise determination of the complete Meso-Neoproterozoic chronostratigraphic frame. However, there still are three problems in the new Proterozoic time scale: ① How can the stratigraphic hiatus of 1.32–1.0 Ga be filled up in the YFDZ? According to the palaeontology information, the most probably developed region(s) of corresponding strata are at western Henan Province (i.e., southern NCC; Gao et al. 2011a, b, c, d, e, f) and/or at Jiaodong Peninsula-Liaodong Peninsula-Xuzhou City-Huaihe River Valley (JLXH) region in eastern NCC (Xing 1979; Xing and Liu 1982; Xing et al. 1985, 1989). ② At present, both the stratigraphic positions of Qingbaikouan ($Pt_3{}^1$) Luotuoling and Jing'eryu Formations in the stratigraphic column are in need of accurate dating (Gao et al. 2011e, 2013b). ③ Within the tectono-stratigraphic framework, there is a suit of obviously banded- distributed epimetamorphic sedimentary strata and a series of magmatic rock belts in between the Yangtze and Cathaysia Blocks, which is geologically called "Jiangnan Orogenic Belt". Its stratigraphic position would be very important for the correlation of Precambrian tectono-stratigraphic frame in South China, and a series of tuffaceous zircon SHRIMP U–Pb ages in the epimetamorphic clastic rocks is also significant for Neoproterozoic stratigraphic correlation in YC (Gao et al. 2008a, 2009a, 2010a, b, c, d, 2011a, b c, d, 2012a, b, c, d, e, 2014a, b, c).

Both stratotype sections in Jixian and in Yangtze Gorges have still been as the Precambrian reference sections and the mapping standards for geological survey in China, particularly for the precise calibration of Meso-Neoproterozoic stratigraphic sequence during the establishment of Chinese stratigraphic chronology.

1.2 New Progress on the Meso-Neoproterozoic Chronostratigraphiy in NCC

The bottom boundary of Mesoproterozoic strata in Jixian stratotype section is a main theme for the approaches of numerous Chinese geologists. In recent year, some geologists pay more concern at the rapakivi granitic pluton or granitic dike, intruding into the Archean Miyun Complex due to their unconformable overlap by the overlying Changchengian Changzhougou Formation (Fig. 1.2), a zircon SHRIMP U–Pb age 1685 ± 15 Ma of the granitic pluton (Gao et al. 2008a, b, c, d) or laser ablation-multiple collector-inductively-coupled plasma mass spectrometer (LA-MC-ICPMS) U–Pb 1673 ± 10 Ma and 1669 Ma of granite dike (Li et al. 2011) are obtained. However, the rapakivi granitic pluton in Miyun Complex could not only reflect the mechanism of geodynamics, but also contain the distinguishable indicators of the Changchengian basal conglomerate and its underlying old weathering crust for the approach to recalibration of the Mesoproterozoic bottom age in NCC (Fig. 1.2). Recently, the discovery of the Changchengian basal conglomerate and the determination of zircon U–Pb age of 1682 Ma (He et al. 2011a, b) and 1673 Ma (Li et al. 2011) in the old weathering crust would provide a new and effective chronostratigraphic dating in YFDZ, NCC, which would give a challenge to the original age 1.6 Ga of Mesoproterozoic bottom boundary assigned in the International Chronostratigraphic Chart (International Commission on Stratigraphy 2013; Ogg et al. 2016). However, the stratigraphic relationship between the Changchengian and the epimetamorphic Palaeoproterozoic Hutuo Group (Table 1.1) remains to be further studied (Gao et al. 1996).

Fig. 1.2 The contact relationship between the old weathering crust of rapakivi granitic pluton and its overlying Changchengian strata

1.3 Stratigraphic Hiatus of 1.2–1.0 Ga (Pt_2^4) in Jixian Stratotype Section, NCC

1.3.1 Stratigraphic Position of the Precambrian Strata in Stratigraphic Column of Western Henan

How can the sedimentary hiatus of 1.2–1.0 Ga be filled up in current stratigraphic column? According to the biostratigraphic information (Gao et al. 2011a, b, c, d, e, f), the sets of missing strata may be found at the junctions among western Henan, southern Shanxi and eastern Shannxi, mainly within the western Henan, and also tectonically on the southern margin of NCC, where the Ruyang and Louyu Groups most probably make up for the stratigraphic hiatus, especially for geologists to position the Luoquan glacial stage in the stratigraphic column (Table 1.2) which is the only diamictitic layer in the NCC, and also acts as a bridge for the correlation of glacial stage or diamictitic layer between the Tarim Block and the Yangtze Craton (Table 1.2).

Since the Jixian stratotype section is as the standard of the Meso-Neoproterozoic reference section in China, the optimization of its chronostratigraphic frame will importantly promote the precision of Precambrian geological mapping and stratigraphic correlation, and also be significant for the research of Precambrian biostratigraphic evolution and the calibration of evolutionary time limit for macroscopic algae in NCC (Table 1.2).

Based on micropalaeoflora (microplants), stratigraphic sequences and contact relation with a few glauconitic isotopic dating (Guan et al. 1988) had first referred the Huanglianduo, Dongjia, Luoquan and Dongpo Formations to the Neoproterozoic "Sinian" (equivalent to the Nanhuan and Sinian in current stratigraphic column Table 1.2). Nevertheless, there always are different cognitions in the Neoproterozoic stratigraphic division and correlation on the southern margin of NCC. By way of multidisciplinary methodologies, the numerous researches of Meso-Neoproterozoic strata have been achieved, dealing with isotope chrono (Xing et al. 1989; Yan and Zhu 1992; Xiao et al. 1997; Yin and Gao 1997; Yin and Yuan 2003; Yin et al. 2004), stratigraphic sequences and sedimentary facies (Cui et al. 1996; Zhou et al. 1999; Gao et al. 2002). As concerns chronostratigraphy, however, only one detrital zircon age 1126 ± 36 Ma was available for the Mesoproterozoic strata (Gao et al. 2006), so that the precise chronostratigraphic position is still a "mystery" in whole the stratigraphic column of western Henan (Table 1.2). Since then, a LA-MC-ICPMS U–Pb age 1611 ± 8 Ma of tuffaceous zircon is first determined at the top of Luoyukou Formation, thus the Luoyukou Formation positioned into the range of the Mesoproterozoic Changchengian (Su et al. 2013). Consequently, the Ruyang and Luoyu Groups in western Henan are attributed to the Early Mesoproterozoic strata, but there remains a large contradiction between the zircon U–Pb age and the acanthomorphic acritarch fossils found in the Ruyang Group (Gao et al. 2011a, b, c, d, e, f; Li et al. 2012). Since the radiation of acanthomorphic acritarchs evolved up to a very high

Table 1.2 Meso-neoproterozoic comprehensive stratigraphic column of China (Gao et al. 2009a, b; Qiao et al. 2007)

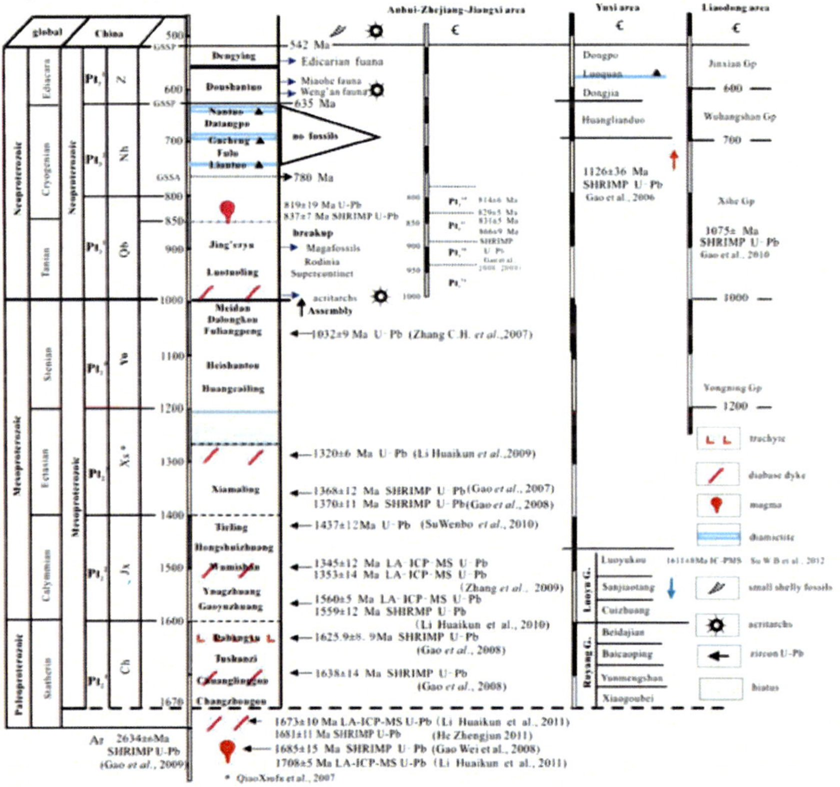

level in the Ruyang Group, even if the tuffaceous zircon U–Pb dating is correct, the evolutional history of early acanthomorphic acritarchs should be re-acquainted. Recently, another chronology progress of the Guandaokou and Luoyu Groups in the southern margin of NCC, the tuff zircon U–Pb dating 1594 ± 12 Ma and 1541 ± 8 Ma have been achieved in the Longjiayuan Formation just above the Guandaokou Group. A series of SHRIMP U–Pb dating have been gain from the Louyu Group similar as the Guandaokou Group, which are 1596 ± 7 Ma, 1596 ± 15 Ma, 1608 ± 17 Ma and 1620 ± 16 Ma. However, the column of the strata frame in western Henan and Shanxi Provinces should be change to the Changchengian (Zhang et al. 2019).

1.3.2 The Stratigraphic Position of Precambrian Strata in the JLXH Region, NCC

Geographically, the abbreviation "JLXH" specifically refers a local name to the Jiaodong Peninsula (eastern Shandong)-Liaodong Peninsula (eastern Liaoning)-Xuzhou City (northwestern Jiangsu)-Huaihe River Valley (centro-northern Anhui), which are geologically in the east and south of NCC.

Within the JLXH region, Liaodong is characterized by the most complete and systematic Neoproterozoic strata, such as Yongning, Xihe, Wuhangshan and Jinxian Groups in ascending order (Table 1.2). As a set of clastic sediments directly overlying on the Qingbaikouan Yongning Group, the Xihe Group consists of Diaoyutai, Nanfen and Qiaotou Formations with a large amount of micropalaeoflora fossils (Xing and Liu 1973). Over the Xihe Group, the Wuhangshan Group is composed of Changlingzi, Nanguanling and Ganjingzi Formations from bottom up, and mainly distributed on the north of Fuxian County, Liaoning Province, among which the lower interval of the Changlingzi Formation is shale intercalated by thin-bedded sandstone and upward gradually by limestone lens and thin-bedded limestone with liquefied veins (so-called "molar tooth" structures) containing a lot of macroscopic algal and worm fossils in siltstone (Xing et al. 1985); the overlying Nanguanling Formation is limestone (with liquefied veins/"molar tooth" structures) interbedded with thin-bedded shale bearing metazoan fossils (Xing et al. 1985); and the Ganjingzi Formation is dolomite. While the Jinxian Group mainly is carbonate rocks, and including Yingchengzi, Shisanlitai, Majiatun, Cuijiatun and Xingmincun Formations. The carbonate rocks in Nanguanling, Ganjingzi, Yingchengzi and Xinmincun Formations contain a lot of liquefied veins/"molar tooth" structures (Qiao et al. 1994, 2001; Liu et al. 2003; Kuang et al. 2004; Meng et al. 2006; Qiao and Gao 2007).

Currently, Neoproterozoic biostratigraphy, lithostratigraphy and stratigraphical sequences in Liaodong Peninsula have achieved some reliable progresses, especially on biostratigraphy (Xing and Liu 1973; Hong et al. 1991; Tang and Gao 1998; Gao et al. 1999; Qiao et al. 2001). As concerns chronostratigraphic dating, however, only shale Rb–Sr age 723 ± 43 Ma in the Changlingzi Formation as well as shale Rb–Sr age 650 Ma and glauconite K–Ar age 579 Ma in the Xinmincun Formation are available (Xing et al. 1985). Even though a complete stratigraphic sequence with numerous fossils developed in the southern Liaodong Peninsula, its stratigraphic position in stratigraphical column is still uncertain due to the lack of diamictitic layer and reliable zircon dating. Even if a detrital zircon SHRIMP U–Pb age 1075 Ma of the Diaoyutai Formation was first reported in the southern Liaodong Peninsula (Gao et al. 2011e), it seems to show that the Late Precambrian strata may be around 1.0 Ga. While more precise zircon SHRIMP U–Pb ages of 928 ± 8 Ma (Liu et al. 2005) and 930 ± 10 Ma (Gao et al. 2009c) are obtained from the diabase sills intruding into carbonate rocks of the Niyuan Formation, resulting in wall-rock optical alteration and remaining a deep imprint for its stratigraphic position in the Xuzhou City (northwestern Jiangsu) to Huai River (centro-northern Anhui) region. The wall-rock alteration would imply the stratigraphic age earlier than that of diabase sill (ca. 1.0 Ga). If the above dating is believable, thus, whole the stratigraphic age would be moved earlier on 1.0 Ga, however, it must have caused a conflict with the traditional biostratigraphic record (Xing et al. 1985). Therefore, the Precambrian chronostratigraphy is still need of further study in the JLXH region.

1.4 The Stratigraphic Position of Qingbaikouan in the Stratigraphic Column

1.4.1 Qingbaikouan in NCC

Qingbaikouan (1000–780 Ma) is a stratigraphic unit at the system-level attributed to the Neoproterozoic Era in the National Stratigraphic Chart of China and equivalent to the Tonian in the International Chronostratigraphic Chart. Based on the biotic characteristics, it serves as an important connecting role for biostratigraphic evolution from the Mesoproterozoic to Palaeozoic, and a significant stratigraphic unit for Precambrian correlation between NCC and YC.

Originally, the Xiamaling Formation belongs to the basal formation of Neoproterozoic Qingbaikouan. Since a zircon SHRIMP U–Pb age of 1368 ± 12 Ma was gained from the bentonitic zircon in the Xiamaling Formation (Gao et al. 2007a, b), the Xiamaling Formation. has been attributed to Mesoproterozoic and overlies on the Jixianian. Therefore, so far only Luotuoling and Jing'eryu formations remain in the Qingbaikouan without a qualified isotopic dating in YFDZ, and thus the current Qingbaikouan is still an "enigma" for its exact positions in the stratigraphic column. Furthermore, it is really in need of confirmation that whether the Qingbaikouan strata are synchronous sediments between YFDZ and Liaodong Peninsula.

The biostratigraphy shows that stratigraphic differences in both regions are the emergency of macroscopic-algae or metazoan. Macroscopic-algae appear at the Luotuoling Formation in Beijing and northern Hebei (YFDZ), while some metazoan fossils in Liaodong. So far the Qingbaikouan stratigraphic division is based on the lithostratigraphic correlation of both regions. By means of detrital zircon dating, the present authors try to approach the upper limit of the Qingbaikouan clastic sedimentary rocks in YFDZ and in Liaodong, and then to judge if both sediments are congenetic strata (Gao et al. 2011e).

1.4.2 Qingbaikouan Stratigraphic Correlation in Liaoning Peninsula, NCC

Sino-Korean Craton (or called NCC within the China territory) consists of two parts, i.e., North China Block (NCB) and Jiao-Liao-Korean Block (JLKB), both of which seem to contain a common Achaean basement, in between the Tan-Lu Fault Belt

(TLFB) is situated (Fig. 1.3). After the Lüliang Movement (ca. 1800 Ma), the Yanliao Faulted-Depression Zone (YFDZ) was developed with Meso-Neoproterozoic sediments (1800–800 Ma) in NCC (Fig. 1.3). Since 800 Ma, the NCC was uplifted and suffered from denudation, so that the Qingbaikouan Jing'eryu Formation. shows discomformable contact with the overlying Cambrian Changping Formation without Nanhuan and Sinian strata (780–541 Ma), while the Palaeo-Tanlu Faulted Belt (P-TLFB) appeared as a new rift, and thus the more complete Neoproterozoic strata were developed in eastern side of JLKB (Fig. 1.3). Even though the biostratigraphic study has shown that the Neoproterozoic strata contain a Sinian fossil assemblage in the Jiao-Liao-Korean Craton (JLKC), its position and dating in stratigraphic column is still a controversial issue due to the lack of diamictitic layer in the Sino-Korean Craton.

Based on the biostratigraphic studies and in view of the difference in fossil assemblages, there should be an old and new relationship of Neoproterozoic strata between NCB and JLKB. Although numerous macroscopic algae are developed either in NCB or in JLKB, the metazoan fossils only appear in the Neoproterozoic strata and implying a new century of Neoproterozoic bio-evolution in JLKB (Xing et al. 1985, 1989), where the lower interval of Neoproterozoic strata is clastic rocks, while the upper interval mainly consists of carbonate plus small amount of clastic rocks, and a large amount of seismites are found along the Tan-Lu Fault Belt (Qiao et al. 1994, 2001). Accordingly, there are some stratigraphically evolutional differences

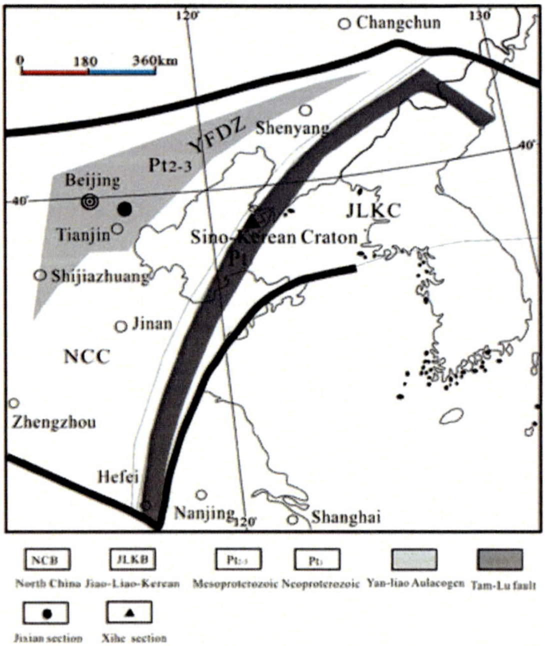

Fig. 1.3 Meso-neoproterozoic tectonic framework and stratigraphic section distribution in Sino-Korean Craton (Qiao et al. 2001, modified). NCC. North China Craton; JLKC. Jiao-Liao-Korean Craton; YFDZ. Yanliao faulted-depression zone

between NCB and JLKB. Therefore, so far the Neoproterozoic stratigraphic correlation between both the blocks is only distinguished by lithostratigraphy and biostratigraphy, and the precise chronostratigraphic dating has still no broken though as an unresolved problem. According to the detrital zircon U–Pb age 1075 Ma of the Diaoyutai Formation in the Lower Xihe Group, the stratigraphic position of the Xihe Group can be basically calibrated in Meso-Neoproterozoic stratigraphic column (Gao et al. 2010e), and thus, the sedimentary age of lower clastic rocks in the Xihe Group should be younger than 1.0 Ga in JLKB as the response of Sino-Korean Craton to the assembly of supercontinent Rodinia.

As the reference section of the Late Precambrian stratigraphy, moreover, the optimization of Meso-Neoproterozoic chronostratigraphic frame in the Jixian stratotype section would be significant for the geological mapping and accurate stratigraphic correlation in China as well as for the study of Precambrian biostratigraphic evolution, especially on the macroscopic algae development time (Gao et al. 1996). The new dating (1.4–1.32 Ga) and reposition of the Xiamaling Formation have made it move from Neoproterozoic Era to Mesoproterozoic Era, and resulting in a new research trend of the Meso-Neoproterozoic stratigraphic boundary in Sino-Korean Craton and dealing with the conversion from Colombia to supercontinents Rodinia. Accordingly, the Qingbaikouan Luotuoling and Jing'eryu Formations are still in need of further discrimination and reconfirmation. Their intensive study will be favorable to the understanding of the Meso-Neoproterozoic tectonic evolution in Sino-Korean Craton (or NCC plus JLKC) as well as to the determination of regional stratigraphic correlation.

1.5 Chronostratigraphic Progress in the Jiangnan Orogenic Belt

1.5.1 Jiangnan Orogenic Belt (JOB)

As a Chinese geographical term, "Jiangnan" means a region on the south of Yangtze River, involving Guizhou, Guangxi, Hunan, Jiangxi, Anhui and Zhejiang Provinces, etc. from west to east in South China, and geologically in between the Yangtze Craton and Cathaysia Block (cf. the inset in Fig. 1.1), and it consists of an old metasedimentary rock belt, and attracts the research interests of numerous Chinese geologists, so that significant research progress has been made since 1930s. Tectonically, this metasedimentary rock belt was initially named "Jiangnan Old Land" (Huang 1945), and then "Jiangnan Axis" (Huang 1954); Hereafter Chen (1956) referred it to an old folded-belt according to geosyncline and platform doctrine. Based on the plate-tectonics, Guo et al. (1985, 1996) defined it as "Jiangnan Proterozoic Mobile Belt", and/or "Meso-Neoproterozoic Island-Arc Collision Orogenic Belt" (Fig. 1.4).

It is generally considered that the Jiangnan late-metasedimentary rock belt is an orogenic belt of late-ocean-continent collision driven by the Cathaysia Block

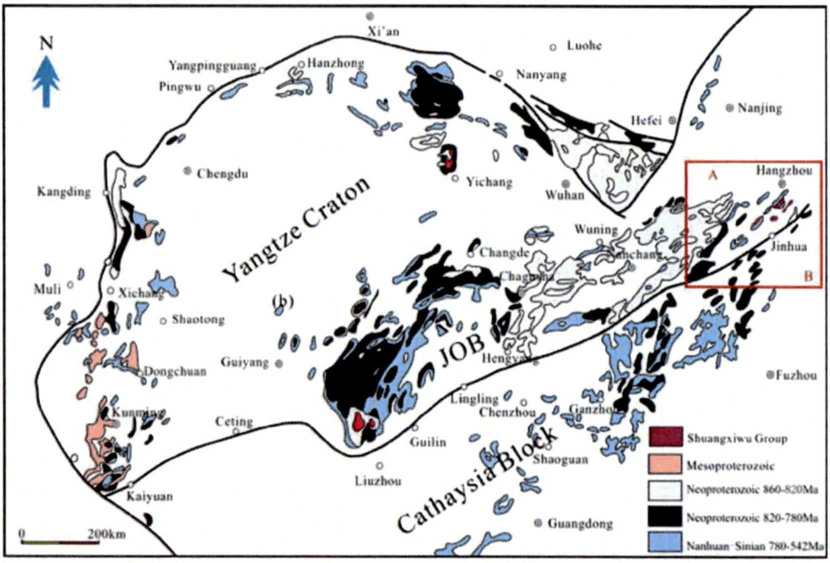

Fig. 1.4 Meso-neoproterozoic tectonic division and stratigraphic distribution in south China including Yangtze Craton, Jiangnan Orogenic Belt and Cathaysia Block. JOB. Jiangnan Orogenic Belt; "Area A" in red frame shows the outcrop area of Shuangxiwu Group in JOB

subducting towards Yangtze Craton (Guo et al. 1980, 1985, 1996; Xu et al. 1992a, b; Xu 1994). Moreover, Wang (1994) further proposed that the Jiangnan Orogenic Belt (JOB) would be divided into two parts roughly along boundary between Hunan and Jiangxi Provinces, the west part is mainly a rift on the passive continental margin as a "new plate" with a 780 Ma-old folded basement, while the east one is a continuously developed active continental margin (Fig. 1.4; Wang et al. 1990, 1994). Until the beginning of 2000 s, based on the study of granite intruding into the Sibao Group (or equivalent strata) and being unconformably superposed by the Banxi Group (or equivalent strata; Table 1.3), the granitic rock mass is mainly constituted by the muscovite-bearing peraluminous granitoids (MPG) and the cordierite-bearing peraluminous granitoids (CPG) with zircon SHRIMP U–Pb age of 840–820 Ma (Gao et al. 2010a), in which the unconformity is named Sibao Orogeny as a geological event. However, so far some geologists still compared the Sibao Orogeny with the Grenville Movement, attributed it to the global tectonic system of supercontinent Rodinia and considered the Banxi Group (or equivalent strata) to be the product of Rodinia breakup at ca. 1000 Ma, and thus recently they inferred that the breakup is caused by the activity of "super mantle plume" (Li et al. 2003, 2007; Wang et al. 2003). Nevertheless, Zhou is the first one who queried that whether the JOB could be related to the Grenville Movement? Consequently, numerous bentonitic zircon SHRIMP U–Pb ages were measured, which would provide reliable chronostratigraphic dating for the recognition of the metasedimentary basement formation and evolution in Jiangnan Orogenic Belt (Gao et al. 2013b).

Table 1.3 Regional stratigraphic table in the Jiangnan orogenic belt

Geological time		System \ Province	Guangxi	Guizhou	Hunan	Jiangxi & partial Anhui		Tectonic movement
			northern	southeastern	northweastern	northweastern	northeastern	
Palaeozoic		Cambrian		Meishucun Fm				
Neoproterozoic	542Ma	Sinian	Dengying Fm			Dengying Fm		Tongwan mov
			Doushantuo Fm			Doushantuo Fm		
	635Ma	Nanhuan	Nantuo Fm Darangpo Fm Gucheng Fm Fulu Fm Changan Fm	Liangjiehe Fm	Nantuo Fm Darangpo Fm Gucheng Fm Fulu Fm Changan Fm	Nantuo Fm	Leigongwu Fm	
				Tiesiao Fm		Liantuo Fm	Xiuning Fm	
	760Ma	Qingbaikou	Danzhou Gp	Xiajiang Gp	Banxi Gp	Majiangqiao Gp	Tantou Gp	Xuefeng or Jinning Oro
	780Ma						Dengshan Gp	
	820Ma		Shibao Gp	Fanjingshan Gp	Lengjiaxi Gp	Shaoqiwa Gp	Shuangqiaoshan Gp	Wuling or Sibao Oro
	850Ma				Cangxi Complex Gp	Yifeng Complex Gp		
	870Ma						TianliSchit Shuangxiwu Gp	Chengjiang Oro or Grenville Mov
Mesoproterozoic	1000Ma						Tieshajie Fm	

Note Oro. orogeny; Mov. movement

1.5.2 New Recognition on the JOB

As concerns so-called "Jiangnan Old Land" or "Island-Arc Collision Orogenic Belt" and Jiangnan Orogenic Belt, the controversy of geologists was mainly focused on one issue: whether there should be bilayered-synthems in its metasedimentary basement beneath the Nanhuan sedimentary cover, i.e., one simple-folded cover-like basement and another complex-folded basement? Which would deal with the starting age of orogenic epoch, the boundary range of each synthem and the determination of unconformable contacts between both synthems in Jiangnan Orogenic Belt. In this case, the bentonitic zircon SHRIMP U–Pb dating of the metasedimentary synthems would play a significant role in the chronostratigraphy of JOB (Gao et al. 2008a, 2010a, b, 2011a, b, c, d, 2012a, b, c, 2013a, b). As the zircon ages of metasedimentary strata are continuously measured, the original so-called "Mesoproterozoic strata" in Jiangnan Orogenic Belt were suspected (Wang et al. 2006, 2008), even Zhou et al. (2008) doubted that whether the JOB could be equivalent to the "Grenville Orogenic Belt"? Since new evidences have provided a series of volcanic rocks developed on the south or east side of Yangtze Craton (Fig. 1.4), all of which implies the tecto-dynamical transformation of unconformable contact between the Sibao and Banxi Groups during Wuling/Sibao Orogeny at 820 Ma ago (Gao et al. 2010b), and thus the Banxi Group can act as a single-folded transitional cover-like synthem under the Nanhuan sedimentary cover, while the Sibao Formation would be a complex synthem. This interpretation of bilayered-synthems for the metasedimentary basement would be reasonable to understand the tectonic setting, metallogenic epoches and stratigraphic division in the east part of Jiangnan Orogenic Belt. First of all, the bentonitic zircon SHRIMP U–Pb ages of 831 ± 6 Ma 829 ± 6 Ma are gained from the Shuangqiaoshan Group (Table 1.3; Gao et al. 2010a), which would greatly delay the chronostratigraphic age of original "Mesoproterozoic" metasedimentary synthem in Jiangnan Orogenic Belt. Second, two bentonitic zircon SHRIMP U–Pb ages of 842 Ma and 814 Ma respectively from the Sibao Group on the border between Guangxi and Guizhou Provinces (Gao et al. 2010a) and from the Jialu Formation of the Xiajiang Group (Table 1.3; Gao et al. 2010b) as well as a zircon age of 827 Ma from the Motianling granite intruding into the Sibao Group, which is underlain by the Xiajiang Group, are measured (Table 1.3; Gao et al. 2010a). Consequently, the upper limit of sedimentary age for Sibao Group and equivalent strata can be determined, and thus the Sibao Orogeny is basically limited to the interval of 827–814 Ma (Table 1.3). Particularly, additional two measured bentonitic zircon SHRIMP U–Pb ages, i.e., one age 822 Ma from of the top of Lengjiaxi Group (Table 1.3) and another age 802 Ma from the Banxi Group could constrain the Sibao/Wuling Orogeny as a tectonic event within 20 Ma (Table 1.3; Gao et al. 2011c).

In the Jiangnan Orogenic Belt, the complex-folded synthem involves the metasedimentary strata of Sibao, Fanjingshan, Lengjiaxi, and Shuangqiaoshan Groups. It is confirmed that above bentonitic zircon SHRIMP U–Pb ages should indicate the original sedimentary ages of these metasedimentary strata, and thus these metasedimentary strata should be attributed to the Neoproterozoic (Table 1.3). Therefore, the age of Wuling Movement, limited by zircon SHRIMP U–Pb dating, should be independent of the Grenville Orogeny (ca. 1000 Ma; Table 1.3), especially for the detrital zircon SIMS (Secondary Ion Mass Spectrometer) U–Pb age 834 ± 4 Ma in the Xingzi Group, the age 830 ± 5 Ma in its overlying Shaoqiwa Formation in Lushan (Table 1.3; Guan et al. 2010) and the metarhyolitic zircon SHRIMP U–Pb ages 840 ± 6 Ma, 833 ± 4 Ma and 831 ± 3 Ma in the Shaoqiwa Formation (Gao et al. 2012d) as well as the chronostratigraphic dating of the Xiushui Formation in Shuangqiaoshan Group and Majianqiao Formation (Table 1.3; Gao et al. 2012e), all these strata are the coeval Neoproterozoic formations, which would basically shake the original concept about the oldest "Mesoproterozoic" stratigraphic age for the metasedimentary folded basement in JOB.

Previously, many treatises on "tectonic event" always regarded one stratigraphic unconformity between the Banxi Group (or equivalent strata) and its underlying Sibao Group (or equivalent strata), i.e., the tectogenetic plane of Sibao/Wuling/Shuangqiaoshan/Shengong Orogeny, as the behavior of global "Grenville Orogeny" around 1.0 Ga ago (Table 1.3), while another regional unconformity between the Banxi Group (or equivalent strata) and its overlying Nanhuan (corresponding to the original "Lower Sinian"; Lu 2002), i.e., the plane of Jinning/Xuefeng Orogeny, was generally limited to 780 Ma (Table 1.3). However, numerous chronostratigraphic dating data show that there is no any stratum older than 1.0 Ga in age is available within the metasedimentary basement in the Jiangnan Orogenic Belt, the major metamorphism and deformation should happen during 830–780 Ma. Therefore, it is believed that the formation of metasedimentary basement as well as the marginal accretion and the final patterning of old land would be independent of the Grenville Orogeny (ca. 1.0 Ga ago), but the products of Jinning/Xuefeng Orogeny at 800 Ma ago in the JOB.

1.5.3 The Stratigraphic Position of Shuangxiwu Group in Western Zhejiang

There is a set of Neoproterozoic epimetamorphic strata developed at the Pingshui Town in Shaoxing City, Zhejiang Province, which is tectonically distributed along the Jiangshan-Shaoxing Fault (JSF) of Qinzhou-Pingxiang-Jiangshan-Shaoxing Collision Belt (QPJSCB; Fig. 1.5), and attributed to the island-arc type of marine spilite-keratophyre formation named "Pingshui Group" by the Zhejiang Geological Survey Team① (Zhejiang Geological Survey Team. 1990. The report of geological survey of Pingshui area (in Chinese).). While Yu et al. (1995) revised the "Pingshui Group" into Pingshui Formation within the Shuangxiwu Group (Tables 1.3 and 1.4), and Chen et al. (2009) provided its keratophyric zircon laser ablation inductively-couple plasma mass spectrometer (LA-ICP-MS) U–Pb ages 904 ± 3 Ma and 906 ± 10 Ma as well as the Hf isotopic and whole rock compositions, thus referred the stratigraphic age of Pingshui Formation to the Early Neoproterozoic. While Ye et al. (2007) measured the zircon SHRIMP U–Pb ages 913 ± 15 Ma and 905 ± 14 Ma from both Taohong and Xiqiu granitic plutons within the Pingshui Formation (Table 1.5). In addition, Li reported that the Pingshui copper ore mostly consists of quartz-pyrite-sphalerite with the metallogenetic age of 899 ± 21 Ma based on several sets of zircon LA-ICP-MS dating data of S compound-bearing quartz veins, However, above reported ages are obviously contradictory each other, and as "unharmonious data", two dating ages 819 Ma and 818 Ma were eliminated by Ye et al. (2007). In such a case, it has needed further to verify the tuffaceous zircon age of Pingshui Formation and its geological significance.

Since the sedimentary age of Pingshui Formation is around 905 Ma, a zircon SHRIMP U–Pb age 902 ± 7 Ma for the Beiwu Formation and two ages 899 ± 8 Ma and 877 ± 9 Ma for the Zhangcun Formation in Shuangxiwu Group, while two zircon U–Pb ages 901 ± 9 Ma and 893 ± 6 Ma have been obtained from the basal conglomerate of the Luojiamen Formation within the Heshangzhen Group (Table 1.4; Gao et al. 2014a, b, c), Consequently, the stratigraphic age of Shuangxiwu Group should be calibrated to 905–878 Ma in stratigraphic column (Tables 1.3 and 1.4), indicating an Early Neoproterozoic sequence.

1.5.4 Dating of Tieshajie Formation and the South Boundary of JOB

Jiangxi Province has covered the major part of Jiangnan Orogenic Belt, where the EW to NEE trending Pingxiang-Jiangshan Fault (PJF) of the Qinzhou-Pingxiang-Jiangshan-Shaoxing Collision Belt (QPJSCB) is very significant for tectonic and stratigraphic division (Fig. 1.5). With a length up to 400 km, the PJF extends along Pingxiang–Yichun–Xingyu–Linchuan–Qianshan–Shangyao–Jiangshan line,

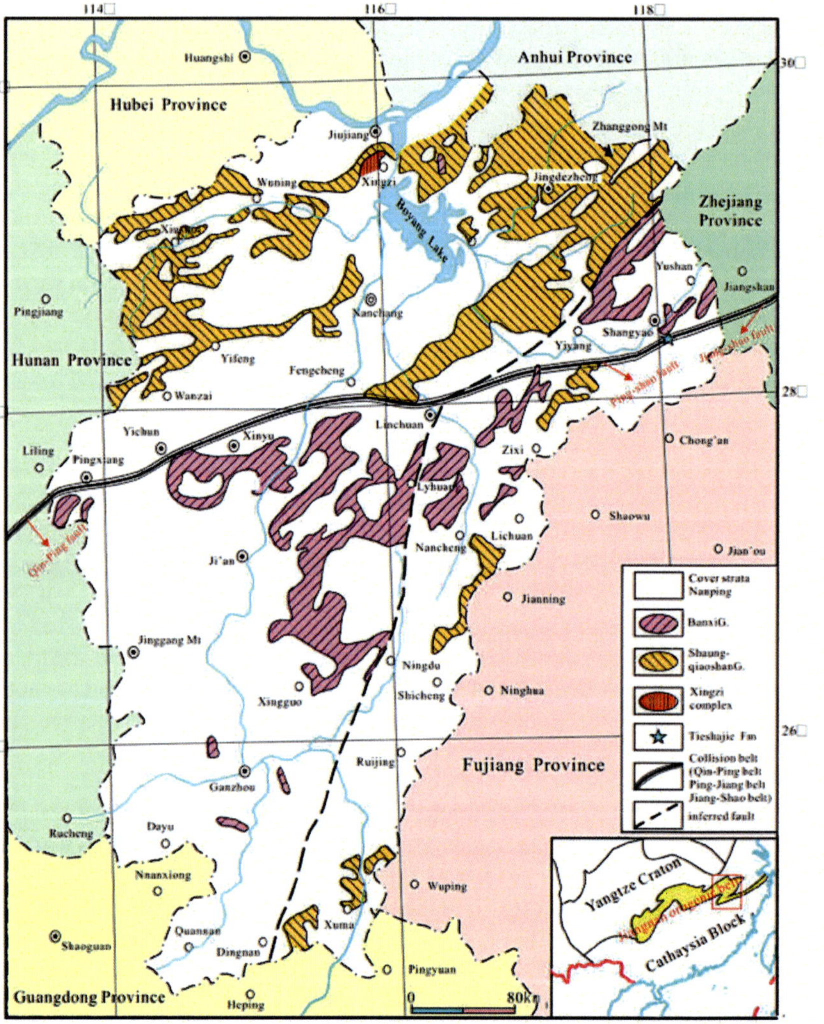

Fig. 1.5 Distribution of Qinzhou-Pingxiang-Jiangshan-Shaoxing collision belt (QPJSCB) and Meso-neoproterozoic strata in Jiangxi territory (Jiangxi Geological and Mineral Bureau 1997, modified). QPF. Qinzhou-Pingxiang Fault; PJF. Pingxiang-Jiangshan Fault; JSF. Jiangshan-Shaoxing fault

running across the central of Jiangxi Province. Its east end is connected to Jiangshan-Shaoxing Fault (JSF) in Zhejiang territory and its southwest end stretches to Qinzhou-Pingxiang Fault (QPF) in western Jiangxi to Guangxi territory (Fig. 1.5). As a long-term regionally chronical active discordogenic fault belt, it acts as a first-order tectonic boundary line between the Jiangnan Orogenic Belt and Cathaysia Block (Figs. 1.1 and 1.5).

Table 1.4 Meso-Neoproterozoic stratigraphic sequence and chronostratigraphy in Jiangnan orogenic belt (Gao et al. 2014b, c).

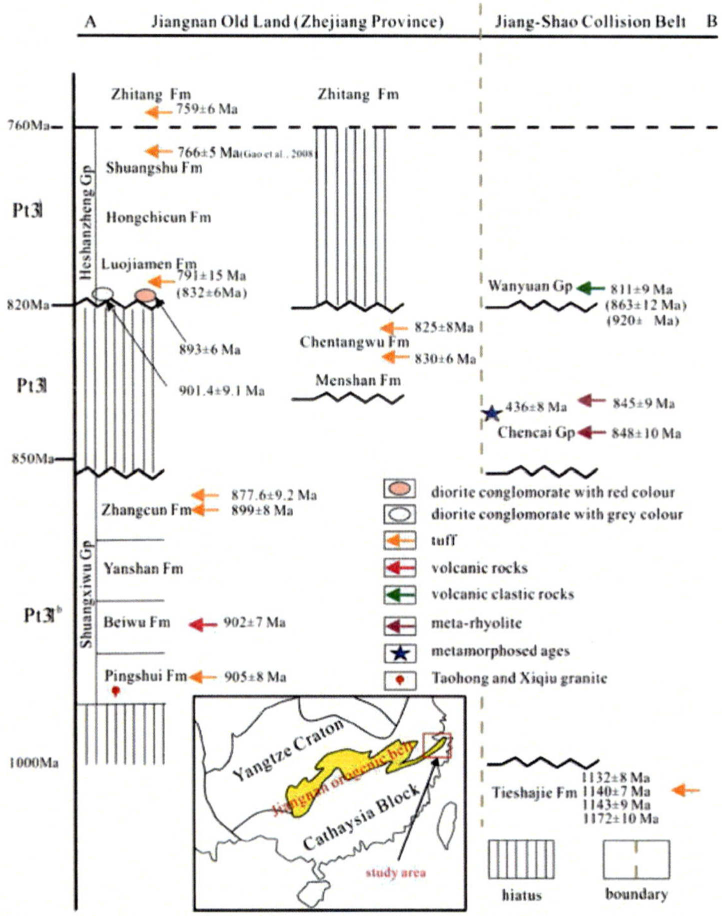

The Orogenic belt also separates the Precambrian strata of northern Jiangxi from that of southern Jiangxi province. The "lower green schist" is equivalent to the Neoproterozoic Shuangqiaoshan Group (831–824 Ma) and Dengshan Group (<820 Ma) as original sedimentary cover beneath the unmetamorphosed Nanhuan (<780 Ma; Table 1.3), mainly distributed in the vast area of northern Jiangxi. While only the mild-deformed and metamorphic sedimentary cover of the Banxi and Tantou Groups and the Nanhuan Yangjiaqiao Group in the southern Jiangxi (Table 1.3 and Fig. 1.5).

Table 1.5 The Stratigraphic scale and chronostratigraphic dating of the YC and JOB

Era.	Sys.	Ser.	Gp.	Fm.	Sym.	Lith. *1	Age(Ma)	Reference
Neoproterozoic	Sinian	Upper	--	Dengying	$Pt_3^2dy(Z_2)$	Tuff	549±5	Yin C Y *et al.*, 2005b
		Lower	--	Doushantuo	$Pt_3^2ds(Z_1)$	--	635±5	Yin C Y *et al.*, 2005b
	Nanhuan		--	Nantuo	Pt_3^2nt	--	--	--
		Middle	--	Datangpo	Pt_3^2dt	Tuff	661±7	Gao L Z *et al.*, 2013b
						Tuff	667±10	Yin C Y *et al.*, 2005b
			--	Gucheng	Pt_3^2gc	--	--	--
			--	Fulu	Pt_3^2fl	Tuff	669±13	--
		Lower	--	Chang'an/Liantuo	Pt_3^2ca	Tuff	778±5	Gao L Z *et al.*, 2013b
	Qingbaikouan	Upper	--	Gongdong	Pt_3^1gd	Tuff	786±6	Gao L Z *et al.*, 2013c
			--	Hetong	Pt_3^1ht	Tuff	801±3	Gao L Z *et al.*, 2013c
			Xiajiang	--	Pt_3^1xj	--	--	--
			Banxi/Danzhou	--	Pt_3^1bx/Pt_3^1dz	--	--	--
		Wuling Movement						
		Middle	Lengjiaxi Sibao Fanjingshan	Lengjiaxi	$Pt_3^1lj/Pt_3^1sb/Pt_3^1fj$	Tuff	822±6	Gao L Z *et al.*, 2012a-c
				Igneous body		Igneous rock	834±4	Gao L Z *et al.*, 2013c
				--		Igneous rock	835±5~837±3	Gao L Z *et al.*, 2014a
				Yuxi		Tuff	842±13	Gao L Z *et al.*, 2013c
		Shengong Movement						
		Lower	Shuangxiwu	Zhangcun	Pt_3^1sx	Tuff	878±6	Gao L Z *et al.*, 2014c
				Igneous body		Igneous rock	878±4	Gao L Z *et al.*, 2014c
				Beiwu		Tuff	878±7	
				Zhangcun		Tuff	901±5	
				Pingshui		Kerato-phyre	904±3	Chen Z H *et al.*, 2009
							906±10	
						Granite	905±14	Ye M F *et al.*, 2007
							913±15	
						Tuff	908±7	Gao L Z *et al.*, 2014c
Mesoproterozoic	--		Guanyang / U. Shennongjia	Tieshajie	Pt_2^3ts	Tuff	1132±6	Gao L Z *et al.*, 2013e
						Tuff	1140±7	
						Tuff	1143±9	
						Tuff	1172±10	

Note *1. lithology of dating sample

A set of meso- to epimetamorphic strata are distributed at the junctions among Jiangxi, Zhejiang and Fujian Provinces, including the Tieshajie Formation (1132–1172 Ma), Tianli Schist (923 Ma), Zhoutan Formation and Wanyuan Group (930–811 Ma; Tables 1.3, 1.4 and 1.5), among which, the Tieshajie Formation only outcrops at the eastern PJF segment of the Qinzhou-Pingxiang-Jiangshan-Shaoxing Collision Belt (QPJSCB; cf. the blue star in Fig. 1.5 for the location of Tieshajie outcrop area), and consists of metamorphic gneiss, schist and slate interbedded with a large amount of volcanic rock fragments (Fig. 1.6). However, Chen et al. (1991) considered that the Tieshajie Formation is not a stratigraphic unit, but the mélange block, and Xu (1990) referred its tectonic setting to a trench-arc-basin system in the northeastern Jiangxi to southern Anhui. Because the collision belt belongs to the ductile fault in regional chronical active discordogenic fault belt, and also a multiperiodic tectonics affected one, in which, its hypometamorphic strata have always been an object for geologists to approach the tectonic setting of JOB continually by way of the stratigraphic dating.

The initial reported isotopic chronostratigraphic dating of the Tieshajie Formation had provided a metaspilite Rb–Sr isochron age 1159 Ma, and a single-grain zircon U–Pb age 1201–1091 Ma of meta-quartz-keratophyre as well as a metarhyolitic zircon U–Pb age 1196 ± 6 Ma (Chen et al. 1991). Thus, it was tentatively clear that the stratigraphic position of the Tieshajie Formation should be the sole known Late Mesoproterozoic stratum in Jiangnan Orogenic Belt (Table 1.3), where the controversy is focused on that whether there really is the older metamorphic rock within the JSF? In this case, a highly precise zircon SHRIMP U–Pb age can confirm the stratigraphic age of Tieshajie Formation.

The present authors tried to explore the Late Mesoproterozoic tectonic event and stratigraphic significance for the Jiangnan Orogenic Belt between the YC and the Cathaysia Block. Along the stratigraphic section of Tieshajie Formation, there outcrops a series of metavolcanics interbedded with clastic rocks, its chronostratigraphic dating would be a reliable means for approach to its tectonic background and metallogenic conditions. Some rock samples of stratigraphic dating (Fig. 1.6a–d) have be collected from the Tieshajie section (Fig. 1.6a–e) at Tieshajie copper mine as follows.

(1) Sample T22-3 (coordinate of sampling site: N28°15.43′, E117°24.35′):

Phyllitic metarhyolitic porphyry contains quartz and feldspar pseudomorphous phenocrysts with 0.2–1.3 mm in size, and showing scattered orientation (Fig. 1.6f). The quartz phenocryst appears as subhedral-anhedral crystal grains with wavy or bound extinction and some corroded border; the feldspar shows subhedral tabular crystal grains and quartz- or sercite-metasomatic pseudomorph; while the rock matrix consists of felsic and neogenic minerals, in which the felsic components appear mostly as felsitic-microcrystal and obvious orientation with <0.1 mm in size, and the neogenic mineral is flaky sericite with <0.2 mm in size and head-to-tail connecting orientation.

There totally are 24 measuring points gained on Sample T22-3, among which, 20 points are plotted on the concordant line (Fig. 1.7, left-upper) with a $^{206}Pb/^{207}Pb$ age 1172.3 ± 9.7 Ma and a corresponding MSWD (mean standard weighted deviation) value 1.2, representing the petrogenetic age of rhyolite.

(2) Sample T22-5 (coordinate of sampling site: N28°15.44′, E117°24.36′):

Metarhyolitic prophyry. The phenocryst is composed of subhedral-anhedral feldspar with 0.5–1.5 mm in size, scattered orientation and the alteration of sericitization, ferritization, etc. While the rock matrix consists of felsic and neogenic minerals, in which the felsic mainly shows microcrystal grains, with <0.1 mm in size and obvious orientation; the neogenic mineral is flaky sericite with <0.1 mm in size (Fig. 1.6g).

Fig. 1.6 Outcrop and microscopic photos of metarhyolite in the Tieshajie formation. **a–d** Outcrop of metarhyolite; **e** sampling section of Tieshajie Formation at Tieshajie copper mine; **f–l** microscopic photos of metarhyolite

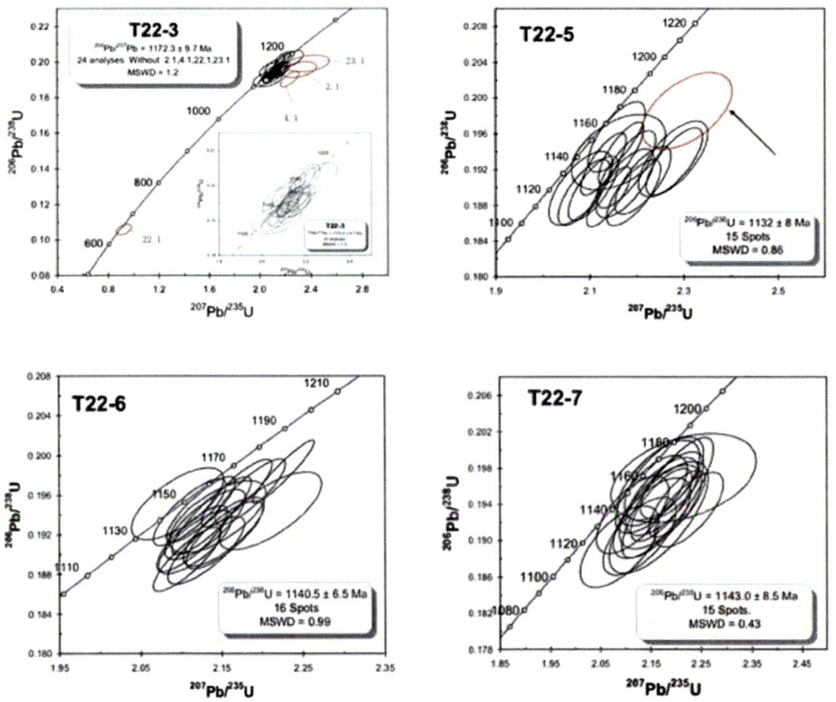

Fig. 1.7 Metarhyolitic zircon U–Pb concordant diagrams in the Tieshajie formation

There are 16 measuring points gained on sample T22-5 and 15 points fallen on the concordant line (Fig. 1.7, right-upper) with a $^{206}Pb/^{207}Pb$ age 1132 ± 8 Ma and a MSWD value 0.86, representing the petrogenetic age of rhyolite.

(3) Sample T22-6 (coordinate of sampling site: N28°15.45′, E117°24.36′):

Metarhyolitic prophyry. Its phenocryst consists of feldspar and quartz generally with 0.2–1.3 mm in size and scattered orientation. The feldspar is near subhedral tabular to anhedral grain with sericitization, silicification, ferritization, etc., but the feldspar type is undistinguishable; the quartz shows subhedral-anhedral grain (Fig. 1.6h), wavy or bound extinction and some corrosive action. While the rock matrix consists of felsic and neogenic minerals; the felsic components appear as felsitic-microcrystal grains with <0.15 mm in size, obvious orientation; and partial quartz shows discontinuous linear and lenticular aggregations.

Totally 16 measuring points are obtained on sample T22-6, all the points are fallen on or around the concordant line (Fig. 1.7, left-lower) with a $^{206}Pb/^{207}Pb$ age 1140 ± 5 Ma and a MSWD value 0.99, representing the petrogenetic age of rhyolite.

(4) Sample T22-7 (coordinate of sampling site: N28°15.47′, E117°24.36′):

Metrhyolite consists of felsic and new minerals. The felsic shows mainly felsitic-microcrystal grains with <0.15 mm in size and obvious orientation (Fig. 1.7, right-lower); partial quartz shows discontinuous linear and lenticular aggregations; while the neogenic mineral is flaky sericite with <0.1 mm in size, head-to-tail connecting orientation and linear aggregation.

There are 15 measuring points fallen on or around the concordant line (Fig. 1.7, right-lower) with a $^{206}Pb/^{207}Pb$ age 1143.0 ± 8.5 Ma and a MSWD value 0.43, representing the petrogenetic age of metarhyolite.

Tectonically, the junctions among Anhui, Zhejiang, Jiangxi and Fujian Provinces belong to the orogenic belt of Meso-Neoproterozoic arc-continent collision type with well-developed Neoproterozoic volcanic-sedimentary rock series. Due to the location in between PJF and JSF (Fig. 1.5), each medium mass has individual name as follows (from west to east): the Wanyuan Terrain, Zhoutan Terrain, Tieshajie Formation, Tianli Schist and Chencai Group. However, both chronostratigraphic position and relationship are still uncertain, and their tectonic significance has always been the controversial focus for Chinese and international geologists. The prominent issue has involved the stratigraphic correlation of Meso-Neoproterozoic volcanic-sedimentary rock series, including the correlation of contemporaneous heteropic facies, even extending to the major lithology of the Jiangnan Orogenic Belt (Tang et al. 1998; Gao et al. 2012a, b, c, d, e). Owing to the traditional cognition based on the early chronostratigraphic methodology, late metamorphism, geochemical index chart (Hu et al. 2011) and various interpretation of metallogenic tectonic background (Yang et al. 2009), the stratigraphic dating of Neoproterozoic volcanic-sedimentary rock series always has multiplicity at the junctions among Zhejiang, Anhui and Jiangxi Provinces, and resulting in a huge impact on tectonic interpretation on Yangtze Craton and Cathaysia Block as well as the discussion on the JSF (Shui et al. 1986; Xing et al. 1992; Li and Mu 1999; Hu 2001; Deng et al. 2005; Wan et al. 2007; Li et al. 2007; Xu et al. 2007; Hu and Deng 2009; Xue et al. 2010; Shu et al. 2011; Yang et al. 1999, 2012).

Currently, the stratigraphic correlation of tectonic belts mainly relies on the precise stratigraphic dating. As for Jiangnan Orogenic Belt, a series of stratigraphic dating data, i.e., the bentonitic zircon U–Pb age 831–829 Ma in the Shuangqiaoshan Formation (Gao et al. 2008a), the metarhyolitic zircon age 860 ± 3 Ma in Zhangcun and Dexing Formations (Liu et al. 2012), the andesitic zircon age 905–878 Ma in the Shuangxiwu Group, and the metarhyolitic zircon SHRIMP U–Pb ages 1132 ± 8 Ma, 1140 ± 7 Ma, 1143 ± 9 Ma and 1172 ± 10 Ma in the Tieshajie Formation, have shown a continuous evolutionary relationship in time and space (Table 1.5 and Fig. 1.8). According to the geochemical feature and petrogenesis of epimetamorphic basalt assemblage, Liu et al. (2012) take this rock assemblage as the arc-back basin tectonic setting of the Early Neoproterozoic Andean active continental margin, which would be favorable to new tectonic interpretation for Jiangnan Orogenic Belt. The Tieshajie Formation is clearly positioned to the Late Mesoproterozoic block, and also the oldest volcanic-sedimentary entity either on the southern margin of the Jiangnan Orogenic Belt or within the JSF. The precise dating of Tieshajie Formation is stratigraphically

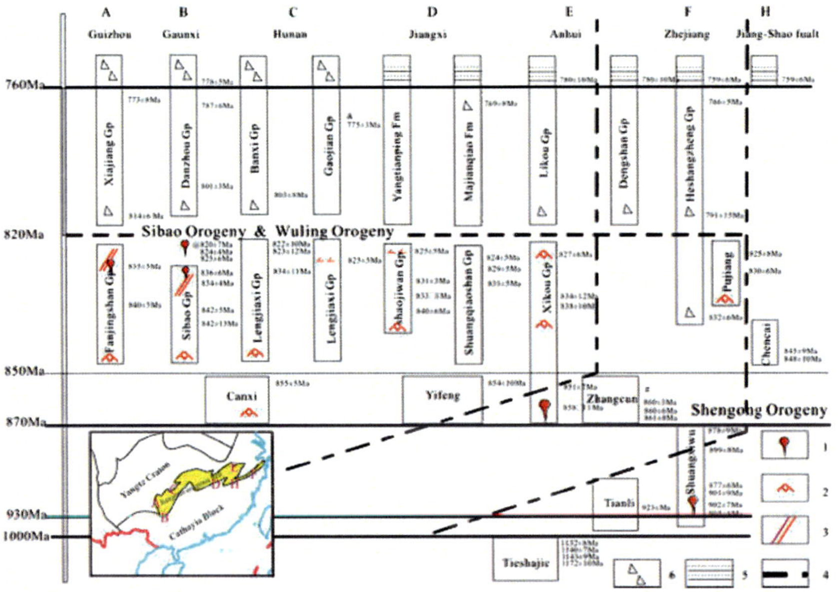

Fig. 1.8 Meso-neoproterozoic stratigraphic correlation (Gao et al. 2014c). *1: Liu et al. 2012; Ma et al. 2013. *2: Yang et al. 2015. Others: Gao et al. 2008a, b, c, 2009a, b, c, 2010a, b, c, d, e, 2011a, b, c, d, e, 2012a, b, c, d, e, 2013a, b, c, d, e. Legend: 1. granite; 2. basalt; 3. diabase dyke; 4. Wuling Orogeny (820 Ma); 5. sandy slate; 6. diamictite or conglomerate

significant for the southern boundary determination of Jiangnan Orogenic Belt and for the searching coeval Tieshajie type of copper metallogenic belt.

1.5.5 The Definition of Tectonic Events in the YC and JOB

The Chinese nomenclature of "tectonic movement" generally means orogeny or epeirogeny. The stratigraphic hiatus shown by orogenic episode would be in different scale, i.e., the geological time interval between the underlying and the overlying strata would be variable. The present authors try to summarize the definition of orogenies in YC and JOB in consideration of new zircon U–Pb age as the constraint factor.

(1) **Sibao Orogeny** (or Sipu Orogeny) named by Li (cf. Yin et al. 1963). It appears as an unconformity between the Banxi Group (or Danzhou Group) and its underlying Sibao Group in Luocheng, northwestern Guangxi. Since the Middle to Upper Sibao Group is roughly equivalent to the Changchengian and Jixianian, while the Banxi Group can be correlated with Qingbaikouan, the Sibao Orogeny would represent the tectonic event and the unconformity between the Jixianian and Qingbaikouan in south China, which occurs at ca. 1.0 Ga ago. Recently, the present authors gained the zircon SHRIMP U–Pb ages 842 ± 13 Ma and 834

± 4 Ma respectively from the tuffaceous sandstone of the Yuxi Formation and from the granodiorite intruding into the Sibao Group, both of which can limit the maximum top age of the Sibao Group. Meanwhile, a SHRIMP zircon U–Pb age 801 ± 3 Ma was gained from the Hetong Formation within the Danzhou Group (Gao et al. 2013e) which is limited the Sibao Orogeny as a tectonic event in between 834 ± 4 Ma and 814 ± 6 Ma.

(2) **Wuling Orogeny** named by the 423 team of Hunan Geological Bureau in 1959. It is attributed to the Early Neoproterozoic folding movement and an unconformity between Lower Guanzhuang Formation of the Banxi and Lengjiaxi Groups (original Lower Banxi Group) with an age of 1.1–0.9 Ga, which is equivalent to the Sibao Orogeny (Gao et al. 2011a). Two zircon SHRIMP U–Pb ages 822 ± 10 Ma and 803 ± 8 Ma were gained respectively from the top of Lengjiaxi Group and its overlying Zhangjiawan Formation, both indicate the Wuling Orogeny as a tectonic event in between 822 ± 10 Ma and 803 ± 8 Ma (Fig. 1.9).

(3) **Shengong Orogeny** named by Ma and Zhang (1977). Based on a remarkable unconformity between the Luojiamen Formation and its underlying Shuangxiwu Group at Fuyang City (Zhejiang) on the north of the JSF. As a folding movement, it happened at ca. 1.0 Ga ago. The Shengong Orogeny leaded the JSF to develop towards foreland basin as well as to deposit the molasse formation of Luojiamen Formation, volcanic rocks of the Hongchicun and Shangshu Formations. The present authors have measured the youngest zircon U–Pb ages 878 ± 9 Ma and 791 ± 15 Ma respectively from the top volcanic rocks of the Zhangcun Formation. Within the Shuangxiwu Group and from the tuff of the Luojiamen

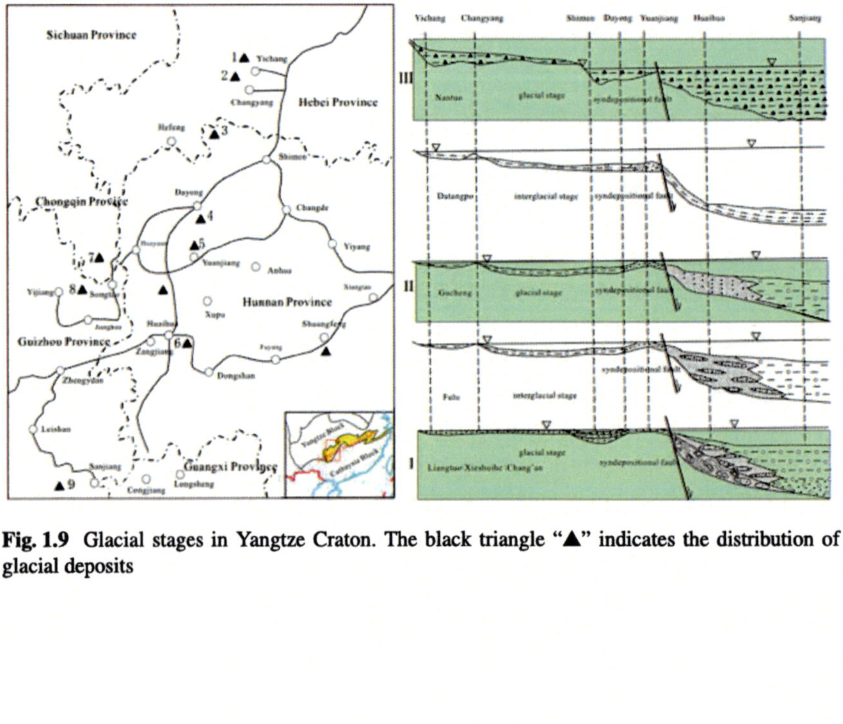

Fig. 1.9 Glacial stages in Yangtze Craton. The black triangle "▲" indicates the distribution of glacial deposits

Formation. Therefore, The Shengong Orogeny should be the known longest tectonic movement or early orogeny in the Jiangnan Orogenic Belt (Fig. 1.9).

(4) **Jinning Orogeny** (or Jinning Movement, Tsinning Movement) named by Misch (1942). As a strong folding movement during Middle Neoproterozoic at Jinning and Yuxi in middle to eastern Yunnan. It is associated remarkable unconformity between Nanhuan Chengjiang Formation and underlying Mesoproterozoic Kunyang Group happened around 800 Ma ago, which is widespread in south China, and resulting in the strongly plicated dislocation of Kunyang Group and the postorogenic molasse formation of Chengjiang Luojiamen Formation. Jinning Orogeny should be equivalent to Prehengjiang, Wannan, Xiuning and Xiufeng Orogenys, which have wide-ranging effect in south China.

(5) **Wannan Orogeny** named by Lee and Hsu (1947), which is a Neoproterozoic tectonic movement in southern Anhui. It is distinguished by an unconformity between the sandstone of Nanhuan Gaoting Formation (late it is called Xiuning Formation) and underlying Lower Likou Group.

(6) **Xuefeng Orogeny** named by Tian (1948). It is a folding movement between Nanhuan and Qingbaikouan, and appearing as an unconformity between the Banxi Group and the diamictitic layer of Nantuo Formation in western Hunan, which occurred about 800 Ma ago.

(7) **Xiuning Orogeny** named by the Geological Department of the Nanjing University (1958). It is a tectonic movement represented by the unconformity between the basal sandstone of Nanhuan Xiuning Formation and its underlying Lower Qingbaikouan Likou Group which is equivalent to the Xuefeng Orogeny.

(8) **Chengjiang Orogeny** is a folding movement within Nanhuan, which is established based on a slight unconformity between the diamictitic layer of Nantuo Formation and its underlying sandstone of Chengjiang Formation, which occurred ca. 750 Ma ago and after the appearance of postorogenic molasse formation during the Jinning Orogeny.

(9) **Pre-Chengjiang Movement** named by Meng and Zhang (1948) based on the remarkable unconformity between the diamictite of Nanhuan Nantuo Formation and its underlying epimetamorphic rocks of Mesoproterozoic Kunyang Group with strong fold and thrust, which occurred at 7.5 Ma ago and also belongs to Early Xingkai (Salair) Orogeny.

1.6 Nanhuan in Yangtze Craton (YC)

The time span of Neoproterozoic Cryogenian is just within the interval of 850–635 Ma on the International Geological Time Scale, and three Cryogenian glacial stages of "Snow Ball" age occurred basically during 720–635 Ma, which is roughly equivalent to the Nanhuan Period (780–635 Ma) in China. At present, the top age of Cryogenian is assigned to 635 Ma, i.e., the bottom age of the Ediacaran as GSSP. Currently, however, the controversial focuses about the GSSP are as follows: ① Get it at the emergence of the earliest diamictitic layer; ② Assign the boundary at the sediments of cold event; ③ Take the most extensive global diamictites as the boundary. Since the Kaigas glacial stage (ca. later than 770 Ma) in South Africa is globally not the most extensive developing one, it would be unsuccessful in the competition. Nevertheless, some geologists proposed that the bottom line (GSSP) of Cryogenian should be assigned at the bottom boundary of the globally most widely developed Sturtian glaciation event at 755 Ma ago in Australia, while other geologists suggest that the Cryogenian GSSP is at the sedimentary age 780 Ma of cold event.

1.6.1 Neoproterozoic Glacial Stage in the YC

The Working Group of International Cryogenian suggests that the boundary of Cryogenian should take following basic features into consideration. ① The existence of diamictitic layer or glacial stage; ② the variational curves of C/O isotopic ratios; ③ chemical index of alteration (CIA); ④ chronostratigraphy U–Pb dating. So far only one completely stratigraphic section contains four stages of Neoproterozoic diamictitic layers and multiperiodic volcanic rocks at Kugluktuk region in Tarim Block, northwestern China, the chronostratigraphic results of which would directly influence the global Cryogenian stratigraphic division and correlation (Gao et al. 2013b, c).

According to the field observed characteristics, there are three Neoproterozoic glacial stages and associated interglacial stages in YC, i.e., Chang'an glacial stage, Fulu interglacial stage, Gucheng glacial stage, Datanpo interglacial stage, Nantuo glacial stage and its overlying interglacial stage in ascending order (Gao et al. 2013b, c; Table 1.5). Due to the vegetational cover of grass *artemisia*, however, the outcrop distribution of diamictites is not so clear in South China, some geologists used to consider the "iron-bearing formation" of Fulu Formation and the "manganese-bearing formation" of the Datanpo Formation to be the same depositional system as one interglacial stage, so that only Chang'an and Nantuo glacial stages remained in South China as compared with the Sturtian and Marinoan glacial stages in Australia. Nevertheless, the present authors and colleagues of the Hunan Geological Survey have first found a continuous stratigraphic section from Fulu, though Gucheng and Datangpo, to Nantuo Formations at the Fenghuangcheng, Hunan Provinces in 2011. Combining the continuous stratigraphic sections of Chang'an to Fulu Formations in Guizhou and Guangxi Provinces, the basic features of the three glacial stages in South China have been confirmed (Fig. 1.9).

1.6.2 Starting Time of the Bottom-Most Glacial Stage in the Nanhuan

The initial emergence of diamictitic layer in Chang'an Formation is usually considered to be the indicator for the starting time of the earliest Nanhuan glacial stage in South China, about which, however, there still are two arguments as follows: ① Whether the Chang'an Formation could be correlated with the Liantuo Formation? ② If the age of Nanhuan could be constrained by the basal tuffaceous zircon age of Chang'an Formation, or by the top tuffaceous zircon age of the Banxi Group?

Since the Yangtze Gorges is the stratigraphic development area of Sinian reference section, and the original "Sinian" has been further subdivided into Nanhuan and Sinian by the Commission on Stratigraphy of China (Lu 2002), so that the Liantuo and Nantuo Formations became the Nanhuan stratotype section. Based on the stratotype section, consequently, two different Precambrian stratigraphic correlation schemes have been successively proposed in China. As a sandstone interval, the Liantuo Formation used always to be correlated with the large set of sandstone in the Banxi Group. Therefore, the Banxi isotopic age was often confused with the Liantuo age, thus the Liantuo Formation was attributed to pre-glacial stage. Nevertheless, some geologists in Hunan, Guangxi and Guizhou geological survey preferred to correlate the Liantuo sandstone with the sandstone in Nanhuan Fulu Formation rather than that in Benxi Group. In this case, the bottom age of Nanhuan Chang'an Formation would be one of objects for Chinese and international geologists to explore the starting time of the global Nanhuan/Cryogenian glacial stages. So far many chronostratigraphic dating data are available, but the bottom age of Chang'an Formation still is the target to be looking for. In this case, the stratotype section with continuous stratigraphic boundaries should contain more reliable isotopic ages, thus Gao et al. (2013b, c) published a series of tuffaceous zircon SHRIMP U–Pb ages from the continuous section of Chang'an Formation at Luocheng, Guangxi Province: ① First time measured two ages 842 $\pm$ 13 Ma from the tuff of Yuxi Formation in Sibao Group and 834 $\pm$ 4 Ma from igneous rock intruding into Sibao Group, respectively; ② Tow ages 801 $\pm$ 3 Ma and 786 $\pm$ 6 Ma respectively from the overlying tuffs of Hetong and Gongdong Formations; ③ Finally, two ages of 778 $\pm$ 5 Ma and 787 $\pm$ 6 Ma respectively for the bottom of Chang'an Formation and for the Datanpo Formation (Table 1.5). Therefore, the above dating can basically restrict the bottom age of the Chang'an Formation earlier than 778 $\pm$ 5 Ma. (Table 1.5), it is more reasonable to assign 780 Ma as the Nanhuan bottom age in YC.

1.7 Neoproterozoic Chronostratigraphy in Tarim Block

Besides the NCC and YC, the Precambrian strata are also developed and preserved in the Tarim Block, which is geographically the current Tarim Basin in the south of Xinjiang Uygur Autonomous Region, northwestern China (Fig. 1.4), where

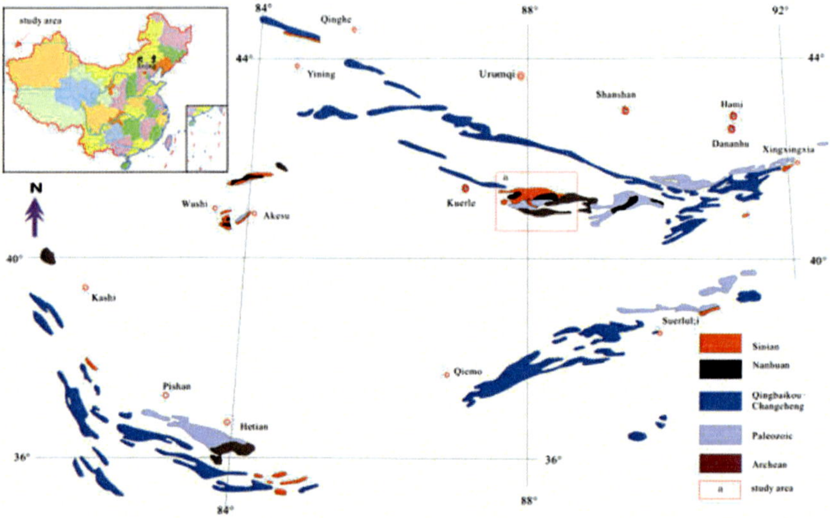

Fig. 1.10 Precambrian stratigraphic distribution in Tarim block

the unmetamorphic Neoproterozoic strata have principally outcropped along the surrounding marginal mountains, especially at the Kuluktag region. The Kuluktag is situated at the northeastern Tarim Block, its Neoproterozoic are mainly composed of the Nanhuan and Sinian sequences (Fig. 1.10).

1.7.1 Neoproterozoic Stratigraphic Frame in Kuluktag

1.7.1.1 Chronostratigraphic Division

Kuluktag region is located in the eastern Tarim Block. Based on the Kuluktag geological survey during 1928–1932, Norm (1935) had initially established the late Precambrian Kuluktag stratigraphic section, and subdivided it into four "series", i.e., Beiyixi, Alatonggou, Teruiaiken and Yukengou (in ascending order). Thereafter, Gao et al. (1980), Gao and Zhu (1984) further reported a perfect Neoproterozoic glacial sequence, and then Cao (1991) named the upper member of Alatonggou Formation as the Huangyanggou Formation at the Kuluktag section. Consequently, the Neoproterozoic glacial sequence involves two systems (i.e., Nanhuan and Sinian) and nine formations at the same stratigraphic section at the Kuluktag region (Fig. 1.11). Based on four Diamictitic layers, four glacial stages plus associated interglacial stages are recognized at the Kuluktag section, and the glacial stages are named in turn, I. Beiyixi, II. Alatonggou, III. Teruiaiken and IV. Hangerqiaok in ascending order. Finally, a systematic division of glacial stages and associated interglacial stages is established on the Neoproterozoic column in Chinese Geological Time Scale, the first three glacial stages are respectively referred to the Nanhuan Beiyixi, Alatonggou and Teruiaiken Formations in turn, and the fourth to Sinian Hangerqiaok Formation (Fig. 1.11; Gao et al. 2010d).

Since then, a new round of global glacial geology research has involved multiple disciplinary directions, e.g., the zircon U–Pb dating (Xu et al. 2005, 2008, 2009; Yin et al. 2007; Gao et al. 2010d, 2013c), stratigraphic C/O ratio (Xu et al. 2003; Xiao et al. 2004) and CIA (Liu et al. 2007) for the Kuluktag stratigraphic section, Tarim

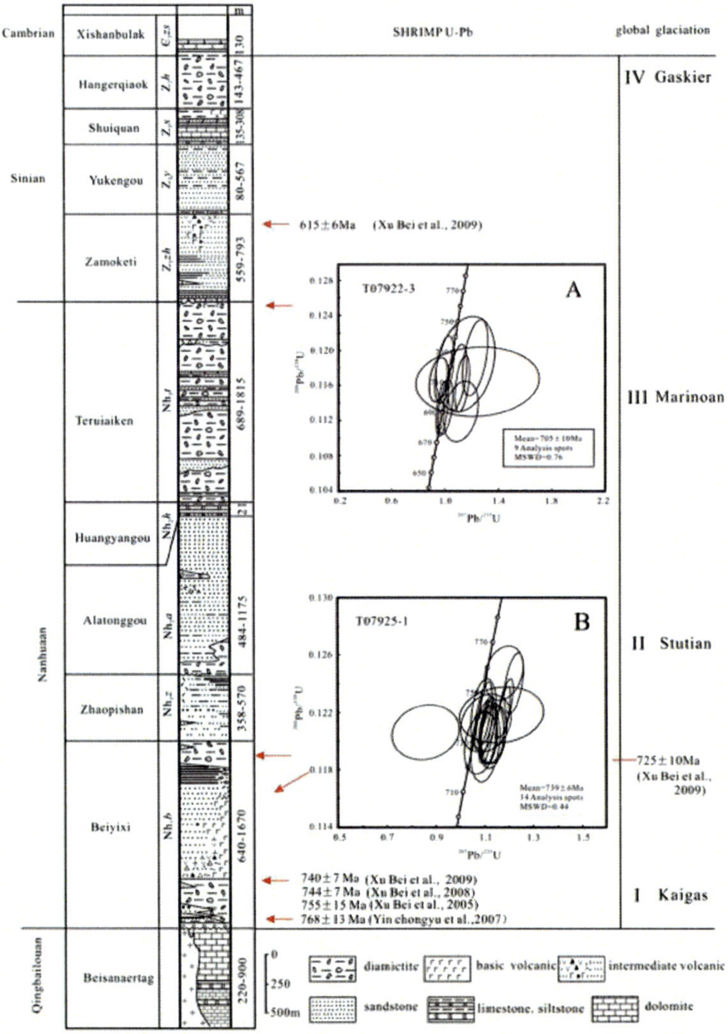

Fig. 1.11 The Nanhuan-Sinian stratigraphic column and zircon dating in Kuluktag, Tarim block (Gao et al. 1980; Cao 1991)

Block, which have an impact on global geology (Xiao et al. 2004; Hoffman and Li 2009). One basic lava zircon SHRIMP U–Pb age 739 ± 6 Ma of Beiyixi Formation and another lava pillow zircon SHRIMP U–Pb age 705 ± 10 Ma of Teruiaiken Formation in Kuluktag area (Fig. 1.11; Gao et al. 2010d) are obtained.

1.7.1.2 Neoproterozoic Kuluktag Glacial Stages

As a complete Neoproterozoic glacial sequence, the Kuluktag Nanhuan and Sinian glacial stages/diamictitic layers could be correlated to the known Chinese and global glacial stages/diamictitic layers as follows (Fig. 1.11).

(1) The I, Beiyixi glacial stage (768 ± 10–740 ± 7 Ma; Xu et al. 2005, 2008, 2009; Yin et al. 2007) may be equivalent to the Nanhuan Chang'an Formation (778 ± 10 Ma) in YC, the Kaigas glaciation in Kalahari Craton, Namibia (780 ± 10–741 ± 6 Ma; Allsopp et al. 1979; Frimmel et al. 1996), Jequitai and Windemere diamictites in Sao Franciscon Craton and Negash diamictite at Ethiopa in Brazil (750–613 Ma; MacGabhann 2005).
(2) The II, Alatonggou glacial stage (<725 ± 10 Ma; Xu et al. 2009) is the same as the Gucheng diamictite in YC and Sturtian diamictite in Australia and Gubrah diamictite in Oman (Allen et al. 2002).
(3) The III, Teruiaiken glacial stage (<705 ± 10 Ma) is the same as the Nantuo tillite in YC and Varanger glacial stage in Baltic, Norway.
(4) The IV, Hangerqiaok glacial stage (580–570 Ma) is correlated with the Luoquan tillite in NCC and Gaskiers diamictite in Baltica, Canadia (ca. 580 Ma; Guerrot and Peucat 1990).

Globally, the above-mentioned four Neoproterozoic glacial stages at one stratigraphic section are never reported, which were separately found at different places, even in different continents, which are respectively named Kaigas (in Africa), Sturtian and Marinoan (in Australia) and Gaskiers (in North American; MacGabhann 2005), among which, the first three global Cryogenian glacial stages (750–635 Ma; Hoffman and Li 2009) are basically comparable with the Chinese Nanhuan glacial stages (780–635 Ma), but their stratigraphic bottom ages are different.

Some geologists tried to assign Cryogenian bottom GSSP to 720 Ma, while the bottom age of Nanhuan is just at 780 Ma. However, the Gaskier glacial stage/diamictitic layer should be equivalent to the Hangerqiaok diamictitic layer to the Hangerqiaok diamictite (580–570 Ma) in Tarim Block (Xiao et al. 2004; Fig. 1.11) and the Luoquan tillite in NCC (Guan et al. 1986; Wu and Guan 1988; Le Heron et al. 2018, 2019). So far Kuluktag region contains solely complete Neoproterozoic section with four glacial stages/diamictitic layers in the globe. Its chronostratigraphic study will affect global Cryogenian division and correlation. So far the global geological events have been concerned dealing with chronostratigraphy (Yin et al. 2003; Hoffman et al. 2004; Zhou et al. 2004; Yin 2005; Zhang et al. 2008; Gao et al. 2013b, c, d), stratigraphic C/O ratio (Burns and Matter 1993; Derry et al. 1994; Kaufman et al. 1997; Kennedy and Christie-Blieck 2001; Chu et al. 2004; Jiang et al. 2003; Xu et al. 2003; Carlos et al. 2004; Xiao et al. 2004), lithochemistry (e.g., CIA; Wang et al. 2002, 2009; Feng et al. 2003) and biotic evolutions (e.g., Weng'an fauna, Miaohe fauna and Edicarian fauna; Knoll 2000; Bowing et al. 2003; Chen et al. 2004; Yin et al. 2007, 2009; Zhou et al. 2007).

The Kuluktag glacial sequence would be very significant for global Neoproterozoic glacial stage correlation (Gao and Chen 2003) as well as for the study of Neoproterozoic sedimentology, glaciology and chronostratigraphy.

1.8 Meso-Neoproterozoic Stratigraphic Frame in China

The Chinese Meso-Neoproterozoic sequences are distributed in three sedimentary basins, i.e., the North China Craton (NCC), Yangtze Craton (YC, incl. the late-epimetamorphogenic Jiangnan Orogenic Belt) in South China, and Tarim Block (TB) in northwest China (cf. the Fig. 11.22 in Chap. 11). In view of the Chronostratigraphic distribution, Mesoproterozoic Changchengian (Pt_2^1), Jixianian (Pt_2^2) and Xiamaling (Pt_2^3) (and equivalent strata) are predominantly developed on northern NCC and southeastern YC, while TB only contains Jixianian (Pt_2^2). Whereas "Yuxian" (Pt_2^4) is mainly deposited on YC (Gao et al. 2015a, b, 2018) and southern NCC (Table 1.6). In addition, Neoproterozoic Qingbaikouan (Pt_3^1) is relatively widespread in all the three basins, but the Nanhuan (Pt_3^2) and Sinian (Pt_3^3) are only distributed in YC, TB and northeastern NCC (Table 1.6). In these cases, however, the starting age (or bottom age), stratigraphic duration and developmental degree of Meso-Neoproterozoic sedimentary sequences are discriminatory.

In NCC, the initial Changchengian (Pt_2^1) strata are referred to Xiong'er Group. Based on a series of measured age of igneous intrusions into the metasedimentary strata of Xiong'er Group (Table 1.7; the references therein), the starting age of the Mesoproterozoic Changchengian (Pt_2^1) strata would be assigned to be 1.8 Ga (Table 1.6; Zhang et al. 2019), which is different from the bottom age 1.6 Ga in the International Stratigraphic Chart (Cohen et al. 2013; Gao et al. 2008b, 2009a, 2010c). While another stratigraphically continuous Mesoproterozoic section is developed in the Yanliao Faulted-Depression Zone (YFDZ) of the northern NCC, where the well-exposed, simple-structured and unmetamorphosed Mesoproterozoic strata with clear top and bottom boundaries (Kao et al. 1934) are the Changchengian (Pt_2^1)-Jixianian (Pt_2^2)-Xiamaling Formation (Pt_2^3x) sequences, and contain a stratigraphic duration of 1670 Ma (e.g., Gao et al. 2008a; Gao et al. 2008a, b, c, d; Li et al. 2011; He et al. 2011a, b) to 1320 Ma (Li et al. 2009a, b; cf. Chap. 10), and total stratigraphic thickness up to 9030 m (cf. Chaps. 2 and 3 and Table 11.12 in Chap. 11). Even though the Changchengian starting age 1670 Ma in YFDZ of northern NCC is later than that (1800 Ma) of Xiong'er Group in southern NCC, the Jixianian (Pt_2^2) Gaoyuzhuang Formation appears as a continuous deposition with underlying Changchengian (Pt_2^1) Dahongyu Formation without hiatus at the depocenter of YFDZ. Moreover, there are no Nanhuan (Pt_3^2) and Sinian (Pt_3^3) sequences deposited in all the NCC except

Table 1.6 The Meso-neoproterozoic chronostratigraphic frame in China

Tectonic unit	Yangtze Craton(YC) plus Jiangnan Orogenic Belt(JOB)		Tarim Block (TB)	North China Craton (NCC)			
	SW.YC	W.YC	NE.TB	SE.NCC	S.NCC	N.NCC	NE.NCC
Lower Cambrian($Є_1$)	Qiongzhusi/Jiulaodong Fm Zhujiaqingi/Midiping Fm		Xishanbulak Fm	Xingji Fm Shangzhangwan Fm	Dongpo Fm	Xiaweidian Fm	Jiacheng Fm Jingxian Gp
541 Ma							
Sinian(Pt_3^3)	Dengying Fm 550 Ma Wangjiawan Fm/	Doushantuo Fm 630±12 Ma	Hankalchough Fm▲▲▲▲ Shuiquan Fm Yukenggou Fm Zamoketi Fm	Louquan Fm▲▲▲▲			
635 Ma							
Nanhuan(Pt_3^2)	Nantuo Fm▲▲▲ Datangpo Fm Gucheng Fm▲▲ Fulu Fm 748±12 Ma Liantuo/Chang'an/Chengjiang Fm▲		Tereeken Fm▲▲▲ Huangyangou Fm Alatungol Fm▲▲ Zhaobishan Fm Baysi Fm▲ 755±15 Ma				Wuhangshan GP
780 Ma							
Qingbaikouan(Pt_3^1)	820Ma Lubiao Fm Huajiaqing Fm Liubatang Fm	Huangling granite/Banxi/Xiajiang/Danzhou Gp Lengjiaxi/Fangjingshan/Sipu Gp Shuangxiwu Gp	Tarim Movement Paergangtag Gp 912±28 Ma		Dongjia Fm Huangliandou Fm	Jinger'yu Fm Luotouling Fm	Xihe Gp Yongning Gp

Tectonic unit	Yangtze Craton(YC) plus Jiangnan Orogenic Belt(JOB)		Tarim Block (TB)	North China Craton (NCC)			
	SW.YC	W.YC	NE.TB	SE.NCC	S.NCC	N.NCC	NE.NCC
1000 Ma							
'Yuxian'(Pt_2^4)	'Yuxian' Chahe Fm Dalongkou Fm Fuliangpeng Fm Heishantuo Fm Chejiacheng Fm Reshuitang Fm	Upper Shennongjia Gp/Tieshanjie Fm*			Huejiazhai Fm Luotupan Fm		
1200 Ma							
Pt_2^3		Lower Shennonjia Gp			Puyu Fm	1320 Ma Xiamaling Fm 1368 Ma	
1400 Ma							
Jixianian(Pt_2^2)	Luzhijiang Fm Heishan Fm 1499.8±3.8Ma		Aler'jigan Gp	Shibeigou Fm Fengjiawang Fm Duguan Fm Xujiansi Fm Longjiayuan Fm		Tieling Fm Hongshuizhuang Fm Wumishan Fm 1483Ma/1487Ma Yangzhuang Fm Gaoyuzhuang Fm 1560Ma/1577Ma	
1600 Ma							
Changchengian (Pt_2^1)	Luoxue Fm Yinmin Fm 1742±13Ma			Gaoshanhe Gp Xiong'er Gp 1751Ma/1800Ma	Luoyu Gp Ruyang Gp	Dahongyu Fm Tuanshanzi Fm Chuanlinggou Fm Changzhougou Fm 1670Ma	
1800 Ma							
Crystalline Basement	Shipingyan Complex (Pt_1)	Shuiyuesi Gp (Pt_1)	Yangjibulak Gp (Pt_1)	Taihua Gp (Ar)		Qianxi Gp (Ar)	

Glacial stages ▲. Chang'an/Beiyixi glacial stage; ▲▲. Gucheng/Alatonggou (Stutian) glacial stage; ▲▲▲. Natuo/Teruiaiken (Marinoer) glacial stage; ▲▲▲▲. Luoquan/Hangerqiao glacial stage

Note S. NCC the southern North China Craton, especially for Luonan in Shaanxi Province; SE NCC the southeastern North China Craton, especially for Lushan in Henan Province; N NCC the northern North China Craton, especially for the Yanliao Faulted-Depression Zone (YFDZ) in Yanshan Mountain; NE NCC the Liaodong Peninsula in eastern Liaoning Province; NE. TB the eastern Tarim Block in the eastern Uygur Autonomous Region of Xinjiang; W. YC the western Yangtze Craton, especially for the current Sichuan Basin, and * only in JOB; SW. YC the southwestern Yangtze Craton especially for the Yunnan and Guangxi Provinces

for the Wuhangshan and Jinxian Groups. The big problem is no tillite and full of indeterminate magefossils and microplants in both groups.

In YC, Mesoproterozoic sequences are only developed in southwestern YC, including Changchengian (Pt_2^1) Yinmin Formation (1742 ± 13 Ma; 1667 ± 13 Ma) and Luoxue Formation, Jixianian (Pt_2^2) Heishen Formation (1499.8 ± 3 Ma) and Luzhijiang Formation, and "Yuxian" (Pt_2^4) Reshuitang, Chejiacheng, Heishan, Fuliangpeng (1032 ± 19 Ma; 1043 ± 7 Ma), Dalongkou and Chahe Formations with a large Pt_3^4 stratigraphic hiatus of 1200 Ma to 1400 Ma and a duration of

Table 1.7 The measured stratigraphic age of igneous intrusions of Changchengian ($Pt_2{}^1$) Xiong'er Group in the southern NCC

No	Formation (Group)	Intrusive body	Location	Age/Ma	References
1	Jidanping formation	Rhyolite, porphyry	Waifangshan	1751 ± 14	He et al., (2009)
2	Xiong'er group	Rhyolite, porphyry	Ruyang	1763 ± 15	Wang et al., (2010)
3	Majiahe formation	Rhyolite	Lushan	1776 ± 20	Zhao et al., (2004)
4	Jidanping formation	Vocanic rock	Xiaoshan	1778 ± 5.5	He et al., (2009)
5	Majiahe formation	Vocanic rock	Xiaoshan	1778 ± 6.5	He et al., (2009)
6	Majiahe formation	Diorite	Shizhaigou	1780 ± 11	Cui et al., (2013)
7	Xushan formation	Vocanic rock	Xiaoshan	1783 ± 20	He et al., (2009)
8	Majiahe formation	A-Granite	Luoning	1786 ± 7	Cui et al., (2013)
9	Majiahe Formation	Diorite	Shizhaigou	1789 ± 3.5	Cui et al. 2013
10	Xushen formation	Diorite	Xiong'ershan	1789 ± 26	Zhao et al., (2004)
11	Jidanping formation	Rhyolite, porphyry	Lushan	1800 ± 16	Zhao et al., (2004)

780–635 Ma, but undeveloped in western YC. However, Neoproterozoic sequences are well developed in YC, especially for Nanhuan ($Pt_3{}^2$) and Sinian ($Pt_3{}^3$). The Nanhuan ($Pt_3{}^2$) consists of three glacial stages and two interglacial stages, i.e., Liantuo/Chang'an/Chengjiang glacial stage (748 ± 12 Ma; 778 ± 5 Ma)-Fulu interglacial stage-Gucheng glacial stage (669 ± 13 Ma)-Datangpo interglacial stage-Natuo glacial stage (667 ± 10 Ma; 663 ± 4 Ma) in ascending order. While the Sinian ($Pt_3{}^3$) is composed of Doushantuo Formation ($Pt_3{}^3ds$) and Dengyin Formation ($Pt_3{}^3dy$) with a duration of 635–541 Ma.

As the Meso-Neoproterozoic sequences in TB, only the Aler'jigan Group ($Pt_2{}^2a$) and Paergangtag Group ($Pt_3{}^1p$; 912 ± 28 Ma) are developed in TB. Whereas the Nanhuan and Sinian ($Pt_3{}^{2-3}$) are well developed four glacial stages, i.e., Nanhuan ($Pt_3{}^2$) Baysi (755 ± 15 Ma; 740 ± 7 Ma; 725 ± 10 Ma; 739 ± 6 Ma), Alatungol (633 ± 23 Ma; < 725 ± 10 Ma), Tereeken (615 ± 6 Ma; 705 ± 10 Ma) and Sinian ($Pt_3{}^3$) Hankalchough glacial stages with a total duration from 615 to 541 Ma.

Acknowledgements This study was supported by the Ministry of Science and Technology (MST) of the People's Republic of China (Grant No. 2015FY310100), and the China Geological Survey (CGS) and IGMA 5000 (Grant No. 12120111200131). The authors would like to thank Professor Liu Dunyi for their help in the SHRIMP laboratory.

References

Allen PA, Bowring S, Leather J, Brasier M, Cozzi A, Grotzinger JP, McCarron G, Amthor JJ (2002) Chronology of neoproterozoic glaciations: new insights from Oman. In: The 16th international sedimentological congress. Abstract vol. 7–8

Allsopp H, Kostlin EO, Welke HJ, Burger AJ (1979) Rb-Sr and U-Pb geochronology of late Precambrian-early palaeozoic igneous activity in the Richtersveld (South Africa) and southern South West Africa. Trans Geol Soc South Afr 82:185–204

Bowing SA, Myrow P, Landing E, Grotzinger J (2003) Geochronological contains on terminal neoproterozoic events and the rise of metazoans. Geophys Res Abstr 5(13):219

Burns SJ, Matter A (1993) Carbon isotopic record of latest proterozoic from Oman. Echlogac Geol Hlvetiae 86:595–607

Cao RG (1991) New observations of the Sinian system in the southern Yardang Mountains, Xinjiang. Reg Geol China 1:30–40 (in Chinese with English abstract)

Carlos JS, de Alvarenga, Roberto V Santos (2004) C-O-Sr isotopic stratigraphy of cap carbonates overlying Marinoan-age glacial diamictites in the Paraguay Belt, Brazil. Precambr Res 131:1–21

Chen GD (1956) Example of "activizing region" in the Chinese platform with special reference to the "Cathaysia" problem. Acta Geol Sin 36(3):239–272

Chen H, Dong WM, Zhou HR (1991) Isotopic dating metamelange block of the Tieshajie Formation. Geol Bull China 2:151–154 (in Chinese with English abstract)

Chen DF, Dong WQ, Zhu BQ, Chen XP (2004) Pb-Pb ages of neoproterozoic Doushantuo phosphorites in South China: constraints on early metazoan evolution and glaciation events. Precambr Res 132:123–132

Chen ZH, Xing GF, Guo KY, Dong YG, Chen R, Zeng Y, Li LM, He ZY, Zhao L (2009) Petrogenesis of keratophyres in the Pingshui group, Zhejiang: constraints from zircon U-Pb ages and Hf isotopes. Chin Sci Bull 54(5):610–617 (in Chinese with English abstract)

Chen JB, Zhang HM, Zhu SX, Zhao Z, Wang ZG (1980) Research on Sinian Suberathem of Jixian, Tianjin. In: Tianjin Institute of Geology and Mineral Resources (ed) Research on precambrian geology, Sinian Suberathem in China. Tianjin: Tianjin Science and Technology Press, pp 56–114 (in Chinese)

Chu XL, Zhang TG, Zhang QR, Feng LJ, Zhang FS (2004) Carbon isotopic variations of Proterozoic carbonates in Jixian, Tianjin, China. Sci China, Ser D Earth Sci 47(2):160–170

Cohen KM, Finney SC, Gibbard PL, Fan JX (2013) The ICS international chronostratigraphic chart. Episodes 36:199–204

Cui XS, Dong WM, Zhou HR (1996) Preliminary research of the outcrop sequence stratigraphy in Sinian and its significance in western Henan Province. China Earth Sci 21(3):249–253

Cui ML, Zhang LC, Zhang BL, Zhu M T (2013) Geochemistry of 1.78 Ga A-type granites along the southern margin of the North China Craton: implications for Xiong'er magmatism during the break-up of the supercontinent Columbia. Int Geol Rev 55(4):496–509

Deng GH, Liu CG, Feng Y (2005) Tectonic features and evolution of the proterozoic orogenic belt between northeastern Jiangxi and southern Anhui. Acta Geoscientica Sin 26(1):9–16 (in Chinese with English abstract)

Derry LA, Brasier MD, Corfield RM (1994) Sr and C isotopes in lower Cambrian carbonates from the Siberian craton: a palaeoenvironmental record during the "Cambrian explosion." Earth Planet Sci Lett 128:671–681

Feng LJ, Chu XL, Zhang QR, Zhang TG (2003) CIA (chemical index of alteration) and its applications in the Neoproterozoic clastic rocks. Earth Sci Frotiers 10(4):539–543 (in Chinese with English abstract)

Frimmel HE, Klotzli US, Siegfried PR (1996) New Pb-Pb single zircon age constraints on the timing of Neoproterozoic glaciation and continental break-up in Namibia. J Geol 104(4):459–469

Gao ZJ, Chen KQ (2003) The Nanhua system of Xinjiang and some geological issues of Nanhua system in China. Geol Surv Res 26(1):8–13 (in Chinese with English abstract)

Gao ZJ, Zhu CS (1984) Precambrian geology in Xinjiang, China. Xinjiang People's Publishing House, Urumuqi (in Chinese)
Gao LZ, Zhang YX, Wang CS, Tian SG, Peng Y, Liu YY, Dong DZ, He HX, Lei BT, Chen MG, Yang LG (1996) The preliminary study on the Meso-Neoproterozoic sequence stratigraphy in Jixian County, Tianjin. Geol Bull China 1:64–74 (in Chinese with English abstract)
Gao LZ, Yin CY, Xing YS (1999) Neoproterozoic micropalaeophyta and sequence stratigraphy. Prof Pap Stratigr Palaeontol 27:28–36 (in Chinese with English abstract)
Gao LZ, Yin CY, Wang ZQ (2002) New view of the neoproterozoic on the southern margin of the North China platform. Geol Bull China 3:131–136 (in Chinese with English abstract)
Gao LZ, Zhang CH, Shi XY, Zhou HR, Wang ZG (2007a) Zircon SHRIMP U-Pb dating of the ash bed from Xiamaling formation, Qingbaikou group in North China. Geol Bull China 26(3):249–255 (in Chinese with English abstract)
Gao LZ, Zhang CH, Shi XY, Zhou HR, Wang ZG, Song B (2007b) A new SHRIMP age of the Xiamaling Formation in the NCP and its geological significance. Acta Geol Sin (english Edition) 81(6):1103–1109
Gao LZ, Yang MG, Ding XZ, Liu YX, Liu X, Ling LH, Zhang CH (2008a) SHRIMP U-Pb zircon dating of tuff in the Shuangqiaoshan and Heshangshen Groups in South China: constraints on the evolution of the Jiangnan neoproterozoic orogenic belt. Geol Bull China 27(10):1744–1758 (in Chinese with English abstract)
Gao LZ, Zhang CH, Shi XY, Song B, Wang ZG, Liu YM (2008b) Mesoproterozoic age for Xiamaling formation in North China plate indicated by zircon SHRIMP dating. Chin Sci Bull 53(17):2665–2671
Gao LZ, Zhang CH, Yin CY, Shi XY, Wang ZG, Liu YM, Liu PJ, Tang F, Song B (2008c) SHRIMP zircon ages: basis for refining the chronostratigraphic classification of the Meso-neoproterozoic strata in North China old land. Acta Geoscientica Sin 29(3):366–376 (in Chinese with English abstract)
Gao W, Zhang CH, Gao LZ, Shi XY, Liu YM, Song B (2008d) Zircon SHRIMP U-Pb age of rapakivi granite in Miyun, Beijing, China and its tectono-stratigraphic implication. Geol Bull China 27(6):793–798 (in Chinese with English abstract)
Gao LZ, Zhang CH, Frank RE, Shi XY, Wang ZQ (2009a) The Jiangnan orogenic belt between the Yangtze and Cathaysia blocks for neoproterozoic context. Acta Geosinica Sin 30(Supp 1):10–11
Gao LZ, Zhang CH, Liu PJ, Ding XZ, Wang ZQ, Zhang YJ (2009b) Recognition of Meso-neoproterozoic stratigraphic framework in North and South China. Acta Geoscientica Sin 30(4):433–446 (in Chinese with English abstract)
Gao LZ, Zhang CH, Liu PJ, Tang F, Song B, Ding XZ (2009c) Reclassification of the Meso- and neoproterozoic chronostratigraphy of North China by SHRIMP zircon ages. Acta Geol Sin 83(6):1074–1084
Gao LZ, Dai CG, Liu YX, Wang M, Wang XH, Chen JS, Ding XZ, Zhang CH, Cao Q, Liu JH (2010a) Zircon SHRIMP U-Pb dating of tuff bed of the Sibao group and implication for its stratigraphy. Geol Bull China 29(9):1259–1267 (in Chinese with English abstract)
Gao LZ, Dai CG, Liu YX, Wang M, Wang XH, Chen JS, Ding XZ (2010b) Zircon SHRIMP U-Pb dating of tuff bed of the Xiajing group and implication for its stratigraphy. Geol China 37(4):1071–1080 (in Chinese with English abstract)
Gao LZ, Ding XZ, Cao Q, Zhang CH (2010c) New geologic time scale of late Precambrian of China and geochronology. Geol Bull China 34(4):1014–1020 (in Chinese with English abstract)
Gao LZ, Wang ZQ, Xu ZQ, Zhang W (2010d) A new zircon SHRIMP dating of the neoproterozoic glaciation events in Kuluktag area in Tarim Basin, Xinjiang. Geol Bull China 29(2–3):33–41 (in Chinese with English abstract)
Gao LZ, Zhang CH, Chen SM, Liu PJ, Ding XH, Liu YX, Dong CY, Song B (2010e) Detrital zircon SHRIMP U-Pb age from the Diaoyutai formation, Xihe group in Liaodong Peninsula, China and its geological significance. Geol Bull China 29(8):1113–1122 (in Chinese with English abstract)
Gao LZ, Chen J, Ding XZ, Liu YR, Zhang CH, Zhang H, Liu YX, Pang WH, Zhang YH (2011a) Zircon SHRIMP U-Pb dating of the tuff bed of Lengjiaxi and Banxi groups, northeastern Hunan:

constraints on the wuling movement. Geol Bull China 30(7):1001–1008 (in Chinese with English abstract)

Gao LZ, Dai CG, Ding XZ, Wang M, Liu YX, Wang XH, Chen JS (2011b) SHRIMP U-Pb dating of intrusive alaskite in the Fanjingshan group and alaskite basal conglomerates: constraints on the deposition of the Xiajiang group. Geol China 38(6):1413–1420 (in Chinese with English abstract)

Gao LZ, Ding XZ, Pang WH, Liu YX, Lu SN, Liu YR, Chen J, Zhang YH (2011c) SRHIMP zircon U-Pb dating of metamorphic tuff from the Precambrian Cangxi complex-group in northeastern Hunan. Geol Bull China 30(10):1479–1484 (in Chinese with English abstract)

Gao LZ, Ding XZ, Pang WH, Zhang CH (2011d) New geologic time scale of Meso-neoproterozoic of China and geochronologic constraint by SHRIMP zircon U-Pb dating. J Stratigr 35(1):1–7 (in Chinese with English abstract)

Gao LZ, Liu PJ, Yin CY, Zhang CH, Ding XZ, Liu YX, Song B (2011e) Some detrital Zircon SHRIMP dating of Meso-neoproterozoic in North China and implication. Acta Geol Sin (engl Ed) 85(2):801–811

Gao W, Zhang CH, Wang ZG (2011f) The discovery of large-scale acanthomorphic acritarch assemblage on the southern margin of North China old land and an analysis of its palaeogeographic environment. Geol China 38(5):1232–1243 (in Chinese with English abstract)

Gao LZ, Ding XZ, Zhang CZ, Chen J, Liu YR, Zhang H, Liu YX, Pang WH (2012a) Revised chronostratigraphic framework of the metamorphic strata in the Jiangnan orogenic belt, South China and its tectonic implications. Acta Geol Sin (engl Ed) 86(2):339–349

Gao LZ, Ding XZ, Zhang CH, Lu SN, Liu YX, Peng WH (2012b) A revised chronostratigraphic dating of metamorphosed basement strata of Jiangnan old land and its implication for Wuling tectonic movement. Resour Sur Environ 33(2):71–76 (in Chinese with English abstract)

Gao LZ, Ding XZ, Zhang CH, Wang ZQ, Chen J, Liu YR (2012c) SHRIMP dating of Cangshuipu group in the middle part of the Jiangnan Orogen and its implications for tectonic evolutions. Geol China 39(1):13–20 (in Chinese with English abstract)

Gao LZ, Huang ZZ, Ding XZ, Liu YX, Zhang CH, Wang ZQ, Pang JF, Han KY (2012d) The relationship between the Xiaoqiwa Formation and the Xingzi Complex Group, constraints on Zircon SHRIMP U-Pb dating in Northwestern Jiangxi. Acta Geoscienca Sin 34(3):295–304 (in Chinese with English abstract)

Gao LZ, Huang ZZ, Ding XZ, Liu YX, Peng JF, Zhang CH (2012e) Zircon SHRIMP U-Pb dating of Xiushui and Majianqiao formations in northwestern Jiangxi Province. Geol Bull China 32(7):1086–1093 (in Chinese with English abstract)

Gao LZ, Ding XZ, Liu YX, Huang ZZ, Zhang CH, Xu XM, Wu XL, Song ZR, Zhang H (2013a) The revision of the Chentangwu formation in neoproterozoic stratigraphic column: constraints on zircon U-Pb dating of tuff from the Mengshan section in Pujiang County, Zhejiang Province. Geol Bull China 32(7):988–995 (in Chinese with English abstract)

Gao LZ, Ding XZ, Yin CY, Zhang CH, Ettensohn FR (2013b) Qingbaikouan and Cryogenian in South China: constraints by SHRIMP zircon U-Pb dating. Acta Geol Sin 87(6):1540–1553

Gao LZ, Guo XP, Ding XZ, Zong WM, Zhang CH, Wang ZQ (2013c) Nanhuan glaciations and its correlation in Tarim block, China. Acta Geoscientica Sin 34(1):39–57 (in Chinese with English abstract)

Gao LZ, Liu YX, Ding XZ, Song ZR, Huang ZZ, Zhang CH, Zhang H, Shi ZG (2013d) Geochronographic dating of the Tieshajie Formation in the Jiang-Shao fault zone and its implications. Geol Bull China 32(7):996–1005 (in Chinese with English abstract)

Gao LZ, Lu JP, Ding XZ, Wang HR, Liu YX, Li J (2013e) Zircon U-Pb dating of neoproterozoic tuff in Southwest Guangxi Province and implications for stratigraphic correlation. Geol China 40(5):1443–1452 (in Chinese with English abstract)

Gao LZ, Chen JS, Dai CG, Ding XZ, Wang XH, Liu YX, Wang M, Zhang H (2014a) SHRIMP zircon U-Pb dating of tuff in Fanjingshan group and Xiajiang group from Guizhou and Hunan Provinces and its stratigraphic implications. Geol Bull China 33(7):949–959 (in Chinese with English abstract)

Gao LZ, Ding XZ, Liu YX, Zhang CH, Zhang H, Huang ZZ, Xu XM, Zhou ZY (2014b) SHRIMP zircon U-Pb dating of neoproterozoic Chencai complex in Jiangshan-Shaoxing fault zone and its implications. Geol Bull China 35(5):641–648 (in Chinese with English abstract)

Gao LZ, Zhang H, Ding XZ, Liu YX, Zhang CH, Huang ZZ, Xu XM, Zhou ZY (2014c) SHRIMP zircon U-Pb dating of the Jiangshan-Shaoxing faulted zone in Zhejiang and Jiangxi. Geol Bull China 33(6):763–775 (in Chinese with English abstract)

Gao LZ, Yin CY, Ding XZ, Wang ZJ, Zhang H (2015a) Rating data of the Neoproterozoic chronostratigraphy and stratigraphic correlation in South China. Acta Geoscientica Sin 36(5):533–545

Gao LZ, Yin CY, Zhang H, Tang F, Ding XZ, Wang Y, Zhang CH (2015b) SHRIMP zircon U-Pb dating of the Liubatang formation in the jinning area, Yunnan Province, and its implication for the jinning movement. Geol Bull China 34(9):1595–1604 (in Chinese with English abstract)

Gao LZ, Zhang H, Zhang C, Ding XZ, Yin CY, Wu ZJ, Song B (2018) Collate and stipulate the sequences of the mesoproterozoic Kunyang group in Eastern Yunnan and Its position in stratigraphic column of China. Geol Rev 64(2):283–298 (in Chinese with English abstract)

Gao ZJ, Peng CW, Li YA, QIan JX, Zhu CS (1980) The Sinian system and its glacial deposits in Quruqtagh, Xinjiang: Sinian Subearthem in China. Tianjin Science and Technology Press, Tianjin, pp 186–213 (in Chinese)

Gao LZ, Zhang CH, Zhao X, Yan QR (2006) The distribution pattern and geological significance of the detrital zircon in the Proterozoic sedimentary rocks in Jiaozuo Yuntai in Henan Province. In: Proceedings of the first international symposium on development within geoparks: science and management, pp 116–119

Graham A, Zhou S, Porter S, Halverson GP (2016) A new rock-based definition for the Cryogenian Period (circa 720–635 Ma). Episodes 39(1). https://doi.org/10.18814/epiiugs/2016/v39i1/89231

Guan BD, Wu RT, Hambrey MJ, Geng WC (1986) Glacial sediments and erosional pavements near the Cambrian-Precambrian boundary in western Henan Province, China. J Geol Soc 143:311–323

Guan JP, He B, Li DW (2010) SIMS U-Pb dating of the detral zircons from the Xingzi Group in Lushan area and its geological significance. Geotecton Metallog 34(3):402–407 (in Chinese with English abstract)

Guan BD, Geng WC, Wu ZQ Du HY (1988) Middle and upper mesoproterozoic in Northern Slope of Eastern Qinling ranges. Henan Science and Tectonic Press, Zhengzhou

Guerrot C, Peucat JJ (1990) U-Pb geochronology of the upper Proterozoic Cadomina Orogeny in the northern American Massif, France. In: D'Lemos RS et al (eds) The Cadomian Orogeny, vol. 51. Geological Society Special Publication, London, pp 13–26

Guo LZ, Shi YS, Ma RS, Lu HF, Ye SF, Ding ZZ, Chen SZ, Xie B (1985) Plate movement and crustal evolution of the Jiangnan Proterozoic mobile belt, Southeast China. Earth Sci J Assoc Geol Collab Jpn 39(2):156–166

Guo LZ, Lu HF, Shi YS, Ma RS, Sun Y, Shu LS, Jia D, Zhang QL, Charvet J, Faure M (1996) On the Meso-neoproterozoic Jiangnan island arc: its kinematics and dynamics. Geol J China Univ 2(1):1–13 (in Chinese with English abstract)

Guo LZ, Shi YS, Ma, RS (1980) The geotectonic framework and crustal evolution of South China. In: Scientific Paper on Geology for international exchange. Geological Publishing House, Beijing, pp 109–116 (in Chinese)

Haverson GP, Maloof AC, Hoffman PF (2004) The Marinoan glaciation (Neoproterozoic) in northern Svalbard. Basin Res 16:297–324

He YH, Zhao GC, Sun M, Xia XP (2009) SHRIMP and LA-ICP-MS zircon geochronology of the Xiong'er volcanic rocks: implications for the Paleo-mesoproterozoic evolution of the southern margin of the North China Craton. Precambr Res 168(3):213–222

He ZJ, Niu BG, Zhang XY, Zhao L, Liu RY (2011a) Discovery of the palaeo-weathered mantle of the rapakivi granite covered by the Proterozoic Changzhougou formation in the Miyun area, Beijing and their detrital zircon dating. Geol Bull China 30(5):798–802 (in Chinese with English abstract)

He ZJ, Zhang XY, Niu BG, Liu RY, Zhao L (2011b) The palaeo-weathering mantle of the Proterozoic rapakivi granite in Miyun County, Beijing and the relationship with the Changzhougou formation of Changchengian system. Earth Sci Front 18(4):123–130 (in Chinese with English abstract)

Hoffman PF, Li ZX (2009) A palaeogeographic context for neoproterozoic glaciation. Palaeogeogr Palaeoclimatol Palaeoecol 277:158–172

Hoffman KH, Condon DJ, Bowring SA (2004) U-Pb zircon date from the Neoproterozoic Ghaub Formation, Zambia: constraints on Marinoan glaciation. Geology 32:817–820

Hong ZM, Huang ZF, Liu XL (1991) The upper Cambrian geology in the Southern part of Liaodong Peninsula. In: Journal of stratigraphy. Geological Publishing House (in Chinese), Beijing

Hu KM (2001) Initial discussion on the tectonic evolution of Jiangshan-Shaoxing fault zone. Geol Zheijang 17(2):1–11 (in Chinese with English abstract)

Hu ZR, Deng GH (2009) Tectonic characteristics of the Qinzhou-Hangzhou joint belt. J East China Statute Technol 26(2):114–122

Hu YH, Gu MG, Xu Y, Yu SQ, Wang J, He Y (2011) The confirmation of the age of Caledonian Chencai group in Zhuji area of Zhejiang province and its geological significance. Geol Bull China 30(11):1661–1670 (in Chinese with English abstract)

Huang JQ (1954) On major tectonic forms of China. Geological Publishing House, Beijing (in Chinese)

Huang TK (1945) On major tectonic forms of China. Mem Geol Surv China, A(20):1–165

International Commission on Stratigraphy (2013) International chronostratigraphic chart. J Stratigr 37(3):250

Jiang G, Kennedy YMJ, Christie-blick N (2003) Stable isotopic evidence for methane seeps in Neoproterozoic postglacial cap carbonate. Nature 426:822–825

Kao CS, Hsiug YH, Kao P (1934) Preliminary notes on Sinian stratigraphy of North China. Bull Geol Soc China 13:243–288

Kaufman AJ, Knoll AH, Nabornne GM (1997) Isotopes, ice ages and terminal Proterozoic earth history. Geology 94:6600–6605

Kennedy MJ, Christie-Blieck N (2001) Carbon isotopic composition of neoproterozoic glacial carbonates as a test of palaeooceanographic models for snowball earth phenomena. Geology 29:1135–1138

Knoll AH (2000) Learning to tell Neoproterozoic time. Precambr Res 100:3–20

Kuang HW, Liu YX, Meng XH, Ge M (2004) Sedimentary lithofacies and petrological feature of neoproterozoic MT structures-bearing carbonates in Jilin-Liaoning area. Acta Geoscientica Sin 25(6):647–652

Le Heron DP, Vandyk TM, Wu GH, Li M (2018) New perspectives on the Luoquan glaciation (Ediacaran-Cambrian) of North China. Depositional Rec 4:274–292

Le Heron DP, Vandyk TM, Kuang HW, Liu YQ, Chen XS, Wang YC, Yang ZR, Scharfenberg L, Davies B, Shields G (2019) Bird's-eye view of an Ediacaran subglacial landscape. Geology 47(8):705–709

Lee YY, Hsu SWC (1947) Stratigraphy and orogeny of southern Anhui. Publ National Res Inst Geol Mem 6:161–165

Li JH, Mu J (1999) Tectonic constraints from Chinese cratonic blocks for the reconstruction of Rodinia. Sci Geol Sin 34(3):259–272 (in Chinese with English abstract)

Li XH, Li ZX, Ge WC, Zhou HW, Li WX, Liu Y, Michael TD (2003) Neoproterozoic granitoids in south China: crustal melting above a mantle plume at ca. 825 Ma? Precambr Res 122:45–83

Li ZX, Wartho JA, Occhipinti S, Zhang CL, Li XH, Bao CM (2007) Early history of the eastern Sibao Orogen (South China) during the assembly of Rodinia: new mica $^{40}Ar/^{39}Ar$ dating and SHRIMP U-Pb detrital zircon provenance constraints. Precambr Res 159:79–94

Li CH, Xing GF, Jiang YH, Dong YG, Yu XM, Chen ZH, Jiang Y, Chen R (2009a) LA-ICP-MS U-Pb dating of zircons from sulfide-bearing quartz veins in the Pingshui copper deposit, Zhejiang Province, and its geological implications. Geol China 37(2):477–487 (in Chinese with English abstract)

Li HK, Lu SN, Li HM, Sun LX, Xiang ZQ, Gen JZ, Zhou HY (2009b) Zircon anad beddekeyite U-Pb precision dating of basic rock sills intruding Xiamaling formation, North China. Geol Bull China 28(10):1396–1404 (in Chinese with English abstract)
Li HK, Zhu SX, Xiang ZQ, Su WB, Lu SN, Zhou HY, Geng JZ, Li S, Yang FJ (2010) Zircon U-Pb dating on tuff bed from Gaoyuzhuang formation in Yanqing, Beijing: further constraints on the new subdivision of the Mesoproterozoic stratigraphy in the northern North China Craton. Acta Petrol Sin 26(7):2131–2140 (in Chinese with English abstract)
Li HK, Su WB, Zhou HY, Geng JZ, Xiang ZQ, Cui YR, Liu WC, Lu SN (2011) The base age of the Changchengian system at the northern NCC should be younger than 1670 Ma: constraints from zircon U-Pb LA-MC-ICPMS dating of a granite-porphyry dike in Miyun County, Beijing. Earth Sci Frontier 18(3):108–120 (in Chinese with English abstract)
Li M, Liu PJ, Yin CY, Tang F, Gao LZ, Chen SM (2012) Acritarchs from the Baicaoping formation (Ruyang Group) of Henan. Acta Palaeontol Sin 51(1):76–87
Li HK, Zhang CL, Yao CY, Xiang ZQ (2013) U-Pb zircon age and Hf isotope compositions of Mesoproterozoic sedimentary strata on the western margin of the Yangtze massif. Sci China Earth Sci 56:628–639 (in Chinese with English abstract)
Li HK, Su WB, Zhou HY, Tian H, Yang LG, Huff D, Ettensihn FR (2014) The first precise age constraints on the Jixian system of the Meso- to Neoproterozoic standard section of China: SHRIMP zircon U-Pb dating of bentonites from the Wumishan and Tieling formations in the Jixian section, North China Craton. Acta Petrol Sin 30(10):2999–3012 (in Chinese with English abstract)
Li CD, Zhao LG, Chang QS, Xu YW, Wang SY, Xu T (2017) Zircon U-Pb dating of tuff bed from Luoyukou formation in western Henan province on the southern margin of the North China Craton and its stratigraphic attribution discussion. Geol China 44(3):511–525 (in Chinese with English abstract)
Liu YX, Kuang HW, Cai GY, Meng XH, Ge M (2003) Depositional environments of molar-tooth of the neoproterozoic Yingchengzi formation in Southern Liaoning. Geol Bull China 22(6):419–425 (in Chines with English abstract)
Liu YQ, Gao LZ, Liu YX, Song B, Wang ZX (2005) Zircon U-Pb dating of magnesite magma in early neoproterozoic Xuhuai area. Chin Sci Bull 50(22):2514–2521
Liu SW, Yang PT, Wang ZQ, Luo P, Wang YQ, Luo GH, Wang W, Boran G (2012) LA-ICPMS zircon U-Pb ages and geochemistry of neoproterozoic low-grade metavolcanic rocks in Wuyuan-Dexing area of northeastern Jiangxi Province. Acta Petrol Sin 29(2):581–593
Liu B, Xu Bei, Meng XY, Kou XW, HE JY, Wei W, Mi H (2007) Study on the chemical index of alteration of Neoproterozoic strata in the Tarim plate and its implications. Acta Petrol Sin 24(12):1664–1667 (in Chinese with English abstract)
Lu SN (2002) Discussion on the new subdivision of the neoproterozoic in China. Geol Rev 48(3):242–248 (in Chinese with English abstract)
Lu SN, Zhao GC, Wang HC, Hao GJ (2008) Precambrian metamorphic basement and sedimentary cover of the NCC: A review. Precambr Res 160:77–93
Ma RS, Zhang JK (1977) Precambrian division and the discovery of Shengong orogeny in north-eastern Zhejiang—on several methods in the Precambrian research. J Nanjing Univ (nat Sci) 1:68–90 (in Chinese)
Ma HY, Sun HQ, Huang JZ, Ma TQ (2013) U-Pb dating of zircon from the Shangzheng group tuff in central Hunan province and its geological implication. Mineral Explor 4(1):69–74 (in Chinese with English abstract)
MacGabhann BA (2005) Age constraints on Precambrian glaciations and the subdivision of Neoproterozoic time. Galway. IUGS Ediacaran Subcommission Circulat, Ireland, p 13
Meng XM, Zhang XT (1948) Dongchuan Sichuan geology. Western Publication No, Former Central Research Institute of Geology, p 17
Meng XH, Ge M, Liu YX (2006) Study of the neoproterozoic Microspar (molar tooth) carbonate events, sequence stratigraphy of the Sino-Korean plate and the establishment of the Beihuan system. J Stratigr 30(3):211–222

Misch P (1942) Sinian stratigraphy of central eastern Yunnan. Nat Univ Peking, Contr Coll Scil, p 4
National Commission for Stratigraphy of China (2001) China Stratigraphic Guide and China Stratigraphic Guide (Revised). Geological Publishing House, Beijing, p 59
Ogg JG, Ogg GM, Gradstein FM (2016) 4-Cryogenian and Ediacaran. A Concise Geol Time Scale, 29–40
Plumb KA (1991) New Precambrian time scale. Episodes 14:139–140
Qiao XF, Gao LZ (2007) Mesoproterozoic palaeoearthquake and palaeogeography in Yan-Liao Aulacogen. J Palaeogeogr 9(5):337–352
Qiao XF, Song TR, Gao LZ, Peng Y, Li HB, Gao M, Song B, Zhang QD (1994) Seismic sequence in the carbonate rocks by vibrational liquefaction. Acta Geol Sin 26(5):503–509
Qiao XF, Gao LZ, Peng Y (2001) Neoproterozoic in palaeo-Tanlu zone: catastrophes, sequences, and biostratigraphic. Geological Publishing House, Beijing (in Chinese)
Qiao XF, Gao LZ, Zhang CH (2007) New idea of the Meso-neoproterozoic chronostratigraphic chart and tectonic environment in Sino-Korean Plate. Geol Bull China 26(5):503–509 (in Chinese with English abstract)
Shatsky NS (1945) Notes on tectonics of Volga-Ural oil bearing region and adjacent part of the Western slop of the South Ural. Mater Study Geol Struct USSR 26:1–130
Shu LS, Michel F, Yu JH, Jahu BM (2011) Geochronological and geochemical features of the Cathaysia block (south China): new evidence for the Neoproterozoic breakup of Rodinia. Precambr Res 187:263–276
Shui T, Xu BT, Liang RH (1986) Structural framework of continental basement of South China. Science in China (series B) 4:414–422 (in Chinese with English abstract)
Su WB, Li HK, Huff WD, Ettensohn FR, Zhang SH, Zhou HY, Wan YS (2010) Zircon SHRIMP U-Pb geochronology and its geological significance in the Tiling formation. Chin Sci Bull 55(22):2197–2206
Su WB, Li HK, Xu L, Jia SH, Geng JZ, Zhou HY, Wang ZH, Pu HY (2013) Luoyu and Ruyang groups at the south margin of the North China Craton (NCC) should belong in the Mesoproterozoic Chang Chengian System: direct constraints from the LA-MC-ICPMS U-Pb age of the Tuffisite in the Luoyukou Formation, Ruzhou, Henan, China. Geol Surv Res 35(2):96–108
Tang F, Gao LZ (1998) "Sinian Biota" in China. Acta Geol Sin 72(3):193–204 (in Chinese with English abstract)
Tang HF, Li WX, Zhou XM (1998) Comparison of neoproterozoic volcanic-sedimentary rocks from the Zhejiang-Jiangxi-Anhui border area-with a discussion of the problem about synchrono-heteropic facies of volcanism. Acta Geol Sin 72(1):34–41 (in Chinese with English abstract)
Thorkelson DJ, Grant AJ, Mortensen JK, Creaser RA, Villeneuve ME, Menicoll VJ, Layer PW (2005) Early and Middle Proterozoic evolution of Yukon, Can ada. Can J Earth Sci 42(6):1045–1071
Tian QQ (1948) Relationship between the axis of Xuefeng, Hunan Province and the transgression of Palaeozoic Era. Geol Rev 13(3–4):203–210 (in Chinese with English abstract)
Vignersse JL (2005) The specific case of the Mid-Proterozoic rapakivi granites and associated suited with the context of the Columbia super continent. Precambr Res 137:1–34
Wan YS, Zhang QD, Song TR (2003) SHRIMP ages of detrital zircons from the Changchengian in the Mingtombs area, Beijing: constraint on the protolith nature and maximum depositional age the mesoproterozoic cover of the North China Craton. Chin Sci Bull 48(18):1970–1975 (in Chinese with English abstract)
Wan YS, Liu DY, Xu MH, Zhang JM, Song B, Shi YR, Du LL (2007) SHRIMP U-Pb zircon geochronology and geochemistry of metavolcanic and metasedimentary rocks in Northwestern Fujian, Cathaysia block, China: tectonic implications and the need to redefine lithostratigraphic units. Gondwana Res 12:166–183
Wang XF (2015) the geochronology of the Meso-Neoproterozoic strata in the southern margin of North China and its geological significance. China University of Geosciences, Wuhan (in Chinese with English abstract)

Wang HZ, Liu BP, Li ST (1990) Geotectonic units and tectonic development of China and adjacent region. In: Wang HZ et al (eds) Tectonopalaeogeography and palaeogeography of China and adjacent regions. Geological Publishing House, Beijing, pp 246–262 (in Chinese)

Wang HZ, Wang ZQ, Zhang LH (1994) The tectonic development of the continental margin of the China in the proterozoic. In: Wang HZ et al (eds) The neoproterozoic mesoproterozoic and palaeozoic. Geological Publishing House, Beijing, pp 15–37 (in Chinese)

Wang HZ (1994) The continental margin of the China and geotectonic noun system, the tectonic development of the continental margin of the China. In: Wang HZ et al (eds) The neoproterozoic, mesoproterozoic and palaeozoic. Geological Publishing House, Beijing, pp 1–14 (in Chinese)

Wang ZQ, Yin CY, Gao LZ, Liu YQ (2002) Chemostratigraphic characteristics and correlation of the Sinian stratotype in the Eastern Three Gorges Area, Yichang, Hubei Province. Geol Rev 48(4):197–204 (in Chinese with English abstract)

Wang J, Li X, Duan T, Liu DY, Song B, Li ZX, Gao YH (2003) Zircon SHRIMP U-Pb dating for the Cangshuipu volcanic rocks and its implications for the lower boundary age of the Nanhuan strata in South China. Chin Sci Bull 48(22):2500–2506

Wang XL, Zhao GC, Qiu JS, Zhang WL, Liu XM, Zhang GL (2006) LA-ICPMS U-Pb zircon geochronology of the Neoproterozoic igneous rocks from Northern Guangxi, South China: implications for petrogenesis and tectonic evolution. Precambr Res 145:111–130

Wang XL, Zhao GC, Zhou JC, Liu YS, Hu J (2008) Geochronology and Hf isotopes of zircon from volcanic rocks of the Shuangqiaoshan Group, South China: implications for the Neoproterozoic tectonic evolution of the eastern Jiangnan orogen. Gondwana Res 14:355–367

Wang XL, Wang F, Chen FK, Zhu XY, Xiao P, Siebel W (2010) Detrital zircon ages and Hf-Nd isotopic composition of neoproterozoic sedimentary rocks in the Yangtze block: constraints on the deposition age and provenance. J Geol 118(1):79–94

Wang ZQ, Gao LZ, Ding XZ, Huang ZZ (2012) Tectonic environment of the metamorphosed basement in the Jiangnan Orogen and its evolutional features. Geol Rev 58(3):401–413 (in Chinese with English abstract)

Wang ZQ, Yin CY, Gao LZ, Tang F (2009) Chemostratigraphic characteristics of the Nanhua System in Southern Guizhou-Northern Guangxi area. Acta Geosci Sin 30(4):465–474

Wu RT, Guan BD (1988) Glacigenic characteristics of the Luoquan formation and sediment gravity flow reworking on it. Acta Geol Sin (engl Ed) 1:325–339

Xiao SH, Knoll AH, Yin LM, Zhang Y (1997) Neoproterozoic fossils in Mesoproterozoic rocks? A stratigraphic conundrum from the North China platform. Precambr Res 8:197–220

Xiao SH, Bao H, Wang H, Kaufman AJ, Zhou CM, Li G, Yuan XL, Ling H (2004) The neoproterozoic quruqtagh group in eastern Chinese Tianshan: evidence for a post-Marinoan glaciation. Precambr Res 130:1–26

Xing YS, Liu GZ (1973) The Sinian micropalaeoflora and its geological significance in Yanliao region. Acta Geol Sin 47(1):1–64 (in Chinese with English abstract)

Xing YS, Liu GZ (1982) Chinese late Cambrian microflora and its geological significance. Bull Chin Acad Geol Sci 4:55–64 (in Chinese with English abstract)

Xing YS, Liu GZ, Qiao XF, Gao ZJ, Wang ZQ, Zhu H, Chen YY (1989) Precambrian of China, Strata of China 3. Geological Publishing House, Beijing (in Chinese)

Xing FM, Xu X, Chen JF, Zhou TX (1992) The history of the hyperplasia in the southeast of jiangnan continent. Acta Geol Sin 66(1):59–72 (in Chinese with English abstract)

Xing YS, Duan CH, Liang YZ, Cao RG (1985) Late Precambrian palaeontology of China. Geological Memoirs Series 2, Member 2, Geological Publishing House, Beijing (in Chinese)

Xing YS (1979) Sinian of China. In: International journal of communication and geology (2), Stratigraphic, Palaeontological. Geological Publishing House, Beijing, pp 1–2 (in Chinese)

Xu B (1990) The Late Proterozoic trench-basin-arc system in northeastern Jiangxi-southern Anhui provinces. Acta Geol Sin 1:33–42 (in Chinese with English abstract)

Xu B (1994) The basic characteristics of Proterozoic tectonic evolution in the southeastern continental margin of the Yangtze Craton. In: Wang HZ (ed) Tectonic evolution of mesoproterozoic,

neoproterozoic and palaeozoic in Chinese continent. Geological Publishing House, Beijing, pp 189–201 (in Chinese)
Xu B, Guo LZ, Shi YS (1992a) Proterozoic terranes and poly-stages collisional orogeny in Anhui-Zhejiang-Jiangxi Area. Geological Publishing House, Beijing (in Chinese)
Xu B, Guo LZ, Shi YS (1992b) Proterozoic land mass and multiple collision orogenic belts in Anhui. Beijing; Geological Publishing House, Zhejiang and Jiangxi Area (in Chinese)
Xu B, Zhang HF, Yao HT, Li YA (2003) C-isotope composition and significance of the Sinian on the Tarim plate. Chin Sci Bull 48(4):385–389 (in Chinese with English abstract)
Xu B, Jian P, Zhang HF, Zou HB, Zhang LF, Liu DY (2005) U-Pb zircon geochronology and geochemistry of Neoproterozoic volcanic rocks in the Tarim Block of northwest China: implications for the breakup of Rodinia supercontinent and Neoproterozoic glaciations. Precambr Res 136(2):107–123
Xu B, Kou XW, Song B, Wei W, Wang Y (2008) SHRIMP dating of the upper Proterozoic volcanic rocks in the Tarim plate and constraints on the neoproterozoic glaciation. Acta Petrol Sinica 24(12):2857–2862 (in Chinese with English abstract)
Xu B, Xiao SH, Zou HB, Chen Y, Li ZX, Song B, Liu DY, Zhou CM, Yuan XL (2009) SHRIMP zircon U-Pb age constraints on Neoproterozoic Quruqtagh diamictites in NW China. Precambr Res 168:247–258
Xu XS, O'Reilly SY, Griffin WL, Wang XL, Pearson NJ (2007) The crust of Cathaysia: age, assembly and reworking of two terranes. Precambr Res 158:51–78
Xue HM, Ma F, Song YQ, Xie YJ (2010) Geochronology and geochemisty of the neoproterozoic granitoid association from eastern segment of the Jiangnan orogen, China: constraints on the timing and process of amalgamation between the Yangtze and Cathaysia blocks. Acta Petrol Sin 26(11):3215–3244 (in Chinese with English abstract)
Yan YZ, Zhu SX (1992) Discovery of acanthomorphic acritarchs from the Baicaoping formation in Yongji, Shanxi and its geological significance. Acta Micropalaeontologica Sin 9(3):267–282 (in Chinese with English abstract)
Yang MG, Liao RJ, Liu YG (1999) The metamorphic basement types and the division and correlation of metamorphosed strata in Jiangxi. Jiangxi Geol 12(3):201–208 (in Chinese with English abstract)
Yang JH, Wu FY, Liu XM, Xie CW (2005) Zircon U-Pb and Hf isotopes and their geological significance of the Miyun rapakivi granites from Beijing, China. Acta Petrol Sinica 21(6):1633–1644 (in Chinese with English abstract)
Yang SF, Gu MG, Lu CZ, Lu CZ (2009) Geochemical characteristics and tectonic implications of the island-arc volcanic rocks of Mesoproterozoic in Zhangcun, Zhejiang Province. J Jilin Univ Earth Sci Ed 39(4):689–698 (in Chinese with English abstract)
Yang MG, Zhu PJ, Xiong QH (2012) Framework and evolution of the neoproterozoic-early palaeozoic South-China rift system. Acta Geol Sin 86(9):1367–1375 (in Chinese with English abstract)
Yang C, Li XH, Wang XC, Lan ZW (2015) Mid-Neoproterozoic angular unconformity in the Yangtze Block revisited: insights from detrital zircon U-Pb age and Hf-O isotopes. Precambrian 266:165–178
Ye MF, Li XH, Li WX, Li ZX (2007) SHRIMP zircon U-Pb geochronological and whole-rock geochemical evidence for an early neoproterozoic Sibaoan magmatic arc along the southeastern margin of the Yangtze block. Gondwana Res 12:144–156
Yin LF (1920) The geology of His-Shan or the western hills of Peking. Mem Geol Surv China, Series A 1:115
Yin CY (2005) Advance and tendency of the Neoproterozoic chronostratigraphic studies. J Stratigr 29(2):178–180
Yin CY, Gao LZ (1997) New discovery of megafossils in neoproterozoic Luoyukou formation at Lushan in Western Henan. Geol Rev 4:355 (in Chinese with English abstract)

Yin LM, Yuan XL (2003) Review of the microfossil assemblage from the late Mesoproterozoic Ruyang group in Shanxi, China. Acta Micropalaeontologica Sinica 20(1):39–46 (in Chinese with English abstract)

Yin CY, Liu DY, Gao LZ, Wang ZQ, Xing YS, Jian P, Shi YR (2003) Lower boundary age of the Nanhuan and the Gucheng glacial stage: evidence from SHRIMP U-Pb dating. Chin Sci Bull 48(16):1657–1662

Yin LM, Yuan XL, Bian LZ, Hu J (2004) Late mesoproterozoic microfossil assemblage on northern slope of eastern Qinling mountains, China: a new window on early eukaryotes. Acta Palaeontol Sin 43(1):1–13 (in Chinese with English abstract)

Yin CY, Liu YQ, Gao LZ, Wang ZQ, Tang F, Liu PJ (2007) Phosphatized biota in early sinian (Ediacaran)–Weng'an Biota and its environment. Geological Publishing House, Beijing (in Chinese)

Yin ZX, Xu DY, Pu QY (1963) Information compilation of the name of crustal movement in China. Geol Rev 23(Supp):20–81 (in Chinese)

Yin CY, Tang F, Liu PJ, Gao LZ Wang ZQ, Chen SM (2009) New advances in the study of biostratigraphy of the Sinian (Ediacaran) Doushantuo formation in South China. Acta Geoscientica Sin 30(4):421–432

Yu GH, Bao CM, Fang BX, Ma WP, Song FQ, He SC (1995) Brief introduction to the achievements on lithostartigarphic clearing in Zhejiang. Geol Zhejiang 11(1):1–14 (in Chinese)

Zhang SH, Jiang GQ, Han Y (2008) The age of the Nantuo formation and Nantuo glaciation in South China. Terra Nova 20(4):289–294

Zhang SH, Zhao Y, Yang ZY, He ZF, Wu H (2009) The 1.35 Ga diabase sills from the northern NCC: implications for breakup of the Columbia (Nuna) supercontinent. Earth Planet Sci Lett 288:588–600

Zhang H, Gao LZ, Zhou HR, Song B, Ding XZ, Zhang CH, Liu HG, Gong CQ (2019) Chronology progress of the Guandaokou and Luoyu groups in the southern margin of North China Craton: constraints on zircon U-Pb dating of tuff by means of the SHRIMP. Acta Petrol Sin 35(8):2470–2486 (in Chinese with English abstract)

Zhao TP, Zhai MG, Xia B, Li HM, Zhang YX, Wan YS (2004) Zircon U-Pb SHRIMP dating for the volcanic rocks of the Xiong'er group: constraints on the initial formation age of the cover of the North China Craton. Chin Sci Bull 49(23):2495–2502

Zhao XF, Zhou MF, Li JW, Sun M, Gao JF, Sun WH, Yang JH (2010) Late palaeoproterozoic to early mesoproterozoic Dongchuan group in Yunnan, SW China: implications for tectonic evolution of the Yangtze block. Precambr Res 182(1–2):57–69

Zhou CM, Tucker R, Xiao SH, Peng Z, Yuan XL, Chen Z (2004) New constraints on the ages of neoproterozoic glaciations in South China. Geology 32:437–440

Zhou CM, Xie GW, Kathleen M, Xiao SH, Yuan XL (2007) The diversification and extinction of Doushantuo-Pertatataka acritarchs in South China: causes and biostratigraphic significance. Geol J 42:229–262

Zhou JC, Wang XL, Qiu JS (2008) Is the Jiangnan orogenic belt a Grenvillian orogenic belt: some problems about Precambrian geology of South China. Geol J China Univ 14(1):64–72 (in Chinese with English abstract)

Zhou HR, Wang ZQ, Cui XS, Lei ZY, Dong WM, Sheng Y (1999) The sequential stratigraphic study of mesoproterozoic and neoproterozoic in the Southern of North China. Geological Publishing House, Beijing, pp 1–90 (in Chinese)

Chapter 2
Meso-Neoproterozoic Stratigraphic Sequences in the Yanliao Faulted-Depression Zone, North China Craton

Shixing Zhu, Huaikun Li, Lixin Sun, and Huan Liu

Abstract The Precambrian sedimentary strata in the Yanliao Faulted-Depression Zone (YFDZ) on the North China Craton is the most-developed Meso-Neoproterozoic sequences in China. The sequences consist of the $Pt_2{}^1$ Changchengian, $Pt_2{}^2$ Jixianian, $Pt_2{}^3x$ Xiamaling Formation and $Pt_3{}^1$ Qingbaikouan, which are traditionally subdivided into 12 formations and 43 members. This chapter aims to review the history and the state of art of the studies on the stratigraphic sequences in aspects covering the tempo-spatial distribution, lithostratigraphic correlation, geochronology and palaeontology. Some important aspects have been discussed in detail. As the oldest unmetamorphosed sedimentary sequence in China, the Changchengian (1670–1600 Ma) shows a regional micro-angle unconformity with the underlying Dahongyu Formation of Jixianian (1600–1400 Ma) with a local conformity at the Dahongyu depocenter. Therefore, both Changchengian and Jixianian should be referred to a set of basically continuous sedimentary strata in the Jixian stratotype section. The Changchengian may be attributed to the aulacogen clastic deposition in the early stage of the YFDZ related to early breakup of the Supercontinent Columbia.

Keywords Yanliao faulted-depression zone (YFDZ) · Changchengian · Jixianian · Xiamaling formation · Qingbaikouan · North China craton (NCC)

2.1 Introduction

The Yanliao Faulted-Depression Zone (YFDZ) is geographically situated at the Yanshan Mountain on the west of Liaohe River regions, and geologically located at the middle segment in the northern margin of the North China Craton (Fig. 2.1), which used to be termed as the "Yanliao Subsidence Zone" or the "Yanshan Paraplatform" in literatures.

Tectonically, the YFDZ includes five depressions (i.e., Liaoxi, Jibei, Jidong, Jiangxi and Xuanlong Depressions) and two uplifts (i.e., Shanhaiguan and Mihuai

S. Zhu (✉) · H. Li · L. Sun · H. Liu
Tianjin Institute of Geology and Mineral Resources, Tianjin 300170, China

T. Wang, *Meso-Neoproterozoic Geology and Petroleum Resources in China*, Springer Geology, https://doi.org/10.1007/978-981-19-5666-9_2

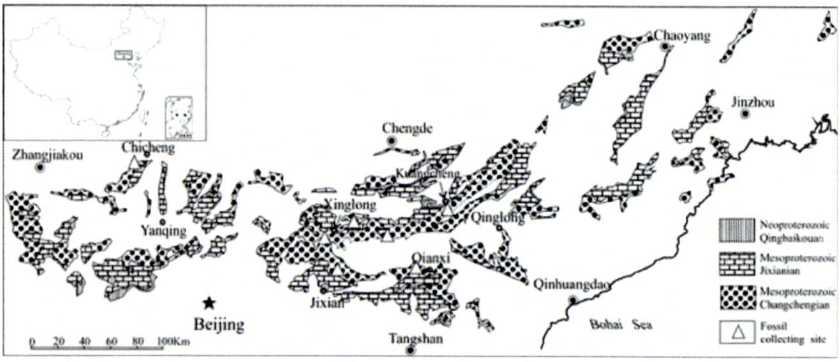

Fig. 2.1 the sketch map showing the distribution of the proterozoic strata in the yanliao faulted-depression zone (YFDZ)

Uplifts; cf. Figure 11.13 in Chap. 11). Stratigraphically, the Precambrian sedimentary strata in the YFDZ consist of the Meso-Neoproterozoic stratigraphic sequences, i.e., $Pt_2{}^1$ Changchengian, $Pt_2{}^2$ Jixianian, $Pt_2{}^3x$ Xiamaling Formation and $Pt_3{}^1$ Qingbaikouan, in the Regional Stratigraphic Chart of China. In the last decade, great advances have been made on research of the Meso-Neoproterozoic strata in the YFDZ.

2.2 Historical Review

The geological studies on the Precambrian sequences in the YFDZ has started since 1922. The Precambrian strata in the area used to be named as "Sinian System", "Sinian Suberathem", "Middle to Upper Proterozoic" and Meso-Neoproterozoic respectively in four different periods, witnessing the research progresses on the Precambrian stratigraphy not only in this area, but also in whole the China.

2.2.1 "Sinian System" Period

The word "Sinian" was initially coined as a tectonic term by Punpelly (1866), and then was used as a stratigraphic term (Richthofen 1882). As a stratigraphic unit, the word "System" was formally defined for "a suite of unmetamorphosed or slightly metamorphosed rocks unconformably overlying the high-grade metamorphosed Wutai System, disconformable underlying the fossil-bearing Lower Cambrian Mantou shales" by Grabau (1922). Although the Nankou section at north Beijing was the earliest investigated "Sinian System" section in the Yanshan Mountain (Tien 1923), while the Jixian section at the southern slope of the eastern Yanshan Mountain has been traditionally used as the stratotype section for the "Sinian System" in China.

The Jixian stratotype section is located in the north of the Jixian District in Tianjin. As a very thick and unmetamorphosed Precambrian sedimentary sequence between the Archean and Cambrian, the Jixian section was initially investigated by Kao C. S., Hsiung Y. H. and Kao P. in 1931 (Kao et al. 1934). It was subdivided into three groups with ten litholostratigraphic members in ascending order (Table 2.1), the Lower Sinian "Nankou Group" (including the "Changcheng quartzite", "Chuanlinggou shale", "Dahongyu quartzite and Lava" and "Gaoyuzhuang limestone"), the Middle Sinian "Jixian Group" (the "Yangzhuang shale", "Wumishan limestone", "Hongshuizhuang shale" and "Tieling limestone") and the Upper Sinian "Qingbaikou Group" (the "Xiamaling shale" and "Jing'eryu limestone"; Table 2.1), and the "Sinian System" unconformably overlies the Archean/Proterozoic Taishan or Wutai Groups, and disconformably underlies the Lower Cambrian "Mantou shale" (Kao et al. 1934). Among the above three groups of "Sinian System", the stratigraphic contacts are disconformable as well. Kao et al. (1934) also noted that there might be a disconformable contact between the "Wumishan limestone" and "Hongshuizhuang shale".

Table 2.1 Subdivision of the Jixian "Sinian system" (Kao et al. 1934)

	Lower-Cambrian		Mantou shale
	Disconformity		
Sinian System	Upper-Sinian Qingbaikou Group		Jingeryu limestone
			Xiamaling shale
	Disconformity		
	Mid-Sinian Jixian Group		Tieling limestone
			Hongshuizhuang shale
			(Disconformity?)
			Wumishan limestone
			Yangzhuang shale
	Disconformity		
	Lower-Sinian Nankou Group		Gaoyuzhuang limestone
			Dahongyu sandstone and lava
		Hangao rock series	Chuanlinggou shale
			Changcheng sandstone
	Unconformity		
	Archean / Proterozoic Taishan Group or Wutai Group		

Since the Gao C. S.'s pioneer work in 1934, the major research advances on the "Sinian System" in the Yanshan Mountain are summarized as follows.

(1) Due to the discovery of trilobite fossils from the upper "Jing'eryu limestone" at the Mingtombs section in the Changping District, north Beijing, Zhang W. Y. and Li T. B. doubted the "Jing'eryu limestone" as Cambrian strata in 1935, and named the trilobite-bearing limestone as the "Changping Formation" belonging to the Early Cambrian in age (Zhang 1935). Sun (1957) confirmed that the upper "Jing'eryu limestone" in the Jixian stratotype section is referred to the Lower Cambrian with an unconformable contact over the "Sinian System". Meanwhile, the tectonic movement corresponding to the unconformity was named as "Jixian Movement" with standard site at the Fujunshan Hill in the north of the Jixian District. While the Cambrian interval of the upper "Jing'eryu limestone" was renamed as the Fujinshan Formation based on local tablet inscription (Wang 1963; Xiang and Guo 1964).
(2) Shen and Liao (1958) considered that the Gaoyuzhuang Formation appears as a conformable contact with its overlying Yangzhuang Formation, and a disconformable contact with its underlying Dahongyu Formation, so that the Gaoyuzhuang Formation should be assigned into the Middle Sinian "Jixian Group" in the stead of the Lower Sinian "Nankou Group" in Table 2.1.
(3) Wang (1963) referred the "Sinian System" of the Jixian section to the Precambrian age in North China, which would be older than the Eocambrian "Sinian System" represented by the Yangtze Gorge section in South China.

2.2.2 "Sinian Suberathem" Period

Owing to the dispute on stratigraphic correlation of the "Sinian System" between South and North China, a National Precambrian Symposium was held in Beijing in 1975, and a National Sinian Stratigraphic Subdivision Chart was tentatively established in the symposium, based on which, it was proposed that the "Sinian System" of South Sinian (represented by the Yangtze Gorges section) was placed above the "Sinian System" of North China (represented by the Jixian section), both constituted the "Sinian Suberathem" of Proterozoic Erathem, which was subdivided into four chronostratigraphic units, i.e., "Sinian System" (in Yangtze Gorge section as the stratotype), "Qingbaikou System", "Jixian System" and "Changcheng System" (in Jixian section as the stratotype) in descending order (Table 2.2; Wang et al. 1980a, b).

Now that the basal interval of the original "Jing'eryu limestone" consists of terrestrial siliciclastic rocks, and thus as a new stratigraphic unit, it used to be alone called as "Longshan sandstone", "Longshan Formation" (Qiao 1976) and "Changlongshan Formation" (Compilation Group of Regional Stratigraphic Table in Beijing 1977) respectively, resulting in the terminologically inconsistent names. Finally, Xing et al. suggested to rename the terrestrial siliciclastic as Luotuoling Formation.

Table 2.2 The subdivision and timescale of the "Sinian Suberathem" in China (Wang et al. 1980a, b)

Paleozoic Erathem				
				570 Ma
Proterozoic Eonothem	Sinian Suberathem	Sinian System (represented by Yangtze Gorge Section)		
				850 Ma
		Qingbaikou System	(Represented by Jixian Section)	
				1050 Ma
		Jixian System		
				1400 Ma
		Changcheng System		
				1950 Ma

Since 1975, more stratigraphic sections of the "Sinian Suberathem" in the YFDZ have been investigated in detail. In addition to the Jixian section (Chen et al. 1980), the results of Mingtombs section in north Beijing (Wang et al. 1980a, b) and sections in the western and eastern segments of the Yanshan Mountains (Du and Li 1980; Xu and Cui 1980) have been reported. The main progresses during the period are summarized in Table 2.3.

Table 2.3 The subdivision and timescale of the "Sinian Suberathem" in Jixian section (Chen et al. 1980)

			Jixian Movement (~850 Ma)
Qingbaikou System	Jingeryu Formation	Jingeryu Subformation: Limestones	
		Changlongshan Subformation: sandstones	
			Yuxian Uplifting
	Xiamaling Formation	Sandstones and shales	
			Qinyu Uplifting (~1050 Ma)
Jixian System	Tieling Formation	Laohuding Subformation	
			Tieling Uplifting
		Daizhuangzi Subformation	
	Hongshuizhuang Formation	shales	
	Wumishan Formation	Various dolomites (from bottom to top: Luozhuang, Mopanyu, Ershilipu and Shanpoling subformations)	
	Yangzhuang Formation	Red mud-dolomites	
			Luanxian Uplifting (~1400 Ma)
Nankou System	Gaoyuzhuang Formation	Various dolomites (from bottom to top: Guandi, Sangshuan, Zhangjiayu and Huanxiusi subformations)	
			Qinglong Uplift
	Dahongyu Formation	Sandstones and volcanic rocks	
			Xingcheng Uplifting (~1700 Ma)
Changcheng System	Tuanshanzi Formation	Dolomites	
	Chuanlinggou Formation	Shales	
	Changzhougou Formation	Conglomerates and sandstones	
			Lvliang Uplifting (~1950 Ma)

2.2.3 "Middle to Upper Proterozoic" Periods

The term of "Sinian Suberathem" had to be abandoned in China, which was officially approved by National Commission on Stratigraphy of China during a special meeting on Precambrian in 1982. The "Sinian System" was specifically redefined to cover the "Sinian System" strata in Yangtze Gorges section as one stratotype in Yangtze Craton (South China), while another stratotype in Jixian section represents original "Sinian System" in North China Craton, and based on the stratigraphic subdivision scheme of Kao et al. (1934; Table 2.4a), which is further divided into "Changchengian Jixianian and Qingbaikouan" systems from bottom to top with a duration of 1800–800 Ma (Table 2.4b; China Commission on Stratigraphy of China 2001).

Hereafter, the standard Middle to Upper Proterozoic Stratigraphy Chart of North China consists of three systems with twelve formations in the Jixian stratotype section, east Yanshan Mountain was established.

In 1998, the National Commission on Stratigraphy of China released an official notice on the recommendation of Geological Time Scale of China. In the timescale, the Proterozoic Eon/Eonothem is subdivided into Palaeoproterozoic, Mesoproterozoic and Neoproterozoic Era/Erathem. The Mesoproterozoic Era/Erathem is further subdivided into the Changchengian (1800–1400 Ma) and the Jixianian (1400–1000 Ma), the Neoproterozoic Era/Erathem was subdivided into the Qingbaikouan (1000–800 Ma) and the Sinian (800–600 Ma). Ever since, the three periods/systems in the Jixian stratotype section was used as the official chronostratigraphic units in

Table 2.4 Brief historical summary of the Proterozoic stratigraphy in the Yanliao Faulted-Depression Zone

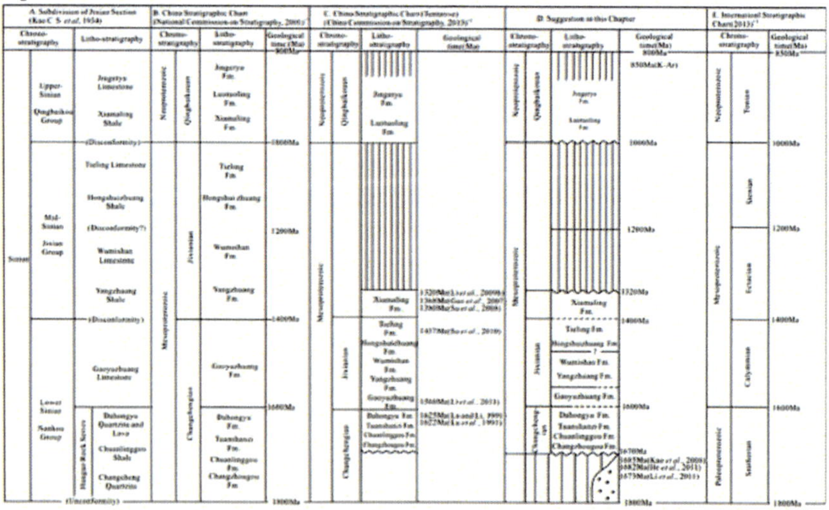

Note: *1. National Commission on Stratigraphy of China (2001); *2. Editorial Board on China Stratigraphic Chart 2013; *3. International Commission on Stratigraphy 2013

the stratigraphic chart of China (China National commission on Stratigraphy 2001; Table 2.4b).

2.2.4 *The Stratigraphic Chart of China in 2013*

Over the last decade, along with application of improved geochronology, great advances have been achieved on chronostratigraphy and stratigraphy, particularly on the new geological age constraining of Xiamaling Formation and Changchengian. The National Commission on Stratigraphy of China issued a new Stratigraphic Chart of China in 2013 (Editorial Board on China Stratigraphic Chart 2013; Table 2.4C). The major revision in the new chart includes:

(1) The Xiamaling Formation (1400–1320 Ma) is placed in the middle of the Mesoproterozoic rather than the lower of the Neoproterozoic Qingbaikouan.
(2) A long period of hiatus (1320–1000 Ma) between the Xiamaling Formation and the Qingbaikouan is recognized.
(3) As the highest horizon of Jixianian, the position of Tieling Formation is moved from the top of the Mesoproterozoic to the middle-lower horizon of the Mesoproterozoic.
(4) The Meso-Neoproterozoic stratotype of Jixian section in YFDZ contains four stratigraphic units: Pt_2^1 Changchengian (1700–1600 Ma), Pt_2^2 Jixianian (1600–1400 Ma), Pt_2^3x Xiamaling Formation (1400–1320 Ma) and Pt_3^1 Qingbaikouan (1000–780 Ma) in ascending order.
(5) The boundary of the Changchengian and Jixianian was changed from the boundary between the Gaoyuzhuang and Yangzhuang Formations to that between the Dahongyu and Gaoyuzhuang Formations, which shows a regional disconformable contact with local conformable contact at the depocenter of Dahongyu Formation.

2.3 Meso-Neoproterozoic Stratigraphic Sequences

In accordance with the China Stratigraphic Chart (tentative; Table 2.4c; Editorial Board on China Stratigraphic Chart 2013), the Meso-Neoproterozoic sequences in the YFDZ are briefly described as follows (Fig. 2.2).

2.3.1 *Changchengian (Pt_2^1)*

The Changchengian in Jixian stratotype section is mainly distributed at the Xiaying area in Jixian Distract, Tianjin, and its lower interval generally is typical aulacogen siliciclastic rocks, and the middle-upper interval mainly comprises muddy and

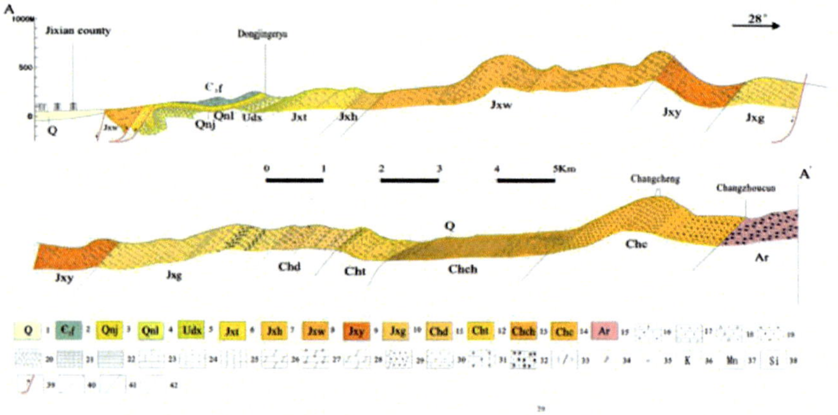

Fig. 2.2 Meso-Neoproterozoic lithostratigraphic sequence of the Jixian Section 2.1. 1. Quaternary; 2. Lower Cambrian Fujunshan Formation; 3. Qingbaikouan Jing'eryu Formation; 4. Qingbaikouan Luotuoling Formation; 5. Xiamaling Formation; 6. Jixianian Tieling Formation; 7. Jixianian Hongshuizhuang Formation; 8. Jixianian Wumishan Formation; 9. Jixianian Yangzhuang Formation; 10. Jixianian Gaoyuzhuang Formation; 11. Changchengian Dahongyu Formation; 12. Changchengian Tuanshanzi Formation; 13. Changchengian Chuanlinggou Formation; 14. Changchengian Changzhougou Formation; 15. Neoarchean Zunhua Group-complex; 16. sandy conglomerate; 17. sandstone; 18. quartzose sandstone; 19. unequal grain sandstone; 20. fine-grained sandstone; 21. silty shale; 22. shale; 23. Limestone; 24. muddy limestone; 25. dolomitic limestone; 26. dolostone; 27. sandy dolostone; 28. silt-bearing muddy dolostone; 29. dolomitic sandstone; 30. carbonate breccia; 31. K-rich trachyte; 32. hornblende plagioclase gneiss; 33. columnar stromatolites; 34. microbial (algae) layer; 35. glauconite; 36. potassium; 37. manganese dolostone; 38. chert nodules and cherty layer; 39. reverse fault; 40.disconformity; 41. unconformity; 42. conformity

carbonate rocks, and the upper interval mainly consists of siliciclastic rocks and potassium-rich volcanic rocks, with a total stratigraphic thickness of 2525 m and an age of 1670–1600 Ma. It includes following four formations, i.e., Changzhougou, Chuanlinggou, Tuanshanzi, and Dahongyu Formations in ascending order.

The Changchengian is distributed at both southern and northern sides of the Shanhaiguan submarine uplift in the middle-eastern segment of YFDZ as well as at the surrounding Changping-Huairou submarine uplift in the western segment of YFDZ. The Changchengian appears as a NE-trending distribution controlled by the Yanliao Faulted-Depressions (Fig. 2.3), and unconformably overlies the Archean metamorphic rocks and unconformably underlies the Jixianian strata. It includes four formations as follows.

2.3.1.1 Changzhougou Formation (Pt_2^1c or Chc)

It was formerly known as "Changcheng quartzite" (Kao et al. 1934), "Nankou Series" (Shen and Liao 1958), "Huangyaguan Formation" (Chen and Lu 1963) and "Changzhoucun Formation" (Yu et al. 1964) respectively. Finally, it was formally

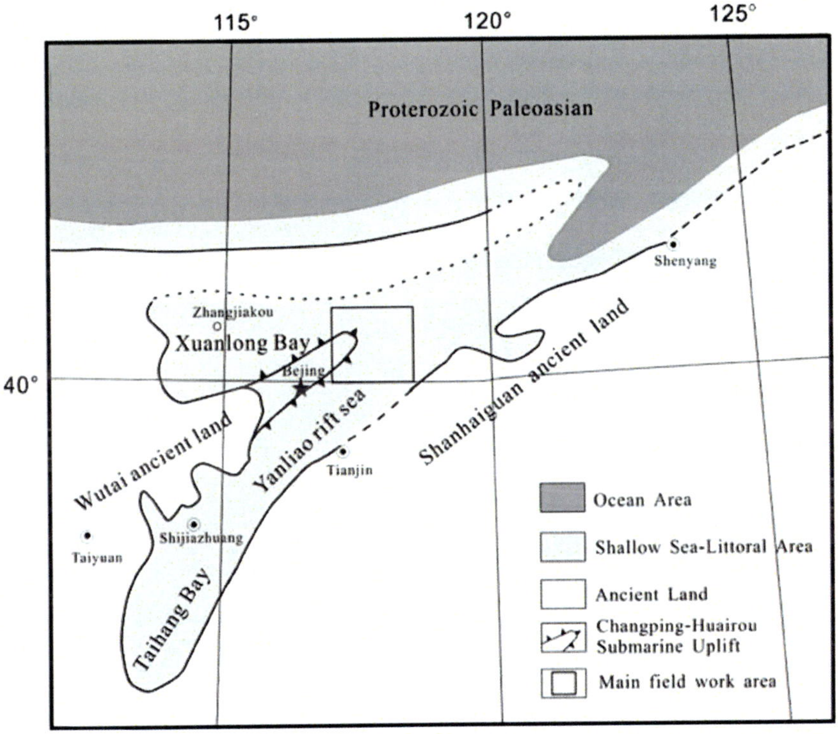

Fig. 2.3 Tectonic-palaeogeographic map of the Changchengian in the YFDZ (Wang 1985)

renamed as the Changzhougou Formation during the field seminar on the "Sinian System" in Jixian section in 1964 (North China Institute of Geosciences 1965), and its stratotype section is established at Xiaying area, Jixian District, Tianjin (China Stratigraphy Code Editorial Board 1999).

Taking the Jixian section as an example, the Changzhougou Formation is composed of siliceous sandstone with stratigraphic thickness of 859 m. It can be subdivided into three members, i.e., the fluvial conglomerate and bebbly coarse feldspathic quartzose sandstone as the lower member (Chang-1 Member), the littoral shoal light-purple quartzified sandstone and white sedimentary quartzite as the middle member (Chang-2 Member) and the tidal zone slaty and wedge feldspathic quartzose sandstone interbedded with lamella silty shale as upper member (Chang-3 Member; Fig. 2.4). The Changzhougou Formation (ca. 1670 Ma) unconformably overlies on the Archean Zunhua Group-complex (2458 Ma) consisting of the garnet-hornblende-plagiogneiss (Fig. 2.5).

The Changzhougou Formation appears as following changes in the YFDZ:

(1) Stratigraphic thickness varies greatly, it can be more than 1000 m thick, e.g. 1065 m at Xinglong, 2048 m at Kuancheng and 1286 m at Pingquan in Jibei Depression from west to east, and it is significantly thinner and only ca.

Fig. 2.4 The lithological characteristics of the Changzhougou formation at Changzhougou Valley in the north of Jixian District, Tianjin. **a**. Conglomerate and gravel-containing coarse feldsparthic quartzose sandstone (Chang-1 Member); **b**. the fluvial cross bedding showing facies (Chang-1 Member); **c**. thick-bedded to massive light-purple quartzified sandstone and white feldspathic quartzose sandstone (Chang-2 Member); **d**. slaty feldspathic quartzose sandstone (Chang-3 Member)

100 m at the edge of the faulted-depression and around the Changping-Huairou submarine uplift (in Jingxi to Xuanlong Depressions).

(2) In the centrality of YFDZ (such as Xinglong, Kuancheng and Pingquan in Jidong to Jibei Depressions), the lower interval of the Changzhougou Formation is generally fluvial coarse deposits predominated by conglomerate and coarse feldspathic quartzose sandstone; while surrounding the Changping-Huairou submarine uplift, the fluvial deposition does not exist except for the basal conglomerate, and it is mainly composed of marine quartzose sandstone with dolostone or dolomitic stromatolites at the top.

(3) In the centrality of YFDZ, there is black fine-grained sediments with carbonaceous shale and sandy shale near the middle interval of the formation; in Pangjiapu area, the formation is only 173.7 m thick and composed of lagoon sandstone and shale in the Chang-1 Member.

(4) At the Xuanhua-Longguan area (in Xuanlong Depression) of YFDZ, there are multiple layers of ferruginous sandstone within the quartzose sandstone of Chang-2 Member, even ferrilites composed of ferruginous stromatolites are enriched to constitute the famous Xuanlong hematite at the top of Chang-2 Member.

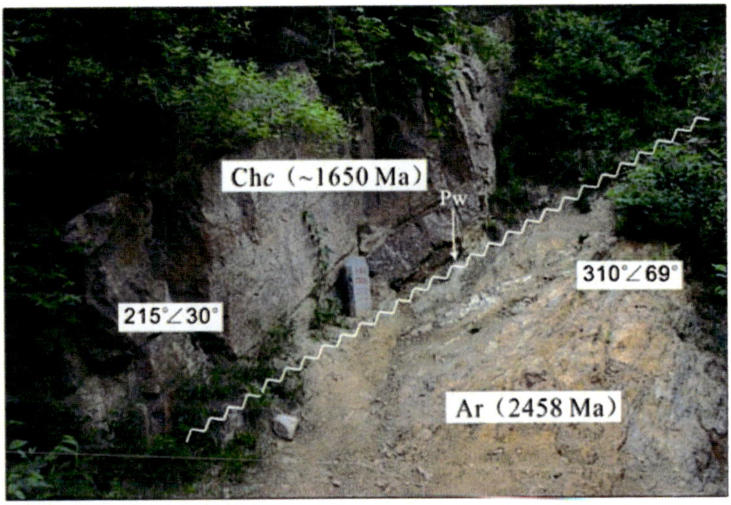

Fig. 2.5 The photograph of unconformable contact (Pw) between the Changzhougou Formation (Chc) and Neoarchean gneiss (Ar). Ar. Neoarchean garnet amphibolite gneiss; Chc. pebbly and feldsparthic course-grained quartzose sandstone at the base of the Changzhougou formation; Pw. palaeosoil layer

2.3.1.2 Chuanlinggou Formation (Pt_2^1ch or Chch)

The "Chuanlinggou shale" was originally named by Kao et al. in (1934), later renamed as the Chuanlinggou Formation in 1959. Geomorphologically, the Chuanlinggou shale forms low hills obviously distinguished from the high mountain terrain of the Changzhougou Formation.

In Jixian section, the Chuanlinggou Formation is composed of the 889 m thick mudstone. It overlies conformably the Changzhougou Formation (Fig. 2.6) and can be subdivided into three members: the lower member (Chuan-1 Member) consists of interbedded yellowish- and greyish-green lenticular sandstone, siltstone and silty illite shales; the middle member (Chuan-2 Member) is characterized by yellowish-green and black shales and silty illite shales; the upper member (Chuan-3 Member) is predominated by black illite shales intercalated by siltstone and fine-grained sandstone as well as some carbonaceous dolostone (Fig. 2.7).

In the Jixian section, magmatic rocks are often found in the Chuanlinggou Formation, which consist of plagioclase-porphyry and hornblende-kersantite with some orthophyre and volcanic breccia. Part of magmatic rocks are subvolcanic rocks of Dahongyu-age or the sills intruded into bedrocks during the Yenshanian (ca. Jurassic–Cretaceous time).

The Chuanlinggou Formation appears as obvious variations of stratigraphic thickness (30–1000 m) and lithofacies in the YFDZ. It can be generally divided into three kinds of lithofacies zones (Fig. 2.8):

Fig. 2.6 The photograph of conformable contact between the Chuanlinggou (Chch) and Changzhougou (Chc) formations at the Mouth of the Changzhougou Valley

Fig. 2.7 The outcrop photographs of the Chuanlinggou formation near Guojiagou and Liuzhuangzi Villages. **a**. sandy shale with intercalations of thin-bedded fine-grained sandstone at Chuan-1 Member; **b**. the "mud-cracks" (?) at the Chuan-1 Member; **c**. the green shale of the Chuan-2 Member; **d**. the black shale intercalated with thin-bedded siltstone and fine-grained sandstone at Chuan-3 Member

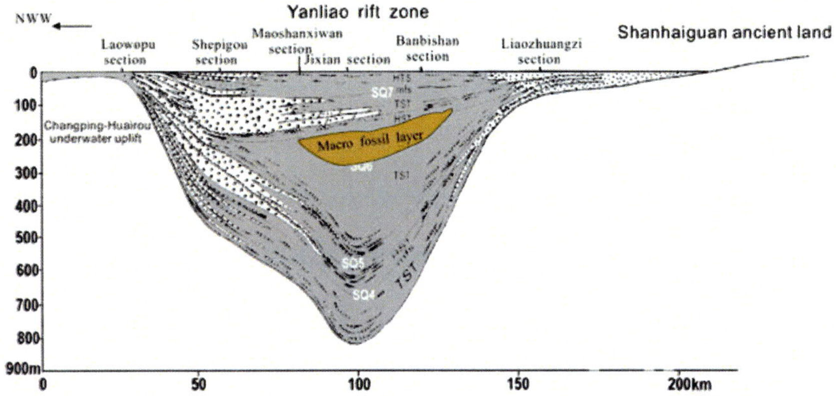

Fig. 2.8 The sketch showing facies changes of the Chuanlinggou formation in YFDZ (Huang et al. 2000)

(1) Barrier island sand-bodies of lagoon basin facies, or the sediments interbedded by multi-bed barrier sand-bodies, each bed is 10-dozens meters thick, but the total stratigraphic thickness would be up to 800–1000 m, especially for the sediments of lagoon basin facies mainly developed in Chuan-2 Member.
(2) With the thinnest stratigraphic thickness, the Chuanlinggou Formation of the Laowopu section isonly 30 m thick at Mingtombs and Xuanlong on the west of the Changping-Huairou submarine uplift (in Jiangxi-Xuanlong Depressions). It is mainly composed of dark-grey, light-purple and greyish-green thin-platy to liminal fine-grained sandstone, siltstone and silty shales with lenticular- and wavy-beddings, and locally developed shrinkage cracks.
(3) In the Mingtombs section and Xuanhua-Longguan zone on the west of the Changping-Huairou submarine uplift zone (in Jingxi-Xuanlong Depressions), and even in the southern segment of the Taihang Mountains (on the shorth of Jingxi Depression), the Chuanlinggou Formation is just dozens meters thick and characterized by the black carbonaceous shale in Chuan-1 Member, the emerald-green potassium-rich shale in the Chuan-2 Member, and the reef-like stromatolitic dolostone dominated by *Eucapsiphora*, etc. in the Chuan-3 Member (Zhu and Chen 1992; Zhu 1993).

The Chuanlinggou Formation is rich-in Acritarches at Jixian, Kuancheng and Pangjiabao, including various unique eukaryotic algal fossils, such as *Leioarachnitum, Trachyarachnitum, Diplomembrana, Schizospora, Goniocystis, Qingshania, Foliomorpha* etc. (Fig. 2.9).

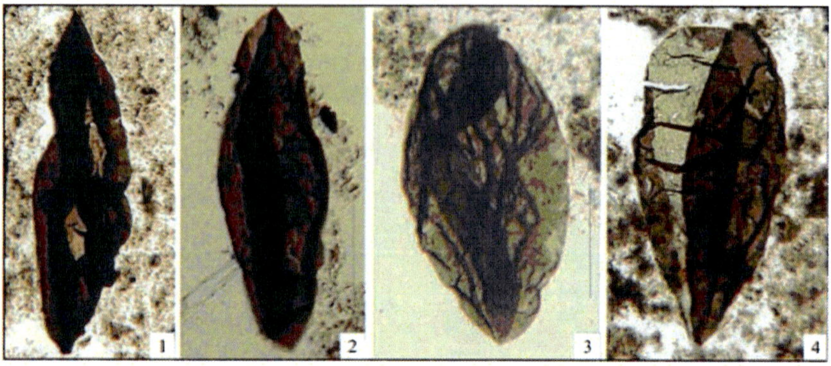

Fig. 2.9 The Microphotographs of Spindle or olive-like acritarch fossils from the Chuanlinggou formation. 1. *Leioarachnitum* sp.; 2. *Leioarachnitum apertum*; 3. *Scapha rugosa*; 4. *Leioarachnitum sinitum*

2.3.1.3 Tuanshanzi Formation (Pt_2^1t or Cht)

It is formerly known as the upper carbonate rocks of the original "Chuanlinggou shale" named by Kao et al. (1934). The regional geological surveying team of Hebei Province found the corresponding carbonate rocks in Jixian section at the Tuanshanzi Village, and proposed a new name of the Tuanshanzi Formation in 1960, which was accepted as a formal lithostratigraphic unit during the symposium on the "Sinian System" in Jixian section in 1964.

The Tuanshanzi Formation in Jixian section is 518 m thick and has conformable contacts with the overlying and underlying strata. It is subdivided into two members. The lower member (Tuan-1 Member) is a 269 m thick greyish-black muddy and silty dolostones interbedded with dolomitic mudstone, which contains straight rhythmic layers due to its variable muddy or sandy contents in dolostones (Fig. 2.10a). Since higher iron content in ankerite, the weathering surface of dolostone often appears as yellowish-brown colour. Abundant carbonaceous fragments and macroscopic algal fossils have often been found in the muddy dolostones. The lithofacies analysis indicates a restricted subtidal and lower energy sedimentary environment, i.e., freshening lagoon facies, in Tuan-1 Member.

While the upper member (Tuan-2 Member) is a 146 m thick dolostone interval with interbedded dolostone, sandy dolostones and dolomitic sandstone, intercalated by quartzose sandstones and sandstone. The Tuan-2 Member mainly shows as medium- to thin-bedded, and the bedding would be gradually thinning upwards (Fig. 2.10b), and ripple marks and mud cracks are common at this interval (Fig. 2.10d), even salty pseudomorphic crystals as well as chute- and scour-molds are sometimes observed on the basal surface of the thin-bedded salty-dolostone. In addition, the stromatolitic bioherms occur in the middle of the Tuan-2 Member (Fig. 2.10c). Therefore, the Tuan-2 Member should be attributed to the sediments of salty intertidal-supratidal environments.

Fig. 2.10 Field photographs showing sedimentary features of the Tuanshanzi formation near Tuanshanzi Village and Dahongyu Valley at Xiaying in Jixian. **a**. The massive and thick-bedded dark-grey ferruginous dolostone in Tuan-1 Member; **b**. the thin-bedded sandy dolostone intercalated by thin-bedded sandstone in the lower Tuan-2 Member; **c**. a stromatolitic bioherm in the middle Tuan-2 Member; **d**. the mud cracks in the upper Tuan-2 Member

The Tuanshanzi Formation exhibits following stratigraphic changes in the YFDZ:

(1) The thickest sequences could be up to a few hundred meters respectively at Jixian, Xinglong and Kuancheng areas (in Jidong-Jibei Depressions).
(2) On the Changping-Huairou submarine uplift (in Jingxi-Xuanlong Depressions) and its west side as well as in Qianxi County (in Jidong Depression), its stratigraphic thickness is significantly thinner, even less than 100 m. In these areas, the Tuan-1 Member is mainly composed of purple ferriferous dolostone, muddy and sand-bearing dolostones and stromatolitic dolostone; while the Tuan-2 Member is chert-stripped dolostone and siliceous stromatolitic dolostone.
(3) In the north of Pinggu (in Jingxi Depression), two layers of ca. 1 m thick K-rich trachyandesite with amygdaloidal structure are intercalated at the upper and middle parts of the Tuanshanzi Formation.
(4) Rhyolitic tuffs are often interbedded within the lower Tuan-1 Member at the Hongqidian (in Jingxi Depression) and Xinglong (in western Jibei Depression), while the interbeds of green K-rich shale occurs at Kuancheng (in central Jibei Depression).

Relatively speaking, the Tuanshanzi Formation is poor in microflora, but the megascopic algal fossils and fossil fragments are abundant in the Tuan-1 Member, especially for the algae of *Tuanshanzia* and *Changchengia* (Fig. 2.11; Zhu and Chen

1995; Yan and Liu 1997), which may mark the first radiation of megascopic multicellular organisms for the early life on the Earth. Regionally, the stromatolitic lithoherms are commonly distributed in the Tuanshanzi Formation mainly for *Gruneria*, *Xiayingella* etc. (Zhu et al. 1978; Tianjin Institute of Geology and Mineral Resources 1980; Zhu and Chen 1992).

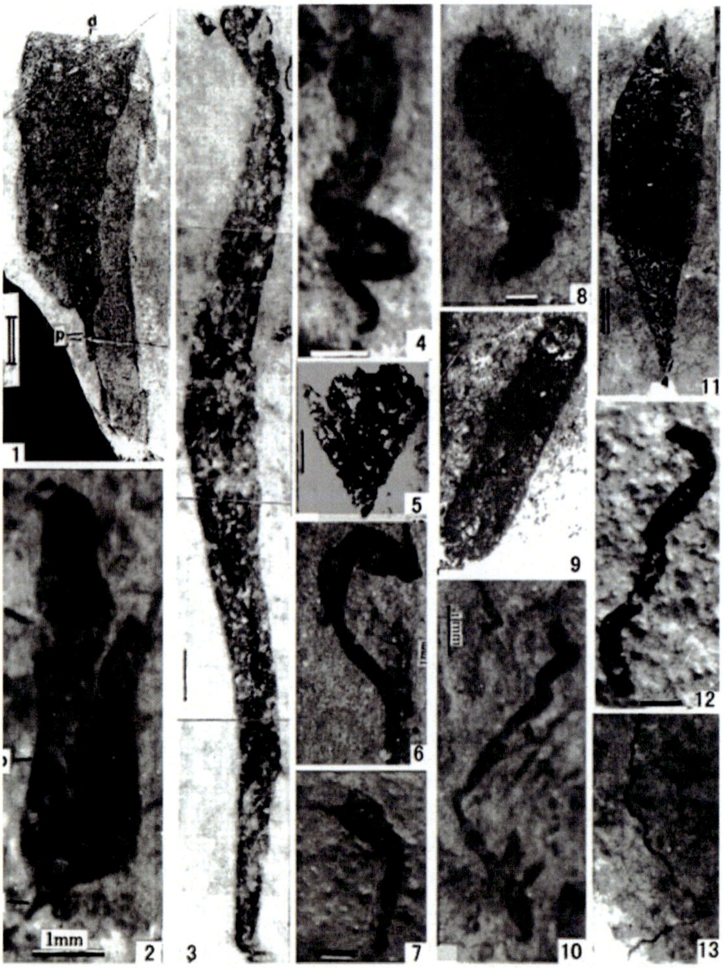

Fig. 2.11 The microphotographs of macroalgal fossils in the Tuanshanzi Formation①. 1, 2. *Changchengia stipitata* (Yan); 3, 4. *Tuanshanzia fasciaria* (Yan); 5. *Eopalmaria pristina* (Yan); 6, 11. *Lanceoforma* sp.; 7. *Longfengshania* sp.?; 8. *Glossophyton* sp.?; 9. *Tawuia* sp. ?; 10. *Vendotaenia* sp.?; 12. *Grypania* sp.; 13. *Tyrasotaenia* sp.?. Scale bars: single lines:1 mm long; double line: 10 mm long. ① Zhu S X, Sun S F, Huang X G. 2006. The multicellular biota from the Mesoproterozoic Changcheng System in Yanshan Mountain. Research Report (in Chinese), slightly modified

The distributional area of Tuanshanzi Formation generally appears as the geomorphological features of low hills, which shows a slightly higher terrain than the distributional area of Chuanlinggou Formation andesitic lava' by Kao et al. (1934) and the "Dahongyu bed" by Shen and Liao (1958).

2.3.1.4 Dahongyu Formation (Pt_2^1d or Chd)

It was named after the "Dahongyu quartzite", and it was renamed as Dahongyu Formation by the First National Conference on Stratigraphy in 1959.

The Dahongyu Formation in Jixian section is 408 m thick and dominated by quartzose sandstone, and intercalated by the volcanic rocks and dolostone (Fig. 2.12). It can be subdivided into three members. The lower member (Da-1 Member) is mainly composed of thick-bedded white quartzose sandstone with the interbeds of purplish-red siltstone, light-green siliceous stripped sandy dolostone, dolomitic quartzose sandstone and bright green K-rich shale. The middle member (Da-2 Member) consist of K-rich mafic volcanic lava (Fig. 2.13b), breccia and agglomerates intercalated by a small amount of quartzose sandstone and tuff. The upper member (Da-3 Member) is the thick-bedded to massive black and white cherty stromatolitic dolostone and chert beds. In above sandstone, the cross beddings, ripple marks and mud cracks are in common. The Dahongyu Formation is conformable contact with the underlying Tuanshanzi Formation in Jixian section and adjacent areas (Fig. 2.13a).

The stromatolites of the Dahongyu Formation are mainly developed in the cherty dolostone of Da-3 Member, most of which are silicified. The main types of stromatolites composed of conical stromatolites, similar to the stromatolitic assemblage in the overlying Gao-1 Member with lager individual and siliceous basic layers, such as *Conophyton dahongyuensis* (Tianjin Institute of Geology and Mineral Resources 1980). In the Dahongyu Formation, microfossils are relatively poor and only found within the muddy interbeds of the Da-1 Member. However, 14 species of 8 genera of Acritarchs have been recently identified (Zhu et al. 1994) and abundant unicellular eukaryotic microfossils have been found in the chert beds of the Da-3 Member.

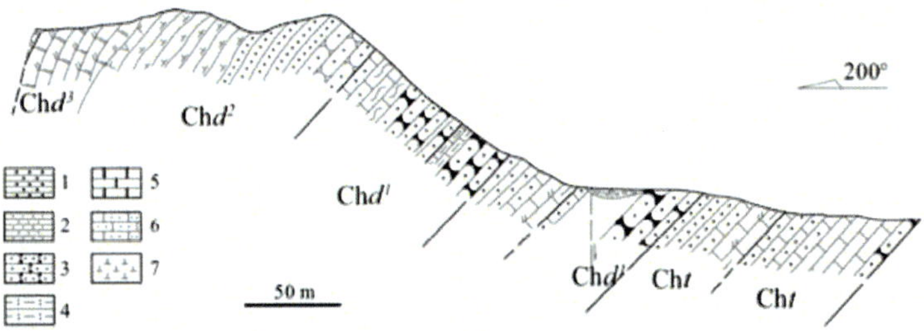

Fig. 2.12 Measured section of the Dahongyu formation at the Dahongyu Valley in the north of Jixian District. Cht. Tuanshanzi formation; Chd. Dahongyu formation. 1. medium-grained sandstone; 2. fine-grained sandstone and siltstone; 3. quartzose sandstone; 4. silty illite shale; 5. dolostone; 6. sandy limestone; 7. K-rich mafic volcanic rocks

Fig. 2.13 The photographs of sandstone and the volcanic rocks in the Dahongyu formation. **a.** The Dahongyu formation (Chd) contacts conformably with the Tuanshanzi formation (Cht); **b.** the K-rich mafic volcanic rocks with amygdaloidal structure in the Da-2 member

A zircon U–Pb age of 1625 ± 6 Ma was reported at the top of the K-rich volcanic lava in the Da-2 Member, providing age constrain (ca.1600 Ma) for the top of the Dahongyu Formation (Lu and Li 1991). The age is recently confirmed by the new U–Pb SHRIMP ages of 1622 Ma and 1625 Ma for the samples from the same outcrop (Lu et al. 2008; Gao et al. 2008a, b, c).

The regional stratigraphic characteristics of the Dahongyu Formation are as follows:

(1) The volcanic rocks are mainly found at Jixian section and its adjacent area, e.g., Pinggu and Xinglong (Jibei Depression) in the middle segment of YFDZ, and at the local area in the eastern segment of YFDZ, e.g., Luanxian (Jibei Depression).
(2) Similar to the underlying formations, the Dahongyu Formation is thicker in Jixian, Kuancheng and Xinglong (in Jidong-Jibei Depressions) where the stratigraphic thickness can be generally up to several hundred meters (e.g., 450 m thick at Kuancheng). While the thickness of Dahongyu Formation would be significantly thinning on the west of Jixian, e.g., only 81 m at the Mingtombs section and 112 m at Pangjiapu section (in Jingxi Depression).
(3) The lithology of the Dahongyu Formation is consistent in YFDZ, which usually forms mountain terrain with medium altitude.

2.3.2 *Jixianian* (Pt_2^2)

The Jixianian is equivalent to the Mesoproterozoic Calymmian of the International Stratigraphic Chart (1600–1400 Ma; Table 2.2d). It consists of the Gaoyuzhuang, Yangzhuang, Wumishan, Hongshuizhuang and Tieling Formations in ascending order. Except for the Hongshuizhuang Formation which is mainly composed of shale, Jixianian is mainly characterized by carbonate sequence and regional disconformable contacts with the underlying Dahongyu Formation and with the overlying Xiamaling

Formation except for the local conformity at the depocenter of Dahongyu Formation (Table 2.4c).

2.3.2.1 Gaoyuzhuang Formation ($Pt_2{}^2g$ or Jxg)

It was formerly known as "Gaoyuzhuang limestone" by Kao et al. (1934), and then renamed as Gaoyuzhuang Formation by the 1st National Conference on Stratigraphy in 1959. Since then it have subdivided into four subformations, namely the Guandi, Sangshuan, Zhangjiayu and Huanxiusi Subformations, by Chen et al. (1980), or ten members (Gao-1 to Gao-10 Members) by Zhu et al.[1]. Totally its stratigraphic thickness in Jixian section would be up to 1596 m.

(1) **Guandi Subformation** (i.e., Gao-1 and Gao-2 Members): The ca. 3 m thick quartzose sandstone and 4 m thick greyish-purple sandy shale with mud cracks are at its base interval. Upwardly, it is chert- stripped, -knobby and -stromatolitic dolostones of the intertidal-supratidal subfacies (Fig. 2.14a). It can be subdivided into two members: the lower member (Gao-1 Member) is rich-in siliceous clasts, and the upper member (Gao-2 Member) is rich-in manganese. Totally the subformation is 267 m thick.
(2) **Sangshu'an Subformation** (i.e., Gao-3 Member): Its lower interval is composed of the manganese-rich siltstone or silty-shale of the intertidal-subtidal subfacies (Fig. 2.14b), locally containing small-size manganese ore, and the upper interval is characterized by the thick-bedded manganese-rich limy-dolostones. The subformation is 282 m thick.
(3) **Zhangjiayu Subformation** (i.e., Gao-4 to Gao-8 Members): It is about 700 m thick and mainly composed of dark micritic-dolostone, limy-dolostone and dolomitic-limestone of subtidal subfacies. Knobby-limestone, gliding and molar-tooth structures are often observed (Fig. 2.14c, d). Organic fragments are enriched within the laminated-bedding and sometimes preserved as megascopic algal fossils such as *Grypania* Walter and *Parachuaria* Sun (Du and Tian 1985; Walter et al. 1990; Sun 2006; Sun et al. 2006; Du et al. 2009). The Gaoyuzhuang macrofossil assemblage is reported in this subformation (Zhu et al. 2016).

    In this subformation, the greyish-black micritic-dolostone rich-in organic matter can be attributed to over-mature hydrocarbon source rock. The subformation can be further subdivided into six members, i.e., Gao-3 to Gao-8 Members in ascending order.
(4) **Huanxiusi Subformation** (i.e., Gao-9 and Gao-10 Members): The subformation is 347 m thick. It has two members. The lower member (Gao-9 Member) consists of the medium- to thick-bedded coarse-crystal dolomitic limestone, with bituminous and bituminous brecciated dolostones at the base.

[1] Zhu S X, Sun S F, Sun L X, Liu H. 2009. Mesoproterozoic carbonate palaeontological study. Tianjin: Tianjin Institute of Geology and Mineral Resources (in Chinese).

Fig. 2.14 Field photographs of the Gaoyuzhuang formation in Jixian section. **a.** Silicic stromatolite dolostone in the Guandi subformation; **b.** the boundary between the chert-stripped dolostone (g1) at the top of the Guandi subformation and manganese silty shale (g2) at the base of the Shangshuan subformation; **c.** knobby-limestone in the lower interval of the Zhangjiayu subformation; **d.** the "molar tooth structure" in the upper interval of the Zhangjiayu subformation

While the upper member (Gao-10 Member) is thick cherty coarse-crystal dolostone with large concentric- and interlocking-nodules and multiple layers of karstic-breccia.

Based on regional geological observation and isopach map, the distribution of Gaoyuzhuang Formation is as follows (Fig. 2.15):

(1) The stratigraphic thickness of Gaoyuzhuang Formation varies between 80 and 1990 m with a depocenter at the Jixian-Qianxi (in Jidong Depression), where its stratigraphic thickness would be more than 1900 m (at Qianxi).
(2) While the stratigraphic thickness is gradually thinning along the SE-direction and tending to zero line towards Shanhaiguan Uplift as well as along NW-direction up to the stratigraphic denudation boundary towards Chongli-Longhua Faulted-Belt along the northern boundary of YFDZ.
(3) The stratigraphic isopach shows a NW-striking in the west segment, a EW-striking in the middle segment and a NE-striking in the east segment of YFDZ.
(4) In comparison with the distribution of other Changchengian formations, the Gaoyuzhuang Formation not only has larger distributional range, but also there is limestone sedimentation within the middle interval of Zhangjiayu Subformation,

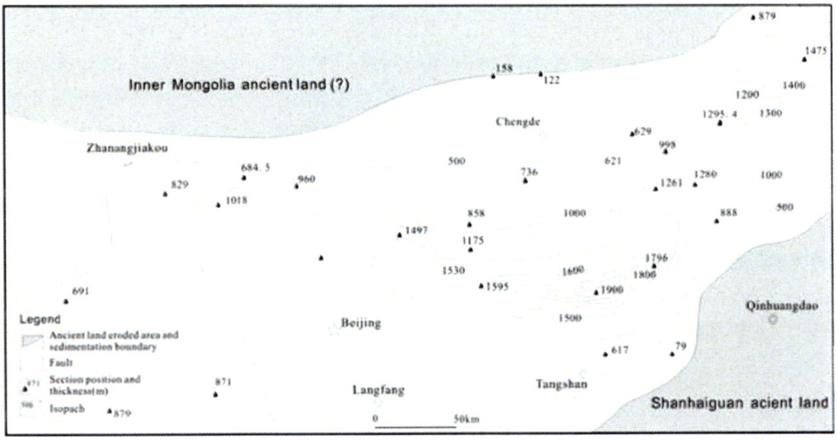

Fig. 2.15 Isopach map of the Jixianian Gaoyuzhuang formation in YFDZ

even both knobby-limestone of basin facies and slump structures of clinothem are observed.

The dolomitic strata of Gaoyuzhuang Formation often appear as mountain terrain with medium elevation.

2.3.2.2 Yangzhuang Formation (Pt_2^2y or Jxy)

It was originated from the "Yangzhuang shale" named by Kao et al. (1934) after the name of Yangzhuang Village in the north of Jixian. The Yangzhuang Formation is characterized by a striking purplish-red or brick-red silty dolostone (Fig. 2.16).

The Yangzhuang Formation can be subdivided into three members. In addition to the purplish-red and greyish-white muddy dolostones (so called "soft dolostones") in Yang-2 Member, its upper and lower members (Yang-1 and Yang-3 Menbers) are interbedded with cherty-nodule and stripped stromatolitic dolostones, dark-grey bituminous dolostone and siliceous dolostone (so called "hard dolostones") so as to become the rhythmic layering of soft- and hard-dolostones in Jixian section.

Based on regional geological observation and stratigraphic isopach map, the distributional characteristics of Yangzhuang Formation are as follows (Fig. 2.17):

(1) The stratigraphic thickness of the Yangzhuang Formation varies from 12 to 770 m. Its depocenter is also located at the Jixian-Qianxi in Jidong Depression with the thickness of up to 500 m and more.
(2) Its thickness is thinning from a few hundred meters to zero meter towards NE- and NWW-directions so that its isopach distribution along NE- and NWW-strikes respectively in the east and west segments of YFDZ, which was constrained by the Shanhaiguan and Inner Mongolia ancient lands (Fig. 2.17).

Fig. 2.16 Outcrop photograph of the purplish-red and greyish-white muddy dolostones in Yangzhuang formation at Jixian section

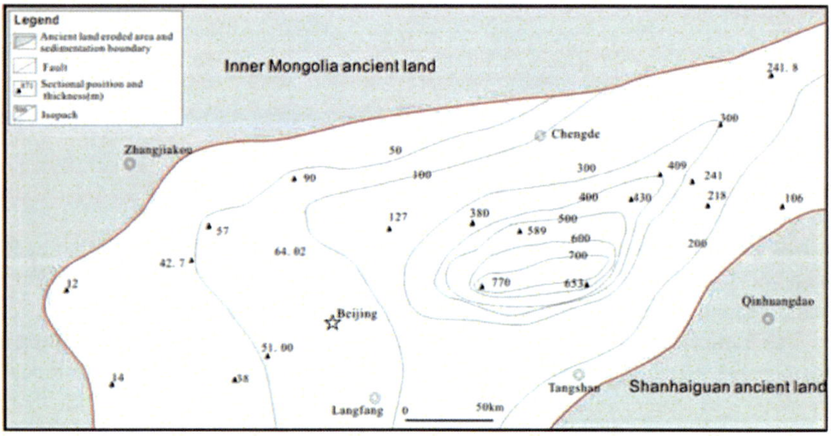

Fig. 2.17 Isopach map of the Jixianian Yangzhuang formation in YFDZ

(3) In the west segment of YFDZ, its thickness is mostly less than 100 m on the west of Miyun (in Jingxi Depression), and it would be thinner further westwards, even the Yangzhuang Formation was no longer deposited on the west of Xuanhua-Yangyuan (in the west of Xuanlong Depression).

(4) The purplish-red strata of Yangzhuang Formation are mainly distributed around the Jixian and Qianxi (in Jidong Depression) as well as Kuangcheng (in Jibei Depression) and Xinglong (in Xuanlong Depression) from east to west in the YFDZ, while the purplish-red strata disappear in the far-east and far-west

segments of YFDZ so that it is hard to distinguish the Yangzhuang Formation from its overlying Wumishan Formation.

(5) The Yangzhuang Formation has conformable contact with the underlying Gaoyuzhuang Formation, but unconformable contact in some areas, e.g., Luanxian (in Jibei Depression).

2.3.2.3 Wumishan Formation (Pt_2^2w or Jxw)

It was named after "Wumishan limestone" of Kao et al. (1934). The Wumishan Formation is composed of various dolostones and contains following characteristics:

(1) The Wumishan Formation is huge-thick up to 3416 m so as to be the thickest lithostratigraphic unit of the Meso-Neoproterozoic sequence in Jixian stratotype section.
(2) Its carbonate rocks are predominated by dolomitic microbiolites accounting for 80–90% of the total thickness for the formation. The microbiolites often show as brown pisiform, sphaeroideous and spotted laminations which are stringed into the basic rhythmic layer of some stromatolites.
(3) The sedimentary rhythms are extremely well developed. In fact, the huge-thick Wumishan Formation is superposed by sedimentary rhythms and sedimentary cycles with different orders (Fig. 2.18), among which, the most basic sedimentary rhythmic layer comprises five rhythmic units (Units A to E) in ascending order (Fig. 2.19):

Fig. 2.18 Photograph of rhythmic dolostones in the Wumishan formation near the Sangyuan Village in the north of Jixian district

Fig. 2.19 Photograph shows five rhythmic units of a basic rhythmic layering in the Wumishan formation near the western Wangzhuang Village in the north segment of Jixian section. Units A to E constitute one basic sedimentary rhythmic layering

① Unit A (basal zone), the sandy muddy or micrite dolostone with mud cracks and salty pseudomorphic crystals of upper supratidal subfacies;

② Unit B (lower zone), lamellated siliceous-stripped microcrystal dolostone of intertidal subfacies, containing stratiform and arched stromatolitic types of microbolites, commonly known as "algal-mat dolostone" or "lower algal-mat layer";

③ Unit C (medium zone), thick-bedded to massive dolosparlite of subtidal subfacies, containing microbiolite of cotted and conical stromatolites.

④ Unit D (upper zone), lamellated siliceous-stripped microcrystal dolostone of intertidal subfacies, containing stratiform and arched stromatolitic types of microbolites, which is referred to the lower part of "upper algal-mat layer";

⑤ Unit E (top zone), light siliceous-stripped microcrystal dolostone of supratidal subfacies (freshwater leached zone), also containing light silicified stratiform and arched stromatolite, commonly attributed to the upper part of "upper algal-mat zone".

Based on incomplete statistics, it is indicated that the Wumishan Formation in Jixian section is totally composed of more than 400 sedimentary rhythms. According to the assemblage of the basic rhythmic units and different sedimentary cycles, the Wumishan Formation can be further subdivided into four subformations with eight members. The basic characteristics of these subformations are as follows.

(1) **Luozhaung Subformation** (i.e., Wu-1 and Wu-2 Members): It is totally 860 m thick. The lower interval is characterized by the rhythmic layers of grey thromboltic-dolostone, algal-mat dolostone and dolomitic shale; the middle

interval comprises the rhythmic layer of stromatolitic dolostone, algal-mat dolostone and dolomicrite; the upper interval consists of the rhythmic layer of algal-mat dolostone, dolomicrite, silty muddy dolostone and dolomitic shale; the top interval contains dolomitic breccia and silicalite. The subformation is rich-in ministromatolites *Pseudogymnosolen* and *Scyphus*, etc.

(2) **Mopanyu Subformation** (i.e., Wu-3 and Wu-4 Members): With stratigraphic thickness of 766 m, the Subformation consists mainly of thick-bedded to massive thrombolite-dolostone, stromatolitic- dolostone, chert-thrombolite or banded micritic-dolostone, algal-mat dolostone and dolomitic shale. Generally, there is a layer of siliceous crust or red-bed on the top of each rhythm. The subformation is characterized by the big to huge conical stromatolites, such as *Conophyton lituum, Jacutophyton furcatum,* etc.

(3) **Ershilipu Subformation** (i.e., Wu-5 and Wu-6 Members): It is 963 m thick. The basal interval is the purplish-red sandy muddy dolostone, dolomitic sandstone and sparite-dolorudite. The lower interval consists of the rhythmic layer of the greyish-white dolomicrite, algal-mat dolostone and dolomitic shale intercalated with oolitic silicalite and sparite-dolorudite. The upper interval is the rhythmic layer of the grey massive thrombolite-dolostone, algal-mat dolostone, stromatolitic dolostone, muddy dolostone and dolomitic shale. The subformation contains abundant conical and columnar types of stromatolites, such as *Conophyton lituum, C. shanpoulingense, Colonnella* cf. *discreta*, etc.

(4) **Shanpoling Subformation** (i.e., Wu-7 and Wu-8 Members): It is 827 m thick. The basal interval is grey dolomitic quartzose sandstone. The lower interval is mainly composed of greyish-while lime-dolostone intercalated by chert-stripped dolomicrite. The upper interval is light-grey chert-banded limy-dolostone, microbial mat dolostone and thick-bedded stromatolitic dolostone. The stromatolites are abundant in the subformation, representing respectively by a few columnar *Colonnella* near the bottom, the conical *Conophyton* and *Jacutophyton* in the middle, and the medium-sized columnar *Pseudochihisienella inconspicua, Wumishanella changzilingensis* and *Paraconophyton inconspicum* at the top. Glauconite infillings in the spaces between of stromatolitic columns are common at top of the subformation.

Regionally, the variation of stratigraphic thickness and lithofacies in Wumishan Formation are as follows:

(1) The total stratigraphic thickness of the Wumishan Formation varies from 650 m to 3330 m (Fig. 2.20). Its thickest one is up to 3368 m at Qinglong (i.e., in Jidong Depression).
(2) In the east segment of YFDZ, its isopach line appears as NE-striking extension with the maximal depocenter distributed around Jixian (in Jidong Depression), where the Wumishan Formation thicker than 3000 m and the stratigraphic thickness is gradually thinning from 910 to 43 m along SE-direction from the Shanhaiguan-Tangshan towards the Shanhaiguan ancient land.
(3) In the west segment of YFDZ, its isopach line appears as EW-striking extension, the stratigraphic thickness is also gradually thinning along W-direction, even

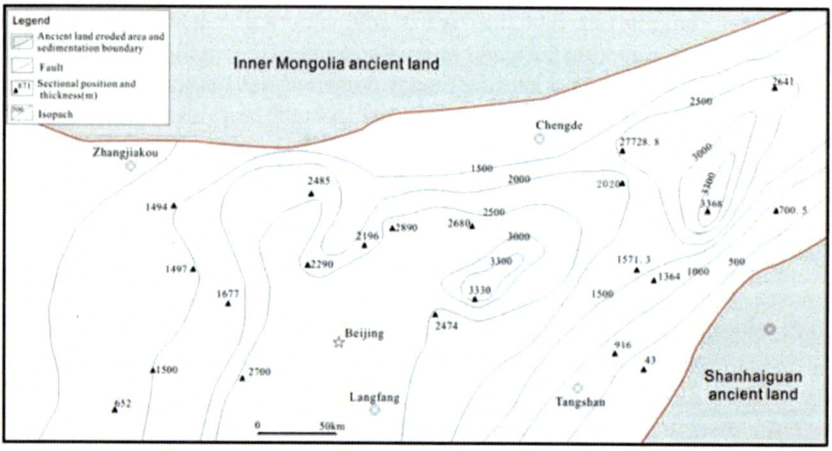

Fig. 2.20 Isopach map of the Jixianian Wumishan formation in YFDZ

reaching a zero isopach line on the west of Xuanhua-Yangyuan adjacent to the Taihang ancient land (i.e., in the Xuanlong-Jingxi Depressions).

(4) In the northern marginal area of YFDZ, the variation of stratigraphic thickness is still uncertain, probably due to denudation and the influence of Chongli-Longhua palaeo-faults.

(5) As a whole, the isopach map of Wumishan Formation shows a palaeo-structural framework of two depressions with larger settlement range and two uplifts with smaller settlement range from west to east in YFDZ, which were alternately distributed probably due to the superposition of the NE-trending faulted-depressions on the NW-trending faulted-depressions. Among both depressions, one depocenter is at Qinglong with the maxima stratigraphic thickness of 3368 m, another at Jixian with maxima thickness of 3330 m.

The Wumishan Formation has conformably contact with the underlying Yangzhuang Formation.

2.3.2.4 Hongshuizhang Formation (Pt_2^2h or Jxh)

It was renamed from the "Hongshuizhuang shale" (Kao et al. 1934) after the name of the Hongshuizhuang Village in the north of Jixian District.

The Hongshuizhuang Formation is mainly composed of black, dark-grey and yellowish-green shales with the stratigraphic thickness is 131 m in the Jixian section, and can be subdivided into two members. The lower member (Hong-1 Member) is dominated by yellowish-green thin-bedded muddy dolostone intercalated by dark-grey thin mudstone rich-in organic matter. The upper member (Hong-2 Member) consists mainly of black, grey and yellowish-green shales with simatic- and pyrite-nodules and lenticular dolostone (Fig. 2.21).

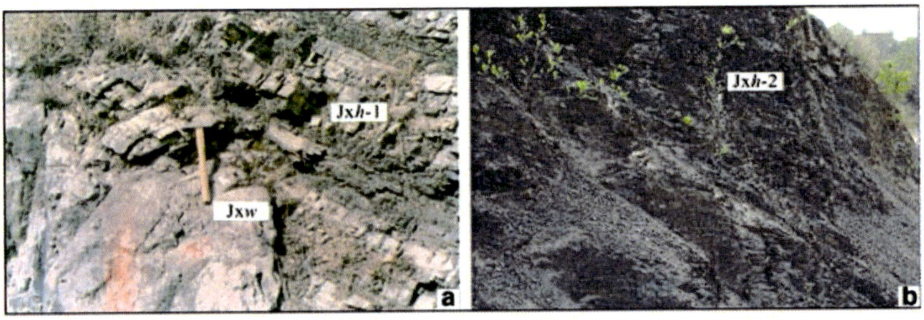

Fig. 2.21 The photographs show the lithology of the Jixianian Hongshuizhuang formation in the Jixian section near the Hongshuizhuang Village in north of Jixian District. **a**. The Hong-1 Member (Jxh-1) composed of thin bedded muddy dolostone with interbeds of black shale and its conformable contact with underlying Wumishan formation (Jxw); **b**. the Nong-2 member mainly consisting of black shale (Jxh-2)

Based on the analyses of stratigraphic isopach map, the Hongshuizhuang Formation shows following variations (Fig. 2.22):

(1) The extensional orientation of its isopach lines is along the NEE-direction in the east Segment of YFDZ and the NWW-direction in the west segment, showing the stratigraphic thickness varies from 40 to 140 m and a depocenter distributed from Kuancheng (in Jibei Depression) through Jixian to Qinglong (in Jidong Depression) with maximal thickness of 130–140 m at Qinglong.
(2) Owing to the constraint of Shanhaiguan ancient land, the stratigraphic thickness tends to vary from 130 to 0 m along the SE-direction from Shanhaiguan to Tangshan (in Jidong Depression) in the east segment.

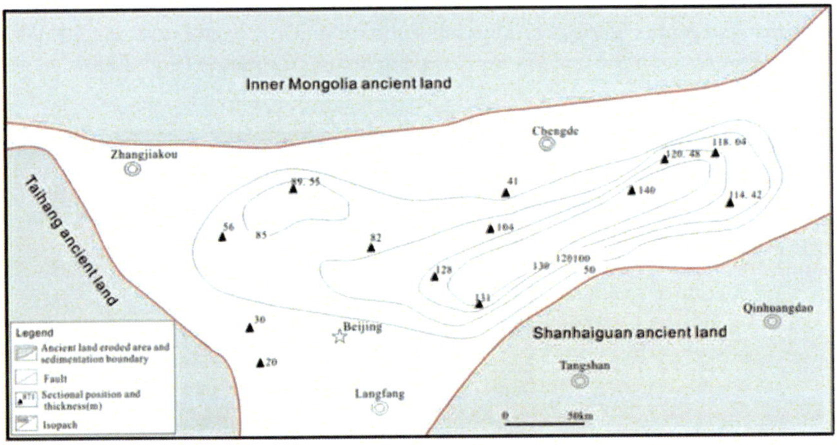

Fig. 2.22 Isopach map of the Jixian Hongshuizhuang formation in YFDZ

(3) In the west segment, the stratigraphic thickness would thin out westwards with the thickness from 130 to 0 m, even it has been denuded out on the west of Xuanhua-Yangyuan (in Xuanlong Depression) obviously due to the control of Taihang ancient land.
(4) Along the northern border of YFDZ, the stratigraphic variation would be uncertain, probably showing a zero isopach line, due to the influences of denudation and the Chongli-Longhua palaeo-faults.

The dark-grey muddy dolostone of the lower Hongshuizhuang Formation directly overlap on the underlying stromatolitic dolostone of the Wumishan Formation with a disconformable contact of mutational change.

2.3.2.5 Tieling Formation ($Pt_2{}^2t$ or Jxt)

It is mainly a carbonate sequence. Kao et al. (1934) had originally called it the "Tieling limestone" after the name of Tieling Village in the north of Jixian District. In view of the lithological variation between its two stratigraphic intervals, Chen et al.(1980) presented a clear depositional hiatus and the reversal of palaeomagnetic poles. Accordingly, the Tieling Formation has been subdivided into two subformations or members, i.e., the Daizhuangzi and Laohuding Subformations (or Tie-1 and Tie-2 Members) in ascending order.

(1) **Daizhuangzi Subformation** (i.e., Tie-1 Member): With total stratigraphic thickness of 153 m, its basal interval is marked by the greyish-white thin-bedded or lenticular quartzose sandstone, and the lower interval consists of brown manganiferous stromatolitic intradolostone and manganiferous dolomicrite with rudaceous and sandy clasts intercalated by thin greyish-green shale. The upper interval is characterized by the variegated shales intercalated by manganiferous dolostone. This sub-Formation (Tie-1 Member) is conformably in contact with the underlying Hongshuizhuang formation (Fig. 2.23).

Fig. 2.23 The conformable contact between Tieling formation (Jxt) and Hongshuizhuang Formation (Jxh) in the north of Jixian District

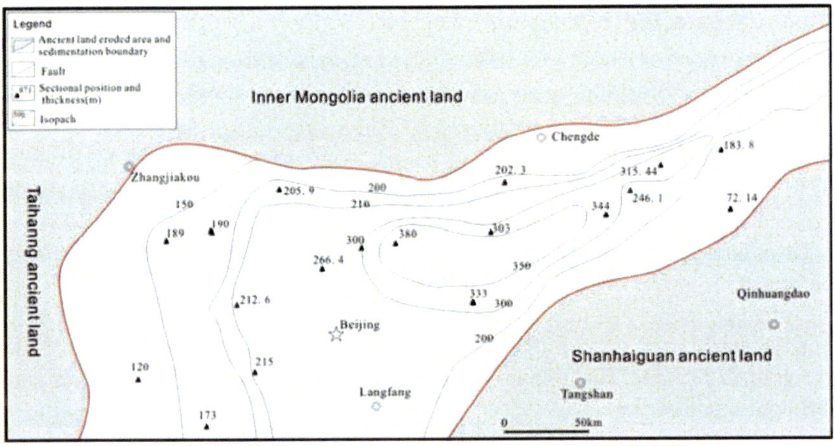

Fig. 2.24 Isopach map of the Jixian Tieling formation in YFDZ

(2) **Laohuding Subformation** (i.e., Tie-2 Member): It is a limestone sequence with a thickness of 181 m in the Jixian section. Its lower interval mainly consists of lime-dolostone, dolomitic limestone and wormkalk-calcirudite, and the upper interval is almost composed of stromatolitic limestone except for the muddy and dolomitic limestone at its top.

Based on the analyses of stratigraphic isopach map, the Tieling Formation shows following variations (Fig. 2.24):

(1) The stratigraphic thickness of Tieling Formation varies from 120 to 380 m. Its depocenter with thickness of 330–380 m is located in the Miyun-Kuancheng-Jixian area, where the maximum thickness is up to 380 m in Miyun County (in Jingxi-Jibei-Jidong Depressions; Fig. 2.24).
(2) The distribution range of Tieling Formation would be similar to that of the underlying Hongshuizhuang Formation (cf. Figures 2.22 and 2.24). But the Tieling Formation is completely absent due to the long-term erosion before the deposition of the overlying Luotuoling Formation on the east of Zunhua (in Xuanlong Depression).
(3) In the east segment of YFDZ, its isopach line shows a NE-Orientational extension, and the stratigraphic thickness is thinning SE-towards from 300 m to zero m along the Shanhaiguan to Tangshan (in Jidong Depression) because of the constraint of Shanhaiguan ancient land.
(4) In the west segment, the isopach line extends along EW-direction, the thickness gradually thins out westwards, and it would tend to zero line on the west of Xuanhua-Yangyuan (in Xuanlong Depression) due to the influence of the Taihang ancient land.

2.3.3 *Xiamaling Formation* ($Pt_2{}^3x$)

The Xiamaling Formation was named after the "Xiamaling shale" reported by Ye (1920) based on the outcrop section near Xiamaling Village in the Mentougou, Beijing (in Jingxi Depression).

The Xiamaling Formation in the Jixian section is a suite of fine-grained clastic sediments with the largest thickness of 168 m at Luotouling in Jixian (Fig. 2.25). At its basal interval, granule conglomerates can be seen. The lower interval is mainly composed of grey or greyish-purple coarse sandstone, greyish-black silty-shale and siltstone commonly with cross-beddings and ripple marks. The upper interval is predominated by the greyish-black and yellowish-green silty-shale intercalated with fine-grained siltstone. The formation is in disconformable contact with the underlying Tieling Formation.

According to the field investigation and isopach map analysis, the stratigraphic characteristics of the Xiamaling Formation are summarized as follows (Fig. 2.26).

(1) The stratigraphic thickness of Xiamaling Formation varies generally from 100 to 537 m. Its distributional orientation is nearly EW-direction, which is controlled by the Shanhaiguan ancient land on the Southeast side, the Inner Mongolia ancient land on the Northwest-North side, and the Taihang ancient land on the West side (Fig. 2.26).
(2) The isopach center/depocenter is located geographically at the Zhaojiashan section at Huailai (in Xuanlong Depression) in west segment of YFDZ, where the maximum stratigraphic thickness of Xiamaling Formation is up to 545 m (Fig. 2.26).

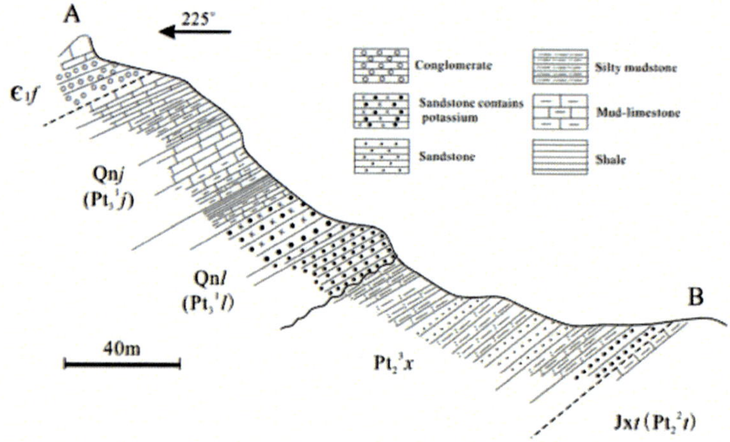

Fig. 2.25 the measured Luotuoling to Laoguading geological cross-section in Jixian. $Pt_2{}^2t$. Jixianian Tieling formation; $Pt_2{}^3x$. Xiamaling Formation; $Pt_3{}^1l$. Qingbaikouan Luotuoling Formation; $Pt_3{}^1j$. Jing'eryu Formation; $\epsilon_1 f$. Lower Cambrian Fujunshan formation

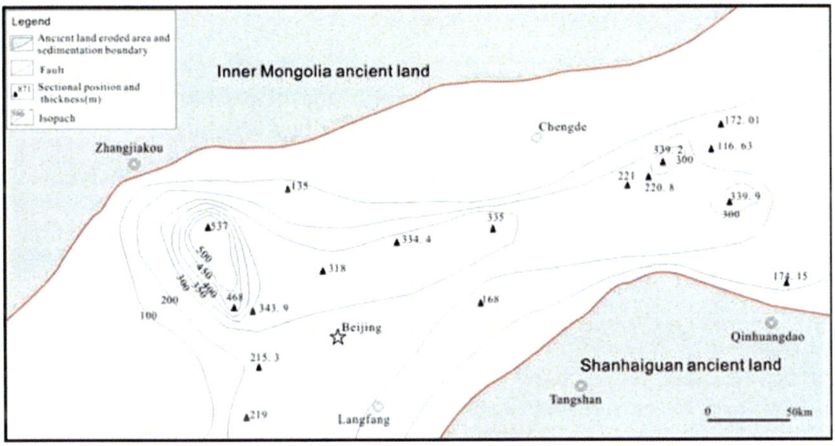

Fig. 2.26 Isopach map of the Xiamaling formation in the YFDZ

(3) The Xiamaling Formation in Xuanlong Depression can be subdivided into four members in ascending order: ① Xia-1 Member ($Pt_2{}^3x^1$) sandy shale; ② Xia-2 Member ($Pt_2{}^3x^2$) greyish-green shale; ③ Xia-3 Member ($Pt_2{}^3x^3$) black shale; and ④ Xia-4 Member ($Pt_2{}^3x^4$) variegated shale intercalated with marl (Fig. 2.27).

(4) In comparison, the Xiamaling Formation at Jixian section in Jidong Depression (the east segment of YFDZ) is only equivalent to the Xia-1 Member plus the basal interval of Xia-2 Member of the Zhaojiashan section in Xuanlong Depression. At Kuancheng section in Jibei Depression (the middle segment of YFDZ), the Xiamaling Formation only represents part of the Xia-1 Member of the Zhaojiashan section.

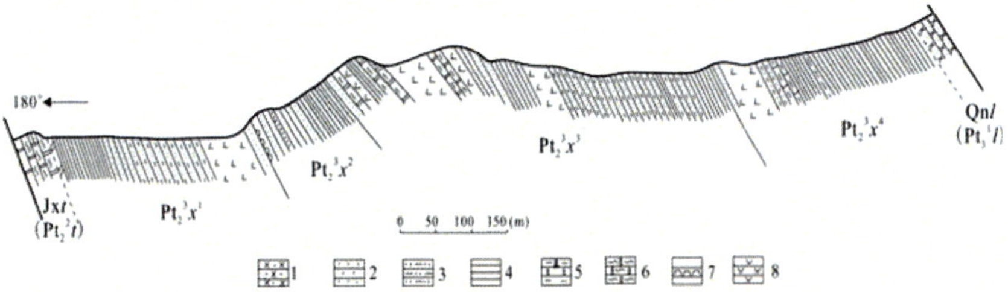

Fig. 2.27 The measured Zhaojiashan cross-section of Xiamaling formation ($Pt_3{}^1x$) at Huailai County in west segment of YFDZ (Du and Li 1980, modified). 1. Quartzose sandstone; 2. sandstone; 3. muddy siltstone; 4. shale; 5. muddy dolostone; 6. siliceous banded and striped dolostone; 7. sandstone with iron nodules. $Pt_2{}^2t$. Tieling formation; $Pt_2{}^3x$. Xiamaling formation; $Pt_3{}^1l$. Luotuoling formation

2.3.4 *Qingbaikouan ($Pt_3{}^1$ or Qn)*

The Qingbaikouan is named after the original "Qingbaikou Group" from Kao et al. (1934), and includes two Formations, i.e., Luotuoling and Jing'eryu Formations (Table 2.4).

2.3.4.1 Luotuoling Formation ($Pt_3{}^1l$or Qn*l*)

It unconformably overlays the Xiamaling Formation (Fig. 2.28). On a regional scale, the stratigraphic contact relationship seems to be a generally micro-angular unconformity. In Jixian stratotype section, the Luotuoling Formation is mainly composed of siliciclastic rocks with stratigraphic thickness of 118 m. It can be subdivided into two members in ascending order:

(1) **Luo-1 Member**: It consists mainly of a sandstone interval which includes:

① Yellowish-brown medium- to thick-bedded, bebbly feldspathic quartzose sandstone and lenticular granule conglomerate as the basal interval.
② Feldsparthic quartzose sandstone intercalated by greyish-yellow muddy siltstone as the lower interval, in which the large tabular-, wedge-, chevron- and herringbone-cross-beddings, ripple mark and mud cracks on bedding surface of muddy siltstone are well-developed.

Fig. 2.28 The micro-angular unconformable contact between Luotuoling (Qnl) and Xiamaling formations ($Pt_3{}^1x$) at the Xiazhuangzi Village in the north of Jixian district). $Pt_2{}^3x$. shales at the top of the Xiamaling formation; Qn*l* ($Pt_3{}^1l$). basal conglomerate and bebbly feldsparthic coarse sandstone of the Luotuoling formation

③ Greyish-white thick-bedded to massive glauconitic quartzose sandstone intercalated by light-grey medium- to thin-bedded quartzose sandstone and grey shaly silty-shale as the middle interval.

④ Greyish-white thick-bedded to massive quartzose sandstone intercalated by greyish-green silty-shale as the upper interval, containing megascopic carbonaceous fossils represented by *Longfengshania* (Du and Tian 1985; Du et al. 2009) and other microscopic acritarchs.

(2) **Luo-2 Member**: It is characterized by the variegated shale consisting mainly of greyish purple, greyish-black and greyish-green shales.

2.3.4.2 Jing'eryu Formation (Pt_3^1j or Qnj)

It is equivalent to the lower part of the original "Jing'eryu limestone" (Kao et al. 1934).

The Jing'eryu Formation in Jixian section is mainly a set of marine carbonate deposits with a total thickness of 112 m. Its lower interval consists of grey or greyish-purple thin-bedded marl with a basal granule- and glauconitic coarse sandstone of 10–20 cm thick. The middle interval comprises grey and egg-cyanic thick-bedded micritic limestone intercalated by marl. The upper interval is characterized by grey thin-bedded muddy and limy dolostone, dolomitic limestone intercalated by purplish-red shale.

The stratigraphic contact between the Jing'eryu Formation and its underlying Luotuoling Formation is generally regarded as conformable, but it could not be excluded that there still is short-term depositional discontinuity between both formations according to the presence of a thin-bedded glauconitic bebbly coarse sandstone on the contact plane (Fig. 2.29). The Jing'eryu Formation is overlain by the Lower Cambrian Fujunshan Formation with a micro-angular unconformity that marks the famous "Jixian Movement" (Fig. 2.30).

A few of megascopic carbonaceous compressions have been found in the greyish-green thin limestone near the base of the Jing'eryu Formation at the Dongjingyu Village in Jixian, such as *Chuaria circularis* and *Shouhsienia shouhsiensis* (Zhu et al. 1994).

At the north ridge to the west of Jing'eryu Village, the Lower Cambrian Fujunshan Formation is composed of ① brow to red palaeo-weathering crust, ② breccia-bearing coarse sandstone or carbonate-cemented breccia, ③ bituminous oncolitic limestone, and ④ massive leopard limestone in ascending order. The leopard limestone contains trilobites represented by *Redlichia chinensis* Walcott and *Megapalaeolenus fengyangensis* Chu.

Fig. 2.29 The Jing'eryu formation (Qn*j*) and its contact relationship with the underlying Luotuoling formation (Qn*l*)

Fig. 2.30 The lower Cambrian Fujunshan formation (**b**) and its underlying Jing'eryu formation (**a**) at the Beiling in Jixian District. The white dash-line shows the micro-angular unconformity

2.4 Geochronology and Chronostratigraphy

In comparison with the "Chinses Stratigraphic Chart 2001" (Table 2.4b; China Commission on Stratigraphy 2001), the "Chinese Stratigraphic Chart 2013" (Table 2.4b) has been updated to reflect the advances on the Meso-Neoproterozoic geochronology and lithostratigraphy in the YFDZ during the past decades, which are summarized as follows.

2.4.1 Stratigraphic Boundary and Dating Between the Xiamaling Formation and Qingbaikouan

Previous stratigraphic studies indicated that the Xiamaling Formation in YFDZ is bounded by the unconformities with both underlying Ticling and overlying Luotuoling Formations and a geological age about 1000–900 Ma. Therefore, the Xiamaling Formation used to be referred to the base of the Neoproterozoic Qingbaikouan. The previous dating includes the following date:

(1) A K–Ar age of 1050 Ma from the glauconite in the upper Tieling Formation (Zhong 1977);
(2) An age of 956 Ma from the illitic shale in the Xiamaling Formation (Yu and Zhang 1984);
(3) A Pb-Pb age of 879 ± 18 Ma from the lower Xiamaling Formation at Xishan, west Beijing (Qiao 1976).

However, a number of new dating results have been published from the Xiamaling Formation in recent dozen years, which are summarized as follows.

(1) The SHRIMP zircon U–Pb ages of the bentonitic beds within the Xia-3 Member, i.e., 1368 ± 12 Ma and 1370 ± 11 Ma at western Beijing in the Jingxi Depression (Gao et al. 2007a, b), and 1366 ± 9 Ma at the Zhaojiashan section of the Xuanlong Depression (Gao et al. 2008a, b, 2009).
(2) The SHRIMP zircon U–Pb ages 1379 ± 12 Ma and 1380 ± 36 Ma of K-bentonite from Xiamaling Formation in above two locations (Su et al. 2008).
(3) The SIMS baddeleyte U–Pb ages of 1327.5 ± 2.4 Ma and 1327.3 ± 2.3 Ma for the gabbro-diabase sills intercalated within the Xiamaling Formation at Kuancheng and Pingquan in Jibei Depression[2](cf. Chap. 10).
(4) The SHRIMP zircon U–Pb age of 1320 ± 6 Ma for the diabase sill of the Xiamaling Formation at the Kuancheng in Jibei Depression (Li et al. 2009).
(5) A SHRIMP zircon U–Pb age of 1372 ± 18 Ma of the K-bentonite in the Xiamaling Formation at the border area between Jibei and Liaoxi Depressions was also reported by Su et al. (2010).

[2] Wang T G, Zhong N N, Zhu S X, et al. 2009. Petroleum prospectivity and reginal play predication of the lower stratigraphic assemblage in North China Platform (internal report). Beijing: China University of Petroleum-Beijing.

(6) A SHRIMP zircon U–Pb age of 1437 ± 21 Ma of the K-bentonites in the underlying Tieling Formation is dated (Su et al. 2010).

To sum up, the above geochronologic results of Xiamaling Formation are on the basis of isotopic dating using different minerals such as baddeleyte, zircon and bentonite, and all the ages just in between 1400 and 1320 Ma. According to this duration of age, therefore, the Xiamaling Formation should be attributed to the Middle Mesoproterzoic, rather than the Early Neoproterozoic Qingbaikouan as considered previously. If the glauconite K–Ar age of ca. 1000 Ma or 900 Ma from the basal Luotuoling Formation is reliable (Yu and Zhang 1984; Qiao and Gao 1997), and then it should have a long-term stratigraphic break of 320–420 Ma between the Mesoproterozoic Xiamaling and the Neoproterozoic Luotuoling Formations.

2.4.2 Stratigraphic Boundary Between the Changchengian and Jixianian

As shown in Table 2.4a and b, the boundary between the Changchengian and Jixianian was originally designated at the stratigraphic boundary between the Gaoyuzhuang and Yangzhuang Formations (Kao et al. 1934; China Commission on Stratigraphy 2001), Because Kao et al. (1934) considered that the stratigraphic contact relationship between the Gaoyuzhuang and overlying Yangzhuang Formations is paraconformable and the contact between the Gaoyuzhuang and underlaying Dahongyu Formations is conformable at Jixian stratotype section.

In the early 1950s, Shen and Liao first pointed out that the Gaoyuzhuang Formation has a disconformable contact with the underlying Dahongyu Formation, and a conformable contact with the overlying Yangzhuang Formation. Accordingly, they initially proposed that Gaoyuzhuang Formation should be assigned to the Jixianian in the stead of Changchengian (Shen and Liao 1958).

However, the above significant point of view did not cause wide-spread concern. In view of its importance, the present author has focused on the stratigraphic contact relationship, and carried a review and regional reinvestigation. The results of new observation are emphasized as follows.

2.4.2.1 The Contact Relationship of Gaoyuzhuang Formation with Its Overlying and Underlying Strata

The regional disconformable contact between the Jixianian Gaoyuzhang and its underlying Changchengian Dahongyu Formations (Table 2.4d) is confirmed by field geological evidences in most areas of the YLDZ except for the depocenter of Jibei Depression at Kuancheng (north Hebei) where a local conformable contact is observed (Fig. 2.31d). The following geological observations could provide evidences for the above-mentioned stratigraphic contact relationship.

Fig. 2.31 The basal contact of the Gaoyuzhuang formation. **a**. the basal quartzose sandstone of the Gaoyuzhuang formation truncates the conical stromatolites on the top surface of the Dahongyu formation, Xiaohongyugou Valley, Jixian section; **b**. the weathering crust on the top of the Dahongyu formation, Xiazhuangzi Village, Qianxi County; **c**. the basal conglomerate at the base of Gaoyuzhuang formation, Matiyu Village, Qianxi County; **d**. the thick-bedded basal sandstone of the Gaoyuzhuang formation conformably overlays the thin-bedded sandstone at the top of the Dahongyu formation, Yamenzi Villagy, Kuancheng County

(1) The conical stromatolite Conophyton dahongyuense was truncated at the top surface of the Dahongyu Formation by the basal quartzose sandstone of its overlying Gaoyuzhang Formation in Jixian stratotype section (in Jidong Depression; Fig. 2.31a).
(2) The ferruginous weathering crust is often seen at the top of the Dahongyu Formation in the eastern and western YLDZ (in Jidong and Xuanlong Depressions; Fig. 2.31b).
(3) A 2–5.6 m thick basal conglomerate bed is found at the base of the Gaoyuzhuang Formation in Qinglong and Qianxi (in Jidong Depression; Fig. 2.31c).
(4) The Gaoyuzhang Formation often overlaps the Palaeoproterozoic or Archean metamorphic strata at the marginal areas of the YLDZ such as the Taihang and Wutai Mountains, showing an overlap sequence.

In a word, the above evidences indicate that the stratigraphic boundary between Gaoyuzhuang Formation and its underlying strata should be a regional disconformity or overlap unconformity, indicating a crustal movement before the sedimentation of Gaoyuzhuang Formation previously called as "Qinglong Uplifting", which is characterized by one Gaoyuzhang basal conglomerate bed at the south of Qinglong (in Jidong Depression; Chen et al. 1980).

Although another Yangzhuang basal conglomerate bed is also found at Taoyuan (in Jidong Depression) and somewhere else, its gravels mainly consist of gneiss, basic and acid volcanic rocks with the maximal grain-size up to 40 cm, showing an obvious local disconformable contact relationship between the Yangzhuang and its underlying Gaoyuzhuang Formations, and marking the crustal movement called as "Luanxian Uplifting". As contrast with the "Qinglong Uplifting", however, the "Luanxian Uplifting" only results in a local-scale disconformity between the Mesoproterozoic Yangzhuang and Gaoyuzhuang Formations.

Moreover, the Yangzhuang Formation with the interbeds of the purplish-red muddy dolostone also shows a comformable contact relationship with the underlying Gaoyuzhuang Formation in the Jixian section (Figs. 2.31d and 2.32) where the "Luanxian Uplifting" would be inexistent so that the Yangzhuang Formation can only be distinguished by the occurrence of purplish-red sandy and muddy dolostone as the sole identification marker.

In addition, even the Yangzhuang Formation lacks the purplish-red sandy and muddy dolostone, and its lithology is similar to that of the upper Gaoyuzhang Formation in the western and northern YLDZ, in this case, the Yangzhuang Formation can hardly be distinguished from the Gaoyuzhuang Formations, in between sometimes there only is a layer of 1–2 m thick white silicalite (or siliceous crustation) at the base of Yangzhuang Formation.

From the point of view of the lithostratigraphy in the YLDZ, the contact relationship between Yangzhuang and Gaoyuzhang Formations is principally regional conformable with a few local exceptions (Fig. 2.32; Table 2.4d). Therefore, it is

Fig. 2.32 Conformable contact between the Gaoyuzhuang formation (Jxg) and its overlying Yangzhuang formation (Jxy) in the Jixian section

very reasonable to assign the stratigraphic boundary between the Jianxianian and its underlying Changchengian at basal surface of the Gaoyuzhang Formation.

2.4.2.2 The Age for the Boundary Between Jixianian and Changchengian

A single grain zircon U–Pb age of 1625 ± 6 Ma from the volcanic lava intercalated within the middle of Dahongyu Formation in Jixian has provided an age constraining on the top boundary of the Dahongyu Formation to ca. 1600 Ma (Lu and Li 1991). This age was confirmed by new zircon SHRIMP U–Pb ages of 1622 Ma and 1625 Ma sampled from same outcrop (Lu et al. 2008; Gao et al. 2008a, 2009).

As described above, although there is a regional disconformity between Gaoyuzhang and Dahongyu Formations, a nearly conformable contact is also observed at the depocenter of sedimentary depression in the YLDZ (in Jibei Depression), suggesting that the depositional break between the Gaoyuzhang and Dahongyu Formations continued not so long. Therefore, the age of 1600 Ma is roughly equivalent to the age for the Jixianian basal boundary. The basal age of the Jixianian is supported by a SHRIMP zircon U–Pb age of 1559 ± 12 Ma and a LA-MC-ICPMS zircon U–Pb age of 1560 ± 5 Ma from the volcanic ask bed at the upper Gaoyuzhang Formation in the west YLDZ (Li et al. 2010).

Since the Jixianian spans a time from 1600 Ma to 1400 Ma (Table 2.4d), it is equivalent to the Calymmian in the International Stratigraphic Chart (Table 2.4e).

2.5 Conclusions and Perspectives

(1) The Meso-Neoproterozoic strata in the YLDZ are subdivided into three systems plus one individual formation, i.e., Pt_2^1 Changchengian (incl. Changzhougou, Chuanlinggou, Tuanshanzi and Dahongyu Formations), Pt_2^2 Jixianian (incl. Gaoyuzhuang, Yangzhuang, Wumishan, Hongshuizhuang and Tieling Formations), Pt_2^3x Xiamaling Formation as well as Pt_3^1 Qingbaikouan (incl. Luotuoling and Jing'eryu Formations) in ascending order.

(2) The Xiamaling Formation is newly attributed to the Middle Mesoproterozoic strata based on the ages (e.g., 1368 Ma from the ash beds in its lower interval and 1320 Ma from the diabase sill in its middle-upper interval).

(3) There is a long-lasting depositional break (from 1320 Ma to 1000 Ma, ca. 320 Ma,) between the Mesoproterozoic Xiamaling and Neoproterozoic Luotuoling Formations. The regional micro-angular unconformity at the base of the Qingbaikouan may reflect a tectonic movement related to the Greenville Orogeny that led to formation of the Supercontinent Rodinia.

(4) The Changchengian (1670–1600 Ma) is the oldest unmetamorphosed sedimentary sequence in China. It is predominated by siliciclastic sequence which represents the deposition in the early stage of the Yanliao Faulted-Depression Zone

(YFDZ) related to the early breakup of Supercontinent Columbia and also to the earliest radiation of megascopic multicellular organisms for the early life on the Earth.

(5) The stratigraphic contact relationship between the Changchengian Dahongyu Formation and its overlying Jixianian Gaoyuzhuang Formation appears as a regional disconformity with local conformity at the Dahongyu depocenter. Both Changchengian and Jixianian should be referred to a basically continuous sedimentary sequences in the Jixian stratotype section, and as a very complete Calymmian sequence it should be attributed to the Mesoproterozoic Erathem.

(6) As the stratotype, the completeness of Meso-Neoproterozoic strata from Changchengian through Jixianian to Qingbaikouan at the Jixian section must be reassessed, it is fully qualified as the candidate for the global stratotype of the Calymmian System.

Acknowledgements This research has been supported by the National Natural Science Foundation of China (41272015), China Geological Survey (1212010611802) and China Petroleum and Chemical Corporation (YPH08086). The authors greatly appreciate Professors Lu Songnian, Huang Xueguang and Sun Shufen from the Tianjin Institute of Geology and Mineral resources for their helps both in the field and labs, valuable information and advices. In addition, we thanks Yang Ligong and others from the "Tianjin Protection Office on Jixian section of the Meso- and Neoproterozoic sequences" for their assistance in working on the Jixian section.

References

Chen JB, Zhang HM, Zhu SX, Zhao Z, Wang ZG (1980) Research on Sinian suberathem of Jixian, Tianjin. In: Tianjin Institute of Geology and Mineral Resources (ed). Research on precambrian geology, Sinian suberathem in China. Tianjin Science and Technology Press, Tianjin. 56–114 (in Chinese)

Chen RH, Lu ZB (1963) Sinian standard geological profile in Jixian of Hebei. Geol Ser Precambr Geol 1:99–127

China Commission on Stratigraphy (2001) Guide to Stratigraphy in China and introduction to the guide, revised. Geological Publishing House, Beijing (in Chinese)

China Stratigraphy Code Editorial Board (1999) China stratigraphic code in the mesoproterozoic. Geological Publishing House, Beijing (in Chinese)

Compilation Group of Regional Stratigraphic Table in Beijing (1977) North China Stratigraphic table: Beijing City Fascicle. Geological Publishing House, Beijing (in Chinese)

Du RL, Tian LF (1985) The discovery and study of longfengshan macroalgae in Qingbaikou system, Yanshan. Geol Sin 3:183–190 (in Chinese with English abstract)

Du RL, Tian LF, Hu HB, Sun LN, Chen J (2009) The neoproterozoic qingbaikou period longfenshan biota. Science Press, Beijing (in Chinese)

Du RL, Li PJ (1980) Sinian suberathem in the western Yanshan ranges. In: Tianjin Institute of Geology and Mineral Resources (ed) Research on precambrian geology: Sinian suberathem in China. Tianjin Science and Technology Press, Tianjin, pp 341–357 (in Chinese)

Editorial Board on China Stratigraphic Chart (2013) China stratigraphic chart (Tentative). Geological Publishing House, Beijing (in Chinese)

Gao LZ, Zhang CH, Shi XY, Zhou HR, Wang ZQ (2007a) Xiamaling tuff SHIRIMP U-Pb zircon dating of Qingbaikou system in Northern China. Geol Bull 26(3):249–255

Gao LZ, Zhang CH, Shi XY, Zhou HR, Wang ZQ (2007b) A new SHRIMP age of the Xiamaling formation in the North China plate and its geological significance. Acta Geol Sin 81(6):1103–1109

Gao LZ, Zhang CH, Shi XY, Song B, Wang ZQ, Liu YM (2008a) Mesoproterozoic age for Xiamaling formation in North China plate indicated by zircon SHRIMP dating. Chin Sci Bull 53(17):1665–2671

Gao LZ, Zhang CH, Yin CY (2008b) New evidence for the age of SHRIMP zircon in the mesoproterozoic and neoproterozoic strata in north China. Earth J 20(3):366–376

Gao W, Zhang CH, Gao LZ (2008c) Zircon SHRIMP U-Pb ages and tectonic implications of Rapakivi granite, Beijing Miyun. Geol Bull 27(6):793–798 (in Chinese with English abstract)

Gao LZ, Zhang CH, Liu PJ, Tang F, Song B, Ding XZ (2009) Reclassification of the Meso- and Neoproterozoic chronostratigraphy of North China by SHRIMP zircon ages. Acta Geol Sin 83(6):1074–1084

Grabau AW (1922) The Sinian system. Bull Geol Soc China 1:44–88

Huang XG, Zhu SX, He YZ,Chen HN, Liu WX (2000) Research on the meso-neoproterozoic sequence-stratigraphy in Chengde Area, North China. Tianjin Institute of Geology and Mineral Resources, Tianjin (in Chinese)

International Commission on Stratigraphy (2013) International chronostratigraphic chart. J Stratigr 37(3):250

Kao CS, Hsiung YH, Kao P (1934) Preliminary notes on Sinian stratigraphy of North China. Bull Geol Soc China 13(2):243–288

Li HK, Lu SN, Li HM, Sun LX, Xiang ZQ, Gen JZ, Zhou HY (2009) Zircon and beddeleyite U-Pb precision dating of basic rock sills intruding Xiamaling formation, North China. Geol Bull China 28(10):1396–1404 (in Chinese with English abstract)

Li HK, Zhu SX, Xiang ZQ, Su WB, Lu SN, Zhou HY, Gen JZ, Li S, Yang FJ (2010) Zircon U-Pb dating of the tuff of Gaoyuzhuang formation in Yanqing, and its further constraints on the northern part of North China middle proterozoic subdivision of the new program. Acta Petrol Sinica 26(7):2131–2140 (in Chinese with English abstract)

Lu SN, Li HM (1991) Single particle zircon U-Pb method accurate measurement of Dahongyu formation volcanic rock of Changcheng system in Jixian. Bull Chin Acad Geol Sci 22:137–145 (in Chinese with English abstract)

Lu SN, Zhao GC, Wang HC, Hao GJ (2008) Precambrian metamorphic basement and sedimentary cover of the North China Craton: a review. Precambr Res 160:77–93

North China Institute of Geosciences (1965) Proceeding of the field seminar on the Jixian section. North China Institute of Geosciences, Tianjin (in Chinese)

Punpelly R (1866) Geological researches in China, Japan and Mongolia. Smithson Contrib Knowl 202:38–39

Qiao XF (1976) Investigation on stratigraphy of the Qingbaikou group of the Yanshan Mountains, North China. Sci Geol Sinica 3:246–265 (in Chinese with English abstract)

Qiao XF, Gao M (1997) Carbonate Pb-Pb isotopic dating of Qingbaikou system in North China and its significance. Earth Sci-J China Univ Geosci 22(1):1–7 (in Chinese with English abstract)

Richthofen FV (1882) China vol. 2. Verlay von Dietrich Reimer, Berlin, p 244

Shen QR, Liao DC (1958) Stratigraphy and sedimentary minerals of Sinian in Yanshan mountains. Chin J Geol 38:263–278 (in Chinese with English abstract)

Su WB, Zhang SH, Warren DH (2008) SHRIMP U-Pb age of K-bentonite beds in Xiamaling Formation: Implications for revised subdivision of the neoproterozoic history of the Noeth China. Gondwana Res 14:543–553

Su WB, Li HK, Huff WD, Ettensohn FR, Zhang SH, Zhou HY, Wan YS (2010) Zircon SHRIMP U-Pb chronology and its geological significance in Tieling potassium bentonite rock. Chin Sci Bull 55(22):2197–2206

Sun YZ (1957) Cambrian lower boundary problem. Geol Knowl 4:1–2 (in Chinese)

Sun SF (2006) Microplants of the mesoproterozoic-neoproterozoic erethem in Jixian, China. Geological Publishing House, Beijing (in Chinese)

Sun SF, Zhu SX, Huang XG (2006) Discovery and geological significance of macrofossils in the Gaoyuzhuang formation of mesoproterozoic in Jixian County, Tianjin, China. Acta Paleontol Sinica 45(2):207–220 (in Chinese with English abstract)

Tianjin Institute of Geology and Mineral Resources (1980) Research on precambrian geology: sinian suberathem in China. Tianjin Science and Technology Press, Tianjin (in Chinese)

Tien CC (1923) Stratigraphy and palaeontology of the Sinian rocks of Nankou. Bull Geol Soc China 2(1–2):105–110

Walter MR, Du RL, Horodyski RJ (1990) Coiled carbonaceous megafossils from the Middle Proterozoic of Jixian (Tianjin) and Montana. Am J Sci 290-A:133–148

Wang YL (1963) The boundary problem of Sinian and Cambrian in Northern China. Geologica Sinica 43(2):116–140 (in Chinese with English abstract)

Wang HZ (1985) China ancient geography atlas. Map Publishing House, Beijing, pp 1–85

Wang CQ, Xiao ZZ, Shi FM, Xu HF, Li ZC (1980a) Sinian suberathem in the Ming Tombs, Beijing. In: Tianjin Institute of Geology and Mineral Resources (ed) Research on precambrian geology: sinian Suberathem in China. Tianjin Science and Technology Press, Tianjin, pp 332–340 (in Chinese)

Wang YL, Lu ZB, Xing YS, Gao ZJ, Lin WX, Ma GG, Zhang LY, Lu SN (1980b) Subdivision and correlation of the upper precambrian in China. In: Tianjin Institute of Geology and Mineral Resources (ed) Research on precambrian geology: sinian suberathem in China. Tianjin Science and Technology Press, Tianjin, 29 (in Chinese)

Xiang LW, Guo ZM (1964) Trilobite fossils and stratigraphic significance of limestone in the Changping Formation, Hebei. Acta Paleontol Sinica 12(4):622–625

Xu ZC, Cui BZ (1980) Sinian suberathem in the eastern Yanshan ranges. In: Tianjin Institute of Geology and Mineral Resources (ed) Research on precambrian geology: sinian suberathem in China. Tianjin Science and Technology Press, Tianjin, pp 358–369 (in Chinese with English contents)

Yan YZ, Liu ZL (1997) The macro-algae of the Tuanshanzi formation, the changcheng system of Jixian County, China. Acta Paleontol Sin 36(1):18–41

Ye LF (1920) Beijing Xishan geological record. Geological Bulletin No. 1, former Ministry of Agriculture Geological Survey

Yu RB, Zhang XQ (1984) Study on the late Cambrian isotopic geochronology of Yanshan area. Tianjin Inst Geol Mineral Res Published 11:1–24

Yu JZ, Cui SQ, Qiu GL (1964) After a comparative study of the Sinian stratigraphy in Liaodong area and its relationship with the Yanshan area, China. Geol J 44(1):1–12

Zhang WY (1935) The boundary between the Sinian and the Precambrian strata in Beijing, China. Academia Sin Fac Rep 6(2):30–50

Zhong FD (1977) On Sinian stratum chronology China Yanshan region from Sinian stratum isotopic age. Sci China Ser D 2:151–161 (in Chinese)

Zhu SX (1993) China stromatolites. Tianjin University Press, Tianjin (in Chinese)

Zhu SX, Chen HN (1992) Characteristics o palaeoproterozoic stromatolites in China. Precambr Res 57:135–163

Zhu SX, Chen HN (1995) Megascopic multicellular organisms from the 1700 Ma Tuanshanzi formation in the Jixian area, North China. Sci New Ser 270(5236):620–622

Zhu SX, Liang YZ, Cao RJ, Zhao WJ (1978) Summary of stromatolites study of sinian Suberathem Jixian stratotype section. Acta Geol Sin 3:209–221

Zhu SX, Xing YS, Zhang PY, Yanzhong LQL, Sun SF, Cao F, Niu SW, Chen HN, Liu GZ, Yin CY, Gao LZ, Yue Z, Du RL, Li PJ, Bu DA, Zhang ZY, Shi GJ (1994) Biostratigraphic sequence of the middle-upper proterozoic on North China platform. Geological Publishing House, Beijing (in Chinese)

Zhu SX, Zhu MY, Knoll AH, Yin ZJ, Zhao FC, Sun SF, Qu YG, Shi M, Liu H (2016) Decimetre-scale multicellular eukaryotes from the 1.56-billion-year-old Gaoyuzhuang formation in north China. Nat Commun 7:1–8

Chapter 3
Meso-Neoproterozoic Sequence Stratigraphy, Sedimentary Facies and Source-Reservoir-Seal Bed Assemblage in Jibei Depression, Yanliao Faulted-Depression Zone

Shunshe Luo, Zhenzhong Gao, Hongwei Kuang, Yuan Shao, and Mingli Xi

Abstract The Meso-Neoproterozoic sedimentary basin in Yanliao Faulted-Depression Zone (YFDZ) is famous for its complete sedimentary strata, continuous outcrop, well stratigraphic preservation, abundant flora and algae fossils, etc. Based on the field investigation results in the Jibei Depression, YFDZ, including sedimentary petrology, sequence stratigraphy, etc., the Meso-Neoproterozoic sequence can be divided into 13 2-order sequences and 79 3-order sequences. It is believed that the sedimentary facies in Jibei Depression are referred to a carbonate rock system of epicontinental sea and a clastic rock system of pericontinental sea. The carbonate platform facies are mainly distributed in the Jixianian Gaoyuzhuang, Yangzhuang, Wumishan, Hongshuizhuang and Tieling Formations as well as Qingbaikouan Jing'eryu Formation; the organic reef facies mainly limited to the Gao-9 to Gao-10 Members of Gaoyuzhuang Formation at the western margin; The neritic shelf facies and baffle-free coastal facies are mainly distributed in Xiamaling and Luotuoling Formations. The research results show that there are 3 source-reservoir-seal bed assemblages, i.e., the Hongshuizhuang Formation (source)-Tieling Formation (reservoir)-Xiamaling Formation (seal), Hongshuizhuang Formation (source)-Xiamaling sandstone (reservoir)-Xiamaling shale (seal), and Gaoyuzhuang Formation (source)-Wumishan Formation (reservoir)-Hongshuizhuang Formation (seal).

Keywords Meso-neoproterozoic sequence · Yanliao faulted-depression zone (YFDZ) · Jibei depression · Sequence stratigraphy · Source-reservoir-seal bed assemblages

S. Luo (✉) · Z. Gao · H. Kuang
Key Laboratory of Education Ministry for Petroleum Resources and Exploratory Technologies, Yangtze University, Jingzhou 434023, Hubei, China

S. Luo · Z. Gao · H. Kuang · Y. Shao · M. Xi
School of Geosciences, Yangtze University, Jingzhou 434023, Hubei, China

T. Wang, *Meso-Neoproterozoic Geology and Petroleum Resources in China*, Springer Geology, https://doi.org/10.1007/978-981-19-5666-9_3

3.1 Regional Geology

The Yanliao Faulted-Depression Zone (YFDZ) ranges from Zhangjiakou (Hebei Province) at the west to Beipiao and Fuxin (Liaoning Province) at the east, and crosses 4 provinces/cities of Hebei, Beijing (City), Tianjin (City) and Liaoning. As an oldest potential oil- and gas-bearing tectonic unit in China, the west-central segment of YFDZ shows near EW-trending distribution, and then turns to NE-orientation at its east part. Tectonically, YFDZ belongs to an active structural unit in the North China Craton (NCC), which used to be called as "Yanliao Subsidence Zone" (Chen et al. 1980). Its north side is adjacent to the Inner Mongolia Axis, and its south side is geographically connected with the North China Plain. As for its internal secondary tectonic unit, there are 2 uplifts and 5 depressions i.e., Shanhaiguan and Mihuai Uplifts as well as Liaoxi, Jidong, Jibei, Jingxi and Xuanlong Depressions from east to west (Fig. 3.1).

During the Meso-Neoproterozoic period, YFDZ was a rift-depression zone at the northern margin of NCC. Basically, faulting and subsidence are the major tectonic activities and resulting in a set of huge-thick and laterally stable Meso-Neoproterozoic marine carbonate sequences intercalated with partial clastic rocks, and the total stratigraphic thickness would be up to 8000–9000 m and more. This unmetamorphosed sedimentary sequences are wide-spread and well outcropped, which can be divided into 3 systems and 12 formations, i.e., the Mesoproterozoic $Pt_2{}^1$ Changchengian Changzhougou, Chuanlinggou, Tuanshanzi and Dahongyu Formations, $Pt_2{}^2$ Jixianian Gaoyuzhuang, Yangzhuang, Wumishan, Hongshuizhuang

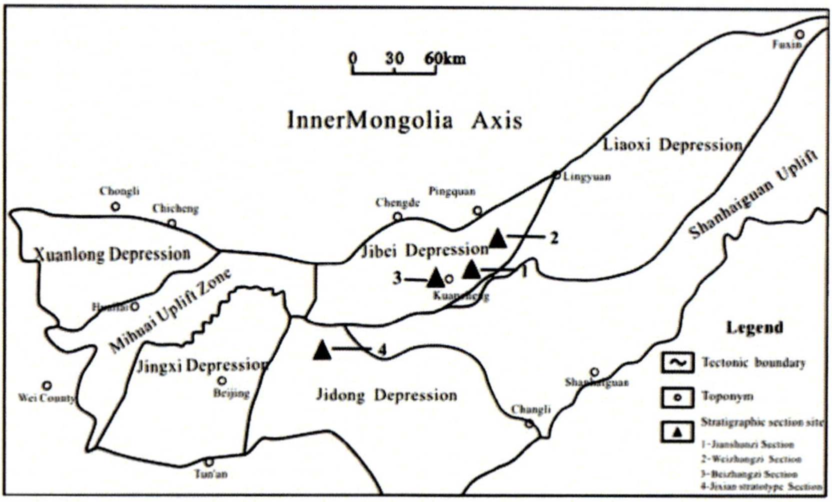

Fig. 3.1 Tectonic division of YFDZ and the position of measured Meso-Neoproterozoic stratigraphic sections in Jibei and Jidong depression (Wang 1980, modified). 1. Jianshanzi section; 2. Weizhangzi section; 3. Beizhangzi section; 4. Jixian stratotype section

and Tieling Formations, and one extra Pt_2^3x Xiamaling Formation as well as the Neoproterozoic Pt_3^3 Qingbaikouan Luotuoling and Jing'eryu Formations.

This chapter only tries to discuss the Meso-Neoproterozoic stratigraphy of Jibei Depression in the north-central segment of YFDZ based on the investigation of a measured stratigraphic section continuously outcropped at Jianshanzi, Weizhangzi and Beizhangzi in turn (Fig. 3.1), and the Meso-Neoproterozoic stratigraphy deals with measured sequences of Pt_2^1 Changchengian, Pt_2^2 Jixianian, Pt_2^3x Xiamaling Formation and Pt_3^1 Qingbaikouan.

By comparison with the Jixian stratotype section in Jidong Depression (Fig. 3.1), the measured stratigraphic thicknesses list in Table 3.1 (cf. Chap. 2), while the Pt_2^1 Changchengian is unmeasured and thus not included here.

3.2 Sequence Stratigraphic Framework

Based on the investigation of three outcropped stratigraphic sections, in combination with the chronostratigraphic, biostratigraphic and chemostratigraphic research results, the Meso-Neoproterozoic sequence of Gaoyuzhuang to Jing'eryu Formations in Jibei Depression has been studied in detail as follows.

3.2.1 Basis of Sequence Division

3.2.1.1 Chronostratigraphy

The chronostratigraphy is to determine the stratigraphic age of sedimentary sequence, look for the isotopic mutation on the sequence boundaries and providing the basis for stratigraphic division. So far the published isotopic chronostratigraphic data are summarized in Fig. 3.2.

3.2.1.2 Petrostratigraphy

Various lithology or lithofacies assemblages could indicate the different types of sedimentary sequences, while the variation of sedimentary sequences results from sea level fluctuation and marine transgression-regression processes, especially within the parasequence. The superposition of different lithofacies appear as the various superposed types of micro-cyclic sequences, and their further combination constitutes a specific sedimentary system tract or higher-order sequence. Therefore, the petrostratigraphic unit is the most basic and direct object for sequence stratigraphic study, Moreover, the appearance of some special lithofacies will often be of obvious indicative significance for sequence division, such as the composition of basic cycles in Wumishan Formation, the appearance of siliceous incrustation, the formation of

Table 3.1 The meso-neoproterozoic stratigraphic sequences and thicknesses in Jibei depression by comparison with the Jixian stratotype section in Jidong depression

Era	System	Formation	Member	Stratigraphic thickness/m			
				Jibei depression		Jidong depression (Jixian stratotype section)	
Cambrian		Fujunshan	–	–		–	
Neo-proterozoic	Qingbaikouan	Jing'eryu	–	9.2		94	
		Luotuoling	Luo-2	2.6		138	45
			Luo-1				93
Mesoproterozoic		Xiamaling	Xia-4	369.5	369.5	198	198
			Xia-3				
			Xia-2				
			Xia-1				
	Jixianian	Tieling	Tie-2	211.1	123.5	290	145
			Tie-1		87.6		145
		Hongshuizhuang		101.7		114	
		Wumishan	Wu-8	2.947.4	547.6	2902	763
			Wu-7		313.7		
			Wu-6		260.2		835
			Wu-5		483.0		
			Wu-4		93.6		870
			Wu-3		462.5		
			Wu-2		452.7		434
			Wu-1		334.1		
		Yangzhuang	Yang-3	322.2	61.0	1048	300
			Yang-2		154.1		547
			Yang-1		107.2		201
		Gaoyuzhuang	Gao-10	938.9	64.3	1543	277
			Gao-9		44.4		
			Gao-8		105.3		706
			Gao-7		49.9		
			Gao- 6		93.8		
			Gao-5		96.6		
			Gao-4		104.2		
			Gao-3		124.4		245
			Gao-2		124.7		315
			Gao-1		131.3		
	Changchengian		–	–	–	–	–

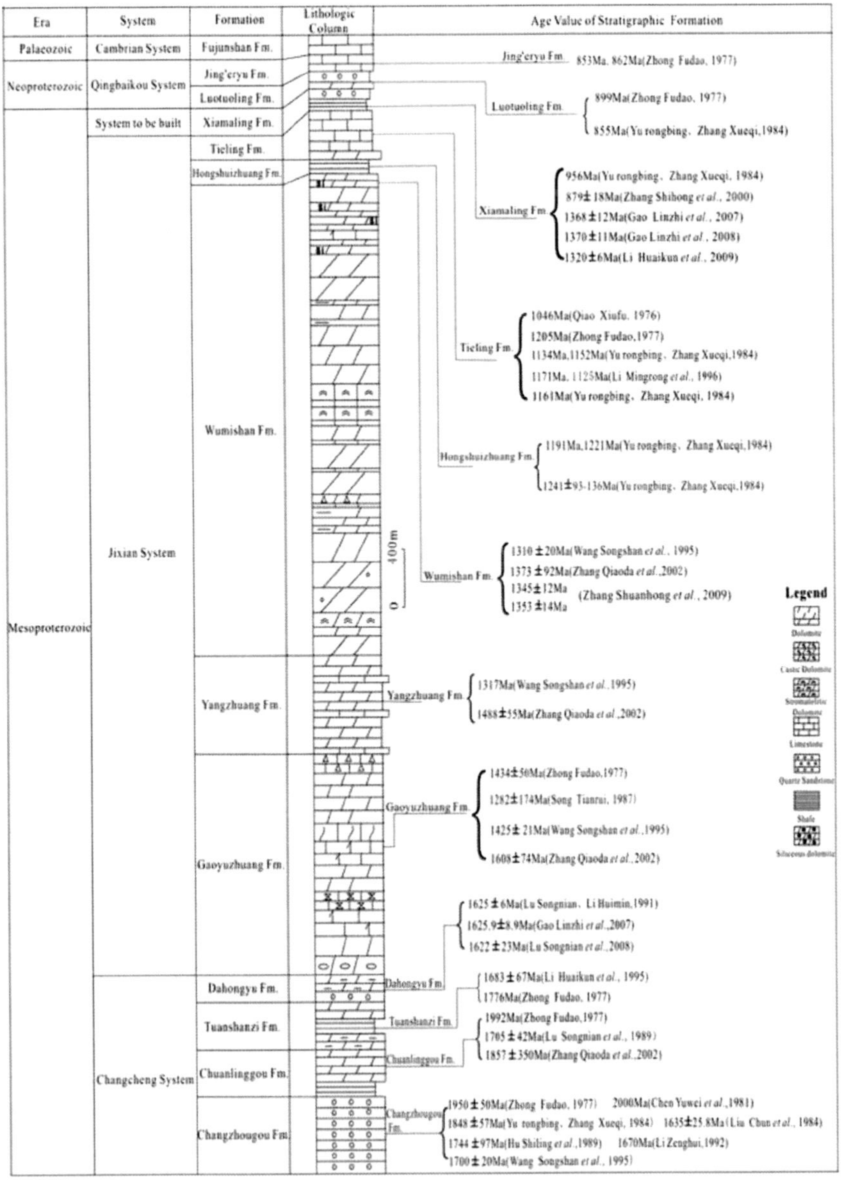

Fig. 3.2 Compilation of measured isotopic ages of Meso-neoproterozoic sequences in YFDZ depression

karst breccia and associated sedimentary structures of lithofacies significance (Wu 2002).

3.2.1.3 Bio-sequence Stratigraphy

Due to the very old geologic age and lack of hard bodied palaeontological fossils, the Meso- Neoproterozoic sequences mainly contain cryptalgal fossils, stromatolites and other non-skeletal carbonate rocks (i.e., thrombolite, oncolite, etc.) in YFDZ, which have been called as microbiolites. Since stromatolites of numerous genera and species are well developed in YFDZ (Table 3.2), various microbiolitic occurrence or morphology could be applied to the division of stratigraphic sequence, i.e., different stromatolites and their assemblage types have been used to distinguish the variation of seawater depth, so that the sea level change could be recognized, and providing the basis for the sequence division (Min et al. 2002). As major species, the laminated, mild- wave-shaped and conic-columnar stromatolites, thrombolites and oncolites are developed in YFDZ. It is commonly known that the conical stromatolite was deposited within the environment of the deepest waters, followed by the columnar stromatolite, thrombolites and oncolites, while the mild-wave-shaped and lamellated stromatolites were in the environment of shallower waters. When different microbialites and their assemblages occur within same sedimentary sequence, it would be helpful and effectively to divide its parasequence, and to distinguish the fluctuation of sea level, which are the critical markers for the determination of system tract and for the dividing and correlating of stratigraphic sequences.

3.2.1.4 Chemostratigraphy

Along with sedimentation, various chemical signals of sedimentary environment are preserved in the sedimentary sequence, providing significant information for petrologic, biologic and chronostratigraphic processes. The chemostratigraphy deals with the application of chemical signals to stratigraphy, mainly using petrochemical element composition, content and variation for the stratigraphic division and correlation as well as for the characterization of geochemical environment and evolutional rules in the sedimentary period (Wu 1999).

In YFDZ, the Meso-Neoproterozoic chemostratigraphy had mainly been focused on the Jixian stratotype section in Jidong Depression, while the investigation of field outcropped Meso-Neoproterozoic sequences were also conducted simultaneously with systematically sampling, analyses and detections of C, O and Sr isotopes (Liu et al. 2007), macro- and micro-elements and X-ray diffraction (Liu et al1984a; b) respectively at the Jianshanzi (at Kuancheng), Weizhangzi (at Lingyuan) and Beizhangzi (at Kuancheng) sections (section Nos. 1, 2, 3 in Fig. 3.1) in Jibei Depression. Their chemical analytic data have provided new evidence for the accurate division of sequence boundaries and for the correction and optimization of 3-order

Table 3.2 Division of the meso-neoproterozoic biostratigraphy

Stratigraphy			Biostratigraphy			
Era	System	Formation	Organic-walled microfossils		Macroscopic algal fossil assemblage	Stromatolitic assemblage
			Shale facies	Chert facies		
Neoproterozoic	Qingbaikouan	Jing'eryu	*Nucellosphaeridium, Tasmanites*			
		Luotuoling				
Mesoproterozoic	Xiamaling		*Microconcentrica, Jixiania*		*Chuaria-Shouhsienia*	
	Jixianian	Tieling	*Trachysphaeridium acis*	*Eomycetopsis, Bigeminococcus*	*Chuaria cicularis*	*Tielingella-Chihsienella*
		Hongshui-zhuang	*Orygmatosphaeridium, Quadratimorpha*			
		Wumishan	*Asperatopsophosphaera umishanensis*		*Wumishania bifurcata*	*Conophyton-Pseudogymnosolen*
		Yangzhuang	*Asperatopsophosphaera, Kildinella*		*Grypania spiralis*	*Conophyton cylindricium*
		Gaoyuzhuang	*Pseudofavososphaera, Gunflintia*			
	Changchengian	Dahongyu	*Leiosphaeridia parvula, Stictosphaeridium*	*Oscillatoriopsis, Myxococcoides*	*Tuanshanzia-Changchengia*	*Conophyton dahongyuense*
		Tuanshanzi	*Trachysphaeridium attenuatum, Eomycetopsis*	*Gunflintia, Eomycetopsis*		
		Chuanlinggou	*Trachysphaeridium, Diplomembrana , Foliomorpha*		*Chuaria-Tyrasotaenia*	*Gruneria, Xiayingella*
		Changzhougoun	*Leiopsophosphaera, Schizofusa, Foliomorpha*		*Chuaria-Shouhsienia-Tawuia*	

sequences, these contents of petrochemical components and elements would show a mutation at the boundaries of stratigraphic sequence (e.g., Figs. 3.3 and 3.4).

The late erosion and epigenetic diagenesis often happened and resulting in some obvious variation of isotopic compositions on the sedimentary boundaries. According to the isotopic mutation, 3-order sequence boundaries can be clearly recognized (Tian et al. 2006). For instance, based on the isotopic data of the Yangzhuang Formation at Jianshanzi section (section No. 1 in Fig. 3.1), the distributional curves of C and O isotopic composition synchronously show a mutation at the boundaries between SQ. 10 and SQ. 11 (Fig. 3.5), by which the recognition of sequence boundaries and the division of 3-order sequences in field outcropped section could be calibrated and optimized.

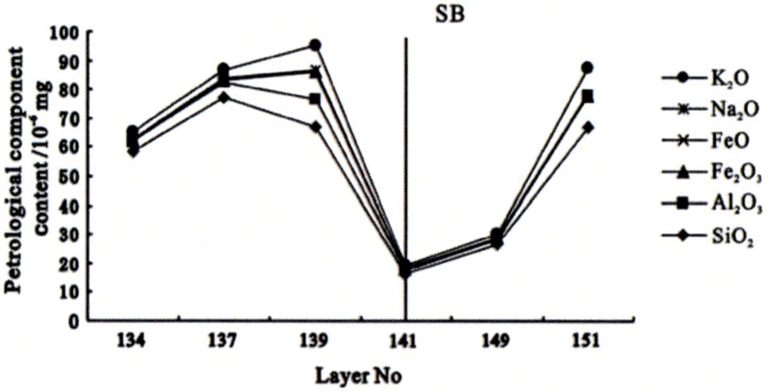

Fig. 3.3 Determination of 3-order sequence boundary based on the petrochemical components (SQ. 5 and SQ. 6 in Gaoyuzhuang Formation, Jibei Depression). SQ. sequence; SB.sequence boundary

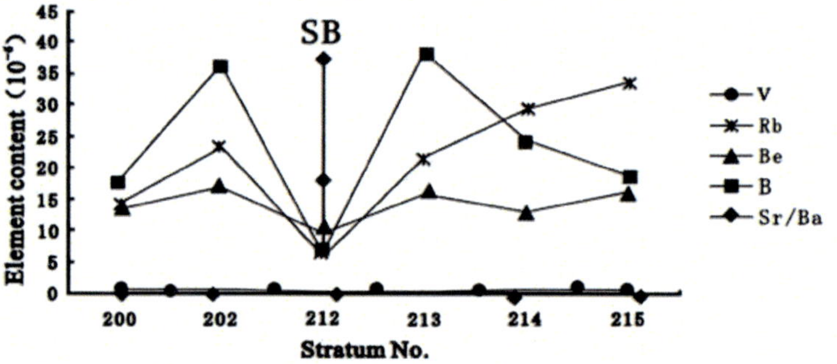

Fig. 3.4 Division of 3-order sequence boundary based on the contents of macro-elements (SQ. 10 and SQ. 11 in Yangzhuang Formation, Jibei Depression). SQ: Sequence, SB: Sequence boundary

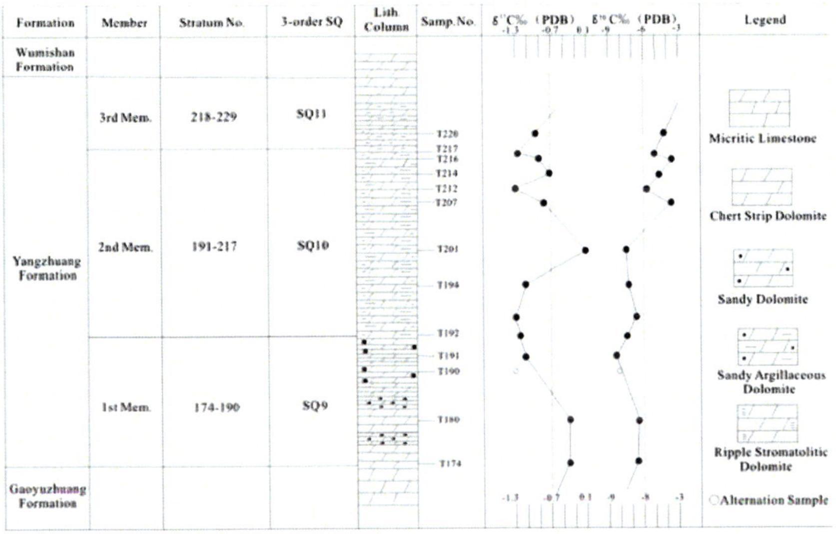

Fig. 3.5 The recognition of 3-order sequence boundary based on C and O isotopic composition in Yangzhuang Formation, Jibei Depression

3.2.1.5 Sequence Stratigraphy

In field outcrop study, the major work of sequence stratigraphy is to recognize the sequence units of different orders, which should be distinguished by the geologic boundaries of different orders. Therefore, the recognition of sequence-associated geological boundary, i.e., bedding plane, sea-flooding surface and unconformity contact, in the outcropped stratigraphic section would be an important basic work (Meng and Ge 2002).

During the processes of outcrop investigation and precise stratigraphic division, the first and foremost task is to recognize sequence boundary, it may be an unconformity contact or conformity interface corresponding to the unconformity contact, which should be laterally continuous, globally comparable and covered in whole the basin. The sequence boundary would appear in the subsidence stage of global sea level. Owing to the distinct subsidence ranges of sea level, meanwhile, there are two types of sequence boundaries, i.e., type-I and type-II boundaries, which show different types of sedimentary characteristics easy to be recognized (Wang et al. 2011).

The incised valley is one of the important markers for identification of type-I sequence boundary. On the continental shelf, type-I sequence boundary would act as a bottom boundary of incised valley, while the first maximum sea-flooding surface (MSF) or transgression surface as the top boundary. In the region between the incised valleys, the unconformity contact would be a corresponding epicontinental exposed surface, and thus the fossil soil or rootlet bed could be taken as a distinguishable marker.

Since the geohistorical type-II sequence boundary is difficult to be preserved, it has seldom been studied. So far the known distinguishable markers of type-II sequence boundary include the onlap of overlying stratum, the downward drifting of the coastal onlap and the slightly cutting of epicontinental exposed surfaces. However, it is still hard to recognize these boundaries on the slop break towards continental side along the sedimentary strand line.

3.2.2 *Recognition of Sequence Boundary*

3.2.2.1 Markers of Sequence Boundary

The previous analyses on the tectonic setting of sedimentary basin and the division of sedimentary sequences in YFDZ have revealed that the individual formations are basically within the sedimentary environments above continental shelf, and generally for lack of the low-stand system tract (LST) in their sedimentary systems, while the transgressive system tract (TST) mainly turns upward into a high-stand system tract (HST). Meanwhile, their sedimentary environments are rarely within the deep-water basin. The condensation level of condensed interval is hardly to meet the theoretical requirements.

According to the characteristics of cyclic sequence superposition in carbonate rocks at circum tidal-flat (Liu et al. 1997), therefore, the sediments and sedimentary structures of shallow- water environment, especially for these with exposed-markers, could be applied to distinguish the sedimentary boundary. While those sediments and sedimentary structures of deeper-water environment would be the products during the maximum sea-flooding period (MSF), which can act as a marker of sea-flooding surface. Thus, the recognition of above two kinds of boundary markers would be significant for stratigraphic sequence division in YFDZ.

In the chapter, the following twelve markers observed in YFDZ are used as the basis for sequence boundary division, among which, seven are the markers for shallow-waters, while five for deep-waters (Zhu et al. 1993; Gao et al. 1996; Huang et al. 2001; Zhou et al. 2006; Fig. 3.6).

(1) **Markers of shallow-water environmemt**:

① Weathering crust and disconformable boundary, i.e., the strata beneath the boundary would have stratigraphic hiatus to different extent (Fig. 3.6a);

② Desiccation phenomenon in the supratidal or shoreland zone, especially for mud cracks (Fig. 3.6b);

③ Siliceous encrustation with a variable thickness, the commonly exposure marker of carbonate rock and mainly found in the tidal-flat carbonate rocks of Wumishan, Yangzhuang and Gaoyuzhuang Formations in YFDZ, which is created by silica aggregation of carbonate rocks exposed to the surface under the leaching effect of fresh water (Fig. 3.6c);

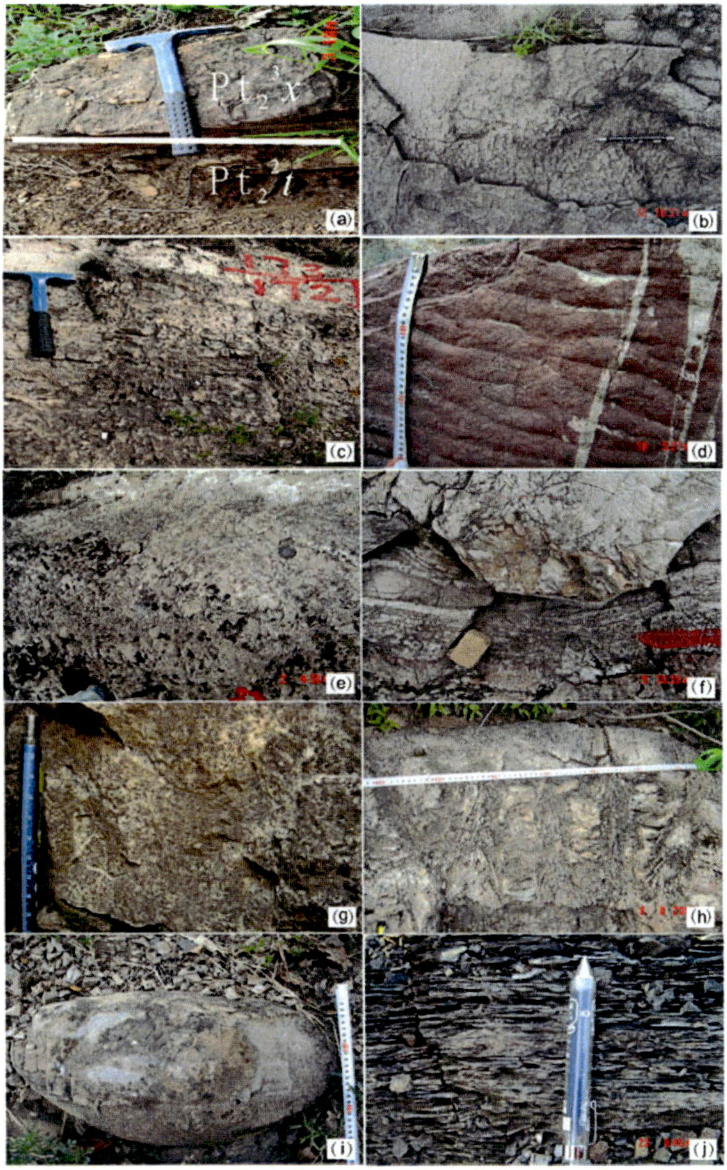

Fig. 3.6 The distinguishable markers of sequence boundaries in Jibei Depression. **a**. Ferruginous weathering crust/disconformable boundary between Tieling ($Pt_2{}^2t$) and Xiamaling ($Pt_2{}^3x$) Formations; **b**. mud cracks of Layer No. 358, Wumishan Formation; **c**. silliceous encrustation of Layer No. 172, Gaoyuzhuang Formation; **d**. ripple markers of Layer No. 209, Yangzhuang Formation; e. karst breccia of Layer No. 8, Wumishan Formation; **f**. erosional surface at the top of Layer No. 48, Wumishan Formation; g. thrombolitic dolostone of Layer No. 246, Wumishan Formation; **h**. conical stromatolites of Layer No. 160, Wumishan Formation; **i**. large knobby-limestone of Layer. No. 141, Gaoyuzhuang Formation; **j**. blackish-grey shale of Layer No. 8, Hongshuizhuang Formation

④ Ripple marks, it generally indicates a shallow water environment (Fig. 3.6d);
⑤ Bird'eye structures, which generally acts as a marker of supratidal zone, and indicating a very shallow-waters, even exposed on surface;
⑥ Fossil karst surface and karst breccia, which are created by exposed carbonate rock, the filled karst breccia are very good markers of exposure (Fig. 3.6e);
⑦ Erosional surface (Fig. 3.6f).

(2) **Markers of deep-water environment**:

① Thick-bedded to massive thrombolytic dolostone, mainly deposited by thrombolite and oncolite as the products in the high energy turbulent environment of subtidal-zone (Fig. 3.6g);
② Conical stromatolite, which is deposited in deeper-waters of the upper subtidal-zone (Fig. 3.6h);
③ Knobby-limestone (Fig. 3.6i);
④ Dark-color foliaceous shale (Fig. 3.6j);
⑤ "Molar-tooth" structure (Fairchild et al. 1997; Frank and Lyons 1998).

3.2.2.2 Markers of Sedimentogeochemistry

The contents of macro- and micro- elements and the range of related specific ratios in strata are closely associated with the variation of sea level during sedimentary period. In the sedimentary environmental study, so far the most commonly used micro-elements include B, Sr, Ba, V and Ni and their specific ratios not only for the distinction of fresh-water and sea-water sediments and the determination of palaeo-salinity and palaeo-climate (Wang et al. 1997), but also for recognition of the sedimentary environment and its relationship with the sea level fluctuation.

In this chapter, all the rock samples of target stratum have been systematically collected by 10 m sampling interval for micro-element, 20 m for macro-element, and 40 m for C and O isotopic analyses in YFDZ. The fresh powdered rock samples were analyzed by X-ray spectrometer at the Tianjin Geology and Mineral Institute. The analyzed items include the macro-elements Al, Si, Fe, Mg, Ca, K, Mn, P and ignition loss as well as the micro-elements V, Rb, Sr, Ba, B. In addition, the C, O, Sr elements were also analyzed for isotopic data (Yan et al. 2005).

Based on previous data in YFDZ, the correlation between the variation of macro- and micro- element contents or isotopic values and the sea level fluctuation are summarized in Table 3.3. For instance, field stratigraphic investigation shows that the Yangzhuang Formation has been divided into 3 members based on the lithologic characteristics. The Yang-1 Member consists of greyish-white and purplish-red muddy dolostones intercalated with greyish-white siliceous dolostone, which turn upward into a thick-bedded to massive siliceous and stromatolitic dolostones. The Yang-2 Member has a set of purplish-red muddy siliceous dolostone at its bottom, but the thick-bedded to massive greyish-white and purplish-red muddy dolostones become

Table 3.3 The effect of macro- and micro-elements as well as isotopic data on sea level fluctuation

Sea level rising	Sea level decline
Decreases of Al_2O_3, SiO_2, Fe_2O_3, K_2O, TiO_2	Increases of Al_2O_3, SiO_2, Fe_2O_3, K_2O, TiO_2
Rising of MnO	Decrease of MnO
Increases of V, Rb, Be, B, Sr/Ba, Ca/Mg	Decreases of V, Rb, Be, B, Sr/Ba, Ca/Mg
Positive drifting of $\delta^{13}C$	Negtive drifting of $\delta^{13}C$
Negtive drifting of $\delta^{18}O$	Positive drifting of $\delta^{18}O$
Decrease of $^{87}Sr/^{86}Sr$	Increase of $^{87}Sr/^{86}Sr$

major lithology upwards. The Yang-3 Member is mainly composed of greyish-white and purplish-red thin-bedded micrite or muddy dolostones and sandy dolostone, showing a gradually shallowing course of sedimentary waters upwards. The above stratigraphic division is also verified by the analytic results of micro-element content (Fig. 3.7), each member shows a larger cycle of sea level variation, i.e., the micro-element contents of V, Rb, Sr/Ba, B, Be etc. appear as a low–high-low fluctuation within each member, and a mutation from low to high contents occurs at the interface between two adjacent members, which would characterizes a transitional surface from regression to transgression (Fan et al. 1977). In view of the general variational trend of micro-elements, whole the Yangzhuang Formation has shown an evolutional process from the higher sea level to the lower sea level.

Along the stratigraphic sequence from bottom up in the Jibei Depression, the variation of C and O isotopic composition obviously shows a rule of cyclicity and a positive correlation in the most cases. The $\delta^{13}C$ value appears as a low amplitude and high frequency fluctuation with a variable range from −3 to 3‰, while $\delta^{18}O$ value is in a high amplitude and high frequency oscillation at a range from −8 to −2‰. The rising of $\delta^{13}C$ is mostly associated with the evolution of sedimentary environment from intertidal to subtidal zones, sea level rising, sea-water desalination and the biomass increasing. The declining of $\delta^{13}C$ represents the environments of intertidal to supratidal zones, which is mostly corresponds to the sequence boundary.

While the Gaoyuzhuang knotty limestone and Hongshuizhuang shale show the particularity, their $\delta^{13}C$ has negative low value, which represents the sedimentation during the maximum sea-flooding period (MSF) and indicating a non-sequence boundary (Chu et al. 2003). The variation of O isotope has a salting environment in YFDZ, and the sea water salinity reached the maximum level at the late Yangzhuang and the early Wumishan sedimentary periods. Subsequently the salinity would be gradually decreasing until the late Wumishan sedimentary period, and then it was going up again.

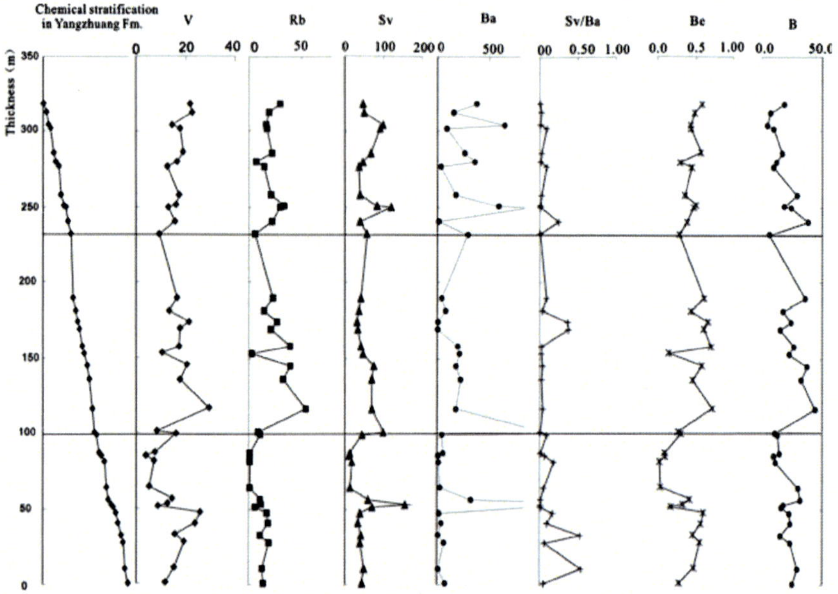

Fig. 3.7 The relationship between micro-element distributional curves and stratigraphic sequence of Yangzhuang Formation at the Jianshanzi section (Kuancheng) in Jibei Depression

3.2.3 *Stratigraphic Sequence Division*

According to above listed principles and basis for sequence division, the strata from the Mesoproterozoic Pt_2^2 Jixianian Gaoyuzhuang Formation to Neoproterozoic Pt_3^1 Qingbaikouan Jing'eryu Formation in Jibei Depression are divided into 13 2-order sequences and 39 3-order sequences (Figs. 3.8, 3.9, and 3.10), which are a set of epicontinental sea carbonate sediments in general (Table 3.1). It is found out that each formation contains different rock units, and the superimposition of these rock units would constitute the most basic micro-cyclic sequence of each formation, while the different superimposition modes would of micro-cyclic sequences provide a foundation for the division of system tract of 3-order sequences.

The characteristics of the 3-order sequences are described for stratigraphic section division in Jibei Depression of YFDZ as follows.

3.2.3.1 Gaoyuzhuang Formation (Pt_2^2g)

There is a 13.2 m thick quartzose sandstone at the bottom of Gaoyuzhuang Formation, which appears as a transitional conformable contact with underlying thick-bedded quartzose sandstone of shoreland facies in the Dahongyu Formation at Jianshanzi section, Kuancheng (section No. 1 in Fig. 3.1). By means of field investigation and

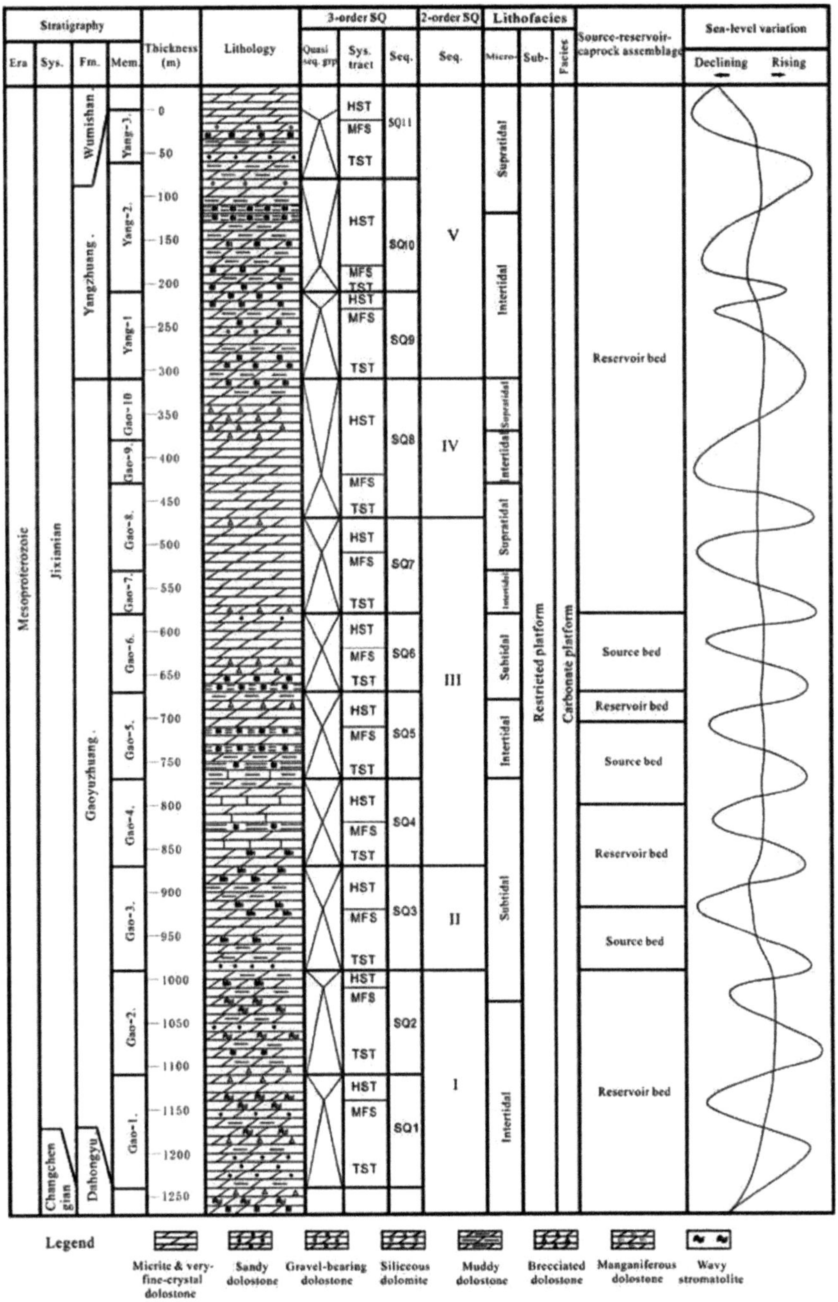

Fig. 3.8 Sequence column of Gaoyuzhuang-Yangzhuang Formations in Jibei Depression. HST. highstand system tract; TST. transgressive system tract; MFS. Maximum flooding surface

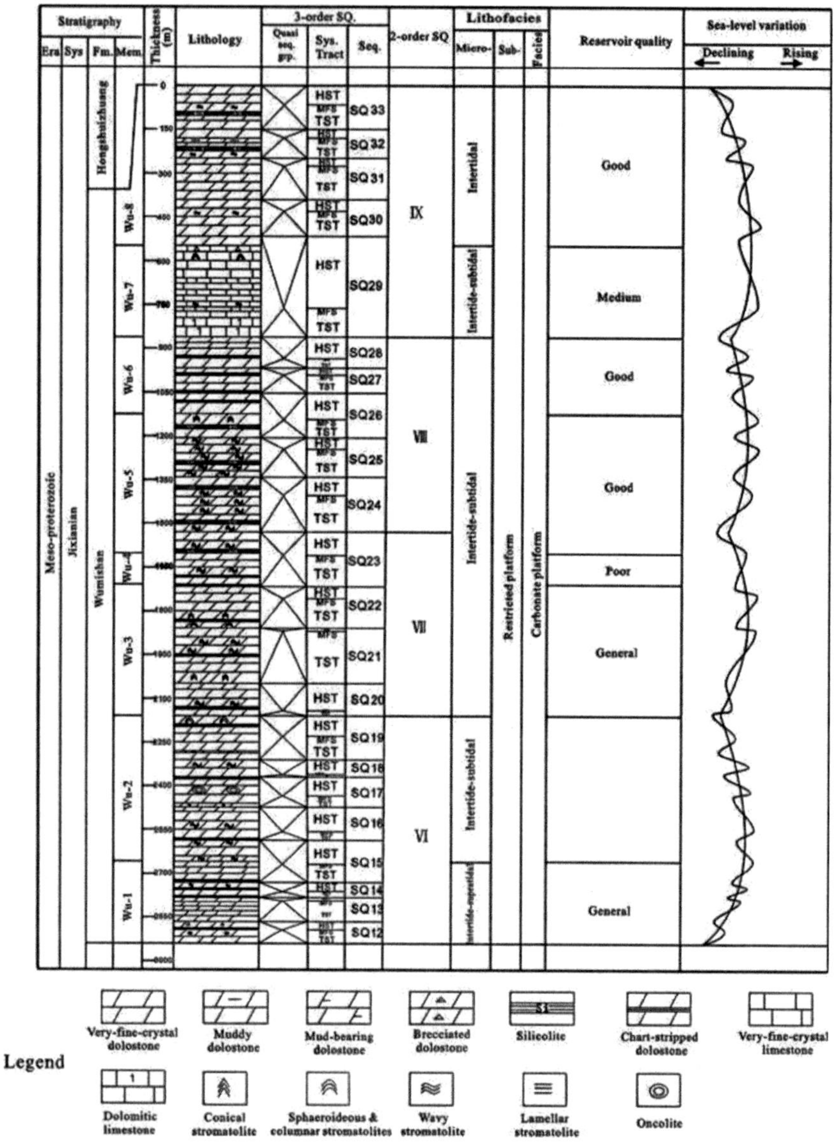

Fig. 3.9 Sequence column of Wumishan formation in Jibei depression. HST. highstand system tract; TST. transgressive system tract; MFS. Maximum flooding surface

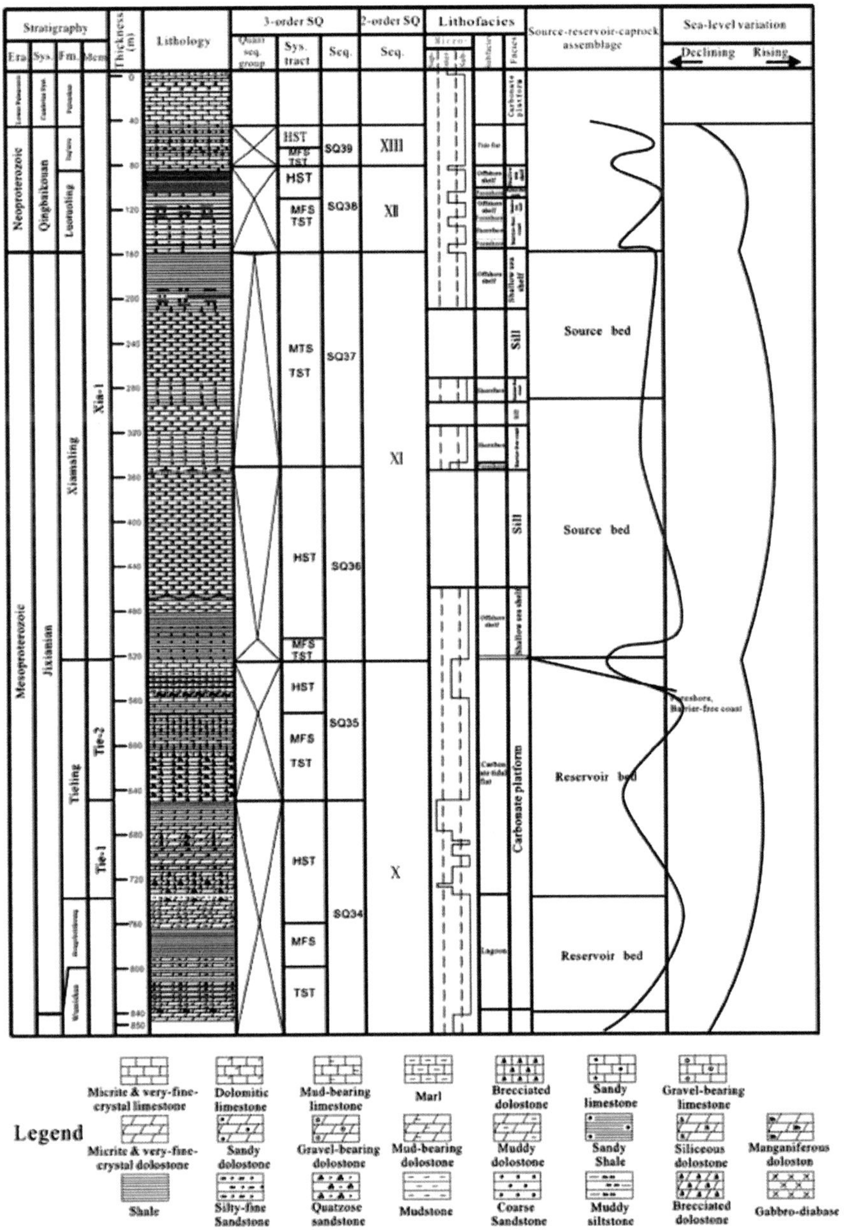

Fig. 3.10 Sequence column of Hongshuizhuang-Jing'eryu Formations in Jibei Depression. HST. highstand system tract; TST. transgressive system tract; MFS. Maximum flooding surface

laboratory analysis, especially for the analyses of macro- and micro-elements as well as the C, O and Sr isotopes, Gaoyuzhuang Formation is divided into 173 layers, 10 members and 8 3-order sequences (Fig. 3.8 and Table 3.4). Based on the micro-cyclic superimposition mode, different system tracts, i.e., HST, TST and MFS, are divided (Fig. 3.8 and Table 3.4).

3.2.3.2 Yangzhuang Formation (Pt_2^2y)

Based on field investigation of sedimentary facies, petrochemical component and its isotope, the Yangzhuang Formation is divided into 56 layer, 3 members and 3 3-order sequences (Fig. 3.8 and Table 3.5). According to the divergence in superimposition modes of micro-cycles, different system tracts are identified (Fig. 3.8 and Table 3.5). In Yangzhuang Formation, the greyish-white and purplish-red thin-bedded muddy dolostone occurs at the bottom, it turns upward into a grey siliceous-stripped or nodular micritic dolostone, which constitutes a transgressive system tract (TST), the emergency of a grey thick-bedded micritic dolostone with turbulent conical stromatolite or brownish-red thick-bedded micritic dolostone represents the maximum sea-flooding period (MSF), while the high-stand system tract (HST) is composed of greyish-white muddy or siliceous dolostone and silicalite.

Overall, Yangzhuang Formation shows a conformable contact with the underlying Gaoyuzhuang Formation at the measured Jianshanzi section (Kuancheng; section No. 1 in Fig. 3.1), but a local disconformity contact is also found at Jixian in Jidong Depression (cf. Chap. 2).

3.2.3.3 Wumishan Formation (Pt_2^2w)

Its measured stratigraphic section of Wumishan Formation is situated at the Weizhangzi section, Lingyuan (section No. 2 in Fig. 3.1). After the arid sedimentary period of Yangzhuang Formation, the palaeo-climate has gradually turned to moisture in YFDZ, and the main sedimentary environment and sediments of Wumishan Formation are attributed to the tidal flat facies of shallow-waters. The micro-cyclic superimposition modes and sedimentary environments of Wumishan Formation in Jibei Depression are shown in Figs. 3.9, and 3.12 genetic rock units are recognized as the fundamental rock units of the micro-cyclic zones in Wumishan Formation. It is the various superimposition modes that constitute the sedimentary assemblages of transgressive system tracts (TST), maximum sea flooding surface (MSF) and highstand system tract (HST) respectively in Wumishan Formation. By way of petrochemical, micro-elemental and C, O and Sr isotopic analyses, the recognized 3-order sequences are calibrated and optimized, and thus totally 411 layers, 8 Members and 22 3-order sequences have been divided in Wumishan Formation of Jibei Depression (Fig. 3.9 and Table 3.6).

Table 3.4 Sequence division and system tract features of Gaoyuzhuang formation in Jibei depression

Formation	3-order sequence	2-order sequence	Lithologic features of system tract	Layer No
Gaoyuzhuang	SQ. 8	SQ. IV	TST: lamellated algal dolomitic limestone, lamellated algal limy- and lime-bearing dolostones	168–173
			MSF: lamellated algal dolomitic limestone and grey siliceous clotted lamellated dolomitic limestone	
			HST: siliceous clotted/nodular and laminated algal limestone and algal dolostone	
	SQ. 7	SQ. III	TST: gray thick-bedded to massive dolostone, medium-bedded to massive lime-bearing dolostone intercalated by thin-bedded dolostone	157–167
			MSF: the interbedding of gray lamellated algal dolostone and thick-bedded to massive lime-bearing dolostone	
			HST: grey massive to thick-bedded mud-bearing dolostone intercalated by algal-fragment micritic to fine-crystal limy dolostone and dark-grey thick-bedded dolostone	
	SQ. 6		HST: lamellated algal limy dolostone and sandy very-fine crystal to micritic limy dolorudite	148–156
			MSF: dark-grey thin-bedded knobby limestone	
			TST: grey thick- to medium-bedded mud-bearing dolostone intercalated by siliceous strips and dark-gray to gray massive limy dolostone	

(continued)

Table 3.4 (continued)

Formation	3-order sequence	2-order sequence	Lithologic features of system tract	Layer No
	SQ. 5		TST: silcolite intercalated by thin-bedded or knobby micritic limestone, which turns upward into an interbedding of micritic limestone	134–147
			MSF: an interbedding of blackish-grey medium- to thin-bedded micritic dolostone, silcolite	
			HST: grey medium-bedded lamellated dolostone	
	SQ.4		TST: the unequally thick interbedding of medium—to thin-bedded manganiferous dolostone and medium- to thin-bedded laminated dolomitic limestone and the interbedding of shaly micritic limestone and equally thick dolomitic limestone	113–133
			MSF: the interbedding of dark-gray knobby limestone	
			HST: an unequally thick interbedding of thick-bedded to massive very-fine-crystal limestone and khaki thin-bedded lime-bearing muddy dolostone	
	SQ. 3	SQ. II	TST: gray manganiferous muddy dolostone, thin-bedded muddy dolostone	92–112
			MSF: the medium-bedded manganiferous micritic dolostone	
			HST: medium- to thick-bedded micritic dolostone, which turn upward into medium- to thin-bedded micritic dolostone	

(continued)

Table 3.4 (continued)

Formation	3-order sequence	2-order sequence	Lithologic features of system tract	Layer No
	SQ. 2	SQ. I	TST: dark-gray thick-bedded siliceous dolostone, gray medium-bedded siliceous dolostone and gray thick-bedded laminated stromatolitic dolostone	56–91
			MSF: the medium- to thick-bedded conical stromatolitic dolostone	
			HST: the interbedding of wavy stromatolitic and muddy dolostones	
	SQ. 1		TST: quartzose sandstone, sandy dolostone and thin-bedded sandstone as main lithology	28–55
			MFS: the dark-gray thick-bedded stromatolitic dolostone and thin-bedded or shaly dolostone	
			HST: thick-bedded to massive and stromatolitc dolostone and thin-bedded muddy dolostone	

3.2.3.4 Hongshuizhuang Formation (Pt_2^2h)

The measured Hongshuizhuang section is situated at the Beizhangzi section, Kuangcheng (section No. 3 in Fig. 3.1). It shows a depositional break of disconformity with underlying Wumishan Formation. Lithologically, thin-bedded muddy dolostone is developed at its bottom, and turning upward into a transgressive system tract (TST) composed of black shale with some muddy dolostone. It is divided into 11 layers, 2 Members and 1 3-order sequence (Fig. 3.10 and Table 3.7). The black shale in layer No. 8 indicates a sedimentary period of the deepest waters, and then the emergency of greyish-white silty muddy dolostone reflects the beginning of a decline in sea level and entering a highstand system tract (HST; SQ. 34 in Fig. 3.10).

3.2.3.5 Tieling Formation (Pt_2^2t)

The measured Tieling section is situated at the Beizhangzi section, Kuangcheng (section No. 3 in Fig. 3.1). It is in conformable contact with the underlying

Table 3.5 Sequence division and system tract features of Yangzhuang formation in Jibei depression

Formation	3-order sequence	2-order sequence	Lithologic features of system tract	Layer no
Yangzhuang	SQ. 11	SQ. V	TST: thin-bedded quartzose sandstone and silt-bearing muddy dolostone, etc	218–229
			MSF: the vertical accretionary bed of brownish-yellow Medium-bedded micritic dolostone intercalated by sandy micritic dolostone	
			HST: arenitic- and rudaceous muddy dolostones etc	
	SQ. 10		TST: undeveloped	191–217
			MSF: shaly muddy dolostone	
			HST: an interbedding of siliceous micritic to very-fine-crystal dolostone and arenitic muddy dolostone	
	SQ. 9		TST: purplish-red muddy dolostone and gray siliceous dolostone	174–190
			MSF: the small conical stromatolitic dolostone and disordered conical stromatolitic dolostone	
			HST: medium- to thick-bedded chert lumps, chert-clot-, chart-nodule-bearing siliceous dolostone	

Hongshuizhuang Formation, and divided into 35 layers, 2 members and 2 3-order sequences (Fig. 3.10 and Table 3.7).

The Tie-1 Member is developed at its lower part, and consists of manganifeous dolostone intercalated by shaly muddy dolostone, and turns upward into silty muddy dolostone and siliceous dolostone, reflecting a sea water gradually shallowing process. While the Tie-2 Member is mainly composed of tempestitic calcirudite and shale, indicating the sea level rising and waters deepening process. In between, a ferruginous weathering crust is developed on the top boundary of Tie-1 Member resulted from the tectonic movement of "Tieling Uplifting".

3.2.3.6 Xiamaling Formation (Pt_2^3x)

Its measured stratigraphic section is situated at the Beizhangzi section, Kuangcheng (section No. 3 in Fig. 3.1). Under normal circumstances, the Xiamaling Formation

Table 3.6 Sequence division and system tract features of Wumishan formation in Jibei depression

Formation	3-order sequence	2-order sequence	Lithologic features of system tract	Layer No
Wumishan	SQ. 33	SQ. IX	HST: siliceous-stripped dark-grey micritic dolostone	393–411
			MSF: large wavy stromatolitc micritic dolostone	
			TST: grey micritic dolostone turning upward into a grey silicalite, grey lamellated and large wavy stromatolitc micritic dolostone	
	SQ. 32		HST: grey medium-bedded micritic dolostone turning upward into grey siliceous clot- or strip-bearing micritic dolostone	376–392
			MSF: grey medium- to thick-bedded micritic dolostone	
			TST: grey lamellated micritic dolostone turning upward into an interbedding of grey medium- to thick-bedded micritic dolostone and greyish-white silicalite	
	SQ. 31		HST: grey medium- to thick-bedded siliceous clotted or stripped micritic dolostone turning upward into grey lamellated stromatolitic micritic dolostone	355–375
			MSF: the huge thick micritic dolostone	
			TST: grey medium-bedded-layer micritic dolomite or black siliceous stripped micritic dolostone turning upward into black thin-bedded silicalite, etc	
	SQ. 30		HST: grey lamellated micritic stromatolitic dolostone slightly with siliceous clots and strips turning upward into grey lamellated stromatolitic micritic dolostone	331–354
			MSF: grey lamellated larger mildly-wavy stromatolitic micritic to very-fine- crystal dolostone	

(continued)

Table 3.6 (continued)

Formation	3-order sequence	2-order sequence	Lithologic features of system tract	Layer No
			TST: an interbedding of grey slightly with flesh-red medium- to thin-bedded micritic dolomite and grey lamellated stromatolitic dolostone	
	SQ. 29		HST: grey thin-bedded stromatolitic micritic limestone and karst breccia, etc	294–330
			MSF: the interbedding of dark-grey thin-bedded micritic limestone	
			TST: dark-grey medium-bedded limestone turning upward into grey medium-bedded limy dolostone and grey thin-bedded muddy dolostone	
	SQ. 28	SQ. VIII	TST: grey medium-bedded bebbly micritic dolostone, thin-bedded algal lamellated micritic dolostone and thin-bedded silicalite, etc	275–293
			MSF: grey medium-bedded very-fine-crystal algal lamellated thrombolitic dolostone	
			HST: grey medium-bedded micritic dolostone turning upward into medium-to Thick-bedded algal lamellated micritic dolostone	
	SQ. 27		HST: grey medium-bedded micritic dolostone turning upward into thin-bedded silicalite	260–274
			MSF: grey medium- to thin-bedded thrombolitic very-fine-crystal dolostone	
			TST: grey thin-bedded algal lamellated micritic dolostone intercalated by siliceous strips or clots, etc	
	SQ. 26		HST: grey medium-bedded algal lamellated micritic dolostone	230–259
			MSF: grey thick-bedded thrombolitic dolosparlite	

(continued)

Table 3.6 (continued)

Formation	3-order sequence	2-order sequence	Lithologic features of system tract	Layer No
			TST: Grey algal lamellated micritic dolostone intercalated by mediun- to thick-bedded stromatolitic very-fine-crystal powder dolostone with siliceous strips	
	SQ. 25		HST: grey lamellated siliceous-stripped dolostone turning upward into grey and grey silicalite	213–229
			MSF: massive fine-crystal dolostone	
			TST: medium-bedded algal lamellated siliceous-strip-bearing micritic dolostone turning upward into thick-bedded wavy stromatolitic micritic to very-fine-crystal dolostone	
	SQ. 24		HST: siliceous bulk or strip contained crystal powder dolomite turning upward into a thick-bedded mildly-wavy stromatolitic micritic dolostone	191–212
			MSF: thick-bedded wavy and arched stromatolitic micritic dolostone	
			TST: arched stromatolitic micritic dolostone turning upward into grey medium-bedded algal lamellated dolostone	
	SQ. 23	SQ. VII	HST: grey medium-bedded micritic dolostone turning upward into siliceous stripped micritic dolostone	170–190
			MSF: huge thick-bedded to massive mildly-wavy and conical stromatolitic dolostone	
			TST: grey medium-bedded micritic dolostone turning upward into thick-bedded mildly-wavy, arched and a little conical stromatolitic dolostone	

(continued)

Table 3.6 (continued)

Formation	3-order sequence	2-order sequence	Lithologic features of system tract	Layer No
	SQ. 22		HST: grey thick-bedded micritic dolostone turning upward into laminated siliceous stripped micritic dolostone	158–169
			MSF: dark-grey thick-bedded oncolite-bearing dolostone	
			TST: grey siliceous stripped micritic dolostone turning upward into dark-grey thick-bedded oncolite-bearing dolostone	
	SQ. 21		HST: grey medium-bedded micritic dolostone turning upward into siliceous- stripped micritic dolomite	129–157
			MSF: grey thick-bedded very-fine-crystal stromatolitic dolostone	
			TST: grey medium- to thick-bedded siliceous-clotted or stripped micritic dolostone and thin-bedded laminated stromatolitic dolostone	
	SQ. 20		HST: grey algal laminated micritic dolostone turning upward into siliceous stripped micritic dolomite or silicalite	121–128
			MSF: grey thick-bedded to massive conical stromatolitic micritic dolostone	
			TST: grey medium- to thick-bedded micritic dolostone turning upward into wavy stromatolitic micritic dolostone	
	SQ. 19	SQ. VI	HST: grey siliceous-thrombolitic or stripped micritic dolostone turning upward into grey siliceous-encrust	100–120
			MSF: grey thick-bedded to massive very-fine-crystal dolostone with basal mildly-wavy and small-conical stromatolites	
			TST: Micritic and very-fine-crystal dolostone, rich-in arched stromatolite	

(continued)

Table 3.6 (continued)

Formation	3-order sequence	2-order sequence	Lithologic features of system tract	Layer No
	SQ. 18		HST: grey flat-laminated stromatolitic micritic dolostone turning upward into siliceous-encrust	87–99
			MSF: large mildly-wavy stromatolite and small conical stromatolite dolostone	
			TST: grey siliceous-thrombolitic micritic dolostone, greyish-white siliceous-stripped dolostone or silicalite	
	SQ. 17		HST: grey silicalite turning upward into karst breccia	75–86
			MSF: thick-bedded oncolitic dolostone	
			TST: grey medium-bedded micritic dolostone and siliceous-stripped micritic dolostone	
	SQ. 16		HST: grey medium-bedded siliceous-stripped micritic dolostone	53–74
			MSF: grey thick-bedded to massive mildly-wavy or hemispheric stromatolitic dolostone	
			TST: grey medium-bedded micritic dolostone and laminated stromatolitic dolostone	
	SQ. 15		HST: grey medium- to thick-bedded micritic dolostone turning upward into siliceous-thrombolitic or -stripped dolostone	33–52
			MSF: dark-grey thick-bedded mildly-wavy and hemispheric stromatolitic micritic dolostones	
			TST: dark-grey gravel- to sand-bearing micritic dolostone	
	SQ. 14		HST: grey medium-bedded micritic dolostone or siliceous-stripped dolostone	24–32

(continued)

Table 3.6 (continued)

Formation	3-order sequence	2-order sequence	Lithologic features of system tract	Layer No
			MSF: dark-grey mildly-wavy, laminated and small-conical algal micritic dolostone	
			TST: thin-bedded karst breccia turning upward into dark-grey medium- to thick-bedded micritic to very-fine-crystal dolostone	
	SQ. 13		HST: karst breccia turning upward into yellowish-grey sand-bearing muddy dolostone and dolomitic sandstone	14–23
			MSF: vertically accreted grey thick-bedded dolostone	
			TST: yellowish-grey medium- to thin-bedded silt-bearing micritic dolostone and laminated micritic dolostone	
	SQ. 12		HST: grey siliceous-stripped dolostone	1–13
			MSF: thick-bedded laminated stromatolitic dolostone	
			TST: light-grey medium- to thin-bedded sand- and mud-bearing dolostone, grey medium- to thick-bedded micritic dolostone and siliceous-stripped micritic dolostone	

involves 4 members, i.e., from Xia-1 to Xia-4 Members in ascending order, especially in the Xuanlong Depression at the west segment of YFDZ (cf. Chap. 2). Owing to the effect of tectonic movement called "Yuxian Uplifting", however, most strata of the Xiamaling Formation had been denuded, only the Xia-1 Member remained, which can be divided into 17 layers and 2 3-order sequences in the Jibei Depression (Fig. 3.10 and Table 3.7).

In addition, the basal bituminous bebbly coarse quartzose sandstone of Xia-1 Member appears as a disconformable contact with the underlying thin-bedded laminated micritic limestone on the top of Tie-2 Member of Tieling Formation, indicating the effect of another tectonic movement named "Qinyu Uplifting".

According to the different micro-cyclic superimposition modes from Xiamaling Formation to Luotuoling Formation, different system tracts are identified (Fig. 3.10). In Xiamaling Formation, the basal bituminous coarse quartzose sandstone turns

Table 3.7 Sequence division and system tract features of Hongshuizhuang-Jing'eryu formations in Jibei depression

Formation	3-order sequence	2-order sequence	Lithologic features of system tract	Layer No
Jingeryu	SQ. 39	SQ. XIII	HST: the interbedding of micritic limestone and leopard limestone on its top	70–74
			MSF: black-grey medium- to thick-bedded micritic limestone	
			TST: thin-bedded micritic limestone	
Luotuoling	SQ. 38	SQ. XII	HST: shale and glauconite-bearing calcirudite	64–69
			MSF: yellowish-grey (weathered) thin-bedded muddy siltstone intercalated by greenish-grey shale	
			TST: greenish-yellow silty shale intercalated by thin-bedded fine-medium-grained quartzose sandstone	
Xiamaling	SQ. 37	SQ. XI	HST: greyish-yellow medium- to thick-bedded bebbly coarse quartzose sandstone	55–63
			MSF: dark-grey shale	
			TST: thin-bedded fine sandstone to siltstone and thin-bedded fine-medium quartzose sandstone	
	SQ. 36		HST: medium- to thin-bedded shale and quartz sandstone etc	47–54
			MSF: dark-grey shale intercalated by muddy-silty lenticles	
			TST: bebbly quartzose sandstone intercalated by sand-stripped shale	

(continued)

Table 3.7 (continued)

Formation	3-order sequence	2-order sequence	Lithologic features of system tract	Layer No
Tieling	SQ. 35		HST: both thin interbedding of algal-mat limestone and micritic limestone and the unequal thick interbedding of thin-bedded micritic limestone and shale	28–46
			MSF: yellowish-green shale intercalated by knobby or lenticlar limestone	
			TST: dolomitic limestone or limy dolostone intercalated by shale	
Hongshui-zhuang	SQ. 34	SQ. X	HST: silty-bearing muddy dolostone and manganifeous micritic dolostone	12–27 (Tie-2 Member)
			MSF: huge thick black shale	1–11 (Hongshuizhuang Formation)
			TST; the unequal thick interbedding of black siliceous shale and black-grey thin-bedded muddy dolostone	

upward into a thin-bedded shale locally with sandy strips, which constitute a transgressive system tract (TST), where the appearance of dark-grey shale would represent the maximum sea flooding (MSF).

In addition, there are 2–4 layers of gabbro-diabase sills intruding into the Xiamaling Formation and resulting in regional wall-rock thermo-optical alteration, the Xiamaling shale was altered into carbonaceous shale, hornfels and slate, in Jibei Depression (cf. Chaps. 10 and 12).

3.2.3.7 $Pt_3{}^1l$ Luotuoling Formation

The measured Luotuoling stratigraphic section is situated at the Beizhangzi section, Kuangcheng (section No. 3 in Fig. 3.1). It consists of variegated clastic rocks, and can be divided into 6 layers and 1 3-order sequence (Fig. 3.10 and Table 3.7). Its basal greyish-yellow ferriferous bebbly coarse quartzose sandstone is in disconformable contact with the underlying Xiamaling shale. The quartzose sandstone contains pinnate cross-beddings, reflecting the coastal deposition of shallow water.

3.2.3.8 $Pt_3{}^1j$ Jing'eryu Formation

The measured Jing'eryu section is situated at the Beizhangzi section, Kuangcheng (section No. 3 in Fig. 3.1). It is in conformable contact with the underlying Luotuoling Formation, and divided into 4 layers and 1 3-order sequence (Fig. 3.10 and Table 3.7). It occurs as a lithological assemblage of sandy calcirudite, thin-bedded muddy siltstone and micritic limestone, showing the waters deepening process, and the interbedding of purplish-grey thin-platy marl and micritic limestone as the product of MSF. Consequently, earth crust was rising on a large-scale and sea level was declining until the occurrence of layer No. 76 leopard-skin limestone of the Lower Cambrian Fujunshan Formation, implying the termination of Meso-Neoproterozoic deposition almost 400 Ma. In between Jing'eryu and Fujunshan Formations, there is an unconformable surface.

As a whole, the Meso-Neoproterozoic Gaoyuzhuang to Jing'eryu Formations are totally divided into 39 3-order sequences in Jibei Depression (Fig. 3.11 and Tables 3.4, 3.5, 3.6 and 3.7) by comparison with the Jixian stratotype section in Jidong Depression, YFDZ.

3.3 Sedimentary Environments and Lithofacies

The geochemical study of Meso-Neoproterozoic stratigraphic column in YFDZ was started since the late 1970s on, its early work was about the geochemical study of petrochemical components. In the late 1970s, the research target was concentrated on petroleum geology and geochemistry, while organic geochemical study was performed on Meso-Neoproterozoic and Lower Palaeozoic carbonate rocks in YFDZ, NCC. Along with the data accumulation of Meso-Neoproterozoic sedimentary geochemical analysis and the progress of mathematical geologic and geochemical researches, these research results of new disciplines have been introduced into the stratigraphic division and correlation in YFDZ. The petrochemical components and micro-elements as well as the stable isotopes of Meso-Neoproterozoic strata are significant for the analyses of palaeosalinity, palaeoclimate and modern sea level variations and etc. (Deng and Qian 1993; Li et al. 1999).

In this chapter, most of the analytical samples for petrochemical components, micro-elements and strontium isotopes are dolostones with a few limestone samples, which are fresh rocks free from alteration, mineralization and secondary weathering, and sampled from the Meso-Neoproterozoic carbonate strata in Jibei Depression. The stratigraphic distribution of rock samples is listed in Table 3.8.

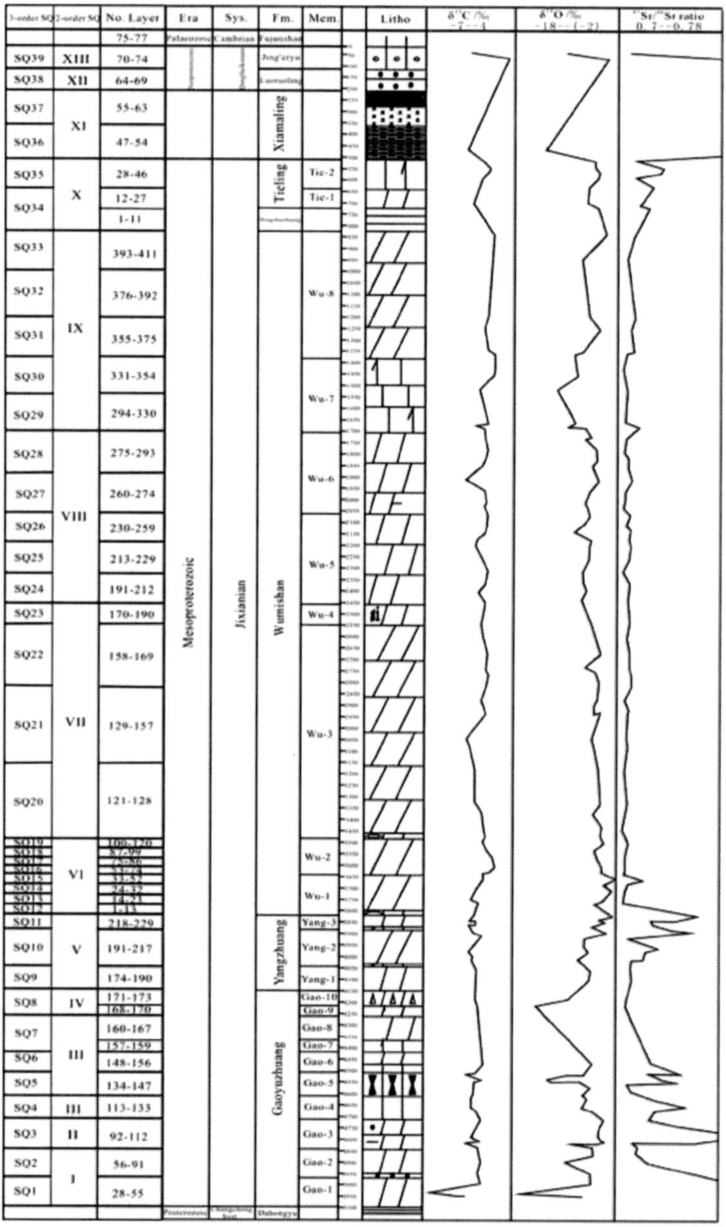

Fig. 3.11 Meso-Neoproterozoic stratigraphic sequence column in Jibei depressions

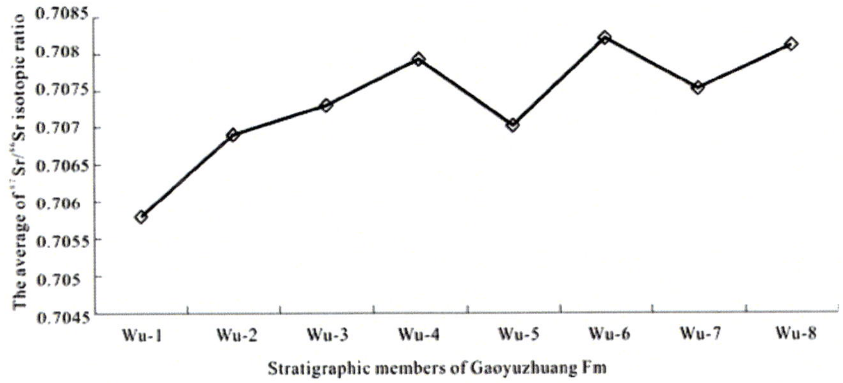

Fig. 3.12 Distribution of the average $^{87}Sr/^{86}Sr$ isotopic ratio in each member of Wumishan formation in Jibei depression

Among the rock samples, the petrochemical components are tested and analyzed using TRITON mass spectrometer by Tianjin Institute of Geology and Mineral Resources. The micro-elements are tested using PW4400/40 X-ray fluorescence spectrometer, the tested elements involve Ca, Mg, Si, Al, Fe, Mn, Na, K, B, Sr, Ba and Sr isotopes, from which a certain number of elements with high content and larger changes are chosen as the object of study, the tested rock sample numbers are listed in Table 3.8.

3.3.1 *Variation of Palaeo-Seawater Depth*

According to research, the Sr remaining time in seawater would be about 2.5×10^6 a, which is three orders of magnitude longer than the completely seawater mixing time of 10^3 years (McArthur et al. 1992). Therefore, the Sr isotopic composition of global seawater should be uniform in any time, which is confirmed by the observed $^{87}Sr/^{86}Sr$ isotopic ratios in modern seawater (Yang et al. 2001). However, the seawater $^{87}Sr/^{86}Sr$ ratio changes over time since Phanerozoic on. Generally, the initial value of $^{87}Sr/^{86}Sr$ ratio in sedimentary carbonate rocks ranges from 0.706 to 0.710 (Huang and Zhou 1997). It is commonly considered that the Sr isotopic composition in seawater originates mainly from the weathering product of continental crust and the mantle-sourced materials of hydrothermal carrying from mid-oceanic ridge. As compared to the global mantle-sourced $^{87}Sr/^{86}Sr$ average of 0.70 (Palmer and Elderfield 1985), the crust-sourced one has a higher global average of 0.711,9 and a very low $^{87}Sr/^{86}Sr$ mass difference (Palmer and Edmond 1989). so that the Sr isotopic fractionation can be negligible in the precipitation of carbonate minerals, the Sr isotopic characteristics of normalized minerals can be directly calibrated by the $^{87}Sr/^{86}Sr$ ratio of sea-water. When the plates collision, tectonic uplifting and accompany sea level fall, the exposed

Table 3.8 Stratigraphic distribution of Meso-neoproterozoic rock sample numbers for microelements and oxide analyses in Jibei depressions

Formation	Gaoyuzhuang	Yangzhuang	Wumishan	Hongshuizhuang	Tieling	Xiamaling	Luotuoling
Sample number	84	23	251	5	15	11	5

area of old land is expanded, the crust-sourced Sr from continental weathering would increase in ocean, resulting in the $^{87}Sr/^{86}Sr$ ratio is relatively raising in seawater (Lan 2001). While the submarine volcanic activity, sea-floor spreading and accompany sea level rise would cause the increase of vast dissolved mantle-sourced Sr in seawater on the one hand, and on the other hand the decrease of crust-sourced Sr duo to the old land shrinking. The superimposed effect of both factors leads to the smaller $^{87}Sr/^{86}Sr$ ratio in seawater. Thus, it follows that the fluctuation of $^{87}Sr/^{86}Sr$ ratio shows a negative correlation with contemporaneous sea level eustacy. During the geological history, the positive drift of $^{87}Sr/^{86}Sr$ ratio in marine carbonate rocks means sea level decline and old land expansion, while the negative drift reflects sea level rise and old land shrinking. Therefore, if no large-scale submarine volcanic activities, the global sea level variation would be the most important constraining factor for the Sr isotopic composition in seawater (Pratt 1998a, b). In China, a certain progress has been made on the Sr isotopic study of Phanerozoic marine carbonate strata, and its evolutional curves become more and more perfect. However, the Precambrian Sr isotopic study is still less studied, especially for the sporadic Mesoproterozoic Sr isotopic research. Since there is no large-scale submarine volcanism, tectonic activity only shows simple up-down motion during the sedimentary period of Gaoyuzhuang Formation in YFDZ, the sea level variation can be distinguished based on the Sr isotopic characteristics.

As to the Gaoyuzhuang Formation, it can be seen from Table 3.9 that the $^{87}Sr/^{86}Sr$ ratio of carbonate rocks appears as an overall getting smaller trend from Gao-1 to Gao-3 Members, while the ratio trends to increase relatively in the middle to upper Gao-3 Member, with a minimum of 0.7022 at the lower to middle Gao-3 Member, which illustrates that the sea level shows a relative rising trend from Gao-1 to Gao-3 Members, and then a decline trend since the middle Gao-3 Member, with the highest sea level during the early to middle period of Gao-3 Member. In this case, the higher $^{87}Sr/^{86}Sr$ ratio of Gao-1 Member indicates its relatively low sea level, it is consistent with previous research conclusion, i.e., the Gaoyuzhuang Formation has basically inherited the shallow-water sedimentary environment of Dahongyu Formation in the initial sedimentary period, but the fluctuation of its $^{87}Sr/^{86}Sr$ ratio is notable as a whole, which reflects the frequent variation of sea level during the Gaoyuzhuang sedimentary period (Fig. 3.12).

Based on the longitudinal change of $^{87}Sr/^{86}Sr$ ratio, and in combination with its sedimentary characteristics, it can be known that the shallow-water carbonate sediments of tidal flat facies had been widely developed at the initial period of Gaoyuzhuang Formation; and then the regional sea level rising and basin waters expanding resulted in the deeper-water carbonate sediments of platform basin facies at the mid period; while the sea level turned to decline so that the sedimentary waters shown a shallower tendency in the final period. Moreover, the $^{87}Sr/^{86}Sr$ ratio of Yangzhuang Formation shows an overall rising trend, which indicated an overall recovery process (Liu et al. 2005).

From the variation of $^{87}Sr/^{86}Sr$ ratio in Fig. 3.13, it can be seen that the sea level still appeared as a declining trend during the sedimentary period of Wumishan Formation, among which the $^{87}Sr/^{86}Sr$ ratios of Gao-5 and -7 Members get smaller,

Table 3.9 Meso-neoproterozoic B content and the Mg/Al, Mg/Ca, $^{87}Sr/^{86}Sr$ ratios in Jibei depression

System	Formation	Member	$B/10^{-6}$	Mg/Al	Mg/Ca	$^{87}Sr/^{86}Sr$
Pt_3^1 Qingbaikouan	Luotuoling	Average	30.5	0.11	2.04	0.80
Pt_2^3	Xiamaling	Average	28.4	0.20	1.01	0.73
Pt_2^2 Jixianian	Tieling	Average	12.1	7.50	0.57	0.75
	Hongshuizhuang	Average	4.5	2.92	1.64	0.78
	Wumishan	Wu-8	9.9	98.7	0.65	0.71
		Wu-7	21.0	5.72	0.08	0.71
		Wu-6	14.1	127.3	0.71	0.71
		Wu-5	10.0	134.6	0.71	0.71
		Wu-4	10.2	154.1	0.73	0.71
		Wu-3	11.7	123.2	0.70	0.71
		Wu-2	6.3	148.9	0.73	0.71
		Wu-1	7.6	155.5	0.72	0.71
		Average	11.4	118.5	0.63	0.71
	Yangzhuang	Yang-3	18.9	9.69	0.73	0.73
		Yang-2	28.9	7.12	0.72	0.73
		Yang-1	30.7	8.41	0.70	0.71
		Average	26.2	8.42	0.72	0.73
	Gaoyuzhuang	Gao-10	15.4	28.3	0.49	–
		Gao-9	8.0	32.2	0.16	–
		Gao-8	18.6	8.55	0.31	–
		Gao-7	15.34	9.20	0.43	–
		Gao-6	23.18	6.01	0.43	–
		Gao-5	52.3	3.95	0.34	0.72
		Gao-4	48.7	4.51	0.35	0.71
		Gao-3	50.9	14.7	0.64	0.71
		Gao-2	51.7	63.3	0.64	0.72
		Gao-1	20.4	13.0	0.64	0.74
		Average	30.5	18.4	0.44	0.72

indicating the sea level rising. While the $^{87}Sr/^{86}Sr$ ratio is totally expanded gradually in Hongshuizhuang to Luotuoling Formations (Fig. 3.14), it reflects a declining process of water depth.

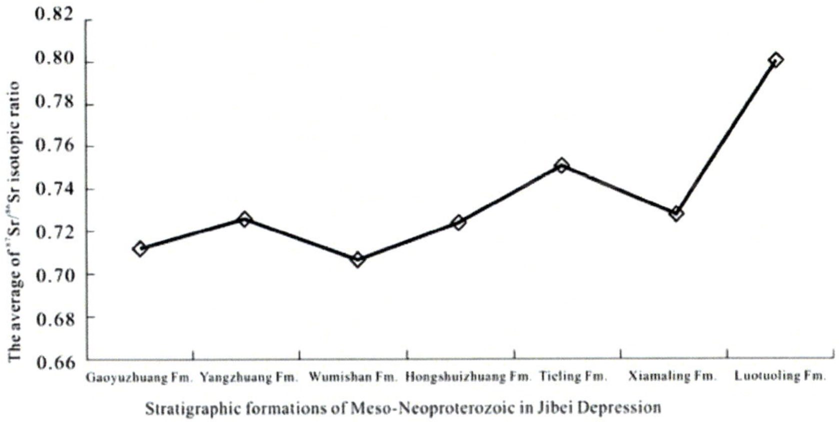

Fig. 3.13 Statistics of average $^{87}Sr/^{86}Sr$ isotopic ratio in each member of Gaoyuzhuang-Luotuoling formations in Jibei depression

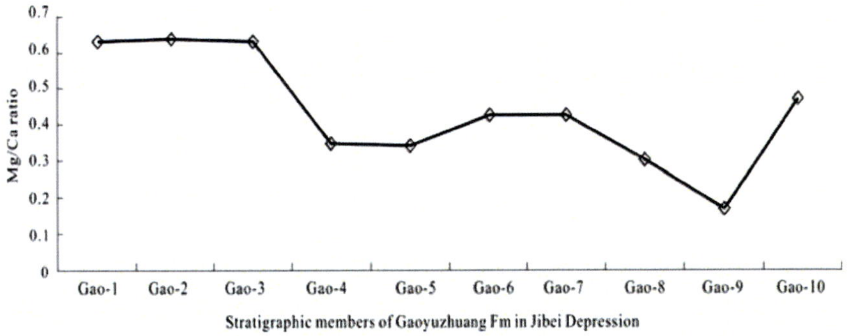

Fig. 3.14 Statistics for the average Mg/Ca ratio of each member in Gaoyuzhuang formation in Jibei depression

3.3.2 Palaeo-Climate and Palaeo-Salinity

The Mg/Ca ratio can be taken as a good marker of climatic changes, its high ratio indicates a arid and hot climate, on the contrary its low ratio shows a wet climate (Kuang et al. 2005). It is because the basic layer consists of sodium carbonate rock, when sodium salt begins to be precipitated, the concentration of Mg^{2+} and Ca^{2+} would be very low in water medium due to their full precipitation; besides, the chemical activity of Mg^{2+} is much worse than that of Ca^{2+}. As compared with each other, Mg^{2+} is almost exhausted, so that the Mg/Ca ratio in strata shows low or very low value (Song 2005). Thus, a necessary supplement should be made for the Mg/Ca ratio as a climate marker, that is, if the soluble salts (e.g., sodium and potassium salts etc.) are not involved in precipitation, a high Mg/Ca ratio would indicate an arid and hot

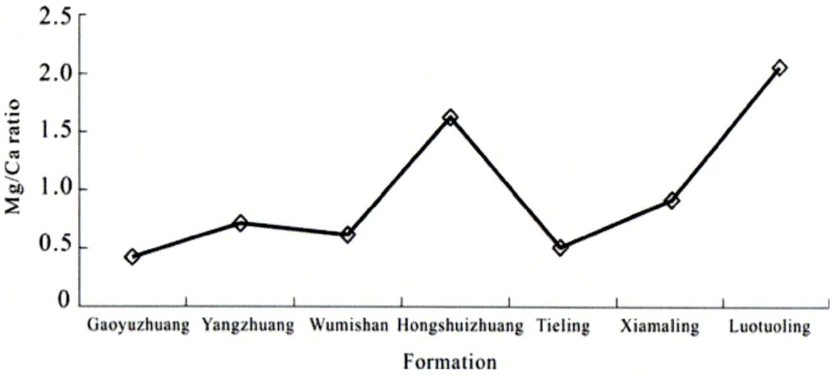

Fig. 3.15 Statistics for the average Mg/Ca ratio in each members of Gaoyuzhuang-Luotuoling formations in Jibei depression

climate, while they are involved, a low Mg/Ca ratio and relatively high K^+ and Na^+ contents indicate an arid and hot climate.

As shown in Fig. 3.14, the variation of Mg/Ca ratio is very characteristic in Gaoyuzhuang Formation, which appears as an overall undulation trend of rising-falling-rising, reflecting a variable process from relative arid and hot, gradually though wet, again to arid and hot palaeo-climates. Meanwhile, the palaeo-salinity of sea-water should also experience an evolutional process of gradually reducing, and then increasing.

From the change tendency of Mg/Ca ratio (Fig. 3.15), it can be known that the Mg/Ca ratio is relatively low, with the averages of 0.4425, 0.6320 and 0.5655 respectively in Gaoyuzhuang, Wumishan and Tieling Formations, indicating relatively wet palaeo-climate and low palaeo-salinity of seawater, among which the Mg/Ca ratio is only 0.11 in Wu-7 Member, reflecting the lowest palaeo-salinity of sea-water and the most humid palaeo-climate, which would result from a maximum transgression. While drier palaeo-climate and higher palaeo-salinity of sea-water would occur in Yangzhuang, Hongshuizhuang, Xiamaling and Luotuoling Formations.

3.3.3 *Sedimentary System and Lithofacies*

The Meso-Neoproterozoic strata contain a variety of sedimentary rocks in YFDZ, which can be attributed to three rock types. Type I is carbonate rocks, mainly to be dolostones with less limestones. Dolostones include crystalline, muddy, stromatolitic, grain and transition dolostones; while limestones involve micritic, very-fine-crystalline, grain, knobby, banded and transition limestones. Type II is terrigenous clastic rocks, mainly to be sandstone, conglomerate and shale/mudstone. Type III is other rock types, such as silicalite, tempestites, baffle-bonded-stone and breccia, etc.

Fig. 3.16 Sedimentary facies markers in Jibei Depression. a. Banding micritic limestone, Gao-2 Member; b. Lamellated stromatolite, Wu-2 Member; c. columnar stromatolitic limestone, Tie-2 Member; d. brownish-grey medium-bedded fine-grained quartzose sandstone with meso-trough- shaped cross-bedding, Luotuoling Formation; e. ripple markers, Yangzhuang Formation; f. bafflestone, Gao-9 Member

As the sedimentary structures, a flat-bedding, parallel bedding, cross-bedding, wavy-bedding and ripple marks, scouring surface, mud cracks and stromatolite, etc.are mainly developed (Fig. 3.16).

A complete stratigraphic section from the Jixianian plus Xiamaling Formation to Qingbaikouan, with total stratigraphic thickness of 5047.5 m, has been composed by the three outcrop sections respectively at Weizhangzi, Jianshanzi and Beizhangzi in Jibei Depression (Fig. 3.1). Based on regional geological setting, sedimentary characteristics and facies markers (e.g., rock colour, authigenic minerals, graininess, composition, texture, sedimentary structure, and palaeontological fossil etc.), in combination with the microscopic observation of more than 1000 rock slices, grading analysis of 20 samples and petrochemical analysis of more than 2000 samples (including componants and micro-elements, Sr, C and O isotopes and X-ray diffraction), two sedimentary systems of the Meso-Neoproterozoic marine carbonate and clastic rocks are established in Jibei Depression (Song 2007). The sedimentary facies, subfacies and micro-facies are listed in Table 3.10 for details.

3.3.3.1 Carbonate Platform Facies

The carbonate platform developed in Jibei Depression, YFDZ should be referred to an epicontinental sea type of carbonate platform, which contains a flat and open terrain, shallower waters, few terrigenous supply, well-developed dolostones, numerous polymorphous stromatolites, various dolorudites without significant changes in

Table 3.10 Meso-Neoproterozoic sedimentary facies division in Jibei Depression

Sedimentary system	Facies	Subfacies	Micro-facies	Distribution in formation and member
Carbonate rock system of epicontinental sea	Carbonate platform	Tide flat	Supratidal zone	Gaoyuzhuang and Wumishan Formations
			Intertidal zone	Yangzhuang and Hongshuizhuang Formations
			Subtidal zone	Tieling and Jing'eryu formations
		Lagoon	Lagoon mud	Gaoyuzhuang and Hongshuizhuang formations
	Organic reef	Baffle-bonding reef	Baffle-bonding rock	Gao-9 and -10 members
Clastic rock system of pericontinental sea	Neritic shelf	Transitional zone	–	Xiamaling and Luotuoling formations
		Offshore shelf		
	Barrier-free coast	Foreshore	–	Xiamaling and Luotuoling formations
		Shoreface	–	Xiamaling and Luotuoling formations

lithology and facies. Its remarkable sea level fluctuation results in the obviously periodic and rhythmic deposition.

The sedimentary markers of phreatic turbulence and arid evaporation indicate a broad Meso-Neoproterozoic tidal wave action zone in YFDZ. The overall elemental geochemical characteristics show the dominant CaO and MgO contents with relative amounts of $\omega(CaO) > \omega(MgO)$, and less SiO_2, Al_2O_3, Fe_2O_3, FeO and K_2O contents etc., while Sr, Sr/Ba and Sr/Ca ratios increase gradually with the deepening sedimentary waters.

The carbonate platform facies are mainly distributed in the Jixianian Gaoyuzhuang, Yangzhuang, Wumishan, Hongshuizhuang and Tieling Formations (Wang and Chen. 1993) as well as Qingbaikouan Jing'eryu Formation, which consist of tidal flat and lagoon subfacies (Table 3.10).

1. **Tidal Flat Subfacies**

Its geological agency is constrained by stronger tidal action and weaker wavy action in Jibei Depression, the range of influence is very wide laterally up to 100–1000 km. Due to the periodic flux and reflux, the lateral distribution of facies zone sediments appears as zonation. According to the intensity of waters energy, it can be subdivided into three microfacies, i.e., the supratidal, intertidal and subtidal zones (Zhao 1988).

(1) **Supratidal zone**: It is in between the average high tidal level and the maximum high tidal level, often exposed to the atmosphere, and flooded by seawater only during big storm and large tide. It usually shows a transitional relationship with the upper part of intertidal zone. In the zone, dolomitization would be stronger so as to forming penecontemporaneous dolostone. Its rock types are mainly of slight-grey, grey and brownish-grey lamellated or thin- to thick-bedded micritic- to very-fine-crystalline dolostones, micritic limestone and muddy dolostone. Since there often keeps arid environment in the zone, algal stromatolitic deposition is not well developed. Sometimes a few doped ferruginous and often mixed clastic materials would result in the yellowish-grey and -green, purplish-red shales and the transitional type of carbonate and clastic rocks, such as muddy and sandy dolostones etc. Occasionally the arid evaporative sedimentary marks, i.e., mud cracks, gypsum pseudomorphic crystal, bird'eye structure and etc. can be seen. The petrochemical components and elemental isotopic characteristics appear as the lowest V, Rb, Be, P_2O_5 and MnO contents and the higher SiO_2 and Al_2O_3 contents because the supratidal zone is within the nearest distance to the coast, and more sufficient terrigenous material supply is available, so that it has the minimum $\delta^{13}C$ average, only around −0.78‰, and the maximum $\delta^{18}O$ average up to −5.10‰.

(2) **Intertidal zone**: It is located between the average high and low tidal levels as the widest zone in the tidal flat environment, characterized by intermittent energy deposition. In general, muddy and silty dolostones/limestones would be formed by the doping of mud and silts. The wavy and bedded stromatolites are often developed, while conical-columnar stromatolite occasionally seen, and generally turning upwards from columnar through wavy to bedded stromatolites. Lithologically, it mainly includes very-fine and fine-crystalline dolostones, endoclastic and oolitic dolostones intercalated by dark-grey siliceous dolostone, siliceous shale with siliceous thrombolite, locally manganese and sometimes terrigenous silts. Usually the sediments appear as lighter colour, mud cracks, bird'eye structure and pinnate cross-bedding, etc.

The intertidal zones can be divided into intertidal low and high energy zones. The low energy zone is mainly characterized by lamellated and micro-wavy stromatolitic dolostones, edgewise dolostone and low-amplitude ripple markers, while the high energy zone by large-wavy and medium to small conical-columnar algal stromatolitic dolostones, sparite-grain-dolostone, grain-dolostone (including endoclastic, oolitic, algal oolitic and pelletal dolostones), brecciated dolostone and cross-bedding, etc. The petrochemical components and elemental isotopic characteristics present that V, Rb, Be, P_2O_5 and MnO contents are in between these of supratidal and subtidal zones, and the SiO_2 and Al_2O contents are smaller, because the intertidal zone is far from the coast and terrigenous supply is less. Moreover, both $\delta^{13}C$ average 0.17‰ and $\delta^{18}O$

average -5.61‰ in the intertidal zone are also in between the those of supratidal and subtidal zones.

(3) **Subtidal zone**: It is below the average lower tidal level, and can be further divided into subtidal high and low energy zones.

The high energy zone is below the average low tidal level, but above the wave base with stronger hydrodynamic force. Its rock types include dolarenite-dolorudite or calcarenite- calcarrudite, oolitic and thrombolitic dolostones, etc., while large conical or columnar algal stromatolitic dolostone and algal reef are often seen, locally with manganifeous and siliceous stripes, tempestites as well as the sedimentary structures, e.g., various cross- and flat-beddings and ripple markers etc.

The low energy zone is below the wave base. Owing to its deeper waters, weaker light and less algal activity, its rock types are mainly chemically-precipitated, algal-relic-barren and mostly thick-bedded micritic dolostone/limestone, knobby limestone, shale, and sometimes with more manganous sediments. In the case of high muddy content, the muddy dolostone with flat-bedding would be often seen. The petrochemical and elemental characteristics show higher V, Rb, Be, P_2O_5 and MnO contents and lower SiO_2 and Al_2O_3 contents due to its furthest distance from coast and almost lack of terrigenous supply. In addition, the value $\delta^{13}C$ average is the highest one up to 0.59‰, but the $\delta^{18}O$ average the lowest only -6.36‰ in the subtidal zone.

2. **Lagoon Subfacies**

The oolite, algal oolite and algal thrombolite etc. are constantly transported from the subtidal high energy zone to the mid-low position of the intertidal zone by water current, resulting in the shoal type of deposit. The shoal berrier effect would isolate the nearshoal waters from the open-sea. To a certain extent, the current is constrained and seawater in a restricted or semi-restricted circulation, so that tide becomes major agent on the landward side, and wave action is unobvious in the euxinic lagoon environment. Due to its low energy, the sediments are mainly of dark shale intercalated by thin-bedded dolostone with flat-bedding, ferromanganese nodules and pyrite in lagoon.

3.3.3.2 Organic Reef Facies

Organic reef facies is not well developed in YFDZ. It is only discovered within the Gao-9 and Gao-10 Members of Gaoyuzhuang Formation from the Panjiadian section (in Xinglong County) on the west margin of Jibei Depression to the Guzifang section (in Zhangjiakou City) in Xuanlong Depression (Fig. 3.17). The sedimentary waters in both Jibei and Xuanlong Depressions were connected each other during Gaoyuzhuang sedimentary period.

It appears as a light-grey huge-thick massive very-fine-crystalline algal dolostone, within which, the inside algal filaments or macroscopic algae can be dimly seen with vertically and branching sessile growth. Lots of pores and vugs are developed with extension vertical to bedding plane. While very-fine- to fine-crystalline dolostones

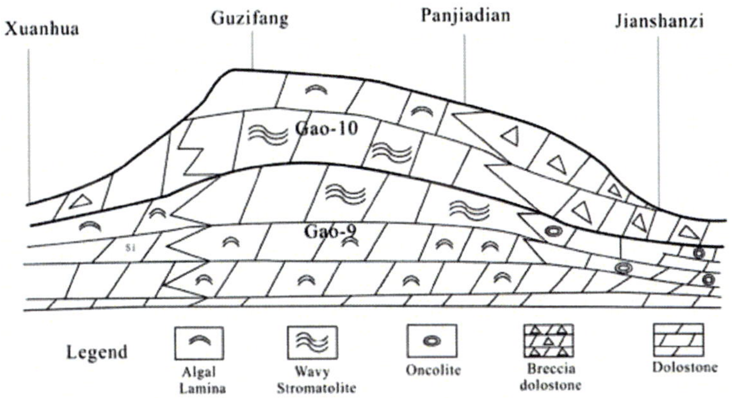

Fig. 3.17 Schematic diagram of organic reef distribution in Gao-9 and Gao-10 members of Gaoyuzhuang formation in YFDZ

are filled in between algae, and also filled or semi-filled by crystalline dolomite or chert and quartz. The reef-building organisms of organic reef are referred to blue-green algae and macroscopic algae. While the baffle-bonding reef is formed due to the baffling and bonding of sediments in the sea-water (Xiao et al. 2001).

Based on the regional distribution of organic reefs, since the algal lime-nodular or oncolitic limestone and karstic breccia in the Gao-9 and Gao-10 Members are found at the Jianshanzi section (in Kuancheng County) in Jibei Depression on the eastern side of Fig. 3.17, while the siliceous stripped dolostone, stromatolitic dolostone and thick-bedded brecciated dolostone are developed at Xuanhua section in Xuanlong Depression on the midwestern side of Fig. 3.17, the occurrence of organic reef looks like a connected hummocky drum.

3.3.3.3 Neritic Shelf Facies

The shelf facies zone is above wave base, it contains flat terrain, small slope and weaker hydrodynamic force. Its waters could keep up well exchange with external circulation, and mainly resulting in suspended sediments, which consist of large number of shale sandwiched with thin-bedded muddy siltstones, fine sandstone or limestone lenticle. The overall petrochemical and elemental characteristics show that the predominant SiO_2 and lower CaO and MgO contents, while the Al_2O_3, Fe_2O_3, FeO, K_2O contents, etc. are obviously higher than those in carbonate platform, Sr content, Sr/Ba and Sr/Ca ratios increase gradually with the deepening waters. The neritic shelf facies discovered in YFDZ is distributed in the Xiamaling Formation and Qingbaikouan Luotuoling Formation, which can further divided into 2 subfacies of offshore shelf and transitional zone.

1. **Offshore shelf subfacies**

   It is situated in neritic area from the outside of transition zone to the internal edge of continental slope. Its lithology is mainly of dark-grey shale with thin-bedded siliceous mudstone, ferrian limestone lenticles and yellowish-brawn (weathering color) muddy dolostone, and its bedding is undeveloped under weaker hydrodynamic condition. The darker colour shale implies a reduction sedimentary environment of deeper waters. Due to it far away from the coast zone and less supply of terrigenous material, carbonate rocks is only slightly developed in the offshore shelf subfacies. Compared with the other subfacies in the sedimentary system of epicontinental clastic rocks, the petrochemical and elemental characteristics appear as the highest SiO_2 content, the higher micro-elements contents (e.g., V, Rb, Be, etc.), and the lowest contents of most petrochemical components (e.g., Al_2O_3, Fe_2O_3 and K_2O, etc.).
2. **Transitional zone subfacies**

   It is located below wave base at a transitional zone between the nearshore and offshore shelfs. Its lithology is mainly of greyish-green shale sandwiched with thin-bedded muddy siltstone and fine sandstone. The shale contains higher content of terrigenous material, micro-flat-bedding and ripple-bedding due to weaker hydrodynamic conditions, locally with authigeneous minerals of ooide-containing chlorite and glauconite and often with perlimonite and siderite nodules and lenticle. The nodules appear as 10–50 cm in size and bedding distribution, while the lenticle is up to 60 cm in length. The siderite nodules and lenticle indicate an alkalescent reducing sedimentary environment of relative calm waters. The petrochemical and elemental characteristics show higher SiO_2 content, higher contents of micro-elements (e.g., V, Rb and B), and lower contents of petrochemical components (e.g., Al_2O_3, Fe_2O_3, K_2O, etc.).

3.3.3.4 Baffle-Free Coastal Facies

The baffle-free coastal facies is well connected with ocean, and affected by evident wave and offshore current, where seawater can fully flow and circulate, therefore it is also called a coast of open sea type. Based on hydrodynamic conditions as well as sedimentary features (i.e., rock composition, texture, structure, etc.), the overall petrochemical and elemental characteristics show higher SiO_2 content, lower CaO and MgO contents, and obviously higher Al_2O_3, Fe_2O_3, FeO and K_2O contents, etc. compared to these in carbonate platform facies, while Sr content, Sr/Ba and Sr/Ca ratios are gradually increasing with the waters deepening. The baffle-free coastal facies is distributed in Xiamaling Formation and Qingbaikouan Luotuoling Formation, which can be subdivided into shoreface and foreshore subfacies.

1. **Shoreface subfacies**

   It is also called nearshore subfacies, and situated between average low tidal line and normal wave base, while its geographical location is equivalent to the subtidal zone, where the lithology of sediments is mainly of grey to dark-grey ferruginous fine-sandstone to siltstone and yellowish-green shale. In sandstone, sand grains change from fine to coarse and sandstone thickness from thin to thick ascendingly, generally with grain size of 0.05–5 mm, better sorting and roundness, and appearing as oval- to subrounded-shapes. As for mineral composition, quartz content accounts for ca. 80% and more, which is mainly muddy- and siliceous-cemented, and often containing pyrite, siliceous and glauconitic authigenic minerals. Due to stronger hydrodynamic condition, flat-bedding, lenticular-bedding and small ripple- bedding are developed in sandstone. Its higher MnO content would indicates an oxidizing environment of shallower waters. The petrochemical and elemental characteristics present the lowest contents of micro-elements (e.g., V, Rb, Be, etc.), and lower contents of petrochemical components (e.g., Al_2O_3, Fe_2O_3, K_2O, etc.) in the clastic rock subfacies.
2. **Foreshore subfacies**

   The foreshore is located between the average high-tidal and low-tidal levels, geographically equivalent to the intertidal zone, but different from the intertidal zone in sedimentary mechanism and sediment characteristics. The foreshore hydrodynamic condition is characterized by wavy-washing, thus, the sediments have high abrasivity and sufficient elutriation. Its lithology is mainly of greyish-yellow medium- to thick-bedded granule-bearing medium- to coarse-grained quartzose sandstones, locally with asphalt. In sandstone, quartz content accounts for 90%, sand grain size are 2–5 mm with better sorting and rounding, which are cemented by ferriferous-muddy, siliceous matters and glauconite. Moreover, clear flat-, flushing- and cross-beddings are developed, the flushing-bedding appears as low-angle cutting each other by multiple sets of flat-bedding with 2–8 mm thick fine layer and 6–14 cm-thick measures. The petrochemical and elemental characteristics show the lower contents of micro-elements (e.g., V, Rb, Be, etc.), and relatively lower contents of petrochemical components (e.g., Al_2O_3, Fe_2O_3, K_2O, etc.).

3.3.4 Facies Analysis of Stratigaphic Sections

In the context, only the facies analysis of measured Pt_2^2 Jixianian, Xiamaling Formation (Pt_2^3x) and Pt_3^1 Qingbaikouan stratigraphic sections are involved and the unmeasured Pt_2^1 Changchengian section is not included.

3.3.4.1 Pt_2^2 Jixianian

1. **Gaoyuzhuang Formation (Pt_2^2g)**

Along with the sea level fluctuation, the Gaoyuzhuang Formation has been through an evolutional process of intertidal-subtidal-intertidal-supratidal zones. Stratigraphically, it is totally divided into 10 members (Gao-1 to Gao-10 Members), 146 layers (layer Nos. 28–173) and five 2-order stratigraphic sequences (SQs. I–IV) in ascending order at the Jianshanzi section in Jibei Depression as follows (Tables 3.1 and 3.4; Figs. 3.8 and 3.11; cf. Chap. 2).

Gao-1 Member: It covers Nos. 28–55 layers (Table 3.4; the same below). There is a *ca.* 13 m thick basal quartzose sandstone, turning upwards into siliceous-tripped, -nodular and stromatolitic dolostones sandwiched with muddy dolostone, thin-bedded quartzose sandstone, medium- to thick-bedded dolostones and bebbly dolomitic sandstone, which is characterized by terrigenous clast- and manganese-containing as well as flat-bedding and local wavy, arched and columnar stromatolites. It is referred to the sediments of supratidal to intertidal zones and equivalent to the lower portion of 2-order SQ. I (Table 3.4).

Gao-2 Member: It covers Nos. 56–91 layers, and mainly consists of dark-grey to grey siliceous dolostone, muddy and manganiferous dolostones and dark-grey to grey thin-bedded stromatolitic dolostone, in which laminated, wavy, conical and columnar stromatolites are developed, which belong to the sediments of intertidal to subtidal zones, and is equivalent to the upper portion of 2-order SQ. I.

Gao-3 Member: It covers Nos. 92–112 layers. As the main lithology, thin-bedded muddy siltstone and manganiferous shale appear at the lower interval, while manganiferous dolostone, silty and fine-crystal dolostones are at the upper interval. Which belong to the sediments of intertidal to subtidal zones, and is equivalent to the 2-order SQ. II.

Gao-4 Member: It covers Nos. 113–133 layers. It is mainly of dark- and light-grey micritic dolostone, dolomitic and muddy limestones, and characterized by manganiferous-, siliceous-strip- and nodule-containing. A few knotty-limestones are developed in its medium to low intervals indicating a MFS in the transgression process. After the occasionally emerged intertidal environment, the Gao-4 Member has entered a low-energy environment of stagnant waters, and resulting in subtidal sediments unsuitable for algal thriving, which is equivalent to the basal portion of 2-order SQ. III.

Gao-5 Member: It covers Nos. 134–147 layers. Its basal interval has an interbedding of black siliceous and dolomitic shales sandwiched with knotty-limestone, and turning upwards into an interbedding of thick-bedded knobby-limestone, silicalite and dark-grey medium- to thick-bedded dolostone. While a layer of brecciated dolostone, locally with flat-bedding, is developed on its top interval. It belongs to the sediments of subtidal to intertidal zones, and is equivalent to the lower portion of 2-order SQ. III.

Gao-6 Member: It covers Nos. 148–156 layers. The dark-grey medium-bedded to massive calcarenite, sandy limestone and massive dolarenite, with limy-strip and lenticle, occur at its basal interval, and local cross-bedding is dimly visible in the basal calcarenite. While it turns upwards into an interbedding of dark-grey thin-bedded kobby-limestone, dolomitic limestone and thick- to medium-bedded limy dolostone, which also sandwiched with shale, and making up multi-rhythms. Since the Gao-6 Member on, sea level has short rises, and then falling down. Which are referred to the sediments of intertidal to subtidal zones, and equivalent to the middle portion of 2-order SQ. III.

Gao-7 Member: It covers Nos. 157–159 layers. It mainly appears as an interbedding of limestone and dolostone with shale interbeds, resulting in multi-rhythms. There are more shale interbeds at its top and basal intervals, and thus each individual layer of dolostone is thinner. However, the shale interbed is thinner, and dolostone is thicker in the middle interval. Flat-lamellas, wavy lamination and local limy-strip, lenticle and ripple marks, etc. are developed in dolostone, while cross-bedding can be dimly visible at the bottom of calcarenite. Which belong to the sediments of intertidal zone, and are equivalent to the middle to upper portions of 2-order SQ. III.

Gao-8 Member: It covers Nos. 160–167 layers. Its basal interval is mainly composed of dolomitic limestone, and the limy component is gradually reduced upward. The thin-bedded limy dolostone with limy-thrombolite or strip occurs at its lower to middle intervals. While the upper interval is of dark-grey thick-bedded micritic dolostone, limy dolostone and dolomitic limestone with siliceous-strip, thrombolite, and occasionally visible simatic nodule. The larger ripple marks and mud cracks are exposed on some bedding planes. Due to the continuous shallowing of waters, Gao-8 Member is attributable to a sedimentary environment of supratidal zone, and equivalent to the middle to upper portions of 2-order SQ. III.

Gao-9 Member: It covers Nos. 168–170 layers. Its lower interval is mainly of dark- to black-grey medium- to thick-bedded and massive micritic limestones and dolomitic limestone, but dolomitic limestone contains lamellas. While dark-grey medium- to thick-bedded micritic to very-fine-crystalline limestones, with a little dolomitic limestone, visible dolomitic laminae, a few siliceous-thrombolites and nodules, in its upper interval. As a whole, micritic limestone plus local oncolite is more developed, and very-fine-crystalline limestone less developed. Moreover, limestone often contains sandy and silty matters, occasionally with dolomitic matter. Which belong to the sediments of intertidal zone, and are equivalent to the lower portion of 2-order SQ. IV.

Gao-10 Member: It covers Nos. 171–173 layers. Its lower interval is mainly composed of grey to dark-grey medium- to thick-bedded and massive micritic to very-fine-crystalline limestones and dolomitic limestones, which contain more thromobolitic-, stripped- and nodular-silicalites. While its upper interval is of thick-bedded to massive breccia, the brecciated component includes dolostone and limy dolostone, which appear as angular-shape and a large disparity in gravel sizes, i.e., the large one can be up to 10 cm, but generally it is 1–5 cm. A few siliceous-thrombolites or strips occur at top 2 m of the upper interval. Which indicate a shallowing waters again. It is referred to the sediments of intertidal to supratidal zones and equivalent to the middle to upper portions of 2-order SQ. IV.

2. **Yangzhuang Formation ($Pt_2{}^2y$)**

It inherits the regressive tendency of late sedimentary period of Gaoyuzhuang Formation, and belongs to a nearshore sediments of epicontinental sea with sufficient supply of terrigenous clastic material under the aridic climate. It is principally characterized by purplish-red silty-muddy micritic dolostone, mainly constituting a set of intertidal to supratidal sediments, and appearing as the alternating strata of red and white colors and the red stratum is predominant in the Yang-3 Member, which reflects a periodic arid climatic condition and oxidized medium environment. Stratigraphically, it is totally divided into 3 members (Yang-1 to Yang-3 Members), 56 layers (layer Nos. 174–229) and one 2-order stratigraphic sequences (SQ. V) in ascending order at the Jianshanzi section in Jibei Depression as follows (Tables 3.1 and 3.5; Figs. 3.8 and 3.11; cf. Chap. 2).

Yang-1 Member: It covers Nos. 174–190 layers (Table 3.5; the same below), which is dominated by purplish-red and grey medium-bedded micritic dolostone locally with small conical-columnar, wavy- and mildly-wavy-stromatolites. In addition, mud cracks, shrinkage cracks, bird's-eye structure, low-amplitude and asymmetrical ripple markers and pinnate cross-bedding, etc. are visible. It is referred to the sediments in intertidal to supratidal zones and equivalent to the lower portion of 2-order SQ. V.

Yang-2 Member: It covers Nos. 191–217 layers, and is dominated by purplish-red and brawnish- red sandy and muddy dolostones with the typical features of the increasing terrigenous mud and sands, and constituting a sedimentary cycle of thin-bedded quartzose sandstone and sandy to muddy or micritic dolostones, occasionally sandwiched with lamellated- and wavy-stromatolitic dolostones, and shallow water ripple markers are often visible, sometimes the arid markers (mud crack, gypsum and halitic pseudocrystals, etc.) can be seen. It belongs to the sediments of the supratidal zone, and is equivalent to the middle to upper portion of 2-order SQ. V.

Yang-3 Member: it covers Nos. 218–229 layers. Its lower interval is primarily light-grey siliceous-tripped micritic dolostone intercalated by sandy dolostone. While its middle to upper intervals are the interbedding of red and white micritic dolostones. It is referred to the sediments of the intertidal to supratidal zones, and equivalent to the top portion of 2-order SQ. V.

3. **Wumishan Formation ($Pt_2{}^2w$)**

It is remarkably characterized by huge stratigraphic thickness, very developed sedimentary rhythmic layers (Fig. 3.18) and numerous carbonatic biolite, which are resulted from the frequent eustatic variations and changeable seawater depth during its sedimentary period.

In Jibei Depression, the total stratigraphic thickness of Wumishan Formation is up to 2947.2 m, which can be subdivided into more than 400 different rhythms, almost

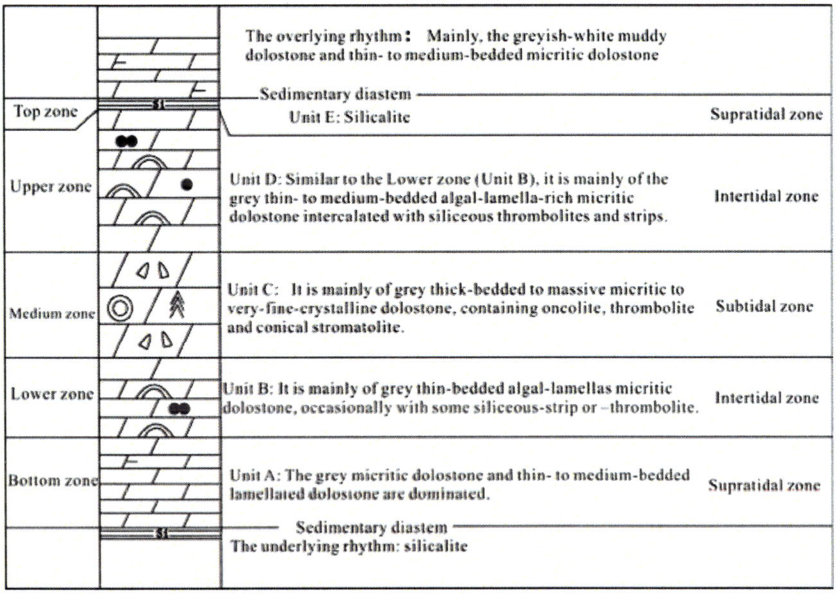

Fig. 3.18 The complete assemblage of five basically rhythmic units for the sedimentary rhythms in Wumishan Formation

all the individual unit is composed of 5 basically rhythmic units from Units. A to E in ascending order (Fig. 3.18), most of the adjacent units are separated each other by a depositional diastem. These rhythms would reveal a low-amplitude but very frequent oscillation of sea level, which obviously shows the features of vibrating subsidence.

Wumishan Formation can be divided into 8 members (from Wu-1 to Wu-8 Members), 411 layers (layer Nos. 1–411) and 4 2-order stratigraphic sequences (SQs. VI to IX) in ascending order at the Weizhangzi section in Jibei Depression (Tables 3.1 and 3.6; Figs. 3.9 and 3.11; cf. Chap. 2).

Taking Jibei Depression as an example, the lithology-lithofacies characteristics in each member of Wumishan Formation are briefly described in ascendant order as follows.

Wu-1 Member: It covers Nos. 1–52 layers, and mainly consists of grey thrombolitic, stromatolitic and algal-lamellated (or called algal-mats) dolostones, and dolomitic shale, commonly appearing as an incomplete rhythmic assemblage of Units A-B-D-E. With less thickness, the Unit C is seldom seen or generally disappears, indicating a shallower water environment of supratidal or intertidal zone. With larger thickness and deeper sedimentary waters, however, the Unit C emerges more frequently in 3-order SQ. 14 (Table 3.6), showing deeper waters.

Overall, the Wu-1 Member is equivalent to the lower to middle portions of 2-order SQ. VI.

Wu-2 Member: It covers Nos. 53–120 layers, is mainly of grey algal-lamellated and silty-muddy dolostones and dolomitic shale, which constitute three kinds of incomplete assemblages of basically rhythmic units, i.e., the major rhythmic assemblage of Units A-B-D-E in a shallow water environment, the minor assemblages of Units B-C-D in a relatively deepening waters and of Units B-D-E in a relatively shallowing waters, in ascending order. In general, the Wu-2 Member is mainly referred to the sediments of intertidal zone and equivalent to the upper portion of 2-order SQ. VI.

Wu-3 Member: It covers Nos. 121–169 layers, and mainly involves grey thick-bedded to massive thrombolitic and stromatolitic dolostones, siliceous-thrombolite or -strip micritic dolostones, algal-lamellated dolostone and dolomitic shale. The Wu-3 Member is equivalent to the most portion of the 2-order SQ. VII, and including the 3-order SQs. 20 to 22. The SQ. 20 is composed of the incomplete assemblage of Units A-B-D-E in the shallow waters, while the SQ. 21 and SQ. 22 are characterized by the complete assemblage of Units A-B-C-D-E (Fig. 3.18) and the dominant Units B, C and D, even the SQ. 21 contains large-size conical and columnar stromatolites, which indicate the sediments of deep water environment as the products of transgressive process at the subtidal zone.

Wu-4 Member: It covers Nos. 170–190 layers, and mainly consists of grey medium- to thick-bedded micritic and very-fine-crystalline dolostones. The Wu-4 Member inherits the sedimentary environment of the Wu-3 Member. Its rhythmic units B, C, D are dominated, there basically is no obvious change of sea level, and it is mainly referred to the sediments of subtidal zone and equivalent to the top portion of 2-order SQ. VII.

Wu-5 Member: It covers Nos. 191–259 layers, and is constituted by the rhythmic layer of grey micritic dolostone, algal-lamellated dolostone and dolomitic shale sandwiched with ooid silicalite, sparry-dolorudite. On the whole, the rhythms are characterized by frequent rhythmic assemblage of Units B-C-D in deeper water environment of intertidal zone, in between a few assemblages of Units A-B and/or B-E are occasionally sandwiched. Even though the Unit C of subtidal zone still exists, its thickness is reduced. However, the assemblage of Units A-B-E starts to emerge frequently on the top of Wu-5 Member, which reflects a shallowing sedimentary environment of waters. As a whole, the Wu-5 Member has presented a regressive process, the sediments of supratidal zone are obviously more developed than those of its underlying strata, and thus the sedimentary environment of Wu-5 Member is mainly in the subtidal to intertidal zones, and equivalent to the lower portion of 2-order SQ. VIII.

Wu-6 Member: It covers Nos. 260–293 layers. Its rhythms are composed of grey massive thrombolitic, algal-lamellated and stromatolitic dolostones, muddy dolostone and dolomitic shale, and mainly appear as the rhythmic assemblage of Units A-B-E in the shallow waters of supratidal-intertidal-supratidal zones, while its minor assemblage of Units A-E, accompanied by bird'eye structure, is in the shallower waters of supratidal zone at the bottom of Wu-6 Member. In general, by comparison with the Wu-5 and Wu-7 Members, the sedimentary waters of Wu-6 Member would be shallower, and obviously with a lower $\delta^{13}C$ value. The Wu-6 Member is equivalent to the upper portion of 2-order SQ. VIII.

Wu-7 Member: It covers Nos. 294–354 layers. Its lower interval is mainly constituted by the interbedding of grey and dark-grey thin-bedded micritic limestone, muddy and limy dolostones. Locally, the limestone presents flesh-pink and purplish-red with lamellated- and mildly-wavy- stromatolites as well as flat-bedding, micro-scour-surface. While the limestone turns into darker colour, thin and locally shaly-beddings, and contains arched-, columnar- and conical-stromatolites and abnormally developed molar tooth structure in its upper interval. The Wu-7 Member is equivalent to the lower portion of 2-order SQ. IX.

In general, the ^{13}C values in Wu-7 Memberber is higher than those in other members of the Wumishan Formation, indicating the deepest sedimentary waters and a tendency of ^{13}C value rising and sea-water deepening during the sedimentary period of Wu-1 to Wu-7 Members, resulting in the deepest waters and the sedimentary environment of subtidal zone at Wu-7 Member.

Wu-8 Member: It covers Nos. 355–411 layers, and consists of light-grey siliceous-stripped limy-dolostone and algal-lamellated and thick-bedded stromatolitic dolostones with conical and columnar stromatolites, in which the assemblage of Units B-D-E is dominated, locally with volcanic rocks. Due to the effect of more active crust activity and more shallow seawater, the sediments of Wu-8 Member is referred to the sedimentary environment of the intertidal to subtidal zones, and equivalent to the middle to upper portions of 2-order SQ. IX.

4. **Hongshuizhuang Formation ($Pt_2{}^2h$)**

After the large-scale seawater transgression in the sedimentary period of Wumishan Formation, the Earth's crust rises into land, causing seawater regression and a great reduced oceanic area during the period of Hongshuizhuang Formation.

The Hongshuizhuang Formation is mainly constituted by a set of greyish-black and yellowish-green shale. It is divided into 2 members (from Hong-1 to Hong-2 Members), 11 layers (layer Nos. 1–11) in ascending order as well as one 3-order stratigraphic sequences (SQ. 34) together with the overlying Tie-1 Member, belonging to the 2-order SQ. X at the Jianshanzi section in Jibei Depression (Tables 3.1 and 3.7; Figs. 3.10 and 3.11; cf. Chap. 2).

The Hong-1 Member (lower one) is mainly of an interbedding of greyish-black siliceous and muddy shales and medium- to thick-bedded muddy dolostone, in which the greyish-black shale is rich in organic matter. While the Hong-2 Member (upper one) consists of yellowish-green, greyish-black and black shale and muddy dolostone with simatics, pyritic nodule and lenticle. The Hongshuizhuang Formation belongs to the lagoonal deposit and is equivalent to the lower portion of 2-order SQ. X at the Kuangcheng section in Jibei Depression (Tables 3.1 and 3.7; Figs. 3.10 and 3.11; Zhu et al. 2005).

5. **Tieling Formation (Pt_2^2t)**

A small-scale transgression happened again and the waters is gradually deepened and the sedimentary environment has transited from intertidal to supratidal zones to subtidal zone during the sedimentary period of Tieling Formation. The Tieling Formation is divided into 2 members (from Tie-1 to Tie-2 Members), 35 layers (layer Nos. 12–46) in ascending order, and two third of the 2-order SQ. X at the Kuangcheng section in Jibei Depression (Tables 3.1 and 3.7; Figs. 3.10 and 3.11; cf. Chap. 2).

Tie-1 Member: It is composed of brown manganiferous endoclastic stromatolitic dolostone and sandy rudaceous-arenitic micritic dolostone sandwiched with greyish-green, and variegated shales at the lower interval and sandwiched with manganiferous dolostone at the upper interval, which indicate a set of alternating sediments of low and high energy sedimentary environments in intertidal to supratidal zones. It is equivalent to the lower portion of 2-order SQ. X, and in conformable contact with the underlying Hongshuizhuang Formation.

Tie-2 Member: There are one set of slightly equal-thick interbedding of grey thin-bedded micritic limestone and greyish-green shale, locally with tempestitic calcirudite in the lower interval, and another interbedding of thin-bedded limy-dolostone and shale often with dolomitic lenticle, cross-bedding and occasionally siliceous-nodule in the middle interval, while an interbedding of mildly-wavy-stromatolitic limestone and greyish-green shale, sandwiched with knotty-limestone is in the upper interval. It is referred to the sediments of subtidal zone, and equivalent to the middle to upper portions of 2-order SQ. X.

3.3.4.2 Xiamaling Formation (Pt_2^3x)

Owing to the influence of the "Qinyu Uplifting" at the end of Tieling Formation, the YFDZ rises into land again. Until the Late Mesoproterozoic, or rather the sedimentary period of Xiamaling Formation, the seawater transgression entered the Northern of North China from the northeast direction, and a small inland bay was created. Its seawater was shallower and the seabed is under the weak-reduction to weak-oxidation conditions during the early stage of Xiamaling Formation. Due to the differential Earth's crust subsidence and the gradually seawater deepening, however, the sea area was much broadened and the oxidation condition of seabed strengthened on the SW of Kuancheng in Jibei Depression during the middle to late stages. Therefore, the sediments of middle to lower Xiamaling Formation are rich-in organic matter, so that the dark colour shale was dominated (cf. This chapter).

Xiamaling Formation can be typically divided into 4 members (from Xia-1 to Xia-4 Members) with a maximum thickness up to 545 m in the Xuanlong Depression (cf. Chap. 2). Due to the "Qinyu Uplifting" and consequently the stratigraphic denudation, however, only the Xia-1 Member remains in Jibei Depression. The residual Xia-1 Member is divided into 17 layers (layer Nos. 47–63) in ascending order and equivalent to the 2-order SQ. XI at the Kuangcheng section (Tables 3.1 and 3.7; Figs. 3.10 and 3.11; cf. Chap. 2).

The residual Xiamaling shale mainly appears as greyish-black or dark-grey colour with greenish-grey, greyish-brown or yellowish-green weathering colour, which is attributed to the sediments of shallow sea shelf environment. Sandstone can only be visible at its basal and the middle intervals, the lithology of basal sandstone is white and yellowish-grey medium-bedded bebbly quartzose sandstone with 1–3.8 m in individual layer thickness. Its granularity varies from fine-grain to coarse sandstone, which are well-sorting and rounded with subround- to round-shapes, and the quartz sands account for about 90% or more, and are cemented by silica. Moreover, the sandstone contains flat- or cross-bedding and corrosion border, indicating a stronger hydrodynamic sedimentary environment and a set of foreshore sediments (cf. This chapter).

It is noticeable that the intergranular opening of Xiamaling basal sandstone is often filled with black solid bitumen so as to be bituminous sandstone found at the Shuangdong (at Pingquan County), Lujiazhuang (at Kuancheng County, Hebei) and Longtangou (at Lingyuan County, Liaoning) in the Jibei Depression, where the outcrops of bituminous sandstone can be intermittently traced up to 5–8 km along with the bottom boundary of Xiamaling Formation, even a paragenetic occurrence of pure bitumen- and silica-cemented sandstones can be found at the Longtangou section (cf. Chaps. 11 and 12).

There generally are 2–4 gabbro-diabase sills (named as $\beta\mu$-1 to $\beta\mu$-4 in ascending order) intruding into the Xiamaling Formation in Jibei Depression, especially the sill $\beta\mu$-1. Based on the statistics of 11 stratigraphic sections, the total sill thickness ranges from 117.5 m to 312.3 m and the ratio of sills to stratum thickness in Xiamaling Formation is from 0.5 to 1.6. The intrusion of gabbro-diabase sills generally results in a regional wall-rock alteration of the Xiamaling shale to different extent, even the original Xiamaling basal oil reservoir is altered into fossil-oil-reservoir (cf. Chaps. 10–12).

3.3.4.3 $Pt_3{}^1$ Qingbaikouan

1. **Luotuoling Formation ($Pt_3{}^1l$)**

   It is a set of siliceous clastic rock rich-in glauconite, including 6 layers (layer Nos. 64–69) and being equivalent to the 2-order SQ. XII at the Kuangcheng section in Jibei Depression (Tables 3.1 and 3.7; Figs. 3.10 and 3.11; cf. Chap. 2).

   The greyish-yellow medium-grained to coarse bebbly quartzose sandstone, with numerous pinnate-, flushing-, large tubular-, wedge- and chevron-cross-beddings, appears at the basal interval, which indicates the repeated scouring by tidal action and belongs to the foreshore sediment.

While the greyish-green fine- to medium-grained glauconitic quartzose sandstone constitutes the lower interval, in which the well-sorting and rounded quartz sands account for 80% or more, generally with grain size of 1–2 mm, appearing as subround- to round-shapes and siliceous and muddy cementations, often contains pyrite and silica, the fine grains, small cross- and flat-beddings reveal the weaker hydrodynamic condition and the shoreface sediments.

An interbedding of purplish-red silty shale and fine-grained glauconitic quartzose sandstone occurs in the upper interval, in which quartz sands are dominated (more than 85%) with ferruginous plus a little siliceous mosaic cementations and flat-bedding. It is referred to the sediment of offshore shelf.

2. **Jing'eryu Formation ($Pt_3{}^1j$)**

It includes 5 layers (layer Nos. 70–74) and is equivalent to the 2-order SQ. XIII at the Kuangcheng section in Jibei Depression (Tables 3.1 and 3.7; Figs. 3.10 and 3.11; cf. Chap. 2).

The yellow medium-bedded coarse quartzose sandstone, sandy calcirudite and thin-bedded muddy siltstone constitute its basal interval. The sandy calcirudite has coarser gravels with the maximum grain size up to 2.0 cm × 1.5 cm, generally showing bedding distribution and indicating the stronger hydrodynamic condition. Its lower interval is composed of grey thin-bedded micritic limestone or marl and shale, turning upward into thin-bedded marl and purplish-red shale, the shale is thickening and the dolomitization of marl is enhanced. While the thin-bedded micritic limestone, marl, silty limestone and small amount of purplish-red shale appear in the upper interval, where the limestone is mostly thick-bedded with larger thickness, which present a quiet sedimentary environment as the sediment of subtidal zone.

3.3.5 *Sedimentary Mode*

During the Meso-Neoproterozoic period, the lithological type of marine sediments includes clastic rock and carbonate rock in YFDZ. With the purpose of research, the sedimentary modes of both clastic rock and carbonate rock should be respectively studied.

The lithology and lithofacies of the carbonate strata are more monotonous without evident variation in YFDZ, but it has very strong rhythmicity. A lot of micritic and very-fine-crystalline dolostones, seldom with grain dolostone, are developed in the stratigraphic section with several thousand meters in thickness. As sedimentary structures, algal stromatolites of morphological diversity and abundant quantity are very-well developed, but the wave-genetic cross-bedding is occasionally visible. The above characteristics indicate that the Meso-Neoproterozoic sedimentary environment is mainly controlled by tidal flat in YFDZ, with a flatter palaeogeographical setting, and belonging to the broadly extended, low sloped and shallow water carbonate sedimentary environment of epicontinental sea.

In this chapter, based on the measured Meso-Neoproterozoic section, the data collection of numerous facies marks and facies analysis in YFDZ, following the rules of facies sequence and facies change, referring to the methodology of facies zone division by sea-water energy and tidal action (Irwin 1965; Young et al. 1972), the Meso-Neoproterozoic sedimentary modes of carbonatic rocks (Fig. 3.19) and clastic rocks (Fig. 3.20) have been established in YFDZ, which mainly show the rules of lateral facies change.

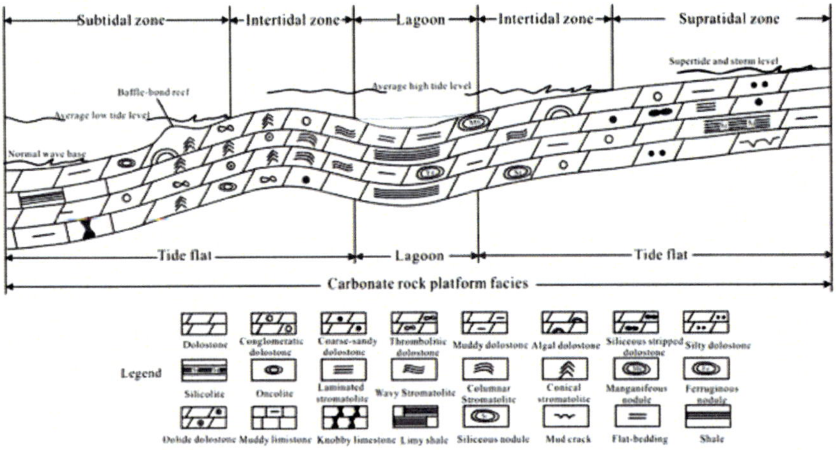

Fig. 3.19 The sedimentary mode of Meso-Neoproterozoic carbonatic rocks in YFDZ

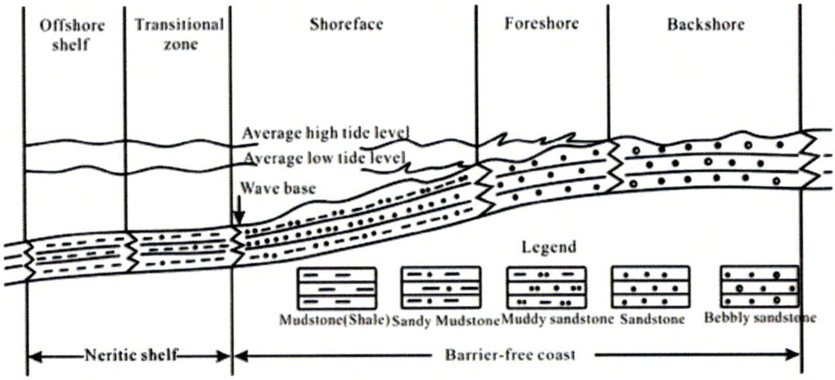

Fig. 3.20 The Meso-neoproterozoic sedimentary mode of clastic rocks in YFDZ

3.4 Source-Reservoir-Seal Bed Assemblage Within the Sequence Stratigraphic Framework

The base level cycle change would cause the increase and decrease of sedimentary accommodating space. As the geography shifts, the migration of accommodating space within the base level cycle would make the sediments to be accumulated at different proportions in different sedimentary environments, which would induce the facies changes in genetic types, geometric morphology and special distribution, even in the specific source-reservoir-seal bed assemblage (Zhao et al. 1977).

3.4.1 Control of Sequence on Source and Reservoir Beds

Generally speaking, the large-scale hydrocarbon source bed and seal bed are mainly developed at the maximum sea-flooding period (MSF) of each 3-order sequence within the sequence stratigraphic framework, and the favorable zones of reservoir bed development are mainly found at the early transgression period of each sequence and in the late period of high-stand system tract (HST), especially near the sequence boundary. A more complete source-reservoir-seal bed assemblage can be created in a same sequence or between the adjacent sequences (Zou et al. 2004).

3.4.1.1 Sedimentary Sequence Control on Source Beds

Generally, hydrocarbon source bed is developed at the turning position of long-term base level cycle from up to down, which is situated at the location of the maximum accommodating space in base level cycle.

Gaoyuzhuang sedimentary period is one of the Meso-Neoproterozoic maximum transgression periods and also the continuity and development of transgression during Dahongyu Period in YFDZ when the sea area is greatly expanded. In Jibei Depression, the major lithologies of lower to middle Gao-3 Member (SQ. 3), upper Gao-4 Member (SQ. 4), lower Gao-5 Member (SQ. 5) and Gao-6 Member (SQ. 6) are mainly grey and dark-grey muddy dolostone and limestone with less shale, among which stromatolite is very well developed with various biological forms, showing a dominant reducing environment of subtidal zone, which could be attributed to the maximum flooding surface (MSF) and transgression system tract (TST) of a 3-order sequence. With good organic type, this Gaoyuzhuang Formation contains the average of total organic carbon content (TOC) 1.16%, and the source beds above TOC 0.5% can be up to 164 m in cumulative thickness, with the equivalent vitrinite reflectance (VR_o) average 1.59%, which is referred to a favorable potential over-mature source bed in Jibei Depression (cf. Chap. 11).

As for the Hongshuizhuang Formation, the interval of black or blackish-brown shale is developed at its lower portion (the lower SQ. 34), belonging a strong reducing environment of lagoon facies and the MSF and TST of a 3-order sequence. Its source bed shows higher organic abundance and moderate maturity with TOC average of 4.65%, chloroform extract of 0.265% and VR_o of 1.19%, which is attributed to a favorable high mature source bed (cf. Chap. 11).

Even though the black shale of Xiamaling Formation is considered as the third source bed, especially for its Xia-3 Member at Xuanlong Depression in YFDZ, due to the strong denudation of "Qinyu Uplifting" (cf. Chap. 2) and the thermo-optical alteration of gabbro-diabase sills (Fig. 3.21; cf. Chaps. 10–12), only the Xia-1 Member remains but the black shale/mudstone is mostly altered into hornfels and/or slate, so that its hydrocarbon potential is exhausted and the source-reservoir-seal bed assemblage unavailable.

In a word, the source beds are mainly developed in the reducing sedimentary environments of subtidal zone, lagoon and offshore shelf, where the sedimentary environments of sea waters are relatively deeper and occluded for the Meso-Neoproterozoic strata in YFDZ (Liu et al. 1985). The large-scale source beds are mainly deposited during the maximum sea-flooding period (MSF). or rather, the maximum sea transgression period, which is constrained by each 3-order sequences in the sequence framework.

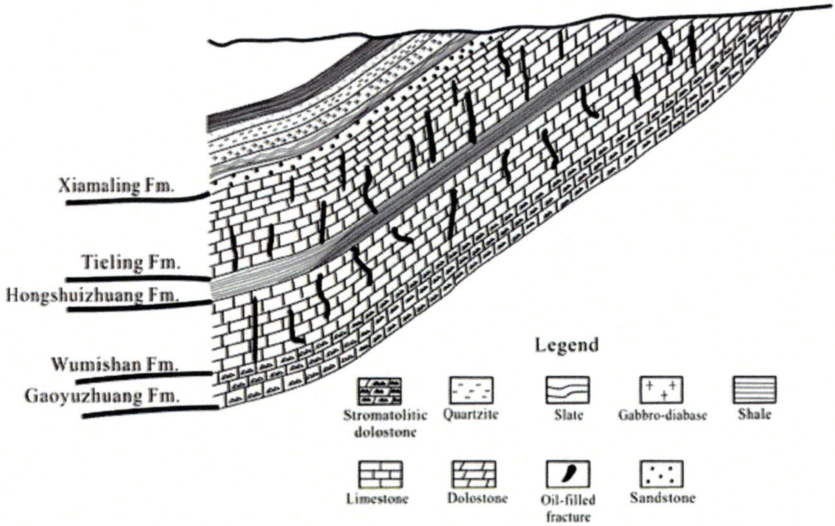

Fig. 3.21 Configuration and preservation conditions of source-reservoir-seal bed assemblage in Jibei depression

3.4.1.2 Sedimentary Sequence Control on the Reservoir Beds

The distribution, lithology, rock texture and sedimentary structures of reservoir beds are controlled by favorable sedimentary facies belt, and thus, to a considerable degree, the development of primary porosity and the evolution of diagenesis are also constrained in reservoir rocks.

Based on the statistics of drilling and outcrop data, the lithological types of Meso-Neoproterozoic reservoirs beds are dominantly carbonate rocks and sandstone, while the lithology of oil-bearing layer is mainly algal carbonate rocks, crystalline dolostone and sandstone, among which the rock types of algal carbonate reservoir bed are mainly of thrombolitic and stromatolitic dolostones and limestones as well as oncolitic and lamellated dolostones. In addition, the palaeo-karstic zone and structural fractures are also important reservoir spaces.

It can be seen from Table 3.11 that the favorable reservoir beds are mainly distributed in Gaoyuzhuang, Wumishan and Xiamaling Formations in Jibei Depression. In Gaoyuzhuang Formation, the Gao-1 (SQ. 1), Gao-2 (SQ. 2), Gao-9 and Gao-10 (SQ. 8) Members contain favorable reservoir property, their lithologies are mainly of micritic to very-fine-crystalline dolostone, algal stromatolitic and epigenetic palaeo-karstic dolostones, which are attributed to the high energy environment in intertidal to subtidal zones, and also the early TST and the late HST. Their macroscopic reservoir spaces are mainly structural and dissolved fractures, solution cavities/pores and less interbedding fractures with the average of palaeo-surface porosity about 10% on the thin section.

In Wumishan Formation, the Wu-1 (SQ. 12 to SQ. 15), Wu-3 (SQ. 20 to SQ. 22) and Wu-8 (SQ. 30 to SQ. 33) Members have favorable reservoir property, their lithologies are mainly of grain dolostone, dolomitic sandstone, algal stromatolitic dolostone, micritic and very-fine-crystalline dolostones, which are referred to the high energy environment in the intertidal zone, and the early TST and the late HST. Their macroscopic reservoir spaces are mainly structural and dissolved fractures, solution cavities, etc., with the average of palaeo-surface porosity about 10% and the maximum up to 15% on the thin section.

In Tieling Formation, the Tie-1 (upper SQ. 34) and Tie-2 Members (SQ. 35) possess favorable reservoir property, their lithologies are mainly of grain dolostone, very-fine-crystalline dolostone and limestone. Which belong to a high energy environment in the intertidal to subtidal zones, and the MFS and TST in a 3-order sequence. Their major macroscopic reservoir spaces are structural joint, interbedding and dissolved fractures, etc., with the average of palaeo-surface porosity about 5–8% on the thin section.

In Xiamaling Formation, the basal sandstone of Xia-1 Member has favorable reservoir property. It is of quartzose sandstone, and belongs to an offshore shelf environment, and the MFS and TST in the 3-order sequence SQ. 36. Its major reservoir spaces are intergranular and dissolved pores, with a palaeo-surface porosity 12–18% on the thin section.

Table 3.11 Stratigraphic position, lithological property and sedimentary environment of favorable reservoir beds in Jibei depression

Formation	Stratigraphic position	Lithology	Sedimentary facies zone	Reservoir evaluation
Xiamaling	Xia-1 member (SQs. 36–37)	Siliceous sandstone	Supratidal to intertidal zones	Favorable
Tieling	Tie-2 member (SQ. 35)	Clastic dolostone, very-fine-crystalline dolostone, and limestone	Intertidal and subtidal zones	
	Tie-1 member (upper SQ. 34)			
Wumishan	Wu-8 member (SQs. 30–33)	Micritic and very-fine-crystalline dolostones	Supratidal and subtidal zones	
	Wu-3 member (SQs. 20–22)	Algal stromatolitic dolostone, micritic and very-fine-crystalline dolostones	Intertidal to subtidal zones, supratidal zone and subtidal zones	
	Wu-1 member (SQs. 12–15)	Grainy dolostone dolomitic sandstone	Intertidal to subtidal zones, supratidal and intertidal zones	
Gaoyuzhuang	Gao-9 to Gao-10 Members (SQ. 8)	Palaeo-karstic dolostone	Supratidal exposure zone	
	Gao-2 Member (SQ. 2)	Micritic and very-fine-crystalline dolostones, algal stromatolitic dolostone, dolomitic sandstone	Supratidal to subtidal zones, intertidal to subtidal zones, supratidal to intertidal zones	
	Gao-1 member (SQ. 1)	Micritic and very-fine-crystalline dolostones, dolomitic sandstone, silty dolostone	Supratidal and subtidal zones	

In short, on the view of sequence stratigraphy, the large-scale reservoir beds are mainly developed in the early period of TST and the late period of HST for each sequence, especially at the vicinity of sequence boundary.

3.4.2 *Configuration of Source-Reservoir-Seal Bed Assemblage*

Generally, the high-quality reservoir bed appears at the early stage of base-level rising and at the late stage of it falling down in the long-term base level cycle, while the favorable source bed and seal bed at the maximum sea-flooding period (MFS) of a long-term base level cycle. Moreover, a more complete source-reservoir-seal bed assemblage would be created in the interior of a long-term base level cycle, or between the adjacent long base-level cycles.

The Meso-Neoproterozoic strata in YFDZ contain favorable source-reservoir-seal bed assemblages, and possess the basic petroleum geological conditions of indigenous oil and gas reservoir formation. In Jibei Depression, the Hongshuizhuang black and dark-grey shale/mudstone is the major mature to high-mature source bed, while the Gaoyuzhuang black and dark-grey stromatolitic dolostone the potential over-mature source bed. There vertically are 2 kinds or 6 sets of potential source-reservoir-seal bed assemblages (Fig. 3.21; cf., Chaps. 10, 11, 12).

(1) Vertical source-reservoir-seal bed assemblages resulted from oil vertical migration and entrapment:

① Gaoyuzhuang black dolostone (source bed)-Wumishan dolostone of fracture and karst types (reservoir bed)-Wumishan/Yangzhuang carbonatic rocks (seal bed), e.g., the Shuangdong fossil- oil-reservoir in Pingquan.

② Gaoyuzhuang black dolostone (source bed)-Tieling limestone of fracture and karst types (reservoir bed)-Hongshuizhuang shale (seal bed), e.g., Shuangdong fossil-oil-reservoir in Pingquan.

③ Gaoyuzhuang black dolostone (source bed)-Xiamaling basal bituminous sands/sandstone (reservoir bed)-Xiamaling shale (seal bed), e.g., Shuangdong, Lujiazhuang and Longtangou fossil- oil-reservoirs in Pingquan, Kuancheng and Lingyuan, respectively.

④ Hongshuizhuang black and dark-grey shale/mudstone (source bed)-Tieling limestone of fracture and karst types (reservoir bed)-Hongshuizhuang shale (seal bed), e.g., Shuangdong and Huapi fossil-oil-reservoirs in Pingquan and Kuangcheng, respectively.

⑤ Hongshuizhuang black and dark-grey shale/mudstone (source bed)-Xiamaling basal bituminous sands/sandstone (reservoir bed)-Xiamaling shale (seal bed), e.g., Shuangdong, Lujiazhuang and Longtangou fossil-oil-reservoirs in Pingquan, Kuancheng and Lingyuan, respectively.

(2) Lateral source-reservoir-seal bed assemblages resulted from oil lateral migration and entrapment:

⑥ Hongshuizhuang black and dark-grey shale (source bed)-Wumishan dolostone of fracture and karst types (reservoir bed)-Wumishan/Yangzhuang carbonatic rocks (seal bed), e.g., the Shuangdong fossil-oil-reservoir in Pingquan.

3.5 Conclusion

(1) Based on the characteristics of petrostratigraphy, following the sequence stratigraphic principle, and combined with the geochemical markers of C, O, Sr isotopes, petrochemical components and micro-elements, etc., the Meso-Neoproterozoic stratigraphic section in Jibei Depression, YFDZ is divided and investigated in details, and its basic framework of the sequence stratigraphy is established. It is considered that during the long Meso-Neoproterozoic Gaoyuzhuang to Jing'eryu Periods, the Jibei Depression has been through the multiple high-level tectonic movements, deposition events and sea level eustacy. Consequently, 13 2-order sequences and 39 3-order sequences as well as the sedimentary system tracts have been established.

(2) The Meso-Neoproterozoic sequences in YFDZ are a set of huge-thick marine carbonate strata intercalated by clastic rocks, and mainly attributed to 2 kinds of sedimentary systems, i.e., epicontinental sea and marginal sea sedimentary systems, in which the marine carbonate sedimentary system contains 2 sedimentary facies, 3 subfacies and 5 microfacies:

   ① Sedimentary facies: carbonate platform and organic reef;
   ② Subfacies: carbonate tidal flat, lagoon and baffle-bonding reef;
   ③ Microfacies: supratidal zone, intertidal zone, subtidal zone, lagoonal mud and baffle-bonding rock.

The marine clastic rock sedimentary system can be divided into 2 sedimentary facies and 4 subfacies:

① Sedimentary facies: shallow shelf and baffle-free coast;
② Subfacies: offshore shelf, transition zone, foreshore and shoreface.

(3) The carbonate platform facies are widely developed in Gaoyuzhuang, Yangzhuang, Wumishan, Hongshuizhuang, Tieling and Jing'eryu Formations in Jibei Depression. The organic reef facies is only found within the Gao-9 and Gao-10 Members of Gaoyuzhuang Formation in the western of Jibei Depression which extend from Jibei to Xuanlong Depressions (cf. Figure 3.17). While the clastic rock shallow shelf and baffle-free coast facies are mainly developed in Xiamaling and Luotuoling Formations.

(4) In view of the distribution of favorable source beds and favorable reservoir beds within the stratigraphic sequences in Jibei Depression, the source beds of shallow shelf facies are mainly distributed at the maximum sea-flooding period (MSF) of each 3-order sequence in Gaoyuzhuang and Hongshuizhuang Formation, while the favorable reservoir beds in the early period of transgressive system tract (TST) and the late period of high-stand system tract (HST) in each sequence of Gaoyuzhuang, Tieling and Wumishan Formations, especially near the sequence boundary.

(5) The Meso-Neoproterozoic sequences in Jibei Depression possess basic petroleum geologic conditions for the formation of indigeneous oil and gas reservoirs with potential source-reservoir-seal bed assemblage. In Jibei Depression, major source beds are the Hongshuizhuang black and dark-grey shale/mudstone as well as the Gaoyuzhuang black and dark-grey stromatolitic dolostone. There are 6 sets of potential source-reservoir-seal bed assemblages, i.e., 5 vertical assemblages from Hongshuizhuang source bed to Tieling and Xiamaling reservoir beds respectively, and from Gaoyuzhuang source bed to Wumishan, Tieling and Xiamaling reservoir beds respectively as well as laterally 1 assemblage from Hongshuizhuang source bed to Wumishan reservoir bed.

Acknowledgements Thanks for the full support for the the analyses of micro-elements by Tianjin Geological Survey Center of China Geological Survey; Thanks for the financial aid and full support for the research work provided by National Natural Science Foundation of China (Grant No.: 40772078) and SINOPEC Marine Facies Forward Looking Project (Grant No: YPH08025).

References

Chen JB, Zhang HM, Zhu SX, Zhao Z, Wang ZG (1980) Research on precambrian geology: sinian suberathem of Jixian, Tianjin. In: Tianjin Institute of Geology and Mineral Resources (ed) Research on precambrian geology, Sinian Suberathem in China. Tianjin Science and Technology Press, Tianjin, pp 56–114 (in Chinese with English contents)

Chu XL, Zhang TG, Zhang QR, Feng LJ, Zhang FS (2003) The variation of carbon isotope of carbonate rocks of Proterozoic Era in Jixian County. Sci China 33(10):951–959 (in Chinese)

Deng HW, Qian K (1993) Sedimentary geochemistry and environmental analysis. Gansu Science and Technology Publishing House, Lanzhou (in Chinese)

Fairchild IJ, Einsele G, Song TR (1997) Possible seismic origin of molar-tooth structures in Neoproterozoic carbonate ramp deposits north China. Sedimentology 44:611–636

Fan DL, Yang H, Dai YD, Zhang YN, Wang LC, Zhang RF (1977) Sedimentary geochemistry of Sinian strata in the Chihsien region, north China. Geochemistry 6(3):161–172 (in Chinese with English abstract)

Frank TD, Lyons TW (1998) "Molar-tooth" structures: a geochemical perspective on a Proterozoic enigma. Geology 26:683–686

Gao LZ, Zang YX, Wang CX, Tian SJ, Peng Y, Liu YY, Dong DZ, He HX, Lei BT, Chen ME, Yang LG (1996) Meso- and neoproterozoic sequence stratigraphy in Jixian. Reg Geol China 1:64–74 (in Chinese)

Huang XG, Zhu SX, Hu YZ (2001) Some basic problems in research on sequence stratigraphy of the Meso- and Neoproterozoic Strata in Jixian area. Prog Precambr Res 24(4):201–219 (in Chinese with English abstract)

Huang SJ, Zhou SH (1997) Carbon and strontium isotopes of Late Palaeozoic marine carbonates in the Upper Yangtze platform, southwest China. Acta Geol Sinica 71(3):282–292 (in Chinese with English abstract)

Irwin ML (1965) General theory of epeiric clear water sedimentation. AAPG 49:445–459

Kuang HW, Liu YX, Meng XH, Ge M (2005) The geochemical features and its environmental significance of the Sinian carbonates in Jilin-Liaoning area. Nat Gas Geosci 16(1):54–58 (in Chinese with English abstract)

Lan XH (2001) Research progress of marine strontium isotope. Geol Sur Res 17(10):1–3 (in Chinese)
Li RW, Chen JS, Zhang SK (1999) Stable carbon and oxygen isotopic composition of carbonates in middle mesoproterozoic Wumishan formation and sea-level change. Chin Sci Bull 23:2130–2136
Liu C, Wang QC, Zhang JC, Chen BY, Yuan XG, Chen HX, Liu HS (1984a) The preliminary palaeomagnetic results of geologic sections of Early and Middle Proterozoic from the Taihan Mountain area and its geologic implication. Sci Geol Sin 19(4):455–460 (in Chinese with English abstract)
Liu YJ, Cao LM, Li ZL, Wang HL, Chu TQ, Zhang JR (1984b) Elemental geochemistry. Science Press, Beijing (in Chinese)
Liu BQ, Liang DG, Fang J, Jia RF, Fu JM (1985) Organic matter maturity and oil-gas prospect in midder-upper proterozoic and lower palaeozoic carbonate rocks in Northern China. Geochemistry 14(2):150–162 (in Chinese with English abstract)
Liu YQ, Liu XW, Li Y (1997) Tectonic cyclic sequences in the mesoproterozoic and neo-proterozoic aulacogen of Yanshan–a concept of aulacogen tectonic cycle and its hierarchy. Acta Geosci Sin 18(2):142–149 (in Chinese with English abstract)
Liu PJ, Wang CW, Sun YW, Zhang BF, Wang LH, Yue SF (2005) The geochemical characters of the Gaoyuzhuang and Yangzhuang formations of mesoproterozoic in Pingquan County, Hebei Province. J Jilin Univ (earth Sci Ed) 35(1):1–6 (in Chinese with English abstract)
Liu JQ, Jia BJ, Yang P, Chen YL, Peng B, Li ZJ (2007) The application of carbon, oxygen and strontium isotopes to the study the middle-upper Jurassic sequence stratigraphy in Longwei area, Qiangtang Basin. Acta Petrol Sinica 28(3):253–260 (in Chinese with English abstract)
McArthur JM, Burnett J, Hancock JM (1992) Strontium isotopes at K/T boundary, discussion. Nature 355(6355):28
Meng XH, Ge M (2002) Research on cyclic sequence, events and formational evolution of the Sino-Korean Plate. Earth Sci Front (china Univ Geosci Beijing) 9(3):125–140 (in Chinese with English abstract)
Min LR, Chi ZQ, Zhu GX, Yao PY, Niu PS (2002) Analysis of the palaeoenvironment of quaternary stromatol ites at Dongmulian, Yangyuan County of Hebei Province. Acta Geosci Sin 76(4):446–453 (in Chinese with English abstract)
Palmer MR, Elderfield H (1985) Sr isotope composition of seawater over the past 75 Myr. Nature 314(6011):526–528
Palmer MR, Edmond JM (1989) The strontium isotope budget ofthe modern ocean. Earth Planet Sci Lett 92(1):11–26
Pratt BR (1998a) Gas bubble and expansion crack origin of "molar tooth" calcite structures in the middle proterozoic belt supergroup, western Montana-discussion. J Sediment Res 68:1136–1140
Pratt BR (1998b) Molar-tooth structure in Proterozoic carbonate rock: origin from synsedimentary earthquake and implications for the natural and evolution of basins and marine sedimentary. GSA Bullitin 110(8):1028–1045
Song MS (2005) Sedmentary environment geochemistry in the Shasi section of southern ramp, Dongying depression. Mineral Petrol 25(1):67–73 (in Chinese with English abstract)
Song TR (2007) Sedimentary facies indicators and sedimentary environments models of the Changcheng system of mesoproterozoic in ming Tombs district, Beijing. J Palaeogeogr 9(5):461–472 (in Chinese with English abstract)
Tian JC, Chen GW, Zhang X, Nie YS, Zhao Q, Wei DX (2006) Application of sedimentary geochemistry in the analysis of sequence stratigraphy. J Chengdu Univ Technol (sci Technol Edn) 33(1):30–35 (in Chinese with English abstract)
Wang TG (1980) The original property of oil show and its petroleum geologic significance in the Sinian Suberathem of Yanshan area. Pet Explor Dev 7(2):34–53 (in Chinese)
Wang KF, Chen JS (1993) Constaints on the stable isotopic composition of sedimentary carbonates from the Tielino formation in the Yanshan region. Geochimica 22(1):10–17 (in Chinese with English abstract)

Wang SJ, Huang XZ, Tuo JC, Shao HS, Yan CF, Wang SQ, He ZR (1997) Evolutional characteristics and their palaeo-climate significance of trace elements in the Hetaoyuan formation, Biyang depression. Acta Sedimentol Sinica 15(S1):65–69 (in Chinese with English abstract)

Wang F, Chen HD, Zhao JX, Chen AQ, Su ZT, Li H (2011) Sequence boundary and petroleum geology of the Cambrian-Permian strata in the Ordos Basin. Sediment Geol Tethyan Geol 31(1):6–12 (in Chinese with English abstract)

Wu ZY (1999) Chemostratigraphy and its development. J Stratigr 23(3):234–240 (in Chinese with English abstract)

Wu TS (2002) Late precambrian (Meso-to neoproterozoic) lithostratigraphic units in North China and their multiple division and correlation. Geol China 29(2):147–154 (in Chinese with English abstract)

Xiao CT, Li YB, Hu MY, Gong WP, Xiao AC, Lin KX, Zhang CS (2001) Discovery of the middle Jurassic Liostrea barrier reefs at Baqen County, Northern Tibet. Geol Bull China 20(1):90–93 (in Chinese with English abstract)

Yan ZB, Guo FS, Pan JY, Guo GL, Zhang YJ (2005) Degree of effects of the main uncertain factors on investment income rate in mining production. Contrib Geol Mineral Resour Res 20(1):53–56 (in Chinese with English abstract)

Yang JD, Zheng WW, Wang ZZ, Tao XC (2001) Age determination of the uppe precambrian system of Northern Jiangsu-Anhui by using Sr and C isotopes. J Stratigr 25(1):44–47 (in Chinese with English abstract)

Young LM et al (1972) Carbonate facies in ordovician of Northern Arkansas. AAPG 56:68–80

Zhao Z (1988) A sedimentary mode of the tidal flat of epicontinental sea. Acta Sedimentol Sin 6(2):68–75 (in Chinese with English abstract)

Zhao CL, Li RF, Zhou JS (1977) Sedimentology and petroleum geology of the meso- and neoproterozoic in North China. Geological Publishing House, Beijing (in Chinese)

Zhou HR, Mei MX, Luo ZQ, Xing K (2006) Sedimentary sequence and stratigraphic framework of the neoproterozoic qingbaikou system in the Yanshan region, North China. Earth Sci Front 13(6):280–290 (in Chinese with English abstract)

Zhu SX, Huang XG, Sun SF (2005) New progress in the research of the mesoproterozoic Changcheng system (1800–1400 Ma) in the Yanshan Range, North China. J Stratigr 29(11):437–449 (in Chinese with English abstract)

Zhu SX et al (1993) The stromatolites of China. Tianjin University Press, Tianjin, pp 146–186 (in Chinese)

Zou CN, Chi YL, Li M, Xue SH (2004) The analytical technology of non-marine sequence stratigraphy–a guidance for the industrialization of oil and gas exploration. Petroleum Industry Press, Beijing (in Chinese)

Chapter 4
Research Progress of Ediacaran (Sinian) Biostratigraphy in South China

Pengju Liu, Chongyu Yin, and Feng Tang

Abstract Abundant micro- and macrofossils have been found from the Ediacaran (Sinian) deposits in South China. These fossils offer diverse information for understanding the evolution of early lives before the Cambrian Explosion, and also provide fundamental evidences for the chronological division and correlation of Ediacaran System. Until now all the Ediacaran microfossils were found from the middle and lower part of Doushantuo Formation in South China with two assemblages established, i.e., the lower *Tianzhushania spinosa* assemblage in the Dou-2 Member of Doushantuo Formation and the upper *Hocosphaeridium anozos-H. scaberfacium* assemblage in Dou-3 Member. *Tianzhushania spinosa* assemblage can be correlated with the microfossil assemblage in the Infrakrol Formation in Lesser Himalaya of India, and the *Hocosphaeridium anozos-H. scaberfacium* assemblage can be correlated with the ECAP (Ediacaran complex acritarch palynoflora) assemblage in South Australia. In other aspects, several exceptionally preserved Ediacaran macrofossil biotas have been reported from South China, including the Lantian biota from the lower part of Doushantuo Formation, Wenghui and Miaohe biotas from the upper part of Doushantuo Formation, as well as Xilingxia (Shibantan), Gaojiashan, Wulingshan and Jiangchuan biotas from the middle and upper part of the Dengying Formation. Among these macrofossil biotas, Miaohe and Wenghui biotas can be correlated with the White Sea biota in Russia and the well-known Ediacara assemblage from the western Flinders Ranges, South Australia; Xilingxia (Shibantan), Gaojiashan, Wulingshan and Jiangchuan biotas can be correlated with the Nama biota in Namibia. According to the biostratigraphy and carbon isotope stratigraphy, the Ediacaran system in South China is suggested to be divided into two series and six stages.

Keywords South China · Ediacaran · Fossil assemblages · Biostratigraphy

P. Liu (✉) · C. Yin · F. Tang
Institute of Geology, Chinese Academy of Geological Sciences, Beijing 100037, China

T. Wang, *Meso-Neoproterozoic Geology and Petroleum Resources in China*,
Springer Geology, https://doi.org/10.1007/978-981-19-5666-9_4

4.1 Introduction

The Ediacaran system, which lies on the uppermost of the Neoproterozoic, was formally established in 2004 by International Commission on Stratigraphy. The initial global standard stratotype section and point (GSSP) of the Ediacaran Period lies at the base of a texturally and chemically distinctive carbonate layer that overlies glaciogenic rocks (diamictite) in an exposure along Enorama Creek in the Flinders Ranges, South Australia; the period's end coincides with the beginning of the Cambrian Period, which is defined by its own initial GSSP residing in Newfoundland, Canada, and the formally defined base (GSSP) of the Ediacaran Period in Enorama Creek, Australia, is located at the contact of Marinoan glacial rocks and overlying Ediacaran cap carbonates (Knoll et al. 2004). The uppermost of the Neoproterozoic is called Sinian System in China, and the lower boundary of the redefined Sinian is located at the contact of the Nantuo diamictite and overlying the Doushantuo cap carbonate (Xing et al. 1999). The redefined Sinian is equivalent to that of the global Ediacaran since the Nantuo glacial deposits have been correlated with Marinoan glacial deposits (Zhou et al. 2004). Thus, many researchers recommend that the Ediacaran System should be adopted to replace the "Sinian System" despite the popularity of the latter term among Chinese stratigraphers (Peng et al. 2012). Here we adopt Ediacaran.

At present, with the discovery of the more and more macro- and microfossils in the world (Grey 2005; Narbonne 2005; McCall 2006; Fedonkin et al. 2007; Sergeev et al. 2011; Moczydłowska and Nagovitsin 2012; Liu et al. 2014b, 2015; Xiao et al. 2014; Joshi and Tiwari 2016; Liu and Moczydłowska 2019 and therein), the further subdivision and intercontinental correlation of the Ediacaran strata have been one of the major focuses of the Ediacaran studies (Liu et al. 2012a, 2014a; Narbonne et al. 2012; Xiao et al. 2016). During the past decades, the well exposed and continuous Ediacaran sequences in South China, consisting of richly fossiliferous Doushantuo and Dengying Formations, including microfossils preserved in chert nodules and phosphorites, and macrofossils preserved in black shale and muddy thin-bedded carbonate (Yin and Li 1978; Yin and Liu 1988; Ding et al. 1996; Zhang et al. 1998; Xiao et al. 2002 2014; Zhao et al. 2004; Yuan et al. 2011; Chen et al. 2014a, b; Liu et al. 2014b; Liu and Moczydłowska 2019; Shang et al. 2019), have been intensively investigated. Notwithstanding the significance of these fossils for understanding the evolution of early multicellular life, some fossil species are widely distributed, showing potential for global biostratigraphic correlation (Zhou et al. 2007; Zhu et al. 2008; Liu et al. 2013, 2014a, b; Xiao et al. 2013; Chen et al. 2014a, b; Liu and Moczydłowska 2019; Shang et al. 2019). Next, we will introduce the Ediacaran macro- and microfossils in South China and their application to biostratigraphy in detail.

4.2 Lithostratigraphy

The Ediacaran System in South China, which is underlain by the Cryogenian Nantuo diamictite and overlain by Cambrian System, consists of the Doushantuo and Dengying Formations. In the Yangtze Gorges area (typical area) of South China, the Doushantuo Formation is generally divided into four lithostratigraphic members (Dou-1 to Dou-4 Members; Fig. 4.1). Dou-1 Member is characterized by ca. 5 m-thick cap dolostone overlying glacigenic deposits of the Nantuo Formation. Dou-2 Member is composed of 80–120 m-thick black shale intercalated with medium-bedded dolostone and muddy dolostone with abundant chert nodules. Dou-3 Member usually consists of 40–60 m-thick medium-bedded dolostone with chert bands or lenses, banded dolostone and limestone. Dou-4 Member is characterized by ca. 10 m-thick black shale with large dolomitic thrombolites. The Dengying Formation is generally divided into three members (in ascending order): the Hamajing, Shibantan and Baimatuo Members (Fig. 4.1). The Hamajing Member is characterized by 10–200 m-thick light-grey massive dolostone. The Shibantan Member is composed of 100–200 m-thick dark-grey thin-bedded limestone with chert nodules. The Baimatuo Member usually consists of 40–400 m-thick light-grey dolostone with erosional surface at the top and overlain by various Cambrian successions (Zhu et al. 2007).

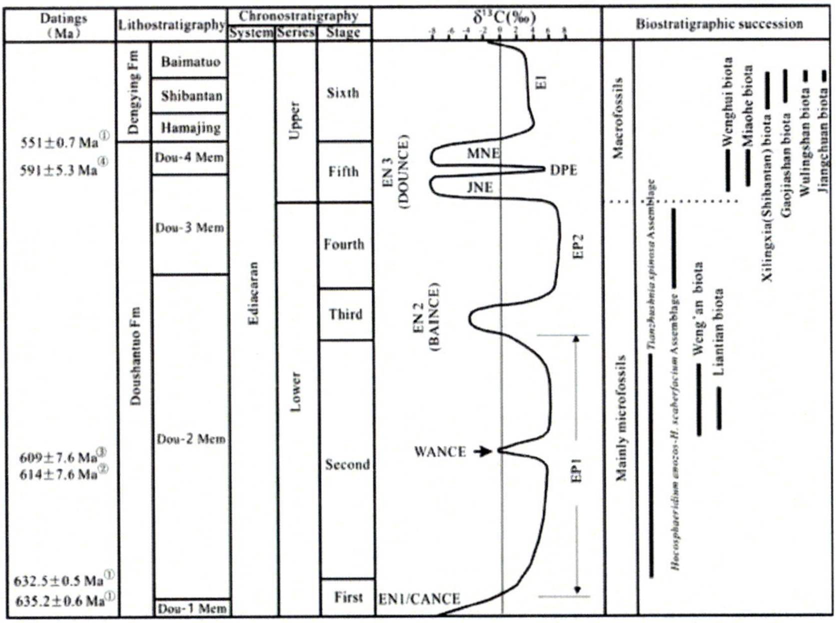

Fig. 4.1 Integrated Ediacaran stratigraphy in South China, showing chronostratrgraphic subdivision, $\delta^{13}C$ evolution, and stratigraphic occurrences of fossil assemblages. ① Condon et al. 2005; ② Liu et al. 2009b; ③ Zhou et al. 2017a; ④ Zhu et al. 2013a, b

4.3 δ^{13}C Chemostratigraphy

A large amount of δ^{13}C chemostratigraphic data from the Ediacaran Doushantuo and Dengying Formations in Yangtze Gorges, South China have been reported during the past 30 years (Lambert et al. 1987; Yang et al. 1999; Wang et al. 2002; Chu et al. 2003; Jiang et al. 2003, 2007, 2011; Shen et al. 2005; Zhou and Xiao 2007; Zhu et al. 2007, 2013a; An et al. 2015; Zhou et al. 2017b), and the composite curve for δ^{13}C from several typical sections shows two positive excursions, four negative excursions and one intermediate values (Fig. 4.1), which are, in ascending order: ① a negative δ^{13}C excursion (EN1/CANCE)[1] in the Dou-1 Member (cap carbonate overlying the Nantuo diamictite) to the lowermost of the Dou-2 Member;②a pronounced positive δ^{13}C excursion (EP1)[2] from lower to upper Dou-2 Member; ③ a mild negative δ^{13}C excursion (EN2/BAINCE)[3] in the uppermost Dou-2 Member; ④ a positive δ^{13}C excursion (EP2) in the lower and middle Dou-3 Member; ⑤ a pronounced negative δ^{13}C excursion (EN3/DOUNCE)[4] from upper Dou-3 and Dou-4 Members; ⑥ an intermediate values (EI)[5] from lower Hamajing to upper Baimatuo Members;⑦ a pronounced negative δ^{13}C at the upper most part of the Baimatuo Member. In addition, a small negative δ^{13}C excursion (WANCE)[6] in the middle part of the EP1 can be observed in some sections (Sawaki et al. 2010; Liu et al. 2013; Zhu et al. 2013a, b), and this event can be determined in about 610 Ma (Liu et al. 2009b; Zhou et al. 2017a). Furthermore, the EN3 consists of two negative δ^{13}C excursion and one positive δ^{13}C excursion in some sections (An et al. 2015; Zhou et al. 2017b), and the positive excursion termed the Diaoyapo excursion (DPE) and two negative excursion termed the Jiuqunao negative excursion (JNE) and the Miaohe negative excursion (MNE) (Zhou et al. 2017b).

4.4 The Characteristics of Microfossils Biota and the Biostratigraphic Succession of Microfossils

The Ediacaran microfossils in South China were preserved in chert nodules and phosphorites of the Doushantuo Formation, the most representative silicified Lagerstätten from the Yangtze Gorges area, whereas the most representative phosphatized Lagerstätten at Weng'an, Guizhou Province. These microfossils provided important material for studying the evaluation of early life, stratigraphic subdivision and international biostratigraphic correlation. Especially, previous studies have indicated

[1] EN. negative carbon isotope excursion; CANCE.cap carbonate negative carbon isotope excursions.

[2] EP. positive carbon isotope excursion.

[3] BAINCE. Baiguoyuan negative carbon isotope excursion.

[4] DOUNCE. Doushantuo negative carbon isotope excursion.

[5] EI. carbon isotope intermediate values.

[6] WANCE. Weng'an negative carbon isotope excursion.

that the large Ediacaran acanthomorphic acritarchs is a useful tool for stratigraphic subdivision and correlation (Grey et al. 2003; Grey 2005; Willman and Moczydlowska 2008; Yin et al. 2009, 2011; McFadden et al. 2009; Liu et al. 2012a, 2013, 2014a, b; Moczydłowska and Nagovitsin, 2012; Joshi and Tiwari, 2016; Liu and Moczydłowska, 2019).

4.4.1 *Characteristics of Microfossil Biota and Biostratigraphic Succession*

4.4.1.1 Silicified Microfossils from the Yangtze Gorges Area

The Ediacaran silicified microfossils preserved in chert nodules of Dou-2 and Dou-3 Members of the Doushantuo Formation in the Yangtze Gorges area, and the palaeontological research on those microfossils is based on petrographic thin section. Abundant silicified microfossils, including large acanthomorphic acritarchs, spherical and filamentous cyanobacteria, multicellar algae, tubular microfossils, vase-shaped fossil and embryos, have been reported during the past 40 years (Yin and Li 1978; Zhang 1981, 1984; Awramik et al. 1985; Yin 1985, 1986, 1987; Yin and Liu 1988; Yin 1990, 1996; Zhang et al. 1998; Xiao, 2004; Zhou et al. 2005, 2007; Yin et al. 2007, 2008, 2011; Xie et al. 2008; Liu et al. 2009a, b, 2013, 2014a, b; McFadden et al. 2009; Chen et al. 2010; Ouyang et al. 2015; Liu and Moczydłowska 2019; Shang et al. 2019). Among all microfossils, the large acanthomorphic acritarchs are the most important for stratigraphic subdivision and correlation. Based on spatial and temporal distribution of the acanthomorphic acritarchs from the Doushantuo Formation in several sections, two assemblages (lower *Tianzhushania spinosa* assemblage and upper *Hocosphaeridium anozos-H. scaberfacium* assemblage) with distinct differences of taxonomic form can be distinguished (Yin et al. 2009, 2011a, b; Liu et al. 2013, 2014a, b). The assemblages are found in the lower (Dou-2 Member) and upper (Dou-3 Member) Doushantuo Formation, respectively, and are separated in this region by a negative δ^{13}C excursion (EN2), in which microfossils have not yet been found (Fig. 4.1).

The lower *Tianzhushania spinosa* assemblage occurs in the Dou-2 Member of the Doushantuo Formation, and all microfossils of this assemblage are found in the stratigraphic intervals defined by first positive δ^{13}C excursion (EP1), which first appeared at ca. 6.8 m above the lower boundary of the Doushantuo Formation (Liu and Moczydłowska 2019). Based on previous investigations, at least 28 species of acanthomorphic acritarchs were identified from this assemblage, including *Appendisphaera grandis, A.? hemisphaerica, A. magnifica, A. tenuis, Apodastoides basileus, Asterocapsoides sinensis, Briareus borealis, Cavaspina acuminate, C. basiconica, Cymatiosphaeroides kullingii, Dicrospinasphaera zhangii, Distosphaera speciosa, Eotylotopalla dactylos , E. delicata, Ericiasphaera magna, E. rigida, E. sparsa, E.*

spjeldnaesii, Meghystrichosphaeridium chadianensis (i.e., *Mengeosphaera chadianensis*), *M. perfectum* (i.e., *Mengeosphaera perfecta*), *Papilomembrana compta, Taedigerasphaera lappacea, Tanarium conoideum, Tianzhushania spinosa, T. polysiphonia, T. tuberifera* (i.e., *Yinitianzhushania tuberifera*), *Variomargosphaeridium lithoschum, Weissiella* cf. *brevis*, etc. (Yin et al. 2007, 2008; Zhou et al. 2007; McFadden et al. 2009; Chen et al. 2010; Liu et al. 2013, 2014a; Ouyang et al. 2015; Liu and Moczydłowska 2019). According to a quantitative evaluation, the species *Tianzhushania spinosa* occurs throughout the assemblage and is the dominant taxon (McFadden et al. 2009; Liu et al. 2013), the taxa *Tianzhushania tuberifera* (i.e., *Yinitianzhushania tuberifera*) is with a higher abundance (Fig. 4.2), whereas all other species are comparatively rare. In addition, many species restricted to the time interval represented by this assemblage in the Yangtze Gorges area, such as *Briareus borealis, Ericiasphaera rigida, E. sparsa, Tianzhushania spinosa, T. polysiphonia, T. tuberifera* (i.e., *Yinitianzhushania tuberifera*), etc. (Liu et al. 2013, 2014a).

The upper *Hocosphaeridium anozos-Hocosphaeridium scaberfacium* assemblage occurs in the Dou-3 Member of the Doushantuo Formation, and all microfossils of this assemblage are found in the stratigraphic intervals defined by second positive $\delta^{13}C$ excursion (EP2). By comparison with the lower *Tianzhushania spinosa* assemblage, acanthomorphic acritarchs of the upper assemblage are more abundant both in terms of diversity and number of individuals (Liu et al. 2013, 2014a, b). Abundant microfossils have been identified by Liu et al. (2014b), including 66 species of acanthomorphic acritarchs, 7 species of spherical microfossils, 12 species of spherical and filamentous cyanobacteria, 4 species of multi-cellar algae, and 2 species of tubular microfossils. Detailed microfossils list exhibited Table 1 in Liu et al. (2014b). Among these fossils, most acanthomorphic acritarchs were new species or first found from South area. The species *Hocosphaeridium anozos* and *Hocosphaeridium scaberfacium* occurs throughout the assemblage and are the dominant taxa (Fig. 4.3). The taxa *Appendisphaera clava, A. ? hemisphaerica, A. longispina, Eotylotopalla delicata, Knollisphaeridium maximum, Mengeosphaera bellula, M. constricta, Schizofusa zangwenlongii, Sinosphaera rupina, Tanarium acus, T. elegans, Variomargosphaeridium floridum, Xenosphaera liantuoensis* are with a relatively large abundance.

4.4.1.2 Phosphatized Microfossils of Weng'an Biota in Weng'an Area

The Weng'an biota preserved within phosphorites of the upper Doushantuo Formation at the Weng'an-Fuquan area, Guizhou Province. All microfossils preserved as phosphate permineralization in three-dimensional detail. These phosphatic microfossils are easily freed from their matrix by acetic acid digestion. Hence the palaeontological research on those microfossils is based on both petrographic thin section and SEM photomicrographs. Abundant phosphatezed microfossils, including large acanthomorphic acritarchs, spherical microfossils, spherical and filamentous cyanobacteria, multicellar algal, tubular microfossils, lichen-like symbiosis, embryos and metazoan, have been reported during the past 30 years (Chen and Liu 1986; Yuan

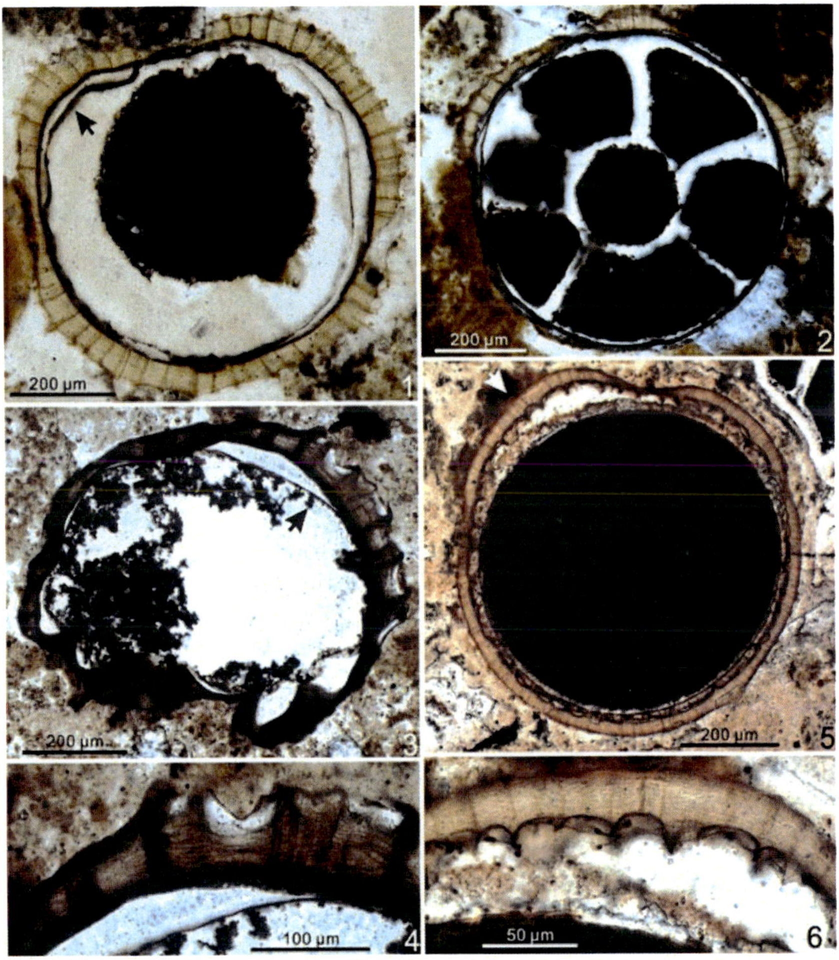

Fig. 4.2 The representative taxa of *Tianzhushania spinosa* assemblage. 1 and 2. *Tianzhushania spinosa*, 2 with daughter cells; 3 and 4. *Tianzhushania polysiphonia*, 4, enlarged view of marked area (arrow) in 3; 5 and 6. *Yinitianzhushania tuberifera*, 6, enlarged view of marked area (arrow) in 5

et al. 1993, 1998, 2005; Xue and Tang 1995; Xiao et al. 1998, 2000, 2007, 2014; Xiao and Knoll 1999, 2000; Zhang et al. 1998; Chen et al. 2000, 2002, 2004; Liu et al. 2006, 2008; Yin et al. 2013, 2015, 2016; Chen et al. 2014a, b). Among those microfossils, the acanthomorphic acritarchs are with the largest diversity, 54 species of acanthomorphic acritarchs (including the genus *Megasphaera*) have been identified (cf. Table 1 in Xiao et al. 2014), and most of those species have been also found from the Dou-2 Member of the Doushantuo Formation in the Yangtze Gorges area, such as *Appendisphaera grandis*, *A. tenuis*, *Cavaspina acuminate*,

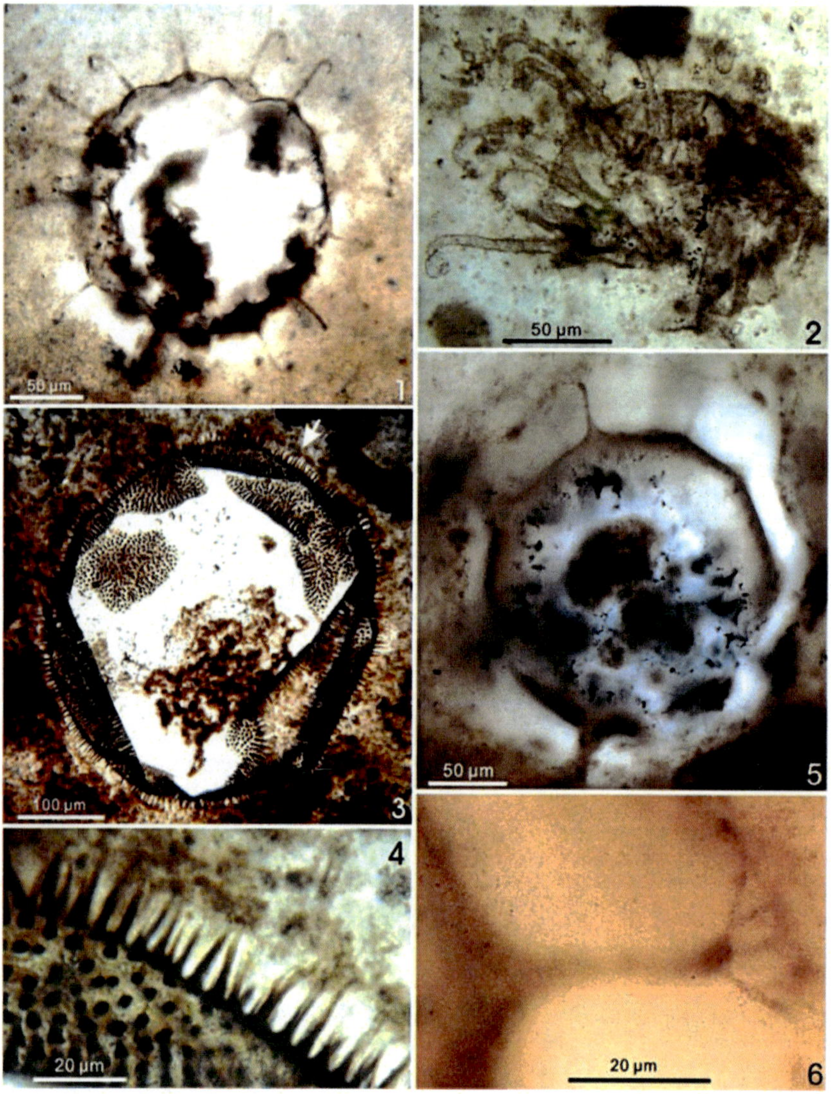

Fig. 4.3 The representative taxa of *Hocosphaeridium anozos-H. Scaberfacium* assemblage. 1. *Hocosphaeridium anozos*; 2. *Hocosphaeridium scaberfecium*; 3 and 4. *Knollisphaeridium maximum*, 4, enlarged view of marked area (arrow) in 3; 5 and 6. *Variomargosphaeridium litoschum*, 6, enlarged view of upper process in 5, pointing to the divarication of the process

Dicrospinasphaera zhangii, Distosphaera speciosa, Eotylotopalla dactylos, Ericiasphaera magna, E. rigida, Mengeosphaera chadianensis, Papillomembrana compta, Tianzhushania spinosa, T. polysiphonia, Yinitianzhushania tuberifera, etc. The spherical microfossils *Megasphaera inornata* and *Megasphaera ornata*, which are interpreted as rest eggs and embryos (Xiao et al. 1998), are the dominant taxa and with the largest abundance in the Weng'an biota. However, based on morphological and taphonomic studies of permineralized microfossils in cherts and phosphorites of the Doushantuo Formation in the Weng'an area, the phosphatized *Megasphaera ornata* and the chert-preserved *Yintianzhushania tuberifera* (i.e., *Tianzhushania tuberifera*) was regarded as representing the same taxon preserved by different mineralized processes, and the chert-preserved *Tianzhushania spinosa* may be a senior synonym of phosphatized *T. inornata* (Yin et al. 2004). We think this explanation is reasonable. Hence the Weng'an biota is similar with the *Tianzhushania spinosa* assemblage in the Yangtze Gorges area although it contains several typical taxa of the *Hocosphaeridium anozos-Hocosphaeridium scaberfacium* assemblage. such as *Appendisphaera grandis, A. tenuis, Cavaspina acuminate, Dicrospinasphaera zhangii, Distosphaera speciosa, Eotylotopalla dactylos, Ericiasphaera magna, E. rigida, Mengeosphaera chadianensis, Papillomembrana compta, Tianzhushania spinosa, T. polysiphonia, Yinitianzhushania tuberifera*, etc. The spherical microfossils *Megasphaera inornata* and *Megasphaera ornata*, which are interpreted as rest eggs and embryos (Xiao et al. 1998), are the dominant taxa and with the largest abundance in the Weng'an biota. However, based on morphological and taphonomic studies of permineralized microfossils in cherts and phosphorites of the Doushantuo Formation in the Weng'an area, the phosphatized *Megasphaera ornata* and the chert-preserved *Yintianzhushania tuberifera* (i.e., *Tianzhushania tuberifera*) was regarded as representing the same taxon preserved by different mineralized processes, and the chert-preserved *Tianzhushania spinosa* may be a senior synonym of phosphatized *T. inornata* (Yin et al. 2004). We think this explanation is reasonable. Hence the Weng'an biota is similar with the *Tianzhushania spinosa* assemblage in the Yangtze Gorges area although it contains several typical taxa of the *Hocosphaeridium anozos-Hocosphaeridium scaberfacium* assemblage.

4.4.2 International Biostratigraphic Correlation of Microfossils

Apart from South China, Ediacaran large acanthomorphic acritarchs have been also found from other continents, including Australia (Zang and Walter 1992; Grey, 2005; Willman et al. 2006; Willman and Moczydlowska 2008, 2011), eastern Europe Craton (Veis et al. 2006; Vorob'eva et al. 2006, 2009a, b), Baltoscan-dian Platform (Vidal 1990), Siberia (Moczydłowsk et al. 1993; Vorob'eva et al. 2008; Golubkova et al. 2010; Moczydłowska and Nagovitsin 2012), northern India (Tiwari and Knoll 1994;

Tiwari and Pant 2004; Shukla and Tiwari 2014; Joshi and Tiwari 2016), and Svalbard (Knoll and Ohta 1988; Knoll 1992). Some globally distributed taxa with short and well-documented stratigraphic ranges provide useful basis for biostratigraphic subdivision and international correlation of Ediacaran strata (Grey et al. 2003; Grey 2005; Willman and Moczydlowska 2008, 2011; Vorob'eva et al. 2009b; Sergeev et al. 2011; Liu et al. 2012a, 2013, 2014a, b; Moczydłowska and Nagovitsin 2012; Joshi and Tiwari 2016; Liu and Moczydłowska 2019).

At present, the dominant and representative taxon *Tianzhushania spinosa* of the lower assemblage is only found from the lower Doushantuo Formation (Dou-2 Member) in South China (Liu et al. 2013, 2014a) and the Infrakrol Formation in Lesser Himalaya, India (Joshi and Tiwari 2016), which suggest that the lower *T. spinosa* assemblage can be correlated with Infrakrol Formation in India. In addition, the lower *T. spinosa* assemblage appears to be missing in Siberia and the East European Craton as well as in Australia since the *T. spinosa* is not found in these continents. However, there is no evidence that the equivalent stratigraphic level to the *T. spinosa* assemblage of South China is missing in those continents because there is a significant thickness of strata between the Marinoan (Nantuo) diamictite and the interval with the acanthomorphic assemblage. Whether the absence of the *T. spinosa* assemblage in those continents is due to preservation problems or for other reasons remains unclear, and certainly requires further work and interpretation.

Previously studies indicated that the upper *Hocosphaeridium anozos-Hocosphaeridium scaberfacium* assemblage can be correlated with the Ediacaran complex acritarch palynoflora (ECAP) of South Australia since both the dominant and representative taxa were also preserved in ECAP (Liu et al. 2013, 2014a). The correlation scheme is further supported by a major negative $\delta^{13}C$ excursion which marks the extinction of the Ediacaran acanthomorphic acritarchs both in South China and Australia. The excursion (DOUNCE) occurs in the upper part of the Doushantuo Formation of South China and is regarded to be equivalent to the Wonoka excursion of Australia (Jiang et al. 2007; Zhu et al. 2007). Similarly, many taxa from the upper acanthomorphic assemblage in the Yangtze Gorges area have been reported from Siberia and the East European Craton (Vorob'eva et al. 2009b; Sergeev et al. 2011; Moczydłowska and Nagovitsin 2012), which demonstrated that the upper assemblage in the Yangtze Gorges area is stratigraphically correlative with these from Siberia and the East European Craton. In addition, almost all reported acanthomorphic acritarchs from the East Siberia were preserved in Ura Formation and equivalent strata and disappeared below the negative $\delta^{13}C$ excursion equivalent to the DOUNCE-Shuram-Wonoka excursion (Vorob'eva et al. 2009b; Sergeev et al. 2011; Moczydłowska and Nagovitsin 2012), which support the correlation scheme. However, this is just a rough correlation scheme that needs to be refined in the future.

4.5 Characteristics of Macrofossils Biota and Biostratigraphic Succession of Macrofossils

The Ediacaran (Sinian) in South China contains several exceptionally preserved macrofossil biotas in its shales and carbonates, including Miaohe biota, Lantian biota, Wulingshan biota, Gaojiashan biota, Wenghui biota, Jiangchuan biota, and Xilingxia (Shibantan) biota. These biotas mainly contain macroalgae and many metazoan fossils. These fossils provide important material for studying the origin and evolution of early metazoan, biostratigraphic subdivision and correlation, and the establishment of the Ediacaran chronostratigraphic framework.

4.5.1 Characteristics of Typical Macrofossil Biota

4.5.1.1 Lantian Biota

Lantian biota, which was first reported in the 1980s (Xing et al. 1985), was preserved within the Early Ediacaran slope-basinal black shales in southern Anhui Province, East China. The Ediacaran System (Sinian) in this area consists of Liantian Formation and the overlying Piyuancun Formation, both formations are respectively correlated with the Doushantuo and Dengying Formations in the Yangtze Gorges area. The Lantian Formation is divided into four members that are similar to and can be correlated with the four members of the Doushantuo Formation. The lowest Dou-1 Member is a ca. 2 m-thick cap dolostone conformably overlying a Marinoan-age glacial diamictite, the Dou-2 Member consists of 35 m-thick black shale, the Dou-3 Member is a 34 m-thick dolostone interbedded with mudstones in the lower part and ribbon rocks in the upper part, the Dou-4 Member consists of a 20 m-thick silty mudstone. The macrofossils were collected from black shale of the Dou-2 Member (Yuan et al. 2011).

Lantian biota is with a considerable degree of taxonomic diversity and morphological differentiation. Previous studies have described numerous macroalgae fossils, including simple vesicles (e.g., *Chuaria*), dichotomously branching forms (e.g., *Doushantuophyton*, Fig. 4.1), loosely bundled filaments (e.g., *Huangshanophyton*), conical structures with densely packed filaments (such as *Anhuiphyton*), fan-shaped forms (e.g., *Flabellophyton*, Fig. 4.2), and chained rings (e.g., *Obisiana*). In additions, at least five morphotypes of putative macrometazoan have been reported (Yuan et al. 2011; Wan et al. 2016 and therein). In general, this biota is dominated by *Anhuiphyton*, *Huangshanophyton* and *Flabellophyton*.

The precise age of the Lantian biota is still unknown due to the lack of isotopic dating. However, we may reasonably conclude that its age should be in Early Ediacaran Period, perhaps shortly after the Marinoan glaciation since its occurred position is just above the Ediacaran cap dolostone.

4.5.1.2 Miaohe Biota

Miaohe biota, which was first reported in the 1980s (Zhu and Chen 1984), was preserved within Late Ediacaran black shales of the Dou-4 Member in Doushantuo Formation in the Yangtze Gorges area, South China (Xiao et al. 2002; An et al. 2015 and therein). All macrofossils preserved as carbonaceous compression and with a considerable degree of taxonomic diversity and morphological differentiation. More than 100 macrofossil species had been described in the 1990s (Ding et al. 1996), but in an examination of published and new materials, Xiao et al. (2002) recognized only about twenty distinct taxa. Most of these fossils can be interpreted unambiguously as colonial prokaryotes or multicellular algae. Of these, the unbranching clavate form *Baculiphyca* (Fig. 4.4) and dichotomously branching form *Enteromorphites* and *Doushantuophyton* (Fig. 4.3) are with considerable abundance. In addition, the taxon *Eoandromeda* is a typical Ediacaran-type animal fossil (Tang et al. 2008, 2011; Zhu et al. 2008), and the taxa *Protoconites* and *Calyprina* are most closely resemble animal remains (Xiao et al. 2002).

The age of the Miaohe biota is constrained by the age 551.1 ± 0.7 Ma obtained from the ash bed at the top of the black shale of Dou-4 Member (i.e., Miaohe Member; Condon et al. 2005). Combined with the eight-armed Ediacara animal fossils *Enteromorphites* also found from the Rawnsley Quartzite in western Flinders Ranges, South Australia (Zhu et al. 2008). It is suggested that the Miaohe biota can correlate with the White Sea biota (560–550 Ma) in Russia and the Ediacara assemblage from the western Flinders Ranges, South Australia.

4.5.1.3 Wenghui Biota

Wenghui biota, which was first reported in 2004 (Zhao et al. 2004), was preserved within upper black shales of the Doushantuo Formation in the Wenghui area, Guizhou Province, Southwest China. This biota is similar to the Miaohe biota. All macrofossils are preserved as carbonaceous compression and the most taxonomic composition are same to the Miaohe biota, e.g., *Baculiphyca, Doushantuophyton, Enteromorphites, Eoandromeda* (Fig. 4.1), *Glomulus, Longifuniculum, Miaohephyton, Sinocylindra*, etc. (Zhao et al. 2004; Tang et al. 2008; Wang et al. 2008). In addition, both biotas occur in black shale of the upper Doushantuo Formation. Hence this biota can correlate with the Miaohe biota.

4.5.1.4 Xilingxia Biota (Shibantan Assemblage)

The Xilingxia biota was preserved within the limestone of the Shibantan Member of the Ediacaran Dengying Formation in the Yangtze Gorges area, South China. Abundant trace fossils and macroalgal fossil *Vendotaenia* as well as Ediacara-type fossil *Paracharnia* and *Sinotubulites* had been reported from the Shibantan Member

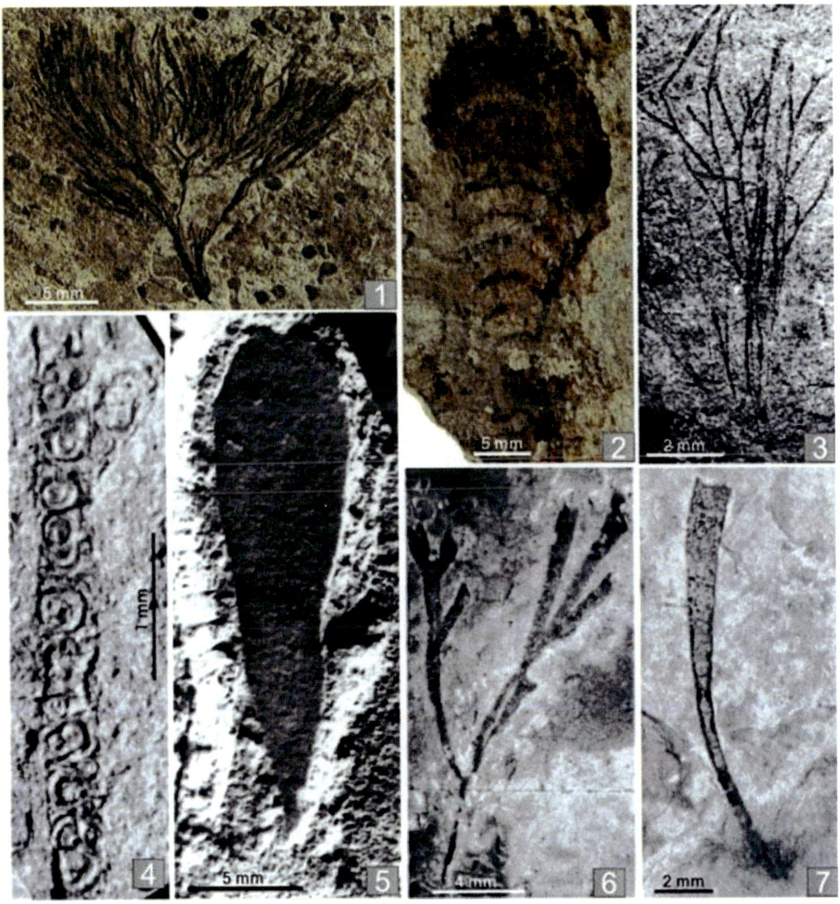

Fig. 4.4 Macroalgal fossils from the Eiacaran Doushantuo and Liantian Formations. 1. ***Doushantuophyton cometa*** from Lantian biota (Yuan et al. 2011); 2. ***Flabellophyton strigata*** from Lantian biota (Yuan et al. 2011); 3. ***Doushantuophyton lineare*** from Miaohe biota (Xiao et al. 2002); 4. ***Miaohephyton bifurcatum*** from Miaohe biota (Xiao et al. 2002); 5. ***Gesinella hunanensis*** from Wulingshan biota (Steiner and Erdtmann 1992); 6. ***Konglingiphyton erecta*** from Miaohe biota (Xiao et al. 2002); 7. ***Baculyphyca taeniata*** from Miaohe biota (Xiao et al. 2002)

in 1980s (Ding and Chen 1981; Sun 1986; Zhao et al. 1988). Recently, abundant iconic Ediacara-type fossils, including *Charniodiscus, Hiemalora, Pteridinium* (Fig. 4.5), *Rangea, Wutubus* (Fig. 4.3), and *Yangtziramulus*, etc. were reported from the Shibantan Member (Xiao et al. 2005; Shen et al. 2009; Chen et al. 2014a, b). These fossils significantly expand the ecological ranges of several key Ediacara taxa and support that they are marine organisms (Chen et al. 2014a, b). Most of these Ediacara-type fossils have wide geographic distribution and long stratigraphic span (Chen et al. 2014a, b). However, the Dengying Formation is constrained between

551 and 541 Ma since the uppermost Doushantuo Formation was dated at 551.1 ± 0.7 Ma (Condon et al. 2005) and overlain Yanjiahe Formation contains Early Cambrian small shelly fossils and microfossils (Guo et al. 2014; Shang et al. 2016; Ahn and Zhu 2017and therein). It suggests that the Xilingxia biota can correlate with Nama biota (550–540 Ma) in Namibia.

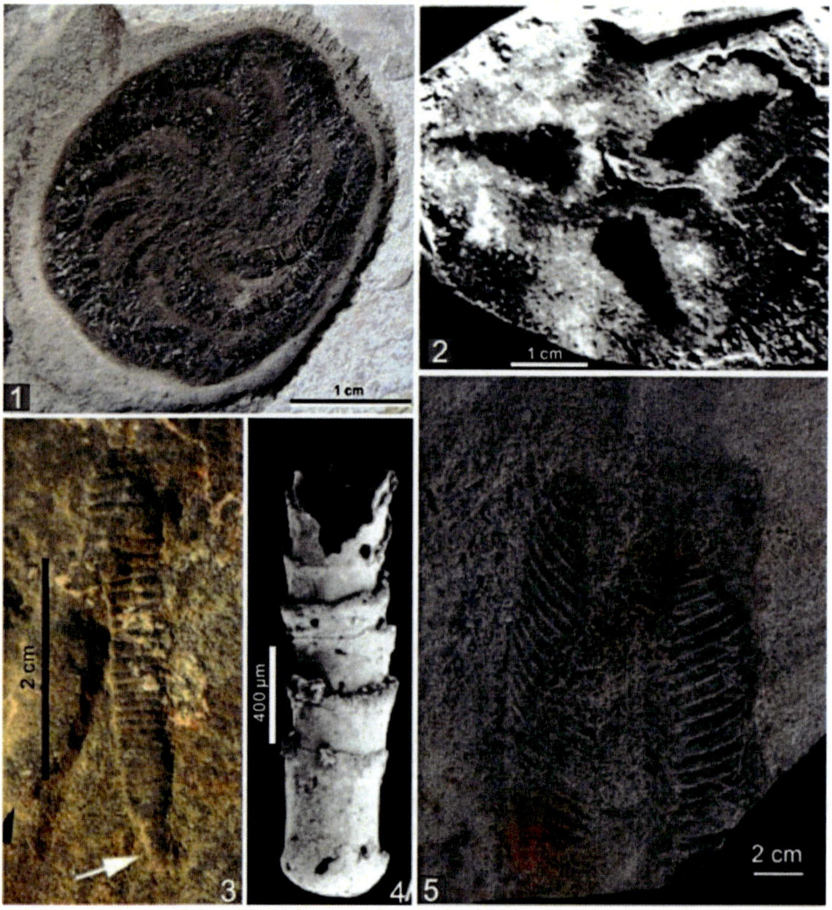

Fig. 4.5 Ediacara-type animal fossils from the Doushantuo and Dengying Formations. 1 *Eoandromeda octobrachiata* from Wenghui biota (Tang et al. 2008); 2 *Triactindiscus sinensis* from Wenghui biota (Zhao et al. 2010); 3 *Wutubus annularis* from Xilingxia biota (Chen et al. 2014a, b); 4 *Cloudina hartmannae* from Gaojiashan biota (Hua et al. 2005); 5 *Pteridinium* from Xilingxia biota (Chen et al. 2014a, b)

4.5.1.5 Wulingshan Biota

Wulingshan biota, which was first reported and named in 1990s (Steiner et al. 1992; Chen et al. 1999), was preserved within the black shales of the Late Ediacaran Liuchapo Formation in western Hunan Province, South China. All macrofossils preserved as carbonaceous compression. 17 macrofossil species have been described in the 1990s (Steiner et al. 1992; Chen et al. 1999), mainly are macroalgal fossils, such as ribbon-like *Chenlidenella, Gesinella* (Fig. 4.5) and *Longifuniculum*, vertical and unbranched form *Longfengshania, Paralongfengshania, Miaohenella* and *Cystoculum*, filamentous with bifurcating end *Setoralga*, and medusoid fossil *Taoyuania, Liaonanella* and *Wulingshania*, etc. (Chen et al. 1999). Recently, a new CA-ID TIMS zircon U–Pb age 545.76 ± 0.66 Ma was reported from an ash bed at the lower Liuchapo Formation in western Hunan (Yang et al. 2017), which suggest that the age of the Xilingxia biota should be less than 545 Ma.

4.5.1.6 Gaojiashan Biota

The Gaojiashan biota were the soft-bodied fossil-Lagerstätte composed of substantial pyritized, three-dimensionally preserved tubular and conotubular animal fossils, trace fossils and macroagal fossils, the fossils preserved in siliciclastic and carbonate rocks in the Gaojiashan Member of the Late Ediacaran Dengying Formation (Hua et al. 2007; Cai et al. 2017 and therein). Three fossil assemblages of the Gaojiashan biota have been established as follows (in ascending orders): ① *Shaanxilithes-Helminthopsis* trace fossil assemblage appears at the base of the Gaojiashan Member; ② *Conotubus-Gaojiashania-Protolagena* assemblage occurs in the middle part of the Gaojiashan Member; ③ *Sinotubulites-Cloudina* assemblage is found from the upper part of the Gaojiashan Member (Hua et al. 2001), among which, the iconic fossil of the *Shaanxilithes, Sinotubulites, Cloudina* (Fig. 4.4) was also found from the Shibantan assemblage in the Yangtze Gorges area, Nama biota in Namibia and many other Late Ediacaran sections in the world (Hua et al. 2007; Yang et al. 2016 and therein); which shows that the Gaojiashan biota can correlate with the Xilingxia (Shibantan) assemblage and the Nama biota.

4.5.1.7 Jiangchuan Biota

Recently, a highly diversified macrofossil assemblage was discovered from the Jiucheng Member of the Late Ediacaran Dengying Formation in the Jiangchuan area, eastern Yunnan Province, Southwest China (Tang et al. 2006a, 2007). All fossils preserved as carbonaceous compression in grey silty shales. In addition to abundant *Vendotaenia* and *Tyrasotaenia*, large mulicellular benthos such as *Tawuia*-like and Longfengshaniaceae fossils with diverse holdfast structures, have a distinct dominance, while there are still some other macrofossils with peculiar configurations as well as indistinct relatives. There is an obvious distinction between this assemblage and the macrofossil assemblages from the Doushantuo Formation and the Shibantan Member of the Dengying Formation in South China (Tang et al. 2007, 2009). The Jiangchuan assemblage may be synchronous with the Shibantan assemblage since the Jiucheng Member in eastern Yunnan can be correlated with the Shibantan Member in the Yangtze Gorges area (Luo et al. 1988).

4.5.2 Biostratigraphic Succession of Macrofossils

According to the stratigraphic spans of all Ediacaran macrofossil biotas, these macrofossils mainly occur in three horizons in South China as follows (in ascending orders).

(1) The lower Dou-2 Member of the Doushantuo Formation. The most important and typical macrofossil assemblage in this horizon is the Lantian biota. In addition, a few macroscopic compression fossils were also reported from this horizon, including simple forms *Chuaria* and *Tawuia,* and dichotomously branching form *Enteromorphites* (Tang et al. 2006b). The macrofossils assemblage is mainly composed of macroalgal fossils and some putative macrometazoan fossils, among which, the popular taxa are *Anhuiphyton, Doushantuophyton* and *Flabellophyton* (Yuan et al. 2011). In the Yangtze Gorges and neighbouring areas, several ages have been reported from ash beds and black shales in the Doushantuo Formation, including ID-TIMS zircon U-Pb age 632.5 ± 0.5 Ma from lower part of the Dou-2 Member (Condon et al. 2005), SHRIMP II zircon U-Pb age 614 ± 7.6 Ma from the middle part of the of the Dou-2 Member (Liu et al. 2009b), and Re-Os age 591.1 ± 5.3 Ma from the base of the Dou-4 Member (Zhu et al. 2013b). Thus, the age of this assemblage is constrained between 632 and 591 Ma, more likely ca. 615 Ma since it occurs in the middle part of the Dou-2 Member, and it should be synchronous with microfossil *Tianzhushania spinosa* assemblage.
(2) The uppermost Dou-4 Member of the Doushantuo Formation. There are two important assemblages occurred in this horizon, Miaohe and Wenghui biotas. All macrofossils preserved as carbonaceous compression and dominated by macroalgae fossils. In addition, a few iconic Ediacara-type fossils also occurs in this horizon, such as eight-armed form *Eoandromeda* (Fig. 4.1) and triridged lobe fossil *Triactindiscus*, etc. (Fig. 4.2; Tang et al. 2008; Wang et al. 2008; Zhu et al. 2008; Zhao et al. 2010). As discussing above, this assemblage can correlate with the White Sea biota and its age is about 560–550 Ma.
(3) The Dengying Formation. The Xilingxia, Gaojiashan, Wulingshan, and Jiangchuan biotas occur in this horizon. The most important feature is that the abundant Ediacaran skeletonized tubular fossils (e.g., *Cloudina*; Fig. 4.4) and other Ediacara-type fossils (e.g., *Charniodiscus, Hiemalora, Pteridinium, Rangea*, etc.) were found from this horizon (Hua et al. 2005, 2007; Chen et al. 2014a, b; Cai et al. 2017). In addition, a lot of macroalgal fossils and trace fossils

were reported from this level (Steiner et al. 1992; Tang et al. 2007; Chen et al. 2013). As discussing above, the age of this level is constrained in 550–541 Ma and can be correlated with the Nama biota.

4.6 Biostratigraphic Succession and Chronostratigraphic Framework

Since ratification of the Ediacaran System by the IUGS in 2004 (Knoll et al. 2004), further subdivision of the system has been the most urgent task in studies of the Ediacaran. In recent years, abundant Ediacaran micro- and macrofossils have been found from many sections in the world, some fossil species are widely distributed, showing potential for global biostratigraphic correlation and subdivision although it was constrained by facies control, taphonomic bias and taxonomic problems (Grey et al. 2003; Grey 2005; Willman and Moczydłowska 2008; Yin et al. 2009; McFadden et al. 2009; Vorob'eva et al. 2009a; Liu et al. 2013, 2014a). In South China, intensive integrated investigations of the Ediacaran sequences during the past years have demonstrated that the sections in the Yangtze Gorges area have potential to establish criteria for chronostratigrapic subdivision of the Ediacaran System. Based on integrated bio- and chemostratigraphic data, two series and five stages have been previously proposed (Zhu et al. 2007; Liu et al. 2012a). Accordingly, four important boundaries/points were tentatively suggested for chronostratigraphic subdivision of the Ediacaran System (Liu et al. 2014a). According to recent advances on the study of the Bio- and Chemostratigraphy, the Ediacaran of South China is further subdivided into two series and six stages (Fig. 4.1). The lower series contains abundant microfossils mainly large acanthomorphic acritarchs, and the upper series yields a lot of macroalgae fossils and Ediacara-type animal fossils. The base boundary of the upper series is marked at nadir of the third negative $\delta^{13}C$ excursion (or DUNCE). The first stage is related to the Nantuo (i.e., Marinoan) glacier, and consists of the cap dolostone and the lowest part of the Dou-2 Member of the Doushantuo Formation. The second stage mainly consists of the lower part of the Dou-2 Member and is restricted to the interval defined by the first positive $\delta^{13}C$ excursion (EP1), and the base boundary is marked by emergence of the dominant and representative taxon of the *Tianzhushania spinosa* assemblage. The third stage consists of the upper part of the Dou-2 Member and the lowest part of the Dou-3 Member, its basal boundary is marked by occurrence of the second negative $\delta^{13}C$ excursion (EN2), and so far no any fossil is found from this level in South China. The fourth stage consists of the middle part of the Dou-3 Member, its basal boundary is marked by emergence of the dominant and representative taxa of the *Hocosphaeridium anozos-H. scaberfacium* assemblage. The fifth stage consists of upper part of the Dou-3 Member and the Dou-4 Member, and it is characterized by rich macroalgae fossils and eight-armed Ediacara-type animal fossil *Enteromorphites*. The sixth stage consists of the Denying Formation, characterized by abundant skeletonized tubular fossils and other Ediacara-type animal fossils, and its basal boundary is marked by the occurrence of the skeletonized tubular fossil *Cloudina* or *Cloudina*-like fossils (e.g., *Conotubus*, *Gaojiashania*, *Sinotubulites*).

4.7 Conclusion

South China is one of the most typical areas in the world for studying the Ediacaran successions since it is with full stratigraphic sections, long research history, abundant micro- and macrofossils, perfect carbon isotopic profile, and important isotopic dating. According to previous study results, we have established two microfossil assemblages and three macrofossil assemblages in South China. In addition, based on biostratigraphic and $\delta^{13}C$ data and U-Pb zircon ages, the Ediacaran System is subdivided into two series and six stages. However, the current Ediacaran chronostratigraphic subdivision is a preliminary groping scheme, which needs more work in the future.

Acknowledgements This study was supported by the National Natural Science Foundation of China (Grant No. 41872024), China Geology Survey (Grant No. DD20190008) and the National Key Research and Development Program of China (Grant No. 2016YFC0601001).

References

Ahn SY, Zhu MY (2017) Lowermost Cambrian acritarchs from the Yanjiahe Formation, South China: implication for defining the base of the Cambrian in the Yangtze Platform. Geol Mag 154(6):1217–1231

An ZH, Jiang GQ, Tong JN, Tian L, Ye Q, Song HY, Song HJ (2015) Stratigraphic position of the Ediacaran Miaohe biota and its constrains on the age of the upper Doushantuo $\delta^{13}C$ anomaly in the Yangtze Gorges area, South China. Precambr Res 271:243–253

Awramik SM, McMenamin DS, Yin CY, Zhao ZQ, Ding QX, Zheng SS (1985) Prokaryotic and eukaryotic microfossils from a Proterozoic/Phanerozoic transition in China. Nature 315(6021):655–658

Cai YP, Cortijo I, Schiffbauer JD, Hua H (2017) Taxonomy of the Late Ediacaran index fossil Cloudina and a new similar taxon from South China. Precambr Res 298:146–156

Chen ME, Liu KW (1986) The geological significance of newly discovered microfossils from the Upper Sinian (Doushantuo age) phosphorites. Sci Geol Sin 1:46–53

Chen XH, Wang XF, Wang CS, Li ZH, Chen LD (1999) A preliminary study on carbonaceous megafossils from the from the Late Sinian Liuchapo Formation of west Hunan. Geol Min Resour South China 2:15–30 (in Chinese with English abstract)

Chen JY, Oliveri P, Li CW, Zhou GQ, Gao F, Hagadorn JW, Peterson KJ, Davidson EH (2000) Precambrian animal diversity: putative phosphatized embryos from the Doushantuo Formation of China. Proc Natl Acad Sci 97(9):4457–4462

Chen JY, Oliveri P, Gao F, Dornbos SQ, Li CW, Bottjer DJ, Davidson EH (2002) Precambrian animal life: probable developmental and adult cnidarian forms from Southwest china. Dev Biol 248(1):182–196

Chen JY, Bottjer DJ, Oliveri P, Dornbos SQ, Gao F, Ruffins S, Chi HM, Li CW, Davidson EH (2004) Small bilaterian fossils from 40 to 55 million years before the Cambrian. Science 305(5681):218–222

Chen SM, Yin CY, Liu PJ, Gao LZ, Tang F, Wang ZQ (2010) Microfossil assemblage from chert nodules of the Ediacaran Doushantuo Formation in Zhangcunping, northern Yichang, South China. Acta Geologica Sinica 84(1):70–77 (in Chinese with English abstract)

Chen Z, Zhou CM, Meyer M, Xiang K, Schiffbauer JD, Yuan XL, Xiao SH (2013) Trace fossil evidence for Ediacaran bilaterian animals with complex behaviors. Precambr Res 224:690–701

Chen L, Xiao SH, Pang K, Zhou CM, Yuan XL (2014a) Cell differentiation and germ-soma separation in Ediacaran animal embryo-like fossils. Nature 516:238–241

Chen Z, Zhou CM, Xiao SH, Wang W, Guan CG, Hua H, Yuan XL (2014b) New Ediacara fossils preserved in marine limestone and their ecological implications. Sci Rep 4(1):4180–4190

Chu XL, Zang QR, Zhang TG, Feng LJ (2003) Sulfur and carbon isotopic variations in Neoproterozoic sedimentary rocks from southern China. Prog Nat Sci 13(11):875–880

Condon D, Zhu MY, Bowring S, Wang W, Yang AH, Jin YG (2005) U-Pb ages from the neoproterozoic Doushantuo Formation. China. Science 308(5718):95–98

Ding QX, Chen YY (1981) Discovery of soft metazoan from the Sinian System along eastern Yangtze Gorge, Hubei. J Wuhan College of Geology 6(2):53–56 (in Chinese with English abstract)

Ding LF, Li Y, Hu XS, Xiao YP, Su CQ, Huang JC (1996) Sinian Miaohe Biota. Geological Publishing House, Beijing, p 221 (in Chinese)

Fedonkin MA, Simonetta A, Ivantsov AY (2007) New data on Kimberella, the Vendian mollusc-like organism (White Sea region, Russia): palaeoecological and evolutionary implications. Geol Soc Spec Publ Lond Geol Soc 286(1):157–179

Golubkova EY, Raevskaya EG, Kuznetsov AB (2010) Lower Vendian microfossil assemblages of East Siberia: significance for solving regional stratigraphic problems. Stratigr Geol Correl 18(4):353–375

Grey K (2005) Ediacaran palynology of Australia. Mem Assoc Australas Palaeontol 31:1–439

Grey K, Walter MR, Calver CR (2003) Neoproterozoic biotic diversification: snowball Earth or aftermath of the Acraman impact? Geology 31(5):459–462

Guo JF, Li Y, Li GX (2014) Small shelly fossils from the Early Cambrian Yanjiahe Formation, Yichang, Hubei, China. Gondwana Res 25(3):999–1007

Hua H, Zhang LY, Zhang ZF, Wang JP (2001) Assemblage zones of Gaojiashan biota and their characteristics. J Stratigr 25(1):13–17 (in Chinese with English abstract)

Hua H, Chen Z, Yuan XL, Zhang LY, Xiao SH (2005) Skeletogenesis and asexual reproduction in the earliest biomineralizing animal Cloudina. Geology 33(4):277–280

Hua H, Chen Z, Yuan XL (2007) The advent of mineralized skeletons in Neoproterozoic Metazoa—new fossil evidence from the Gaojiashan Fauna. Geol J 42(3–4):263–279

Jiang GQ, Sohl LE, Christie-Blick N (2003) Neoproterozoic stratigraphic comparison of the Lesser Himalaya (India) and Yangtze block (south China): paleogeographic implications. Geology 31(10):917–920

Jiang GQ, Kaufman AJ, Christie-Blick N, Zhang SH, Wu HC (2007) Carbon isotope variability across the Ediacaran Yangtze platform in South China: implications for a large surface-to-deep ocean $\delta^{13}C$ gradient. Earth Planet Sci Lett 261(1–2):303–320

Jiang GQ, Shi XY, Zhang SH, Wang Y, Xiao SH (2011) Stratigraphy and paleogeography of the Ediacaran Doushantuo Formation (ca. 635–551 Ma) in South China. Gondwana Res 19(4):831–849

Joshi H, Tiwari M (2016) Tianzhushania spinosa and other large acanthomorphic acritarchs of Ediacaran Period from the Infrakrol Formation, Lesser Himalaya, India. Precambr Res 286:325–336

Knoll AH (1992) Vendian microfossils in metasedimentary cherts of the Scotia Group, Prins Karls Forland, Svalbard. Palaeontology 35:751–774

Knoll AH, Ohta Y (1988) Microfossils in metasediments from Prins Karls Forland, western Svalbard. Polar Res 6(1):59–67

Knoll AH, Walter MR, Narbonne GM, Christie-Blick N (2004) A new period for the geologic time scale. Science 305(5684):621–622

Lambert IB, Walter MR, Zang WL, Lu SN, Ma GG (1987) Palaeoenvironment and carbon isotope stratigraphy of Upper Proterozoic carbonates of the Yangtze Platform. Nature 325:140–142

Liu PJ, Moczydłowska M (2019) Ediacaran microfossils from the Doushantuo Formation chert nodules in the Yangtze Gorges area, South China, and new biozones. Fossils Strata 65:1–172

Liu PJ, Yin CY, Tang F (2006) Microtubular metazoan fossils with multi-branches in Weng'an biota. Chin Sci Bull 51(5):630–632

Liu PJ, Xiao SH, Yin CY, Zhou CM, Gao LZ, Tang F (2008) Systematic description and phylogenetic affinity of tubular microfossils from the Ediacaran Doushantuo Formation at Weng'an, South China. Palaeontology 51:339–366

Liu PJ, Xiao SH, Yin CY, Tang F, Gao LZ (2009a) Silicified tubular microfossils from the upper Doushantuo Formation (Ediacaran) in the Yangtze Gorges area, South China. J Paleontol 83(4):630–633

Liu PJ, Yin CY, Gao LZ, Tang F, Chen SM (2009b) New material of microfossils from the Ediacaran Doushantuo Formation in the Zhangcunping area, Yichang, Hubei Province and its zircon SHRIMP U-Pb age. Chin Sci Bull 54(6):1058–1064

Liu PJ, Yin CY, Chen SM, Li M, Gao LZ, Tang F (2012a) Discussion on the chronostratigraphic subdivision of the Ediacaran (Sinian) in the Yangtze Gorges area. South China. Acta Geologica Sinica 86(6):849–866 (in Chinese with English abstract)

Liu PJ, Yin CY, Chen SM, Tang F, Gao LZ (2012b) Discovery of Ceratosphaeridium (Acritarcha) from the Ediacaran Doushantuo Formation in Yangtze Gorges, South China and its biostratigraphic implication. Bull Geosci 87(1):195–200

Liu PJ, Yin CY, Chen SM, Tang F, Gao LZ (2013) The biostratigraphic succession of acanthomorphic acritarchs of the Ediacaran Doushantuo Formation in the Yangtze Gorges area, South China and its biostratigraphic correlation with Australia. Precambr Res 225:29–43

Liu PJ, Chen SM, Zhu MY, Li M, Yin CY, Shang XD (2014a) High-resolution biostratigraphic and chemostratigraphic data from the Chenjiayuanzi section of the Doushantuo Formation in the Yangtze Gorges area, South China: implication for subdivision and global correlation of the Ediacaran System. Precambr Res 249:199–214

Liu PJ, Xiao SH, Yin CY, Chen SM, Zhou CM, Li M (2014b) Ediacaran acanthomorphic acritarchs and other microfossils from chert nodules of the upper Doushantuo Formation in the Yangtze Gorges area, South China. J Paleontol 88(Sp 72):1–139

Liu AG, Kenchington CG, Mitchell EG (2015) Remarkable insights into the paleoecology of the Avalonian Ediacaran macrobiota. Gondwana Res 27(4):1355–1380

Luo HL, Wu XC, Ouyang L, Jiang ZW, Song XL (1988) New correlation opinions on the sections of Sinian-Cambrian boundary in the Yangtze Platform. Yunnan Geol 7(1):13–27 (in Chinese with English abstract)

McCall GJH (2006) The Vendian (Ediacaran) in the geological record: enigmas in geology's prelude to the Cambrian explosion. Earth Sci Rev 77(1–3):1–229

McFadden KA, Xiao S, Zhou C, Kowalewski M (2009) Quantitative evaluation of the biostratigraphic distribution of acanthomorphic acritarchs in the Ediacaran Doushantuo Formation in the Yangtze Gorges area, South China. Precambrian Res 173(1–4):170–190

Moczydłowsk M, Vidal G, Rudavskaya VA (1993) Neoproterozoic (Vendian) Phytoplankton from the Siberian Platform, Yakutia. Palaeontology 36:495–521

Moczydłowska M, Nagovitsin KE (2012) Ediacaran radiation of organic-walled microbiota recorded in the Ura Formation, Patom Uplift, East Siberia. Precambr Res 198–199:1–24

Narbonne GM (2005) The ediacarabiota: Neoproterozoic origin of animals and their ecosystems. Annu Rev Earth Planet Sci 33:421–442

Narbonne G, Xiao SH, Shields G (2012) The Ediacaran Period. In: Gradstein F et al (eds) The geologic time scale 2012, vol 1. Elsevier, Amsterdam, pp 413–435

Ouyang Q, Zhou CM, Guan CG, Wang W (2015) New microfossils from the Ediacaran Doushantuo Formation in the Yangtze Gorges area, South China, and their biostratigraphic implications. Acta Palaeontologica Sinian 54(2):207–229

Peng SC, Wang XF, Xiao SH, Tong JN, Hua H, Zhu MY, Zhao YL (2012) A call to replace the chronostratigraphic unit Sinian System (Period) with the global Ediacaran System (Period). J Stratigr 36(1):55–59 (in Chinese with English abstract)

Sawaki Y, Ohno T, Tahata M, Komiya T, Hirata T, Maruyama S, Windley B, Han J, Shu D, Li Y (2010) The Ediacaran radiogenic Sr isotope excursion in the Doushantuo Formation in the Three Gorges area, South China. Precambrian Res 176(1–4):46–64

Sergeev VN, Knoll AH, Vorob'Eva NG (2011) Ediacaran microfossils from the Ura Formation, Baikal-Patom Uplift, Siberia: taxonomy and biostratigraphic significance. J Paleontol 85(5):987–1011

Shang XD, Liu PJ, Yang B, Chen SM, Wang CC (2016) Ecology and phylogenetic affinity of the Early Cambrian tubular microfossil Megathrix longus. Palaeontology 59(1):13–28

Shang XD, Liu PJ, Moczydłowska M (2019) Acritarchs from the Doushantuo Formation at Liujing section in Songlin area of Guizhou Province, South China: implications for Early-Middle Ediacaran biostratigraphy. Precambr Res 334:105453

Shen YA, Zhang TG, Chu XL (2005) C-isotopic stratification in a Neoproterozoic postglacial ocean. Precambr Res 137(3–4):243–251

Shen B, Xiao SH, Zhou CM, Yuan XL (2009) Yangtziramulus zhangi new genus and species, a carnonate-hosted macrofossil from the Ediacaran Dengying Formation in the Yangtze Gorges area, south China. J Paleontol 83(4):575–587

Shukla R, Tiwari M (2014) Ediacaran acanthomorphic acritarchs from the Outer Krol Belt, Lesser Himalaya, India: their significance for global correlation. Palaeoworld 23:209–224

Steiner M, Erdtmann BD, Chen JY (1992) Preliminary assessment of new Late Sinian (Late Proterozoic) large siphonous and Filamentous "megaalgae" from eastern Wulingshan, North-central Hunan, China. Berhner Geowissenschafiiche Abhandlungen (e) 3:305–319

Sun WG (1986) Late precambrian pennatulids (sea pens) from the eastern Yangtze Gorge, China: paracharnia gen. nov. Precambrian Res 31(4):361–375

Tang F, Song XL, Yin CY, Liu PJ, Awramik SM, Wang ZQ, Gao LZ (2006a) Discoveries of Longfengshaniaceae from the uppermost Ediacaran (Sinian) in eastern Yunnan, South China and significances. Acta Geol Sin 80(11):1643–1649 (in Chinese with English abstract)

Tang F, Yin CY, Bengtson S, Liu YQ, Wang ZQ, Liu PJ, Gao LZ (2006b) A new discovery of macroscopic fossils from the Ediacaran Doushantuo formation in the Yangtze Gorges area. Chin Sci Bull 51(12):1487–1493

Tang F, Yin CY, Liu PJ, Wang ZQ, Gao LZ (2007) Discovery of diverse macrofossil assembalges from the Jiucheng Member of uppermost Ediacaran in eastern Yunnan. J Palaeogeogr 9(5):533–540 (in Chinese with English abstract)

Tang F, Yin CY, Bengtson S, Liu PJ, Wang Z, Gao LZ (2008) Octoradiate spiral organisms in the Ediacaran of South China. Acta Geologica Sinica Engl Ed 82(1):27–34

Tang F, Yin CY, Liu PJ, Gao LZ, Wang ZQ (2009) Neoproterozoic macrofossil records in South China and biostratigraphic successions and correlations. Acta Geoscientica Sinica 30(4):505–522 (in Chinese with English abstract)

Tang F, Bengtson S, Wang Y, Wang XL, Yin CY (2011) Eoandromeda and the origin of Ctenophora. Evol Dev 13(5):408–414

Tiwari M, Knoll A (1994) Large acanthomorphic acritarchs from the Infrakrol Formation of the Lesser Himalaya and their stratigraphic significance. J Himal Geol 5(2):193–201

Tiwari M, Pant C (2004) Neoproterozoic silicified microfossils in Infra Krol Formation, Lesser Himalaya. India. Himalayan Geology 25(1):1–21

Veis AF, Vorob'eva NG, Golubkova EY (2006) The early Vendian microfossils first found in the Russian Plate: taxonomic composition and biostratigraphic significance. Stratigr Geol Correl 14(4):368–385

Vidal G (1990) Giant acanthomorph acritarchs from the upper proterozoic in Southern Norway. Palaeontology 33(2):287–298

Vorob'eva NG, Sergeev VN, Semikhatov M (2006) Unique lower Vendian Kel'tma microbiota, Timan ridge: new evidence for the paleontological essence and global significance of the Vendian system. Doklady Earth Sci 410(3):366–371

Vorob'eva NG, Sergeev VN, Chumakov N (2008) New finds of early Vendian microfossils in the Ura Formation: revision of the Patom Supergroup age, Middle Siberia. Doklady Earth Sci 419(6):782–787

Vorob'eva NG, Sergeev VN, Knoll AH (2009b) Neoproterozoic microfossils from the northeastern margin of the East European Platform. J Paleontol 83(2):161–196

Vorob'eva NG, Sergeev VN, Knoll AH (2009a) Neoproterozoic microfossils from the margin of the East European Platform and the search for a biostratigraphic model of Lower Ediacaran rocks. Precambrian Res 173(1–4):163–169

Wan B, Yuan XL, Chen Z, Guan CG, Pang K, Tang Q, Xiao SH (2016) Systematic description of putative animal fossils from the Early Ediacaran Lantian Formation of South China. Palaeontology 59(4):515–532

Wang ZQ, Yin CY, Gao LZ, Liu YQ (2002) Chemostratigraphic characteristics and correlationi of the Sinian stratotype in the eastern Yangtze Gorges area, Yichang Hubei Province. Geol Rev 48(7):408–415 (in Chinese with English abstract)

Wang Y, Wang XL, Huang YM (2008) Megascopic symmetrical metazoans from the Ediacaran Doushantuo Formation in the northeastern Guizhou, South China. Earth Sci J China Univ Geosci 19(3):200–206 (in Chinese with English abstract)

Willman S, Moczydlowska M (2008) Ediacaran acritarch biota from the Giles 1 drillhole, Officer Basin, Australia, and its potential For biostratigraphic correlation. Precambr Res 162(3–4):498–530

Willman S, Moczydłowska M (2011) Acritarchs in the Ediacaran of Australia—local or global significance? Evidence from the Lake Maurice West 1 drillcore. Rev Palaeobot Palynol 166(1–2):12–28

Willman S, Moczydlowska M, Grey K (2006) Neoproterozoic (Ediacaran) diversification of acritarchs—a new record from the Murnaroo 1 drillcore, eastern Officer Basin, Australia. Rev Palaeobot Palynol 139(1–4):17–39

Xiao SH (2004) New multicellular algal fossils and acritarchs in Doushantuo chert nodules (Neoproterozoic; Yangtze Gorges, south China). J Paleontol 78(2):393–401

Xiao SH, Knoll AH (1999) Fossil preservation in the Neoproterozoic Doushantuo phosphorite Lagerstatte, South China. Lethaia 32(3):219–240

Xiao SH, Knoll AH (2000) Phosphatized animal embryos from the Neoproterozoic Doushantuo Formation at Weng'An, Guizhou, South China. J Paleontol 74(5):767–788

Xiao SH, Zhang Y, Knoll AH (1998) Three-dimensional preservation of algae and animal embryos in a Neoproterozoic phosphorite. Nature 391(6667):553–558

Xiao SH, Yuan XL, Knoll AH (2000) Eumetazoan fossils in terminal Proterozoic phosphorites? Proc Natl Acad Sci USA 97(25):13684–13689

Xiao SH, Yuan XL, Steiner M, Knoll AH (2002) Macroscopic carbonaceous compressions in a terminal Proterozoic shale: a systematic reassessment of the Miaohe biota, south China. J Paleontol 76(2):347–376

Xiao SH, Shen B, Zhou CM, Xie GW, Yuan XL (2005) A uniquely preserved Ediacaran fossil with direct evidence for a quilted bodyplan. Proc Natl Acad Sci USA 102(29):10227–10232

Xiao SH, Droser M, Gehling JG, Hughes IV, Wan B, Chen Z, Yuan XL (2013) Affirming life aquatic for the Ediacara biota in China and Australia. Geology 41(10):1095–1098

Xiao SH, Zhou CM, Liu PJ, Wang D, Yuan XL (2014) Phosphatized acanthomorphic acritarchs and related microfossils from the Ediacaran Doushantuo Formation at Weng'an (South China) and their implications for biostratigraphic correlation. J Paleontol 88(1):1–67

Xiao SH, Narbonne GM, Zhou CM, Laflamme M, Grazhdankin DV, Moczydłowska-Vidal M, Cui H (2016) Toward an Ediacaran time scale: problems, protocols, and prospects. Episodes 39(4):540–555

Xiao SH, Hagadorn JW, Zhou C, Yuan (2007) Rare helical spheroidal fossils from the Doushantuo Lagerstatte: Ediacaran animal embryos come of age? Geolody 35(2):115–118

Xie GW, Zhou CM, Xiao MKA, S H. Yuan X L. (2008) Microfossils discovered from the Sinian Doushantuo Formation in the Jiulongwan section, east Yangtze Gorges area, Hubei Province, south China. Acta Palaeontologica Sinica 47(3):279–291 (in Chinese with English abstract)

Xing YS, Bi ZG, Wang XF (1985) A discovery of macroscopic fossil algae from Sinian in southern Anhui, China. Bull Inst Geol Chin Acad Geol Sci 12:32

Xing YS, Yin CY, Gao LZ (1999) Boundaries and subdivision of the Sinian System. Geoscience 16(2):202–204 (in Chinese)

Xue YS, Tang TF (1995) Large spheroidal chlorophyte fossils from Doushantuo Formation phosphoric sequence (Late Sinian), central Guizhou, South China. Acta Palaeontologica Sinica 34(6):688–706 (in Chinese with English abstract)

Yang JD, Sun WG, Wang ZZ, Xue YS, Tao XC (1999) Variations in Sr and C isotopes and Ce anomalies in successions from China: evidence for the oxygenation of Neoproterozoic seawater? Precambr Res 93(2–3):215–233

Yang B, Steiner M, Zhu MY, Li GX, Liu JN, Liu PJ (2016) Transitional Ediacaran-Cambrian small skeletal fossil assemblages from South China and Kazakhstan: implications for chronostratigraphy and metazoan evolution. Precambr Res 285:202–215

Yang C, Zhu MY, Condon DJ, Li XH (2017) Geochronological constraints on stratigraphic correlation and oceanic oxygenation in Ediacaran-Cambrian transition in South China. J Asian Earth Sci 140:75–81

Yin LM (1985) Microfossils of the Doushantuo Formation in the Yangtze Gorge District, Western Hubei. Palaeontologia Cathayana 2:229–249

Yin LM (1986) Sinian microfossil plants from the Yangtze region. J Stratigr 4:262–269 (in Chinese with English abstract)

Yin CY (1990) Spinose Acritarchs from the Doushantuo Formation in the Yangtze Gorges and its Geological Signifience. Acta Micropalaeontologica Sinica 7(3):265–270 (in Chinese with English abstract)

Yin CY (1996) New discovery of the Sinian Doushantuo microfossils from Miaohe, Zigui, Hubei. Acta Geosinica Sinica 17(3):322–329 (in Chinese with English abstract)

Yin LM, Li ZP (1978) Precambrian microfossils of southwest China, with reference to their stratigraphical significance. Mem Nanjing Inst Geol Palaeontol Academia of Sinica 10:41–108 (in Chinese with English abstract)

Yin CY, Bengtson S, Yue Z (2004) Silicified and phosphatized Tianzhushania, spheroidal microfossils of possible animal origin from the Neoproterozoic of South China. Acta Palaeontol Pol 49(1):1–12

Yin LM, Zhu MY, Knoll AH, Yuan XL, Zhang JM, Hu J (2007) Doushantuo embryos preserved inside diapause egg cysts. Nature 446(7136):661–663

Yin LM, Zhou CM, Yuan XL (2008) New data on Tianzhushania—an Ediacaran diapause egg cyst from Yichang, Hubei. Acta Palaeontologica Sinica 47(2):129–140 (in Chinese with English abstract)

Yin CY, Liu PJ, Chen SM, Tang F, Gao LZ, Wang ZQ (2009) Acritarch biostratigraphic succession of the Ediacaran Doushantuo Formation in the Yangtze Gorges. Acta Palaeontologica Sinica 48(2):146–154 (in Chinese with English abstract)

Yin CY, Liu PJ, Awramik SM, Chen SM, Tang F, Gao LZ, Wang ZQ, Riedman LA (2011a) Acanthomorph biostratigraphic succession of the Ediacaran Doushantuo Formation in the East Yangtze Gorges. South China Acta Geologica Sinica-English Edition 85(2):283–295

Yin LM, Wang D, Yuan XL, Zhou CM (2011b) Diverse small spinose acritarchs from the Ediacaran Doushantuo Formation, South China. Palaeoworld 20:279–289

Yin ZJ, Zhu MY, Tafforeau P, Chen JY, Liu PJ, Li G (2013) Early embryogenesis of potential bilaterian animals with polar lobe formation from the Ediacaran Weng'an biota, South China. Precambr Res 225:44–57

Yin ZJ, Zhu MY, Davidson EH, Bottjer DJ, Zhao FC, Tafforeau P (2015) Sponge grade body fossil with cellular resolution dating 60 Myr before the Cambrian. Proc Natl Acad Sci 112(12):E1453–E1460

Yin ZJ, Zhu MY, Bottjer DJ, Zhao FC, Tafforeau P (2016) Meroblastic cleavage identifies some Ediacaran Doushantuo (China) embryolike fossils as metazoans. Geology 44(9):735–738

Yin CY, Liu GZ (1988) Micropalaeofloras of the Sinian System of Hubei. In: Zhao Z et al (eds) The Sinian System of Hubei. China University of Geoscience Press, Wuhan, pp 90–100, 170–180 (in Chinese)

Yin LM (1987) Microbiotas of latest Precambrian sequences in China. In: Nanjing Institute of Geology and Palaeontology, Academia Sinica (ed) Stratigraphy and palaeontology of systemic boundaries in China, Precambrian-Cambrian Boundary, vol 1. Nanjing University Publishing House, Nanjing, pp 415–494

Yuan XL, Hofmann HJ (1998) New microfossils from the Neoproterozoic (Sinian) Doushantuo Formation, Wengan, Guizhou Province, southwestern China. Alcheringa 22(3–4):189–222

Yuan XL, Wang QF, Zhang Y (1993) Late Precambrian Weng'an biota from Guizhou, Southwest China. Acta Micropalaeontologica Sinica 10(4):409–420 (in Chinese with English abstract)

Yuan XL, Xiao SH, Taylor TN (2005) Lichen-like symbiosis 600 million years ago. Science 308(5724):1017–1020

Yuan XL, Chen Z, Xiao SH, Zhou CM, Hua H (2011) An Early Ediacaran assemblage of macroscopic and morphologically differentiated eukaryotes. Nature 470(7334):390–393

Zang WL, Walter MR (1992) Late Proterozoic and Cambrian microfossils and biostratigraphy, Amadeus Basin, central Australia. Assoc Australas Palaeontol Mem 12:1–132

Zhang ZY (1981) Precambrian microfossils from the Sinian of South China. Nature 289:792–793

Zhang ZY (1984) A new microphytoplankton species from the Sinian of western Hubei Province. Acta Botanica Sinian 26(1):94–98 (in Chinese with English abstract)

Zhang Y, Yin LM, Xiao SH, Knoll AH (1998) Permineralized fossils from the terminal Proterozoic Doushantuo Formation, south China. J Paleontol 72(4):1–52

Zhao YL, He MH, Chen MW, Peng J, Yu MY, Wang Y, Yang RJ, Wang PL, Zhang ZH (2004) Discovery of a Miaohe-type Biota from the Neoproterozoic Doushantuo formation in Jiangkou County, Guizhou Province, China. Chin Sci Bull 49(20):2224–2226

Zhao YL, Wu MY, Peng J, Yang XL, Yang RJ, Yang YN (2010) Triridged bobe fossils from the Miaohe biota from the Edicaran Doushantuo Formation from Jiangkou County, Guizhou Province, SW China. Acta Micropalaeontologica Sinica 27(4):305–314 (in Chinese with English abstract)

Zhao ZQ, Xing YS, Ding QX, Liu GZ, Zhao YX, Zhang SS, Meng XY, Yin CY, Ning BR, Han PG (1988) The Sinian System of Hubei. China University of Geoscience Press, Wuhan (in Chinese)

Zhou CM, Xiao SH (2007) Ediacaran δ^{13}C chemostratigraphy of South China. Chem Geol 237(1–2):89–108

Zhou CM, Tucker R, Xiao SH, Peng ZX, Yuan XL, Chen Z (2004) New constraints on the ages of Neoproterozoic glaciations in south China. Geology 32(5):437–440

Zhou CM, Xie GW, Xiao SH (2005) New data of microfossils from Doushantuo Formation at Zhangcunping in Yichang, Hubei Province. Acta Micropalaeontologica Sinica 22(3):217–224

Zhou CM, Xie GW, McFadden K, Xiao SH, Yuan XL (2007) The diversification and extinction of Doushantuo-Pertatataka acritarchs in South China: causes and biostratigraphic significance. Geol J 42(3–4):229–262

Zhou CM, Li XH, Xiao SH, Lan ZW, Ouyang Q, Guan CG, Chen Z (2017a) A new SIMS zircon U-Pb date from the Ediacaran Doushantuo Formation: age constraint on the Weng'an biota. Geol Mag 154(6):1193–1201

Zhou CM, Xiao SH, Wang W, Guan CG, Ouyang Q, Chen Z (2017b) The stratigraphic complexity of the Middle Ediacaran carbon isotopic record in the Yangtze Gorges area, South China, and its

implications for the age and chemostratigraphic significance of the Shuram excursion. Precambr Res 288:23–38
Zhu WQ, Chen ME (1984) On the discovery of macrofossil algae from the Late Sinian in the eastern Yangtze Gorges, South China. Acta Botanica Sinica 26(5):558–560 (in Chinese with English abstract)
Zhu MY, Zhang JM, Yang AH (2007) Integrated Ediacaran (Sinian) chronostratigraphy of South China. Palaeogeogr Palaeoclimatol Palaeoecol 254(1–2):7–61
Zhu MY, Gehling JG, Xiao SH, Zhao YL, Droser M (2008) Eight-armed Ediacara fossil preserved in contrasting taphonomic windows from China and Australia. Geology 36(11):867–870
Zhu B, Becker H, Jiang S, Pi D, Fischer-Gödde M, Yang J (2013a) Re-Os geochronology of black shales from the Neoproterozoic Doushantuo Formation, Yangtze platform, South China. Precambr Res 225:67–76
Zhu MY, Lu M, Zhang JM, Zhao FC, Li GX, Yang AH, Zhao X, Zhao MJ (2013b) Carbon isotope chemostratigraphy and sedimentary facies evolution of the Ediacaran Doushantuo Formation in western Hubei, South China. Precambr Res 225:7–28

Chapter 5
Neoproterozoic Stratigraphy, Depositional Environments and Hydrocarbon Source-Reservoir-Seal Bed Assemblage in South China

Maoyan Zhu, Junming Zhang, Aihua Yang, Guoxiang Li, Fangchen Zhao, Miao Lu, Zongjun Yin, Lanyun Miao, and Chunlin Hu

Abstract The well-exposed Neoproterozoic sedimentary successions of South China are rich in fossils and mineral resources, and have proven crucial to understanding Neoproterozoic stratigraphy, biological evolution and palaeoenvironmental changes. Based on analysis of the most complete Ediacaran (Sinian) successions in the Yangtze Gorges area and the pre-Ediacaran (Cryogenian and Tonian) succession in the deep water basin of the southeast Yangtze Craton (YC), a Neoproterozoic stratigraphic framework for South China is established and discussed, and recent advances plus remaining problems in the subdivision and correlation of the Neoproterozoic successions in South China are summarized. Also reviewed are the characteristics and spatiotemporal distribution of hydrocarbon source-reservoir-seal beds in the Neoproterozoic (mainly Ediacaran/Sinian) in this region.

Keywords Stratigraphy · Depositional facies · Neoproterozoic · Cryogenian · Ediacaran/Sinian · South China · Source-reservoir-seal beds

5.1 Introduction

The Neoproterozoic Era (1000–541 Ma) witnessed the transition of the marine biosphere from a primitive system characterized by microbes to a more complex and modern one dominated by multicellular life, including the first megascopic plants

M. Zhu (✉) · J. Zhang · G. Li (✉) · F. Zhao · M. Lu · Z. Yin · L. Miao · C. Hu
State Key Laboratory of Palaeobiology and Stratigraphy, Nanjing Institute of Geology and Palaeontology and Center for Excellence in Life and Palaeoenvironment, Chinese Academy of Sciences, Nanjing, China

M. Zhu · C. Hu
College of Earth and Planetary Sciences, University of Chinese Academy of Sciences, Beijing, China

A. Yang
State Key Laboratory for Mineral Deposits Research, School of Earth Sciences and Engineering, Nanjing University, Nanjing, China

T. Wang, *Meso-Neoproterozoic Geology and Petroleum Resources in China*, Springer Geology, https://doi.org/10.1007/978-981-19-5666-9_5

and animals (Butterfield 2011). Accompanying these revolutionary developments were intensive and wide-spread tectonic activities associated with the assembly and breakup of the supercontinent Rodinia and the subsequent formation of Gondwana (Li et al. 2008, 2013; Zhao et al. 2018). Additional changes at the global scale included dramatic fluctuations in the general climate (e.g., Snowball Earth; Hoffman et al. 1998; Hoffman and Schrag 2002), a rapid increase in oxygen levels in the atmosphere and ocean (e.g., Shields-Zhou and Och 2011; Lyons et al. 2014), and substantial changes in seawater chemistry and nutrient cycling (e.g., Knauth 2005; Komiya et al. 2008). At the same time, these dramatic physical and chemical events were unfolding, complex multicellular life underwent accelerating diversification prior to the exponential radiation of animals during the Early Cambrian (so-called "the Cambrian explosion"; Yin et al. 2007; Love et al. 2009; Yuan et al. 2011; Erwin et al. 2011; Yin et al. 2015; Van Iten et al. 2016; Darroch et al. 2018). Drastic modification of Neoproterozoic near surface systems probably involved complex interactions among the lithosphere, hydrosphere, atmosphere, and biosphere, though at this point extra-terrestrial factors cannot be ruled out entirely (e.g., Gaidos et al. 2007; Erwin and Valentine 2013; Maruyama et al. 2014; Zhang et al. 2014; Shields et al. 2019; Wood et al. 2019). In short, then, the co-evolution of Earth and life during the Neoproterozoic is a critical area of multidisciplinary research.

Establishment of a precise chronostratigraphic framework is fundamental to decipher the history of Earth-life system changes. However, precise subdivision and correlation of Precambrian sedimentary successions has proven difficult to achieve owing to these successions commonly being incomplete and intensely deformed and metamorphosed, and to the scarcity of age-diagnostic Precambrian fossils. For this reason, Global Standard Stratigraphic Ages (GSSAs) were introduced to subdivide Precambrian rocks on the basis of major magmatic-tectonic episodes (Plumb and James 1986; Plumb 1991). Although significant advances in Precambrian geochronology and time scale development have been achieved in recent decades (Gradstein et al. 2012), the present international Precambrian chronostratigraphic framework is poorly suited for global correlation, and thus regional and national chronostratigraphic schemes have been established. The Neoproterozoic Erathem is currently subdivided into the Tonian (1000–720 Ma), Cryogenian (720–635 Ma), and Ediacaran (635–541 Ma) systems (Gradstein et al. 2020); but only the Ediacaran System is defined in accordance with the principle of the global standard stratigraphic section and point (GSSP), which was ratified by the International Union of Geological Science (IUGS) in March 2004 (Knoll et al. 2004). Thus far no additional subdivisions of the three systems have been proposed. Therefore, the current Neoproterozoic chronostratigraphic framework hinders achievement of a better understanding of Earth-life system changes during this critical time interval.

Well-developed Neoproterozoic stratigraphic successions are widely exposed in China, where they have been studied for many decades. In the current Stratigraphic Chart of China (SCC) 2014, issued by the National Commission on Stratigraphy of China (NCSC), the three-fold subdivision of the Neoproterozoic Erathem is adopted, with the Qingbaikouan (1000–780 Ma), Nanhuan (780–635 Ma), and Sinian (635–541 Ma) systems being defined on the basis of stratotype sections in China (Table

5.1). These regional chronostratigraphic names have widely been used in the Chinese literature and appear as well in international publications. However, according to the International Stratigraphic Code (cf. Murphy and Salvador 1999), "chronostratigraphic units, as divisions of rock bodies based on geologic time, are in principle worldwide in extent, and important in providing a worldwide basis for communication and understanding". Since the Nanhuan and Sinian systems are defined by the same criteria as those employed in defining the Cryogenian and Ediacaran systems (Yin and Gao 2013), use of the Chinese regional names can be confusing and thus should be abandoned (Peng et al. 2012), a recommendation with which the present authors are in full agreement (Table 5.1). It should also be noted that the age of 780 Ma assigned to the base of the Nanhuan System in the SCC 2014 should be replaced by an age of 720 Ma.

Prior to the 1950s, the Neoproterozoic stratigraphic succession in South China, then named the Sinian System, was based on sections in the eastern Yangtze Gorges area and comprised the unmetamorphosed strata overlying the metamorphosed rock complex (Sandouping Group) and underlying the Cambrian strata with metazoan fossils (Lee and Chao 1924). However, in the peripheral portions of the YC, in particular the southeastern area, a thick succession underlying the Sinian succession is present (Table 5.1 and Fig. 5.1). The older succession, which is composed of volcanic-sedimentary rocks and known as the Banxi Group and equivalent strata, has generally been interpreted as being pre-Sinian in age (Liu et al. 1999). The unconformities at the base of the Sinian succession in the Yangtze Gorges area and at the base of the pre-Sinian Banxi Group were regarded as erosion surfaces resulting from the collision between the YC and Cathaysia Block (i.e., the Jinning Movement) around 900 Ma (e.g., Liu 1991; Liu et al. 1999; Xing et al. 1999;). The metamorphosed rock complex was treated as Mesoproterozoic or older.

During the past two decades, extensive field and laboratory investigations in South China have brought about improved understanding of the Neoproterozoic successions in this critical region. As shown in Table 5.1 and Fig. 5.1, the Neoproterozoic stratigraphic framework has been revised based on Ediacaran stratotype sections in the Yangtze Gorges area and pre-Ediacaran stratotype sections in the deep basin located along the southeastern margin of the YC.

The present chapter aims to summarize important advances in our knowledge and understanding of Neoproterozoic stratigraphy and sedimentary facies in South China. Also reviewed here are Neoproterozoic source-reservoir-seal bed assemblages which have played and will play important roles in deep hydrocarbon exploration in this region.

Table 5.1 Neoproterozoic stratigraphic chart of South China

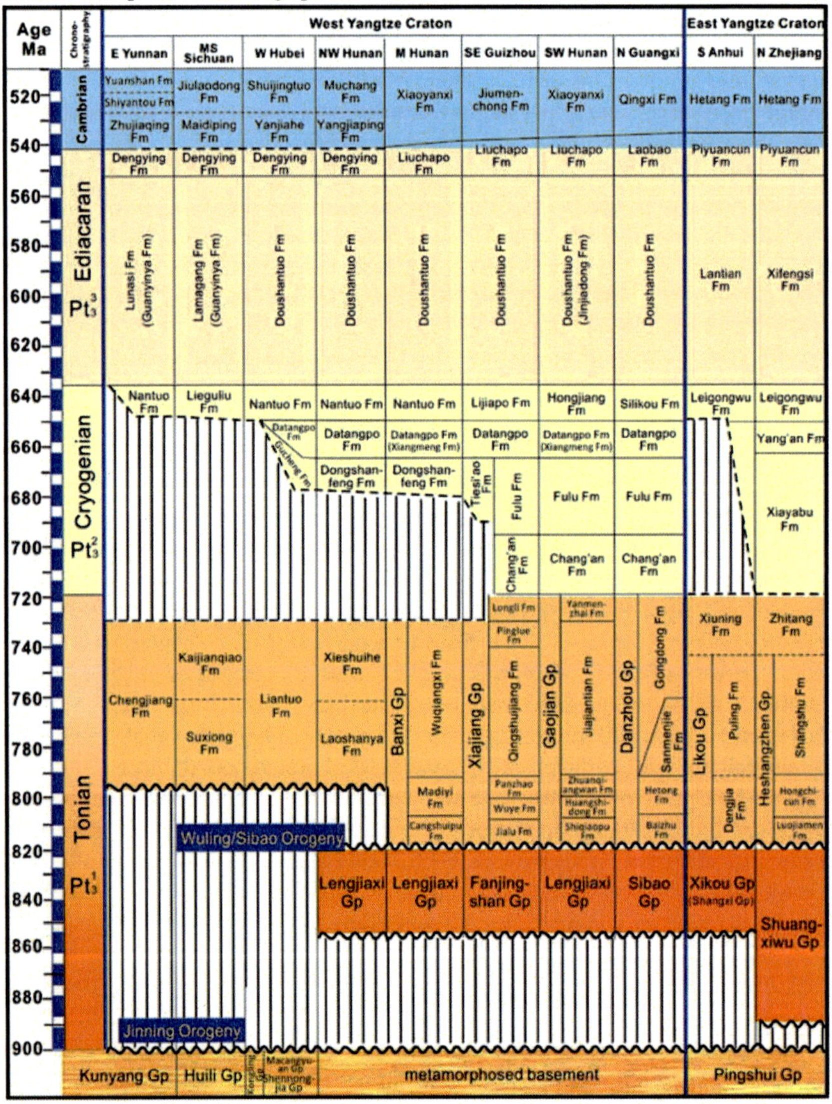

5.2 Neoproterozoic Stratigraphic Framework and Tectonic Background of South China

Owing to the complex tectonic–magmatic history of the South China, which formed during the Early Neoproterozoic through assembly of the YC (or called Yangtze Platform) and Cathaysia Block, Neoproterozoic successions in this region differ

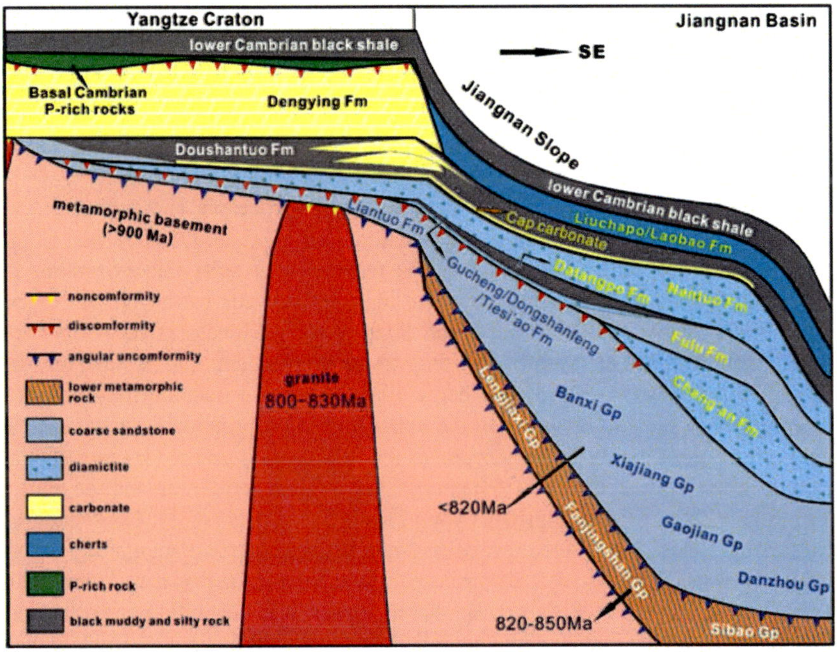

Fig. 5.1 Neoproterozoic tectostratigraphic framework of South China

substantially from each other (Table 5.1 and Fig. 5.1). In general, successions on the YC (or called Yangtze Platform) are less tectonized than others, but nevertheless are incomplete. In the outer shelf area of the YC and the Jiangnan Basin (also coined as the "South China Basin" in literature), the successions are complete stratigraphically but experienced more frequent tectonic disruption such as the mid-Palaeozoic Guangxi Movement and subsequent episodes of Mesozoic and Cenozoic deformation. The Neoproterozoic successions in the area of the Cathaysia Block (also known as the Zhujiang Basin in literature) underwent extensive deformation and metamorphism during multiple tectonic–magmatic events, and consequently they are poorly preserved and have received relatively little attention. For these reasons, the present chapter focuses on the Neoproterozoic successions of the YC and the Jiangnan Basin.

As shown in Table 5.1 and Fig. 5.1, two Early Neoproterozoic angular unconformities exist in South China. The first is associated with the collisional orogeny (i.e., the Jinning Movement) between the YC and Cathaysia Block, which ended at ca. 900 Ma. The metamorphosed rocks underlying the unconformity belong mainly to the Palaeoproterozoic to *c* Mesoproterozoic records and constitute the crystalline basement of the YC. The second unconformity, situated at the base of the Banxi Group and equivalent strata, resulted from another pulse of tectonic activity known as the Sibao or Wuling Movement. The successions between both angular unconformities are composed of weakly metamorphosed volcanic-sedimentary rocks collectively assigned to the Lengjiaxi, Fanjingshan, Sibao and Shuangqiaoshan Groups (Table

5.1). Geochronological data indicate that these rocks formed during 820–850 Ma (Gao et al. 2012), though with the Shuangqiaoshan Group in the eastern part of the YC dating to ca. 890 Ma (Li et al. 2009; Gao et al. 2011). It should be noted that the purported Neoproterozoic age of the Macaoyuan Group in the Shennongjia area of northwest Hubei, which unit was considered to be equivalent in age to the Sibao/Lengjiaxi Group, is supported by results of subsequent research.

At present, there is no consensus regarding the processes and timing of the assembly of the YC and Cathaysia Block (e.g., Li et al. 2009, 2012 and reference therein). The origins and ages of the two major angular unconformities in the basal Neoproterozoic successions are also unresolved. In the literature, the terms Jinning Movement (Orogeny) and Sibao/Wuling Movement commonly are used to refer to the same event. According to the original definitions, however, the Jinning Movement (ca. 900 Ma) is defined by the angular unconformity between the Neoproterozoic successions and the underlying Mesoproterozoic metamorphic rocks, while the Sibao/Wuling Movement (ca. 820 Ma) is marked by the angular unconformity between the Banxi/Xiajiang/Gaojian/Danzhou Group and the underlying Sibao/Lengjiaxi/Fanjingshan Group (Table 5.1 and Fig. 5.1). Consequently, whether the Sibao Group and correlative strata between the two angular unconformities represent deposits of a rift basin, back arc basin or active continental shelf remains uncertain (Li et al. 2012). Be that as it may, it is generally agreed that the Banxi Group and equivalent strata were deposited in a rift basin (i.e., the Jiangnan Basin) within the South China following the final assembly of the YC and Cathaysia Block (e.g. Wang 2000; Wang and Li 2003). The Banxi Group and equivalent strata lack obvious tectonic deformation, and therefore they constitute the early sedimentary cover of South China, which is the focus of the Early Neoproterozoic strata in the present chapter.

Shown in Fig. 5.1 is a conceptual model of the Neoproterozoic tectostratigraphic framework, sedimentary facies evolution and tectonic history of South China. The model highlights the large difference between the stratigraphic successions of the YC and the Jiangnan Basin. The thick pre-Ediacaran sedimentary cover, which is composed of coarse siliciclastic rocks, is developed mainly in the Jiangnan Basin, where it shows a rift basin deposit, as indicated by the presence of volcanic rocks and abundant tuffs in the Banxi Group and equivalent strata. A large stratigraphic gap exists below the Cryogenian or Ediacaran in most parts of the YC (Table 5.1 and Fig. 5.1), though with incomplete thin successions occurring in some areas. This gap has traditionally been interpreted as reflecting uplift of the YC (i.e., the Chengjiang or Xuefeng Movement); however, it now seems more likely that the gap is associated with the buildup of Cryogenian ice sheets and associated subglacial erosion (Wang 2000). Therefore, the hypothesis of a Chengjiang or Xuefeng Movement cannot explain the absence of a Cryogenian succession on the YC.

During the Ediacaran Period, the Jiangnan Basin became a deep water trough following the termination of rifting, while a carbonate platform formed on the YC. In contrast to the pre-Ediacaran successions, the Ediacaran succession in the Jiangnan Basin is composed of condensed fine-grained siliciclastic rocks, shales, and cherts.

The Ediacaran succession on the YC is therefore much thicker than that in the Jiangnan Basin.

5.3 Tonian Sedimentary Cover of South China

As conceived by the present authors, the Tonian sedimentary cover, the oldest in South China, comprises the stratigraphic succession between the Cryogenian succession and the unconformity at the top of the Sibao Group and equivalent strata. Tonian strata were deposited in rift basins bordering the margins of the YC (Li et al. 2009). As shown in Fig. 5.1 and discussed above, the Tonian successions exhibit dramatic changes along transects extending from the YC to the Jiangnan Basin. The stratigraphy and distribution of these successions may be summarized with reference to the following eight sections (Fig. 5.2): the Liantuo-Wangfenggang section in the Yangtze Gorges, western Hubei (Wang et al. 1980); the Yangjiaping section in northwest Hunan (Yin et al. 2004); the Guzhang section in western Hunan; the Yiyang-Taojiang section in Yiyang, central Hunan (Wang 2000); the Dengbao section in northeastern Guizhou (Zhu 1976); the Jinping Section in southeast Guizhou (Yang et al. 2012); the Huangshidong section in Qianyang (Hongjiang), western Hunan (Huang et al. 1996); and the Huangjin section in Luocheng, northern Guangxi (Yang et al. 2012).

5.3.1 General Characteristics and Spatial Distribution of the Banxi Group

Spatio-temporal variation in the Banxi Group and correlative successions (hereafter termed the Banxi successions) in South China are summarized in Fig. 5.2.

5.3.1.1 Stratigraphic Thickness

In general, the total thickness of the successions increases from the YC to the Jiangnan Basin. For example, the succession in the Yangtze Gorges area is less than 100 m thick and represented by the Liantuo Formation, which is composed of purplish conglomerates and sandstone overlying the Huangling Granite (Zhao et al. 1980). In contrast, the thickness of the succession in the Jiangnan Basin can exceed 10,000 m, as for example in the Jinping area of southeast Guizhou (Yang et al. 2012). The stratigraphic thickness decreases southeastward within the Jiangnan Basin, as for example in southwest Hunan and northern Guangxi, where the strata range from about 1000–2000 m-thick.

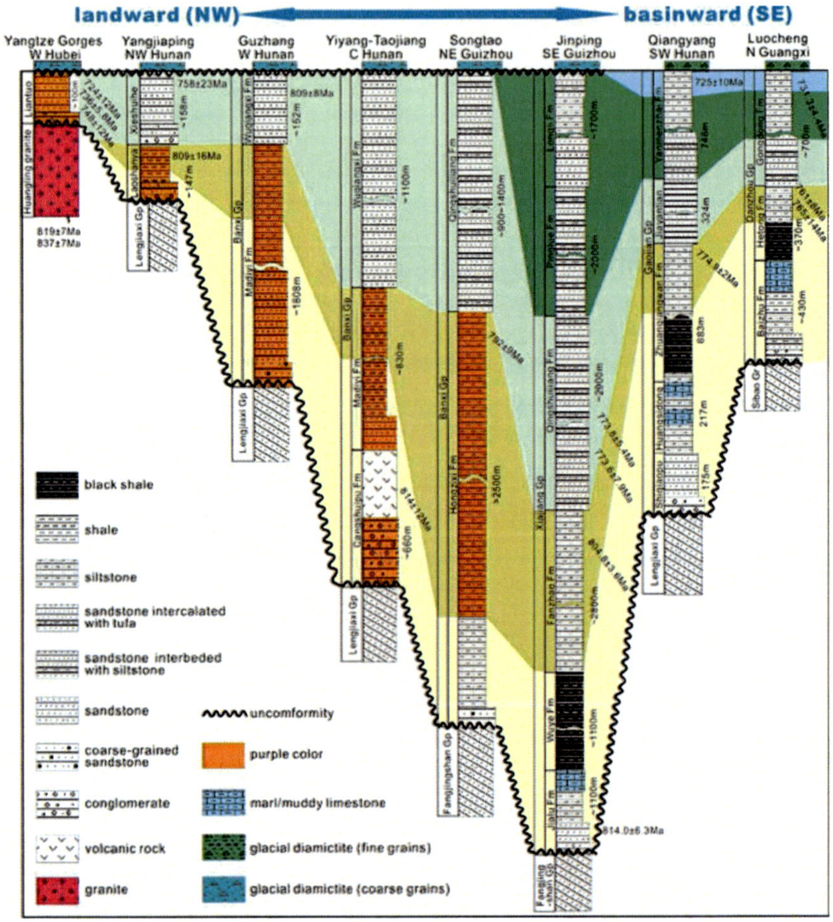

Fig. 5.2 Variation and correlation of the Tonian reference sections from the different sedimentary facies of South China

5.3.1.2 Sediment Grain Size and Color

Moving from the YC to the Jiangnan Basin, sediment grain size decreases, and the color changes from reddish or pinkish to grey or dark grey. These changes reflect deepening of sedimentary waters toward the Jiangnan Basin, with the YC as the sediment source area. The organic-rich muddy siltstone interval, which-measures from several hundreds to thousands of meters thick in the lower part of the Banxi/Xiajiang/Danzhou Group in southeast Guizhou, southwest Hunan and northern Guangxi, and was deposited in a deep water basin.

5.3.1.3 Two Sedimentary Cycles

Together, the Banxi successions comprise two fining upward sequences (Figs. 5.2 and 5.3). Although the successions become incomplete from the deep basin towards the YC, where the top of the successions is capped by an unconformity, both the sequences can still be recognized. In the Hupingshan area of northwest Hunan, both the sequences are represented in the Yangjiaping section by the Lhanya and Xieshuihe Formations, respectively (Liu et al. 1999). In western Hunan, the lower sequence is represented by the Madiyi Formation, which is composed of conglomerate and coarse sandstone in the lower part and purplish mudstone in the middle to upper parts. Volcanic conglomerates occur at the base of the Madiyi Formation in Yiyang and Taojiang in central Hunan. In central and western Hunan, the Wuqiangxi Formation records only the lower half of the upper sequence. Like the succession in western Hunan, the succession in northeastern Guizhou consists of the Hongzixi and Qingshuijiang Formations, which together represent both sequences. The Hongzixi Formation is composed mostly of purplish mudstone and tuffs with coarse sandstone in its basal part.

The Qingshuijiang Formation is composed predominantly of greywacke overlying a basal quartz sandstone. In the deep basin, in southeast Guizhou, southwest Hunan and northern Guangxi, the lower sequence starts with a basal conglomerate and sandstone grading upward into fine sediments. A carbonate interval occurs in the topmost part of the Jialu Formation (in Xiajiang Group), the Huangshidong Formation (in Gaojian Group) and the Baizhu Formation (in Danzhou Group), all of Huang

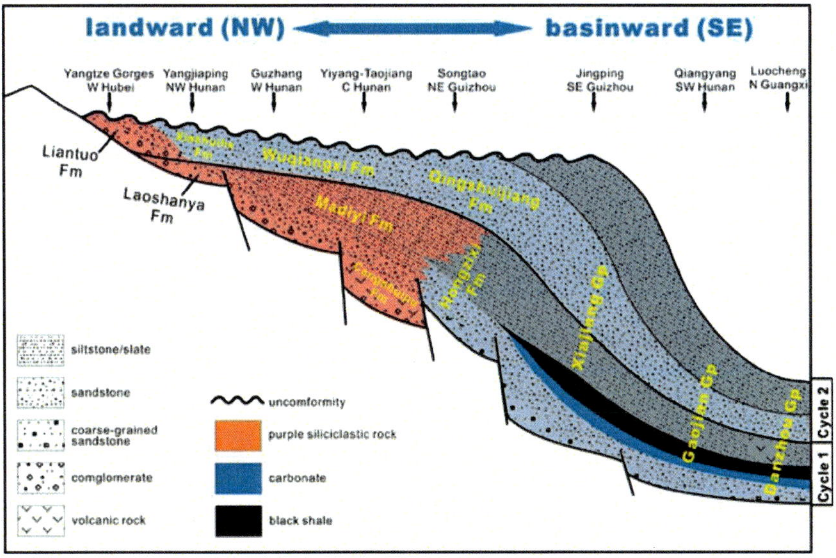

Fig. 5.3 Schematic diagram of the Tonian sedimentary facies model of South China

which represent the middle portion of the lower sequence. Overlying the carbonate-rich interval is a dark organic-rich interval of silty mudstone. The upper sequence begins with the sandstones of the Qingshuijiang Formation (in Xiajiang Group), the Jiayantian Formation (in Gaojian Group) and the Gongdong Formation (in Danzhou Group), and shows upward changes in grain size. Bouma sequences are common in the upper part of the sequence, and coarse sediments appear again near its top.

5.3.1.4 Volcanic Rocks and Tuffs

Volcanic conglomerates, basic and intermediate-acidic extrusive rocks, and hydrothermal features are common in the Jiangnan Basin. However, whereas volcanic rocks in the lower sequence of the upper part of the Cangshuipu Formation are terrestrial and intermediate-acidic, while those in the upper sequence in the Hetong Formation are bimodal and marine.

5.3.2 Sedimentary Facies and Depositional Environments

Liu et al. (1999) interpreted the thick Banxi successions as shallow water deposits, based on the absence of features diagnostic of flysch sequences. However, sedimentary structures indicative of deep water conditions are actually common in the Banxi successions in the Jiangnan Basin. For example, diagnostic Bouma sequence features are present in the upper part of the Gaojian and Danzhou Groups, suggesting deposition from deep water turbidity currents (Yang et al. 2012). During the past two decades, the nature of the Jiangnan Basin has been a hot research topic, and two alternative models, summarized below, have been proposed.

5.3.2.1 Rifting Model

According to this model, the two sedimentary cycles in the Banxi successions were deposited within a rift basin. The lower cycle represents the early rifting stage, when the basin was relatively small and characterized by terrestrial volcanism producing intermediate-acidic extrusive rocks. The upper cycle was formed during the intensive rifting stage, during which the basin grew larger, resulting in stratigraphic onlapping of the YC. The basin also underwent deepening, as indicated by the common occurrence of turbidites. The presence of marine bimodal volcanic rocks provide further support of the rifting model, as does the presence within the basin of active mutually aligned grabens, indicated by the spatio-temporal distribution of the Tonian successions (Wang 2000; Jiang et al. 2012; Wang and Li 2003).

5.3.2.2 Remnant Ocean Model

Under this model, the Jiangnan Basin was a remnant ocean basin opening to the west between the YC and Cathaysia Block. A corresponding volcanic arc system and deep forearc basin existed in the Jiangnan Basin. The spatio-temporal distribution of the Banxi successions was controlled by faulting associated with subduction and collision. The thickest part of the Xiajiang Group, located in southeast Guizhou between the Huaihua-Xinhuan and Huaihua-Jinzhou-Liping Faults, was deposited in a backarc basin. The area between the Huaihua-Jinzhou-Liping and Sanjiang-Rong'an Faults was an arc uplift, termed the Sibao Arc Uplift. The area on the east of the Sangjiang-Rong'an Fault and in the south of Shuangfeng, Hunan, where the succession is represented by the Gaojian and Danzhou Groups, was situated on a deep slope. The deepest part of the forearc basin was located in the area on the east of the Xinghua-Chengbu and Longsheng Faults, where the succession is represented by dark slates and cherts of the Dajiangbian Group (at Xintian) (e.g., Xu et al. 2012).

In the authors' opinion, the remnant ocean model suffers from serious deficiencies. In addition to uncertainty on the tectonic significance of the volcanic rocks, there is no sedimentary evidence to support the existence of the Sibao Arc Uplift.

Clearly, detailed studies of sedimentary basins and their tectonic framework are crucial for understanding the origins of sedimentary facies and successions as well as their stratigraphic subdivision and correlation. We believe that a simple sedimentary model best explains the spatio-temporal distribution of the Banxi successions in South China (Fig. 5.3). Specifically, in the first sedimentary cycle, the sediments are reddish or purplish in color and show an increase in average grain size toward the landward side. Within the basin, sediment color changes to grey or dark grey, and the average grain size and total thickness of the strata decrease. The dark organic-rich fine-grained interval represents the condensed sequence close to the maximum flooding surface. In the second sedimentary cycle, the succession on the YC consists of purplish coarse sediments; moving toward the basin, the sediments become darker and finer. The thickest succession was deposited on the continental shelf, where sedimentation rates were generally high. The existence of a continental shelf in southeast Guizhou is indicated by the common occurrence of turbidites in corresponding basinal deposits of the Gaojian and Danzhou Groups in southwest Hunan and northern Guangxi (Fig. 5.3).

5.3.3 *Stratigraphic Subdivision and Correlation*

The Banxi Group and equivalent successions in South China are readily discernible in the field. The base of the group coincides with the unconformity on top of the Sibao Group and equivalent successions, while its top is marked by the first occurrence of the glacial diamictite at the base of the Cryogenian System. However, it is difficult to further subdivide and correlate the Banxi successions because of the absence of reliable biostratigraphic markers. Key problems of interpretation include: ① the precise ages of the base and top of the Banxi and equivalent intervals; ② whether these surfaces are time planes or diachronous; and ③ whether there are any regional chronostratigraphic markers. Available geochronological data suggest that the base of the Banxi successions is younger than 820 Ma and that the upper surface is no older than 718 Ma (Ma et al. 1984; Yin et al. 2003; Wang et al. 2003; Gao et al. 2010; Ma et al. 2013; Du et al. 2013; Lan et al. 2014, 2015).

Given the apparent lack of bio- and chemostratigraphic markers, further subdivision and correlation of the Banxi successions must be based on sequence stratigraphy in combination with geochronological data. As discussed above, two major depositional cycles may be distinguished, and thus sequence boundaries and maximum flooding surfaces may be employed for these purposes (Figs. 5.2 and 5.3). The first marker horizon is the sequence boundary between the two cycles, which surface is situated at the base of multiple rock units in Jiangnan Basin, namely the Qingshuijiang Formation (in Xiajiang Group), the Jiayantian Formation (in Gaojian Group), the Gongdong/Sanmenjie Formation (in Danzhou Group) and the Xieshuihe Formation in northwest Hunan. A radiometric age of 800–760 Ma from bimodal volcanic rocks near the base of the second sequence provides an age constraint for the base of the second cycle. Because this cycle formed during the peak stage of rifting, thus it is recorded on the YC, where it is represented by the well-known and precisely dated Liantuo and Chengjiang Formations (Ma et al. 1984; Gao and Zhang 2009; Du et al. 2013; Lan et al. 2015). The Fanzhao Formation (in Xiajiang Group) was originally equated with the second sequence (Jiang et al. 2012), but this hypothesis is not supported by geochronological data (ca. 805 Ma), and therefore the Fanzhao Formation probably belongs the first sequence.

It should be noted here that Yin and Gao (2013) proposed an age of 780 Ma for the base of the Nanhuan System (Cryogenian System). This proposal was adopted by the Stratigraphic Chart of China 2014. Additionally, Yin and Gao (2013) assigned the Liantuo Formation in the Yangtze Gorges area to the Cryogenian, based on chemical index of alteration (CIA) data (Feng et al. 2004; Wang et al. 2006). However, given the absence of any evidence of glaciation in these rocks as well as a reported age of ca. 714 Ma for the top of the formation (Gao and Zhang 2009; Lan et al. 2015), it is highly doubtful that it is Cryogenian in age. Moreover, proper use of CIA data generally requires careful analysis of lithofacies and comparison of CIA data from different facies (e.g. Dobrzinski et al. 2004; Bahlburg and Dobrzinski 2011).

5.4 Cryogenian System of South China

In China, the Cryogenian System is termed the Nanhuan System, which was first defined and adopted as a system-level chronostratigraphic unit in the Stratigraphic Chart of China (SCC) in 2002. When this system was named, its upper boundary was placed at the base of the Ediacaran (Sinian) System, but the position of its basal boundary is uncertain because it was placed at the base of the Liantuo Formation in the Yangtze Gorges area, where Ediacaran strata rest nonconformably on the Huangling Granite. Also, an age of 800 Ma, rather than 850 Ma (the age of the base of the global Cryogenian System at that time) for the base of the Nanhuan System was adopted in the SCC of 2002. Because of this ambiguity, the position of the base of the Nanhuan System and thus its regional and global correlation have been widely debated (Lu 2002; Zhang et al. 2003, 2011; Yin et al. 2003, 2004; Peng et al. 2004; Wang 2005; Zhang and Chu 2006; Lin et al. 2010). However, in 2012 the International Commission on Stratigraphy (ICS) redefined the base of the Cryogenian System using the rock-based GSSP concept, placing it at the first occurrence of the oldest and globally recognizable glaciogenic deposits in agreement with the proposal of the International Subcommission on Neoproterozoic Stratigraphy (ISNS; Shields-Zhou et al. 2012). Shortly thereafter, it was decided that the base of the Nanhuan System in China should be placed at the bottom of the first glacial deposits (i.e., the Chang'an Formation) in South China with an age of 780 Ma (Yin and Gao 2013). For this reason, the term Cryogenian System instead of Nanhuan System is used here, as discussed above.

Updated geochronological data indicate that the base of the Cryogenian successions of South China is no older than 718 Ma (Lan et al. 2014) in agreement with ages from Oman and Laurentia (Bowring et al. 2007; Macdonald et al. 2010). Together these age data support the hypothesis of rapid initiation of the Cryogenian global glaciation at about 720 Ma (Hoffman et al. 1998, 2017). This date has replaced the previously accepted age of 850 Ma for the base of the Cryogenian System in the International Chronostratigraphic Chart (ICC; Shield-Zhou et al. 2016). Additional geochronological data for the Cryogenian successions of South China (Zhou et al. 2004; Liu et al. 2015; Wang et al. 2019; Rooney et al. 2020) indicate that the base of the Datangpo Formation is 660.98 ± 0.74 Ma and that the Nantuo Formation is no older than 657.17 ± 0.78 Ma. However, cyclostratigraphic data suggest that the Nantuo Formation may be younger than 651 Ma (Bao et al. 2018). Moreover, even though currently available radiometric age data aid in correlating the Cryogenian succession of South China with successions in other parts of the world, regional subdivision and correlation are complicated by substantial spatio-temporal changes, and thus remain unresolved (Zhang et al. 2011; Yin and Gao 2013; Lin et al. 2013).

Eight reference sections along a transect extending from the YC to the Jiangnan Basin are selected here to summarize major spatio-temporal variation in the Cryogenian successions of South China (Fig. 5.4). The sections are: the Yangtze Gorges section in Yichang, western Hubei; the Yangjiaping section in northwest Hunan (Liu et al. 1999); the Longbizui section near Guzhang, western Hunan; the composite section in Songtao, northeastern Guizhou (Wang et al. 1984; Xu et al. 1991; He 1997; Huang et al. 2010); the Hongjiang-Qiangyang section in western Hunan (Peng et al. 2004); the Lijiapo section in Congjiang, southeast Guizhou (Lu et al. 2010); and the Shiyan section in Sanjiang, northern Guangxi (Zhang et al. 2011). The Cryogenian successions on the eastern YC are poorly exposed and understood (Shi et al.

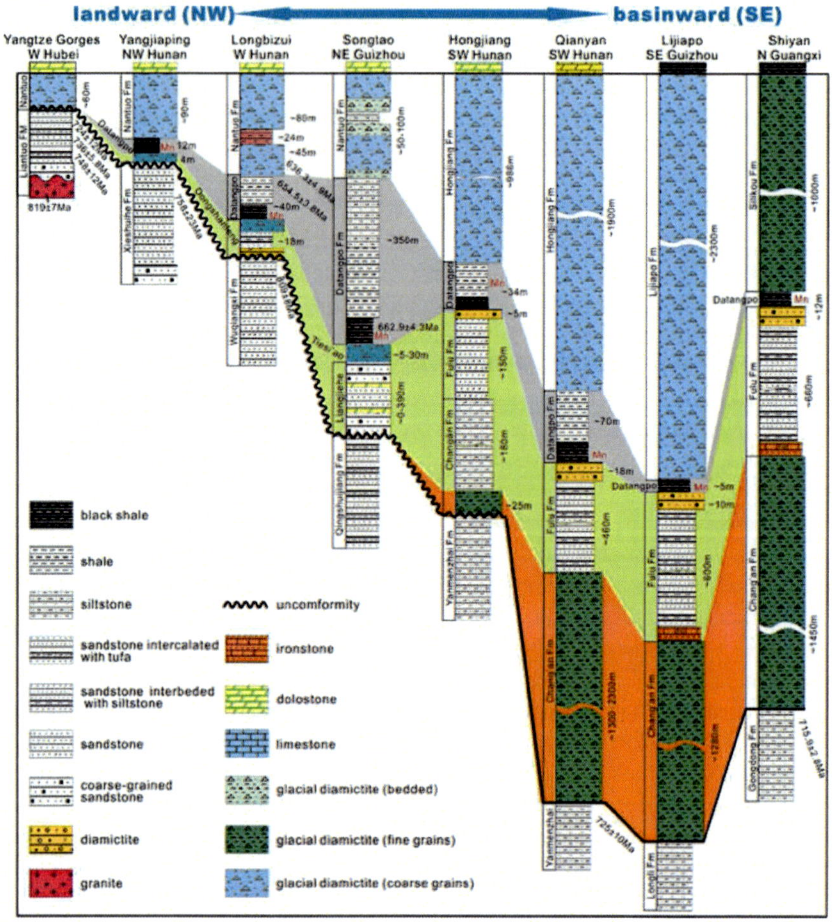

Fig. 5.4 Variation and correlation of the Cryogenian reference sections from different sedimentary facies of South China

1985; Wang and Bi 1985; Guan et al. 2012; Qian et al. 2012) and thus are not reviewed here.

5.4.1 *General Characteristics and Spatial Distribution*

The Cryogenian succession in South China is complete and consists of two sets of glacial diamictites and an interglacial interval. In the Jiangnan Basin, the total thickness of this succession exceeds 4000 m. In the deep basin area, the Cryogenian succession conformably overlies Tonian strata, as shown in the Zhaoxing section at

Liping, southeast Guizhou (Zhang and Chu 2006). However, the Cryogenian succession is incomplete, resting upon a sharp basal unconformity, in the area extending from the Jiangnan Basin to the YC, where the lower glacial diamictite is absent (e.g., in the Yangtze Gorges area). On the western YC, the Cryogenian succession usually consists of a few meters of diamictite or is absent.

In the Jiangnan Basin, the lower glacial deposits are represented by the Chang'an Formation, which consists of over 2200 m of massive clast-poor fine-grained and muddy diamictites. The rare clasts are predominantly millimetric in size. Stratified sandstone, siltstone and slate occur in the Chang'an Formation, while the overlying Fulu Formation consists of well-stratified coarse siliciclastic rocks. Also in the Fulu Formation, turbidites with complete Bouma sequence, as well as slump and fluid escape structures are common, and banded iron formation (BIF) is present in the basal part of the unit. A carbonate interval measuring 0.5–23 m in thickness is present in the Tongdao area of southwest Hunan (Lin et al. 2010). Present near the top of the Fulu Formation is an interval (<10 m), termed the Gucheng Member or "Gucheng Formation", that is composed of coarse-grained sandstone or granular diamictite showing high textural maturity (Peng et al. 2004; Lu et al. 2010; Zhang et al. 2012). The clasts in this interval are mostly millimetric. The same interval is discontinuous or absent in some of the sections in the deep basin. Within the Jiangnan Basin, the overlying Datangpo Formation is usually thin (ranging from a few meters to tens of meters thick), consisting of black shale in the basal part and greenish siltstone in the upper part. Manganese ore layers, including those exploited in mines in southeast Guizhou and southwest Hunan, are common near the base of the Datangpo Formation (Liu and Zhou 2002; Yang et al. 2010). The upper glacial diamictites in the Jiangnan Basin generally are clast-poor and muddy with siliceous mudstone beds. The sedimentary features of the upper glacial diamictites differ from those in the Nantuo Formation in the Yangtze Gorges area, with different lithostratigraphic unit names (e.g., the "Hongjiang Formation" in southwest Hunan, the "Lijiapo Formation" in southeast Guizhou and the "Silikou Formation" in northern Guangxi) were used by local geological surveys.

Present in the transitional area between the Jiangnan Basin and the YC are two sets of glacial diamictites and an interglacial interval. However, the lower diamictite shows substantial variation in thickness (ranging from tens of centimeters- to tens of meters-thick) and sedimentary features, and has been given different lithostratigraphic names, including the Gucheng Formation in Changyang of Hubei, the Dongshanfeng Formation in northwest Hunan and the Tiesi'ao Formation in northeastern Guizhou. The Tiesi'ao Formation, composed of massive diamictites with intercalations of sandstone and carbonate, rests unconformably on the Tonian Xiajiang Group (Wang et al. 1986; He 1998; Huang et al. 2010). In some localities, for example northeastern Guizhou, a coarse-grained sandstone (i.e., the Liangjiehe Formation) occurs between the Tiesi'ao Formation and the Xiajiang Group. However, recognition of the Liangjiehe and Tiesi'ao Formations in the field is difficult, and thus both the formations together have been referred to as the Tiesi'ao Formation. The Datangpo Formation consists of a manganese-rich lower black shale member and an upper member composed of greenish or muddy siltstone. The thickness of the Datangpo Formation and its manganese ore layers varies dramatically, with the formation attaining a maximum thickness of 300 m and the thickness of the manganese ore layers correlating positively with the total thickness of the formation (Xu et al. 2005; Huang et al. 2010). The Nantuo Formation diamictite is more clast-rich and contains larger clasts than that in the deep basin. The thickness of the Nantuo diamictite in the transitional area usually is less than 100 m. Finely laminated clast-free reddish shale or silt beds occur within the Nantuo diamictite. The base of the Nantuo diamictite is conformable in some areas, for example in the Wuhe section in eastern Guizhou, but unconformable in other areas such as Guzhang in western Hunan.

The Cryogenian succession is absent over much of the YC. Where present, it is usually incomplete and thin (<100 m), consisting entirely of the upper glacial deposits of the Nantuo Formation, which are composed of terrestrial or proximal clast-rich massive glacial diamictites. Glaciofluvial interbeds consisting of pebbly sandstone graded diamictite, or sandstone or siltstone are common.

5.4.2 Sedimentary Facies and Environments

The Cryogenian depositional basin in South China was similar to that of Late Tonian times (Fig. 5.5). Cryogenian sedimentation and facies evolution were controlled mainly by the tectonic evolution of the rift basin, with climatically controlled fluctuation in sea level as another governing factor. In recent decades, research on the Cryogenian succession of South China has focused on the following set of topics: the dynamics and timing of glaciations (Huang et al. 2016; Lang et al. 2018; Hu and Zhu 2020; Rooney et al. 2020; Yan et al. 2020), the redox evolution of the Jiangnan Basin (Wei et al. 2016, 2018; Cheng et al. 2018; Peng et al. 2019), the genesis of the banded iron formation (BIF) in the basal Fulu Formation (Busigny et al. 2018; Zhu et al. 2019) and the genesis of the manganese ores in the basal Datangpo Formation (An et al. 2014; Yu et al. 2016; Zhou et al. 2016). Because available data were acquired through various approaches applied to a limited number of sections and localities, the evolution of Cryogenian facies in South China remains poorly understood. Current understanding of these topics is summarized below.

Characteristic glaciomarine sedimentary features are common in the Chang'an Formation, which consists predominantly of massive clast-poor muddy diamictite with ice-rafted debris lenses and lonestones. The rare small clasts comprise a variety of lithologies. Turbidites and sandstones within the diamictite can be interpreted as underflow fan lobes. Together, all these features suggest that the Chang'an Formation diamictite is ice distal in origin.

Unlike the widespread Late Tonian succession, the Chang'an glacial diamictite is limited to the Jiangnan Basin. This may reflect the substantial drop in sea-level associated with severe Snowball-Earth-style glaciation during the Early Cryogenian. Recently, we identified distinct subglacial deformation structures underlying the base of the Tiesi'ao/Dongshanfeng Formation in northeastern Guizhou and western

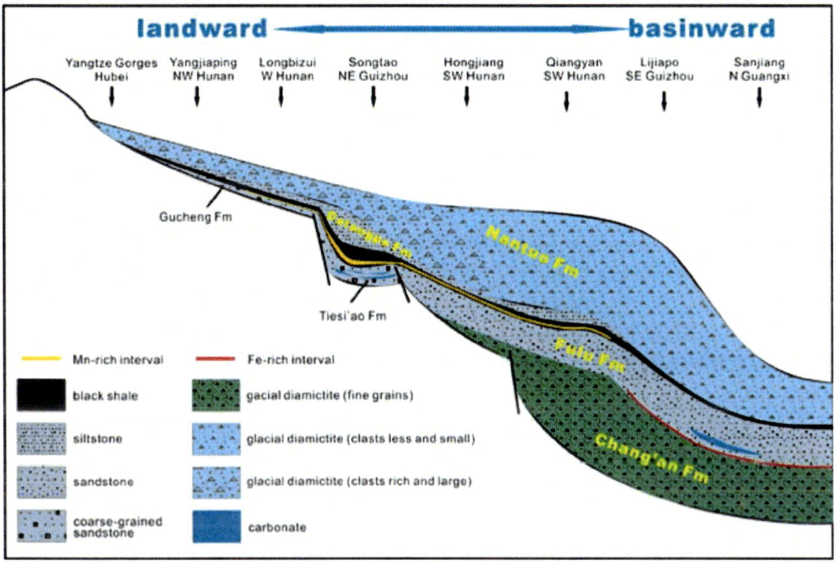

Fig. 5.5 Schematic diagram of the Cryogenian sedimentary facies model of South China

Hunan. This discovery suggests that the ice grounding line reached the slope area of the Jiangnan Basin, and thus provides strong support for the hypothesis of a major drop in global sea level associated with the buildup of an exceptionally thick ice sheet during the Sturtian Glaciation (Hu and Zhu 2020). In addition, facies analysis of the Tiesi'ao/Dongshanfeng Formation indicates that the facies association consisting of diamictite, pebbly sandstone, and sandstone represents the proximal glaciomarine environments of five successive episodes of glacial advance and retreat. This work suggests further that the Early Cryogenian (Sturtian) ice sheets in South China were warm based, rapidly moving, and sensitive to changes in climate.

Glaciogenic sedimentary features are readily discernible in the upper diamictite of the Nantuo Formation. In the proximal area, these features include massive clast-rich diamictite with a low maturity matrix and shattered-, striated- and facetted-clasts, and laminated muddy diamictite and siltstone with dropstones. The facies of the Nantuo Formation show pronounced changes along a transect extending from the proximal area to the deep basin (Fig. 5.5). Thus, moving from the YC to the Jiangnan Basin, the clast-rich coarse-grained diamictite changes to clast-poor muddy diamictite. In the proximal facies, glaciofluvial interbeds are common, while the deep basin hosts massive siliceous mudstones. Also, whereas the Nantuo Formation in the proximal area consists of terrestrial glacial deposits, the basinal Lijiapo or Silikou Formation is composed of glaciomarine deposits (Lin et al. 2013).

The interval between the upper and lower diamictites composed of sandstone, siltstone and black shale, and is generally thought to be interglacial in origin. However, it is unclear whether the sandstone interval in the Fulu Formation between the upper diamictite (Nantuo Formation) and the lower diamictite (Chang'an/Jiangkou Formation) represents interglacial or glacial deposits. Diagnostic sedimentary features such as dropstones and glacial diamictites have not been observed, and putative glacial diamictites have proven to be the Bouma sequence A in turbidites. A thin interval of manganese-rich black shale overlies the Fulu sandstone within the basin, and diamictite/conglomerate beds occur near the top of this unit elsewhere, and so it is likely that the Fulu Formation represents underflow fan lobes and glacial turbidites.

Results of numerous geochemical investigations indicate that the organic- and manganese-rich interval in the basal Datangpo Formation was deposited in a stratified ocean, with the formation of anoxic/euxinic bottom waters resulting from the rapid sea-level rise and high rate of nutrient supply following termination of the Chang'an glaciation (i.e., the Sturtian glaciation; Wei et al. 2016; Yu et al. 2016; Ye et al. 2018; Peng et al. 2019; Ai et al. 2021; Tan et al. 2021). The dramatic thickness variation of the Datangpo Formation in the shelf area was caused by extensional faulting (Zhou et al. 2016), indicating the continuation of rifting activity during the Cryogenian Period.

5.4.3 Stratigraphic Subdivision and Correlation

The Precambrian glaciogenic successions in South China were originally thought to record a single long-term glacial period, termed the "Nanhua great glacial age" (the Chinese term, "Nanhua", meaning South China), from which the name Nanhuan System/Period originated when it was formally adopted by the Stratigraphic Chart of China (SCC) in 2002. Wang et al. (1980) first subdivided the "Nanhua great glacial age" into two glacial periods separated by an interglacial phase, namely the Chang'an glacial, the Fulu interglacial and the Nantuo glacial periods.

This subdivision was revised by Lu et al. (1985), who maintained that the lower half of the Fulu Formation represents the glaciomarine deposits, only its upper manganese-rich interval represents an interglacial stage. Consequently, the revised subdivision comprises the Gucheng glacial, the Datangpo interglacial and the Nantuo glacial periods in ascending order with three designated stratotype sections, namely, the Gucheng section of the Gucheng Formation in Changyang County, Hubei for the Gucheng glacial period, the Datangpo section of the Datangpo Formation in Songtao County, northeastern Guizhou for the Datangpo interglacial period, and the Nantuo Formation in the eastern Yangtze Gorges area for the Nantuo glacial period respectively. Lu et al. (1985) concluded that the Gucheng Formation consists of ice-proximal massive morainic debris and tillite associated with laminated mudstone of a glacial lake. The authors concluded further that the time-equivalent Tiesi'ao Formation of the Gucheng glaciation in northeastern Guizhou consists of proximal glaciomarine deposits, while the Chang'an Formation in the deep basin consists of distal glaciomarine deposits with ice-rafted dropstones.

An alternative hypothesis is that the Gucheng glacial period is not synchronous with the Chang'an glacial period, and a corresponding subdivision consisting of three glacial periods and two interglacial periods was proposed (Peng et al. 2004). Listed in ascending order, these are the Chang'an glacial, the Fulu interglacial, the Gucheng glacial, the Datangpo interglacial and the Nantuo glacial periods. As noted above, the Fulu Formation may not consist of interglacial deposits. The Gucheng glaciation was thought to represent the short final stage of the Jiangkou glaciation, named after the Jiangkou Formation in southwest Hunan (Zhang and Chu 2006). According to Zhang and Chu (2006), the Jiangkou glacial sequence is equivalent to the Chang'an and Fulu Formations in northern Guangxi, the Jiangkou glacial period being an alternative term for the earlier proposed Chang'an glacial period. However, use of the term Chang'an glaciation for the first Cryogenian glaciation in South China has been widespread, and therefore this popular term should be retained.

Based on data derived from extensive geological mapping in Guizhou and adjacent areas, Lin et al. (2010, 2013) proposed a new subdivision of Cryogenian strata in South China. These authors argued that the Fulu Formation and overlying Gucheng and Datangpo Formations record a single long interglacial period with two cooler interludes, i.e., the Sanjiang sub-interglacial, the Longjia sub-glacial, the Lanyang sub-interglacial, the Liangjiehe (Gucheng) sub-glacial, and the Datangpo sub-interglacial stages in ascending order. As discussed above, however, the Fulu Formation is poorly understood, and thus additional work is needed to test Lin et al.'s (2010, 2013) model. Two key problems that must be resolved are whether the Gucheng glaciation is equivalent to the Chang'an glaciation and the nature of the depositional environments of the Fulu Formation.

It is generally agreed that the Gucheng/Dongshanfeng/Tiesi'ao Formations in the shelf area, which units show diagnostic features of glacial deposition, correlate with the top part of the Fulu Formation in the basin (Peng et al. 2004; Zhang et al. 2012; Lin et al. 2013). The succession in northeastern Guizhou is the key reference section, since in that area the Tiesi'ao Formation is underlain by the Liangjiehe Formation (Wang et al. 1984), which is thought to correlate with the Fulu Formation.

It is also unclear whether the Xieshuihe Formation in northwest Hunan is Cryogenian in age as argued by Zhang and Chu (2006). As in the case of the Liantuo Formation in the Yangtze Gorges area, CIA and geochronological data (758 Ma; Yin et al. 2003) for the Liangjiehe Formation must be reevaluated. A new SIMS zircon U–Pb age of 691.9 ± 8.0 Ma (MSWD = 1.3) from the upper Xieshuihe Formation supports an Early Cryogenian age for this unit. Because the age of the Xieshuihe Formation was interpreted as being Early Cryogenian, it was proposed that the Yangjiaping section, which exposes the Xieshuihe, Dongshanfeng and Nantuo Formations in northwest Hunan, be designated the stratotype section for the Cryogenian (Nanhuan) System in China (Yin et al. 2003). However, the Xieshuihe Formation lacks diagnostic glacial features, and thus Zhang and Chu (2006) suggested that the Zhaoxing section at Liping in southwest Guizhou be chosen instead. However, the Zhaoxing section is poorly exposed and disrupted by faulting, and so Lu et al. (2010) proposed that the Lishuping section in Congjiang of southeast Guizhou should be selected.

To summarize, the Cryogenian succession in South China is complete and well exposed. Subdivision and regional correlation of these strata can serve as a basis for global subdivision and correlation. Future integrated investigations of the Cryogenian sequences of South China will no doubt result in improved understanding of Earth system evolution during this critical time interval, and will also benefit exploration for manganese ores and hydrocarbon source rocks in South China.

5.5 The Ediacaran (Sinian) System of South China

The Ediacaran System in China has traditionally been named the Sinian System, which term has a long history featuring multiple revisions of its definition (cf. Zhu et al. 2007 and reference therein). Although the Sinian System is still adopted in the Stratigraphic Chart of China (SCC), its current definition is consistent with that of the Ediacaran System as used since 2001. In South China, the base of the Ediacaran System coincides with the base of the Doushantuo Formation, which surface coincides with the base of the cap carbonate overlying the Nantuo diamictite.

During the past two decades, intensive investigations have resulted in great advances in our understanding of Ediacaran stratigraphy, palaeobiology and palaeoenvironments (e.g. Li et al. 1998; Xiao et al. 1998, 2000; Chen et al. 2000, 2004, 2006, 2009; Jiang et al. 2003, 2011; Condon et al. 2005; Yin et al. 2007; Zhu et al. 2007a, b, 2008, 2013; McFadden et al. 2008; Li et al. 2010, 2016; Bristow et al. 2011; Yuan et al. 2011; Sahoo et al. 2012; Liu et al. 2013, 2014a, b; Chen et al. 2014, 2018, 2019; An et al. 2015). Stratigraphic studies of the Ediacaran System in China have focused on sections in western Hubei and adjacent areas, particularly the Yangtze Gorges, where the stratotype section of the Ediacaran/Sinian System is located. During the Ediacaran Period, the tectonic-palaeogeographic framework of South China changed substantially. Specifically, the Jiangnan Basin became a deep basin, while the YC gradually became a carbonate platform. During formation of the carbonate platform, the transitional area between the YC and the Jiangnan Basin became an unstable slope with highly variable and incomplete stratigraphic sequences. This variation is reflected in the variety of lithostratigraphic unit names applied to the component facies (Table 5.2).

5.5.1 *General Characteristics and Spatial Distribution*

Ten reference sections are chosen here to summarize facies variation in the Ediacaran successions of South China (Fig. 5.6).

Table 5.2 The Ediacaran stratigraphic chart of South China

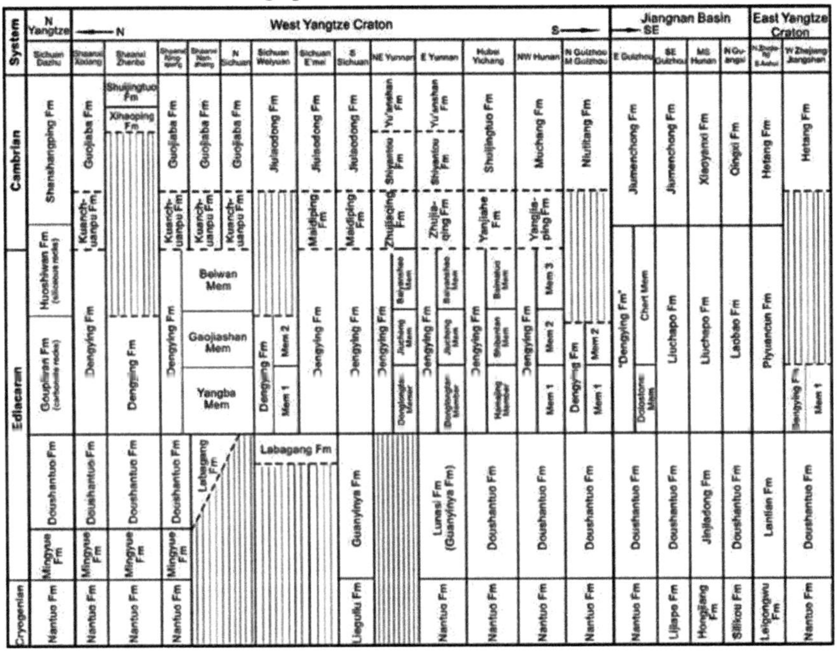

5.5.1.1 Dazhu Section, Wanyuan, Northeastern Sichuan

The slope facies of northwestern YC. Overlying Muchang Formation, the Cryogenian diamictite is a coarse siliciclastic rock interval, the so-called "Mingyue Formation". Above this sequence is a black shale interval with lenticular carbonate interbeds, which is similar to the Dou-2 Member of Doushantuo Formation in the Yangtze Gorges area. In the Dabashan area (northeastern Sichcuan and southeast Shaanxi), the black shale interval contains manganese-rich carbonate and manganese ore layers. The Goupiwan Formation, composed of limestone and dolostone, is overlain by the Huoshiwan Formation, which is composed of thin-bedded black cherts with thin shale interbeds. No cap carbonate has been reported between the coarse siliciclastic interval at the base of the Ediacaran System and the underlying Cryogenian diamictite, neither in the northwest slope area nor in the shallow water facies on the western YC.

5.5.1.2 Yangba Section, Nanjiang, Northwest Sichuan

The first type of nearshore platform facies succession on the western YC, but this Ediacaran succession is incomplete. The basal unit is composed of coarse siliciclastic rocks assigned to the Labagang Formation, which rests unconformably on Neoproterozoic granites or metamorphic basement rocks. The overlying thick

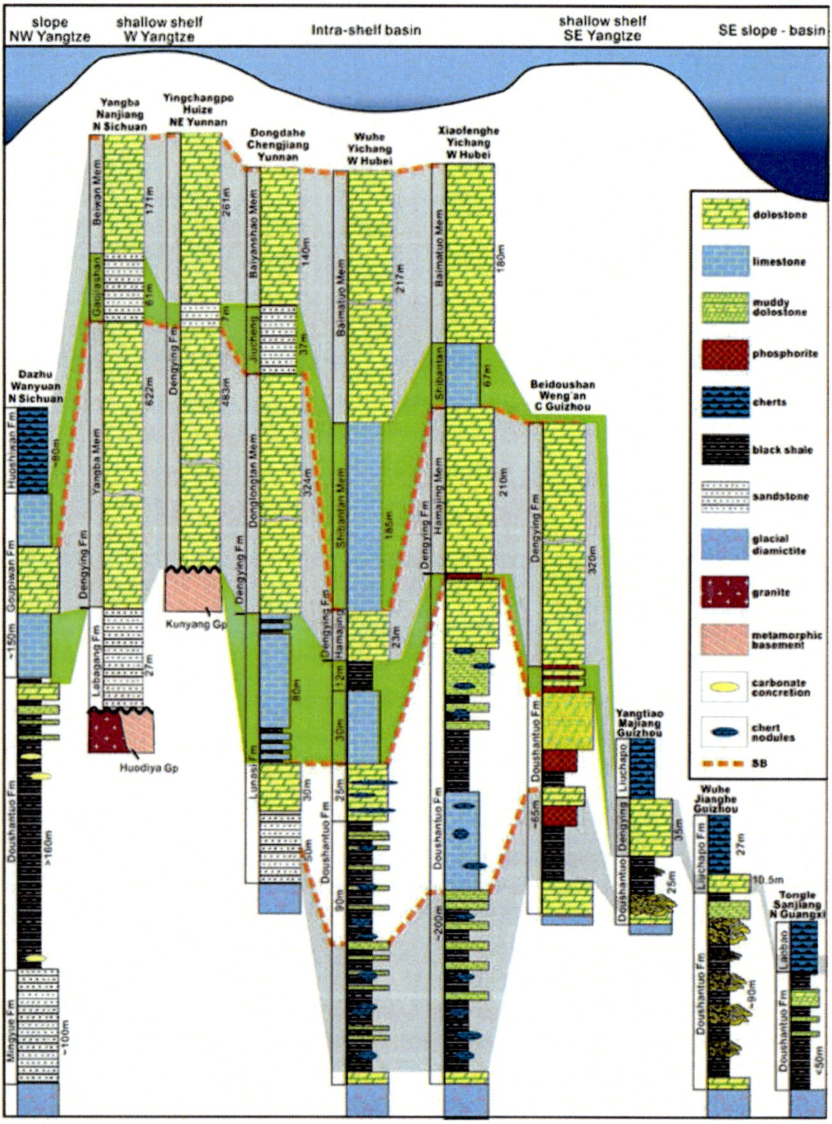

Fig. 5.6 Variation and correlation of the Ediacaran/Sinian reference sections from the different sedimentary facies of South China

carbonate of the Dengying Formation consists of three members, with the lower ("Yangba Member") and upper ("Beiwan Member") being dolostone intervals, and the middle ("Gaojiashan Member") consisting of variegated siliciclastic rocks. A similar Ediacaran succession is present in the western YC.

5.5.1.3 Yinchangpo Section, Huize, Northeastern Yunnan

The second type of nearshore platform facies succession on the western YC. Unlike the Yangba section, the Yinchangpo section only exposes the carbonates of the Dengying Formation above the Neoproterozoic granites and metamorphic basement rocks. Its three parts of the Dengying Formation are lithologically similar to those in the Yangba section, but are named the "Donglongtan", "Jiucheng" and "Baiyanshao" Members in ascending order.

5.5.1.4 Dongdahe Section, Chengjiang, Eastern Yunnan

The third type of nearshore platform facies succession on the western YC. The variably thick Ediacaran succession overlies the Nantuo diamictite. The succession is composed of purplish coarse siliciclastic rocks in the lower part and dolostone and limestone in the upper part, which is named the "Guanyinya Formation" in sorthern Sichuan and the "Lunasi Formation" in eastern Yunnan. No cap carbonate is present. The upper Dengying Formation, described in detail by Zhu et al. (2007b), is similar to the upper Dengying Formation in the Yinchangpo section.

5.5.1.5 Wuhe Section, Yangtze Gorges, Western Hubei

A deep intrashelf basin within the YC. This section has traditionally been regarded as the stratotype section of the Ediacaran/Sinian System of China. The Ediacaran succession consists of the Doushantuo and overlying Dengying Formations. The Doushantuo Formation is subdivided into four members. Dou-1 Member, the lowest member, is a cap carbonate measuring 4 to 5 m-thick and resting upon the Nantuo diamictite. Dou-2 Member is composed of black shale with silty dolostone interbeds and abundant chert nodules with large acanthomorph acritarchs. Dou-3 Member is a carbonate interval consisting of a lower dolostone with lenticular chert beds and nodules and an upper limestone. Dou-4 Member, also termed the "Miaohe Member", is a black shale interval with large carbonate concretions. The Dengying Formation is subdivided into three members. The lower or "Hamajing Member" is composed of dolomitic grainstone and oolitic dolostone. The middle or "Shibantan Formation" consists of finely laminated, dark gray limestone containing abundant soft-bodied Ediacaran fossils and trace fossils. The upper or "Baimatuo Member" is a thick interval composed of massive and thickly bedded dolostone.

5.5.1.6 Xiaofeng Section, Yichang, Western Hubei

The second type of deep intrashelf succession within the YC. Unlike the Wuhe section in the intrashelf basin, this section contains a distinct carbonate interval within Dou-2 Member, and oolitic dolostone is more common in Dou-3 Member. Also, there is no black shale interval near the top of the formation (Zhu et al. 2013).

5.5.1.7 Beidoushan Section, weng'an, Central Guizhou

A shallow water facies on the southeast margin of the YC. This section is famous for its abundant phosphatized embryos (the "Weng'an Biota") and has been described in detail by Zhu et al. (2007b). Overlying the Nantuo diamictite, the lower Doushantuo Formation is a dolostone interval measuring ca. 20 m-thick but lacking the diagnostic cap carbonate features seen elsewhere. Two phosphorite intervals separated by a dolostone interval occur in the middle part of the formation, while oolitic phosphatized dolostone overlies the upper black phosphorite. The upper part consists of interbedded phosphorite and dolostone, which grades upward into oolitic dolostone in the lower Dengying Formation. Notably, the Doushantuo Formation contains two karstic surfaces (Zhu et al. 2007b).

5.5.1.8 Yangtiao Section, Central Guizhou

An upper slope succession on the YC. The Yangtiao section, described by Zhu et al. (2007b), exposes one of the thinnest Ediacaran successions in South China. The Doushantuo Formation is only about 13 m-thick and consists of mixed shale and thin carbonate beds with several slump and brecciated layers. Again the basal dolostone lacks diagnostic features of a cap carbonate. The Dengying Formation is a dolostone interval measuring only 32 m-thick, and it is overlain by Cambrian black cherts and siltstone with abundant phosphatic concretions.

5.5.1.9 Jianhe Section, Central Guizhou

A slope facies on the YC. The Doushantuo Formation is about 90 m-thick and includes a basal cap carbonate as well as more than ten carbonate slump beds. Overlying the Doushantuo Formation are 10 m of massive dolostone of the Dengying Formation, which is overlain by well bedded cherts of the Liuchapo Formation (27 m). Slumps, brecciated layers and undeformed olistostromes are common in the transitional zone between the YC and Jiangnan Basin (Vernhet et al. 2007).

5.5.1.10 Tongle Section, Sanjiang, Northern Guangxi

A deep water succession of the Jiangnan Basin. The total thickness of this condensed interval is only about 50 m. The section consists of black shale and siltstone of the Doushantuo Formation, which lacks a basal cap carbonate, along with bedded black cherts of the Liuchapo/Laobao Formation.

Based on the foregoing ten reference sections, facies variation in the Ediacaran successions may now be summarized. The total thickness of the successions ranges from 50 m to 1000 m. Not surprisingly, sedimentary facies change dramatically as one moves from the shallow platform though the slope to the deep basin, but in spite of this variation a two-stage sedimentary sequence is readily recognizable along the entire transect (Fig. 5.6). On the YC, the lower half of the Ediacaran succession is dominated by black or dark gray, organic-rich shale and siltstone with interstratified carbonate beds and a basal cap carbonate. On the western YC, the Doushantuo Formation consists of coarse siliciclastic rocks and lacks a basal cap carbonate.

The upper half of the Ediacaran succession on the YC consists of the thick carbonates of the Dengying Formation, which contains three members. The middle member is composed of finely laminated limestone in the central portion of the platform and siliciclastic rocks in the western part.

In the deep water Jiangnan Basin, the succession consists of black shale of the Doushantuo Formation, with or without a basal cap carbonate, and black cherts of the Liuchapo/Laobao Formation in ascending order.

Because the slope facies exhibit abundant slump structures, successions in this facies tend to be incomplete and thus are not suitable for high resolution stratigraphic correlation.

5.5.2 Sedimentary Facies and Environments

To a great extent, the distribution and evolution of Ediacaran/Sinian successions in South China was controlled by palaeogeography and tectonics (Figs. 5.7 and 5.8). During Early Ediacaran times, a large portion of the western YC was emergent, with high relief and widespread exposure of basement rocks of different ages. The coarse siliciclastic rocks underlying the Dengying carbonate may represent nearshore or estuarine delta deposits, or even lacustrine sediments. The Ediacaran sequences usually have an unconformable base and thus are incomplete. In the middle and eastern portions of the YC and in the Jiangnan Basin, Ediacaran successions are complete and composed of mudstones and carbonate sediments deposited far offshore. On the margin of the YC and in the outer shelf area, multiple phosphorite and manganese ore layers formed, suggesting upwelling of nutrient-rich water from the open ocean on the northern and southeastern margins of the YC. Finally, during Late Ediacaran times, the YC was almost completely flooded, leading to the formation of a vast carbonate platform.

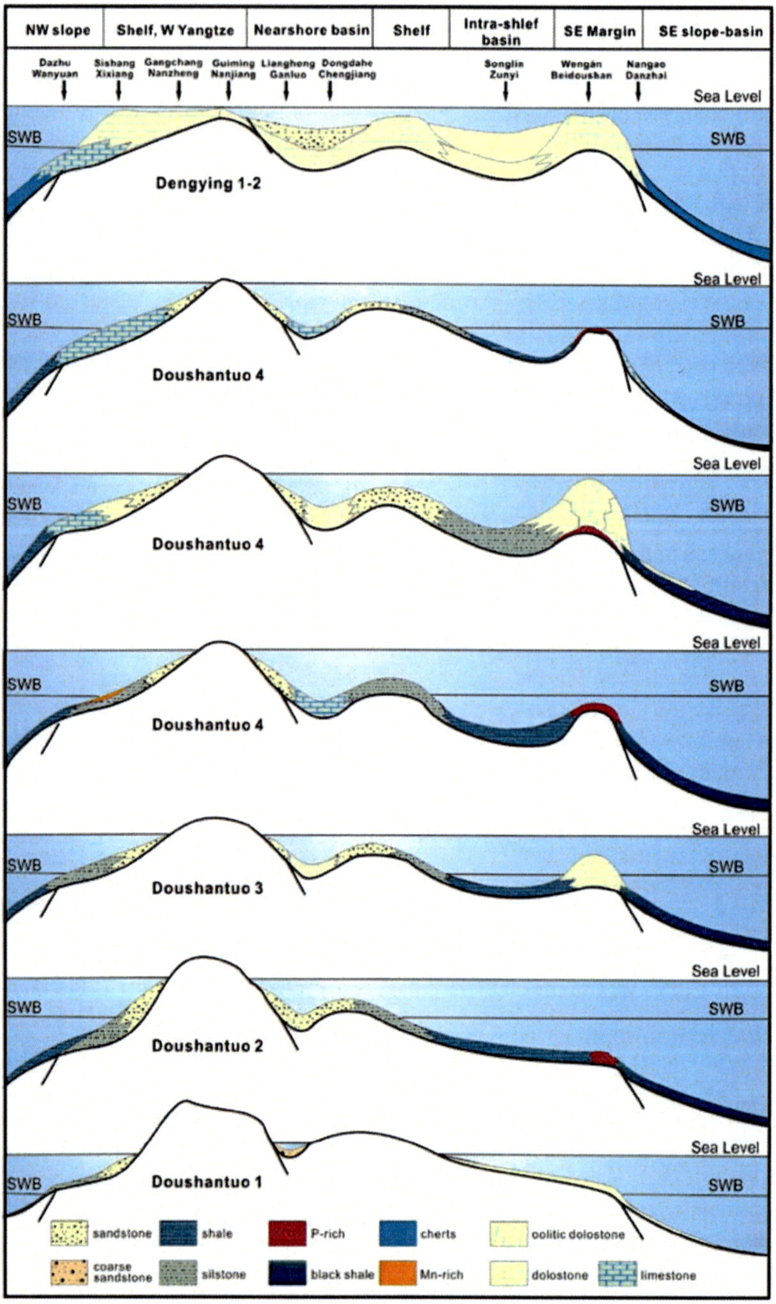

Fig. 5.7 Schematic diagram of the Ediacaran sedimentary facies model and evolution of South China. A transect from southern Shaanxi though Sichuan and eastern Yunnan to southeast Guizhou along the western YC

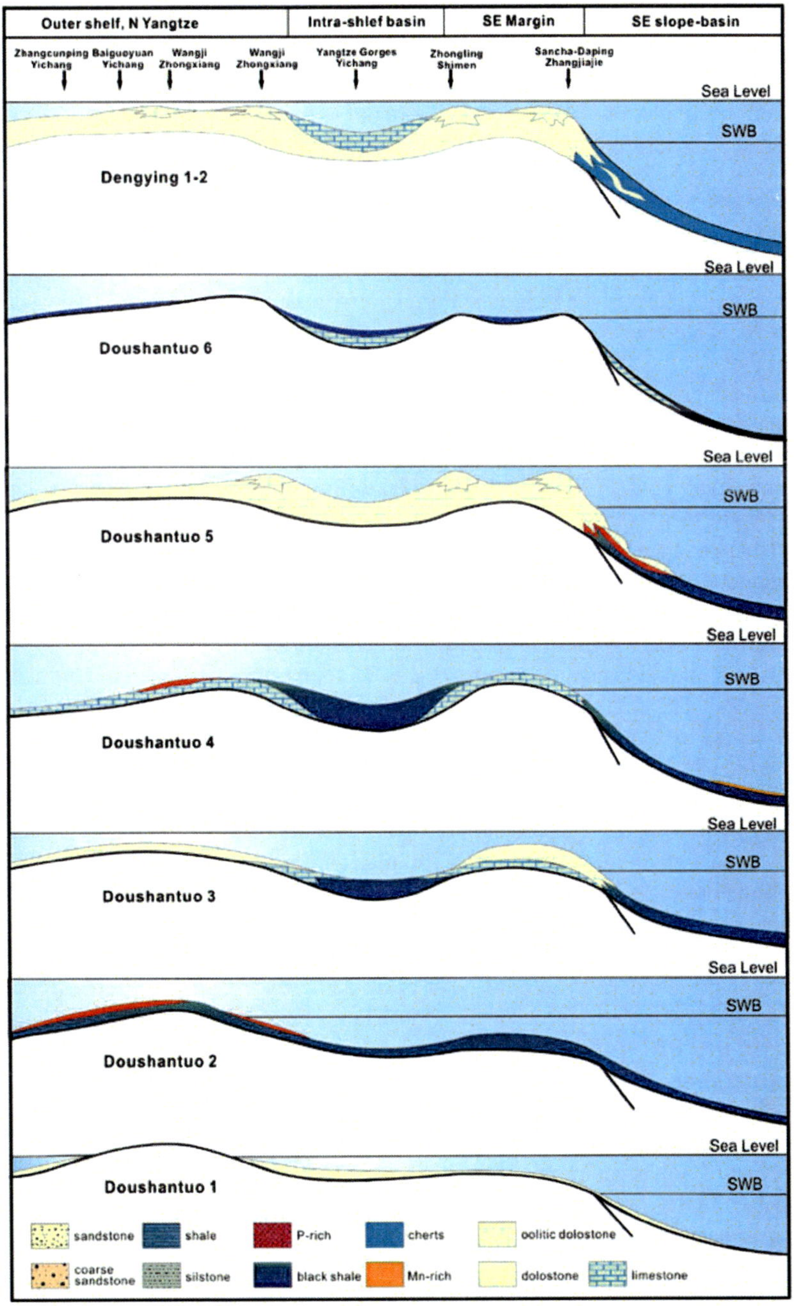

Fig. 5.8 Schematic diagram of the Ediacaran sedimentary facies model and evolution of South China. A transect from northern Yichang to western Hunan along the Control YC

Based on analyses of sedimentary sequences, structures and fabrics, several sedimentary facies can be recognized (Figs. 5.7 and 5.8). It should be noted here that the Yangtze Gorges area was an intrashelf basin, an interpretation that is supported by the following lines of evidence: ① the sequence is complete (lacking obvious erosion surfaces or depositional hiatuses); ② the organic-rich black shale interval with abundant chert nodules in the lower and middle Doushantuo Formation is very thick (up to 100 m); ③ a thick organic-rich black shale member with large carbonate concretions occurs in the upper part of the Doushantuo Formation; ④ slump structures are common in the Hamajing Member of the Dengying Formation; and ⑤ the middle or Shibantan Member of the Dengying Formation consists of finely laminated limestone with abundant chert nodules. Sequences in the intrashelf basin differ substantially from those in the nearshore shallow shelf area of the western YC and on the southeastern margin of the YC, which was fully rimmed (Figs. 5.7 and 5.8). The intrashelf deep basin covered a large area within the YC, including northwest Guizhou, eastern Chongqing, northwest Hunan, and western Hubei, and was interpreted as an interior lagoon (Jiang et al. 2011) or freshwater lake (Bristow et al. 2009). However, these two hypotheses, both of which imply restricted circulation, are not consistent with available geochemical and palaeobiological data (Zhu et al. 2013). As noted above, an active slope was present on the southeastern YC, as shown by variable sequences containing abundant slump structures, large slides, brecciated beds, conglomerate and olistostromes (Vernhet et al. 2007; Zhu et al. 2007b).

Overall, the Ediacaran System on the YC comprises four major depositional sequences (Zhu et al. 2007b, 2013). The first (S1) and second (S2) sequences are represented by the lower and upper Doushantuo Formation, respectively. The third sequence (S3) consists of the uppermost Doushantuo and lower Dengying Formations, and the fourth sequence (S4) is composed of the middle and upper parts of the Dengying Formation. These sequence boundaries are readily discernible in successions deposited on the shallow shelf and platform margin. The sequence boundary (SB2) between S1 and S2 is the karstic surface at the top of the middle dolostone member as displayed in the Beidoushan section, but in the deep intrashelf basin SB2 is difficult to recognize. The boundary (SB3) between S2 and S3 may be detected throughout most of the shelf area. For example, in the Beidoushan section SB3 is the karstic surface at the top of the Doushantuo Formation, while in the Yangtze Gorges area SB3 is located at the top of Dou-3 Member. The boundary (SB3) between S3 and S4 occurs at the base of the middle Dengying member, where it is usually an erosion surface associated with an abrupt facies change. A phosphate-rich horizon may occur near the boundary in some areas. SB3 is located at the base of the "Jiucheng/Gaojiashan Member" on the western YC and at the base of the "Shibantan Member" in the Yangtze Gorges area. The overall history of Ediacaran sequence and facies development on the YC is summarized schematically in Figs. 5.6, 5.7 and 5.8.

5.5.3 Stratigraphic Subdivision and Correlation

The Ediacaran/Sinian System of China was first subdivided into the Doushantuo and Dengyingxia Stages. This subdivision was widely used and formally adopted in the Stratigraphic Chart of China (SCC) (National Commission on Stratigraphy of China 2002). The Tianjiayuanzi section, located near Doushantuo village at Sandouping in the Yangtze Gorges, was designated as the stratotype section of the Doushantuo Stage. The roadcut along the northern side of the Yangtze River between the Liantuo and Shipai Villages was designated as the stratotype section of the Dengyingxia Stage. Because the boundary between these two stages was placed at the lithostratigraphic boundary between the Doushantuo and Dengying Formations, Yin et al. (1993) redefined the base of the Dengyingxia Stage as the base of the "Shibantan Member" of the Dengying Formation, noting that the base of the "Shibantan Member" is a major sequence boundary and that Ediacaran soft-bodied and tubular fossils first appear above this horizon. Based on biostratigraphic data, Wang et al. (2001) proposed a new subdivision consisting of two series and four stages. They retained the base of the "Shibantan Member" as the boundary between the lower and upper series. The lower series was subdivided into the Tianjiayuanzi and Miaohe Stages, with the first occurrence of the "Miaohe biota" defining the base of the Miaohe Stage. The upper series was subdivided into the Sixi and Longdengxia Stages, with the boundary between them being defined by the first occurrence of cloudiniids, which datum occurs near the base of the "Baimatuo Member".

Absolute age estimates for previously erected chronostratigraphic subdivisions were based mainly on the thickness of the depositional successions. However, now that precise ID-TIMS zircon U–Pb ages of volcanic ashes near the base (635 Ma) and top (551 Ma) of the Doushantuo Formation are available (Condon et al. 2005), the Early Ediacaran chronostratigraphic models must be revised. A new chronostratigraphic scheme consisting of two series (the lower or Xiadongian Series and the upper or Yangtzean Series) and five stages was proposed by Zhu et al. (2007b) on the basis of chemo-, bio-, and sequence stratigraphic data. In the scheme, the sequence boundary in the middle part of the Doushantuo Formation is also the boundary between the lower Xiadongian and upper Yangtzean series. Zhu et al. (2007b) believed that this boundary records the eustatic sea level change associated with the Gaskiers glaciation (581 Ma, though later dated at 580 Ma; Pu et al. 2016), noting also that it coincides with a negative $\delta^{13}C$ excursion (WANCE) and a rapid increase in the diversity of acanthomorph acritarchs.

Because the chronostratigraphic scheme of Zhu et al. (2007b) is based largely on $\delta^{13}C$ chemostratigraphy, which is strongly influenced by facies variation, it is problematic for subdividing and correlating sequences in the highly varied depositional facies of South China. Nevertheless, additional chemo- and biostratigraphic data, collected recently from localities throughout the YC, provide useful information for improving the scheme. These data are summarized by category below.

(1) **Sequence stratigraphy**. As discussed above, the Ediacaran System on the YC comprises four major depositional sequences (Zhu et al. 2007b, 2013; Yang et al. 2015). Geochronological data indicate that S4 (middle to upper Dengying Formation) is a third-order sequence spanning fewer than 5 Ma, while S1–S3 are megasequences that each span more than 30 Ma. Since the Doushantuo Formation represents a strongly condensed interval, further sequence stratigraphic subdivision of this rock unit is not really feasible. However, parasequence analysis is feasible for the thick carbonate of the Dengying Formation (e.g. Hu et al. 2019; Ding et al. 2021).

(2) **Biostratigraphy**. The Ediacaran System of South China contains abundant fossils preserved in a variety of taphonomic windows (cf. Zhu 2010 and references therein), including phosphatized metazoan embryos and microscopic adults from the Weng'an biota; macroscopic carbonaceous fossils of the Lantian, Miaohe and Wenghui biotas in the Doushantuo Formation; tubular fossils of the Gaojiashan and Xilingxia biotas (cf. Cai et al. 2010, 2019; Liang et al. 2020); and Ediacara-type fossils of the Shibantan Biota in the Dengying Formation (cf. Chen et al. 2014; Xiao et al. 2020). Both skeletonized tubular fossils and soft-bodied Ediacara-type fossils are potentially useful for biostratigraphic zonation. More significantly, diverse large acanthomorph acritarchs are present within phosphorites and chert nodules in the Doushantuo Formation. Two distinct acritarch assemblage zones, namely the *Tianzhushania spinisa* and *Tanarium conoideum-Hocosphaeridium scaberfacium-Hocosphaeridium anozos* zones, have proven to be very useful for both regional and global correlation (Liu et al. 2013, 2014a, b). Additional, still finer acritarch assemblage zones have been proposed.

(3) **Chemostratigraphy**. Since the publication of the first $\delta^{13}C_{carb}$ chemostratigraphic study of Ediacaran strata in the Yangtze Gorges area (Lambert et al. 1987), additional data published during the past two decades have greatly contributed to the development of a more precise Ediacaran chemostratigraphic scheme (e.g. Jiang et al. 2007, 2011; Zhou et al. 2017b; Zhu et al. 2007a, b, 2013; McFadden et al. 2008; Sawaki et al. 2010; Lu et al. 2013; Tahata et al. 2013; An et al. 2015; Chen et al. 2015; Furuyama et al. 2016; Cui et al. 2017; Gao et al. 2018). In particular, three distinct negative $\delta^{13}C$ excursions are present in the Doushantuo Formation (Jiang et al. 2007; Zhu et al. 2007a). However, whereas the positions of both the excursions near the base and top of the Doushantuo Formation, respectively, are well defined, geographical variation in the magnitude and stratigraphic position of the negative $\delta^{13}C$ excursion (termed the N2, EN2, or WANCE, i.e., Weng'an negative carbonate isotope excursion) in the middle Doushantuo Formation has hindered precise correlation.

Based on additional data from successions ranging in origin from shallow shelf to intrashelf basin, and located mainly in western Hubei, Zhu et al. (2013) determined that there are two negative $\delta^{13}C$ excursions in the middle Doushantuo Formation. In order to avoid terminological confusion, Zhu et al. (2007a, 2013) proposed the following new acronyms for the four negative excursions in the Doushantuo Formation: CANCE (i.e., cap carbonate negative carbon isotope excursion), recorded in the cap carbonate at the base of the Doushantuo Formation and equivalent to N1 of Jiang et al. (2007) and EN1 of Zhou et al. (2007a, b); WANCE, situated near the sequence boundary in the middle Doushantuo Formation (i.e., the Dou-2 Member in Yangtze Gorges area); BAINCE (i.e., Baiguoyuan negative carbon isotope excursion), located near the top of Dou-2 Member in Yangtze Gorges area and equivalent to N2 of Jiang et al. (2007) and EN2 of Zhou et al. (2007a, b); and DOUNCE (i.e., Doushantuo negative carbon isotope excursion), situated at the top of Dou-3 Member and equivalent to N3 of Jiang et al. (2007) and EN3 of Zhou et al. (2007a, b).

Even though the WANCE excursion has not attracted wide attention, it is present in several sections in western Hubei (cf. Tahata et al. 2013; Chen et al. 2015; Gao et al. 2018). Also, and as suggested by examination of sections in the western Huangling Anticline near Yichang, western Hubei, there may be an additional short-lived negative excursion above the DOUNCE near the top of the Doushantuo Formation (Zhou et al. 2017b) or in the lower Dengying Formation (An et al. 2015). This possibility makes the termination time of the DOUNCE (equiv. Shuram) excursion uncertain, as an age of 550.5 ± 0.8 Ma from the basal Dengying Formation has already been widely accepted for this event (Condon et al. 2005; Bowring et al. 2007).

The Ediacaran strontium isotope ($^{87}Sr/^{86}Sr$) profile for the Yangtze Gorges area increases in value upsection from 0.7080 to 0.7085 (Sawaki et al. 2010). A major positive shift occurs near the base of Dou-3 Member, with $^{87}Sr/^{86}Sr$ values peaking at 0.709,0 in the DOUNCE interval.

(4) **Geochronology and cyclostratigraphy.** Following the publication of ID-TIMS zircon U–Pb ages from the base (635 Ma) and top (551 Ma) of the Doushantuo Formation (Condon et al. 2005), two zircon U–Pb ages, namely a SHRIMP zircon U–Pb age (614.0 ± 7.6 Ma) from a horizon below SB2 and a SIMS zircon U–Pb age (609 ± 5 Ma) from immediately above SB2, were reported from the middle Doushantuo Formation in the Zhangcunping area of northern Yichang, western Hubei (Liu et al. 2009; Zhou et al. 2017a). In addition, two SIMS zircon U–Pb ages (553.6 ± 2.7 Ma and 546.3 ± 2.7 Ma) were obtained from ash beds in the lower and middle Jiucheng Member of the Dengying Formation in eastern Yunnan (Yang et al. 2017a). Zircon U–Pb ages have also been obtained from ash beds in the slope sections, including an age of 557 ± 3 Ma in the lower Dengying Formation; an age of 550 ± 3 Ma in the lower Liuchapo Formation in the Fanglong section near Jianhe, Guizhou (Zhou et al. 2018); and an age of 545.76 ± 0.66 Ma in the lower Liuchapo Formation in the Longbizui section in western Hunan (Yang et al. 2017b). Together with cyclostratigraphic data (Gong et al. 2017; 2019; Sui et al. 2018, 2019), the foregoing zircon U–Pb ages provide better age constraints for key Ediacaran stratigraphic boundaries, biotas and events.

Based then on the latest advances in sequence stratigraphy, biostratigraphy, chemostratigraphy and geochronology, the Ediacaran chronostratigraphic scheme of Zhu et al. (2007a, b) may be revised as shown here in Table 5.3. Like the earlier model, the new scheme consists of two series and five stages. Except for the base of the Ediacaran System (equiv. the base of Stage 1) and the base of the terminal stage (Stage 5), the bases of the other three stages are defined by carbon isotope chemostratigraphic markers, as summarized below.

(1) **Ediacaran Stage 1**. The base of this stage coincides with the GSSP of the Ediacaran System, which occurs at the base of the cap carbonate overlying the Nantuo diamictite. The age of the base is constrained by an age of 635 Ma from the top of the cap carbonate (Condon et al. 2005; Zhou et al. 2019).

Table 5.3 Integrated stratigraphic chart of the Ediacaran System of South China

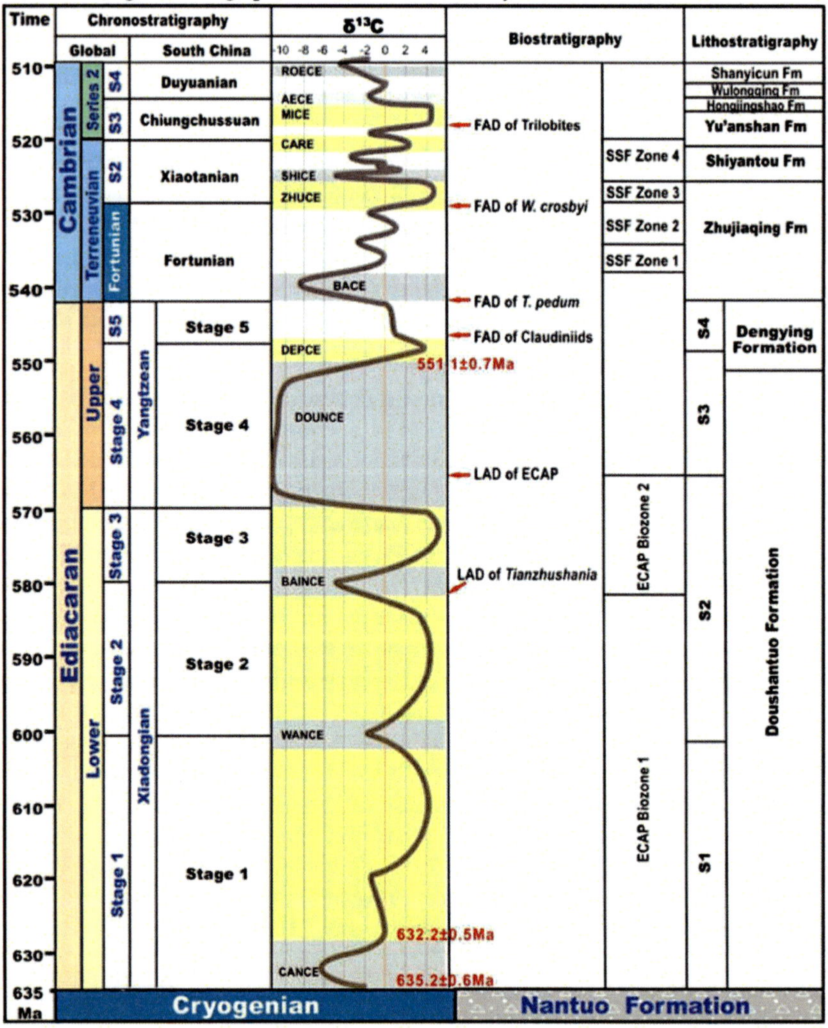

(2) **Ediacaran Stage 2.** The base of this stage is defined by the onset of the WANCE, which is situated near SB2 in the middle Doushantuo Formation. An approximate age of 600 Ma for the base of Stage 2 is suggested by two ages, one from below (614.0 ± 7.6 Ma) and another from above (609 ± 5 Ma) this level in northern Yichang (Liu et al. 2009; Zhou et al. 2017a).

(3) **Ediacaran Stage 3.** The base of this stage is defined by the onset of the BAINCE, which coincides with the boundary between the *Tianzhushania spinisa* and *Tanarium conoideum-Hocosphaeridium scaberfacium-Hocosphaeridium anozos* acritarch assemblage zones. The BAINCE event probably is related to the Gaskiers glaciation, as indicated by oxygen isotope data and the occurrence of glendonite within the BAINCE interval (Tahata et al. 2013; Furuyama et al. 2016; Wang et al. 2020). Therefore, the age of the base of Ediacaran Stage 3 is about 580 Ma (Pu et al. 2016).

(4) **Ediacaran Stage 4.** The base of this stage is defined by the onset of the DOUNCE, which coincides with the mass extinction of large acanthomorph acritarchs recorded in the upper part of Dou-3 Member in the Yangtze Gorges area (Lu et al. 2013). The DOUNCE, the largest negative carbon isotope excursion in Earth history, may have resulted from oxygenation of a large dissolved organic carbon reservoir in the deep ocean (Condon et al. 2005; Shields et al. 2019). The DOUNCE also coincides with the initial diversification of macroscopic complex organisms (e.g. Zhu et al. 2008), corroborating the hypothesis of deep ocean oxygenation. Given the global importance of the base of Stage 3, we recommend that it can be recognized as a series boundary within the Ediacaran System. Finally, and based on cyclostratigraphy, the age of the onset of the DOUNCE (i.e., the base of Stage 4) is approximately 570–571 Ma (Sui et al. 2019; Gong and Li 2020).

(5) **Ediacaran Stage 5.** The base of this stage is defined by the first occurrence of cloudiniids, a characteristic group of terminal Ediacaran tubular fossils reported nearly from all continents. The boundary may be situated close to the basal Gaojiashan and Shibantan Members of the Dengying Formation, but this hypothesis requires further investigation (cf. Cai et al. 2010, 2019; Liang et al. 2020). Integrated correlations indicate that the age of the base of Stage 5 is approximately 551–546 Ma (Condon et al. 2005; Yang et al. 2017a).

Except in the slope and deep basin areas, for which chemo- and biostratigraphic data are not available, the foregoing chronostratigraphic scheme should be useful for subdividing and correlating Ediacaran successions throughout the YC (Fig. 5.6). It should be noted, however, that the bases of Stages 2 and 4 are associated with sequence boundaries, and therefore the records of the WANCE and DOUNCE may be incomplete or complicated by unconformities or syndepositional slides (Lu et al. 2013; Zhu et al. 2013). Also, because the Ediacaran succession in the Yangtze Gorges area was deposited in an intrashelf basin, the short lived WANCE event may not be readily detectable there owing to the lack of carbonate strata. Be that as it may, available chemostratigraphic data indicate that the base of Stage 2 is close to the sharp contact between a thick carbonate bed and an overlying condensed interval with abundant chert nodules in the Yangtze Gorges (e.g. Tahata et al. 2013; Zhu et al. 2013; Chen et al. 2015; Gao et al. 2018). The abrupt change in depositional facies across this contact may have resulted from eustatic sea level change, a hypothesis that is corroborated by a substantial redox change through these strata (Li et al. 2010).

Of course, our updated Ediacaran chronostratigraphic scheme differs from the two series/six stages model recently proposed by Zhou et al. (2019). In their scheme, the base of Stage 2 is defined by the FAD (first appearance datum) of acanthomorph acritarchs, which level is situated about 2.8 m above the top of the cap carbonate in the Jiulongwan section. This definition was originally recommended in the Ediacaran Subcommission on Stratigraphy (e.g. Xiao et al. 2016). The problem with this definition is that it makes Stage 1 a very short interval lasting fewer than 3 Ma (Condon et al. 2005), though clearly the most significant difference between the two schemes is the additional stage placed by Zhou et al. (2019) within Stage 3 of our model. These authors maintained that a short-lived, negative carbon isotope excursion, supposedly present above the DOUNCE near the top of the Doushantuo Formation (Zhou et al. 2017b) in the western Huangling Anticline in Yichang, Hubei can be used to define a new stage (Stage 5) characterized by the Miaohe Biota and situated below the terminal stage (Stage 6). In addition, Zhou et al. (2019) proposed an age of 580 Ma for the base of the DOUNCE (equiv. EN3, Stage 3). Importantly, the points of disagreement between the two alternative chronostratigraphic schemes reflect persistent uncertainties in chemostratigraphy and the lack of age constraints for certain key events, and therefore additional study of the problem is required.

5.6 Source-Reservoir-Seal Bed Assemblages of South China

Actual and potential hydrocarbon source-reservoir-seal bed assemblages in Neoproterozoic strata of South China have been known for some time, and include fossil-oil-reservoirs in the Dengying dolostone in the Nanshanping area of Hunan and the Taishan area of Zhejiang as well as bitumen-rich portions of the Dengying dolostone in northern Sichuan and sorthern Shaanxi. Although the source rocks have generally experienced thermal over-maturation, the recent discovery of the Weiyuan and Anyue Gasfields in the western YC indicates that Neoproterozoic marine source and reservoir beds of the YC are important targets for deep zone hydrocarbon exploration, as summarized briefly below.

5.6.1 Tonian Source Beds

Tonian source rocks are composed of dark gray silty mudstones with a thickness up to hundreds of meters, and occur mainly in the lower part of the first sedimentary sequence of the Banxi Group and equivalent strata in the deep Jiangnan Basin, including the Wuye Formation of the Xiajiang Group in southeast Guizhou, the lower

Zhuanqiangwan Formation of the Gaojian Group in southwest Hunan, and the lower Hetong Formation of the Danzhou Group in northern Guangxi (Figs. 5.1 and 5.2).

5.6.2 *Cryogenian Source Beds*

The black shale interval in the lower part of the Datangpo Formation is a condensed interglacial transgressive sequence with a high TOC content (1.6–4.3%, avg. 2.8%; Ai et al. 2021). The black shale is generally associated with manganiferous carbonate or manganese ore, and its distribution is limited to the southeast margin of the YC and the Jiangnan Basin, with a depocenter in the border area of northeastern Guizhou and western Hunan (Figs. 5.3 and 5.4; Xu et al. 1991; He 1997; Liu and Zhou 2002; Huang et al. 2010; Yang et al. 2010). The thickness of the black shale ranges from a few meters to tens of meters, and was controlled by NE-trending extensional faulting associated with rifting along the southeast edge of the YC (e.g., Zhou et al. 2016; Qu et al. 2020).

5.6.3 *Ediacaran Source Beds*

The thick and widespread black shales and carbonates of the Doushantuo Formation are the main Ediacaran source beds. These strata constitute the transgressive system tracts of three depositional sequences having a total thickness of 30–379 m (Figs. 5.6, 5.7 and 5.8). According to Ai et al. (2021), the TOC content of black shales in the lower and middle parts of the Doushantuo Formation ranges from 1.2 to 3.5% (avg. 2.1%), especially for the black shales with the highest TOC content (7.7–9.4%) in Dou-4 Member (the "Miaohe Member").

The distribution, thickness, and TOC content of the Doushantuo source beds were controlled by depositional facies (e.g. Qu et al. 2020; Xiao et al. 2020). The source beds formed predominately in the intrashelf basins on the YC as well as in the outer shelf areas and deep basins along the southeast and northern margins of YC (Figs. 5.7 and 5.8). In most cases, the source beds are less than 60 m-thick, but they measure up to 389 m-thick in the intrashelf basin near Hefeng in southwest Hubei; 579 m thick near Hujiaba in Ningqiang; 310 m thick near Dazhu in Wanyuan and 309 m thick near Mingyue in Chengkou, in the outer shelf area on the northern margin of the YC. In the western YC including a large part of the current Sichuan Basin, where the Doushantuo Formation is poorly developed or composed mainly of thin coarse siliciclastic beds, Doushantuo source rocks are of limited extent or absent.

In addition to the Doushantuo Formation, source beds also occur in upper Ediacaran strata. In the intrashelf basin in the Yangtze Gorges area of western Hubei, organic-rich black and dark gray muddy limestone with a thickness of 65–185 m occurs in the middle Dengying Formation. Potential source beds in the slope deposits and Jiangnan Basin consist of organic-rich thin-bedded cherts and black shale in the

Liuchapo/Laobao/Piyuancun Formations. These strata vary from 20 to 400 m in total thickness, with the thickest deposits occurring in Dexing and Hengfeng in western Zhejiang (201–393 m), and in the Quanzhou-Linggui-Hechi area of northern Guangxi (78–230 m).

Intrashelf (rift) basins also formed during middle Dengying times on the western YC. These include a NE-trending basin in the Wanyuan-Dazhou area of northwest Sichuan and the NW-trending Deyang-Anyue area in central Sichuan (e.g., Zhao et al. 2017). Siliciclastic-dominated portions of the Gaojiashan/Jiucheng members of the middle Dengying Formation may be potential source rocks.

5.6.4 Potential Neoproterozoic Reservoir and Seal Beds

The widespread massive dolostone of the Dengying Formation on the YC contains important hydrocarbon reservoir beds, e.g., those in the Weiyuan and Anyue Gasfields in the western YC (i.e., the current Sichuan Basin). The reservoir beds consist of microbial carbonate mounds and shoals deposited along the platform margin and interior uplifts. The reservoir space consists of solution pores, vugs, karst caves, cracks and intercrystalline pores. Not surprisingly, the reservoir beds exhibit variable porosity and permeability (Liu et al. 2008). For example, the dolostone reservoir beds have a porosity of 1.39% and a permeability of 0.61×10^{-3} μm^2 in southwest Sichuan, a porosity of 1.86–2.05% and a permeability of 0.01×10^{-3}–8.02×10^{-3} μm^2 in central Sichuan, and a porosity of 1.78% and a permeability of $0.074{,}6 \times 10^{-3}$ μm^2 in the Dingshan Well in southeast Sichuan.

The development and distribution of the Dengying reservoir beds were controlled by depositional facies and late karstification, particularly at the top of the Dengying Formation, further improved the reservoir quality. For example, widespread deposition of mound-shoal facies plus late karstification along the margins of the NW-trending Deyang-Ziyang Faulted-Sag may have controlled the development of the extensive and continuous Dengying reservoir beds in the Gaoshiti-Moxi area of the western YC (Luo et al. 2017; Li et al. 2020).

The best Dengying reservoir beds occur mainly in the lower member (equiv. Member 1, Donglongtan and Hamajing Members in Table 5.2 and Fig. 5.6, and Deng-2 Member in other chapters) and upper member (equiv. Member 3, Baiyanshao and Baimatuo Members in Table 5.2 and Fig. 5.6, and Deng-4 Member in other chapters). The Deng-2 reservoir beds are composed predominantly of dolomitic grainstone and microbiolite and crystalized dolostone, and range in thickness from 140 to 1100 m. The Deng-2 dolostone occurs on the southwestern YC, where it measures up to 1100 m-thick at Changning in sorthern Sichuan and 939 m thick in the Luquan-Dongchuan area of northeastern Yunnan. In western Hunan and the Hubei area of the central YC, the Deng-2 dolostone is only 151–355 m-thick. The shoal facies, represented by dolomitic grainstone and oolite, occurs on the platform margins in the Zhangjiajie-Chili area of central Hunan, the Yangtze Gorges, and the Fangxian area of western Hubei. Intraclastic, oncolitic and oolitic dolostones are common in the Nanjing-Zhenjiang area of Jiangsu on the eastern YC.

The Deng-4 reservoir beds are similar in distribution to the Deng-2 reservoir beds. The Deng-4 reservoir beds measure 162–410 m-thick in northeastern Yunnan and southwest Sichuan on the southwestern YC, and 146–627 m-thick in sorthern Shaanxi on the northern YC. In the central and southern portions of the current Sichuan Basin, the upper Dengying Formation is incomplete, owing possibly to uplift and erosion at the end of the Ediacaran Period. In western Hunan and Hubei, the upper Dengying Formation reaches a thickness of 555 m. Fossil-oil-reservoirs have been discovered in the upper Dengying Formation at Nanshanping of Chili (western Hunan) and Taishan of Yuhan (Zhejiang).

The Dengying reservoir beds are overlain throughout the YC by a thick Cambrian sequence consisting of black shale, mudstone, and siltstone. The reservoirs underlie the Lower Cambrian Qiongzhusi, Guojiaba, and Niutitang Formations on the western YC; the Shuijingtuo and Xiaoyanxi Formations on the central YC; and the Hetang Formation on the eastern YC. Fine grained rocks in all these formations are rich in organic matter, and together with the underlying Dengying reservoirs and Doushantuo source beds they form excellent source-reservoir-seal bed assemblages for deep hydrocarbon exploration.

Acknowledgements This is a summary of our research on the Neoproterozoic strata of South China based on field work conducted during the past two decades. We thank our colleagues from all over the world, as well postgraduate students Xinglian Yang, Meijuan Zhao and Xin Zhao for their help in the field and collaboration, Prof. Heyo Van Iten (Hanover College and Cincinnati Museum Center) for language editing, and Prof. Tieguang Wang for careful editing proof. Our research was supported by the Strategic Priority Research Program (B) of the Chinese Academy of Sciences (Grant No. XDB26000000, 18000000), the National Natural Science Foundation of China (Grant No. 41921002), and the Ministry of Science and Technology (MST) of the People's Republic of Chinay (Grant No. 2013CB835000).

References

Ai JY, Zhong NN, Zhang TG, Zhang Y, Wang TG, George SC (2021) Oceanic water chemistry evolution and its implications for post-glacial black shale formation: insights from the Cryogenian Datangpo Formation, South China. Chem Geol 120083

An ZH, Jiang GQ, Tong JN, Tian L, Ye Q, Song HY, Song HJ (2015) Stratigraphic position of the Ediacaran Miaohe biota and its constrains on the age of the upper Doushantuo δ^{13}C anomaly in the Yangtze Gorges area, South China. Precambr Res 271:243–253

An ZZ, Zhang RB, Chen JC, Qin Y, Pan W, Wu GW, Peng QY, Zheng C, Zhang FF, Zhu XK, Wang HB (2014) Geological and geochemical characteristics of Daotuo super large manganese ore deposit in Songtao County of Guizhou Province: constraint on formation mechanism of Mn-carbonate ores. Mineral Deposits 33(4):870–884 (in Chinese with English abstract)

Bahlburg H, Dobrzinski N (2011) A review of the chemical index of alternation (CIA) and its application to the study of neoproterozoic glacial deposits and climate transitions. In: Arnaud E, Halverson G P, Shields-Zhou G (eds) The geological record of neoproterozoic glaciations. Geol Soc Lond Mem 36:81–92

Bao XJ, Zhang SH, Jiang GQ, Wu HC, Li HY, Wang XQ, An ZZ, Yang TS (2018) Cyclostratigraphic constraints on the duration of the Datangpo formation and the onset age of the Nantuo (Marinoan) glaciation in South China. Earth Planet Sci Lett 483:52–63
Bowring SA, Grotzinger JP, Condon DJ, Ramezani J, Newall MJ, Allen PA (2007) Geochronologic constraints on the chronostratigraphic framework of the Neoproterozoic Huqf Supergroup, Sultanate of Oman. Am J Sci 307(10):1097–1145
Brstow TF, Kennedy MJ, Derkowski A, Droser ML, Jiang GQ, Creaser RA (2009) Mineralogical constraints on the paleoenvironments of the Ediacaran Doushantuo Formation. Proc Natl Acad Sci 106:13190–13195
Brstow TF, Bonifacie M, Derkowski A, Eiler JM, Grotzinger JP (2011) A hydrothermal origin for isotopically anomalous cap dolostone cements from South China. Nature 274:68–71
Busigny V, Planavsky NJ, Goldbaum E, Lechte MA, Feng LJ, Lyons TW (2018) Origin of the Neoproterozoic Fulu iron formation, South China: insights from iron isotopes and rare earth element patterns. Geochim Cosmochim Acta 242:123–142
Butterfield NJ (2011) Animals and the invention of the Phanerozoic Earth system. Trends Ecol Evol 26:81–87
Cai YP, Hua H, Xiao SH, Schiffbauer JD, Li P (2010) Biostratinomy of the Late Ediacaran pyritized Gaojiashan Lagerstätte from southern Shaanxi, south China: importance of event deposits. Palaios 25(8):487–506
Cai YP, Xiao SH, Li GX, Hua H (2019) Diverse biomineralizing animals in the terminal Ediacaran Period herald the Cambrian explosion. Geology 47(4):380–384
Chen JY, Oliveri P, Li CW, Zhou GQ, Gao F, Hagadorn JW, Peterson KJ, Davidson EH (2000) Precambrian animal diversity: putative phosphatized embryos from the Doushantuo Formation of China. PNAS 97:4457–4462
Chen JY, Bottjer DJ, Oliveri P, Dornbos SQ, Gao F, Ruffins S, Chi HM, Li CW, Davidson EH (2004) Small bilaterian fossils from 40 to 55 million years before the Cambrian. Science 305:218–222
Chen JY, Bottjer DJ, Davidson EH, Dornbos SQ, Gao X, Yang YH, Li CW, Li G, Wang XQ, Xian DC, Wu HJ, Hwu YK, Tafforeau P (2006) Phosphatized polar lobe-forming embryos from the Precambrian of Southwest China. Science 312:1644–1646
Chen JY, Bottjer DJ, Li GX, Hadfield MG, Gao F, Cameron AR, Zhang CY, Xian DC, Tafforeau P, Liao X, Yin ZJ (2009) Complex embryos displaying bilaterian characters from Precambrian Doushantuo phosphate deposits, Weng'an, Guizhou, China. PNAS 106:19056–19060
Chen XH, Zhou P, Zhang BM, Wang CS (2015) Stable isotope records of the Ediacaran Doushantuo Formation in the eastern Yangtze Gorges and its significance for choronostratigraphy. Geol China 42:207–223 (in Chinese with English abstract)
Chen Z, Zhou CM, Xiao SH, Wang W, Guan CG, Hua H, Yuan XL (2014) New Ediacara fossils preserved in marine limestone and their ecological implications. Sci Rep 4(1):1–10
Chen Z, Chen X, Zhou CM, Yuan XL, Xiao SH (2018) Late Ediacaran trackways produced by bilaterian animals with paired appendages. Sci Adv 4(6):eaao6691
Chen Z, Zhou CM, Yuan XL, Xiao SH (2019) Death march of a segmented and trilobate bilaterian elucidates early animal evolution. Nature 573(7774):412–415
Cheng M, Li C, Chen X, Zhou L, Algeo TJ, Ling HF, Feng LJ, Jin CS (2018) Delayed Neoproterozoic oceanic oxygenation: evidence from Mo isotopes of the Cryogenian Datangpo formation. Precambr Res 319:187–197
Condon D, Zhu MY, Bowring S, Wang W, Yang AH, Jin YG (2005) U-Pb ages from the Neoproterozoic Doushantuo Formation, China. Science 308:95–98
Cui H, Kaufman AJ, Xiao SH, Zhou CM, Liu XM (2017) Was the Ediacaran Shuram Excursion a globally synchronized early diagenetic event? Insights from methane-derived authigenic carbonates in the uppermost Doushantuo Formation, South China. Chem Geol 450:59–80
Darroch SAF, Smith EF, Laflamme M, Erwin DE (2018) Ediacaran extinction and Cambrian explosion. Trend in Ecology and Evolution 33:653–663

Ding Y, Li ZW, Liu SG, Song JM, Zhou XQ, Sun W, Zhang XH, Li SJ, Ran B, Peng HL, Li ZQ, Wang H, Chen DZ (2021) Sequence stratigraphy and tectono-depositional evolution of a Late Ediacaran epeiric platform in the upper Yangtze area, South China. Precambr Res 354:106077

Dobrzinski N, Bahlburg H, Strauss H, Zhang QR (2004) Geochemical climate proxies applied to the Neoproterozoic glacial succession on the Yangtze Platform, South China. In: Jenkins G, McMaenamin M, McKay C P, Sohl L (eds) The extreme proterozoic: geology, geochemistry and climate. Am Geophys Union Monogr Ser 146:13–32

Du QD, Wang ZJ, Wang J, Qiu YS, Jiang XS, Deng Q, Yang F (2013) Geochronology and paleoenvironment of the pre-Sturtian glacial strata: evidence from the Liantuo Formation in the Nanhua rift basin of the Yangtze Craton, South China. Precambr Res 233:118–131

Erwin DH, Valentine JW (2013) The Cambrian explosion: the construction of animal biodiversity. Roberts and Company Publisher, Greenwood Village, Colorado

Erwin DH, Laflamme M, Tweedt SM, Sperling EA, Pisani D, Peterson KJ (2011) The Cambrian conundrum: early divergence and later ecological success in the early history of animals. Science 334:1091–1097

Feng LJ, Chu XL, Zhang QR, Zhang TG, Li H, Jiang N (2004) New evidence of deposition under cold climate for the Xieshuihe formation of the Nanhua system in northwestern Hunan, China. Chin Sci Bull 49(1):420

Furuyama S, Kano A, Kunimitsu Y, Ishikawa T, Wang W (2016) Diagenetic overprint to a negative carbon isotope anomaly associated with the Gaskiers glaciation of the Ediacaran Doushantuo Formation in South China. Precambr Res 276:110–122

Gaidos E, Dubuc T, Dunford M, Mcandrew P, Padilla-Ganino J, Studer B, Weersing K, Stanley S (2007) The Precambrian emergence of animal life: a geobiological perspective. Geobiology 5:351–373

Gao LZ, Dai CG, Liu YX, Wang M, Wang XH, Chen JS, Ding X (2010) Zircon SHRIMP U-Pb dating of the tuffaceous bed of Xiajiang Group in Guizhou Province and its stratigraphic implication. Geol China 37:1071–1080 (in Chinese with English abstract)

Gao LZ, Dai CG, Ding XZ, Wang M, Liu YX, Wang XH, Chen JS (2011) SHRIMP U-Pb dating of intrusive alaskite in the Fanjingshan Group and alaskite basal conglomerate constraints on the deposition of the Xiajiang Group. Geol China 38(6):1413–1420 (in Chinese with English abstract)

Gao LZ, Liu YX, Ding XZ, Zhang CH, Wang ZQ, Chen J, Liu YR (2012) SHRIMP dating of Cangshuipu Group in the middle part of the Jiangnan Orogen and its implications for tectonic evolutions. Geology in China 39(1):12–20 (in Chinese with English abstract)

Gao W, Zhang CH (2009) Zircon SHRIMP U-Pb ages of the Huangling granite and the tuff beds from Liantuo Formation in the Three Gorges area of Yangtze River, China and its geological significance. Geol Bull China 28(1):45–50 (in Chinese with English abstract)

Gao YP, Zhang XL, Zhang GJ, Chen KF, Shen YA (2018) Ediacaran negative C-isotopic excursions associated with phosphogenic events: Evidence from South China. Precambr Res 307:218–228

Gong Z, Li MS (2020) Astrochronology of the Ediacaran Shuram carbon isotope excursion, Oman. Earth Planet Sci Lett 547:116462

Gong Z, Kodama KP, Li YX (2017) Rock magnetic cyclostratigraphy of the Doushantuo Formation, South China and its implications for the duration of the Shuram carbon isotope excursion. Precambr Res 289:62–74

Gong Z, Kodama KP, Li YX (2019) Paleomagnetism and rock magnetic cyclostratigraphy of the Ediacaran Doushantuo Formation, South China: constraints on the remagnetization mechanism and the encoding process of Milankovitch cycles. Palaeogeogr Palaeoclimatol Palaeoecol 528:232–246

Gradstein FM, Ogg JG, Schmitz MD, Ogg GM (2012) The geologic time scale 2012. Elsevier, Oxford

Gradstein FM, Ogg JG, Schmitz MD, Ogg GM (2020) The geologic time scale 2020. Elsevier, Oxford

Guan CG, Wan B, Chen Z, Fu Q (2012) A revisit of the Neoproterozoic glacial deposit in southern Anhui. J Stratigr 36:611–629 (in Chinese with English abstract)
He MH (1997) Event deposition and stratigraphic correlation of Sinian Datangpo time in eastern Guizhou and its neighbouring regions. Guizhou Geol 14(1):21–29 (in Chinese with English abstract)
He MH (1998) Discussion on sedimentary facies, environmental evolution and tectonic setting during the Tiesiao-Nantuo Period of Sinian in the eastern Guizhou and its neighborhood. Guizhou Geol 15(1):26–31 (in Chinese with English abstract)
Hoffman PF, Schrag DP (2002) The snowball earth hypothesis: testing the limits of global change. Terra Nova 14:129–155
Hoffman PF, Kaufman AJ, Halverson GP, Schrag DP (1998) A neoproterozoic snowball earth. Science 281(5381):1342–1346
Hoffman PF, Abbot DS, Ashkenazy Y, Benn DI, Brocks JJ, Cohen PA, Cox GM, Creveling JR, Donnadieu Y, Erwin DH, Fairchild IJ, Ferreira D, Goodman JC, Halverson GP, Jansen MF, Le Hir G, Love GD, Macdonald FA, Maloof AC, Partin CA, Ramstein G, Rose BEJ, Rose CV, Sadler PM, Tziperman E, Voigt A, Warren SG (2017) Snowball Earth climate dynamics and Cryogenian geology-geobiology. Sci Adv 3:e1600983
Hu CL, Zhu MY (2020) Lithofacies and glacio-tectonic deformation structures of the Tiesi'ao/Dongshanfeng Formation on the Yangtze Craton, South China: implications for Sturtian Glaciation dynamics. Palaeogeogr Palaeoclimatol Palaeoecol 538:109481
Hu MY, Gao D, Wei GQ, Yang W, Xie WR (2019) Sequence stratigraphy and facies architecture of a mound-shoal-dominated dolomite reservoir in the Late Ediacaran Dengying Formation, central Sichuan Basin, SW China. Geol J 54(3):1653–1671
Huang DG, Mu J, Wang AH (2010) Manganese bearing series and early sedimentary environment in Nanhua Period in Yinjiang-Songtao area of Guizhou. Guizhou Geology 27(1):13–21 (in Chinese with English abstract)
Huang KJ, Teng FZ, Shen B, Xiao SH, Lang XG, Ma HR, Fu Y, Peng YB (2016) Episode of intense chemical weathering during the termination of the 635 Ma Marinoan glaciation. Proc Natl Acad Sci 113(52):14904–14909
Huang JZ, Tang XS, Zhang XY, Guo LQ (1996) On the correlation between the Liantuo Formation of the Yangtze Gorge and the Banxi Group of Hunan. J Stratigr 20:232–236 (in Chinese with English abstract)
Jiang GQ, Kennedy MJ, Christie-Blick N (2003) Stable isotopic evidence for methane seeps in Neoproterozoic postglacial cap carbonates. Nature 426:822–826
Jiang GQ, Kaufman AJ, Christie-Blick N, Zhang SH, Wu HC (2007) Carbon isotope variability across the Ediacaran Yangtze platform in South China: implications for a large surface-to-deep ocean $\delta^{13}C$ gradient. Earth Planet Sci Lett 261(1–2):303–320
Jiang GQ, Shi XY, Zhang SH, Wang Y, Xiao SH (2011) Stratigraphy and paleogeography of the Ediacaran Doushantuo Formation (ca. 635–551 Ma) in South China. Gondwana Res 19:831–849
Jiang XS, Wang J, Cui XZ, Zhuo JW, Xiong GQ, Lu JZ, Liu JH (2012) Zircon SHRIMP U-Pb geochronology of the Neoproterozoic Chengjiang Formation in central Yunnan Province (SW China) and its geological significance. Sci China Earth Sci 42:1496–1507 (in Chinese)
Knauth LP (2005) Temperature and salinity history of the Precambrian ocean: implications for the course of microbial evolution. Palaeogeogr Palaeoclimatol Palaeoecol 219:53–69
Knoll AH, Walter MR, Narbonne GM, Christie-Blick N (2004) A new period for the geologic time scale. Science 305:621–622
Komiya T, Komiya T, Hirata T, Kitajima K, Yamamoto S, Shibuya T, Sawaki Y, Ishikawa T, Shu DG, Li Y, Han J (2008) Evolution of the composition of seawater through geologic time, and its influence on the evolution of life. Gondwana Res 14:159–174
Lambert IB, Walter MR, Zhang WL, Lu SN, Ma GG (1987) Paleoenvironment and carbon isotope stratigraphy of upper Proterozoic carbonates of the Yangtze platform. Nature 325:140–142
Lan ZW, Li XH, Zhu MY, Chen ZQ, Zhang QR, Li QL, Lu DB, Liu Y, Tang GQ (2014) A rapid and synchronous initiation of the wide spread Cryogenian glaciations. Precambr Res 255:401–411

Lan ZW, Li XH, Zhu MY, Zhang QR, Li QL (2015) Revisiting the Liantuo Formation in Yangtze Craton, South China: SIMS U-Pb zircon age constraints and regional and global significance. Precambr Res 263:123–141

Lang XG, Chen JT, Cui H, Man L, Huang KJ, Fu Y, Zhou CM, Shen B (2018) Cyclic cold climate during the Nantuo glaciation: evidence from the Cryogenian Nantuo Formation in the Yangtze Craton, South China. Precambr Res 310:243–255

Lee LS, Chao YT (1924) Geology of the Gorge district of the Yangtze (from Ichang to Tzekuei) with special reference to the development of the Gorges. Bull Geol Soc China 3(3–4):351–391

Li C, Love GD, Lyons TW, Fike DA, Sessions AL, Chu XL (2010) A stratified redox model for the Ediacaran Ocean. Science 328:80–83

Li C, Zhu MY, Chu XL (2016) Preface: Atmospheric and oceanic oxygenation and evolution of early life on Earth: new contributions from China. J Earth Sci 27(2):167–169

Li CW, Chen JY, Hua TE (1998) Precambrian sponges with cellular structures. Science 279:879–882

Li XH, Li WX, Li ZX, Lo CH, Wang J, Ye MF, Yang YH (2009) Amalgamation between the Yangtze and Cathaysia Blocks in South China: constraints from SHRIMP U-Pb zircon ages, geochemistry and Nd-Hf isotopes of the Shuangxiwu volcanic rocks. Precambr Res 174:117–128

Li XH, Li WX, He B (2012) Building of the South China Block and its relevance to assembly and breakup of Rodinia supercontinent: observations, interpretations and tests. Bull Mineral Petrol Geochem 31(6):543–559 (in Chinese with English abstract)

Li YQ, He DF, Li D, Li SJ, Wo YJ, Li CX, Huang HY (2020) Ediacaran (Sinian) palaeogeographic reconstruction of the Upper Yangtze area, China, and its tectonic implications. Int Geol Rev 62(12):1485–1509

Li ZX, Bogdanova SV, Collins AS, Davidson A, De Waele B, Ernst RE, Fitzsimons ICW, Fuck RA, Gladkochub DP, Jacobs J, Karlstrom KE, Lu S, Natapov LM, Pease V, Pisarevsky SA, Thrane K, Vernikovsky V (2008) Assembly, configuration, and break-up history of Rodinia: a synthesis. Precambr Res 160:179–210

Li ZX, Evans DAD, Halverson GP (2013) Neoproterozoic glaciations in a revised global palaeogeography from the breakup of Rodinia to the assembly of Gondwanaland. Sed Geol 294:219–232

Liang DD, Cai YP, Nolan M, Xiao SH (2020) The terminal Ediacaran tubular fossil Cloudina in the Yangtze Gorges area of South China. Precambr Res 351:105931

Lin SJ, Xiao JF, Lu DB, Liu AM, Mou SY, Chen R, Yi CX, Wang XL (2010) Re-division about Fulu Formation and Fulu interglacial epoch in Hunan-Guizhou-Guangxi border area in South China. Geol Bull China 29:195–204 (in Chinese with English abstract)

Lin SJ, Lu DB, Xiao JF, Xiong XH, Li YT (2013) Stratigraphy of the Nanhuan system in Guizhou Province. J Stratigr 37:542–557 (in Chinese with English abstract)

Liu HY (1991) The Sinian system in China. Science Press, Beijing (in Chinese)

Liu HY, Hao J, Li YJ (1999) The late precambrian strata and their geological evolution of Eastern China. Science Press, Beijing (in Chinese)

Liu PJ, Yin CY, Gao LZ, Tang F, Chen SM (2009) New material of microfossils from the Ediacaran Doushantuo Formation in the Zhangcunping area, Yichang, Hubei Province and its zircon SHRIMP U-Pb age. Chin Sci Bull 54(6):1058–1064

Liu PJ, Yin CY, Chen SM, Tang F, Gao LZ (2013) The biostratigraphic succession of acanthomorphic acritarchs of the Ediacaran Doushantuo Formation in the Yangtze Gorges area, South China and its biostratigraphic correlation with Australia. Precambr Res 225:29–43

Liu PJ, Chen SM, Zhu MY, Li M, Yin CY, Shang XD (2014a) High-resolution biostratigraphic and chemostratigraphic data from the Chenjiayuanzi section of the Doushantuo Formation in the Yangtze Gorges area, South China: implication for subdivision and global correlation of the Ediacaran System. Precambr Res 249:199–214

Liu PJ, Xiao SH, Yin CY, Chen SM, Zhou CM, Li M (2014b) Ediacaran acanthomorphic acritarchs and other microfossils from chert nodules of the upper Doushantuo Formation in the Yangtze Gorges area, South China. J Paleontol 88(SP72):1–139

Liu PJ, Li XH, Chen SM, Lan ZW, Yang B, Shang XD, Yin CY (2015) New SIMS U-Pb zircon age and its constraint on the beginning of the Nantuo glaciation. Science Bulletin 60:958–963

Liu SG, Ma YS, Wang GZ, Cai XY, Huang WM, Zhang CJ, Xu GS, Yong ZQ, Pan CL (2008) Formation and conservation mechanism of the high-quality reservoirs in Sinian-lower Palaeozoic in Sichuan Basin. Petrol Geol Recovery Effi 15:1–5 (in Chinese with English abstract)

Liu TS, Zhou XL (2002) Characteristic phase and metallogenic model of sediment facies during Xiangmeng Period Early Sinian in southwest Hunan. Hunan Geology 21:30–34 (in Chinese with English abstract)

Love GD, Grosjean E, Stalvies C, Fike DA, Grotzinger JP, Bradley AS, Kelly AE, Bhatia M, Meredith W, Snape CE, Bowring SA, Condon DJ, Summons RE (2009) Fossil steroids record the appearance of Demospongiae during the Cryogenian. Nature 457(7230):718–723

Lu DB, Xiao JS, Lin SJ, Liu AM, Mou SY, Chen R, Yi CX, Wang XL (2010) The Nanhuaan section at Lijiapo Village, Congjiang County, Guizhou Province in Hunan-Guizhou-Guangxi adjacent region, China—a good section with sedimentary record in Nanhuaan large glaciation age. Geol Bull China 29:1143–1151 (in Chinese with English abstract)

Lu M, Zhu MY, Zhang JM, Shields GA, Li GX, Zhao FC, Zhao X, Zhao MJ (2013) The DOUNCE event at the top of the Ediacaran Doushantuo Formation of South China: Wide stratigraphic occurrence and non-diagenetic origin. Precambrian Res 225:86–109

Lu SN (2002) Discussion of the new subdivision of the Neoproterozoic in China. Geol Rev 48(3):242–248 (in Chinese with English abstract)

Lu SN, Ma GG, Gao ZJ, Lin WX (1985) Primary research on glacigenous rocks of Late Precambrian in China. In: Precambrian Geology Editorial Committee (ed) Precambrian geology, vol 1. The Collected Works of Late Precambrian Glacigenous Rocks in China. Geological Publishing House, Beijing. (in Chinese)

Luo B, Yang YM, Luo W, Wen L, Wang WZ, Chen K (2017) Controlling factors of Dengying Formation reservoirs in the central Sichuan paleo-uplift. Petrol Res 2(1):54–63

Lyons TW, Reinhard CT, Planavsky NJ (2014) The rise of oxygen in Earth's early ocean and atmosphere. Nature 506:307–315

Ma GG, Li HQ, Zhang ZC (1984) An investigation of the age limits of the Sinian in South China. Bull Yichang Inst Geol Miner Resour Chin Acad Geol Sci 8:1–29 (in Chinese with English abstract)

Ma HY, Sun HQ, Huang JZ, Ma TQ (2013) U-Pb dating of zircon from the Gaojian Group tuff in central Hunan Province and its geological implication. Miner Explor 4:69–74 (in Chinese with English abstract)

Macdonald FA, Schmitz MD, Crowley JL, Roots CF, Jones DS, Maloof AC, Strauss JV, Cohen PA, Johnston DT, Schrag DP (2010) Calibrating the cryogenian. Science 327(5970):1241–1243

Maruyama S, Sawaki Y, Ebisuzaki T, Ikoma M, Omori S, Komabayashi T (2014) Initiation of leaking Earth: an ultimate trigger of the Cambrian explosion. Gondwana Res 25(3):910–944

McFadden KA, Huang J, Chu XL, Jiang GQ, Kaufman AJ, Zhou CM, Yuan XL, Xiao SH (2008) Pulsed oxidation and biological evolution in the Ediacaran Doushantuo Formation. PNAS 105:3197–3202

Murphy MA, Salvador A (1999) International subcommission on stratigraphic classification of IUGS international commission on stratigraphy-international stratigraphic guide-an abridged version. Episodes 22(4):255–271

National Commission on Stratigraphy of China (2002) Explanatory memorandum of the regional chronostratigraphic chart (geological time table) of China. Geological Publishing House, Beijing (in Chinese)

Peng SC, Wang XF, Xiao SH, Tong JN, Hua H, Zhu MY, Zhao YL (2012) A call to replace the chronostratigraphic unit Sinian System (Period) with the global Ediacaran System (Period). J Stratigr 36(1):55–59 (in Chinese with English abstract)

Peng X, Zhu XK, Shi FQ, Yan B, Zhang FF, Zhao NN, Peng PA, Li J, Wang D, Shields GA (2019) A deep marine organic carbon reservoir in the non-glacial Cryogenian ocean (Nanhua Basin, South China) revealed by organic carbon isotopes. Precambr Res 321:212–220

Peng XJ, Liu YR, Wu NJ, Chen JC, Li JQ (2004) Correlation of the Nanhuan Strata on the southern margin of the Yangtze Landmass. J Stratigr 28(4):354–359 (in Chinese with English abstract)
Plumb KA (1991) New Precambrian time scale. Episodes 14(2):139–140
Plumb KA, James HL (1986) Subdivision of Precambrian time: recommendations and suggestions by the Subcommission on Precambrian Stratigraphy. Precambr Res 32(1):65–92
Pu JP, Bowring SA, Ramezani J, Myrow P, Raub TD, Landing E, Mills A, Hodgin E, Macdonald FA (2016) Dodging snowballs: Geochronology of the Gaskiers glaciation and the first appearance of the Ediacaran biota. Geology 44(11):955–958
Qian MP, Zhang ZY, Jiang Y, Yu MG, Yan YK, Ding BL (2012) The Neoproterozoic tillite in southeastern China. J Stratigr 36(3):587–599 (in Chinese with English abstract)
Qu HJ, Li P, Dong YP, Yang B, Chen S, Han X, Wang K, He M (2020) Development and distribution rules of the main Neoproterozoic source and reservoir strata in the Yangtze Craton, Southern China. Precambrian Res 350:105915
Rooney AD, Yang C, Condon DJ, Zhu MY, Macdonald FA (2020) U-Pb and Re-Os geochronology tracks stratigraphic condensation in the Sturtian snowball Earth aftermath. Geology 48(6):625–629
Sahoo SK, Planavsky NJ, Kendall B, Wang XQ, Shi XY, Scott C, Anbar AD, Lyons TW, Jiang GQ (2012) Ocean oxygenation in the wake of the Marinoan glaciation. Nature 489:546–549
Sawaki Y, Ohno T, Tahata M, Komiya T, Hirata T, Maruyama S, Windley B, Han J, Shu DG, Li Y (2010) The Ediacaran radiogenic Sr isotope excursion in the Doushantuo Formation in the Three Gorges area, South China. Precambrian Res 176(1–4):46–64
Shi SF, Jiang CR, Zhang JK (1985) A study on the Sinian tillites in western Zhejiang Province. In: Precambrian Geology Editorial Committee (ed) Precambrian geology, vol 1. The Collected Works of Late Precambrian Glacigenous Rocks in China. Geological Publishing House, Beijing, pp 261–282. (in Chinese)
Shields GA, Mills BJ, Zhu MY, Raub TD, Daines SJ, Lenton TM (2019) Unique Neoproterozoic carbon isotope excursions sustained by coupled evaporite dissolution and pyrite burial. Nat Geosci 12(10):823–827
Shields-Zhou GA, Och L (2011) The case for a Neoproterozoic oxygenation event: geochemical evidence and biological consequences. GSA Today 21:4–11
Shields-Zhou GA, Hill AC, Macgabhann BA (2012) Chapter 17: the Cryogenian Period. In: Gradstein FM, Ogg JG, Schmitz MD et al (eds) The geologic time scale 2012. Elsevier, Oxford, pp 399–411
Sui Y, Huang CJ, Zhang R, Wang ZX, Ogg J (2019) Astronomical time scale for the middle-upper Doushantuo Formation of Ediacaran in South China: implications for the duration of the Shuram/Wonoka negative $\delta^{13}C$ excursion. Palaeogeogr Palaeoclimatol Palaeoecol 532:109273
Sui Y, Huang CJ, Zhang R, Wang ZX, Ogg J, Kemp DB (2018) Astronomical time scale for the lower Doushantuo Formation of Early Ediacaran, South China. Sci Bull 63(22):1485–1494
Tahata M, Ueno Y, Ishikawa T, Sawaki Y, Murakami K, Han J, Shu DG, Li Y, Guo JF, Yoshida N (2013) Carbon and oxygen isotope chemostratigraphies of the Yangtze platform, South China: decoding temperature and environmental changes through the Ediacaran. Gondwana Res 23(1):333–353
Tan Z, Jia W, Li J, Yin L, Wang S, Wu J, Song J (2021) Geochemistry and molybdenum isotopes of the basal Datangpo formation: implications for ocean-redox conditions and organic matter accumulation during the Cryogenian interglaciation. Palaeogeogr Palaeoclimatol Palaeoecol 563:110169
Van Iten H, Leme JM, Pacheco ML, Simões MG, Fairchild TR, Rodrigues F, Galante D, Boggiani PC, Marques AC (2016) Origin and early diversification of phylum Cnidaria: key macrofossils from the Ediacaran System of North and South America. In: Goffredo S, Dubinsky Z (eds) The Cnidaria, past, present and future. Springer, Switzerland, pp 31–40
Vernhet E, Heubeck EC, Zhu MY, Zhang JM (2007) Stratigraphic reconstruction of the Ediacaran Yangtze platform margin (Hunan Province, China) from margin-originated large-scale olistolith. Palaeogeogr Palaeoclimatol Palaeoecol 254:123–139

Wang D, Zhu XK, Zhao NN, Yan B, Li XH, Shi FQ, Zhang FF (2019) Timing of the termination of Sturtian glaciation: SIMS U-Pb zircon dating from South China. J Asian Earth Sci 177:287–294
Wang J (2000) Neoproterozoic rifting history of South China significance to Rodinia breakup. Geological Publishing House, Beijing (in Chinese)
Wang J (2005) New advances in the study of "the Nanhuaan System"—with particular reference to the stratigraphic division and correlation of the Nanhuaan System, South China. Geol Bull China 24:491–495 (in Chinese with English abstract)
Wang J, Li ZX (2003) History of Neoproterozoic rift basins in South China: implications for Rodinia break-up. Precambr Res 122:141–158
Wang J, Li XH, Duan TZ, Liu DY, Song B, Li ZX, Gao YH (2003) Zircon SHRIMP U-Pb dating for the Cangshuipu volcanic rocks and its implications for the lower boundary age of the Nanhua strata in South China. Chin Sci Bull 48:1663–1669 (in Chinese)
Wang XF, Bi ZG (1985) The Sinian tillites of south Anhui Province. In: Precambrian Geology Editorial Committee (ed) Precambrian geology, vol 1. The Collected Works of Late Precambrian Glacigenous Rocks in China. Geological Publishing House, Beijing, pp 245–260. (in Chinese with English abstract)
Wang XF, Chen X, Wang C, Chen L (2001) Basal boundary and its inner chronostratigraphic subdivision of the Sinian System in China. J Stratigr SP25:370–376. (in Chinese with English abstract)
Wang YG, Yin CY, Zheng SF, Qin SR, Chen YL, Luo QL, Zhu SX, Wang F, Qian Y (1984) The Upper Precambrian and Sinian-Cambrian Boundary in Guizhou. Guizhou People's Publishing House, Guiyang (in Chinese)
Wang YG, Xie ZC, Wang LX, Chen DC, Zhu SC (1986) The stratigraphic sequence and origin of the depositional environment of the Tiesi'ao Formation in eastern Guizhou and its adjacent areas. Reg Geol China 4:341–348 (in Chinese with English abstract)
Wang YL, Lu ZB, Xing YS, Gao ZJ, Lin WX, Ma GG, Zhang LY, Lu SN (1980) Subdivision and correlation of the Upper Precambrian in China. In: Tianjin Institute of Geology and Mineral Resources (ed) Research on Precambrian Geology-Sinian Suberathem in China. Tianjing Science and Technology Press, Tianjin, pp 1–32. (in Chinese with English abstract)
Wang Z, Chen C, Wang JS, Suess E, Chen XH, Ma XC, Wang GZ, Xiao SH (2020) Wide but not ubiquitous distribution of glendonite in the Doushantuo Formation, South China: Implications for Ediacaran climate. Precambr Res 338:105586
Wang ZQ, Yin CY, Gao LZ, Tang F, Liu YQ, Liu PJ (2006) The character of the chemical index of alteration and discussion of subdivision and correlation of the Nanhua system in Yichang area. Geol Rev 52(5):577–585 (in Chinese with English abstract)
Wei W, Wang D, Li D, Ling HF, Chen X, Wei GY, Zhang FF, Zhu XK, Yan B (2016) The marine redox change and nitrogen cycle in the Early Cryogenian interglacial time: evidence from nitrogen isotopes and Mo contents of the basal Datangpo formation, northeastern Guizhou, South China. J Earth Sci 27(2):233–241
Wei W, Frei R, Klaebe R, Li D, Wei GY, Ling HF (2018) Redox condition in the Nanhua Basin during the waning of the Sturtian glaciation: a chromium-isotope perspective. Precambr Res 319:198–210
Wood R, Liu AG, Bowyer F, Wilby PR, Dunn FS, Kenchington CG, Cuthill JFH, Mitchell EG, Penny A (2019) Integrated records of environmental change and evolution challenge the Cambrian explosion. Nat Ecol Evol 3:528–538
Xiao D, Cao J, Luo B, Tan XC, Xiao WJ, He Y, Li KY (2020) Neoproterozoic postglacial paleoenvironment and hydrocarbon potential: a review and new insights from the Doushantuo Formation Sichuan Basin, China. Earth Sci Rev 212:103453
Xiao SH, Zhang Y, Knoll AH (1998) Three-dimensional preservation of algae and animal embryos in a Neoproterozoic phosphorite. Nature 391:553–558
Xiao SH, Yuan XL, Knoll AH (2000) Eumetazoan fossils in terminal Proterozoic phosphorites? PNAS 97:13684–13689

Xiao SH, Narbonne GM, Zhou CM, Laflamme M, Grazhdankin DV, Moczydłowska-Vidal M, Cui H (2016) Towards an Ediacaran time scale: problems, protocols, and prospects. Episodes 39(4):540–555

Xing YS, Yin CY, Gao LZ (1999) Boundary and subdivision of the Sinian System. Earth Sci J China Univ Geosci 13(2):202–204 (in Chinese with English abstract)

Xu XS, Huang HQ, Liu BJ, Wang YG (1991) The sedimentology and origin of Early Sinian manganese deposits from the Datangpo formation, South China. Acta Sedimentologica Sinica 9(1):63–71 (in Chinese with English abstract)

Xu XS, Liu W, Men YP, Zhang HQ (2012) Probe into the tectonic nature of Neoproterozoic southern Hunan-northern Guangxi marine basin. Acta Geol Sin 86(12):1892–1904 (in Chinese with English abstract)

Yan B, Shen WB, Zhao NN, Zhu XK (2020) Constraints on the nature of the Marinoan glaciation: cyclic sedimentary records from the Nantuo Formation, South China. J Asian Earth Sci 189:104137

Yang AH, Zhu MY, Zhang JM, Zhao FC, Lu M (2015) Sequence stratigraphic subdivision and correlation of the Ediacaran (Sinian) Doushantuo Formation of Yangtze Plate, South China. J Palaeogeogr 17(1):1–20 (in Chinese with English abstract)

Yang C, Li XH, Zhu MY, Condon DJ (2017a) SIMS U-Pb zircon geochronological constraints on Upper Ediacaran stratigraphic correlations, South China. Geol Mag 154(6):1202–1216

Yang C, Zhu MY, Condon DJ, Li XH (2017b) Geochronological constraints on stratigraphic correlation and oceanic oxygenation in Ediacaran-Cambrian transition in South China. J Asian Earth Sci 140:75–81

Yang F, Wang ZJ, Wang J, Du QD, Deng Q, Wu H, Zhou XL (2012) An analysis on property and dynamics of the Middle Neoproterozoic sedimentary basin in the western South China: constraint from the sedimentary data of Danzhou Group in Northern Guangxi. Geol Rev 58(5):854–864 (in Chinese with English abstract)

Yang RD, Gao JB, Cheng ML, Wei HR, Xu LQ, Wen XF, Wei X (2010) Sedimentary geochemistry of manganese deposit of the neoproterozoic Datangpo formation in Guizhou Province,China. Acta Geologica Sinica 84(12):1781–1790 (in Chinese with English abstract)

Ye YT, Wang HJ, Zhai LN, Wang XM, Wu CD, Zhang SC (2018) Contrasting Mo-U enrichments of the basal Datangpo formation in South China: implications for the Cryogenian interglacial ocean redox. Precambr Res 315:66–74

Yin CY, Gao LZ (2013) Deposition, time limit and stratigraphic subdivision of the Nanhua system in China. J Stratigr 37(4):534–541 (in Chinese with English abstract)

Yin CY, Liu DY, Gao LZ, Wang ZQ, Xing YS, Jian P, Shi YR (2003) Lower boundary age of the Nanhua system and the Gucheng glacial stage: evidence from SHRIMP II dating. Chin Sci Bull 48:1657–1662

Yin CY, Gao LZ, Xing YS, Wang ZQ, Tang F (2004) Advances in the study on the Nanhua system of the Neoproterozoic and its stratotype in South China. Prof Pap Stratigr Palaeontol 28:1–10 (in Chinese with English abstract)

Yin JC, He TG, Li SL, Cai XL, Wen CQ, Yuan HH, Ye XH (1993) Geological evolution and mineralization from the surrounding areas of Sichuan Basin and its vicinal regions during the Sinian Subera. Press of Chengdu University of Science and Technology, Chengdu (in Chinese)

Yin LM, Zhu MY, Knoll AH, Yuan XL, Zhang JM, Hu J (2007) Doushantuo embryos preserved inside diapause egg cysts. Nature 446:661–663

Yin ZJ, Zhu MY, Davidson EH, Bottjer DJ, Zhao FC, Tafforeau P (2015) Sponge grade body fossil with cellular resolution dating 60 Myr before the Cambrian. PNAS 112(12):E1453–E1460

Yu WC, Algeo TJ, Du YS, Maynard B, Guo H, Zhou Q, Peng TP, Wang P, Yuan LJ (2016) Genesis of Cryogenian Datangpo manganese deposit: hydrothermal influence and episodic post-glacial ventilation of Nanhua Basin, South China. Palaeogeogr Palaeoclimatol Palaeoecol 459:321–337

Yuan XL, Chen Z, Xiao SH, Zhou CM, Hua H (2011) An Early Ediacaran assemblage of macroscopic and morphologically differentiated eukaryotes. Nature 470:390–393

Zhang FF, Zhu XK, Yan B, Kendall B, Peng X, Li J, Algeo TJ, Romaniello S (2015a) Oxygenation of a Cryogenian ocean (Nanhua Basin, South China) revealed by pyrite Fe isotope compositions. Earth Planet Sci Lett 429:11–19

Zhang SG, Zhang YB, Yan HJ (2015b) Introduction to the stratigraphic chart of China (2014). J Stratigr 39(4):359–366 (in Chinese with English abstract)

Zhang QR, Chu XL (2006) The stratigraphic classification and correlation of the Jiangkou glaciation in the Yangtze Craton and the stratotype section of the Nanhuan system. J Stratigr 30(4):306–314 (in Chinese with English abstract)

Zhang QR, Chu XL (2007) Problems in defining the Nanhuan Period. J Stratigr 31(3):222–226 (in Chinese with English abstract)

Zhang QR, Chu XL, Bahlburg H, Feng LJ, Dobrzinski N, Zhang TG (2003) The stratigraphic architecture of the Neoproterozoic glacial rocks in "Xiang-Qian-Gui" region of the central Yangtze Craton, South China. Progress in Natural Science 13(10):783–787

Zhang QR, Chu XL, Feng LJ (2008a) A correlation of the "Xieshuihe formation", Nanhuan system, with a discussion to its glacial sedimentary structures. J Stratigr 32(3):246–252 (in Chinese with English abstract)

Zhang QR, Li XH, Feng LJ, Huang J, Song B (2008b) A new age constraint on the onset of the Neoproterozoic glaciations in the Yangtze Platform, South China. J Geol 116:423–429

Zhang QR, Chu XL, Feng LJ (2011) Neoproterozoic glacial records in the Yangtze Region, China. In: Arnaud E, Halverson G P, Shields-Zhou G (eds) The geological record of neoproterozoic glaciations, vol 36. Geological Society, Memoirs, London, pp 357–366

Zhang QR, Huang J, Chu XL (2012) The Nanhuan system in Xinluhe township of Huaihua, Hunan Province. J Stratigr 36(4):761–763 (in Chinese with English abstract)

Zhang SH, Jiang GQ, Dong J, Han YG, Wu HC (2008c) New SHRIMP U-Pb age from the Wuqiangxi Formation of Banxi Group: implications for rifting and stratigraphic erosion associated with the Early Cryogenian (Sturtian) glaciation in South China. Sci China, Ser D Earth Sci 51:1537–1544

Zhang SH, Jiang GQ, Han YG (2008d) The age of the Nantuo Formation and Nantuo glaciation in South China. Terra Nova 20:289–294

Zhang XL, Shu DG, Han J, Zhang ZF, Liu JN, Fu DJ (2014) Triggers for the Cambrian explosion: hypotheses and problems. Gondwana Res 25(3):896–909

Zhao GC, Wang YJ, Huang BC, Dong YP, Li SZ, Zhang G, Yu S (2018) Geological reconstructions of the East Asian blocks: From the breakup of Rodinia to the assembly of Pangea. Earth Sci Rev 186:262–286

Zhao WZ, Wei GQ, Yang W, Mo WL, Xie WR, Su N, Liu MC, Zeng FY, Wu SJ (2017) Discovery of Wanyuan-Dazhou Intracratonic Rift and its significance for gas exploration in Sichuan Basin, SW China. Petrol Explor Dev 44(5):697–707

Zhao ZQ, Xing YS, Ma GG, Yu W, Wang ZQ (1980) The Sinian System of eastern Yangtze Gorges, Hubei. In: Tianjin Institute of Geology and Mineral Resources (ed) Research on precambrian geology-sinian suberathem in China. Tianjin Science and Technology Press, Tianjin, pp 31–55. (in Chinese)

Zhou CM, Tucker R, Xiao SH, Peng ZX, Yuan XL, Chen Z (2004) New constraints on the ages of Neoproterozoic glaciations in South China. Geology 32:437–440

Zhou CM, Xie GW, McFadden K, Xiao SH, Yuan XL (2007a) The diversification and extinction of Doushantuo-Pertatataka acritarchs in South China: causes and biostratigraphic significance. Geol J 42:229–262

Zhou JB, Li XH, Ge WC, Li ZX (2007b) Age and origin of Middle Neoproterozoic mafic magmatism in southern Yangtze Craton and relevance to the break-up of Rodinia. Gondwana Res 12:184–197

Zhou CM, Li XH, Xiao SH, Lan ZW, Ouyang Q, Guan CG, Chen Z (2017a) A new SIMS zircon U-Pb date from the Ediacaran Doushantuo Formation: age constraint on the Weng'an biota. Geol Mag 154(6):1193–1201

Zhou CM, Xiao SH, Wang W, Guan CG, Ouyang Q, Chen Z (2017b) The stratigraphic complexity of the Middle Ediacaran carbon isotopic record in the Yangtze Gorges area, South China, and its

implications for the age and chemostratigraphic significance of the Shuram excursion. Precambr Res 288:23–38
Zhou CM, Yuan XL, Xiao SH, Chen Z, Hua H (2019) Ediacaran integrative stratigraphy and timescale of China. Sci China Earth Sci 62(1):7–24
Zhou MZ, Luo TY, Huff WD, Yang ZQ, Zhou GH, Gan T, Yang H, Zhang D (2018) Timing the termination of the Doushantuo negative carbon isotope excursion: evidence from U-Pb ages from the Dengying and Liuchapo formations, South China. Sci Bull 63(21):1431–1438
Zhou Q, Du YS, Yuan LJ, Zhang S, Yu WC, Yang ST, Liu Y (2016) Rift basin structure and its control function in Nanhua Period of Guizhou-Hunan-Chongqing Border area. Earth Sci J China Univ Geosci 41:177–188 (in Chinese with English abstract)
Zhu JL (1976) Pre-Sinian in Guizhou, summary of the stratigraphy in Guizhou. Geological Bureau of Guizhou Province, Guiyang. (in Chinese)
Zhu MY (2010) The origin and Cambrian explosion of animals: fossil evidence from China. Acta Palaeontologica Sinica 49(3):269–287 (in Chinese with English abstract)
Zhu MY, Strauss H, Shields GA (2007a) From snowball earth to the Cambrian bioradiation: calibration of Ediacaran-Cambrian earth history in South China. Palaeogeogr Palaeoclimatol Palaeoecol 254:1–6
Zhu MY, Zhang JM, Yang AH (2007b) Integrated Ediacaran (Sinian) chronostratigraphy of South China. Palaeogeogr Palaeoclimatol Palaeoecol 254:7–61
Zhu MY, Gehling JG, Xiao SH, Zhao YL, Droser ML (2008) An eight-armed Ediacara fossil preserved in contrasting taphonomic windows from China and Australia. Geology 36:867–870
Zhu MY, Lu M, Zhang JM, Zhao FC, Li GX, Zhao X, Zhao MJ (2013) Carbon isotope chemostratigraphy and sedimentary facies evolution of the Ediacaran Doushantuo Formation in western Hubei, South China. Precambr Res 225:7–28
Zhu XK, Sun J, Li ZH (2019) Iron isotopic variations of the Cryogenian banded iron formations: a new model. Precambr Res 331:105359

Chapter 6
Characterization and Developmental Background of Global Precambrian Hydrocarbon Source Beds

Ping'an Peng, Wanglu Jia, and Jian Chen

Abstract Precambrian source beds have been discovered worldwide. The sedimentary ages of these strata are mainly 2.7–2.6 Ga, ca. 2.0 Ga, 1.6–1.4 Ga, ca. 1.0 Ga, 0.7–0.6 Ga, and 0.6–0.5 Ga respectively. The characteristics of these source beds are closely related to the evolution of microbe, from aquatic cyanobacteria to eukaryotes as well as the preservation conditions for organic matter. The occurrence and thriving of eukaryotic algae in ecosystems may increase the possibility of oil-prone source bed formation. Geological processes, including crustal weathering, volcanic activity and glaciations, are also essential for the development of large-scale source beds. Thermal degradation of organic matter has been regarded as one of the most important factors controlling Precambrian source bed quality, and thus detailed and precise research on maturity would be very important for the evaluation of oil and gas potential. As macro fossils are rare in Precambrian source beds, the biomarkers, i.e., molecular fossils, present powerful tools for tracing the Precambrian microbial communities. For example, based on some specific biomarkers, the first appearance of sulfate-reducing bacteria and green sulfur bacteria can be dated to 1.64 Ga. The diversity of biomarkers in Precambrian strata provides a basis for oil-source rock correlation. For example, the source beds of bituminous sandstones in the Xiamaling Formation as well as oils in Oman have been successfully confirmed using both the specific biomarkers, 13α(*n*-methyl)-tricyclic terpanes and C_{19} A-norsterane. The researchers also reveal that carbon isotopic composition of organic matter in Precambrian strata remarkably differs from those in the Phanerozoic. Although the tremendous progress has been made in the discovery of Precambrian biomarkers, trace amounts of those compounds in the strata also provoked the debate regarding the indigeneity of Precambrian organic matter. Nevertheless, it is believed that this problem can be solved as analytical technology progresses in the future.

Keywords Precambrian · source bed/rock · Biomarkers · Indigeneity · Developmental mechanisms

---

P. Peng (✉) · W. Jia · J. Chen
State Key Laboratory of Organic Geochemistry, Guangzhou Institute of Geochemistry, Chinese Academy of Sciences, Guangzhou 510640, China

T. Wang, *Meso-Neoproterozoic Geology and Petroleum Resources in China*, Springer Geology, https://doi.org/10.1007/978-981-19-5666-9_6

6.1 Introduction

It is widely accepted that the global sedimentary environment in Precambrian was markedly different from that in Phanerozoic (Li et al. 2010; Craig et al. 2013). Precambrian geological research has been ongoing for more than a century and a great deal of progress has been made (Mossman et al. 2005; Zhang et al. 2007; Wang et al. 2008; Frolov et al. 2011; Liu et al. 2011; Dutta et al. 2013).

As for Precambrian source beds, several key topics, concerning the source and evolution of organic matter, remain to be investigated due to its long historical span and special geological background (Brocks et al. 2003a, 2005; Bao et al. 2004; Sherman et al. 2007; Wang et al. 2008; Grosjean et al. 2009). Although potential source beds rich in organic matter are widely distributed in Precambrian strata, the developmental mechanisms of effective source beds would be also different from those in the Phanerozoic. In view of the preservation of organic matter, the relatively low content of dissolved oxygen or sulfate in Precambrian oceans during early diagenesis (Li et al. 2010) may have suppressed the oxidation of organic matter in the water column and leading active organic matter (e.g., cellulose, protein etc.) to be well preserved, and facilitating some biochemical processes (e.g., fermentation, methane generation etc.). As the most important hydrocarbon source material, lipids may contain higher oxygen content and poorer organic type due to the incomplete remove of active substances. Regarding productivity of organic matter, since the Precambrian oceanic water has higher dissolved iron content than that in modern oceans, the main constraint factor of organic productivity may be referred to phosphorus instead of nitrogen. Crustal weathering and volcanic activity on different temporal and spatial scales would be the major processes for the enrichment of nutrients in waters (Condie et al. 2001). In addition, prior to the development of eukaryotic algae, cyanobacteria would be the main body in Precambrian microbial system, and thus the property of organic matter would be different from that in Phanerozoic algae. To date, all these inferences on the Precambrian source beds are mainly speculative and should be confirmed by future work.

The Precambrian oil and gas may have vast promising prospects (Craig et al. 2013), since a batch of large oil/gas fields have been discovered, especially in the Neoproterozoic strata. For example, the Dhahaban oil/gas system in Oman has produced about 30 million tons of oil annually, while the larger Anyue Gasfield has also been discovered in the Neoproterozoic Sinian Dengying and Lower Cambrian Longwangmiao Formations in the western Yangtze Craton, SW China. Thus, it is very significant for oil/gas exploration to strengthen the research on the characteristics and developmental background of indigenous organic matter in Precambrian source beds, meanwhile, the characteristics of Precambrian organic matter are also closely related to the marine ecosystems at that time. Therefore, the research of Precambrian source beds can also provide important evidence for the early Earth life evolution.

In this chapter, so far reported organic geochemical characteristics of Precambrian source beds in the literature have been summarized, the developmental mechanisms of these source beds discussed, and thus, a number of critical issues on Precambrian source bed studies are outlined here. It is hopeful that this work will provide useful information to the field of Precambrian geochemistry and petroleum exploration.

6.2 Global Records on Precambrian Hydrocarbon Source Beds

6.2.1 *Africa*

6.2.1.1 Archean Source Beds in the Kaapvaal Craton, South Africa

Buick et al. (1998) reported the occurrence of pyrobitumen nodules within the Black Reef Formation (2.59 Ga) and the Witwatersrand Group (2.85 Ga) in the Kaapvaal Cratonic Basin. The presence of these high-mature pyrobitumens suggests that they are hydrocarbons that migrated from neighboring shales. The fact that the shales of the Black Reef Formation and the Witwatersrand Group have total organic carbon contents (TOC) of 9.1% and 0.28%, respectively, which are suggestive of source bed for the pyrobitumen nodules. Additional research by Dutkiewicz et al. (1998) indicated that both strata contained oil-bearing inclusions as convincing evidence for the ancient oil accumulation. In addition, within the same basin, the Nauga and Klein Naute Formations that are also part of the Transvaal Group, with 10–100 m thick black shales, containing TOC 0.1–12% (avg. 3%; Kendall et al. 2010).

6.2.1.2 Palaeoproterozoic Source Beds in the Franceville Basin, Gabon

A molecular compositional analysis on liquid inclusions reported by Dutkiewicz et al. (2007) suggested that oils trapped in inclusions within the FA Formation (2.1 Ga) of the Francevillian Series (dated to 2.1–1.7 Ga in the Franceville Basin) are derived from organic matter of black shales in the overlying FB Formation (1.9 Ga). The FB Formation is 600–1000 m thick and contains TOC up to 15%, which is considered to comprise a set of high quality source bed (Mossman et al. 2005), while its extremely low H/C (0.5) and O/C (0.3) atomic ratios in kerogen are indicative of a considerably high mature phases. Thus, it is generally accepted that these inclusions could be indigenous, and preventing organic matter from contamination of younger hydrocarbons, which are suitable for the study of Precambrian indigenous organic matter.

From oil-bearing inclusions in Precambrian strata, Dutkiewicz et al. (2007) have identified some biomarkers, including low-carbon-numbers *n*-alkanes, regular steranes, C_{30} *n*-propyl cholestanes, hopanes, rearranged hopanes, 2α-methylhopanes, and gammacerane. These biomarkers are indicative of a major source input from cyanobacteria.

6.2.2 North America

6.2.2.1 Palaeoproterozoic Source Beds

Limited records of Palaeoproterozoic source beds are known from North America. Mancuso et al. (1989) reported pyrobitumen present in strata (dated to 1.9–2.0 Ga) in the Great Lakes region, its $\delta^{13}C$ values ranges −35 to −31‰, which could be hydrocarbons generated from Palaeoproterozoic source bed.

6.2.2.2 Mesoproterozoic Source Beds in the Nonesuch Formation, the Midcontinental Rift (1.1 Ga)

Lacustrine source beds are widely distributed across North America and have been studied for many years (Imbus et al. 1988). As a typical case, the Rice Formation in northeastern Kansas is enriched in organic matter (TOC up to 2.5%) with T_{max} values 440–460 °C indicative of mature to over-mature phases (Newell et al. 1993). Organic petrographic studies have referred its organic matter to Type I/II kerogen with $\delta^{13}C$ values −34 to −30‰. Abundant biomarkers, including hopanes, 2α-methylhopanes, 3β-methylhopanes and trace amounts of regular and rearranged steranes, have been identified from the source bed using metastable reaction monitoring gas chromatography-mass spectrometry (MRM GC–MS or GC–MS/MS; Pratt et al. 1991).

6.2.2.3 Neoproterozoic Source Bed in the Walcott Member of the Chuar Group, Western America

Source bed of the Walcott Member is common in Utah and Arizona, western North America (Summons et al. 1988), and have been dated to 900–850 Ma. Organic geochemical analyses show that the source rock contains TOC up to 9% and HI up to 255 mg HC/g TOC, its T_{max} values 433–449 °C indicates its maturity within the oil window (Uphoff 1997). The Walcott Member is considered the hydrocarbon source bed in the overlying Tapeats sandstones. Its detected biomarkers include rearranged hopanes and gammacerane, steranes and hopanes.

6.2.3 Australia

6.2.3.1 Archean Source Beds in the Pilbara Craton

Bitumen nodules preserved in lacustrine sediments of the Fortecue Group (2.75 Ga), the LallaRookh Formation (3.0 Ga), the Mosquito Creak Formation (3.25 Ga) and the

Warrawoona Group (3.46 Ga) were all likely generated from Archean-aged shales (Buick et al. 1998). However, because these shales are at extremely high maturity and have low TOC (ranging 0.21–0.32%), it is still unclear that whether, or not, they could actually act as the bituminous source. The oil-bearing inclusions in the LallaRookh Formation and Warrawoona Group reported by Dutkiewicz et al. (1998) represent migrated petroleum from Archean shales.

Brocks et al. (1999, 2003a, b, c) reported the presence of organic-rich shales in drill cores from the Fortecue Group and overlying Hamersley Group (2.5 Ga). The former contains TOC up to 11.4%, while the latter has values up to 7.9%, and both units are candidate source beds for Archean-aged bituminous nodules. Their extremely high equivalent vitrinite reflectance eqR_o values up to 2.6% in both units have also led to a lower H/C ratio down to 0.1, and the abundant biomarkers recognized in both units include regular steranes, hopanes, 2α- and 3β-methylhopanes. Certainly, 2α-mehtylhopane are characteristic of cyanobacteria, while 3β-methylhopanes indicative of methanotrophic bacteria. Rasmussen et al. (2008) reassessed the biomarkers found in those strata and realized the steranes and hopanes was entered the strata after 2.2 Ga. The new results eliminate the evidence of biomarker indigeneity and assumption of oxygenic photosynthesis 2.7 Ga ago (Brocks et al. 1999, 2003a, b, c).

6.2.3.2 Palaeo-Mesoproterozoic Source Beds in the McArthur Basin

Two sets of source beds have been documented in the McArthur Intracratonic Basin. The first set of source bed is the Palaeoproterozoic shale of the Barney Creek Formation (1.64 Ga), which has been collected from drill cores and shows TOC values up to 8%, HI as high as 500 mg HC/g TOC, and T_{max} values 435–450 °C. The shale is an organic-rich source bed at moderate mature phase (Lee and Brocks 2011), while the associated oil-seeps are suggestive of a palaeo-oil reservoir. Organic geochemical analyses of the shale have abundant biomarkers, including branched alkanes, hopanes, rearranged hopanes, and steranes. It is also noteworthy that the aryl-isoprenoids detected in the Barney Creek Formation by Brocks et al. (2005) support the prevalence of bacterial sulfate reduction (BSR) and the generation of H_2S which diffused into the euphotic layer at that time.

The second set is the Mesoproterozoic shales of the Velkerri Formation (1.43 Ga; Volk et al. 2003). The source bed has TOC values up to 8%, a maxima HI value at 600 mg HC/g TOC, and T_{max} values 435–470 °C. The biomarkers detected in the source bed are similar to those identified in the Barney Creek Formation, although gammacerane is less abundant. Detailed organic petrographic and geochemical researches have shown that oil trapped in inclusions of the Bessie Creek Formation (1.2 Ga) was emplaced from overlying source bed of the Velkerri Formation and comprises a mixture of high-matured condensate and normal oil (Dutkiewicz et al. 2003, 2004; Volk et al. 2005).

6.2.4 *Russia and Europe*

6.2.4.1 Palaeoproterozoic Source Bed in Northwestern Russia

The 600 m-thick Zaonezhskaya Formation (2.0 Ga) which is located near to Lake Onega and contains abundant mature shungite (Melezhik et al. 1999) with TOC values 0.1–50%, trace amounts of N, O, S and H elements. In addition, $\delta^{13}C$ values of this source bed range −45 to −17‰, appearing as a bimodal distribution with a maximum at −34‰. Research has shown that the Zaonezhskaya Formation was deposited in a brackish lacustrine environment, while the shungite is interpreted as an indigenous product of the Zaonezhskaya Formation.

6.2.4.2 Meso-Neoproterozoic Source Beds on the East European Craton

Bazhenova and Arefiev (1996) reported three sets of source beds, including upper sections of lower (ca. 1.4 Ga), middle (ca. 1.2 Ga) and upper (ca. 0.9 Ga) Riphean strata in the Moscow Basin/Syneclise, East European Craton (or called Russia Craton). Their TOC contents range up to 3.0%, 3.2%, and 1.2%, respectively, while their T_{max} values around 435 °C demonstrate a moderate mature phase. Moreover, Vendian strata in this region comprise two sets of source beds, including its lower (ca. 0.65 Ga) and middle sections (ca. 0.63 Ga). TOC values for these beds range up to 3.0% and 1.1%, respectively, while average T_{max} values around 430 °C also demonstrate moderate mature phase.

6.2.4.3 Meso-Neoproterozoic Source Beds in the Siberia Craton, East Russia

Siberia region is a globally well-known oil and gas enrichment area, where its Precambrian hydrocarbons are mainly distributed in the Siberia Craton (Frolov et al. 2011; Kelly et al. 2011; Ulmishek 2001a, b). As consistent with those in Moscow Basin, East European Craton, three sets of Riphean source beds are also developed respectively at the upper section of lower (ca. 1.4 Ga) as well as at middle (ca. 1.2 Ga) and upper Riphean strata (ca. 0.9 Ga) in Siberia Craton. These source beds contain TOC respectively up to 3.0, 0.7 and 1.2%, HI below 100 mg HC/g TOC and T_{max} around 500 °C which is indicative of a high-mature phase.

Kelly et al. (2011) demonstrated the presence of abundant biomarkers in above source beds, including regular and rearranged steranes, C_{30} *n*-propylcholestanes and *i*-propylcholestanes, regular hopanes, 2α- and 3β-methylhopanes, and gammacerane. The C_{30} *i*-propylcholestanes are diagnostic biomarkers of sponges. Based on the above characteristics, the Precambrian oil in Siberia Craton is most probably derived from Riphean and Vendian strata, although no further oil-source correlation analysis has yet been reported.

6.2.5 *Asia*

6.2.5.1 The Neoproterozoic to Lower Cambrian Source Beds in Oman

The Huqf Supergroup (dated to 0.81–0.53 Ga) contains three sets of organic-rich shale source beds in the Ghaba, Fahud, and South Oman Salt Basins of Oman (Hold et al. 1999; Terken and Frewin 2000; Terken et al. 2001; Grosjean et al. 2009). The first set of shale in the Masirah Bay Formation (0.635 Ga) of Nafun Group has TOC up to 4.9%, T_{max} around 435 °C, and HI 300–400 mg HC/g TOC. The second set in the Shuram and Buah Formations (dated to 0.56–0.55 Ga) of the Nafun Group contains TOC up to 11%, T_{max} around 435 °C, and HI 300–600 mg HC/g TOC. The third set in the U and Thuleilat Formations (0.54 Ga) of the Ara Group possess TOC up to 11%, variable T_{max} 425–430 °C, and HI up to 600 mg H/g TOC.

Infracambrian oil- and gas-fields found in above Oman basins have been termed the "Dhahaban petroleum system". Indeed, the Oman oil and gas geological reserves can be up to 16×10^8 m^3 and 1×10^{12} m^3, respectively, in which, 30% of the former and 70% of the latter would be considered as recoverable reserve. It is confirmed by detailed oil-source correlations that the Oman oil was sourced mainly from the shales of the U and Thuleilat Formations and migrated into reservoirs 50 Ma ago, which has made the Oman salt basins to be one of the famous Precambrian oil and gas enrichment regions in the world.

Detailed research has revealed the abundant biomarkers in these source beds, including medium-chain alkanes, gammacerane, tricyclic terpanes, hopanes, 2α- and 3β-methylhopanes, regular steranes, and rearranged hopanes.

Grosjean et al. (2009) indicated he C_{19} A-norsterane reported by provided key information in local oil-source correlation, while Love et al. (2009) applied the catalytic hydropyrolysis method to confirm the occurrence of 24 *i*-propylcholestane in the Masirah Bay Formation (0.635 Ga), and interpreted as the first appearance of sponges.

6.2.5.2 Neoproterozoic Source Beds on the Krol Platform, Northern India

Neoproterozoic source beds in this region were deposited between 635 and 541 Ma ago, subsequent to the Marinoan glaciation. Thus, source beds on the Krol Platform include Krol Group as well as the Infra Krol and Blaini Formations (Kaufman et al. 2006). Certainly, the shales of B and C layers in the Krol Group exhibit the highest TOC values up to 1.85%, while TOC values are relatively lower in the shales of the Infra Krol Formation. Limited research has been carried out to date on the petroleum geochemistry of these units; Dutta et al. (2013) detected the presence of a number of common biomarkers, including regular and rearranged steranes, gammacerane, tricylic terpanes, and methyl chromans. Of these, methyl chromans are indicative of a stratified water column, while C_{29} regular steranes are also most abundant and imply an algal contribution to the source beds. Data show that $\delta^{13}C$ values of phytanes, pristanes, and *n*-alkanes range between −37 and −33‰, towards the light end of the spectrum, while the molecular and $\delta^{13}C$ characteristics of petroleum compounds are very similar to those detected in the Huqf Group, Oman.

6.2.5.3 Neoproterozoic-Lower Cambrian Source Beds in the Tarim Basin, NW China

The breakthrough of deep to ultra-deep oil exploration in the Tarim Basin (Jin et al. 2017; Yang et al. 2020) has greatly stimulated the re-evaluation of the Late Neoproterozoic-Lower Carmbrian source beds recently (Peng 2015; Zhu et al. 2018; Deng et al. 2021; Li et al. 2021). To date, four units of source beds have been shown to be developed in the Tarim (Fig. 6.1a), mainly based on the outcrop investigations. Nevertheless, the details of two Late Neoproterozoic units are still poorly understood when compared with the Lower Cambrian units.

Two old organic-rich units were developed in the interglaciation periods of the Cyrogenian (Nanhuan in China). The oldest unit is observed in the top Aletonggou Formation (Fig. 6.1a), which overlies the diamictites of this formation and underlies the Teruiaiken Formation. The total thickness of organic-rich black mudstone in this unit is ca. 10 m, and the TOC content ranges from 0.97–2.83% (avg. 1.65%; Peng 2015). The latest study further shows that a very thick, organic-rich unit (ca.

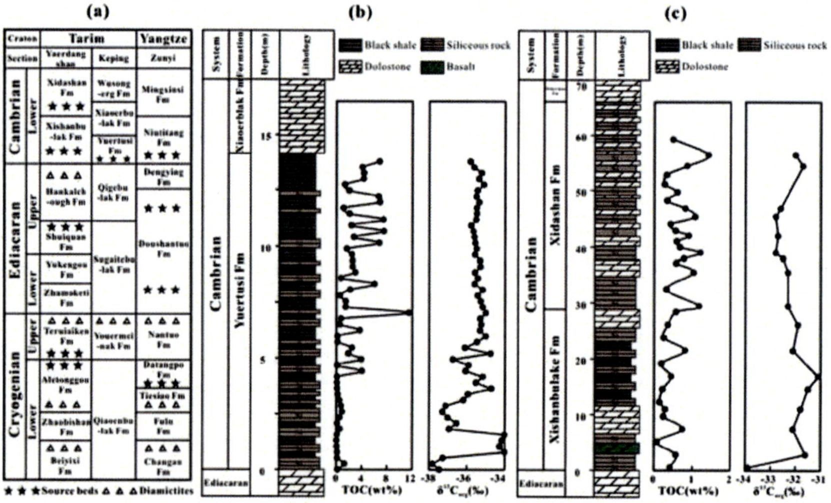

Fig. 6.1 **a** Summary of the distribution of Neoproterozoic-Lower Cambrian source beds in the Tarim Basin (Jia et al. 2003; Peng 2015; Li et al. 2021); **b** the content and carbon isotope values of bulk organic matter in the Lower Cambrian source beds from the Shiairik section, Keping area of western Tarim (Zhu et al. 2018); and **c** the content and carbon isotope values of bulk organic matter in the Lower Cambrian source beds from the Yaerdangshan section, eastern Tarim

320 m) is also encountered in the basal Teruiaiken Formation (Fig. 6.1a), which has the TOC content of 1.0–2.6% (Li et al. 2021). This unit overlies the dolostones in the lowermost Teruiaiken Formation and is overlain by thick diamictites of this formation.

The third organic-rich unit is located in the upper Shuiquan Formation (Sinian in China) deposited during the Middle-Late Ediacaran Period (Fig. 6.1a). This unit has been previously shown to be organic-lean (TOC content of 0.06–0.61%, the Yaerdangshan section; Peng 2015), which may be caused by the weathering effect. By contrast, a recent study has shown that this unit is ca. 60 m thick and has relatively high TOC content ranging from 0.92 to 1.75% (Li et al. 2021).

The youngest organic-rich unit is the well-known Lower Cambrian source beds, which have been increasingly thought to make great contribution to oil in deep reservoirs of the Tarim Basin (Huang et al. 2016). Compared with the Neoproterozoic source beds, the Lower Cambrian source beds have much greater TOC content (up to 22%, Gu et al. 2012). Though generally being characterized by the black shale interbedded with siliceous rocks, this unit shows large variations both in the thickness and in the content and carbon isotopes of bulk organic matter (Fig. 6.1b, c). In a typical section of the western Tarim (Fig. 6.1b), this unit located in the Yuertusi Formation is relatively thin and has very high TOC content (0.05–11.5% with an average of 3.72%) that generally increased upward (Zhu et al. 2018). By contrast, this unit is distributed in two Lower Cambrian formations of the section from the eastern Tarim, i.e. the Xishanbulak Formation and Xidashan Formation (Fig. 6.1c), which are much thicker than their western counterpart. However, the TOC content of the two formations varies from 0.08 to 1.42% (avg. 0.58%), which is much lower than that of the Yuertusi Formation. In addition, the bulk organic matter in the Yuertusi Formation is much more depleted in ^{13}C relative to that in the Xishanbulak Formation and Xidashan Formation. These results are consistent with those summarized by a recent study (Chen et al. 2020). Large variations in both geological and geochemical characteristics displayed by this unit from different locations in the basin could be related to the terrestrial material input, biological types as well as the sedimentary environment on carbon isotopic signatures of carbon pools (Huang et al., 2016; Deng et al. 2021).

6.2.5.4 Neoproterozoic Source Beds in Western Yangtze Craton, China

Two sets of organic-rich source beds have been reported in Nanhuan and Sinian strata. One comprises over-mature black shales of the Xiangmeng Formation or Datangpo Formation (663 Ma) with TOC up to 10% (avg. 3.3%; Tan et al. 2021), and another is the black shales of the Doushantuo Formation or Lantian Formation (<635 Ma) with TOC up to 20% (avg. 3.9%; Wang et al. 2017). Since both are in over-mature phase, few biomarkers could be detected (Wang et al. 2008). Meng et al. (2003) only found dinosterane in very low abundance on mass chromatograph, while Wang et al. (2005), Han et al. (2016; cf. Chap. 15) considered the black shale rich-in organic matter of Neoproterozoic Sinian Doushantuo Formation as the source bed for above

Longmenshan asphaltic veins based on their initial correlation of pregnane, regular steranes, tricyclic terpanes etc. (cf. Chap. 15).

6.2.5.5 Meso-Neoproterozoic Source Beds in Yanliao Faulted-Depression Zone (YFDZ), North China Craton

Three sets of Mesoproterozoic hydrocarbon source beds have been identified in the Yanliao Faulted-Depression Zone (YFDZ), northern China (Liu et al. 2000; Wang et al. 2002; Bao et al. 2004; Zhang et al. 2007; Luo et al. 2013; cf. Chap. 11). The first comprises the over-mature Gaoyuzhuang Formation (ca. 1.6 Ga) which has high TOC values of 0.5–4.29% (avg. 1.16%) as well as very high equivalent vitrinite reflectance eqR_o values 1.38–1.75% (avg. 1.59%) and low HI values 11–45 mg HC/g TOC (avg. 21 mg HC/g TOC). The second is the mature to high-mature shale in Hongshuizhuang Formation (ca. 1.42–1.5 Ga) that has high TOC values 0.5–7.21% (avg. 4.65%), medium to high eqR_o values 0.9–1.42% (avg. 1.19%), and high HI values 97–311 mg HC/g TOC (avg. 233 mg HC/g TOC). While the third is the mature or over-mature shale in the Xia-3 Member of Xiamaling Formation (1.32 ~ 1.4 Ga), with high TOC values up to 2.20–4.71% (avg. 3.16%), variable eqR_o values 0.96–7.53% (avg. 1.87%), and HI values up to 500 mg HC/g TOC. However, the Xiamaling source bed is only limited to the Xuanlong Depression and probably to the Jingxi Depression in YFDZ, North China Craton (cf. Chap. 11).

A good deal of previous researches has been conducted on the basal bituminous sandstone of Mesoproterozoic Xiamaling Formation in the YFDZ. Wang (1980) and Wang et al. (1988) discovered and reported the Xiamaling bituminous sandstone as the oldest fossil-oil-reservoir (1.4–1.327 Ga) in China (cf. Chaps. 10 and 12). Liu et al. (2011) considered that the fossil-oil-reservoir was resulted from the thermo-alteration of original oil-reservoir consequent on magmatic intrusion of gabbro-diabase in Jibei Depression, YFDZ.

Wang (1991), Wang and Simoneit (1995) detected a novel biomarker series of 13α-(*n*-alkyl)-tricyclic terpanes from the Xiamaling bituminous sandstone, and indicated that original liquid oil of the fossil-oil-reservoir would have a microbial source input based on the novel biomarker series, while Wang (2009) suggested that eukaryote algae may be the major microbial source.

The 13α-(*n*-alkyl)-tricyclic terpane series has been detected not only from the bituminous sandstone, but also from the Mesoproterozoic shales (Zhang et al. 2007; Wang et al. 2011) and the related oil-seeps in YFDZ (cf. Chap. 11) as well as from the asphaltic veins of Lower Cambrian Guojiaba Formation in the Longmenshan Nappe Zone, western Yangtze Craton (Huang and Wang 2008; cf. Chap. 15), and that Wang et al. (2005) has suggested that the Guojiaba asphaltic veins would be sourced by the black shale source bed of Neoproterozoic Sinian Doushantuo Formation. So far 13α-(*n*-alkyl)-tricyclic terpanes have never been reported from Phanerozoic sediments. Therefore, 13α-(*n*-alkyl)-tricyclic terpanes may act as a kind of diagnostic biomarkers especially for Precambrian hydrocarbons.

By means of artificially catalytical hydrocracking and GC–MS techniques, the correlation of 13α-(*n*-alkyl)-tricyclic terpanes between the bituminous sandstone or oil-seeps and kerogen-degraded products of source rocks indicates that Mesoproterozoic Hongshuizhuang shale should be the source bed for Xiamaling fossil-oil-reservoir and oil-seeps in YFDZ (cf. Chaps. 12 and 13).

In addition, Peng et al. (1998) and Li et al. (2003) have detected numerous biomarkers, including steranes, rearranged steranes, tricyclic terpanes, gammacerane and rearranged hopanes, from Changchengian, Jixianian and Qingbaikouan strata in the Jidong Depression, YFDZ.

6.3 Developmental Mechanisms of Precambrian Source Beds

Geohistorically, the developmental process of source bed occurs by phases, while Precambrian source beds have presented mainly in six geological periods of time: 2.76–2.67 Ga, ca. 2.0 Ga, 1.6–1.4 Ga, ca. 1.0 Ga, 0.7–0.6 Ga and 0.6–0.5 Ga (Fig. 6.2). The phased development of source beds is positively correlated to crustal weathering, their correlation coefficient is up to 70% (Fig. 6.3; Condie et al. 2001), while the crust weathering process is associated with the supercontinental unification and breakup. It is commonly accepted that the Precambrian supercontinents may have four significant developmental periods, dealing with the Palaeoproterozoic (2.5–2.1 Ga), Mesoproterozoic (ca. 1.8 Ga) and Rodinia (1.3–1.0 Ga) Supercontinents as well as Neoproterozoic Vendian Oldland (ca. 0.6 Ga). The transitional period from supercontinental breakup to reunification would result in a strong crustal weathering stage on land, which would transport a large amount of nutrients from continents into water column for the propagation of aquatic organism, and creating favorable conditions for the development of source bed.

The unification and breakup of supercontinents had a long-term impact on crustal weathering, while the fluctuation between ice and warm ages and volcanic activity would make the nutrients to be rapidly increased for the aquatic ecological system during a shorter time, which would be favorable for the eutrophication of waters, and promoting the phased development of source beds. As for the Precambrian time, the crust physical weathering at Sturtian and Marinoan ice ages leads nutrients to be enriched on land, while a mass of nutrients has been inputted into oceanic waters at subsequent warm age, and facilitating the development of source bed such as the black shales of Datangpo and Doushantuo Formations. Moreover, the hydrolysis of volcanic lava and ash could provide a lot of nutrients to promote source bed development. Volcanic activity could play another role is to bring the ice age to an end. However, the relationship between Precambrian volcanism and source bed development is still not very clear due to its inadequate study, however, it seems to the present authors that C_{30} rearranged hopanes might be the indicator of volcanic

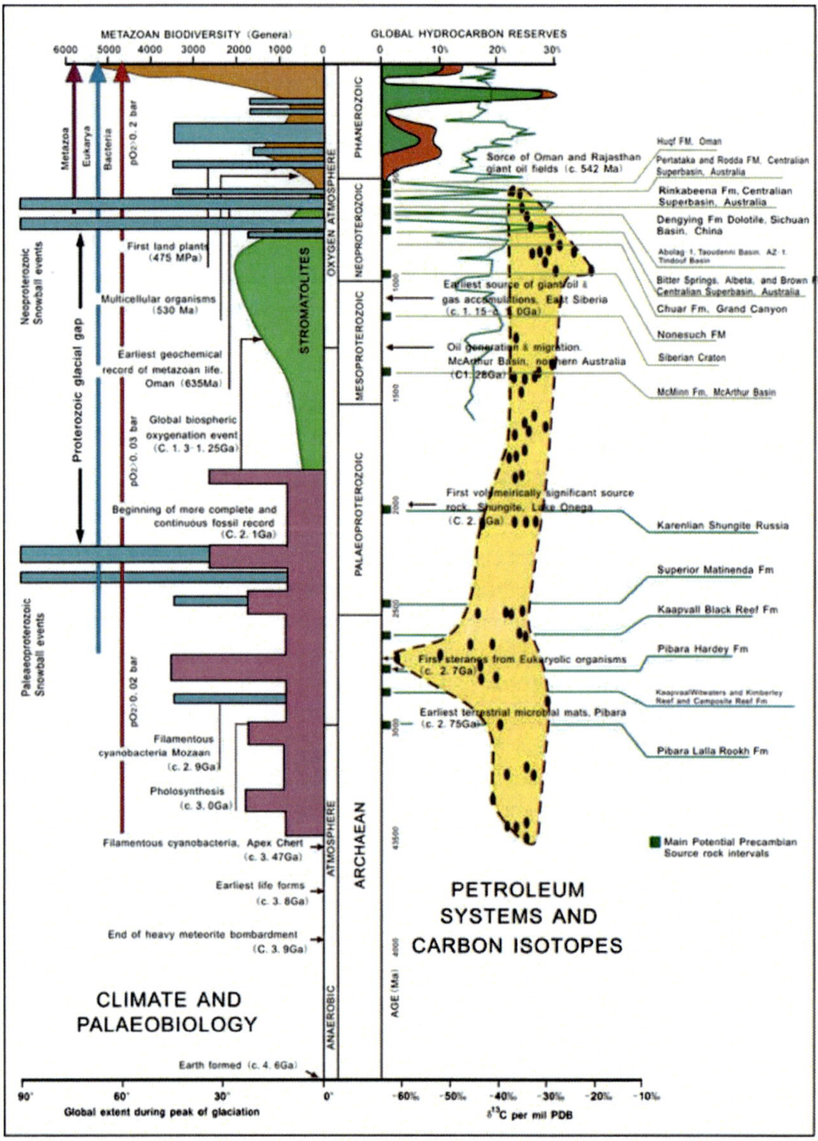

Fig. 6.2 Summary of source bed distribution and biological evolution during the Precambrian Craig et al. (2013)

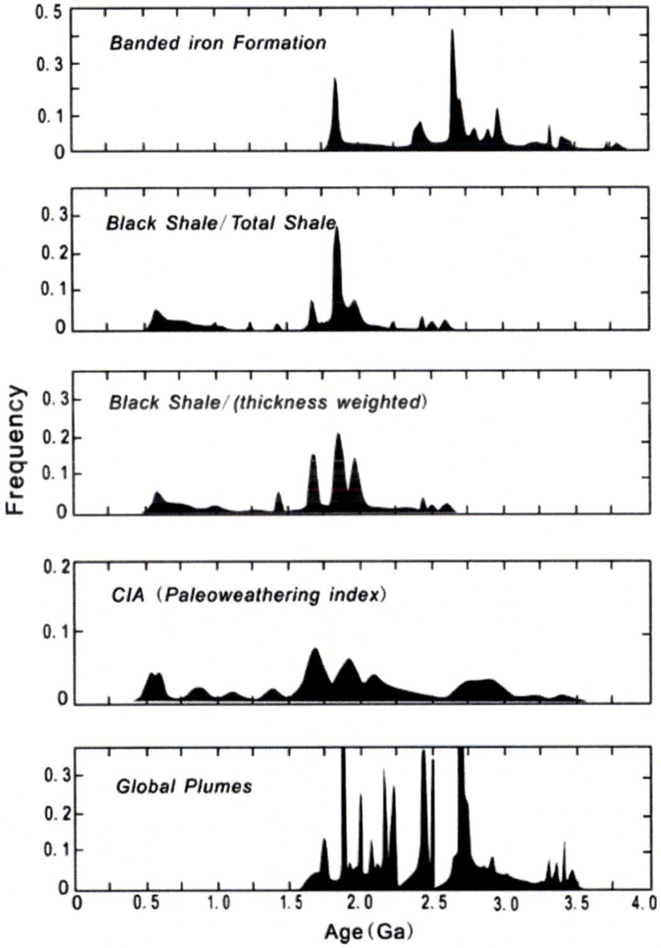

Fig. 6.3 Summary of black shale distribution and the chemical index of alteration (CIA) during the Precambrian (Condie et al. 2001). Most black shales were deposited subsequent on the end of banded iron formation (BIF), i.e., after the great oxidation event (GOE)

activity, if so, the widespread rearranged hopanes in Precambrian source beds could indicate the effect of volcanic activity on source bed development.

The property of Precambrian organic matter can hardly to be clearly recognized generally due to its high maturity, however, in terms of biological evolution, the early Earth organic matter should be likely derived from cyanobacteria during the Early Precambrian, while eukaryotic algae make a greater contribution in later periods. Timing emergence of eukaryotic algae is still a question in debate. According to above-mentioned biomarkers, eukaryotic algae would emerge earlier, but might not be the major component in organic matter. A pyrolytic experiment conducted by Wu et al. (1996) has confirmed that cyanobacteria are more likely to be the main algae prone to gas generation, while eukaryotic algae are probably prone to oil generation. In view of the stratigraphic distribution of Precambrian solid asphalt, oil sands and oil, 1.6–1.4 Ga is probably a key boundary, only solid asphalt emerged before then, while the possibility of oil generation and preservation would be increasing afterwards. This time limit of hydrocarbon phase distribution may be associated with that whether or not the ecological system is under the domination of eukaryotic algae beside the maturity of organic matter.

The distribution of biomarkers of in Precambrian strata is also likely to reflect the evolution of microbial system. For example, since methyl-alkanes and 2α-methyl hopanes are indicative of cyanobacteria, both have been detected from 1.64 Ga-old mid-Proterozoic strata (Brocks et al. 2005), which is later than the earliest occurrence of fossil cyanobacteria (ca. 1.9 Ga, Fischer et al. 2016). Meanwhile, as the biomarker of methane-oxidizing archaea, 3β-methyl hopanes are detected from the contemporaneous strata, providing the first convincing evidence for the processes of methane formation and oxidation in the very early Earth surface system, and revealing the oxygenation in surface sea water and numerous methanogenic bacteria reproduction at the bottom of waters or sediments, which is understandable, since the organic matter of early Earth life could not be entirely oxidized, a considerable number of protein and cellulose may remain and leading the low weight organic molecules and H_2 to be accumulated at bottom waters or in sediment, and to generate methane by methanogenic bacteria. In addition, another key biomarker, aryl-isoprenoid, a biomarker for green sulphur bacteria, has been detected in 1.64 Ga-old strata (Brocks et al. 2005); this biomarker evidences the blooming of sulfate-reducing bacteria and the growth of oceanic sulfur sinks. The discovery of 24 *i*-propylcholestane in the Huqf Supergroup also indicates the emergence of sponges as early as 0.81 Ga (Love et al. 2009), while other specific biomarkers detected in Precambrian strata, including 13α(*n*-methyl)-tricyclic terpanes and C_{19} A-norsterane, have been widely used for oil-source correlations (Wang and Simoneit 1995; Grosjean et al. 2009).

6.4 Discussion

6.4.1 Low HI Values

The above discussion on the characteristics of Precambrian source beds shows that HI values are extremely low with the exception of Huqf Supergroup, which used to be attributed to their high maturity as one optimum explanation. However, there still is another possible reason, i.e., as the dominant Precambrian alga and cyanobacteria have lower H/C atomic ratios. Therefore, it follows that the closer to the Phanerozoic, the more possible to find the source bed prone to oil generation in Precambrian strata, and thus, the source beds after Cryogenian and Ediacaran (Sinian) should be specially drawn to people's attention due to the larger possibility to generate and preserve oil and/or gas as current petroleum exploration focus.

6.4.2 Carbon Isotope Reversals

One significant characteristic of Precambrian source beds is that their carbon isotopic composition $\delta^{13}C$ values of low molecular weight compounds are heavier (enriched with ^{13}C) as comparable to associated kerogen (Fig. 6.4). Previous research tried to explain the phenomena by way of the mixing of hydrocarbons from different source inputs as follows: ① $^{13}C_{n\text{-alkane}} > {}^{13}C_{kerogen}$: some photosynthetic *n*-alkanes could be degraded by heterotrophic archaea and replaced by ^{13}C-enriched *n*-alkanes (Logan et al. 1997; Li et al. 2003). ② $^{13}C_{isoprenoids} > {}^{13}C_{kerogen}$: ^{13}C-enriched halophilic archaea lipids (with heavier carbon isotopic composition) are partly mixed into photosynthetic isoprenoids (Grice et al. 1998; Peng et al. 2000). ③ There is a roughly linear relationship between $\delta^{13}C_{n\text{-alkane}}$ and $\delta^{13}C_{isoprenoid}$: halophilic archaea might be the major heterotrophic entities (Li et al. 2003).

Since the alkyl chain of alginite is difficult to be degraded in water column, the degradation effect of heterotrophic archaea could hardly to result in the above abnormal $\delta^{13}C$ isotopic relationship. Thus, above explanations would remain controversial, and some researchers have considered that the low molecular weight

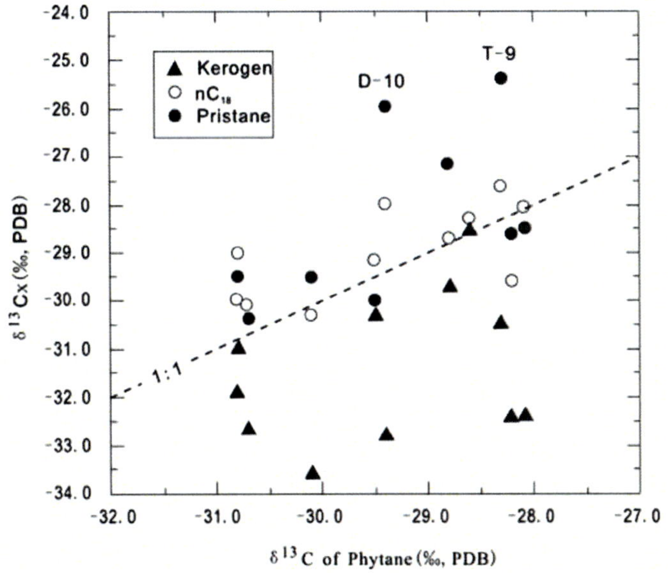

Fig. 6.4 Isotopic compositions of kerogen, n-alkanes versus isoprenoids from Precambrian strata (>1.4 Ga) in the Jixian stratotype section Li et al. (2003)

compounds might consist of migrating hydrocarbons from younger strata (Brocks et al. 2003a; Sherman et al. 2007).

It is also a reasonable explanation that carbon isotope fractionation constrained by hydrocarbon cracking is responsible for carbon isotopic reversal (Liu et al. 2013). In a high-mature phase, low molecular weight compounds are constantly cracked and then expulsed, and leading the carbon isotopic composition to become heavier in residual system (Jia et al. 2014). However, kerogen of macromolecular structure is so stable as to be free from thermal degradation, this kind of difference in the structural stability might result in the carbon isotope reversal.

6.4.3 Indigeneity of Biomarkers

As above, biomarkers in source beds can provide useful biological information, especially to trace microbial evolution. However, it must be on the basis of the indigeneity of biomarkers. The indigeneity of Precambrian biomarkers is a tough challenge faced by researchers.

Laboratory contamination is the first question. Owing to the primitive biomarkers are commonly present at low concentrations, the contamination in laboratory is universal. So long as the solvents are clean enough, and operations conform to the standard, the contamination should be avoidable, so far many laboratories in the world have successfully controlled the contamination to a negligible level (French et al. 2015). For this end, the addition of blank experiments with standard materials in parallel with sample procedures would be the optimum method to control or examine laboratory contamination.

The second one is the field sample pollution. Both field weathering and plant growth could result in sample contamination. So far a common consensus is using drilling core samples to minimize the risk of field pollution.

The third is the contamination of biomarker composition from migrating hydrocarbons of younger strata, which would happen as a geohistorical event. It is the most difficult problem for the detection and exclusion of pollutants. It is also well known that free hydrocarbons are easy to migrate and so the cross-contamination of hydrocarbons between different strata is entirely possible. Some effort has been made using the inclusion-trapped hydrocarbons as indigenous biomarkers, but it is still questionable because most of the trapped hydrocarbons are exogeneous migrating hydrocarbons. One of the methods to exclude the problem is the study of kerogen-bound biomarkers, which has less possibility to be transferred. Nevertheless, it has a certain difficult to obtain useful information from the bound hydrocarbons due to their lower quantity and less information.

In spite of these problems, the desire to understand the evolution of early Earth life in the Precambrian is driving scientists to develop a range of new methods. One case of technical progress is to remove pollutants on the exterior surface of rock samples by way of repeatedly reagent rinsing rock surface, and GC–MS/MS analysis. Another is to cut off the pollutants on exterior surficial layer, just the indigenous hydrocarbons in the interior of rock sample are chosen for further analysis (Sherman et al. 2007). A potential method would be only to choose low maturity and high TOC rock sample, or using kerogen-degraded product of artificially catalytical hydrocracking, for the study of free hydrocarbons (Love et al. 2009; cf. Chap. 12). Thus, taking advantage of these technical advances, indigenous Precambrian biomarkers can be studied.

References

Bao ZD, Chen JF, Zhang SC, Zhao HW, Zhang QH, Li Y (2004) Sedimentary environment and development controls of the hydrocarbon source beds: Middle and Upper Proterozoic in northern North China. Sci China (Ser D-Earth Sci) 47(S2):133–140

Bazhenova OK, Arefiev OA (1996) Geochemical peculiarities of Pre-Cambrian source rocks in the East European Platform. Org Geochem 25(5–7):341–351

Brocks JJ, Logan GA, Buick R, Summons RE (1999) Archean molecular fossils and the early rise of eukaryotes. Science 285(5430):1033–1036

Brocks JJ, Buick R, Logan GA, Summons RE (2003a) Composition and syngeneity of molecular fossils from the 2.78 to 2.45 billion-year-old Mount Bruce Supergroup, Pilbara Craton, Western Australia. Geochimica et Cosmochimica Acta 67(22):4289–4319

Brocks JJ, Love GD, Snape CE, Logan GA, Summons RE, Buick R (2003b) Release of bound aromatic hydrocarbons from Late Archean and Mesoproterozoic kerogens via hydropyrolysis. Geochim Cosmochim Acta 67(8):1521–1530

Brocks JJ, Summons RE, Buick R, Logan GA (2003c) Origin and significance of aromatic hydrocarbons in giant iron ore deposits of the Late Archean Hamersley Basin, Western Australia. Org Geochem 34(8):1161–1175

Brocks JJ, Love GD, Summons RE, Knoll AH, Logan GA, Bowden SA (2005) Biomarker evidence for green and purple sulphur bacteria in a stratified Palaeoproterozoic sea. Nature 437(7060):866–870

Buick R, Rasmussen B, Krapez B (1998) Archean oil: evidence for extensive hydrocarbon generation and migration 2.5–3.5 Ga. AAPG Bull 82(1):50–69

Chen J, Jia WL, Xiao ZY, Peng PA (2020) Catalytic hydropyrolysis of asphaltenes in marine oil from the Tarim Basin, NW China: Implications to complicated oil charging histories in an old composite basin. Mar Pet Geol 114:104232

Condie KC, Marais DJD, Abbott D (2001) Precambrian superplumes and supercontinents: a record in black shales, carbon isotopes, and Palaeoclimates? Precambr Res 106(3–4):239–260

Craig J, Biffi U, Galimberti RF, Ghori KAR, Gorter JD, Hakhoo N, Le Heron DP, Thurow J, Vecoli M (2013) The Palaeobiology and geochemistry of Precambrian hydrocarbon source rocks. Mar Pet Geol 40:1–47

Crick IH, Boreham CJ, Cook AC, Powell TG (1988) Petroleum geology and geochemistry of Middle Proterozoic Mcarthur Basin, northern Australia, 2. Assessment of source rock potential. AAPG Bull 72(12):1495–1514

Deng Q, Wang HZ, Wei ZW, Li SD, Zhang HZ, Liu H, Faboya OL, Cheng B, Liao ZW (2021) Different accumulation mechanisms of organic matter in Cambrian sedimentary successions in the western and northeastern margins of the Tarim Basin, NW China. J Asian Earth Sci 207:104660

Dutkiewicz A, Rasmussen B, Buick R (1998) Oil preserved in fluid inclusions in Archaean sandstones. Nature 395(6705):885–888

Dutkiewicz A, Volk H, Ridley J, George SC (2003) Biomarkers, brines, and oil in the Mesoproterozoic, Roper Superbasin, Australia. Geology 31(11):981–984

Dutkiewicz A, Volk H, Ridley J, George SC (2004) Geochemistry of oil in fluid inclusions in a Middle Proterozoic igneous intrusion: implications for the source of hydrocarbons in crystalline rocks. Org Geochem 35(8):937–957

Dutkiewicz A, George SC, Mossman DJ, Ridley J, Volk H (2007) Oil and its biomarkers associated with the Palaeoproterozoic Oklo natural fission reactors, Gabon. Chem Geol 244(1–2):130–154
Dutta S, Bhattacharya S, Raju SV (2013) Biomarker signatures from Neoproterozoic-Early Cambrian oil, Western India. Org Geochem 56:68–80
Frolov SV, Akhmanov GG, Kozlova EV, Krylov OV, Sitar KA, Galushkin YI (2011) Riphean basins of the central and Western Siberian Platform. Mar Pet Geol 28(4):906–920
Grosjean E, Love GD, Stalvies C, Fike DA, Summons RE (2009) Origin of petroleum in the Neoproterozoic-Cambrian South Oman Salt Basin. Org Geochem 40(1):87–110
Gu Y, Zhao YQ, Jia CS, He GY, Luo Y, Wang B, Lu QH (2012) Analysis of hydrocarbon resource potential in Awati Depression of Tarim Basin. Pet Geol Exp 34(3):257–266 (in Chinese with English abstract)
Fischer WW, Hemp J, Johnson JE (2016) Evolution of oxygenic photosynthesis. Annu Rev Earth Planet Sci 44:647–683
French KL, Hallmann C, Hope JM, Schoon PL, Zumberge JA, Hoshino Y, Peters CA, George SC, Love GD, Brocks JJ, Buick R, Summons RE (2015) Reappraisal of hydrocarbon biomarkers in Archean rocks. PNAS 112(19):5915–5920
Grice K, Schouten S, Nissenbaum A, Charrach J, Sinninghe Damsté JS (1998) Isotopically heavy carbon in the C21 to C25 regular isoprenoids in halite-rich deposits from the Sdom Formation, Dead Sea Basin, Israel. Org Geochem 28(6):349–359
Han KY, Wang GL, Wang TG, Wang LS (2016) Petroleum geological characteristics of asphaltic veins at the Longmenshan Fronthill Belt, West Sichuan Province. In: Sun S, Wang TG (eds) Meso-neoproterozoic geology and petroleum resources in East China. Science Press, Beijing, pp 542–559 (in Chinese)
Hold IM, Schouten S, Jellema J, Damste JSS (1999) Origin of free and bound mid-chain methyl alkanes in oils, bitumens and kerogens of the marine, Infracambrian Huqf Formation (Oman). Org Geochem 30(11):1411–1428
Huang DF, Wang LS (2008) Geochemical characteristics of bituminous dike in Kuangshanliang area of the Northwestern Sichuan Basin and its significance. Acta Petrolei Sinica 29(1):23–28 (in Chinese with English abstract)
Huang HP, Zhang SC, Su J (2016) Palaeozoic oil-source correlation in the Tarim Basin, NW China: a review. Org Geochem 94:32–46
Imbus SW, Engel MH, Elmore RD, Zumberge JE (1988) The origin, distribution and hydrocarbon generation potential of organic-rich facies in the Nonesuch Formation, Central North-American rift system-a regional study. Org Geochem 13(1–3):207–219
Jia CZ, Zhang SB, Wu SZ (2003) Stratigraphy of the tarim basin and adjacent areas (Summary). Science Press, Beijing (in Chinese with English abstract)
Jia WL, Wang QL, Liu JZ, Peng PP, Li BH, Lu JL (2014) The effect of oil expulsion or retention on further thermal degradation of kerogen at the high maturity stage: a pyrolysis study of type II kerogen from Pingliang shale, China. Org Geochem 71:17–29
Jin ZJ, Liu QY, Yun JB, Tenger (2017) Potential petroleum sources and exploration directions around the Manjar Sag in the Tarim Basin. Sci China Earth Sci 60:235–245
Kaufman AJ, Jiang GQ, Christie-Blick N, Banerjee DM, Rai V (2006) Stable isotope record of the terminal Neoproterozoic Krol platform in the Lesser Himalayas of Northern India. Precambr Res 147(1–2):156–185
Kelly AE, Love GD, Zumberge JE, Summons RE (2011) Hydrocarbon biomarkers of Neoproterozoic to Lower Cambrian oils from Eastern Siberia. Org Geochem 42(6):640–654
Kendall B, Reinhard CT, Lyons T, Kaufman AJ, Poulton SW, Anbar AD (2010) Pervasive oxygenation along Late Archaean ocean margins. Nat Geosci 3(9):647–652
Lee C, Brocks JJ (2011) Identification of carotane break down products in the 1.64 billion year old Barney Creek Formation, McArthur Basin, Northern Australia. Org Geochem 42(4):425–430
Li C, Peng PA, Sheng GY, Fu JM, Yan YZ (2003) A molecular and isotopic geochemical study of Meso- to Neoproterozoic (1.73–0.85 Ga) sediments from the Jixian section, Yanshan Basin, North China. Precambr Res 125(3–4):337–356

Li C, Love GD, Lyons TW, Fike DA, Sessions AL, Chu X (2010) A stratified redox model for the Ediacaran ocean. Science 328:8–83

Li JZ, Tao XW, Bai B, Huang SP, Jiang QC, Zhao ZY, Chen YY, Ma DB, Zhang LP, Li NX, Song W (2021) Geological conditions, reservoir forming evolution and favorable exploration directions of marine ultra-deep oil and gas in China. Pet Explor Dev 48(1):1–16 (in Chinese with English abstract)

Liu BQ, Qian JZ, Li X (2000) Characteristics of the Middle-Upper Proterozoic source rocks and analysis of the origin of oil seeps from the Jibei Depression. Mar Orig Petrol Geol 5(1–2):35–46 (in Chinese with English abstract)

Liu H, Liao ZW, Zhang HZ, Cheng B, Tian YK (2013) Review of the study on stable carbon isotope reversal between kerogen and its evolution products: implication for the research of the marine oil reservoirs in Tarim Basin, NW China. Bull Mineral Petrol Geochem 32(4):497–502 (in Chinese with English abstract)

Liu Y, Zhong NN, Tian YJ, Qi W, Mu GY (2011) The oldest oil accumulation in China: mesoproterozoic Xiamaling Formation bituminous sandstone reservoirs. Pet Explor Dev 38(4):503–512 (in Chinese with English abstract)

Logan GA, Summons RE, Hayes JM (1997) An isotopic biogeochemical study of Neoproterozoic and Early Cambrian sediments from the Centralian Superbasin, Australia. Geochimica Et Cosmochimica Acta 61(3):5391–5409

Love GD, Grosjean E, Stalvies C, Fike DA, Grotzinger JP, Bradley AS, Kelly AE, Bhatia M, Meredith W, Snape CE, Bowring SA, Condon DJ, Summons RE (2009) Fossil steroids record the appearance of Demospongiae during the Cryogenian Period. Nature 457(7230):718–721

Luo QY, Zhong NN, Zhu L, Wang YN, Qin J, Qi L, Zhang Y, Ma Y (2013) Correlation of burial organic carbon and palaeoproductivity in the Mesoproterozoic Hongshuizhuang Formation, northern North China. Chin Sci Bull 58(11):1299–1309

Mancuso JJ, Kneller WA, Quick JC (1989) Precambrian vein pyrobitumen—evidence for petroleum generation and migration 2 Ga ago. Precambr Res 44(2):137–146

Melezhik VA, Fallick AE, Filippov MM, Larsen O (1999) Karelian shungite—an indication of 2.0-Ga-old metamorphosed oil-shale and generation of petroleum: geology, lithology and geochemistry. Earth-Sci Rev 47(1–2):1–40

Meng FW, Yuan XL, Zhou CM, Chen ZL (2003) Dinosterane from the Neoproterozoic Datangpo black shales and its biological implications. Acta Micropalaeontologica Sinica 20(1):97–102 (in Chinese with English abstract)

Mossman DJ, Gauthier-Lafaye F, Jackson SE (2005) Black shales, organic matter, ore genesis and hydrocarbon generation in the Palaeoproterozoic Franceville Series, Gabon. Precambr Res 137(3–4):253–272

Newell KD, Burruss RC, Palacas JG (1993) Thermal maturation and organic richness of potential petroleum source rocks in Proterozoic Rice Formation, North-American midcontinent rift system, northeastern Kansas. AAPG Bull 77(11):1922–1941

Peng PA (2015) The evolution, simulation and determination of the deep source rocks from the Tarim Basin. Report for the National Oil and Gas Major Project (2011ZX05008–002–30). Guangzhou: Guangzhou Institute of Geochemistry, Chinese Academy of Sciences (in Chinese)

Peng PA, Sheng GY, Fu JM, Yan YZ (1998) Biological markers in 1.7 billion year old rock from the Tuanshanzi Formation, Jixian strata section, North China. Org Geochem 29(5–7):1321–1329

Peng PA, Sheng GY, Fu JM, Jiang JG (2000) Immature crude oils in the salt lake depositional environment are related to organic matter precipitated at stage of carbonate in salt lake sedimentation sequences. Chin Sci Bull 45(Supplement):1–6

Pratt LM, Summons RE, Hieshima GB (1991) Sterane and triterpene biomarkers in the Precambrian Nonesuch formation, North-American midcontinent rift. Geochim Cosmochim Acta 55(3):911–916

Rasmussen B, Fletcher IR, Brocks JJ, Kilburn MR (2008) Reassessing the first appearance of eukaryotes and Cyanobacteria. Nature 455(23):1101–1104

Sherman LS, Waldbauer JR, Summons RE (2007) Improved methods for isolating and validating indigenous biomarkers in Precambrian rocks. Org Geochem 38(12):1987–2000

Summons RE, Brassell SC, Eglinton G, Evans E, Horodyski RJ, Robinson N, Ward DM (1988) Distinctive hydrocarbon biomarkers from fossiliferous sediment of the Late Proterozoic Walcott member, Chuar Group, Grand-Canyon. Arizona. Geochimica Et Cosmochimica Acta 52(11):2625–2637

Summons RE, Jahnke LL, Hope JM, Logan GA (1999) 2-Methylhopanoids as biomarkers for cyanobacterial oxygenic photosynthesis. Nature 400(6744):554–557

Tan ZZ, Jia WL, Li J, Yin L, Wang SS, Wu JX, Song JZ, Peng PA (2021) Geochemistry and molybdenum isotopes of the basal Datangpo Formation: Implications for ocean-redox conditions and organic matter accumulation during the Cryogenian interglaciation. Palaeogeogr Palaeoclimatol Palaeoecol 563:110169

Terken JMJ, Frewin NL (2000) The Dhahaban petroleum system of Oman. AAPG Bull 84(4):523–544

Terken JMJ, Frewin NL, Indrelid SL (2001) Petroleum systems of Oman: charge timing and risks. AAPG Bull 85(10):1817–1845

Ulmishek GF (2001a) Petroleum geology and resources of the Baykit high province, East Siberia. US Geol Surv Bull 2201-F:18

Ulmishek GF (2001b) Petroleum geology and resources of the Nepa-Botuoba High, Angara-Lena Terrace, and Cis-Patom Foredeep, Southeastern Siberian craton, Russia. US Geol Surv Bull 2201-C:16

Uphoff TL (1997) Precambrian Chuar source rock play: an exploration case history in Southern Utah. AAPG Bull 81(1):1–15

Volk H, Dutkiewicz A, George SC, Ridley J (2003) Oil migration in the Middle Proterozoic Roper Superbasin, Australia: evidence from oil inclusions and their geochemistries. J Geochem Explor 78–79:437–441

Volk H, George SC, Dutkiewicz A, Ridley J (2005) Characterization of fluid inclusion oil in a Mid-Proterozoic sandstone and dolerite (Roper Superbasin, Australia). Chem Geol 223(1–3):109–135

Wang CJ (2009) Biomarker evidence for eukaryote algae flourishing in a Mesoproterozoic (1.6–1.5 Ga) stratified sea on the North Choina Craton. Geochimica et Cosmochimica 73(13):1407.

Wang CJ, Wang M, Xu J, Li Y, Yu Y, Bai J, Dong T, Zhang X, Gai H (2011) 13α-(*n*-alkyl)-tricyclic terpanes: a series of biomarkers for the unique microbial mat ecosystem in the Middle Mesoproterozoic (1.45–1.30 Ga) North China Sea. Mineral Mag 75:2114

Wang J, Chen JF, Wang DR, Zhang SC (2002) Study on the characteristics of carbon isotopic composition and hydrocarbon generation potential of organic matter of Middle-Upper Proterozoic in northern part of North China. Pet Explor Dev 29(5):13–15 (in Chinese with English abstract)

Wang LS, Han KY, Xie BH, Zhang J, Du M, Wan MX, Li D (2005) Reservoiring conditions of the oil and gas fields in the north section of Longmen Mountain nappe structural belts. Nat Gas Indus 27(supplementary issue A):1–5 (in Chinese with English abstract)

Wang TG (1980) Primary properties of Sinian Suberathem oil-seep and its petroleum geological significance in Yanshan region. Pet Explor Dev 7(2):34–52 (in Chinese with English abstract)

Wang TG (1991) A novel tricyclic terpane biomarker in the Upper Proterozoic bituminous sandstone, Eastern Yanshan region. Sci China (Ser B-Chem) 34(4):479–489

Wang TG, Simoneit BRT (1995) Tricyclic terpanes in Precambrian bituminous sandstone from the Eastern Yanshan region, North China. Chem Geol 120(1–2):155–170

Wang TG, Song DF (2016) Meso-Neoproterozoic petroleum resources in the world and in East China. In: Sun S, Wang TG (eds) Meso-neoproterozoic geology and petroleum resources in East China. Beijing Science Press, Beijing, pp 371–400 (in Chinese)

Wang TG, Huang GH, Xu ZY (1988) A fossil oil pool on the basement of the Xiamaling Formation of the Upper Proterozoic in Longtangou, West Liaoning. Oil Gas Geol 9(3):278–287 (in Chinese with English abstract)

Wang TG, Li MJ, Wang CJ, Wang GL, Zhang WB, Shi Q, Zhu L (2008) Organic molecular evidence in the Late Neoproterozoic Tillites for a palaeo-oceanic environment during the snowball Earth era in the Yangtze region, Southern China. Precambrian Res 162(3–4):317–326

Wang W, Guan C, Zhou C, Peng Y, Pratt LM, Chen X, Chen L, Chen Z, Yuan X, Xiao S (2017) Integrated carbon, sulfur, and nitrogen isotope chemostratigraphy of the Ediacaran Lantian Formation in South China: Spatial gradient, ocean redox oscillation, and fossil distribution. Geobiology 15:552–571

Wu QY, Zhang B, Sheng GY, Fu JM (1996) Study of molecular organic geochemistry on hydrocarbons thermal degraded from algae. Bull Mineral Petrol Geochem 15(2):75–79 (in Chinese with English abstract)

Yang HJ, Cheng YQ, Tian J, Du JH, Zhu YF, Li HH, Pan WQ, Yang PF, Li Y, An HT (2020) Great discovery and its signifi cance of ultra-deep oil and gas exploration in well Luntan-1 of the Tarim Basin. China Petrol Explor 25(2):62–72 (in Chinese with English abstract)

Zhang SC, Zhang BM, Bian LZ, Jin ZJ, Wang DR, Chen JF (2007) The Xiamaling oil shale accumulated by Rhodophyta over 800 Ma ago. Sci China (Ser D-Earth Sci) 50(4):527–535

Zhu GY, Chen FR, Wang M, Zhang ZY, Ren R, Wu L (2018) Discovery of the Lower Cambrian high-quality source rocks and deep oil and gas exploration potential in the Tarim Basin, China. AAPG Bull 102(10):2123–2151

Chapter 7
Palaeo-Oceanic Geochemistry and Sedimentary Environments of the Xiamaling Formation in the Yanliao Faulted-Depression Zone, North China Craton

Shuichang Zhang, Xiaomei Wang, Huajian Wang, Jin Su, Yuntao Ye, and D. E. Canfield

Abstract The Mesoproterozoic geochemical record is limited, but critical in revealing the relationships between Precambrian eukaryotic ecosystem evolution and source bed formation. In North China Craton, the Xiamaling Formation (1400–1320 Ma) is a set of tens-million-years-old and well-preserved low-mature to mature sedimentary sequence on a passive continental margin, and records the interaction among climate, ecosystem, and dynamic ocean chemistry. Moreover, the low thermal maturity and excellent preservation of the Xiamaling sediments allow organic and inorganic geochemical analysis. Overall, by means of multidisciplinary investigation, including iron speciation, trace elements, biomarkers, and so on, the palaeo-oceanic geochemistry during the Xiamaling deposition is evaluated. Based on both sedimentological and geochemical criteria, the Xiamaling Formation is tentatively divided into six lithological units in the present study. Of the six units, four units are adequately investigated and each unit contains distinct geochemical features, indicating specific water-column environment respectively with oxic, non-sulfidic anoxic, and sulfidic anoxic sedimentary environments. The variation of the palaeo-oceanic chemistry has significant influence on the primary productivity of organic matter and the preservation of hydrocarbons, which is consistent with the regional palaeo-climatic change, particularly for the place where the Xiamaling Formation was related to the intertropical convergence zone and the corresponding Hadley cell fluctuations.

Keywords Mesoproterozoic · Xiamaling Formation · Oxygen minimum zone (OMZ) · Sedimentary environment · Palaeo-oceanic geochemistry

---

S. Zhang (✉) · X. Wang · H. Wang · J. Su · Y. Ye · D. E. Canfield
Key Laboratory of Petroleum Geochemistry, Research Institute of Petroleum Exploration and Development, CNPC, Beijing 100083, China

D. E. Canfield
Institute of Biology and Nordic Center for Earth Evolution, University of Southern Denmark, 5230 Odense M, Denmark

T. Wang, *Meso-Neoproterozoic Geology and Petroleum Resources in China*, Springer Geology, https://doi.org/10.1007/978-981-19-5666-9_7

7.1 Introduction

As the north margin of the North China Craton (NCC), the Yanliao Faulted-Depression Zone (YFDZ; used to be called as "Yanliao Subsidence Zone", "Yanshan Paraplatform"/"aulacogen") has attracted broadly systematic researches on its stratigraphy, petrology, tectonics, and geochemistry (e.g., Kao et al. 1934; Wang and Simoneit 1995; Zhang et al. 2007; Gao et al. 2008; Lu et al. 2008; Shi et al. 2012; Luo et al. 2013; Zhao et al. 2019). However, seldom work has focused on its Mesoproterozoic palaeo-oceanic geochemistry and sedimentary environment. Over the years, it is generally assumed that the Mesoproterozoic palaeo-ocean was filled with a vast or perhaps even global reservoir of hydrogen sulfide (Canfield 1998; Bjerrum and Canfield 2002; Poulton et al. 2004). Sulfidic anoxic conditions are capable to hinder the expansion and diversification of eukaryotes due to the insolubility of bio-essential trace elements (such as Mo) in sulfidic waters (Anbar and Knoll 2002), making life evolution dull and tedious. Since sulfate supply was limited, the ocean was ferruginous prior to 1.8 Ga, and ferruginous conditions might have prevailed again from 0.7 to 0.54 Ga (Canfield et al. 2008). But the redox characteristics of oceanic environments during "Earth's middle age" (1.8–0.8 Ga) are widely controversial, which is an important transition time from anoxic waters to oxic waters, and from prokaryotic biomass to eukaryotic biomass dominated world, and also a miraculous time for the disappearance of banded iron formation on the Earth since then.

The palaeo-oceanic redox environment greatly affects or even plays a vital role in the process of biological evolution. Palaeo-oceanic redox conditions are quite important for the formation of source beds because it provides not only the living environment for marine biomass, but also the sedimentary and burial context for organic matters. Therefore, it is of great significance to recognize and reconstruct the evolution of oceanic redox environment in order to understand the genetic and developmental mechanism of source beds.

Although free oxygen in the atmosphere was already available during the Great Oxygenation Event at 2.4 Ga, oxygenation of the Proterozoic ocean seems to occur much later, especially during the Mesoproterozoic Era. It was hypothesised that the deep ocean did not get completely oxygenation even until the Neoproterozoic Oxygenation Event (Poulton and Canfield 2011). The reconstruction of oceanic redox conditions tends to be either extremely anoxic or fully oxygen-enriched for a long time. As an important transition, scholars hold different views about the oceanic chemical structure during the Mesoproterozoic. For example, some scientists believe that oceanic oxygenation level was still very low during the Mesoproterozoic and the oxygen level in the atmosphere was lower than 0.1% PAL (present atmospheric level; Planavsky et al. 2014). The depth of free-oxygen-containing seawater beyond the sedimentary surface would not exceed 25 m (Brocks et al. 2005) and the deep water persisted to be iron-enriched (Planavsky et al. 2014). However, evidence also shows that dissolved oxygen content in the Mesoproterozoic Ocean was already high. Molybdenum isotopes have revealed a dynamic ocean redox during the Mesoproterozoic (Ye et al. 2021). Analysis of Roper Group in Northern Australia Craton indicated that marine oxygen contents changed with seawater depth during 1.5–1.4 Ga, which transmitted from oxygen-enriched surface waters to sulfidic anoxic deep waters (Shen et al. 2003). However, another case of the Kaltasy Group in central Russia showed the existence of dissolved oxygen in deep water at 1.42 Ga ago (Sperling et al. 2014).

According to previous researches mentioned above, it is inferred that the Mesoproterozoic oceanic redox environment may be unstable, which also exhibited highly temporal and spatial heterogeneity. But is there any correlation between the changes of oceanic redox environment and the atmospheric oxygen content in the Mesoproterozoic Era? It seems very hard to answer this question due to the limited distribution of Mesoproterozoic sedimentary strata over the world, but we may have primary discussions for this concern by studying the Xiamaling Formation. The Xiamaling Formation formed during 1.40–1.32 Ga with an entire time span for near 80 Ma in YFDZ. Its sedimentary continuity and diversity would provide us an ideal opportunity to reveal the oceanic redox environment and dynamic evolution during the Middle Mesoproterozoic Era.

7.2 Geological Background

7.2.1 Stratigraphic Dating and Subdivision

7.2.1.1 Stratigraphic Dating

The Xiamaling Formation in YFDZ is mainly composed of various kinds of shales, which are positioned over the carbonate rocks of the Tieling Formation and beneath the sandstone of Luotuoling Formation. However, there are two sedimentary hiatuses respectively between the Xiamaling Formation and the underlying Tieling Formation, or the overlying Qingbaikouan Luotuoling Formation (Qu et al. 2014), which indicate a sedimentary discontinuity for the former and a microangular unconformity for the latter, respectively (Fig. 7.1a). According to the K–Ar dating ages from glauconitiic sandstone, the Xiamaling Formation was initially dated to 1000–900 Ma (i.e., the Neoproterozoic age; Li et al. 1996). In recent two decades, however, a series of zircon and baddeleyite U–Pb or Pb-Pb isotopic ages ranging from 1320 ± 6 Ma to 1392 ± 1.2 Ma was measured from the tuff or bentonite beds of the Xiamaling Formation in YFDZ (Fig. 7.1a; Gao et al. 2007, 2008; Su et al. 2008; Li et al. 2009, 2013; Zhang et al. 2015; cf. Chaps. 2 and 10), and demonstrating that the Xiamaling Formation should belong to the Mesoproterozoic instead of the Neoproterozoic. Moreover, two zircon U–Pb ages of 1437 ± 21 Ma (Su et al. 2010) and 1439 ± 14 Ma (Li et al. 2014) obtained from the middle of the underlying Tieling Formation, further constrained the bottom age limit of Xiamaling Formation to be around 1400 Ma.

In addition, the Xiamaling Formation is intercalated with 2–4 gabbro-diabase sills wide-spread in the Jibei Depression. One zircon U–Pb dating age of 1320 ± 6 Ma (Li et al. 2009) and two baddeleyite U–Pb ages of 1327.3 ± 2.3 Ma and 1327.5

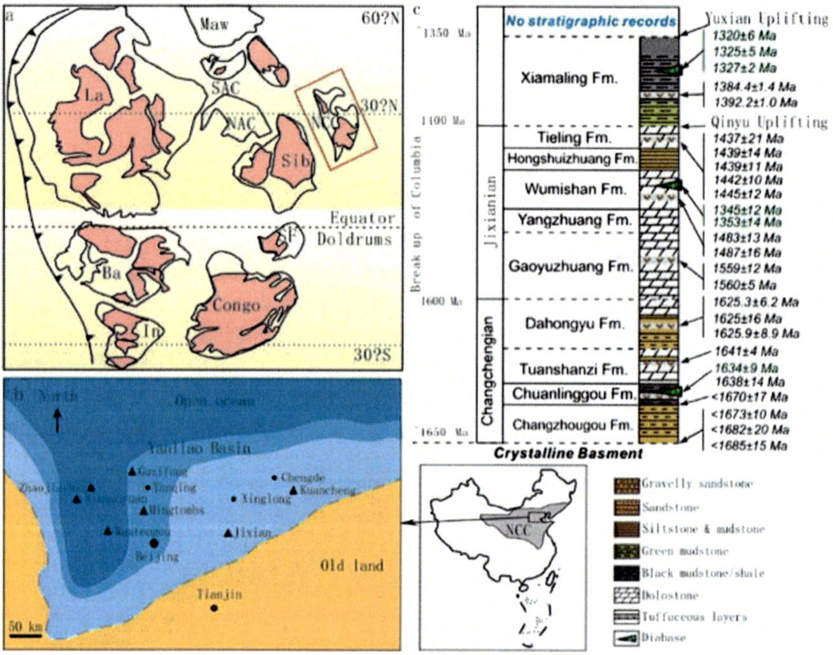

Fig. 7.1 **a** Locality of the NCC at 1400 Ma (Pisarevsky et al. 2014, modified); **b** palaeogeographic map of the middle segment of west YFDZ during the sedimentary period of Xiamaling Formation showing the location of measured stratigraphic sections; and **c** general stratigraphic column of the Mesoproterozoic sequence in the YFDZ, NCC (Wang et al. 2017; Li et al. 2019; Zhang et al. 2019, modified). Ba. Baltica Craton; In. India Craton; La. Laurentia Craton; Maw. Mawson Craton; NAC. North Australian Craton; NCC. North China Craton; SAC. South Australian Craton; SF. São Francisco; Sib. Siberia Craton. ▲ Stratigraphic section location; ● city/county

± 2.4 Ma (cf. Chaps. 10 and 11) respectively obtained from the gabbro-diabase sills, indicate the upper age limit of the Xiamaling Formation should be at 1320 Ma (Fig. 7.1a; cf. Chaps. 2, 10 and 11). Therefore, it is commonly accepted that the age of Xiamaling Formation should be dated between 1400 and 1320 Ma, which is crucial to understand the Mesoproterozoic tectonic evolution in the YFDZ, and even in the entire NCC.

7.2.1.2 Stratigraphic Subdivision

Based on the lithological characteristics, the Xiamaling Formation is traditionally divided into four stratigraphic members in ascending order (Fan 2015 and references therein; cf. Chaps. 2 and 3). The Xia-1 Member is mainly fine sandstone, variegated siltstone and shale. The Xia-2 Member consists of purplish-red mudstone and bright-green silty shale with muddy lens and glauconitic sandstone. The Xia-3 Member is mainly of black shale with paper-oil shale and silicalite interbeds. The Xia-4 Member is mainly composed of variegated silty shale, shale with thin marl lens and stromatolite (Fig. 7.2; cf. Chap. 2).

By comparing the Zhaojiashan, Guzifang (in the Xuanlong Depression), Mentougou (in the Jingxi Depression), Kuancheng (in the Jibei Depression) and other sections, we have made a detailed geochemical and sedimentological investigation of Xiamaling Formation at the Xiahuayuan section (in the Xuanlong Depression; Fig. 7.1b), and then tentatively suggested a stratigraphic division scheme of 6 lithologic units in descending order (Fig. 7.2; Wang et al. 2017; Zhang et al. 2019). Among which, the traditional Xia-3 Member is further subdivided into two units (Units 2 and 3) because the upper black silty shale (Unit 2) is different from the lower black shale and silicalite interbeds (Unit 3) in their geochemical and sedimentological characteristics. Moreover, the original Xia-2 Member is also subdivided into two units (Units 4 and 5) due to the difference between and the upper interbedding of red mudstone and green siltstone (Unit 4) and the lower glauconitic sandstone (Unit 5; Fig. 7.2; Zhang et al. 2015, 2019; Wang et al. 2017).

According to the sedimentary record of the pyroclastic rocks, Meng et al. (2011) suggested the Xiamaling Formation as the transitional product from passive margin to a back-arc basin. On the whole, however, the frequency of pyroclastic occurrence is very low in Xiamaling Formation, so far the Xiamaling pyroclastics is only limited in the Units 2 and 3 (Fig. 7.2), i.e., four tuff beds in Unit 2 and one K-bentonitic bed in Unit 3 (Fig. 7.2), which could be attributed to two episodes of volcanic eruption within the Xiamaling sedimentary period of near 80 Ma. Based on the zircon dating for Unit 3, it could be deduced that the sedimentary rate would be ca. 0.8 m/Ma for the 400 m-thick Xiamaling sedimentary sequence. At that time, moreover, no sedimentary evidence of turbulent or gravitational flow has been recorded as typical sedimentary characteristic of back-arc tectonic environment, and thus the pyroclastics of Units 2 and 3 should be deposited in a deep-water environment instead of the back-arc tectonic environment, which would represent the continuous deposition of passive continental margin based on the sedimentary characteristics of the Xiamaling Formation. In the light of palaeo-magnetic data, it can be determined that the palaeogeographic location of the Xiamaling Formation is just in between 10 and 30°N (Evans and Mitchell 2011; Zhang et al. 2012).

7.2.2 Sedimentary Tectonic Setting and Depocenter

7.2.2.1 Sedimentary Tectonic Setting

According to the zircon U–Pb age of 1673 ± 10 Ma from the granitic porphyry dike of the Changzhougou Formation at the bottom of Changchengian (Li et al. 2011), the basal age limit of the Changchengian should be constrained younger than 1670 Ma. While at the north edge of NCC, the formation and evolution of the YFDZ would be the consequence of the breakup of the supercontinent Columbia during

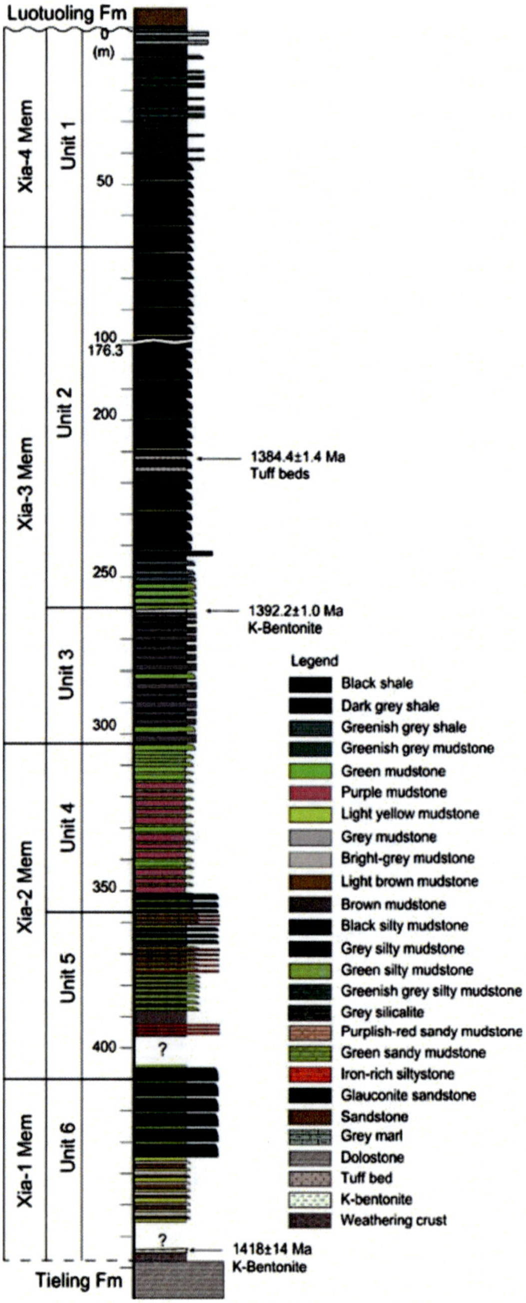

Fig. 7.2 The stratigraphic column of the Xiamaling Formation at the Xiahuayuan section, the Xuanlong Depression (Zhang et al. 2019, modified). Correlation between the stratigraphic divisions of 4 members and 6 lithological units is shown

1670–1320 Ma (Meng et al. 2011; Fig. 7.1a). Due to continued crustal extension and emergence of oceanic crust, the northern NCC was involved into a passive continental margin in the Jixianian sedimentary period (1600–1400 Ma; Qu et al. 2014). After deposition of the Tieling Formation, one crustal movement called "Qinyu Unplifting" has leaded the YFDZ into land, and thus to an erosion process of Tieling Formation (Fig. 7.1c; Qiao 1976). A low-angle subduction of oceanic crust beneath the NCC is considered to be the driving force for the uplifting (Qu et al. 2014). Then a high-angle subduction of oceanic crust might be followed, resulting in the formation of back-arc basin by compression, and causing intense rifting and deposition of the Xiamaling Formation (Qu et al. 2014). However, ash beds are very rare in the Xiamaling Formation, so far only found respectively in two intervals, and the view of back-arc basin is still lack of robust evidence.

Furthermore, there exists another crustal movement after the deposition of Xiamaling Formation, named "Yuxian Uplifting", which resulted in the erosion of upper sediments of the Xiamaling Formation and a basin-wide micro-angle unconformity between the Xiamaling Formation and the overlying Luotuoling Formation in the YFDZ (Fig. 7.1c; Fan 2015; Qu et al. 2014), or an unconformity with the overlying Jurassic in the Xuanlong Depression (Fig. 7.2). This uplifting would be caused by the collision between the NCC and its adjacent block, which is related to an early response to the assembly of supercontinent Rodinia (Qu et al. 2014).

It should be noted that there are numerous gabbro-diabase sills or diabase veins with a uniform age of 1320–1327 Ma not only in the Xiamaling Formation (Fig. 7.1c; Li et al. 2009; Liu et al. 2011; cf. Chap. 10), but also in the underlying Jixianian Wumishan, Gaoyuzhuang, and Tieling Formations, which occurred between 1330–1300 Ma with a peak age of 1320 Ma (Zhang and Zhao 2018). These gabbro-diabase sills/diabase veins are named as the Yanliao large igneous province (Yanliao LIP), which are obviously similar to the Derim-Galiwinku LIP (1320 Ma) in the North Australian Craton (NAC) in age, occurrence, petrographic assemblage, and geochemical compositions (Zhang and Zhao 2018). Besides, the compositions and ages of Late Mesoproterozoic stratigraphic units between both cratons are also very similar. As exemplified by the top stratigraphic horizons with the diabase intrusive bodies, the Xiamaling Formation in the YFDZ, NCC and the Velkerri Formation of Roper Group in NAC are predominated by black shales, indicating both cratons may be connected each other, or adjacent blocks at least during the Late Mesoproterozoic period (Fig. 7.1a). The REE Nb-rich carbonatite in the Bayan-Obo deposits, occurred and dated at 1301 ± 12 Ma mainly in the sill, are also thought to be spatially and temporally linked with the Yanliao LIP, and their origin may be related to the continental rifting and the NCC breakup in the supercontinent Columbia/Nuna (Zhang and Zhao 2018). Therefore, the northern margin of the NCC during the Xiamaling period may still be in an extensional cracking environment, and well connected to the open ocean (Fig. 7.1b) Therefore, the Xiamaling sediments should record the variation of palaeo-oceanic environment at that time.

7.2.2.2 Stratigraphic Depocenter

The stratigraphic depocenter means an area or site of maximum deposition. In comparison with the Jixianian, the Xiamaling depocenter has been shifted from the southeast (the Jidong Depression) to the northwest (the Xuanlong Depression) in the YFDZ (cf. Table 11.12 in Chap. 11). In the Xuanlong Depression, the stratigraphic thickness of the Xiamaling Formation is over 500 m measured at the Zhaojiashan, Xiahuayuan, and Huangtugang sections, Xuanlong Depression (Fig. 7.1b; Fan 2015; Song and Zhang 1983), where sandstone, green or red organic-poor shale, black organic-rich shale, green organic-poor shale and stromatolite were sequentially deposited in ascending order, which can be divided into 4 stratigraphic members (cf. Chaps. 2 and 3) or 6 lithological units. The measured thickness of the Xiamaling Formation is 352 m and 318 m, respectively, at the Mentougou and Mingtombs sections in the Jingxi Depression, and just 168–280 m in the Jibei, Liaoxi, and Jidong Depressions (Fan 2015; Qu et al. 2014), where most strata of the Xiamaling Formation are denuded due to the crustal movement of "Yuxian Uplifting" and only green sandy shale of basal Xia-1 Member remains (Fan 2015; cf. Chaps. 2 and 3). During the Xiamaling sedimentary period of 1400–1320 Ma, the YFDZ experienced a complete sedimentary cycle from crustal subsidence-transgression to uplifting-regression. Consequently, the most complete geological records of palaeo-oceanic chemical environmental variations have been well preserved in the Xuanlong Depression.

7.3 Lithology and Lithofacies of the Xiamaling Formation

Among the six lithological units of the Xiamaling Formation, we have fully investigated the Units 1–4. Each unit possesses distinct geochemical characteristics, and overall reflecting the dynamic variation of sedimentary environments (Wang et al. 2017). The lithology of Units 1–4 is briefly summarized as follows in ascending order (Fig. 7.2).

(1) The basal Unit 6 shows a disconformable contact with the underlying Tieling Formation. From the bottom to up, the lithology varies from conglomerate, through sandy mudstone, to variegated (e.g., grey, greenish-grey, light-brown, light-yellow, etc.) fine clastic shale. Its dominated sedimentary environment should contain high-energy shallow-water environment.

(2) The lower part of Unit 5 dominates with greyish-black thin-bedded shale and siderite concretions or iron-rich sandy mudstones, while the upper part dominates with greyish-green muddy siltstone and glauconitic sandstone interbedded with iron-rich sandy mudstones (Canfield et al. 2018). Its sedimentary environment was transferred from low energy deep-water facies to high energy shallow-water shelf facies.

(3) Unit 4 is composed of the interbedding of variegated thin-bedded mudstone/shale and greyish-green siltstone/sandstone (Fig. 7.3a). The 1–2 cm-thick thin-bedded greyish-green or purplish-red siltstone/sandstone (Fig. 7.3b) contains typically parallel-lamellas, sometimes with cross-bedding (Fig. 7.3c), and the greyish-green siltstone/sandstone vertically turns into purplish-red and/or green mudstones (Fig. 7.3d), among which, the greyish-green siltstone and sandstone are exogenous sediments inputted by distal turbid flow. The purplish-red mudstone should be indigenous sediments, while greyish-green mudstone and siltstone/sandstone occur alternatively. In upper interval of Unit 4, however, purplish-red mudstone disappears, while greyish-green mudstone and siltstone/sandstone occur alternatively (Fig. 7.3d), and the mudstone is regularly thickening. Moreover, the thin-bedded black shale appears at ca. 10 m above the purplish-red mudstone as the top boundary of Unit 4. The sufficient evidence of turbidite and the lack of ripple evidence indicate that Unit 4 was deposited in the same water depth as storm wave base or below the base.
(4) As the most important characteristics of Unit 3, the interbedding of black shale and silicalite frequently occurs, while the purplish-red mudstone is no longer present. The black shale is rich in organic matter with fine-lamellas so as to be paper-oil shale occasionally with thin-bedded silicalite (Fig. 7.3e), but the silicalite is relatively poor-in organic matter. A clearly lithological change can be observed between the black shale and fine-silicalite either on the outcrop (Fig. 7.3e) or on rock slice (Fig. 7.3f). Since siltstone is undeveloped, no sedimentary evidence for waves or water flow modification is observed, and thus the Unit 3 would be attributed to the sediments of deep-water facies. Within the top interval of Unit 3, a ca. 20 m-thick green muddy siltstone bed frequently contains carbonate thin-beds, lens, and nodules as well as occasionally with carbonate-cemented sandstone beds. Once the carbonate and the carbonate-cemented sandstone disappear, black shales would keep on continuously developing.
(5) In Unit 2, black to grey shale shows very clear bedding on its outcrop (Fig. 7.3i), but the bedding is not so obvious in the fresh core samples, i.e., the clear bedding could be resulted from the differential weathering of organic- or sulfide-rich black shales and organic-poor grey shales. Black and grey shales are the major lithology in Unit 2. Microscopic bedding can also be seen from rock slice (Fig. 7.3j). In addition to the top interval, there is no sedimentary evidence for the occurrence of bioturbation in Unit 2 (Fig. 7.3k). Therefore, it can be considered that Unit 2 is also attributed to the sediments of deep-water facies. In general, Units 3 and 2 may represent the maximum sea flooding period, with the deepest water column of the Xiamaling Formation.
(6) Unit 1 mainly consists of black, grey, and greyish-green mudstone/shale, sandy mudstone with siltstone/sandstone interbeds. The dark siltstones/sandstones have cross-bedding, which may reflect that the water was shallow and/or influenced by storm waves. The nodule-bearing carbonate rock occurs near the top interval as the dominant lithology, often with carbonate cross-lamella structure, indicating continuous deposition at the storm wave base. In addition, it has been reported that the uppermost interval contains thin-bedded stromatolitic marl of

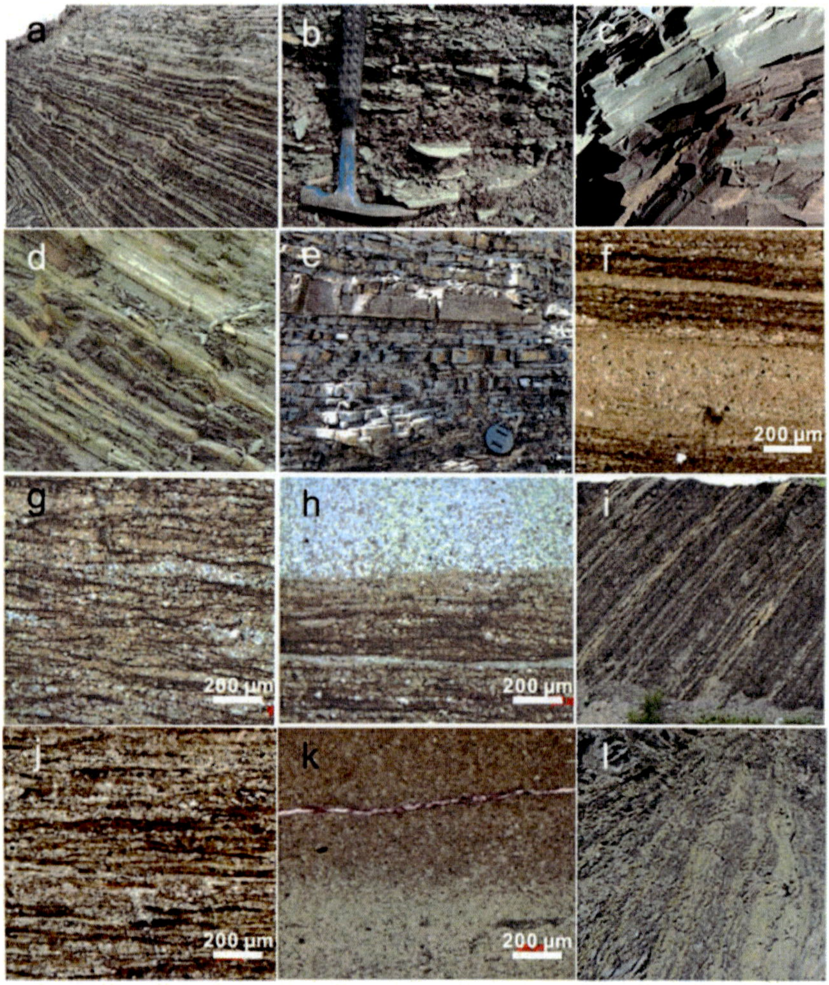

Fig. 7.3 Field outcrop photographs and thin-section micrographs of Units 1–4 in the Xiamaling Formation. Unit 4: **a** outcrop photographs, the typical interbedding of thin-bedded variegated shale/mudstone and greyish-green siltstone/sandstone; **b** outcrop photographs, 1–2 cm-thick greyish-green and purplish-red siltstones/sandstone with parallel-lamella; **c** outcrop photographs, 1–2 cm-thick greyish-green/purplish-red siltstones with cross-bedding; **d** outcrop photographs, greyish-green siltstone turns into purplish-red/green shale/mudstone upwards and downwards. Unit 3: **e** outcrop photograph, typical interbedding of black shale/paper-oil shale and fine silicalite beds; **f** thin-section micrograph, directly contact of black shale and silicalite; **g** thin-section micrograph, typical black shale with dense organic matter beds; **h** thin-section micrograph, directly contact of black shale and silicalite. Unit 2: **i** outcrop photographs, variegated shale. **j** Thin-section micrograph, black shale with dense organic beds; **k** thin-section micrograph, an occurrence of continuous transition from black shale to grey shale by synsedimentary bioturbation. Unit 1: **l** outcrop photograph, the interbedding of greyish-green mudstones and black shales

shallower waters, and the sedimentary evidence of wave disturbance could be observed. Thus, Unit 1 may represent a regression sequence, especially for its upper interval.

7.4 Geochemical Parameters and Profiles

7.4.1 Significance of Palaeo-Oceanic Geochemical Parameters

By means of inorganic, organic and isotopic chemical analyses, a series of geochemical induces, i.e., Fe_{HR}/Fe_T, Fe_{Py}/Fe_{HR}, $\delta^{13}C$, TOC, HI, TMAIs, and trace elements Mo, V, U contents have been acquired to study palaeo-oceanic sedimentary environment of Xiamaling Formation.

(1) **Total organic carbon** (TOC, %) **and hydrogen index** (HI, mg HC/g TOC): TOC is the percentage of total organic carbon in sedimentary rock and HI is the hydrocarbon content in TOC, both of which could be applied to evaluate the abundance of sedimentary organic matter and preservation of hydrocarbons. As the major evaluation parameters, TOC principally reflect the abundance and the primary productivity, while HI would be more sensitive to the preservation of hydrocarbons.
(2) **The content of 2,3,6-trimethylarylisoprenoids** (TMAIs, ng HC/g Rock): Using the analysis of gas chromatography-mass spectrometrography, the biomarker homologous series of C_{13}–C_{23} 2,3,6-trimethylarylisoprenoids (abb. 2,3,6-TMAIs or called 1-alkyl-2,3,6-trimethyl- benzenes; Summons and Powell 1986) have been quantitatively detected from the aromatic fraction of argillaceous rocks in each lithological unit of the Xiamaling Formation, especially for dark shale/mudstone (Fig. 7.4), but the contents of C_{18} and C_{19} 2,3,6-TMAIs is changeable from Units 1 to 4 due to the variation of sedimentary water environment.

As the diagenetic products of aromatic carotenoid pigments isorenieratene and β-isorenieratene in the *Chlorobiaceae* photosynthetic autotrophic green sulfur bacteria (GSB; Summons and Powell 1986, 1987), the quantitative detection of a homologous series of C_{13}–C_{23} 2,3,6-TMAIs, with the low contents of C_{17} and C_{23} homologues (Fig. 7.4), provides a distinctive and significant evidence for their GSB origin (Summons and Powell 1986, 1987). Base on GSB biotope, the successive growth of GSB content in sediments would reflect a salination course in sedimentary water column, while the very high content of 2,3,6-TMAIs indicate an sulfidic anoxic water environment, which is favorable for the breeding of anoxygenic photosynthetic sulfur bacteria *Chlorobiaceae* (Summons and Powell 1987; Brocks et al. 2005).

(3) **Iron speciation:** Iron component analysis is often used to evaluate the oceanic-chemical environment of sedimentary water column (Canfield et al. 1992,

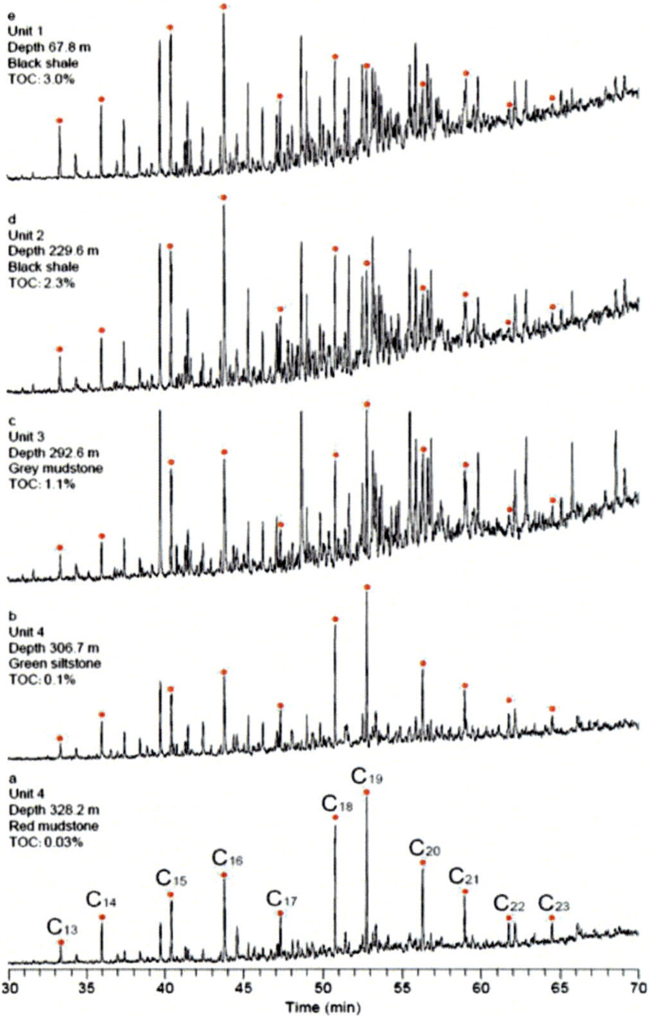

Fig. 7.4 The *m*/*z* 134 mass chromatograms show the distributions of C_{13}–C_{23} 2,3,6-TMAI homologous series detected from the aromatic fractions of core rocks in the Xiamaling Formation. The red dots indicate the homologs of 2,3,6-TMAIs

2008; Raiswell and Canfield 1998; Shen et al. 2002; Poulton et al. 2004; Poulton and Canfield 2005; Raiswell and Anderson 2005; Lyons and Severmann 2006; Planavsky et al. 2011). The total iron composition (Fe_T) is composed of highly reactive iron (Fe_{HR}) and poorly reactive iron. The Fe_{HR} includes iron oxide (Fe_2O_3, e.g., hematite generated under oxic environment), iron carbonate ($FeCO_3$, i.e., siderite generated under weak anoxic environment), and iron sulfide (FeS_2, typically as pyrite Fe_{Py}) generated under non-sulfidic or sulfidic anoxic environment). However, Pyrite (FeS_2) and hematite (Fe_2O_3) cannot be a pair of paragenetic and co-existing minerals. Theoretically, high content of pyrite and highly reactive iron minerals could be the indicators of the sulfidic anoxic water column usually in the euxinic sedimentary environment.

Raiswell and Canfield (1998, 2012) reported that the high Fe_{HR}/Fe_T ratio of sediment indicates the anoxic sedimentary conditions in contemporary oceanic environment, and the Phanerozoic marine sedimentary rocks under anoxic water column have shown the Fe_{HR}/Fe_T ratios exceeding 0.38 (i.e., the highly reactive iron content over 38% of the total iron) would display characteristic sulfidic anoxic or ferruginous sediments in euxinic sedimentary environment. Poulton and Raiswell (2002) has shown that ancient sediments deposited from an oxic water column have average Fe_{HR}/Fe_T ratios of 0.14 ± 0.08. Their observations allow anoxic ($Fe_{HR}/Fe_T > 0.38$) and oxic ($Fe_{HR}/Fe_T < 0.22$) sedimentary conditions to be distinguished (Fig. 7.5).

When ca. 70% or more of the highly reactive iron is bound as sulfide (typically as pyrite), the Fe_{Py}/Fe_{HR} ratio (>0.7) would be indicative of a sulfidic water column in euxinic sedimentary environment, while the ratio < 0.7 would characterize a ferruginous (pyrite and siderite) water-column conditions (Raiswell and Canfield 2012).

Based on the geochemical parameters of Fe_{Py}/Fe_{HR} and Fe_{HR}/Fe_T ratios, Fig. 7.5 shows a sedimentary environment division of water column for the Xiamaling Formation, which can be used to differentiate the palaeo-oceanic environments and the corresponding sedimentary characteristics of Xiamaling palaeo-ocean. The diagram is divided into three areas (Fig. 7.5).

"Area A": Both high reactive iron (Fe_{HR}) and pyrite (Fe_{Py}) are well developed with the Fe_{HR}/Fe_T ratio > 0.38 and the authigenic pyrite predominates over Fe_{HR}

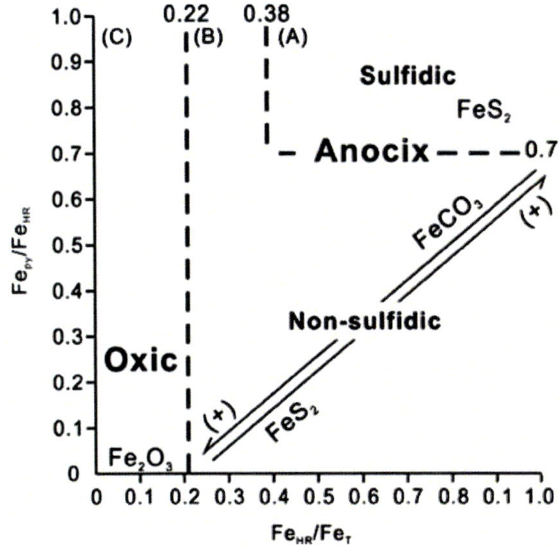

Fig. 7.5 Geochemical environment division of palaeo-oceanic water column. Area A: sulfidic anoxic water column with authigenic pyrite predominance; Area B: non-sulfidic anoxic water column characterized by siderite or siderite plus pyrite; Area C: oxic water column with hematite or hematite plus siderite

with the Fe_{Py}/Fe_{HR} ratio > 0.7. Therefore, the "Area A" would be characterized by authigenic pyrite predominance, and thus reflecting a strong anoxic and sulfidic environment.

"Area B": Outsider the "Area A", the Fe_{HR} is still well developed with the Fe_{HR}/Fe_T ratio > 0.22 in "Area B". In such a case, the "Area B" would show the features of non-sulfidic ferruginous mineral (i.e., pyrite and/or siderite) sedimentation in a non-sulfidic anoxic palaeo-oceanic geochemical environment.

"Area C": Fe_{HR} is poorly developed with the Fe_{HR}/Fe_T ratio < 0.22 (Poulton and Raiswell 2002), and only hematite or hematite plus siderite generated, which reflects an oxic to weak anoxic palaeo-oceanic geochemical environment.

(4) **Trace element content of Mo, U and V** (%): Mo, U and V are all redox-sensitive trace elements. These elements exist in soluble state (e.g., MoO_4^{2-}, HVO_4^{2-}, $UO_2(CO_3)_3^{4-}$) in the oxic waters, which are easy to form insoluble state (e.g., MoS_4^{2-}, V_2O_3, UO_2) when the water turns to be anoxic, resulting in their transformation from water to sediments (Algeo and Lyons 2006; Tribovillard et al. 2006; Algeo and Rowe 2012). Thus, the contents of Mo, U and V are much higher in the sediments of anoxic sedimentary environment than those in oxic environment.

7.4.2 Palaeo-Oceanic Geochemical Profile

The Xiamaling Formation contains not only the different sedimentary environment, but also the distinct geochemical records in its individual lithological unit. The geochemical profiles of the Xiamaling Formation are shown in Fig. 7.6.

In Unit 4 (or the upper Xia-2 Member), the variegated mudstone/shale and siltstone show very low TOC (0.05% to 0.4%, mostly in trace amount) and low HI values (mostly < 50 mg HC/g TOC), trace C_{18}-C_{19} TAMIs (GSB biomarkers), low to medium Fe_{HR}/Fe_T (0.05–0.65, mostly around 0.38) and very low Fe_{Py}/Fe_{HR} (trace amount) ratios, very low contents of Mo (<10 ppm) and U (<5 ppm).

The black shales in Unit 3 (or the lower Xia-3 Member) have very high TOC (half > 5% with the maximum up to 20%) and high HI values (mostly > 200 mg HC/g TOC), low contents of C_{18}–C_{19} TAMIs (individually up to 107.55 ng HC/g Rock), medium Fe_{HR}/Fe_T (mostly > 0.22, around 0.38) and low to medium Fe_{Py}/Fe_{HR} (0.05–0.7, individually > 0.7) ratios, and low to medium contents of Mo (<30 ppm) and U (<15 ppm). It is worth noting that the interval of 290–270 m is characterized by a depletion of V (with the minimum of 7.6 ppm).

The black and green shales in Unit 2 (or the upper Xia-3 Member) are characterized by higher TOC (mostly 2–7%) and very high HI values (up to 800 mg HC/g TOC), high Fe_{HR}/Fe_T (0.2–0.8, mostly > 0.38) and medium to high Fe_{Py}/Fe_{HR} (0.1–0.8, partially > 0.70) ratios, high contents of C_{18}–C_{19} TMAIs (individually up to 269.53 ng HC/g Rock), Mo (up to 47.7 ppm), U (up to 13.6 ppm), and V (up to 819 ppm) especially for its upper interval.

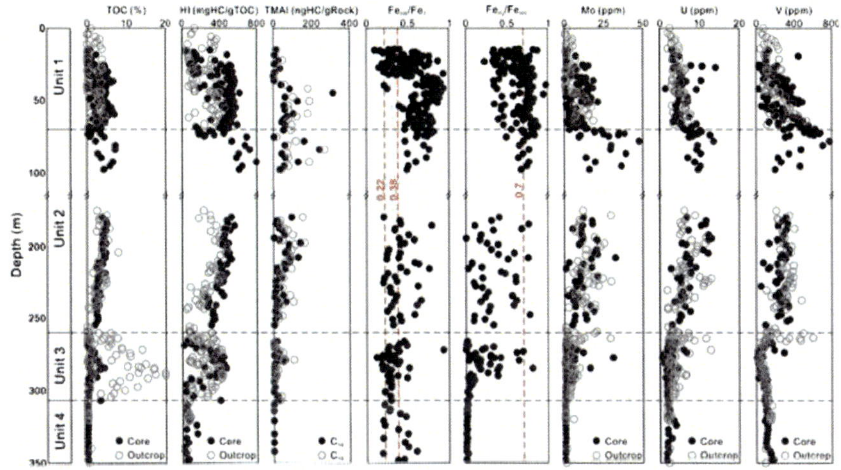

Fig. 7.6 Geochemical profiles of the Xiamaling Formation (Wang et al. 2017, mordified). TOC. total organic carbon content (%); HI. hydrogen index (mg HC/g TOC); TMAI. the content of C_{18}–C_{19} 2,3,6-trimethylarylisoprenoids (ng HC/g Rock); Fe_T. total iron content (%); Fe_{HR}. the content of highly reactive iron; Fe_{Py}. the pyrite content. Mo, U, V. trace elements contents (ppm) of molybdenum, uranium, and vanadium

In Unit 1 (or the Xia-4 Member), the black, grey, and greyish-green mudstones show higher TOC (mostly 2–5%), very high Fe_{HR}/Fe_T (mostly > 0.38, up to 1) and Fe_{Py}/Fe_{HR} (mostly > 0.7) ratios, higher contents of C_{18}–C_{19} TMAIs (partially up to 300 ng HC/g Rock), and medium to high contents of Mo (up to 21.4 ppm), U (up to 14 ppm) and V (up to 622.5 ppm). However, almost all the above-mentioned geochemical parameters show an obvious decrease trend at the upper interval of Unit 1 (Fig. 7.6).

Although these geochemical parameters show dramatic change along the section of Xiamaling Formation in the Xuanlong Depression, YFDZ, the similar geochemical variation has been reported in previous Precambrian literatures, while these parameters are linked with surface runoff rate and water circulation pattern in the sedimentary basin (Rossignol-Strick 1987; Cramp and O'Sullivan 1999; Beckmann et al. 2005) or trade wind intensity, upwelling rate and dynamic trace metals supply (Yarincik et al. 2000; Beckmann et al. 2005; Hofmann and Wagner 2011). As a sedimentary signal below storm wave base, the absence of cross-bedding indicates that the diagenetic process should have happened below storm wave base (Meng et al. 2011). Based on the above geochemical parameters, therefore, the sedimentary environments in four units of the Xiamaling Formation can be evaluated.

7.5 Palaeo-Oceanic Geochemical Environments of Xiamaling Formation

7.5.1 Unit 4: Oxic to Weak Anoxic Environments

Since Unit 4 is composed of variegated mudstone/shale, siltstone and sandstone, among which the highly reactive iron of purplish-red mudstone/shale is rich-in hematite, while the Fe_{HR} of greyish green one contains siderite. Therefore, the variegated strata of Unit 4 typically show low to medium Fe_{HR}/Fe_T (0.2–0.6) and very low Fe_{Py}/Fe_{HR} (trace amount) ratios, very low contents of redox-sensitive elements Mo (<10 ppm) and U (<5 ppm), and the trace GSB biomarkers (C_{18}–C_{19} TAMIs), Overall, it would indicate a water column environment in oxic to weak anoxic conditions.

On the diagram of Fe_{Py}/Fe_{HR} vs. Fe_{HR}/Fe_T ratios, all the rock samples of Unit 4 are plotted along the abscissa (X-axis) within the "Area C" (oxic) and "Area B" (weak anoxic). As the ferruginous minerals (Fe_{HR}), there should be only autogenetic siderite generated in Area B as well as the hematite in Area C without autogenetic pyrite, which would be featured by the ferruginous (hematite- or siderite-containing) mudstones/shales of typical oxic to weak anoxic water column environments, especially for the early typical oxygenating stage of Unit 4 (Fig. 7.7).

Moreover, the mudstone has very low TOC (<0.3%, mostly in trace amount) and low HI value (mostly < 100 mg HC/g TOC; Fig. 7.6), implying that the very low abondance and the very poor preservation of sedimentary organic matter in the Unit 4 as well as the very low primary productivity for the palaeo-ocean during Unit 4 sedimentary period.

7.5.2 Unit 3: Non-Sulfidic Anoxic Environment and Oxygen Minimum Zone (OMZ)

Unit 3 is mainly composed of the interbedding of black shale and silicalite, the black shales have fine-lamellas and are also known as paper-oil shale, but without any indication for waves or water flow modification, representing the sedimentary features of deep-water facies sediments. These shales show medium Fe_{HR}/Fe_T (mostly > 0.22 and around 0.38) and low to medium Fe_{Py}/Fe_{HR} (mostly < 0.7) ratios, low to medium contents of redox-sensitive elements Mo (<30 ppm), U (<15 ppm), and medium amounts of GSB biomarkers (C_{18}–C_{19} TAMIs mostly < 50 ng HC/g Rock, individually up to 107.55 ng HC/g Rock). There is an upward transgression sequence from Unit 4 to Unit 3 (cf., the TMAIs, Fe_{Py}/Fe_T, Mo, and U variation profiles in Fig. 7.6).

On the diagram of Fe_{Py}/Fe_{HR} versus Fe_{HR}/Fe_T ratios (Fig. 7.7), the majority of data points in Unit 3 are plotted mostly into "Area B" (non-sulfidic anoxic), and individually into "Area C" (oxic) and "Area A" (sulfidic anoxic). Overall, Unit 3

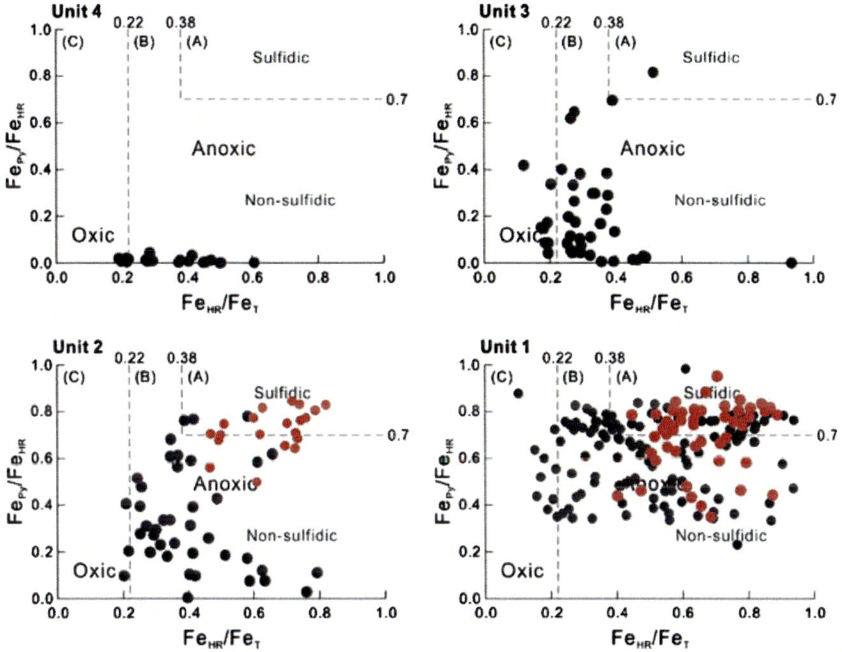

Fig. 7.7 Diagram of Fe_{Py}/Fe_{HR} vs. Fe_{HR}/Fe_T ratios for the differentiation of palaeo-oceanic environments of the Xiamaling Formation. Red dots are mainly attributed to sulfidic anoxic environment from the upper interval of Unit 2 to the lower-middle intervals of Unit 1, while black dots are mostly referred to the non-sulfidic anoxic environment, respectively, in the lower-middle intervals of Unit 2 and the upper interval of Unit 1

would be referred to the non-sulfidic anoxic (pyrite- and/or siderite-containing) deep-water sedimentary environment.

It is noteworthy that shales of Unit 3 are rich in TOC, Mo, and U, but are depleted in V (Fig. 7.6). In modern sedimentary water column, V is commonly released from sediments deposited under low-oxygen conditions and under normal bottom water oxygen levels where oxygen only penetrates a few millimeters into sediment (Emerson and Huested 1991; Nameroff et al. 2002a, b). The profile of V makes a contrast to the Mo and U profiles in Unit 3, which is consistent with OMZ depositional setting as discussed below. Redox-sensitive elements Mo, U, and V are usually co-enriched in sediments deposited in the middle of the anoxic OMZ. However, at the lower part of the anoxic OMZ, where bottom-water oxygen levels begin to rise, V would return to content at or even below the crustal average (Bennett and Canfield 2020). In contrast, Mo and U maintain enrichments above the crustal average in sediments with high TOC values (Zhang et al. 2016; Wang et al. 2017).

OMZ is an oxygen-deficient layer in the contemporary mid-deep ocean. The formation of OMZ is due to the decomposition of surface-derived sinking organic material by aerobic respiration in the subsurface, combined with the introduction of oxygen-rich surface water to the deeper bottom zones. In some cases of modern tropical open ocean (e.g., eastern North Pacific, Arabian Sea, etc.), the OMZ can extend to a few hundred meters deep (Ulloa et al. 2012).

The above OMZ interpretation is also proved by the increased contents of C_{18}–C_{19} TMAIs in Fig. 7.6, which indicate the growth of anoxygenic phototrophic GSB living off the oxidation of either sulfide or iron in the water column (Zhang et al. 2016). The co-occurrence of V depletion and GSB development reveals the emergency of anoxic photic zone with an oxygenated sediment–water interface, which is in agreement with the OMZ setting.

Therefore, in the non-sulfidic anoxic environment, the black shale of Unit 3 contains very high TOC (mostly > 5%, even up to 20%) and elevated HI values (mostly 200 mg HC/g TOC and up to 435 mg HC/g TOC; Fig. 7.6), indicating very high primary productivity and well preservation of sedimentary organic matter.

7.5.3 Unit 2: Non-sulfidic to Partial Sulfidic Anoxic Environments

As euxinic sediments without any sedimentary disturbance, the black to grey shale in Unit 2 displays obvious banding which was resulted from differential weathering. These bandings may not be observable on the surface of fresh core, but micro-scale banding can be observed on rock slice, which would be the features of deep-water facies.

Overall, the shale has higher Fe_{HR}/Fe_T (0.2–0.8, mostly > 0.38) and medium to high Fe_{Py}/Fe_{HR} (0.1–0.8, partially > 0.7) ratios, while higher content of C_{18}-C_{19} TMAIs (individually up to 269 ngHC/gRock) as well as very high contents of Mo (up to 47.7 ppm), U (up to 13.6 ppm) and V (up to 819 ppm), especially for the upper interval of Unit 2.

On the diagram of Fe_{Py}/Fe_{HR} vs. Fe_{HR}/Fe_T ratios (Fig. 7.7), the Unit 2 can also be divided into two parts: red dots for its upper interval are mostly attributed to the sulfidic anoxic "Area A" and partially to the non-sulfidic anoxic "Area B"; while black dots for the lower-middle interval mostly to the non-sulfidic anoxic "Area B" (Fig. 7.7). Both of which indicate an upward palaeo-oceanic environment transformation from non-sulfidic to sulfidic anoxic water column in an enxinic environment.

A gradually upward increasing trend of all these geochemical parameters TOC, HI value, C_{18}–C_{19} TMALs contents, Fe_{HR}/Fe_T, Fe_{Py}/Fe_{HR} and Mo, U, V contents indicate a transgression sequence from non-sulfidic anoxic (the lower-middle intervals) to partial sulfidic (the upper interval) palaeo-oceanic environments (cf. the HI, TMAI, Mo, U and V profiles in Figs. 7.6 and 7.7), which can be resulted from the gradually expanded sea-flooding area and the deepened seawater during the sedimentary period of Unit 2, especially for its upper interval to enter the maximum sea-flooding period.

The sedimentary waters of Unit 2 supported the continuous growth of anoxygenic phototrophic GSB as evidenced by upward increasing contents of C_{18}–C_{19} TMAIs, which are consistent with the enrichment of redox-sensitive trace elements Mo, U and V, especially in the upper intervals of Unit 2 (Fig. 7.6), and the GSB was living in the water column of non-sulfidic anoxic to partial sulfidic anoxic environment.

Overall, the black shale of Unit 2 has higher TOC value (up to 7.2%), higher contents of C_{18}–C_{19} TMAIs, and very high HI value (up to 800 mg HC/g TOC), which demonstrat the higher organic matter abundance and higher primary productivity as well as the very well preservation of sedimentary organic matter, particularly for its upper interval (Fig. 7.6). In addition, the vertical variations of HI, redox-sensitive Mo, U, V and C_{18}–C_{19} TMAIs show the maximum peak values at the upper interval of Unit 2 (Fig. 7.6).

7.5.4 *Unit 1: Sulfidic Anoxic to Non-sulfidic Environments*

Based on multiple geochemical indicators, the lower-middle intervals of Unit 1 have a continuous deep water sulfidic sedimentary environment (Fig. 7.6), its black, grey and greyish-green mudstones/shales contain very high Fe_{HR}/Fe_T (mostly > 0.38, maximum up to 1) and Fe_{Py}/Fe_{HR} (mostly > 0.7, maximum up to 1) ratios, higher content of C_{18}-C_{19} TMAIs (up to 300 ng HC/g Rock) and medium to high contents of Mo (up to 21.4 ppm), U (up to 14 ppm) and V (up to 622.5 ppm). All the above geochemical parameters in the upper interval present an upward decreasing tendency (Fig. 7.6).

On the diagram of Fe_{Py}/Fe_{HR} vs. Fe_{HR}/Fe_T ratios (Fig. 7.7), Unit 1 can also be divided into two parts: red dots for the lower-middle intervals are characterized by the maximum Fe_{HR}/Fe_T and Fe_{Py}/Fe_{HR} ratios, and mainly concentrated in the “Area A”, indicating a typical sulfidic anoxic environment and a maximum sea-flooding period with an euxinic setting. In contrast, black dots for the upper interval are mainly scattered in the “Aera B” with changeable Fe_{Py}/Fe_{HR} and Fe_{HR}/Fe_T ratios, showing a non-sulfidic anoxic environment and a regression sequence. Unit 1 would be a regression sequence from sulfidic anoxic to non-sulfidic palaeo-oceanic environments.

7.6 Sedimentary Cycle and Model of Xiamaling Palaeo-Ocean

The geochemical profiles of the Xiamaling Formation in Fig. 7.6 shows a complete sedimentary cycle, including a continuously steady transgression process (during the sedimentary period from Unit 4 to the middle Unit 2) as well as a rapid regression process (the period of the upper Unit 1), while the upper Unit 2 plus the lower-middle Unit 1 belong to the maximum sea-flooding sequence with the typical sulfidic anoxic sedimentary environment (cf., the TMAIs, Fe_{HR}/Fe_T, Fe_{Py}/Fe_{HR}, Mo, U, V variation profiles in Fig. 7.6).

In Fig. 7.8, we show a suggested palaeo-oceanic geochemical sedimentary model during the Xiamaling sedimentary period. Unit 4 reflects an oxic to weak anoxic environment with periodic upwelling and runoff input under an oxic to non-sulfidic weak anoxic water column. With the highest primary productivity, the black shale of Unit 3 represents a continued non-sulfidic anoxic environment under an OMZ-containing deep water column below storm wave base. Unit 2 is deposited under the non-sulfidic to partial sulfidic anoxic environments, in which its lower-middle interval was non-sulfidic and the upper interval was sulfidic. Unit 1 is in transition from a sulfidic anoxic environment (for the lower-middle intervals) to a non-sulfidic anoxic/oxic environment (for the upper interval).

Even though the sulfidic anoxic sequence of upper Unit 2 to lower-middle Unit 1 to contains the maximum Fe_{HR}/Fe_T and Fe_{HR}/Fe_T ratios, HI value and the highest contents of redox-sensitive TMAL and trace elements (Mo, U and V), however, the non-sulfidic anoxic sequence of Unit 3 has the highest TOC (up to 20%) in the

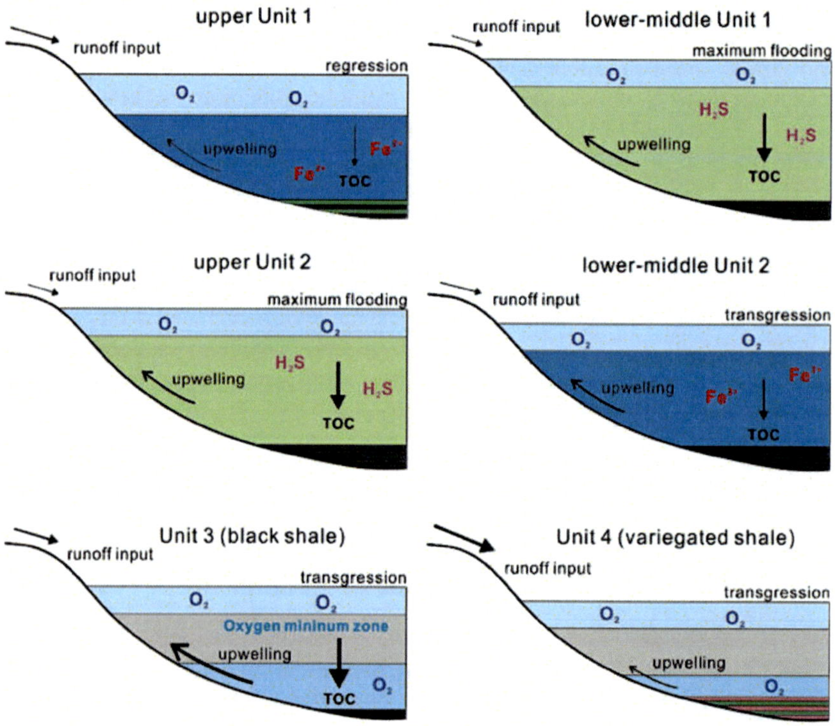

Fig. 7.8 A palaeo-oceanic chemical sedimentary model during the Xiamaling sedimentary period (Wang et al. 2017, modified)

Xiamaling Formation probably due to the constraint of sulfidization on the biodiversity and biomass in Palaeo-oceanic environment. The tropical OMZ is considered to be the hotspots of ocean deoxygenation, mostly related to the robust upwelling of nutrient-rich water that can generate high primary production in the surface and high aerobic respiration in the subsurface (Scholz 2018). Today, surface primary productivity in the Peruvian upwelling region is 1300 g·C/(m^2·a), which is ten times higher than primary productivity in Black Sea (Pennington et al. 2006).

The upward mixing of nutrient-replete deep waters is the key to support high primary productivity. Moreover, enhanced regeneration of phosphorus and iron from mid-depth anoxic waters can act as positive feedback for primary production. Specifically, sedimentary iron and phosphate are released by the recycling of both Fe oxides and organic material, which would fertilize primary productivity and constitute the major sources of dissolved Fe and phosphate to the global ocean (Noffke et al. 2012).

By comparison to the non-sulfidic anoxic sedimentary environment, the sulfidic anoxic environment has monotonous population of aquatic organisms and finite biomass, but minimal oxidation effect for the formation and enrichment of sedimentary organic matter. Consequently, Unit 3 contains the highest TOC in the Xiamaling Formation, while Units 1 and 2 only has higher TOC, but the highest HI value. Therefore, the OMZ-containing non-sulfidic anoxic sequence processes maximum primary productivity, and the sulfidic anoxic sequence owns excellent hydrocarbons preservation condition.

It is suggested that the sedimentary features and palaeo-oceanic geochemical signals of the Xiamaling Formation were probably driven by changes of the intertropical convergence zone and the relationship of the North China Craton (NCC) to Hadley cell circulation, which could have affected water-column productivity, oxygen distribution, as well as sediment source to the basin (Broccoli et al. 2006; Arbuszewski et al. 2013; Wagner et al. 2013; Schneider et al. 2014).

7.7 Conclusions

Palaeo-oceanic chemical sedimentary environments of the water columns were recorded during the deposition of the Neoproterozoic Xiamaling Formation in the YFDZ, NCC. The Xiamaling Formation is traditionally divided into four stratigraphic members in ascending order. In this chapter, however, a stratigraphic division scheme of 6 lithologic units in descending order is tentatively suggested based on a detailed geochemical and sedimentological investigation in order to approach the palaeo-oceanic geochemical environments.

According to the analyses of iron speciation, trace elements, and organic geochemistry, the Unit 4 of the Xiamaling Formation was formed within oxic to weak anoxic sedimentary environments. Notably, runoff source input led to the distinct variegated sediments of this unit. Unit 3 has a non-sulfidic anoxic environment under an OMZ setting with the highest primary productivity of sedimentary organic matter, which seems to be affected by palaeo-climate. Unit 2 was deposited in non-sulfidic to sulfidic anoxic environments below storm wave base. Unit 1 was deposited under sulfidic anoxic to non-sulfidic environments. The OMZ-containing non-sulfidic anoxic sequence processes maximum primary productivity, while the sulfidic anoxic sequence owns the best hydrocarbons preservation condition.

Overall, the Xiamaling Formation constitutes a complete sedimentary cycle with a continuously steady transgression process (Unit 4 to middle Unit 2) and a rapid regression process (upper Unit 1) and its maximum sea-flooding sequence in upper Unit 2 to middle Unit 1.

References

Algeo TJ, Lyons TW (2006) Mo-total organic carbon covariation in modern anoxic marine environments: implications for analysis of paleoredox and paleohydrographic conditions. Paleoceanography 21:PA1016

Algeo TJ, Rowe H (2012) Paleoceanographic applications of trace-metal concentration data. Chem Geol 324:6–18

Anbar AD, Knoll AH (2002) Proterozoic ocean chemistry and evolution: a bioinorganic bridge. Science 297:1137–1142

Arbuszewski JA, Cléroux C, Bradtmiller L, Mix A (2013) Meridional shifts of the Atlantic intertropical convergence zone since the Last Glacial Maximum. Nat Geosci 6(11):959–962

Beckmann B, Flogel S, HoFormationann P, Schulz M, Wagner T (2005) Orbital forcing of Cretaceous river discharge in tropical Africa and ocean response. Nature 437:241–244

Bennett WW, Canfield DE (2020) Redox-sensitive trace metals as paleoredox proxies: a review and analysis of data from modern sediments. Earth Sci Rev 204:103175

Bjerrum CJ, Canfield DE (2002) Ocean productivity before about 1.9 Gyr ago limited by phosphorus adsorption onto iron oxides. Nature 417:159–162

Broccoli AJ, Dahl KA, Stouffer RJ (2006) Response of the ITCZ to Northern Hemisphere cooling. Geophys Res Lett 33(116):L01702

Brocks JJ, Love GD, Summons RE, Knoll AH, Logan GA, Bowden SA (2005) Biomarker evidence for green and purple sulphur bacteria in a stratified Palaeoproterozoic sea. Nature 437(7060):866–870

Canfield DE (1998) A new model for Proterozoic ocean chemistry. Nature 396:450–453

Canfield DE, Raiswell R, Bottrell SH (1992) The reactivity of sedimentary iron minerals toward sulfide. Am J Sci 292(9):659–683

Canfield DE, Poulton SW, Knoll AH, Narbonne GM, Ross G, Goldberg T, Strauss H (2008) Ferruginous conditions dominated Later Neoproterozoic deep-water chemistry. Science 321:949–952

Canfield DE, Zhang SC, Wang HJ, Wang XM, Zhao WZ, Su J, Bjerrum CJ, Haxen ER, Hammarlund EU (2018) A mesoproterozoic iron formation. Proc Nat Acad Sci USA 115:E3895–E3904

Cramp A, O'Sullivan G (1999) Neogene sapropels in the Mediterranean: a review. Mar Geol 153(1–4):11–28

Emerson SR, Huested SS (1991) Ocean anoxia and the concentrations of molybdenum and vanadium in seawater. Mar Chem 34(3–4):177–196

Evans DA, Mitchell RN (2011) Assembly and breakup of the core of Paleoproterozoic-Mesoproterozoic supercontinent Nuna. Geology 39(5):443–446

Fan W (2015) Geological features and research progress of the Mesoproterozoic Xiamaling Formation in the North China Craton: a review after nearly 100 years of study. Geol Rev 61:1383–1406 (in Chinese with English Abstract)

Gao LZ, Zhang CH, Shi XY, Zhou HR, Wang ZQ, Song B (2007) A new SHRIMP age of the Xiamaling Formation in the North China Plate and its geological significance. Acta Geologica Sinica-English Edition 81(6):1103–1109

Gao LZ, Zhang CH, Shi XY, Song B, Wang Z, Liu YM (2008) Mesoproterozoic age for Xiamaling Formation in North China Plate indicated by zircon SHRIMP dating. Chin Sci Bull 53(17):2665–2671

Hofmann P, Wagner T (2011) ITCZ controls on Late Cretaceous black shale sedimentation in the tropical Atlantic Ocean. Paleoceanography 26:PA4223

Kao CS, Hsiung YH, Kao P (1934) Pleliminary notes on Sinian stratigraphy of North China. Bull Geo Soc China 13(2):243–288

Li HK, Lu SN, Li HM, Sun LX, Xiang ZQ, Geng JZ, Zhou HY (2009) Zircon and beddeleyite U-Pb precision dating of basic rock sills intruding Xiamaling Formation, North China. Geol Bull China 28(10):1396–1404 (in Chinese with English abstract)

Li HK, Su WB, Zhou HY, Geng JZ, Xiang ZQ, Cui YR, Liu WC, Lu SN (2011) The base age of the Changchengian System at the northern North China Craton should be younger than 1670 Ma: constraints from zircon U-Pb LA-MC-ICPMS dating of a granite-porphyry dike in Miyun County, Beijing. Front Earth Sci 18:108–120 (in Chinese with English abstract)

Li HK, Lu SN, Su WB, Xiang ZQ, Zhou HY, Zhang Y (2013) Recent advances in the study of the Mesoproterozoic geochronology in the North China Craton. J Asian Earth Sci 72:216–227

Li HK, Su WB, Zhou HY, Xiang ZQ, Tian H, Yang LG, Huff WD, Ettensohn FR (2014) The first precise age constraints on the Jixian System of the Meso- to Neoproterozoic Standard Section of China: SHRIMP zircon U-Pb dating of bentonites from the Wumishan and Tieling formations in the Jixian Section, North China Craton. Acta Petrologica Sinica 30(10):2999–3012 (in Chinese with English abstract)

Li MR, Wang SS, Qiu J (1996) The Ages of Glauconites from Tieling and Jingeryu Formations, Beijing-Tianjin Area. Acta Petrologica Sinica 3:416–423 (in Chinese with English abstract)

Li ZH, Xi SL, Hu JM, Dong XP, Zhang GS (2019) New insights about the Mesoproterozoic sedimentary framework of North China Craton. Geol J 54(1):409–425

Liu Y, Zhong NN, Tian YJ, Qi W, Mu GY (2011) The oldest oil accumulation in China: Mesoproterozoic Xiamaling Formation bituminous sandstone reservoirs. Pet Explor Dev 38(4):503–512

Lu SN, Zhao GC, Wang HC, Hao GJ (2008) Precambrian metamorphic basement and sedimentary cover of the North China Craton: a review. Precambr Res 160(1–2):77–93

Luo QY, Zhong NN, Zhu L, Wang YN, Qin J, Qi L, Zhang Y, Ma Y (2013) Correlation of burial organic carbon and paleoproductivity in the Mesoproterozoic Hongshuizhuang Formation, northern North China. Chin Sci Bull 58(11):1299–1309

Lyons TW, Severmann S (2006) A critical look at iron paleoredox proxies: new insights from modern euxinic marine basins. Geochim Cosmochim Acta 70(23):5698–5722

Nameroff TJ, Balistrieri LS, Murray JW (2002a) Suboxic trace metal geochemistry in the eastern tropical North Pacific. Geochimica Et Coamochimica Acta 66(7):1139–1158

Meng QR, Wei HH, Qu YQ, Ma SX (2011) Stratigraphic and sedimentary records of the rift to drift evolution of the northern North China craton at the Paleo- to Mesoproterozoic transition. Gondwana Res 20(1):205–218

Nameroff T, Balistrieri L, Murray J (2002b) Suboxic trace metal geochemistry in the eastern tropical North Pacific. Geochim Cosmochim Acta 66(7):1139–1158

Noffke A, Hensen C, Sommer S, Scholz F, Bohlen L, Mosch T, Graco M, Wallmann K (2012) Benthic iron and phosphorus fluxes across the Peruvian oxygen minimum zone. Limnol Oceanogr 57:851–867

Pennington JT, Mahoney KL, Kuwahara VS, Kolber DD, Calienes R, Chavez FP (2006) Primary production in the eastern tropical Pacific: a review. Prog Oceanogr 69:285–317

Pisarevsky SA, Elming SÅ, Pesonen LJ, Li ZX (2014) Mesoproterozoic paleogeography: supercontinent and beyond. Precambr Res 244:207–225

Planavsky NJ, McGoldrick P, Scott CT, Li C, Reinhard CT, Kelly AE, Chu X, Bekker A, Love GD, Lyons TW (2011) Widespread iron-rich conditions in the mid-Proterozoic ocean. Nature 477:448–451

Planavsky NJ, Reinhard CT, Wang X, Thomson D, McGoldrick P, Rainbird RH, Johnson T, Fischer WW, Lyons TW (2014) Low Mid-Proterozoic atmospheric oxygen levels and the delayed rise of animals. Science 346:635–638

Poulton SW, Canfield DE (2005) Development of a sequential extraction procedure for iron: implications for iron partitioning in continentally derived particulates. Chem Geol 214(3):209–221

Poulton SW, Canfield DE (2011) Ferruginous conditions: a dominant feature of the ocean through Earth's history. Elements 7(2):107–112

Poulton SW, Raiswell R (2002) The low-temperature geochemical cycle of iron: from continental fluxes to marine sediment deposition. Am J Sci 302:774–805

Poulton SW, Fralick PW, Canfield DE (2004) The transition to a sulphidic ocean approximately 1.84 billion years ago. Nature 431:173–177

Qiao XF (1976) Investigation on stratigraphy of the Qingbaikou Group of the Yanshan Mountains, North China. Chin J Geol 3:246–264 (in Chinese with English abstract)

Qu YQ, Pan JG, Ma SX, Lei ZP, Li L, Wu GL (2014) Geological characteristics and tectonic significance of unconformities in Mesoproterozoic successions in the northern margin of the North China Block. Geosci Front 5:127–138

Raiswell R, Anderson T (2005) Reactive iron enrichment in sediments deposited beneath euxinic bottom waters: constraints on supply by shelf recycling. Geol Soc Lond Spec Publ 248(1):179–194

Raiswell R, Canfield DE (1998) Sources of iron for pyrite formation in marine sediments. Am J Sci 298(3):219–245

Raiswell R, Canfield DE (2012) The iron biogeochemical cycle past and present. Geochem Perspect 1(1):1–210

Rossignol-Strick M (1987) Rainy periods and bottom water stagnation initiating brine accumulation and meal concentrations: 1. The Late Quaternary. Paleoceanography 2(3):333–360

Schneider T, Bischoff T, Haug GH (2014) Migrations and dynamics of the intertropical convergence zone. Nature 513:45–53

Scholz F (2018) Identifying oxygen minimum zone-type biogeochemical cycling in Earth history using inorganic geochemical proxies. Earth-Sci Rev 184:29–45

Shen Y, Canfield DE, Knoll AH (2002) Middle Proterozoic ocean chemistry: evidence from the McArthur Basin, northern Australia. Am J Sci 302(2):81–109

Shen Y, Knoll AH, Walter MR (2003) Evidence for low sulphate and anoxia in a mid-Proterozoic marine basin. Nature 423:632–635

Shi Y, Liu D, Kröner A, Jian P, Miao L, Zhang F (2012) Ca. 1318 Ma A-type granite on the northern margin of the North China Craton: implications for intraplate extension of the Columbia supercontinent. Lithos 148:1–9

Song QC, Zhang ZC (1983) The Xiamaling Formation and its palaeogeographic environment in Xiahuayuan area, Hebei Province. J Stratigr 7(2):104–111 (in Chinese with English abstract)

Sperling E, Rooney A, Hays L, Sergeev V, Vorob'eva N, Sergeeva N, Selby D, Johnston D, Knoll A (2014) Redox heterogeneity of subsurface waters in the Mesoproterozoic ocean. Geobiology 12(5):373–386

Su WB, Zhang SH, Huff WD, Li HK, Ettensohn FR, Chen X, Yang H, Han Y, Song B, Santosh M (2008) SHRIMP U-Pb ages of K-bentonite beds in the Xiamaling Formation: implications for revised subdivision of the Meso- to Neoproterozoic history of the North China Craton. Gondwana Res 14(3):543–553

Su WB, Li HK, Huff WD, Zhang SH, Zhou HY, Wan YS (2010) SHRIMP U-Pb dating for a K-bentonite bed in the Tieling Formation, North China. Chin Sci Bull 55(29):3312–3323

Summons RF, Powell (1986) Chlorobiaceae in Palaeozoic seas revealed by biological markers, isotopes and geology. Nature 319(27):763–765

Summons RF, Powell (1987) Identification of aryl isoprenoids in source rocks and crude oils: Biological markers for the green sulphur bacteria. Geochimica et Coamochimica Acta 51:557–566

Tribovillard N, Algeo TJ, Lyons TW, Riboulleau A (2006) Trace metals as paleoredox and paleoproductivity proxies: an update. Chem Geol 232(1–2):12–32

Ulloa O, Canfield DE, DeLong EF, Letelier RM, Stewart FJ (2012) Microbial oceanography of anoxic oxygen minimum zones. Proc Natl Aca Sci USA 109(40):15996–16003

Wagner T, HoFormationann P, Flögel S (2013) Marine black shale deposition and Hadley Cell dynamics: a conceptual framework for the Cretaceous Atlantic Ocean. Mar Pet Geol 43:222–238

Wang TG, Simoneit BRT (1995) Tricyclic terpanes in Precambrian bituminous sandstone from the eastern Yanshan region, North China. Chem Geol 120(1–2):155–170

Wang XM, Zhang SC, Wang HJ, Bjerrum CJ, Hammarlund EU, Haxen ER, Su J, Wang Y, Canfield DE (2017) Oxygen, climate and the chemical evolution of a 1400 million year old tropical marine setting. Am J Sci 317(8):861–900

Yarincik KM, Murray RW, Peterson LC (2000) Climatically sensitive eolian and hemipelagic deposition in the Cariaco Basin, Venezuela, over the past 578,000 years: results from Al/Ti and K/Al. Paleoceanography 15(2):210–228

Ye Y, Zhang S, Wang H, Wang X, Tan C, Li M, Wu C, Canfield DE (2021) Black shale Mo isotope record reveals dynamic ocean redox during the Mesoproterozoic Era. Geochem Perspect Lett 18:16–21

Zhang SC, Zhang BM, Bian LZ, Jin ZJ, Wang DR, Chen JF (2007) The Xiamaling oil shale generated through Rhodophyta over 800 Ma ago. Sci China Ser D Earth Sci 50(4):527–535

Zhang SC, Wang XM, Hammarlund EU, Wang HJ, Costa MM, Bjerrum CJ, Connelly JN, Zhang BM, Bian LZ, Canfield DE (2015) Orbital forcing of climate 1.4 billion years ago. Proc Nat Acad Sci USA 112(12):E1406–E1413

Zhang SC, Wang XM, Wang HJ, Bjerrum CJ, Hammarlund EU, Costa MM, Connelly JN, Zhang B, Su J, Canfield DE (2016) Sufficient oxygen for animal respiration 1,400 million years ago. Proc Natl Acad Sci USA 113:21731–21736

Zhang SC, Wang XM, Wang HJ, Bjerrum CJ, Hammarlund EU, Haxen ER, Wen HJ, Ye YT, Canfield DE (2019) Paleoenvironmental proxies and what the Xiamaling Formation tells us about the mid-Proterozoic ocean. Geobiology 17(3):225–246

Zhang SH, Zhao Y (2018) The 1.33–1.30 Ga mafic large igneous province and REE-Nb metallogenic event in the northern North China Craton. Earth Sci Front 25(5):34–50 (in Chinese with English abstract)

Zhang SH, Li ZX, Evans DA, Wu H, Li H, Dong J (2012) Pre-Rodinia supercontinent Nuna shaping up: a global synthesis with new paleomagnetic results from North China. Earth Planet Sci Lett 353:145–155

Zhao WZ, Wang XM, Hu SY, Zhang SC, Wang HJ, Guan SW, Ye YT, Ren R, Wang TS (2019) Hydrocarbon generation characteristics and exploration prospects of Proterozoic source rocks in China. Sci China Earth Sci 62(6):909–934

Chapter 8
Meso-Neoproterozoic Multiple Rifting and Magmatism in the North China Craton

Mingguo Zhai, Bo Hu, Peng Peng, Taiping Zhao, and Qingren Meng

Abstract The North China Craton (NCC) was formed around 2.50 Ga. A 2.50–2.35 Ga period involved a time of tectonic quiescence, and then an important orogenic epoch, i.e., the period of Hutuo Orogeny. The tectonic processes of rifting, subduction and collision characterize this tectonic event, and possibly relate to development of the Supercontinents Nuna or Columbia. Three Palaeoproterozoic orogenic belts have been recognized in the NCC, which are termed as the Jiaoliaoji, Jinyü, and Fengzhen Mobile Belts, respectively. The Jiaoliaoji Mobile Belt is located in the eastern NCC, and made up mainly with the Liaohe and Fenzishan Groups. The Jinyü Mobile Belt occurs in the centro-western NCC, and made up of the Lüliang, Hutuo and Zhongtiao Groups. The Fengzhen Mobile Belt developed in the northwestern NCC, and is composed of the Fengzhen and Erdaowa Groups. The Palaeoproterozoic rocks are mostly basic and acid volcanic and sedimentary rocks, but metamorphosed into low-grade amphibolite facies-greenstone facies. Volcanic rocks show bimodal volcanic features in petrology and geochemistry, indicative of an intraplate setting for their occurrence. These orogenic belts apparently recorded the plate tectonic evolution in the Mesoproterozoic time. The NCC underwent extension in the period from 2.3 to 2.0 Ga, as evidenced by the development of rifts and oceanic basins. There happened subduction, collision, and resulting contraction during 2.01–1.95 Ga. The NCC became stabilized after the Hutuo orogeny or during the "Earth's middle age", and remained as a stable platform for more than 1.0 Ga. Meso-Neoproterozoic sedimentary sequences were extensively deposited on the basement. The Xiong'er Rift was formed in the centro-south NCC, and the Yanliao Rift (or called Yanliao Faulted-Depression Zone, YFDZ) occurred in the centro-north

M. Zhai (✉) · P. Peng · Q. Meng
State Key Laboratory of Mineral Resources, Institute of Geology and Geophysics, Chinese Academy of Sciences, Beijing, China

M. Zhai
State Key Laboratory of Continental Geodynamics, Northwest University, Xi'an, China

B. Hu
School of Earth Science and Resources, Chang'an University, Xi'an, China

T. Zhao
Guangzhou Institute of Geochemistry, Chinese Academy of Sciences, Guangzhou, China

T. Wang, *Meso-Neoproterozoic Geology and Petroleum Resources in China*, Springer Geology, https://doi.org/10.1007/978-981-19-5666-9_8

NCC. Two other rifts developed in the northwestern and eastern margins of NCC, i.e., the Zha'ertai-Bayan Obo-Huade Rift and Jiao-Liao-Xu-Huai Rift, respectively. The Meso-Neoproterozoic sequences in the Yanliao Rift record sedimentary development of the NCC, and are divided into the Pt_2^1 Changchengian, Pt_2^2 Jixianian, Xiamaling Formation (Pt_2^3x) and Pt_3^1 Qingbaikouan. The Changchengian started from the Changzhougou Formation with the age of ca. 1.67 Ga. The Xiong'er Rift began with volcanic rocks dated at ca. 1.78 Ma, indicating that the rifting developed earlier than the Yanliao Rift. The Jixianian is corresponding to the Calymmian sequences, and followed by the Xiamaling Formation (Ectasian) and then the Neoproterozoic Qingbaikouan (Tonian). Four magmatic events are recognized in the NCC during Late Palaeo-Neoproterozoic times: ① ca. 1800–1780 Ma Xiong'er magmatism; ② ca. 1730–1620 Ma anorogenic magmatism; ③ ca. 1400–1320 Ma diabase/gabbro-diabase sill/vein swarm; and ④ ca. 925 Ma mafic dyke swarm. These magmatic events suggest that the NCC was in an intraplate/intracraton setting from ca. 1800 Ma to ca. 700 Ma. Late Palaeo-Neoproterozoic ore deposits include magmatic iron deposits and REE-Nb-Fe or Pb–Zn-Cu-Fe deposits, while orogenic metal deposits are absent, and no geologic evidence exists for the Grenville or other orogenic events in the NCC. Accordingly, the NCC was probably far from the Nuna Supercontinent in the Proterozoic. The "Earth's middle age" was a period when the Earth possessed a stable lithosphere and secular warm mantle, which could only result in multiple episodes of magmatism and rifting. It is thus suggested that the Earth experienced an evolution from non-plate tectonics through primitive plate tectonics to modern plate tectonics in the Proterozoic.

Keywords North China Craton (NCC) · Proterozoic · Rifting · Magmatism · Plate tectonics

8.1 Introduction

The North China Craton (NCC) consists mainly of the Precambrian metamorphic basement, Mesoproterozoic-Palaeozoic sedimentary sequences and Mesozoic intrusions. The NCC covers an area over 300,000 km^2, and experienced multi-stage crustal growth (Zhai and Santosh 2011). The early studies of the NCC paid much attention to Early Precambrian geology, its stabilization processes (Zhao 1993; Windley 1995; Liu et al. 2006a, 2006b; Zhao et al. 2001; Kusky et al. 2007a; Zhai 2011) and Mesozoic lithospheric thinning (Fan and Hooper 1991; Fan and Menzies 1992; Menzies et al. 1993; Zhai et al. 2002, 2007; Zhai 2008; Kusky et al. 2014). Figure 8.1 shows distribution of the Archean rocks, Palaeoproterozoic mobile/orogenic belts and a Meso-Neoproterozoic rift system. The Archean rocks are distributed almost all over the NCC, and mostly exposed in metamorphosed supracrustal belts or greenstone belts. The cratonization of the NCC must have been established toward the end of the Archean (Zhao 1993; Zhai et al. 2001; Wan et al. 2011; Geng et al. 2012). Three

Palaeoproterozoic mobile belts developed in the northeastern, central and northwestern parts, termed as the Jiaoliaoji, the Jinyü and the Fengzhen Mobile Belts, respectively (Fig. 8.1; Zhai et al. 2010). The belts are composed of both sedimentary and volcanic rocks, which had been altered into high- to moderate-grade metamorphic rocks. These belts are all underlain by the Archean basement, and evolved within the craton or on the continental margin in ca. 2.35–2.00 Ga. Their Palaeoproterozoic tectonics share some features with the Phanerozoic orogens in some aspects, but are different from the Archean granite-greenstone belts. The NCC experienced a sequential process from rifting, through subduction and accretion, to collision in the Palaeoproterozoic after the 2.5 Ga cratonization event (Zhai and Liu 2003; Kusky and Li 2003; Kröner et al. 2005; Zhai and Santosh 2011; Geng et al. 2012). Drifting of the NCC occurred at 2.3–2.0 Ga, leading to the formation of continental rifts and some oceanic basins (Zhai and Santosh 2011). It was also argued that arc-continent or continent–continent collision was prevailing in the Proterozoic (Zhao et al. 2001; Kusky et al. 2007b; Santosh et al. 2009). Crustal contraction might be operating in 2.01–1.95 Ga possibly due to subduction and collision in the NCC (Wang et al. 2007; Zhai 2011; Stern et al. 2013).

Meso-Neoproterozoic sedimentary basins were relatively poorly studied. Vary thick Meso-Neoproterozoic successions were deposited over the NCC basement, and remain unmetamorphosed (Zhai 2004). The Mesoproterozoic basins include the Xiong'er Rift in the southern NCC (Fig. 8.1), the Yanliao Rift (or called Yanliao

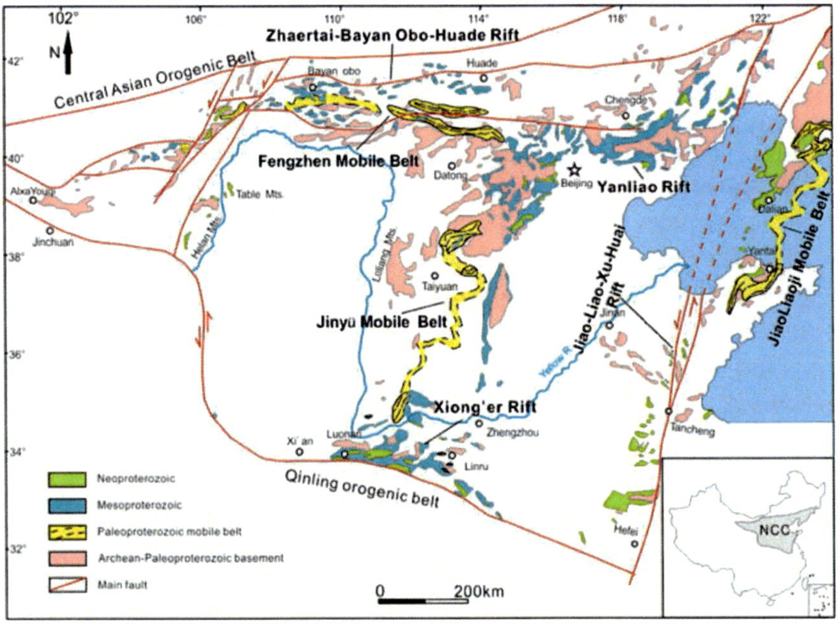

Fig. 8.1 Precambrian sketch map in North China Craton (Hu et al. 2014, modified)

Faulted-Depression Zone, YFDZ) in the northern NCC, the Zha'ertai-Bayan Obo-Huade Rift in the northwestern NCC and the Jiao-Liao-Xu-Huai Rift in the eastern NCC and North Korea, which were considered to have been once connected (Zhao 1993), but they are spatially separated nowadays. Although different in inception and termination times, sedimentary facies associations and stratigraphic thickness, all individual rifts do show something in common. For instance, they all underwent multi-stages of rifting for a quite long time. Four episodes of magmatism existed during the Meso-Neoproterozoic, which was closely related to rifting: ① ca. 1.80–1.78 Ga Xiong'er igneous event; ② ca. 1.72–1.62 Ga anorogenic magmatism; ③ ca. 1.40–1.32 Ga diabase/gabbro-diabase sill/vein swarm; and ④ ca. 925–900 Ma mafic dyke swarm. The four episodes of magmatic events imply that the NCC was in an intraplate setting during a prolonged period from ca. 1.8 Ga to ca. 0.7 Ga. Analyses of sedimentary basins (Meng et al. 2011) and magmatism (Zhai et al. 2014, 2015) suggest that the NCC was in a stable tectonic environment during the "Earth's middle age", or it had involved into a stable platform. Meso-Neoproterozoic strata had remained undeformed until the Late Mesozoic. An extensional setting for the four episodes of magmatism was also supported by geochemical studies (Zhao et al. 2002a, 2002b, 2009a, b; Wang et al. 2004, 2008; Xie 2005; Peng et al. 2006, 2011a, 2011b; Yang et al. 2011; Wang et al. 2011; Wang et al. 2012; Peng 2015), and the magmatic rocks might have been generated as a result of secular warming of the mantle (Korenaga and Jordan 2001; Korenaga 2006; Prokoph et al. 2004).

It is assumed that the NCC might have been located at the remote edge of the Supercontinent Nuna in the Proterozoic if such a supercontinent did exist. No geologic evidence exists in favor of the Grenville or other orogenic events in the NCC (Zhang et al. 2012). The Meso-Neoproterozoic (ca. 1.8 to 0.7 Ga) could be a particular period when the Earth possessed a stable lithosphere and a warm mantle. The lithospheric mantle was warm enough to result in widespread melting of the lower crust and lead to rifting in the overlying continents (Zhai et al. 2015). This paper will address what really happened in the NCC during the "Earth's middle age".

8.2 Rift System and Stratigraphy

Meso-Neoproterozoic strata in the NCC are broadly distributed, and largely unmetamorphosed. Continuous Mesoproterozoic stratigraphic sections are exposed in the Yanliao Rift, and divided into the Pt_2^1 Changchengian, Pt_2^2 Jixianian, Xiamaling Formation (Pt_2^3x) and Pt_3^1 Qingbaikouan (without the Pt_2^4 sequence). Whole the sequences were considered to have developed in rift basins, and started since ca. 1.8 Ga (Chen et al. 1980; Xing 1989; Wang and Li 1990). Contemporaneous strata occurred at the southern, northern and eastern margins of the NCC. Zhai and Peng (2007) named the Proterozoic rifts as the Yanliao, Xiong'er, Zha'ertai-Bayan Obo-Huade, and Jiao-Liao-Xu-Huai Rifts (Fig. 8.1). In practice, all the individual rifts underwent multi-stage extension during Meso-Neoproterozoic times.

8.2.1 Yanliao Rift

Meso-Neoproterozoic stratigraphy in the Yanliao Rift (or called Yanliao Faulted-Depression Zone, YFDZ) has been well established (Fig. 8.2). The Pt_2^1 Changchengian is mainly clastic rocks with minor alkaline volcanic rocks. It is divided into four formations in ascending order (Fig. 8.3): the Changzhougou Formation rests unconformably over the Archean Qianxi Group, mainly comprising marine sandstones, its basal bebbly sandstones might have been deposited in fluvial environments; the Chuanlinggou Formation is dominated by mudstone and fine-grained sandstones with dolostone at its upper part; the Tuanshanzi Formation is mainly composed of dolostone and siltstone with potassiferous basalt and trachybasalt presented at the upper part; the Dahongyu Formation mainly comprises littoral to neritic sandstone and shale, also containing some potassiferous trachyte interlayers. Based on geochemistry studies, the volcanic rocks were formed under the rift setting (Qiu and Liao 1998). The Pt_2^2 Jixianian is mainly composed of dolostone and separated from the underlying Pt_2^1 Changchengian by a local disconformity. Five formations are dividable (Fig. 8.3), stromatolitic dolostone dominates the Gaoyuzhuang, Yangzhuang and Wumishan Formations; whereas the Hongshuizhuang Formation is largely shales; and the Tieling Formation is characterized by manganiferous dolostone, shales and stromatolitic dolostone. The Xiamaling Formation (Pt_2^3x) is mainly shales with some sandy shale, sandstone and diabase/gabbro-diabase sills/veins and overlays disconformably the Tieling Formation (Pt_2^2t). The Pt_3^1 Qingbaikouan includes two formations from bottom upward (Fig. 8.3): the Luotuoling Formation comprises sandstone and variegated shale, which is in micro-angle unconformable contact with the underlying Xiamaling Formation; while the Jing'eryu Formation is dominated by muddy/micritic limestones and marl with basal glauconitic quartzose sandstone.

Ages of individual units in the Meso-Neoproterozoic sequences have been well constrained in recent years. The sedimentary ages of the Changchengian was suggested to be 1670–1600 Ma based on the youngest zircon U–Pb ages of the Changzhougou sandstone (Wan et al. 2003; Li et al. 2011; cf. Chap. 2), zircon U–Pb ages of potassiferous volcanic rocks of the Tuanshanzi and Dahongyu Formations (Li et al. 1995; Lu and Li 1991; Gao et al. 2008a, 2008b; Lu et al. 2008), zircon U–Pb ages of diabase dykes intruding the Chuanlinggou Formation (Gao et al. 2009), and zircon U–Pb ages of granite-porphyry veins that intruding Archean basement and was overlain by the Changzhougou Formation (Li et al. 2011; Fig. 8.3). The Gaoyuzhuang Formation was formed during the period from 1600 Ma to 1400 Ma according to zircon U–Pb ages of tuffs (Fig. 8.3; Li et al. 2010). The Tieling Formation contains K-bentonites that yield zircon U–Pb ages of 1437 Ma (Su et al. 2010). The age of the Wumishan Formation was constrained by zircon and baddeleyite U–Pb ages of 1345–1354 Ma of diabase sills intruding the unit (Zhang et al. 2009, 2017a). Zircon and baddeleyite U–Pb ages of K-bentonites in the Xiamaling Formation and the 1400–1320 Ma basic sills put a tight constraint on the age of the Xiamaling Formation (Fig. 8.3; Gao et al. 2007a, 2007b, 2008a, 2008b; Su et al. 2008, 2010; Li et al. 2009; cf. Chap. 10).

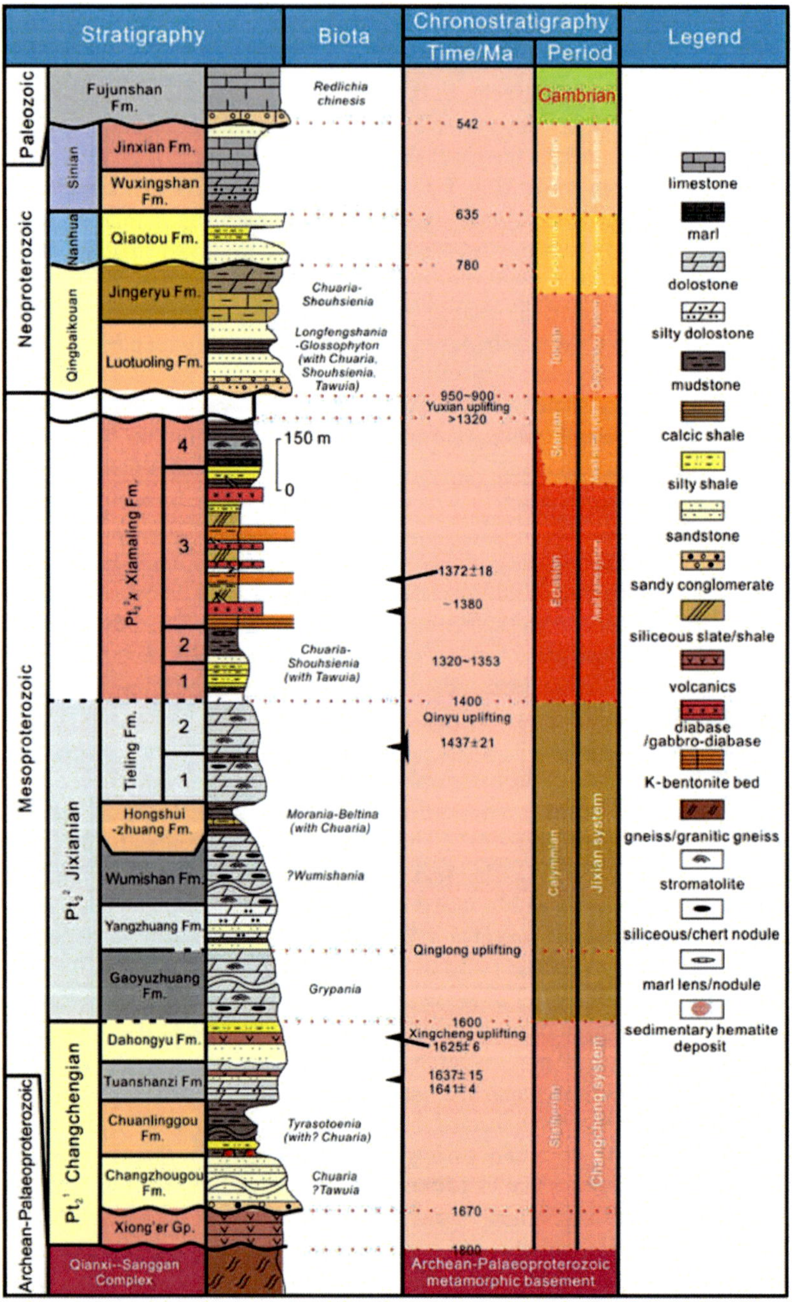

Fig. 8.2 Meso-Neoproterozoic stratigraphy in the NCC

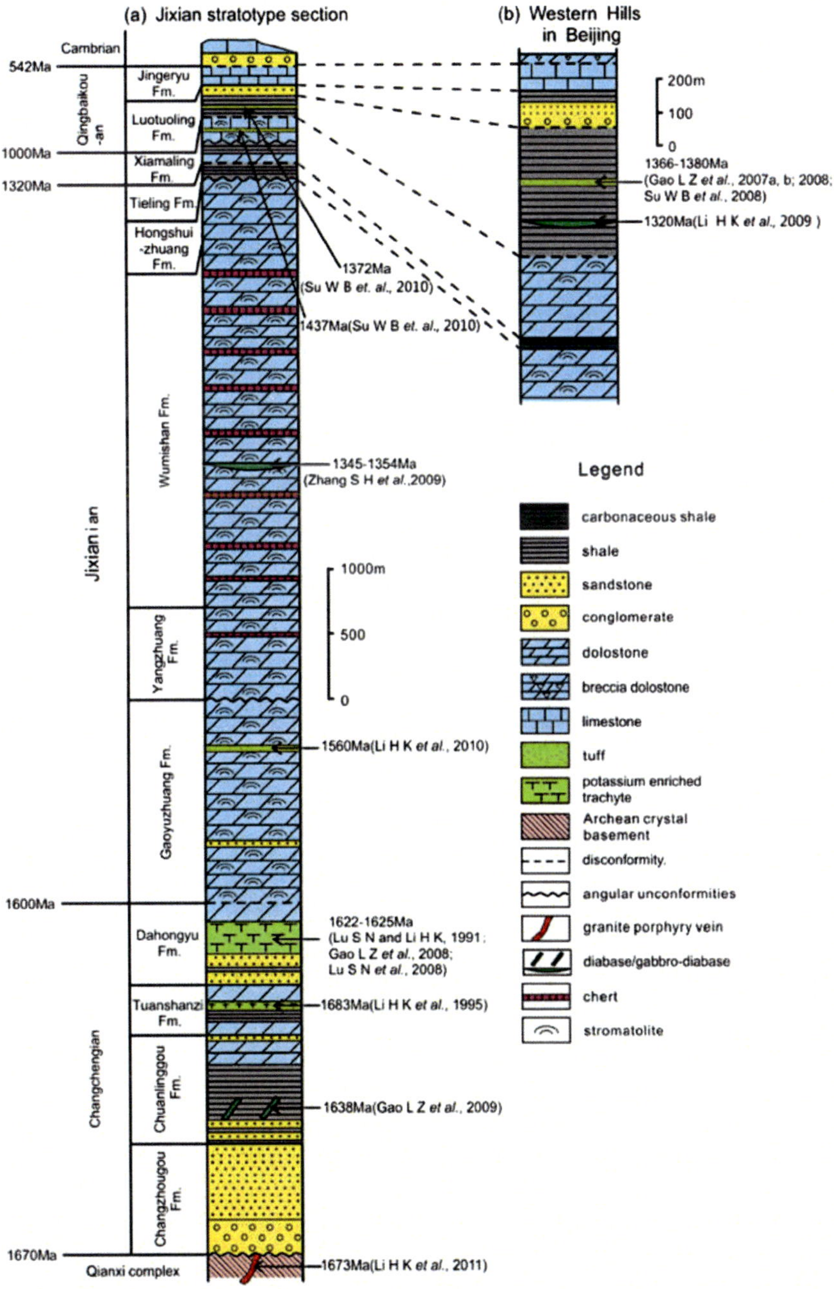

Fig. 8.3 Meso-Neoproterozoic stratigraphic columns in Yanliao Rift. **a** Stratigraphy established at Jixian stratotype section in Tianjin (Gao et al. 2008a, 2008b; modified); **b** stratigraphy established at the Western Hills, Beijing (Institute of Geologic Survey, Beijing Geology and Mineral Resources Bureau 1994, 1:50,000 Yanchi Geological Map, Internal Communication Data in Chinese Geologist, modified)

8.2.2 Xiong'er Rift

The Xiong'er Rift was formed in the south of the NCC, and filled with Mesoproterozoic Xiong'er volcanics and Meso-Neoproterozoic strata. The Xiong'er Group used also to be called "Xiyanghe Group" in literature (Henan Geology and Mineral Resources Bureau 1989). The Xiong'er Group is made up of volcano-sedimentary succession, and divided into four formations in ascending order: the Dagushi Formation, resting unconformably over the Archean-Palaeoproterozoic basement, consists mainly of fluvial and lacustrine sediments; the Xushan Formation is composed of basaltic andesite and andesite; the Jidanping Formation is mostly rhyolite with interlayers of andesitic basalt and andesite; the Majiahe Formation consists chiefly of volcanic rocks, but contains lots of sedimentary and pyroclastic rocks (Fig. 8.4a). The U–Pb ages of Xiong'er Group range from 1.80 Ga to 1.78 Ga, whereas the underlying Taihua complex was dated at ca. 1.84 Ga. It was regarded that the Xiong'er Group developed in an Andean-type continental margin (Jia et al. 1988; Hu et al. 1988; He et al. 2009, 2010; Zhao et al. 2009a, 2009b) or a continental rift (Sun et al. 1985; Zhai et al. 2000; Zhao et al. 2002a, 2002b, 2005, 2007; Cui et al. 2011), but the controversy still remains.

The Xiong'er Group is overlain unconformably by the Ruyang Group at Zhongtiao and Lushan mountains, which consists of four formations, i.e., the Xiaogoubei, Yunmengshan, Baicaoping and Beidajian Formations in ascending order: the Xiaogoubei Formation is mainly conglomeratic facies, and passes upward into the Yunmengshan quartz arenites. The Baicaoping Formation is made up largely with quartz arenites, but contains abundant mudstone; alternated arenite and dolostone are characteristic of the Beidajian Formation. Moreover, the Ruyang Group turns upward continuously into the Luoyu Group that constitutes three formations: the lower one called Cuizhuag Formation composed dominantly of mudstone and its basal conglomerate, the overlying Sanjiaotang Formation is quartz arenites, and passes upward into the Luoyukou Formation characterized by mudstone and dolostone (Fig. 8.4a–c). Su et al. (2012) dated a tuff bed in the upper Luoyukou Formation, which yields a U–Pb age of 1611 ± 8 Ma. In view of the ages of underlying Xiong'er volcanics, the Ruyang and Luoyu Groups might have been formed during the period from 1.75 to 1.6 Ga.

In the Xiong'er mountain, the Guandaokou Group was deposited unconformably over the Xiong'er Group, which consists of the Gaoshanhe, Longjiayuan, Xunjiansi, Duguan and Fengjiawan Formations respectively (Fig. 8.4d). The Gaoshanhe Formation is made up with mudstone and arenites, and the Longjiayuan and Xunjiansi Formations are composed of muddy dolostone and thick-bedded dolostone, while shale and dolostone dominate the Duguan and Fengjiawan Formations. Whereas the Luanchuan Group is divided into the Baishugou, Sanchuan, Nannihu and Meiyaogou Formations, which have been metamorphosed to various degrees (Fig. 8.4d).

A regional disconformity separates the Luoyu Group from the overlying Sinian or Cambrian strata at Zhongtiao and Lushan mountains and Linru area (Fig. 8.4a–c). The Sinian strata include four units, i.e., the Huanglianduo, Dongjia, Luoquan,

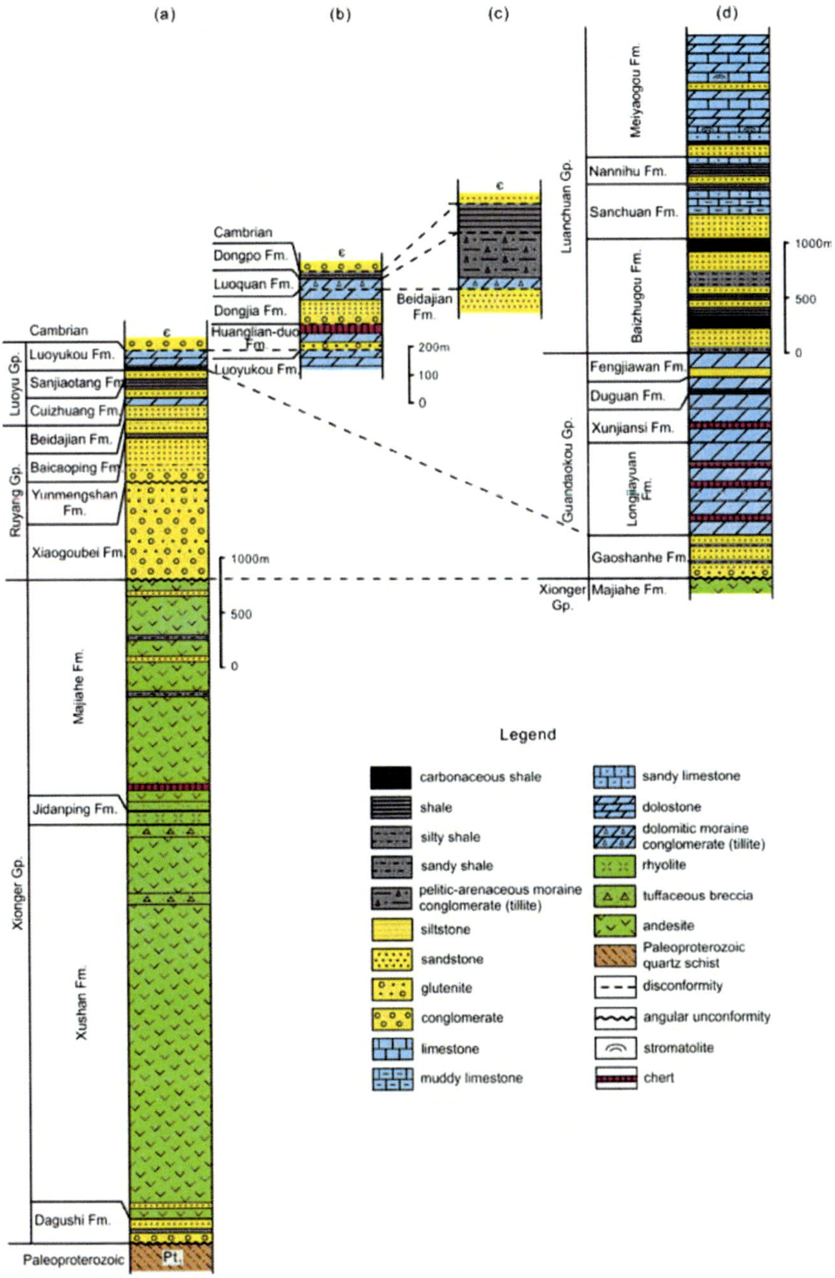

Fig. 8.4 Mesoproterozoic to Neoproterozoic stratigraphic columns in the adjacent area of Shanxi, Shaanxi and Henan Provinces in Xiong'er Rift. **a**. Zhongtiao Mountain at northwestern Xionger Rift (Zhao et al. 2005); **b** Lushan Mountain at eastern Xionger Rift; **c** Linru area at eastern Xionger Rift; **d** Xiong'er Mountain in center Xionger Rift (Henan Geology and Mineral Resources Bureau 1989)

and Dongpo Formations, respectively (Fig. 8.4b, c). The Huanglianduo Formation is mainly dolostone with chert nodules. Arenites and muddy dolostone comprise the Dongjia Formation. The Luoquan Formation is regarded to have resulted from glaciation, as evidence by occurrence of diamictites (Fig. 8.4b, c). The Dongjia and Luoquan Formations can be correlated stratigraphically with the Nantuo and Dengying Formations in south China based on micropalaeoflora fossils (Henan Geology and Mineral Resources Bureau 1989).

8.2.3 *Zha'ertai-Bayan Obo-Huade Rift*

Meta-sedimentary strata also outcropped out at the western segment of the northern margin of the NCC, which is called the Zha'ertai-Bayan Obo-Huade Rift (Fig. 8.1). This rift is mainly composed of three Groups from west to east: the Zha'ertai, Bayan Obo, and Huade Groups, respectively.

The Zha'ertai Group is distributed in Langshan and Zha'ertai mountains. It was deposited over the Guyang Greenstone Belt of Late Archean age, and divided into four units in ascending order (Fig. 8.5a): the Shujigou Formation is composed of conglomerate, arkosic arenites and quartzite; the Zenglongchang Formation is dominated by dolomitic slate, stromatolitic limestone and dolostone; the Agulugou Formation is mainly slate; and the Liuhongwan Formation is largely quartzite. It is worth noting that alkaline volcanics are present in the Shujigou clastic sequence (Wang et al. 1992), giving zircon U–Pb age of 1745 Ma (Li et al. 2007a, 2007b). Alkaline and basaltic volcanics in the Shujigou Formation are indicative of inception of the Zha'ertai-Bayan Obo-Huade Rift. The Zha'ertai Group also contains bi-modal volcanics, and their geochemical characteristics carry implications for continental rifting (Peng and Zhai 1997; Peng et al. 2005a, 2005b, 2007a, 2007b; Wang et al. 1992). Potassic spilite yields Sm–Nd isochronal age of ca. 1824 Ma (Peng 1999); whereas some acid volcanics were dated around 817–805 Ma (Peng et al. 2010). These ages of the volcanics suggest that the Zha'ertai Group was formed during the Meso-Neoproterozoic period.

The Bayan Obo Group outcrops from Bayan Obo to Shangdu area in the Inner Mongolia. This group was deposited over the Late Archean basement, and contain four formations in ascending order (Fig. 8.5b, c): the Dulahala Formation is largely quartzite, conglomerate and coarse-grained sandstone; the Jianshan, Halahuogete and Bilute Formations are dominated by mudstone and turbiditic sandstone; the Baiyinbaolage and Hujitu Formations are also clastic rocks, which are covered by the Ayadeng carbonates. The basement rocks were dated at 1.9 Ma (Wang et al. 2002), thus putting a constraint on the initial time of the Bayan Obo Group. Basaltic volcanics in the lower part of the Bayan Obo Group yields a zircon U–Pb age of 1.73 Ga (Lu et al. 2002), which might represent the onset of sedimentation of the Bayan Obo Group. In addition, igneous carbonate intruded into the Dulahala Formation, and gave a zircon U–Pb age of 1.42 Ga (Fan et al. 2006), indicating that this unit cannot be younger than 1.42 Ga. All the radiometric dates imply that the first

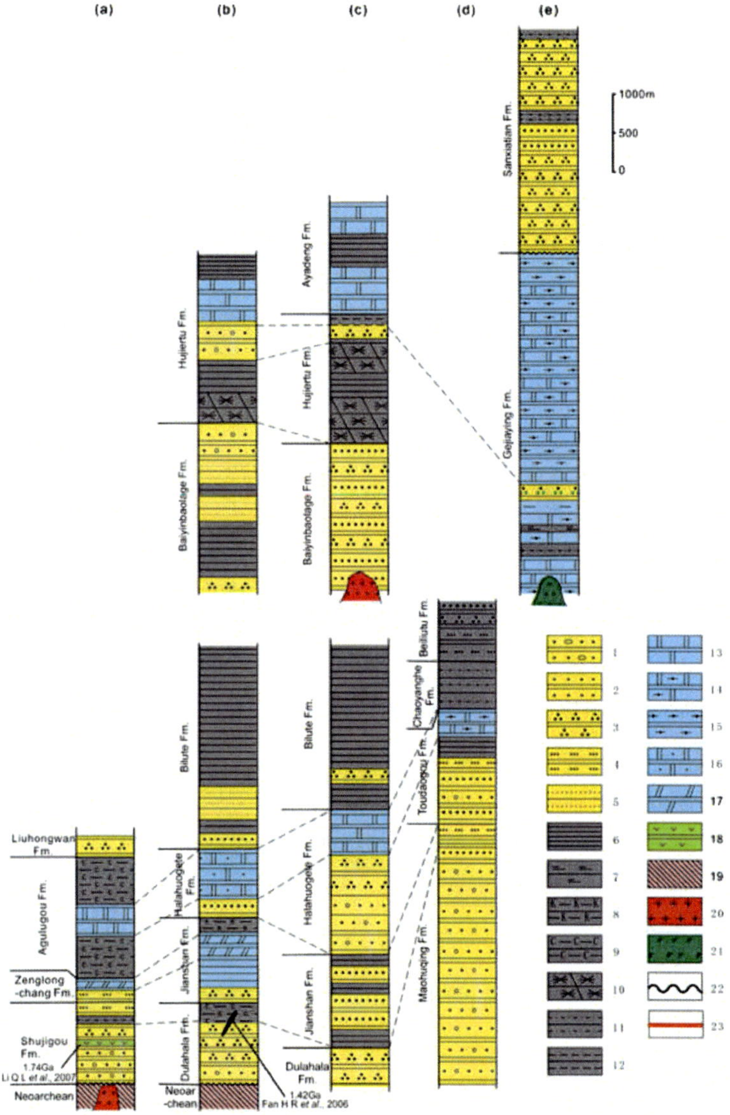

Fig. 8.5 Meso-Neoproterozoic stratigraphic columns of the Zha'ertai-Bayan Obo-Huade Rift in NCC. **a** Zha'ertai Group; **b** Bayan Obo Group (Wang et al. 1989); **c** Bayan Obo Group in Shangdu, Inner Mongolia; **d** Lower Huade Group in Huade, Inner Mongolia; **e** Upper Huade Group in Kangbao, Hebei. Legends: 1. meta-pebbly sandstone; 2. meta-sandstone; 3. quartzite; 4. meta-fine sandstone; 5. meta-siltstone; 6. slate; 7. phyllite; 8. potassic slate; 9. carbonaceous slate; 10. actinolite hornstone; 11. quartz schist; 12. mica schist; 13. marble (crystalline limestone); 14. diopside (tremolite) marble; 15. scapolite diopsidite and calcite diopsidite (tremolite); 16. silty crystalline limestone; 17. dolostone; 18. meta-andesite; 19. archean crystal basement; 20. granite; 21. gabbro; 22. angular unconformity; 23. fault

four units were formed in the period from 1.73 to 1.42 Ga, roughly synchronous with the Changchengian and Jixianian.

The Huade Group spreads over Huade area in Inner Mogolia and Kangbao area in Hebei province. It is considered as the eastward stretching segment of the Banyan Obo Group. The Huade Group. is build up with meta- sedimentary rock, such as siliciclastics and carbonates (Fig. 8.5d, e). No volcanic rocks have been found in the sequence. Ages of detrital zircons from sandstone of the first unit in the Huade Group range from 1.8 to 1.3 Ga, coeval with the Changchengian-Jixianian. The upper two units should be younger, and were deposited from 1.46 to 1.3 Ga (Hu et al. 2009).

8.2.4 *Jiao-Liao-Xu-Huai Rift*

The Jiao-Liao-Xu-Huai Rift (i.e., Jiaodong Peninsula-Liaodong Peninsula-Xuzhou City-Huaihe River Valley, abbs. JLXH; cf. the Sect. 1.3.2 in Chap. 2) includes the Liaodong Depression (or called Fuzhou-Dalian Basin at eastern Liaoning), the Jiaodong Depression (at eastern Shandong) and the Xuhuai Depression (or Xuzhou City-Huaihe River Valley), Pyongnam Basin (North Korea) etc. (Bai et al. 1993).

The Pyongnam Basin was located in the Nangrim Block, and filled with Neoproterozoic rocks resting unconformably on the Archean-Late Palaeoproterozoic basement. The Neoproterozoic succession includes the Sangwon and Kuhyon Systems, which are separated by an angular unconformity and have been metamorphosed into greenschist facies. Figure 8.6 shows the detailed stratigraphic subdivisions of the two systems. Hu et al. (2012) reported that average age of younger detrital zircons from the Jangsusan Formation of the Sangwon System is 984 Ma, indicating that the Sangwon System should not be older than 980 Ma. Peng et al. (2011a) also showed that the Sangwon System was cut by 899 Ma mafic dikes. As a result, the Sangwon System must have been deposited in the period from 1000 Ma to 900 Ma, and can be correlated with the Qingbaikouan in the NCC. The 899 Ma mafic dikes are covered by the Kuhyon System.

The Neoproterozoic successions in the Liaodong Depression comprise the Yushulazi Group, Yongning Formation, Xihe Group, Wuxingshan Group, and Jinxian Group in ascending order (Fig. 8.7a): the Yushulazi Group unconformably rests on the Palaeoproterozoic Liaohe Group, and consists of meta-sedimentary rocks; the Yongning Formation and Xihe Group are composed mainly of arenites, siltstone and limestone; the Wuxingshan Group is made up with siliciclastics and stromatolitic carbonate; while the Jinxian Group is in a conformable contact with the calcite dolostone of Wuxingshan Group in the Dalian Basin (Fig. 8.7b). The Yushulazi Group was regarded to be the Palaeoproterozoic in age. The Yongning Formation and Xihe Group are correlated with the Qingbaikouan, while the Wuxingshan and Jinxian Groups were thought to be the Sinian (Liaoning Geology and Mineral Resources Bureau 1989). Detrital zircons from Yushulazi and Xihe Groups yield the youngest age around 1.1 Ga (Luo et al. 2006; Gao et al. 2010), thus putting a constraint on the timing of these units. Given that mafic dikes are dated at ca. 900 Ma (Peng et al.

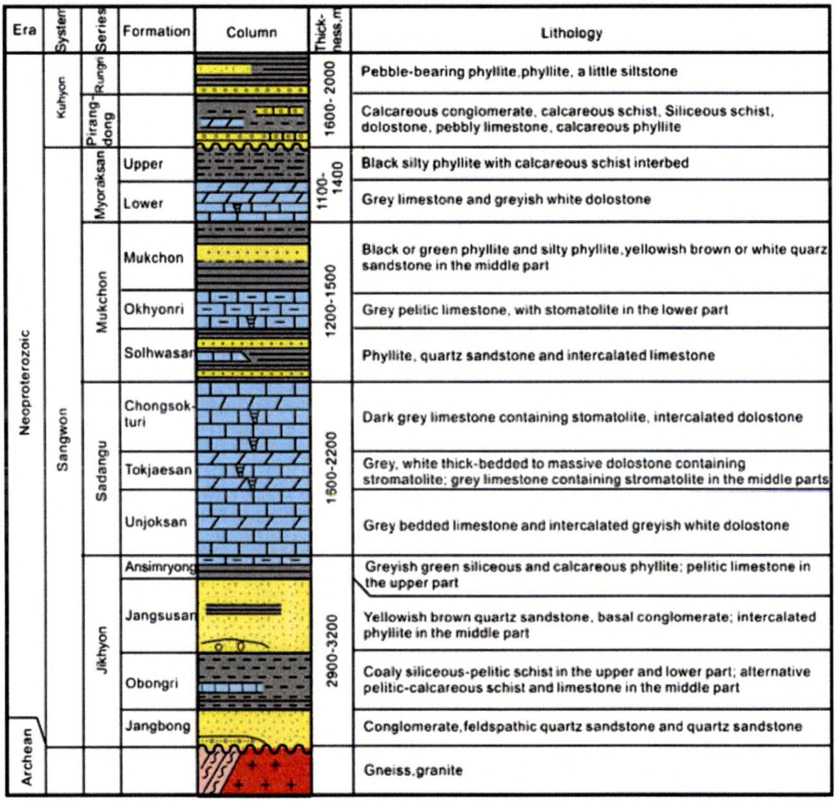

Fig. 8.6 Neoproterozoic stratigraphic column in the Pyongnam Basin, North Korea (Paek et al. 1993, modified)

2011b), the Liaodong Depression (Fuzhou-Dalian basin) might have been initiated during the Middle Mesoproterozoic-Neoproterozoic.

The Penglai Group in Jiaodong Depression is a set of epimetamorphic sequences, and can be divided into four formations (Fig. 8.7c). Li et al. (2007a, 2007b) reported that the youngest peak age of detrital zircon in the Penglai Group is ca. 1.2 Ga, suggesting its starting time of deposition probably from 1.1 to 0.8 Ga. The Tumen Group in Luxi Basin was separated from the Neoarchean Taishan Group by an angular unconformity. It comprises five units and is mostly quartz arenite, limestone and calcareous shale (Fig. 8.7e). The whole-rock isochronal Rb–Sr age of stromatolitic dolostone is ca. 910 Ma (Zhou and Hu 1998; Shandong Provincial No. 4 Institute of Geological and Mineral Survey 2003) and the youngest detrital zircon age is around 1.1 Ga (Hu et al. 2012). Accordingly, the Tumen Group was formed in the Late Meso-Neoproterozoic. Meso-Neoproterozoic successions in Xuhai Depression are divided into 13 Formations from bottom up (Fig. 8.7d): the Lanling, Xinxing and Jushan Formations comprise mainly conglomerates, quartz arenite, fine sandstone, siltstone

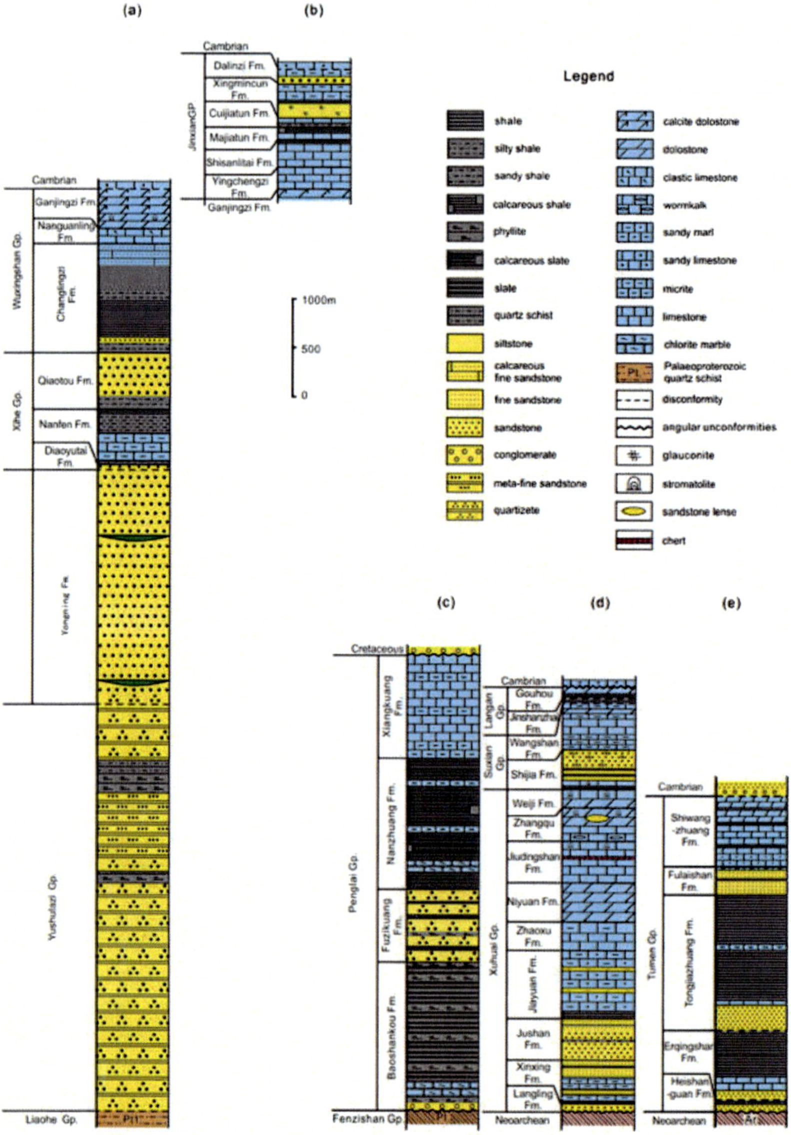

Fig. 8.7 Late Mesoproterozoic-Neoproterozoic stratigraphic column in JLXH Rift of NCC. **a** Fuzhou Basin in Liaodong Depression (Liaoning Geology and Mineral Resources Bureau 1989, modified); **b** Dalian Basin in Liaodong Depression (Liaoning Geology and Mineral Resources Bureau 1989, modified); **c** Jiaodong Depression (Shandong Provincial No. 4 Institute of Geological and Mineral Survey 2003, modified); **d** Xuhuai Depression (Xuzhou-Huaihe River Basin; Geologic Survey Group, Jiangsu Geology Bureau 1977, 1:200,000 Xuzhou Geological Map, Internal Communication Data in Chinese Geologist, modified); **e** Luxi Depression (Western Shandong; Shandong Provincial No. 4 Institute of Geological and Mineral Survey 2003; modified)

and shale; Jiayuan, Zhaoxu, Niyuan, Jiudingshan, Zhangqu and Weiji Formations are carbonatites; Shijia Formation are dominantly by shale, siltstone and glauconitic sandstone; Wangshan Formation is composed by limstone and marlstone; Jinshanzhai and Gouhou Formation are mainly shale and dolostone. Liu et al. (2006a, 2006b) reported zicon U–Pb ages of a diabase sill intruding Zhaoxu, Niyuan and Shijia Formations. Peng et al. (2011a, b) explain that the $^{206}Pb/^{238}U$ average age of 918 Ma represents the crystallization age of this sill. Combing stratigraphic correlation with Liaodong, Jiaodong and Luxi Depresiions, the depositional time of strata in Huhuai Depression may be limited to Late Meso-Neoproterozoic.

Figure 8.8 displays Proterozoic stratigraphic-sedimentary sequences in different rifts of the NCC. The Xiong'er Rift with volcanism early than Changchengian is in the south of the NCC. The Ruyang and Luoyukou Groups, together with the Guandaokou, Luanchuan, Zha'ertai, Bayan Obo and the Huade Groups, were formed roughly coeval with the Changchengian and Jixianian. The Sangwon System in North Korea as well as the Yushulazi, Xihe, Wuxingshan, Peglai and Tumen Groups can be correlated with the Qingbaikouan. The Kuhyon System in North Korea and Huanglianduo-Dongpo Formation in Xiong'er Rift was formed synchronously with the Nanhuan to Sinian.

8.3 Major Igneous Events

The igneous events in NCC took place in the following time intervals, ca. 1780–1730 Ma, 1730–1620 Ma, ca. 1320–1230 Ma and 925–810 Ma, respectively (Peng 2016; Figs. 8.9 and 8.10). The events of ca. 1780 Ma, 1320 Ma and ca. 925 Ma formed three large igneous provinces (LIPs), and the understanding of the LIPs is essential to unravel Proterozoic tectonic evolution of the NCC.

8.3.1 Taihang Igneous Event (ca. 1780 Ma)

The ca. 1780 Ma igneous event resulted in the Taihang dyke swarm, the Lüliang dyke swarm and the Xiong'er volcanic province (Peng et al. 2015; Fig. 8.11). The Miyun and Beitai dyke swarms were possibly formed at ca. 1730 Ma, and presented as the younger event (Peng et al. 2012; Fig. 8.9).

The Taihang and Lüliang dyke swarms are mainly distributed in the central NCC (Fig. 8.11), and composed of diabase and some gabbros. Individual dykes are vertical to subvertical distributed, up to 60 km long and ca. 15 m wide (Peng et al. 2015). The dykes of Taihang swarm are mostly NNW-trending (315°–345°) with a few NE-orientated ones (20°–40°; Fig. 8.11). The NE-orientated dykes mostly occur in the southern Taihang Mountains (Wang et al. 2004), with a few in the southern margin of the NCC (Hou et al. 2008). The dykes of Lüliang swarm are near EW-trending (250°–290°), and are present mainly in the Lüliang, southern Taihang, Huoshan and

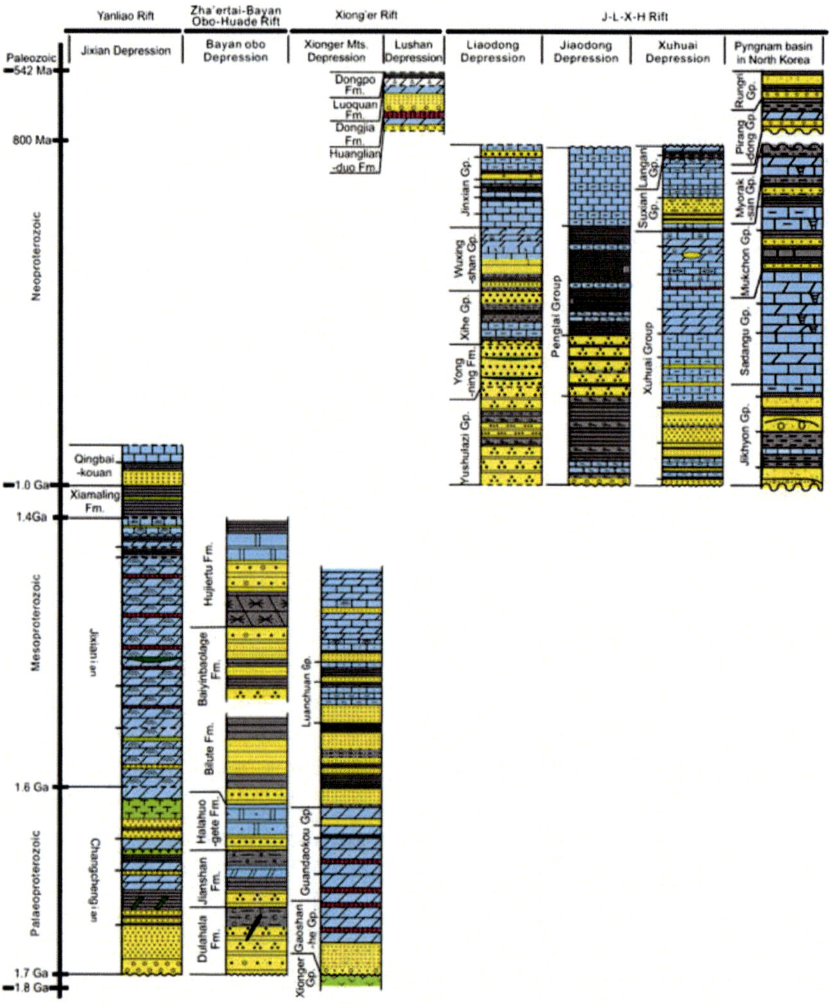

Fig. 8.8 Meso-Neoproterozoic stratigraphic correlation of different rifts in the NCC (the legends are same as Figs. 8.3, 8.4, 8.5, 8.6 and 8.7)

Zhongtiao Mountains (Fig. 8.11). The 250°–270°-striking dykes are dominant in the Lüliang and Taihang Mountains, and the 270°–290°-striking dykes prevail in the Zhongtiao and Huoshan Mountains. The radiating pattern of dykes in the NCC has been well preserved because of less modifications of Mesozoic-Cenozoic intraplate deformation, and magmatic center was likely located in the center of the Xiong'er Rift at the southern margin of the NCC (Peng et al. 2006; Hou et al. 2008; Xu et al. 2014).

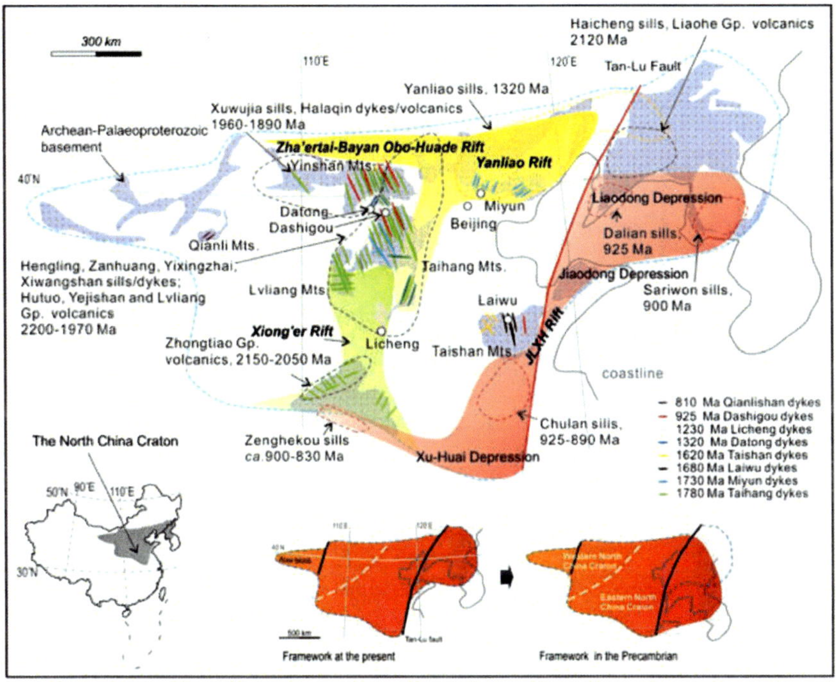

Fig. 8.9 Sketched distribution of Meso-Neoproterozoic dyke swarms in the NCC (Peng 2016; modified). Inset maps show location and reconstructed Precambrian shape of the NCC

The ca. 1730 Ma Miyun dyke swarm appears in the Taihang mountains and Yanliao Rift (Fig. 8.9). The dykes are northeast-oriented, and mostly basaltic and basaltic-andesitic in lithology. Felsic dykes are relatively minor. Trace elements and Sr–Nd isotopes of the dykes are quite similar to those derived from the ancient lithospheric mantle (Peng 2015). The Taihang and Lüliang dykes were dated at ca. 1770–1780 Ma (cf., Halls et al. 2001; Peng et al. 2005a, 2005b, 2006; Wang et al. 2004; Peng 2015).

The Xiong'er Group includes three volcanic sequences in the southern NCC, among which two are within the Lüliang Mountains. These volcanic rocks are 3000–7000 m thick and distributed in a tri-rift system, with one branch extending to the center of the NCC and the other two to the south (Wang 1995; Zhao et al. 2002a, 2002b; Xu et al. 2007; Peng et al. 2008). The Xiong'er Group is made up of the Dagushi, Xushan, Jidanping and Majiahe Formations in an ascending order (Fig. 8.12a): the Dagushi Formation up to 290 m thick is composed of conglomerate, sandstone and shale; the Xushan Formation is mainly comprised of volcanics, including andesite, pyroxene andesite, andesitic basalt, rhyolite and volcaniclastics, and varies from 2400 to 3000 m range from 100 to 1000 m. Pillow lavas are observed in acidic layers, indicative of submarine eruptions; the Majiahe Formation consists of andesite, basaltic andesite, pyroxene andesite, rhyolite, dacite and volcaniclastics, and is ca. 2000 m in total thickness, but up to 3900 m think in the Xiong'er region

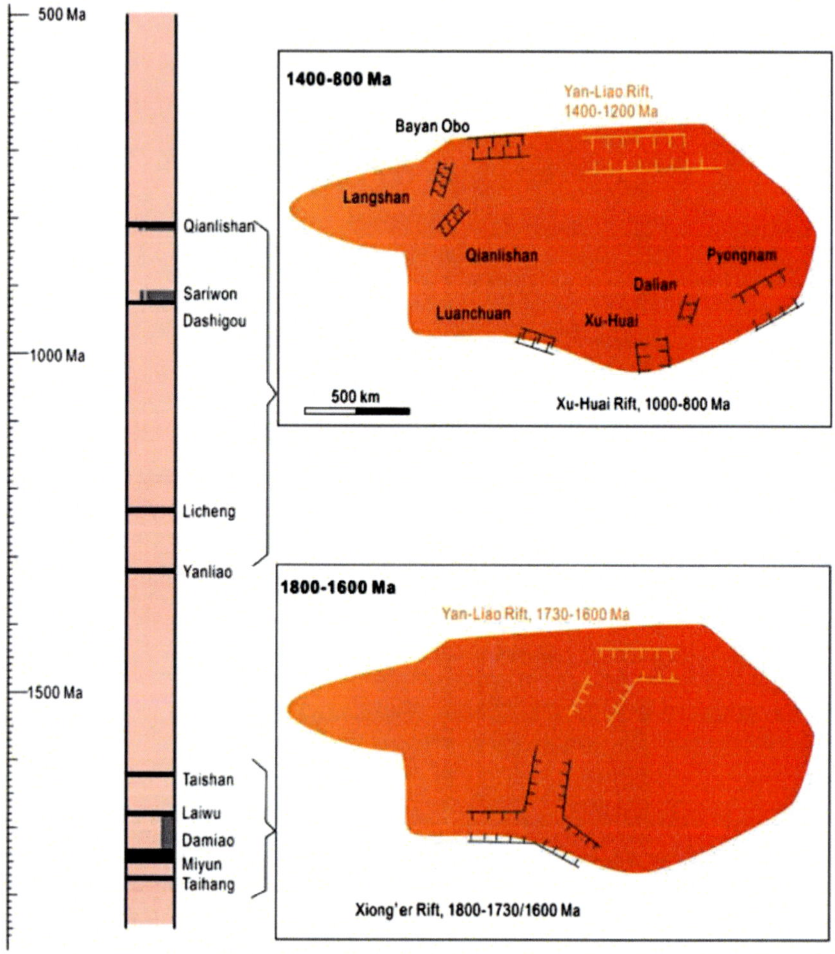

Fig. 8.10 Diagram showing geological "barcode" of the Meso-Neoproterozoic igneous events and distribution of rifts in the NCC (Peng 2016)

(Zhao et al. 2002a, 2002b; Fig. 8.12b). Volcanism appeared to be voluminous during 1770–1790 Ma (Zhao et al. 2004a; He et al. 2009; Cui et al. 2011), and sulfide Pb, Zn and Au deposits presumably borne on volcanism (Weng et al. 2006; Pei et al. 2007).

The Xiong'er Group was considered to be bimodal tholeiitic volcanism in an intracontinental environment (Sun et al. 1985; Zhao et al. 2002a. 2002b; Peng et al. 2008; Wang et al. 2010). However, it was also regarded as arc calc-alkalines (Jia 1987; Hu et al. 1988; He et al. 2009; Zhao et al. 2009a, 2009b). Peng et al. (2015) suggested that the Xiong'er volcanics were related to the Taihang/Lüliang dyke swarms, and basalt and basaltic andesite were actually parts of the dykes. The rhyolite and dacite

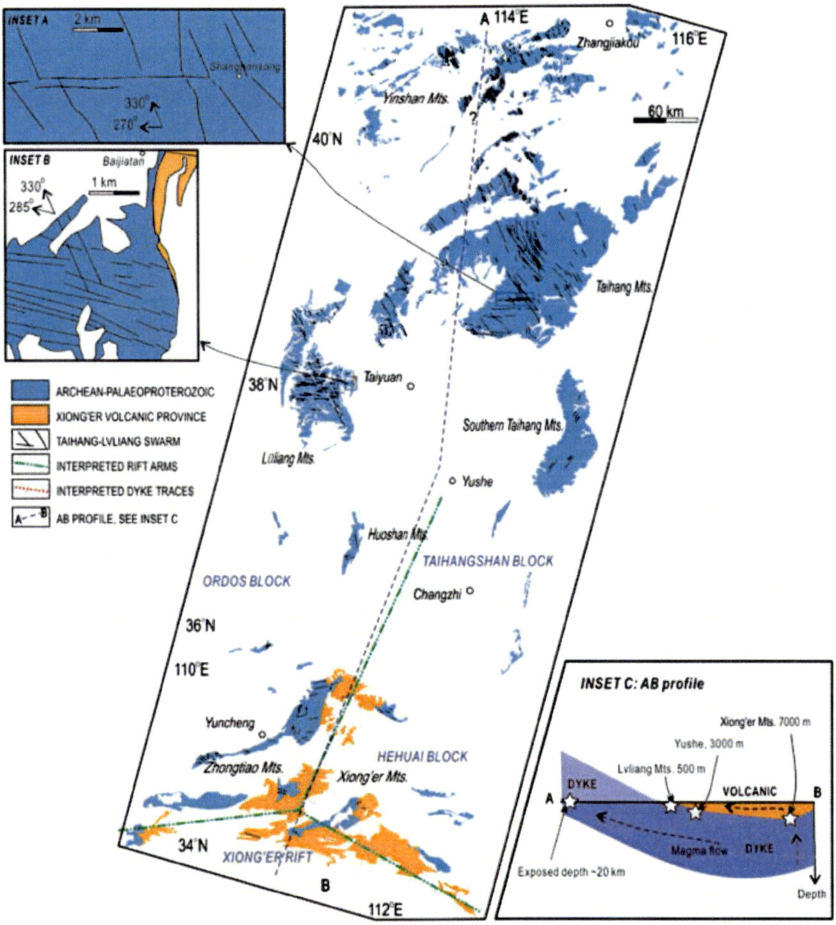

Fig. 8.11 Geological map of the Taihang large Igneous Province (Peng et al. 2008)

could be the immiscible conjugated high-silica portion of high-Ti–Fe mafic dykes (Fig. 8.12c, d; Peng et al. 2007a, 2007b).

The ca. 1780 Ma Taihang-Lüliang dyke swarms and the Xiong'er volcanic province comprise an igneous area of ca. 0.3 Mkm^2, and the magma volume of the dyke swarms and igneous province is estimated to be over 0.02 Mkm^3, and thus comprise a large igneous province (Peng 2015). However, it still remains controversial about its tectonic settings, either syncollisionic (Wang et al. 2004, 2008; He et al. 2009; Zhao et al. 2009a, 2009b) or anorogenic (Zhai et al. 2000; Kusky and Li 2003; Hou et al. 2006, 2008; Wang et al. 2010; Peng et al. 2005a, 2005b, 2008, 2015). Dyke swarms and volcanic province were also thought to be the consequence of drifting of the NCC from adjacent blocks that had been connected with the NCC prior to 1780 Ma (Peng et al. 2005a, 2005b; Hou et al. 2008; Peng 2015).

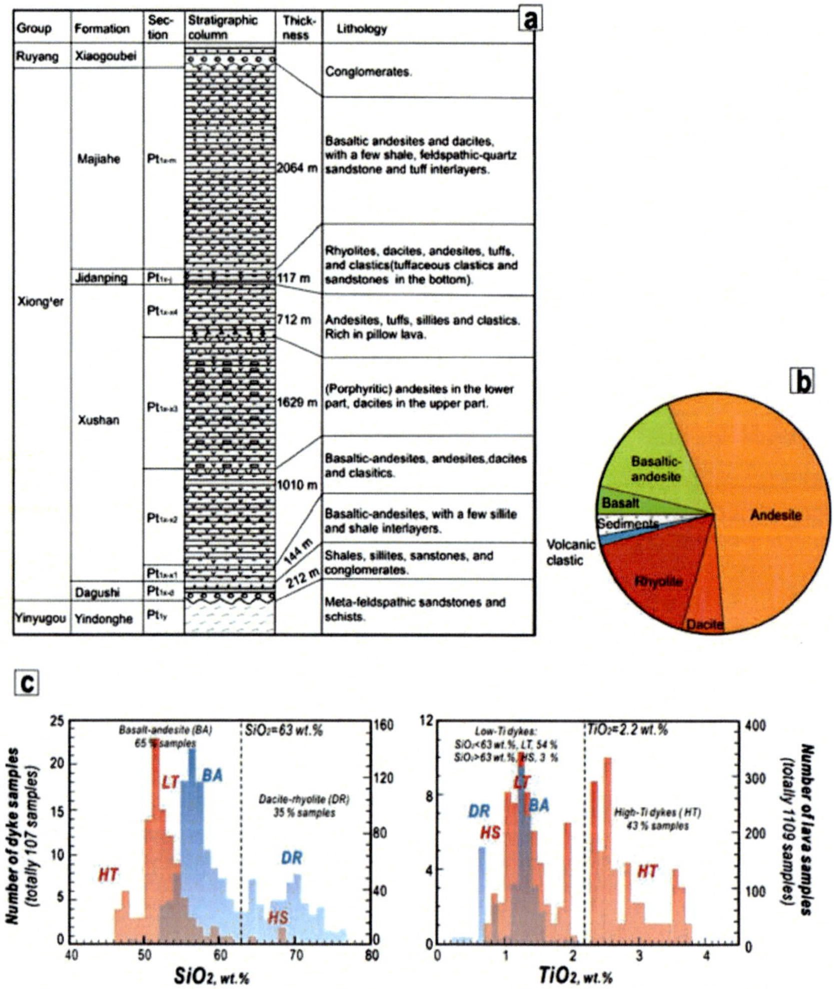

Fig. 8.12 Stratigraphy **a** and pie chart of compositions **b** of the Xiong'er Group (Peng et al. 2015); and **c** comparison histogram of SiO_2 and TiO_2 of the Taihang dyke swarm versus Xiong'er volcanic province (Peng 2016)

8.3.2 Anorogenic Magmatism (ca. 1720–1620 Ma)

There occurred the Damiao anorthosite, the Miyun rapakivi granites and the Dahongyu volcanics in the northern NCC, which are from ca. 1.7 to 1.62 Ga in age. The Longwangzhuang A-type granites, mafic dyke swarms and alkaline rocks also simultaneously developed in the southern NCC. This section will briefly address their geologic and geochemical characteristics, petrogenesis and tectonic settings.

8.3.2.1 Damiao Anorthosite Complex

Massif-type anorthosite complex consists usually of over 90% plagioclase and occurs as a single pluton. Massif-type anorthosite bodies were only formed in the time interval from 2.1 to 0.9 Ga, and commonly hosted abundant Fe–Ti oxide ores. Massif-type anorthosite was considered as an important window for understanding mantle nature, crustal evolution, crust-mantle interaction and mineralization in Proterozoic time. The Damiao anorthosite complex in the northern margin of the NCC is the only massif-type anorthosite intrusion recognized in China. It is exposed in a small area, ca. 100 km^2, and shows a variety of lithologies, including anorthosite (85%), hypersthenfels (10%), mangerite (4%), troctolite (<1%) and minor iron diorite and gabbro veins (Xie 1980, 2005), containing Fe–Ti–P ores.

Crystallization ages of the complex are 1693 ± 7 Ma and 1715 ± 6 Ma, respectively (Zhao et al. 2004b). SHRIMP zircon U–Pb dating gave a crystallization age of 1726 ± 9 Ma (Zhang et al. 2007). Norite and mangerite in complex provide LA-ICP MS and SHRIMP zircon U–Pb ages of 1742 ± 17 Ma and 1739 ± 14 Ma (Zhao et al. 2009a, 2009b). These ages indicate that the Damiao anorthosite complex was formed in the Mesoproterozoic and its emplacement lasted for 10 to 20 Ma.

Exsolutions in plagioclase megacrysts and high Al_2O_3 contents (5.5–9.0%) of orthopyroxene of the Damiao anorthosite complex indicate that these minerals were crystallized under a high pressure (over 10 kbar) and then emplaced under a lower pressure (ca. 4 kbar). Different types of rocks, however, show similar Nd–Hf isotopic compositions, with whole-rock $\varepsilon_{Nd}(t)$ values being −4.0 and −5.4 and zircon $\varepsilon_{Hf}(t)$ values being from -4.7 to -7.5 (Zhang et al. 2007; Zhao et al. 2009a, 2009b). Considering whole-rock characteristics of major and trace elements and variations of mineral compositions, these different types of rocks should have been derived from a single magma. Zhang et al. (2007) argued that recycling of the Archean crust enriched the subcontinental lithospheric mantle, and the primary magma as the result of partial melting then formed the Damiao anorthosite. Based on Sr contents and initial $^{87}Sr/^{86}Sr$ ratios of synchronous mantle-derived mafic dykes and ancient lower crustal xenoliths in the NCC, parental magma of the Damiao anorthosite should have been sourced from a depleted mantle and then assimilated with ca. 30% lower-crustal materials (Al_2O_3 = 15wt%–24wt%; Sr = 800–2000 ppm) or mixed with melts of the lower crust when it ponded at the base of lower crust (Chen et al. 2015). Such a model could account for variable isotopic signatures for different phases in many anorthosite complexes (Chen et al. 2015).

8.3.2.2 Miyun Rapakivi Pluton

The Miyun rapakivi pluton in the north of Beijing is part of the anorthosite-rapakivi granite belt constituted by the Damiao anorthosite complex, Gubeikou K-rich granite, Gudonggou K-rich granite, Lanying quartz-syenite, Xindi anorthosite and Chicheng rapakivi granite (Xie 2005). The Miyun rapakivi granite has been intensively studied (Rämö et al. 1995; Yu et al. 1996; Xie 2005; Yang et al. 2005; Zhang et al. 2007;

Gao et al. 2008a, 2008b). It intruded into the Archean metamorphic series consisting of gneiss, granulite, amphibolite and magnetite quartzite. The rapakivi pluton is exposed in an area of ca. 25 km^2, up to 12 km long and 3 km wide. It is in a fault contact with metamorphic rocks to the north, but intruded into the surrounding rocks to the south (Yang et al. 2005). The Miyun pluton is composed largely of rapakivi hornblende biotite granite and porphyric biotite granite. The porphyritic biotite granite makes up the bulk of the pluton, and mostly outcrops in the center and east of the pluton. Rapakivi granite makes up a quarter of the pluton, and dominates the west of the pluton. In addition, there are minor medium- to fine-crystalline biotite granite, medium-crystalline two-mica granite and light-colored fine-crystalline granite, which are mainly distributed at the edge of the pluton. Many veins also occur in the north and south, inclusive of rapakivi granite dykes, fine-grained biotite granite dykes and diabase dykes (Yu et al. 1996; Xie 2005). More than 30% of K-feldspar phenocrysts have plagioclase rims and typical rapakivi textures.

The Miyun rapakivi pluton were dated by different methods, yielding TIMS zircon U–Pb ages between 1679 Ma and 1735 Ma, SHRIMP zircon U–Pb age of 1685 ± 15 Ma (Gao et al. 2008a, 2008b), and LA-ICP MS zircon U–Pb ages of 1681 ± 10 Ma and 1679 ± 10 Ma (Yang et al. 2005). Yang et al. (2005) proposed that the Miyun granites resulted from partial melting of juvenile crust formed in the Late Archean based on $\varepsilon_{Hf}(t)$ values of zircons in rapakivi granite (-5.0) and two-stage model ages (t_{DM2}) of rapakivi granites from 2.6 Ga to 2.8 Ga. Alternatively, Zhang et al. (2007) suggested that they were formed by fractional crystallization and crustal contamination of basic magma from an enriched mantle (EMI) according to whole-rock characteristics of major and trace elements as well as Sr–Nd-Pb and zircon Hf isotopes. Jiang et al. (2011) reported a formation age of 1697 ± 7 Ma of the Wenquan A-type granite with rapakivi textures in Chicheng country. The Wenquan A-type granites, Miyun rapakivi granites and synchronous granites in the northern NCC have comparable Nd-Hf isotopic (Fig. 8.13) and geochemical features, characterized by similar degrees of oxidation. These facts indicate that they originated from the same source, i.e., the Neoarchean crystalline basement of the NCC.

8.3.2.3 Dahongyu Volcanics

The Changchengian Dahongyu Formation is widely distributed in the northern NCC with an outcropped area of ca. 600 km^2 and a thickness up to 718 m at Pinggu (north Beijing) and 490 m at Jixian (north Tianjin). The Dahongyu Formation is composed mainly of quartzitic sandstone with minor volcanics and dolostone at its middle to upper parts. The volcanics gave zircon U–Pb ages of 1625.3 ± 6.2 Ma (Lu and Li 1991) and 1625.9 ± 8.9 Ma (Gao et al. 2008a, 2008b). The Dahongyu volcanics display variations in rock types of lava, volcanic breccia and tuff from west to east. Magmatism in the west appeared more intense and protracted with the total thickness of volcanic lava up to 500 m. In contrast, volcanic breccia is dominant in the east. In addition, The Tuanshanzi Formation beneath the Dahongyu Formation contains K-rich andesite that yields a zircon U–Pb age of 1683 ± 67 Ma (Li et al.

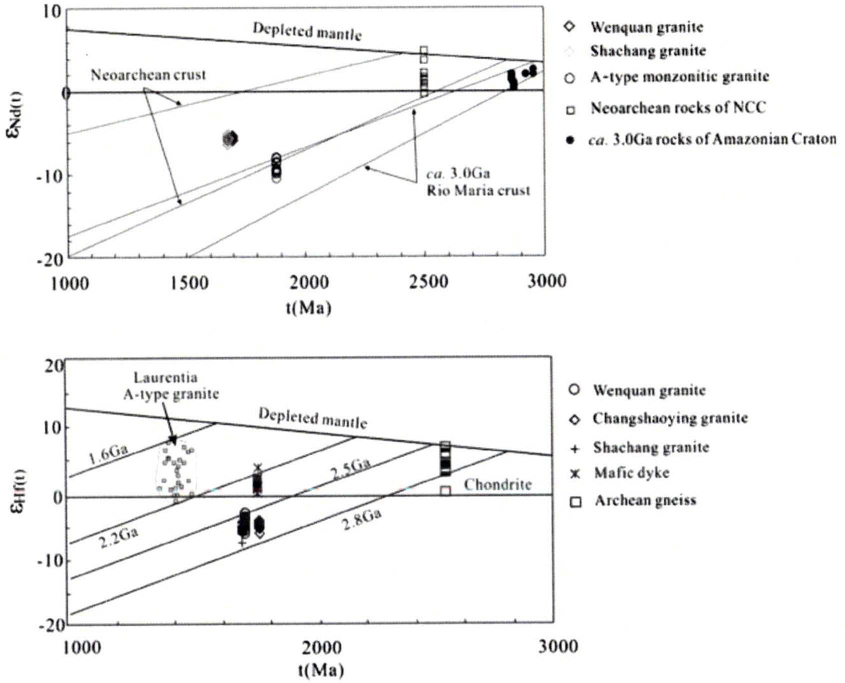

Fig. 8.13 Plots of $\varepsilon_{Nd}(t)$ and $\varepsilon_{Hf}(t)$ values vs. ages for rapakivi granites and related rocks (Jiang et al. 2011)

1995). The Gaoyuzhuang Formation above the Dahongyu Formation also contains tuffs that gave SHRIMP zircon U–Pb age of 1559 ± 12 Ma (Li et al. 2010). These geochronological data put a tight age constraint on the Dahongyu Formation.

Figure 8.14 exhibits stratigraphic columns of the Dahongyu Formation, illustrating spatial variation of volcanic eruption, stratigraphic thickness and lithologies from west to east. Four events of volcanism are recognized (V1, V2, V3, V4), which were punctuated by clastic deposition (I, II, III; Fig. 8.14). The fourth volcanic eruptions were particularly voluminous, and occurred throughout the northern NCC. Volcanic rocks produced by eruptions prior to the fourth event are mostly ultra-potassic, such as high-K alkaline basalts, phonolites and minor volcaniclastics (Hu et al. 2007). Geochemically, volcanic and volcanoclastic rocks are characterized by the enrichment of light rare earth elements (LREE) and large ion lithophilic elements (LILE) and the depletions of high field strength elements (HFSE) and heavy rare earth elements (HREE) with slightly negative anomalies of Nb and Ta and positive Eu anomaly. Constant La/Nb ratios and $\varepsilon_{Nd}(t)$ values indicate that crustal materials had not contaminated the ascending magma. Weak depletion of Nb, Ta and $\varepsilon_{Nd}(t)$ values from −0.66 to 0.63 reflects characteristics of the mantle source. Hu et al. (2007) suggested that parental magma originated from the enriched mantle was both metasomatosed by crustal material and affected by asthenospheric materials.

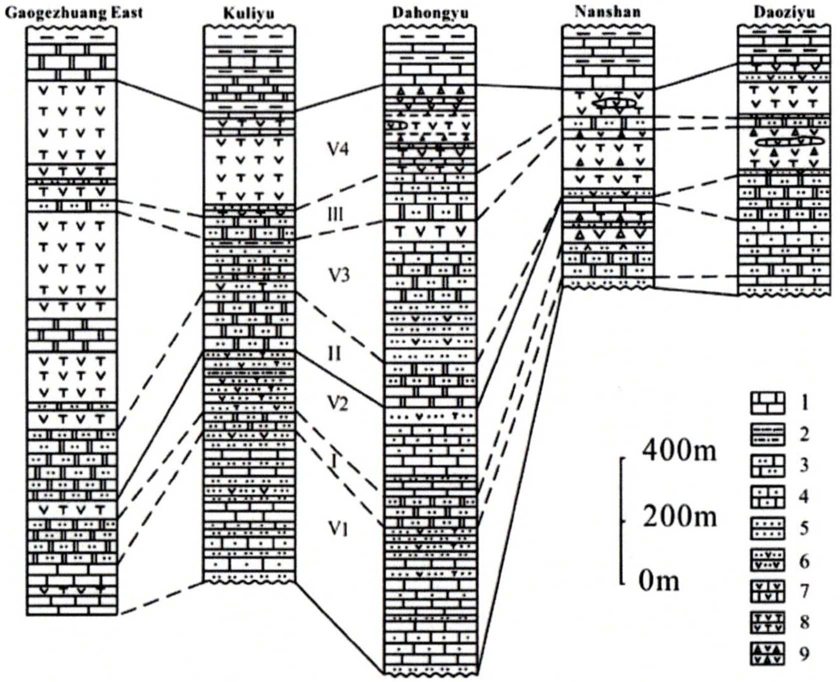

Fig. 8.14 Stratigraphic profiles of the Dahongyu Formation in Jixian stratotype section, Center Yanliao Rift (Hu et al. 2007). 1. Dolomite; 2. silicalite; 3. quartzite; 4. sandy limestone and tuffaceous dolostone; 5. quartz arenite and conglomerate; 6. sandy peperite and tuff; 7. carbonate peperite; 8. potassic lava; 9. volcanic breccia and agglomerate. V1–V4: volcanism events; I–III: clastic deposition

8.3.2.4 Longwangzhuang A-type Granite

Along the southern margin of the NCC, an EW-trending alkaline granite belt stretches over 400 km from Luonan in the west, through Lushi, Luanchuan, Fangcheng, to Wuyang in the east. The Longwangzhuang pluton in Luanchuan is the largest A-type granite in the belt.

The Longwangzhuang pluton is 20 km long from east to west and 10 km wide from south to north, covering an area of 140 km^2. It intruded into the Taihua Group and the Yanshanian Funiushan granite. The pluton is composed of gray arfvedsonite granite, with biotite K-feldspar granite being in the south and north. Late-stage gabbro-diabase, alkaline biotite syenite, granite porphyry and quartz monzonite veins are commonly associated with plutonism.

LA-ICP MS zircon U–Pb ages of the pluton at 1602 ± 6 and 1616 ± 20 Ma (Bao et al. 2009; Wang et al. 2013) are consistent with SHRIMP zircon age of 1625 ± 16 Ma (Lu et al. 2003). The granite is characterized by high silicon and alkaline contents (SiO_2 = 72.2%–76.8%, $K_2O + Na_2O$ = 8.3%–10.2%, $K_2O/Na_2O > 1$), with differentiation index (DI) from 95 to 97, aluminum saturation index (ASI) from 0.96 to 1.13, and very high Fe* number [FeO*/(FeO* + Mg) = 0.90–0.99]. It thus can be classified as the metaluminous to weakly peraluminous, alkalic to calc-alkalic and ferroan A-type granites. The granite is also rich in LILE, especially REE (REE + Y = 854–1572 ppm) with slight enrichment in HFSE (Nb, Ta, Zr, Hf) and strong depletions in Ba, Sr, Ti, and Pb. In addition, it has low $\varepsilon_{Nd}(t)$ values (−4.5 to −7.2) and high Nd model ages (2.3–2.5 Ga) as well as low $\varepsilon_{Hf}(t)$ values (−1.11 to −5.26) and high Hf model ages ($t_{DM}{}^{C}$ = 2.4–2.6 Ga). However, the petrogenesis of Longwangzhuang granite remains controversial. It was proposed that the granite originated from partial melting of the lower crust (Wang et al. 2013), but might be formed through extreme fractional crystallization of alkali basaltic magma due to partial melting of enriched mantle (Bao et al. 2009).

8.3.2.5 Anorogenic Magmatism

The origin of non-orogenic magmatism in the NCC has been a matter of debate (Wang et al. 2013 and references therein). It might be generated by mantle plumes and/or related to breakup of the Columbia supercontinent (Zhao et al. 2002a, b; Hou et al. 2008; Zhang et al. 2007, 2009, 2012, 2017a). Key questions are listed as follows.

(1) The magmatic rocks have geochemical characteristics similar to arc volcanic rocks. For instance, they have enhancd whole-rock $\varepsilon_{Nd}(t)$ and zircon $\varepsilon_{Hf}(t)$ values with depletions in Nb, Ta and other HFSE. Enrichments in LILE possibly originated from an enriched mantle, which had been metasomatized either by palaeo-subduction materials or due to addition of crustal materials (Fig. 8.15).
(2) The magmatic rocks are generally rich in iron and potassium. For example, iron contents in intermediate-basic volcanics of the Xiong'er Group and contemporary intrusives are about 10% (FeO + Fe_2O_3 or total iron as Fe_2O_3), with some up to 15%. In contrast, iron contents in mafic dyke swarms are normally over 10%, and can be up to 20%, with TiO_2 content around 1%. The majority of mafic rocks (including mafic dyke swarms) show an evolution trend and mineralogical characteristics of tholeiitic volcanic rocks. Silica contents are usually more

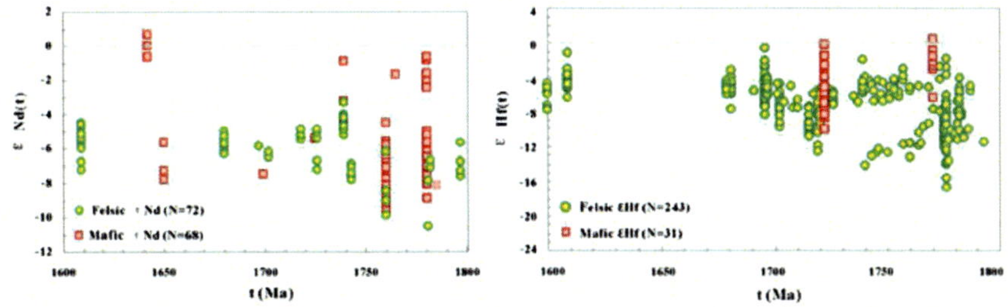

Fig. 8.15 Plots of $\varepsilon_{Nd}(t)$ and zircon $\varepsilon_{Hf}(t)$ ratios vs. ages for anorogenic magmatic rocks

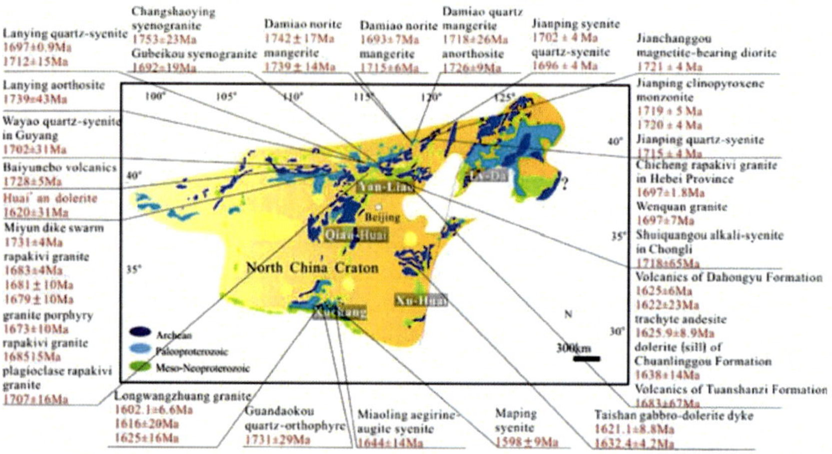

Fig. 8.16 Sketch map showing distribution and ages of Proterozoic anorogenic magmatic rocks in the NCC

than 52%, and ferromagnesian minerals are mainly clinopyroxene and minor hornblende.

(3) No contemporary picrites, OIB (ocean island basalt)-type basalts or other magmatic rocks derived directly from depleted asthenospheric mantle have been found in North China. There did not occur voluminous magmatism during a short period, but happened episodically from 1720 to 1620 Ma.

(4) Compared with 1.78 Ga magmatism, the magmatism in this period lasted for a quite long time, and the resulting magmatic rocks show gradual variations in their geochemical characteristics, which became distinct from plume-related rocks. Given that they occurred in the Xiong'er and the Yanliao Rifts (Fig. 8.16), we thus infer the 1.78 Ga magmatism resulted from rifting related to large igneous event or mantle plume.

8.3.3 *Yanliao Igneous Event (ca. 1320 Ma)*

Many ca. 1320 Ma mafic sills have been identified in the Yanliao Rift (Li et al. 2009; Zhang et al. 2009, 2012, 2017a; cf. Chap. 10; Fig. 8.17). They mostly intruded into the Xiamaling and the Wumishan Formations, but some occur in other formations of the Jixianian (Fig. 8.18). The sills are diabase/gabbro-diabase/dolerite composed of 40–60vol% pyroxene and 40–55vol% plagioclase (Zhang et al. 2009, 2017a), and varied in thicknesses from tens meters to hundred meters, extending to tens of kilometers (Zhang et al. 2009, 2012, 2017a; cf. Chap. 10).

These sills are tholeiitic baslats, and chemically show lithospheric signature (Zhang et al. 2009, 2017a). The Nd t_{DM} (depleted mantle-model age) of whole-rocks and Hf t_{DM} of in-situ zircon and baddeleyite are ca. 1500–2000 Ma (Zhang

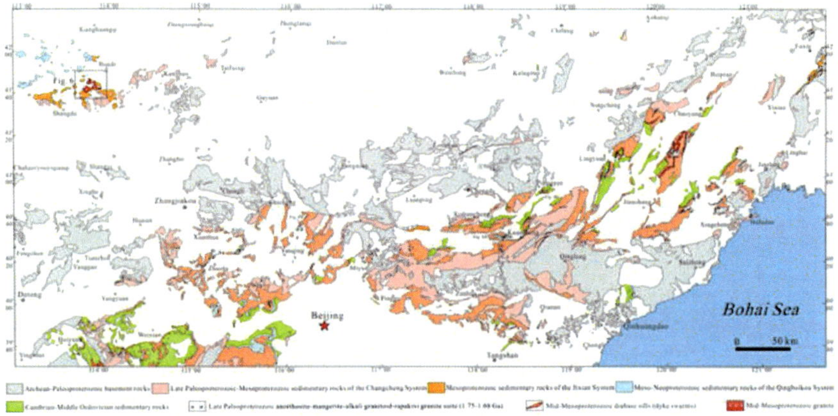

Fig. 8.17 Diagram showing distribution of ca. 1320 Ma sills in the northern margin of the NCC (Zhang et al. 2012)

et al. 2012). The basic sills and coeval granites indicated bimodal magmatism, and might have been related to breakaway of an ancient block from the NCC (Yang et al. 2011; Zhang et al. 2012, 2017b).

Coeval carbonatite dykes were once reported in the Bayan Obo rare earth elements-iron deposits (Yang et al. 2011; Zhang et al. 2017b). The ca. 1230 Ma dykes were also identified in the interior of the NCC (Peng et al. 2015) although they are younger than those in the Yanliao Rift.

8.3.4 Dashigou Igneous Event (ca. 925 Ma)

The Dashigou igneous event happened around 925 Ma, and resulted in the Dashigou large Igneous Province, which is comprised of mafic dykes in the center and sills in the southern and eastern margins of the NCC (Fig. 8.19a–g; Peng et al. 2011b). Some coeval or slightly younger (925–880 Ma) sill complexes are present in the Pyongnam Basin in North Korea (Fig. 8.19a; Peng et al. 2011a), Xuhuai Basin (i.e., Xuzhou City-Huaihe River Valley, Fig. 8.19b; Wang et al. 2012), Dalian Basin in Liaodong Depression (Fig. 8.19c; Zhang et al. 2016) and Luanchuan Basin in Xiong'er Rift (Fig. 8.19d; Wang et al. 2011). Peng et al. (2011b) suggested that these sills were not only developed in the Xuhuai Basin, but also in other two rift basins extending to the southern and eastern margins of the NCC. They together comprised a triple-conjugated rift system (Fig. 8.20).

The Dashigou dykes are individually 10–20 km long and 10–50 m wide, striking mainly NW and NNW (305°–340°). There are also some NNE-oriented gabbro and diabase dykes in the eastern NCC, which are composed of plagioclase (ca. 65vol%),

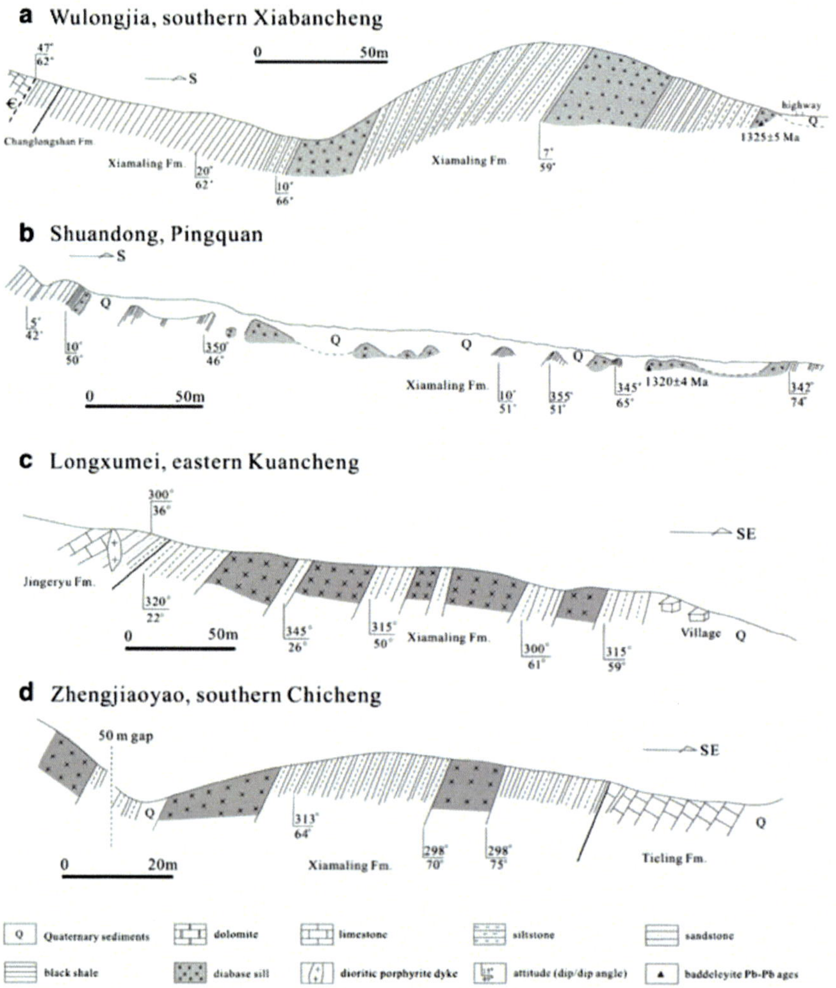

Fig. 8.18 Cross sections show ca. 1320 Ma sills in the Jixianian Xiamaling Formation in the Yanliao Rift (Zhang et al. 2017a)

clinopyroxene (ca. 25vol%) and minor accessory minerals like hornblende and K-feldspar. The dykes gave baddeleyite U–Pb ages from ca. 925–920 Ma (Peng et al. 2011b). These sills are usually several meters to ca. 150 m thick and several kilometers to tens of kilometers long. They are mainly diabase, and display similar petrographic and chemical features. Sills in the Pyongnam Basin and the Dalian Basin experienced low-degree or greenschist facies metamorphism (Peng et al. 2011a; Zhang et al. 2016). Plagioclase and clinopyroxene constitute the bulk of mineral assemblage, but clinopyroxene had been largely altered to hornblende. Metamorphic minerals are common, like epidotite, chlorite, albite and amphibole. One of the sills in the

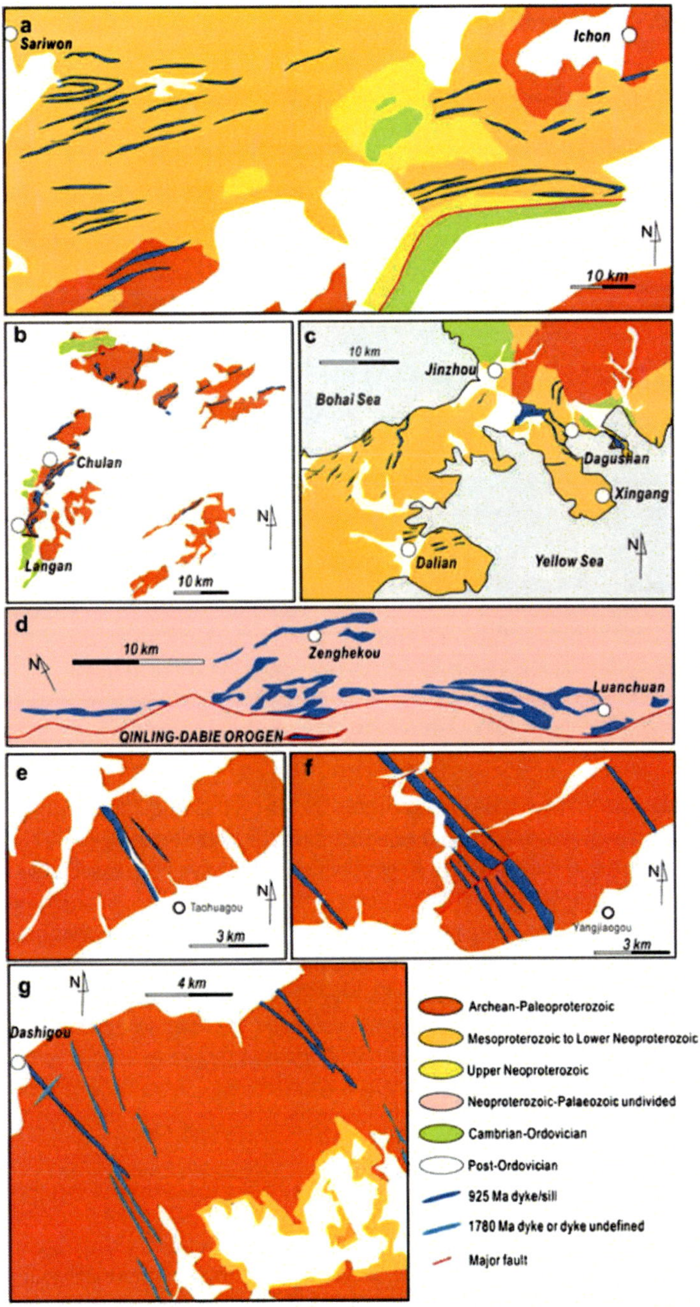

Fig. 8.19 Diagrams showing distribution of dyks and sills in different areas (Peng et al. 2011b). **a.** Sill swarms (925–890 Ma) in the Pyongnam Basin (North Korea); **b** Xuhuai Basin (including the Xuzhou City and Huaihe River Basin); **c** Dalian Basin (or called Liaodong Basin); **d** Luanchuan Basin in Xionger Rift; **e** Dashigou dykes (ca. 925 Ma) in Inner Mongolia; **f** Hebei Province; g. Shanxi Province

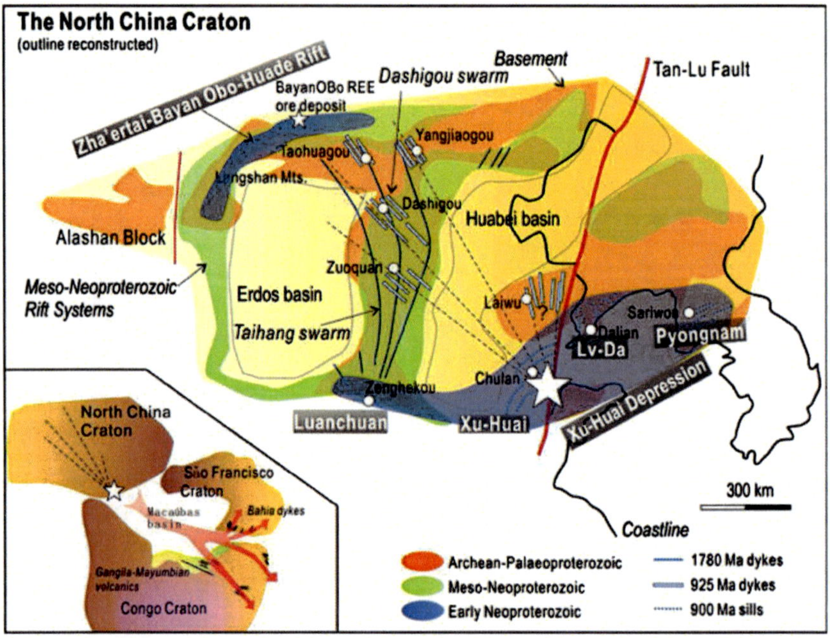

Fig. 8.20 Distribution of Dashigou dyke swarm (ca. 925 Ma) and coeval rift system in NCC (Peng et al. 2011b). Inset map shows a possible connection of the NCC, São Francisco and Congo Cratons. Note that the NCC is modified to its Precambrian shape

Pyongnam Basin yielded a U–Pb age of ca. 400 Ma on baddeleyite from a ca. 900 Ma dolerite sill, which was interpreted as a metamorphic age (Peng et al. 2011a).

It was argued that the ca. 925–880 Ma events might have been associated with the Proterozoic mantle plume (Fig. 8.20; Peng et al. 2011b). Figure 8.21 is a summary of Nd isotopes of 15 major dyke/sill swarms in the NCC, showing that both 1780–1730 Ma and ca. 925–880 Ma dyke events might have resulted from magma related to a mantle plume. In addition, the 1780–1730 Ma event might have modified the sub-continental lithospheric mantle of the eastern NCC, whereas the ca. 925–880 Ma events exerted little influence on the lithospheric mantle (Peng 2015).

8.4 Discussion and Conclusion

There once existed several supercontinents in the Earth's history (Fig. 8.22; Condie 2004). Two tectonic processes are considered responsible for assembly and breakup of the supercontinents, plate tectonics for the assembly and super-plume for the breakup of supercontinents. The proposed supercontinents include Arctica-Kenorland, Nera or Antlantica, Nuna or Columbia, Rodinia and Pangea. Gondwana

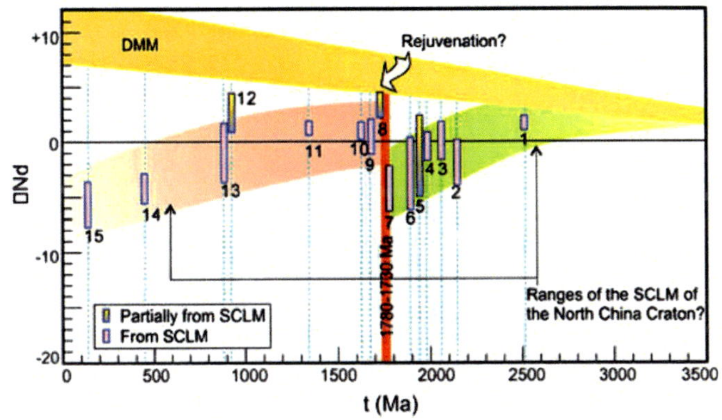

Fig. 8.21 Diagram of ε_{Nd} value ranges of 15 selected mafic dyke/sill swarms in the NCC (Peng 2015 and references therein). 1. Huangbaiyu dykes; 2. Hengling sills; 3. Yixingzhai dykes; 4. Xiwangshan dykes; 5. Xuwujia sills; 6. Halaqin dykes/volcanics; 7. Taihang dykes; 8. Miyun dykes; 9. Laiwu dykes; 10. Taishan dykes; 11.Yanliao sills; 12. Dashigou dykes; 13.Sariwon sills; 14. Shanxi dykes; 15. Mengyin dykes. SCLM. sub-continental lithospheric mantle; DMM. depleted MORB mantle

had existed before Rodinia (Fig. 8.22). Then each supercontinent brokes up into several continents. The reconstruction of supercontinents is based on the studies of ancient orogens and rifts within the continents and the global correlations of geological events. The supercontinent Arctica-Kenorland had some remnants of the NCC, indicating that they were once connected in ca. 2.5 Ga. The NCC then drifted away in 2.0–2.2 Ga (Zhai et al. 2000). The Palaeoproterozoic orogenic belts resulted from subduction and collision processes that led to the formation of Supercontinent Columbia (Zhai and Liu 2003). The Meso-Neoproterozoic NCC was obviously in the stable or extensional tectonic settings, as indicated by the development of multi-stage rifts. No strong crustal contraction happened in this time interval (Zhang et al. 2012; Zhai et al. 2014).

The $\varepsilon_{Hf}(t)$ values of 2.4–2.8 Ga magmatic detrital zircons in both the Huade and Tumen Formations are mostly positive, indicating they were from depleted mantle (Fig. 8.23b–d). The Hf modal age is ca. 2.85 Ga (Fig. 8.23a–c). As implied by the data, the crust grew extensively in ca. 2.85 Ga and was then reworked at ca. 2.5 Ga. This characteristic is consistent with that of the NCC metamorphic basement. The 1.7–2.05 Ga zircons in the Huade Formation have -3 to $+4$ $\varepsilon_{Hf}(t)$ and ca. 2.5 Ga modal ages or -6 to -3 $\varepsilon_{Hf}(t)$ and ca. 2.85 Ga (2.6–3.3 Ga) modal ages, indicating affection of ca. 2.5 Ga and 2.85 Ga crust rocks, respectively. The 1.05–1.7 Ga magmatic detrital zircons in the Tumen and Huade Formations have $\varepsilon_{Hf}(t)$ values with Hf modal ages of 1.6–2.3 Ga, similar to depleted mantle (Fig. 8.23b–d), indicating transformation from the old crust and mantle and corresponding to Mesoproterozoic continental magmatism in the NCC.

Zhai (2011) and Zhai et al. (2014) studied Meso-Neoproterozoic processes of the NCC and proposed that it should be of global implications. Cawood and Hawkesworth (2014) termed the 1.7–0.75 Ga period as "Earth's middle age", and

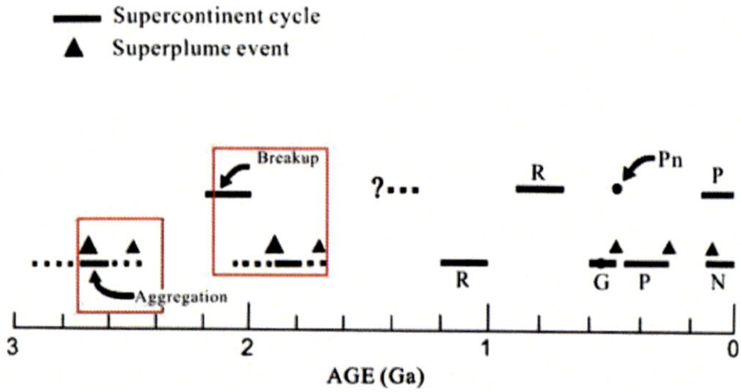

Fig. 8.22 Diagram showing supercontinental cycles (Condie 2004). Black triangles are proposed superplume events; symbol size is proportional to event intensity. R. Rodinia; G. Gondwana; Pn. Pannotia; P. Pangea; N. new supercontinent forming today

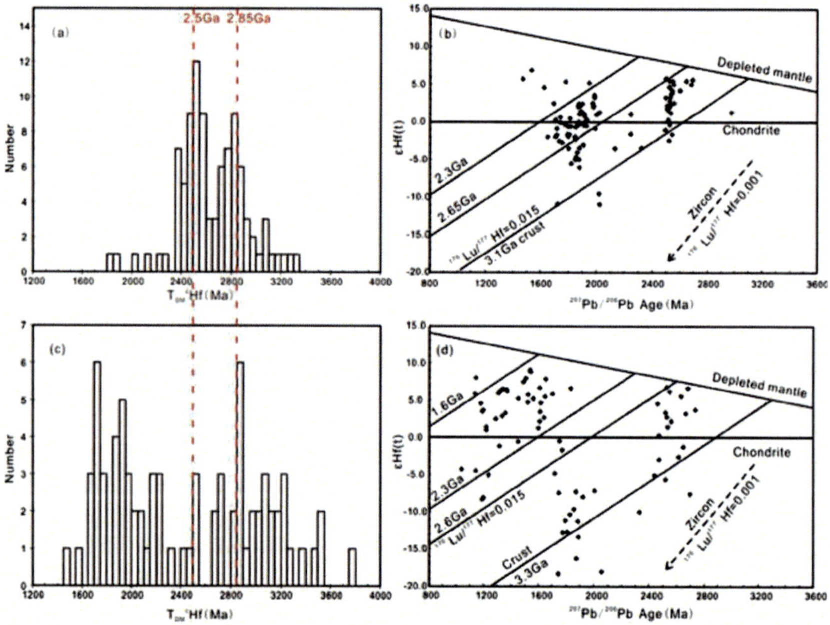

Fig. 8.23 Diagrams showing $\varepsilon_{Hf}(t)$-$^{207}Pb/^{206}Pb$ ages of magmatic detrital zircons from the Huade and Tumen Formations (Hu et al. 2012). **a, b** Huade Formation; **c, d** Tumen Formation

summarized that this time interval was characterized by the following phenomena: few passive margins, absence of significant Sr anomaly in seawater and $\varepsilon_{Hf}(t)$ of detrital zircons, paucity of orogenic gold and volcanics-hosted massive sulfide deposits, and absence of glacial deposits and banded iron formations. By contrast, there widely developed anorthosites and kindred bodies as well as large-scale Mo and Cu mineralization. Cawood and Hawkesworth (2014) ascribed these striking phenomena to the formation of the Supercontinent Nuna since ca. 1.75 Ga, which might have persisted until the breakup of Rodinia at ca. 0.75 Ga. The "Earth's middle age" thus envisioned a period of tectonic stability that contrasted the dramatic changes in preceding and succeeding eras. All the geological phenomena are also characteristic of the Meso-Neoproterozoic NCC, as described by Zhai et al. (2014).

(1) The 1.8–0.7 Ga thick sedimentary sequences are widely distributed in the NCC, which had not experienced strong contraction and metamorphism until Middle Jurassic time. Four rifts developed around the NCC during the Mesooproterozoic-Neoproterozoic, the Xiong'er rift, the Yanliao rift, the Zha'ertai-Banyan Obo-Huade Rift and the Jiao-Liao-Xu-Huai Rift, respectively. Volcanic rocks at the base of the Xiong'er Group are dated at ca. 1.80–1.78 Ga (Zhao et al. 2004a), indicating that rifting had started at the southern edge of the NCC since the Mesoproterozoic. The Yanliao Rift sequences also began developing around 1.67 Ga. The Neoproterozoic is rare in the Yanliao Rift, but well preserved in the northeastern NCC. The Neoproterozoic sequence in the northeastern NCC is divided into the Jing'eryu, Qiaotou, Wuxingshan and Jinxian Formations, which can be correlated with the Qingbaikouan, Nanhuan and Sinian, respectively. The proposed stratigraphic correlation is tentative because of the lack of reliable radiometric dates. A disconformity separates the Proterozoic from the overlying Cambrian, suggesting that the NCC did not experienced strong contraction at the end of the Neoproterozoic. Sedimentation in the NCC took place in an extensional setting from ca. 1.8 Ga to 0.7 Ga, and could have been continuing to ca. 540 Ma after the Hutuo Movement. In other words, the NCC evolved in a stable tectonic setting in the "Earth's middle age", as evidenced by rift basin development and epicontinental sedimentation.
(2) Four-stage magmatism is recognized in the NCC in the Meso-Neoproterozoic: ① ca. 1780–1730 Ma mafic dyke swarms and the Xiong'er volcanism; ② ca. 1720–1620 Ma anorogenic magmatism; ③ ca. 1350–1320 Ma diabase sill swarm; ④ ca. 900 Ma mafic dyke swarm and bimodal volcanism. The first and fourth-stage magmatic events were possibly the result of mantle plume, whereas the second-stage anorogenic magmatism was presumably generated by a thermal mantle. The third-stage diabase sills could have resulted from partial melting of a depleted asthenosphere mantle coupled with slight crustal assimilation. It is thus proposed that the NCC lithosphere should be tectonically stable and underlain by a warmer mantle in the "Earth's middle age". A persistent warmer state of the mantle should have been responsible for multi-stage partial melting of both the mantle and lower crust, and triggered rifting. Meso-Neoproterozoic ore-deposits include magmatic Fe–Ti–P deposits associated with anorthosites and gabbro intrusions, Meso-Neoproterozoic SEDEX (sedimentary exhalative)-type Pb–Zn–Cu deposits, REE–Fe–Nb deposits. All the deposits clearly bore on Mesoproterozoic continental rifting, extension-related magmatism, and mantle upwelling.

(3) The earliest supercontinent had formed prior to 2.5 Ga, and broke up in the 2.35–2.25 Ga period (Condie et al. 2001). Another two supercontinents are assumed to have also existed in the Proterozoic Nuna (or Columbia) and Grenville (Rodinia). The assembly of the Nuna and Grenville Supercontinents was recorded by ca. 2.1–1.95 Ga and ca. 1.1–0.9 Ga metamorphic and magmatic events. The NCC, however, did not undergo any strong compressional deformations and metamorphism during the Late Palaeo-Neoproterozoic interval. Neoproterozoic sedimentary and magmatic evolution offers no support for the Grenville orogeny in the NCC (Zhai et al. 2014). Several magmatic events did occur in Late Mesoproterozoic times in the NCC, but they were related to the breakup of Nuna from ca. 1.78–1.30 Ga and Rodinia from ca. 0.925–0.82 Ga (Zhao et al. 2004a; Zhang et al. 2009; Gao et al. 2009; Peng et al. 2011b, 2014). The duration from ca. 1.78 Ga to 1.30 Ga or 0.82 Ga seems too long for breakup processes to be accepted. The NCC might be located at a remote edge of the Supercontinent Nuna throughout Proterozoic era based on palaeomagnetic studies (Zhang et al. 2000, 2012). This result can well account for the lack of geologic records of Meso-Neoproterozoic orogenies in the NCC. Two possibilities may exist: the NCC was located far from the Grenville orogenic belts, or the Grenville orogeny was not a global tectonic event (Zhai et al. 2014). Therefore, it remains unclear about the existence of Proterozoic supercontinent cycles.

Main conclusions are as follows.

(1) A Palaeoproterozoic tectonic event, the Hutuo Movement, took place in the NCC, involving sequential rifting, subduction, and collision. The NCC then evolved into a stable platform in "Earth's middle age" for more than 1.0 Ga, with Meso-Neoproterozoic sequences being extensively deposited over the metamorphic basement.
(2) There occurred four-episode magmatism during Meso-Neoproterozoic time, which was accompanied by multi-stage rifting. The NCC developed in an intraplate setting from ca. 1.8 Ga to ca. 0.7 Ga.
(3) Magmatic-type iron deposits were closely related to anorthosite-gabbro magmatism and the Meso-Neoproterozoic rifting played an import role in the formation of REE–Nb–Fe and Pb–Zn–Cu–Fe deposits.
(4) No Grenville and other orogenies took place in the NCC. The Proterozoic NCC might be at a remote edge of the Supercontinent Nuna if this supercontinent did exist. The "Earth's middle age" was a particular time when the Earth possessed a stable lithosphere and a warm mantle, as manifested by multiple episodes of magmatism and rifting.

Acknowledgements This work is based on the results of the 973 Program (Grant No. 2012CB4166006) founded by Ministry of Science and Technology (MST) of the People's Republic of China, and other programs (Grant No. 41030316, 41210003) supported by National Nature Science Foundation of China. Thanks go to Peng Runmin, Guo Jinghui, Zhang Yanbin, Li Qiuli, Zhang Xiaohui, Zhao Yue, Hu Jianmin, Sun Yong and Zhang Chengli for their support and valuable data. We also thank Tim Kusky for helpful discussion. Zhong Yan, Zhou Yanyan, Zhao Lei and Wang Haozheng are thanked for preparing illustrations. Two reviewers' comments are great appreciated.

References

Bai J, Huang XY, Dai FY, Wu CH (1993) Precambrian crustal evolution in China. Geological Publishing House, Beijing (in Chinese)

Bao ZW, Wang Q, Zi F, Tang GJ, Du FJ, Bai GD (2009) Geochemistry of the Palaeoproterozoic Longwangzhuang A-type granites on the southern margin of North China Craton: petrogenesis and tectonic implications. Geochimica 38(6):509–522 (in Chinese with English abstract)

Cawood PA, Hawkesworth CJ (2014) Earth's middle age. Geology 42(6):503–506

Chen WT, Zhou MF, Gao JF, Zhao TP (2015) Oscillatory Sr isotopic signature in plagioclase megacrysts from the Damiao anorthosite complex, North China: implication for petrogenesis of massif-type anorthosite. Chem Geol 393–394:1–15

Chen JB, Zhang HM, Zhu SX, Zhao Z, Wang ZG (1980) Studies on Sinian Suberathem in Jixian. In: Tianjin Institute of geology and mineral resources, Chinese Academy of geological sciences (ed) Chinese Sinian Suberathem. Tianjin Science and Technology Press, Tianjin, pp 56–114. (in Chinese with English abstract)

Condie KC (2004) Precambrian superplume event. In: Eriksson PG, Altermann W, Nelson DR, Mueller WU, Catuneanu O (eds) The Precambrian Earth Tempos and events: development in Precambrian Geology. Elsevier, Amsterdam, pp 163–172

Condie KC, Des Marais DJ, Abbot D (2001) Precambrian superplumes and supercontinents: a record in black shales, carbon isotopes and palaeoclimates. Precambr Res 106:239–260

Cui ML, Zhang BL, Zhang LC (2011) U-Pb dating of baddeleyite and zircon from the Shizhaigou diorite in the southern margin of North China Craton: constrains on the timing and tectonic setting of the Palaeoproterozoic Xiong'er Group. Gondwana Res 20:184–193

Fan HR, Hu FF, Chen FK, Yang KF, Wang KY (2006) Intrusive age of No. 1 carbonatite dyke from Bayan Obo REE-Nb-Fe deposit, Inner Mongolia: with answers to comment of Dr. Le Bas. Acta Petrologica Sinica 22(2):519–520. (in Chinese with English abstract)

Fan QC, Hooper PR (1991) The Cenozoic basaltic rocks of Eastern China: petrology and chemical composition. J Petrol 32:765–810

Fan WM, Menzies MA (1992) Destruction of aged lower lithosphere and asthenosphere mantle beneath eastern China. Geotecton Metallog 16:171–179 (in Chinese)

Gao LZ, Zhang CH, Shi XY, Zhou HR, Wang ZQ, Song B (2007a) A new SHRIMP age of the Xiamaling Formation in the North China Plate and its geological significance. Acta Geol Sin 81(6):1103–1109

Gao LZ, Zhang CH, Shi XY, Zhou HR, Wang ZQ (2007b) Zircon SHRIMP U-Pb dating of the tuff bed in the Xiamaling formation of the Qingbaikouan system in North China. Geol Bull China 26(3):249–255 (in Chinese with English abstract)

Gao LZ, Zhang CH, Yin CY, Shi XY, Wang ZQ (2008a) SHRIMP zircon ages: basis for refining the chronostratigraphic classification of the Meso- and Neoproterozoic strata in North China Old Land. Acta Geoscientica Sinica 29(3):366–376 (in Chinese with English abstract)

Gao W, Zhang CH, Gao LZ, Shi XY, Liu YM, Song B (2008b) Zircon SHRIMP U-Pb age of rapakivi granite in Miyun, Beijing, China, and its tectono-stratigraphic implications. Geol Bull China 27(6):793–798 (in Chinese with English abstract)

Gao LZ, Zhang CH, Liu PJ, Ding XZ, Wang ZQ, Zhang YJ (2009) Recognition of Meso- and Neoproterozoic stratigraphic framework in North and South China. Acta Geoscientica Sinica 30(4):433–446 (in Chinese with English abstract)

Gao LZ, Zhang CH, Chen SM, Liu PJ, Ding XZ, Liu YX, Dong CY, Song B (2010) Detrital zircon SHRIMP U-Pb age from the Diaoyutai Formation, Xihe Group in Liaodong Peninsula, China and its geological significance. Geol Bull China 29(8):1113–1122 (in Chinese with English abstract)
Geng YS, Du LL, Ren LD (2012) Growth and reworking of the Early Precambrian continental crust in the North China Craton: constraints from zircon Hf isotopes. Gondwana Res 21:517–529
Halls HC, Campal N, Davis DW, Bossi J (2001) Magnetic studies and U-Pb geochronology of the Uruguayan dyke swarm, Rio de la Plata Craton, Uruguay: palaeomagnetic and economic implications. J S Am Earth Sci 14:349–361
He YH, Zhao GC, Sun M, Xia XP (2009) SHRIMP and LA-ICP-MS zircon geochronology of the Xiong'er volcanic rocks: implications for the Palaeo-Mesoproterozoic evolution of the southern margin of the North China Craton. Precambr Res 168(3–4):213–222
He YH, Zhao GC, Sun M (2010) Geochemical and isotopic study of the Xiong'er volcanic rocks at the southern margin of the North China Craton: petrogenesis and tectonic implications. J Geol 118(4):417–433
Henan Geology and Mineral Resources Bureau (1989) Regional geology of Henan Province. Geological Publishing House, Beijing (in Chinese)
Hou GT, Liu YL, Li JH (2006) Evidence for ~1.8 Ga extension of the Eastern Block of the North China Craton from SHRIMP U-Pb dating of mafic dyke swarms in Shandong Province. J Asian Earth Sci 27:392–401
Hou GT, Santosh M, Qian XL, Lister GS, Li JH (2008) Configuration of the Late Palaeoproterozoic supercontinent Columbia: insights from radiating mafic dyke swarms. Gondwana Res 14:395–409
Hu SX, Lin QL, Chen ZM, Li SM (1988) Geology and metallogeny of the collision belt between the South China and North China Plates. Nanjing University Press, Nanjing (in Chinese)
Hu JL, Zhao TP, Xu YH, Chen W (2007) Geochemistry and petrogenesis of the high-K volcanic rocks in the Dahongyu Formation, North China Craton. J Mineral Petrol 27(4):70–77 (in Chinese with English abstract)
Hu B, Zhi MG, Guo JH, Peng P, Liu F, Liu S (2009) LA-ICP-MS U-Pb geochronology of detrital zircons from the Huade Group in the northern margin of the North China Craton and its tectonic significance. Acta Petrologica Sinica 25(1):193–211 (in Chinese with English abstract)
Hu B, Zhai MG, Li TS, Li Z, Peng P, Guo JH, Kusky TM (2012) Mesoproterozoic magmatic events in the eastern North China Craton and their tectonic implications: geochronological evidence from detrital zircons in the Shandong Peninsula and North Korea. Gondwana Res 22:828–842
Hu JM, Gong WB, Wu SJ, Liu Y, Liu SC (2014) LA-ICP-MS zircon U-Pb dating of the Langshan Group in the northeast margin of the Alxa Block, with tectonic implications. Precambr Res 255:756–770
Jia CZ (1987) Geochemistry and tectonics of the Xiong'er Group in the eastern Qinling mountains of China—a mid-Proterozoic volcanic arc related to plate subduction. Geol Soc Lond Spec Publ 33:437–448
Jia CZ, Shi YS, Guo LZ (1988) Plate tectonics of eastern qinling mountains of China. Nanjing University Press, Nanjing (in Chinese)
Jiang N, Guo JH, Zhai MG (2011) Nature and origin of the Wenquang granite: implications for the provenance of Proterozoic A-type granites in the North China Craton. J Asian Earth Sci 42:76–82
Korenaga J, Jordan TH (2001) Effects of vertical boundaries on infinite Prandtl number thermal convection. Geophys J Int 147(3):639–659
Korenaga J (2006) Archean geodynamics and the thermal evolution of Earth. In: Benn K et al (eds) Archean geodynamics and environments. American Geophysical Union, Washington, pp 7–32
Kröner A, Wilde SA, Li JH, Wang KY (2005) Age and evolution of a Late Archaean to Early Palaeozoic upper to lower crustal section in the Wutaishan/Hengshan/Fuping terrain of northern China. J Asian Earth Sci 24:577–595
Kusky TM, Li JH (2003) Palaeoproterozoic tectonic evolution of the North China Craton. J Asian Earth Sci 22:383–397
Kusky TM, Li JH, Santosh M (2007b) The Paleoproterozoic North Hebei Orogen: North China Craton's collisional suture with the Columbia supercontinent. Gondwana Res 12(1–2):4–28

Kusky T, Windley B, Wang L, Wang ZS, Li XY, Zhu PM (2014) Flat slab subduction, trench suction, and craton destruction: comparison of the North China, Wyoming, and Brazilian Cratons. Tectonophysics 630:208–221

Kusky TM, Windley BF, Zhai MG (2007a) Tectonic evolution of the North China Block: from orogen to craton to orogen. In: Zhai MG et al (eds) Mesozoic Sub-continental Lithospheric Thinning Under Eastern Asia. Geol Soc Lond Spec Publ 280:1–34

Li HK, Li HM, Lu SN (1995) Grain zircon U-Pb ages for volcanic rocks from Tuanshanzi Formation of Changcheng System and their geological implications. Geochimica 24(1):43–48 (in Chinese with English abstract)

Li QL, Chen FK, Guo JH, Li XL, Yang YH, Siebel W (2007a) Zircon ages and Nd-Hf isotopic composition of the Zha'ertai Group (Inner Mongolia): evidence for Early Preoterozoic evolution of the northern North China Craton. J Asia Earth Sci 30:573–590

Li XH, Chen FK, Guo JH, Li QL, Xie LW, Siebel W (2007b) South China provenance of the lower-grade Penglai Group north of the Sulu UHP Orogenic Belt, eastern China: evidence from detrital zircon ages and Nd-Hf isotopic composition. Geochem J 41:29–45

Li HK, Lu SN, Li HM, Sun LX, Xiang ZQ, Geng JZ, Zhou HY (2009) Zircon and beddeleyite U-Pb precision dating of basic rock sills intruding Xiamaling Formation, North China. Geol Bull China 28(10):1396–1404 (in Chinese with English abstract)

Li HK, Zhu SX, Xiang ZQ, Su WB, Lu SN, Zhou HY, Geng JZ, Li S, Yang FJ (2010) Zircon U-Pb dating on tuff bed from Gaoyuzhuang Formation in Yanqing, Beijing: further constraints on the new subdivision of the Mesoproterozoic stratigraphy in the northern North China Craton. Acta Petrologica Sinica 26(7):2131–2140 (in Chinese with English abstract)

Li HK, Su WB, Zhou HY, Geng JZ, Xiang ZQ, Cui YR, Liu WC, Lu SN (2011) The base age of the Changchengian System at the northern North China Craton should be younger than 1,670 Ma: constraints from zircon U-Pb LA-MC-ICP MS dating of a granite porphyry dike in Miyun County, Beijing. Earth Sci Front 18(3):108–120 (in Chinese with English abstract)

Liaoning Geology and Mineral Resources Bureau (1989) Regional geology of Liaoning Province. Geological Publishing House, Beijing (in Chinese)

Liu SW, Zhao GC, Wilde SA, Shu GM, Sun M, Li QG, Tian W, Zhang J (2006a) Th-U-Pb monazite geochronology of the Lüliang and Wutai Complexes: constraints on the tectonothermal evolution of the Trans-North China Orogen. Precambr Res 148:205–225

Liu YQ, Gao LZ, Liu YX, Song B, Wang ZX (2006b) Zircon U-Pb dating for the earliest Neoproterozoic mafic magmatism in the southern margin of the North China Block. Chin Sci Bull 51(19):2375–2382

Lu SN, Li HK (1991) A precise U-Pb single zircon age determination for the volcanics of Dahongyu Formation, Changcheng System in Jixian. Bull Chin Acad Geol Sci 22:137–145 (in Chinese with English abstract)

Lu SN, Yang CL, Li HK, Li HM (2002) A group of rifting events in the terminal Palaeoproterozoic in the North China Craton. Gondwana Res 5(1):123–131

Lu SN, Li HK, Li HM, Song B, Wang SY, Zhou HY, Chen ZH (2003) U-Pb isotopic ages and their significance of alkaline granite in the southern margin of the North China Craton. Geol Bull China 22(12):762–768 (in Chinese with English abstract)

Lu SN, Zhao GC, Wang HC, Hao GJ (2008) Precambrian metamorphic basement and sedimentary cover of the North China Craton: a review. Precambr Res 160:77–93

Luo Y, Sun M, Zhao GC (2006) LA-ICP-MS U-Pb zircon geochronology of the Yushulazi Group in the Eastern Block, North China Craton. Int Geol Rev 48:828–840

Meng QR, Wei HH, Qu YQ, Ma SX (2011) Stratigraphic and sedimentary records of the rift to drift evolution of the northern North China Craton at the Palaeo- to Mesoproterozoic transition. Gondwana Res 20:205–218

Menzies MA, Fan WM, Zhang M (1993) Palaeozoic and Cenozoic lithoprobes and the loss of >120 km of Archean lithosphere, Sino-Korean Craton, China. Geol Soc Lond Spec Publ 76:71–81

Paek RJ, Kan HG, Jon GP, Kim YM, Kim YH (1993) Geology of Korea. Foreign Languages Books Publishing House, Pyongyang

Pei YH, Yan HQ, Ma YF (2007) The relationship between palaeo-volcanic apparatus and mineral resources of Xiong'er Group along Songxian-Ruzhou Zone in Henan Province. Geol Min Resour South China 1:51–58 (in Chinese with English abstract)

Peng RM (1999) Discovery of potassic spilite (or poenite) in Langshan Group of Tanyaokou district, Inner Mongolia, China. Chin Sci Bull 44(6):563–567

Peng P (2015) Precambrian mafic dyke swarms in the North China Craton and their geological implications. Sci China Earth Sci 58:649–675

Peng P (2016) Map of Precambrian dyke swarms and related plutonic/volcanic units in the North China Block (1:2,500,000). Science Press, Beijing (in Chinese)

Peng RM, Zhai YS (1997) The confirmation of the metamorphic double-peaking volcanic rocks in Langshan Group of the Dongshengmiao ore district, Inner Mongolia and its significance. Earth Sci J China Univ Geosci 22(6):589–594 (in Chinese with English abstract)

Peng P, Zhai MG, Zhang HF, Guo JH (2005a) Geochronological constraints on the Palaeoproterozoic evolution of the North China Craton: SHRIMP zircon ages of different types of mafic dikes. Int Geol Rev 47:492–508

Peng RM, Zhai YS, Wang ZG, Han XF (2005b) Discovery of double-peaking potassic volcanic rocks in Langshan Group of the Tanyaokou hydrothermal-sedimentary deposit, Inner Mongolia, and its indicating significance. Sci China Ser D Earth Sci 48(6):822–833

Peng P, Zhai MG, Guo JH (2006) 1.80–1.75 Ga mafic dyke swarms in the central North China Craton: implications for a plume-related break-up event. In: Hanski E et al (eds) Dyke swarms-time markers of crustal evolution. Taylor & Francis, London, pp 99–112

Peng P, Zhai MG, Guo JH, Kusky T, Zhao TP (2007a) Nature of mantle source contributions and crystal differentiation in the petrogenesis of the 1.78 Ga mafic dykes in the central North China Craton. Gondwana Res 12:29–46

Peng RM, Zhai YS, Han XF, Wang ZG, Wang JP, Liu JJ (2007b) Sinsedimentary volcanic activities in the cracking process of the Mesoproterozoic aulacogen of passive continental margin in Langshan-Zhaertai area, Inner Mongolia, and its indicatin significance. Acta Petrologica Sinica 23(5):1007–1017 (in Chinese with English abstract)

Peng P, Zhai MG, Ernst RE, Guo JH, Liu F, Hu B (2008) A 1.78 Ga large igneous province in the North China Craton: the Xiong'er Volcanic Province and the North China dyke swarm. Lithos 101:260–280

Peng RM, Zhai YS, Wang JP, Chen XF, Liu Q, Lü JY, Shi YX, Wang G, Li SB, Wang LG, Ma YT, Zhang P (2010) Discovery of Neoproterozoic acid volcanic rock in the south-western section of Langshan, Inner Mongolia. Chin Sci Bull 55(26):2611–2620

Peng P, Zhai MG, Li QL, Wu FY, Hou QL, Li Z, Li TS, Zhang YB (2011a) Neoproterozoic (~900 Ma) Sariwon sills in North Korea: geochronology, geochemistry and implications for the evolution of the south-eastern margin of the North China Craton. Gondwana Res 20:243–354

Peng P, Bleeker W, Ernst RE, Söderlund U, McNicoll V (2011b) U-Pb baddeleyite ages, distribution and geochemistry of 925 Ma mafic dykes and 900 Ma sills in the North China Craton: evidence for a Neoproterozoic mantle plume. Lithos 127:210–221

Peng P, Liu F, Zhai MG, Guo JH (2012) Age of the Miyun dyke swarm: constraints on the maximum depositional age of the Changcheng system. Chinese Sci Bull 57:105–110

Peng P, Wang X, Lai Y, Wang C, Windley BF (2015) Large-scale liquid immiscibility and fractional crystallization in the 1780 Ma Taihang dyke swarm: implications for genesis of the bimodal Xiong'er volcanic province. Lithos 136–137:106–122

Peng RM, Zhai YS, Wang JP, Liu Q (2014) The discovery of the neoproterozoic rift-related mafic volcanism in the northern margin of North China Craton: implications for Rodinia reconstruction and mineral exploration. In: International conference on continental dynamics, Xi'an, China

Prokoph A, Ernst RE, Buchan KL (2004) Time-series analysis of large igneous provinces: 3500 Ma to present. J Geol 11:1–22

Qiu JX, Liao QA (1998) The acidity, series, tectonic settings and magmatic origin of Middle Proterozoic and Mesozoic volcanic rocks from Beijing area. Acta Petrologica Et Mineralogica 17(2):104–117 (in Chinese with English abstract)

Rämö OT, Haapala I, Vaasjoki M, Yu JH, Fu HQ (1995) 1700 Ma Shachang complex, northeast China: Proterozoic rapakivi granite not associated with Palaeoproterozoic orogenic crust. Geology 23(9):815–818
Santosh M, Sajeev K, Li JH, Liu SJ, Itaya T (2009) Counterclockwise exhumation of a hot orogen: the Palaeoproterozoic ultrahigh-temperature granulites in the North China Craton. Lithos 110:40–152
Shandong Provincial No. 4 Institute of Geological and Mineral Survey (2003) Regional geology of Shandong Province. Shandong Map Press, Jinan. (in Chinese)
Stern RJ, Tsujimori T, Harlow G, Groa LA (2013) Plate tectonic gemstones. Geology 41:723–726
Su WB, Zhang SH, Huff WD, Li HK, Ettensohn FR, Chen XY, Yang HM, Han YG, Song B, Santosh M (2008) SHRIMP U-Pb ages of K-bentonite beds in the Xiamaling Formation: implications for revised subdivision of the Meso- to Neoproterozoic history of the North China Craton. Gondwana Res 14:543–553
Su WB, Li HK, Huff WD, Ettensohn FR, Zhang SH, Zhou HY, Wan YS (2010) SHRIMP U-Pb dating for a K-bentonite bed in the Tieling Formation, North China. Chin Sci Bull 55(29):3312–3323
Su WB, Li HK, Xu L, Jia SH, Gen JZ, Zhou HY, Wang ZH, Pu HY (2012) Luoyu and Ruyang Group at the south margin of the North China Craton (NCC) should belong in the Mesoproterozoic Changchengian System: direct constraints from the LA-MC-ICP MS U-Pb age of the tuffite in the Luoyukou Formation, Ruzhou, Henan China. Geol Surv Res 35(2):96–108 (in Chinese with English abstract)
Sun S, Zhang GW, Chen ZM (1985) Evolution of Precambrian crust in the southern north china fault block. Metallurgical Industry Press, Beijing (in Chinese)
Wan YS, Liu DY, Wang SJ, Yang EX, Wang W, Dong CY, Zhou HY, Du LL, Yang YH, Diwu CR (2011) ~2.7 Ga juvenile crust formation in the North China Craton (Taishan-Xintai area, western Shandong Province): further evidence of an understated event from U-Pb dating and Hf isotopic composition of zircon. Precambr Res 186:169–180
Wan YS, Zhang QD, Song TR (2003) SHRIMP ages of detrital zircons from the Changcheng System in the Ming Tombs area, Beijing: Constraints on the Protolith nature and maximum depositional age of the Mesoproterozoic cover of the North China Craton. Chin Sci Bull 48(22):2500–2506
Wang TH (1995) Evolutionary characteristics of geological structure and oil-gas accumulation in Shanxi-Shaanxi area. J Geol Mineral Resour North China 10(3):283–398 (in Chinese with English abstract)
Wang HZ, Li GC (1990) International stratigraphic time comparison chart. Geological Publishing House, Beijing (in Chinese)
Wang J, Wang BL, Xu CH, Liang YZ, Li JJ, Ma YP, Li SQ (1989) Comparison of sedimentary time between Zhaertai and Bayan Obo groups in inner Mongolia, and their ore-bearing potential. Inner Mongolian People's Publishing House, Hohhot (in Chinese)
Wang J, Li SQ, Wang BL (1992) Langshan-Bayan Obo rift system. Peking University Press, Beijing (in Chinese)
Wang KY, Fan HR, Xie YH, Li HM (2002) Zircon U-Pb dating of basement gneisses in the super-large Bayan Obo REE-Fe-Nb deposit, Inner Mongolia. Chin Sci Bull 47(3):245–248
Wang YJ, Fan WM, Zhang YH, Guo F, Zhang HF, Peng TP (2004) Geochemical, $^{40}Ar/^{39}Ar$ geochronological and Sr-Nd isotopic constraints on the origin of Palaeoproterozoic mafic dikes fromthe southern Taihang Mountains and implications for the ca. 1800 Ma event of the North China Craton. Precambr Res 135:55–77
Wang YJ, Zhao GC, Fan WM, Peng TP, Sun LH, Xia XP (2007) LA-ICP-MS U-Pb zircon geochronology and geochemistry of Palaeoproterozoic mafic dykes from western Shandong Province: implications for back-arc basinmagmatism in the Eastern Block, North China Craton. Precambr Res 154:107–124
Wang YJ, Zhao GC, Cawood PA, Fan WM, Peng TP, Sun LH (2008) Geochemistry of Palaeoproterozoic (~1770 Ma) mafic dikes from the Trans-North China Orogen and tectonic implications. J Asian Earth Sci 33:61–77

Wang XL, Jiang SY, Bai BZ (2010) Melting of enriched Archean subcontinental lithospheric mantle: evidence from the ca. 1760 Ma volcanic rocks of the Xiong'er Group, southern margin of the North China Craton. Precambr Res 182:204–216

Wang QH, Yang DB, Xu WL (2012) Neoproterozoic basic magmatism in the southeast margin of North China Craton: evidence from whole-rock geochemistry, U-Pb and Hf isotopic study of zircons from diabase swarms in the Xuzhou-Huaibei area of China. Sci China Earth Sci 55(9):1461–1479

Wang XL, Jiang SY, Dai BZ, Kern J (2013) Lithospheric thinning and reworking of Late Archean juvenile crust on the southern margin of the North China Craton: evidence from the Longwangzhuang Palaeoproterozoic A-type granites and their surrounding Cretaceous adakite-like granites. Geol J. https://doi.org/10.1002/gj.2464

Wang XL, Jiang SY, Dai BZ, Griffin WL, Dai MN, Yang YH (2011) Age, geochemistry and tectonic setting of the Neoproterozoic (ca. 830 Ma) gabbros on the southern margin of the North China Craton. Precambr Res 190(1–4):35–47

Weng JC, Li ZM, Yang ZQ, Li WZ (2006) Hydrothermally modified Pb-Zn deposit: a new deposit type in volcanic rocks of the Xiong'er Group, Henan, China. Geol Bull China 25(4):502–505 (in Chinese with English abstract)

Windley BF (1995) The evolving continents, 3rd edn. John Wiley and Sons, New York, pp 401–460

Xie GH (1980) Petrochemical characteristics of the anorthosite suit in Damiao, Hebei Province, China. Geochimica 9(3):263–277

Xie GH (2005) Petrology and geochemistry of Damiao anorthosite and Miyun rapakivi granite. The Science Publishing Company, Beijing (in Chinese)

Xing YS (1989) Upper Precambrian system in China: Chinese stratigraphy (3). Geological Publishing House, Beijing (in Chinese)

Xu YH, Zhao TP, Peng P, Zhai MG, Qi L, Luo Y (2007) Geochemical characteristics and geological significance of the Palaeoproterozoic volcanic rocks from the Xiaoliangling Formation in the Lvliang area, Shanxi Province. Acta Petrologica Sinica 23(5):1123–1132

Xu HR, Yang ZY, Peng P, Meert JG, Zhu RX (2014) Palaeo-position of the North China Craton within the supercontinent Columbia: constraints from new palaeomagnetic results. Precambr Res 255:276–293

Yang JH, Wu FY, Liu XM, Xie LW (2005) Zircon U-Pb ages and Hf isotopes and their geological significance of the Miyun rapakivi granites from Beijing, China. Acta Petrologica Sillica 21(6):1633–1644 (in Chinese with English abstract)

Yang KF, Fan HR, Santosh M, Hu FF, Wang KY (2011) Mesoproterozoic mafic and carbonatitic dykes from the northern margin of the North China Craton: implications for the final breakup of Columbia supercontinent. Tectonophysics 498:1–10

Yu JH, Fu HQ, Zhang FL, Wan FX (1996) Anorogenic Rapakivi Granites and related rocks in the Northern of North China Craton. China Science and Technology Press, Beijing (in Chinese)

Zhai MG (2008) Lower crust and lithospheric mantle beneath the North China Craton before the Mesozoic lithospheric disruption. Acta Petrologica Sinica 24(10):2185–2204 (in Chinese with English abstract)

Zhai MG (2011) Cratonization and the Ancient North China Continent: a summary and review. Sci China Earth Sci 54(8):1110–1120

Zhai MG, Liu WJ (2003) Palaeoproterozoic tectonic history of the North China Craton: a review. Precambr Res 122:183–199

Zhai MG, Peng P (2007) Palaeoproterozoic events in the North China Craton. Acta Petrologica Sinica 23(11):2665–2682 (in Chinese with English abstract)

Zhai MG, Santosh M (2011) The Early Precambrian odyssey of North China Craton: a synoptic overview. Gondwana Res 20(1):6–25

Zhai MG, Bian AG, Zhao TP (2000) The amalgamation of the supercontinent of North China Craton at the end of Neo-Archaean and its breakup during Late Palaeoproterozoic and Mesoproterozoic. Sci China (Ser D) 43:219–232

Zhai MG, Guo JH, Zhao TP (2001) Study advances of Neoarchaean-Paleoproterozoic tectonic evolution in the North China Craton. Prog Precambr Res 24(3):17–27 (in Chinese with English abstract)

Zhai MG, Yang JH, Fan HR, Miao LC, Li YG (2002) A Large-scale cluster of gold deposits and metallogenesis in the eastern North China Craton. Int Geol Rev 44:458–476

Zhai MG, Fan QC, Zhang HF, Sui JL (2007) Lower crustal processes leading to Mesozoic lithospheric thinning beneath Eastern North China: underplating, replacement and delamination. Lithos 96:36–54

Zhai MG, Li TS, Peng P, Hu B, Liu F, Zhang YB (2010) Precambrian key tectonic events and evolution of the North China Craton. Geol Soc Lond Spec Publ 338:235–262

Zhai MG, Hu B, Peng P, Zhao TP (2014) Meso-Neoproterozoic magmatic events and multi-stage rifting in the NCC. Earth Sci Front 21(1):100–119 (in Chinese with English abstract)

Zhai MG, Hu B, Zhao TP, Peng P, Meng QR (2015) Late Palaeoproterozoic-Neoproterozoic multi-rifting events in the North China Craton and their geological significance: a study advance and review. Tectonophysics 662:153–166

Zhai MG (2004) 2.1–1.7 Ga geological event group and its geotectonic significance. Acta Petrologica Sinica 20(6):1343–1354. (in Chinese with English abstract)

Zhang SH, Liu SW, Zhao Y, Yang JH, Song B, Liu XM (2007) The 1.75–1.68 Ga anorthosite-mangerite-alkali granitoid-rapakivi granite suite from the northern North China Craton: magmatism related to a Palaeoproterozoic orogen. Precambr Res 155:287–312

Zhang SH, Zhao Y, Yang ZY, He ZF, Wu H (2009) The 1.35 Ga diabase sills from the northern North China Craton: implications for breakup of the Columbia (Nuna) supercontinent. Earth Planet Sci Lett 288:588–600

Zhang SH, Li ZX, Evans DAD, Wu HC, Li HY, Dong J (2012) Pre-Rodinia supercontinent Nuna shaping up: a global synthesis with new palaeomagnetic results from North China. Earth Planet Sci Lett 353–354:145–155

Zhang SH, Zhao Y, Ye H, Hu G (2016) Early Neoproterozoic emplacement of the diabase sill swarms in the Liaodong Peninsula and pre-magmatic uplift of the southeastern North China Craton. Precambr Res. https://doi.org/10.1016/j.precamres.2015.11.005

Zhang SH, Zhao Y, Li XH, Ernst RE, Yang ZY (2017a) The 1.33–1.30 Ga Yanliao large igneous province in the North China Craton: implications for reconstruction of the Nuna (Columbia) supercontinent, and specifically with the North Australian Craton. Earth Planet Sci Lett 465:112–125

Zhang SH, Zhao Y, Liu YS (2017b) A precise zircon Th-Pb age of carbonatite sills from the world's largest Bayan Obo deposit: Implications for timing and genesis of REE-Nb mineralization. Precambr Res 291:202–219

Zhang SH, Li ZX, Wu H, Wang H (2000) New palaeomagnetic results from the Neoproterozoic successions in southern North China block and palaeogeographic implications. Sci China (Series D) 43(Supp):234–244

Zhao ZP (1993) Precambrian crustal evolution of the Sino-Korean paraplatform. Science Press, Beijing (in Chinese)

Zhao GC, Wilde SA, Cawood PA, Sun M (2001) Archaean blocks and their boundaries in the North China Craton: lithological, geochemical, structural and P-T path constraints and tectonic evolution. Precambr Res 107:45–73

Zhao TP, Jin CW, Zhai MG, Xia B, Zhou MF (2002a) Geochemistry and petrogennesis of the Xiong'er Group in the southern regions of the North China Craton. Acta Petrologica Sinica 18(1):59–69 (in Chinese with English abstract)

Zhao TP, Zhou MF, Zhai MG, Xia B (2002b) Palaeoproterozoic rift-related volcanism of the Xiong'er Group in the North China Craton: implications for the break-up of Columbia. Int Geol Rev 44:336–351

Zhao TP, Zhai MG, Xia B, Li HM, Zhang YG, Wan YS (2004a) Zircon U-Pb SHRIMP dating for the volcanic rocks of the Xiong'er Group: constraints on the initial formation age of the cover of the North China Craton. Chin Sci Bull 49(23):2495–2502

Zhao TP, Chen FK, Zhai MG, Xia B (2004b) Single zircon U-Pb ages and their geological significance of the Damiao anorthosite complex, Hebei Province, China. Acta Petrologica Sinica 20(3):685–690 (in Chinese with English abstract)

Zhao TP, Xu YH, Zhai MG (2007) Petrogenesis and tectonic setting of the Palaeoproterozoic Xiong'er Group in the southern part of the North China Craton: a review. Geol J China Univ 13(2):191–206 (in Chinese with English abstract)

Zhao GC, He YH, Sun M (2009a) The Xiong'er volcanic belt at the southern margin of the North China Craton: Petrographic and geochemical evidence for its outboard position in the Palaeo-Mesoproterozoic Columbia Supercontinent. Gondwana Res 16:170–181

Zhao TP, Chen W, Zhou MF (2009b) Geochemical and Nd-Hf isotopic constraints on the origin of the ~1.74 Ga Damiao anorthosite complex, North China Craton. Lithos 113:673–690

Zhao TP, Wang JP, Zhang ZH, Wang FY, Zhang BC, Yuan ZL, Wang SY, Xu YH, Chen W (2005) Proterozoic Geology of Mt. Wangwushan and Adjacent Areas, China. China Land Press, Beijing (in Chinese)

Zhou JB, Hu K (1998) Tectonic activities of the Yihe-Shuhe Fault during the Jinning Movement period. Seismol Geol 20(3):208–212 (in Chinese with English abstract)

Chapter 9
Neoproterozoic Magmatism and Tectonic Evolution in South China

Li Xianhua and Li Wuxian

Abstract Neoproterozoic magmatic rocks are widespread in South China, which can be divided into three major groups based on their formation ages and their genetic relationships with regional tectonics, metamorphism and basin evolution: ① Group 1 formed during the syn-orogenic process (1.0–0.9 Ga); ② Group 2 formed in the early rift phase (0.85–0.80 Ga); and ③ Group 3 formed in the major rift phase (0.79–0.75 Ga). Despite numerous studies during the past two decades, the tectonic settings responsible for the Neoproterozoic magmatism, in particular the intensive granitoid plutonism in South China is still an issue of hot debate. It is noteworthy that the geochemical characteristics of granitoid rocks are reflective of their source compositions as well as the melting and crystallization histories of the melts, thus could not be simplistically used to assess the tectonic regime under which the granitoids were formed. On the other hand, geochemistry of basaltic rocks (including basalts and basaltic dikes) reflects the compositions and thermal structures of their mantle sources that are relevant to different petrotectonic associations. In this chapter we compile the published high-precision isotopic ages and geochemical and isotopic data for the Neoproterozoic basaltic rocks and the pre-Neoproterozoic crystalline basement rocks from South China. These datasets, combined with other aspects of geological records, are used to investigate the petrogenesis of the basaltic rocks, as well as their mantle source compositions and potential temperatures, aiming to shed new lights on the Neoproterozoic petrotectonic evolution in South China. The Group 1 syn-orogenic basaltic rocks occur sporadically along the Yangtze Craton margins, including the ca. 0.9 Ga Yanbian basalts and the ca. 1.0 Ga Huili mafic dykes in western margin, the ca. 0.95–0.89 Ga Xixiang basalts in northwestern margin and the ca. 0.96 Ga Pingshui spilites in southeastern margin. They are closely associated with the calc-alkalinic intermediate-acidic volcanic/granitic rocks; all of the rocks are clearly deformed and metamorphoses to varying degrees. The Xixiang basalts and Pingshui basalts are calc-alkaline in compositions with geochemical characteristics

L. Xianhua (✉)
State Key Laboratory of Lithospheric Evolution, Institute of Geology and Geophysics, Chinese Academy of Sciences, Beijing 100029, China

L. Wuxian
State Key Laboratory of Isotope Geochemistry, Guangzhou Institute of Geochemistry, Chinese Academy of Sciences, Guangzhou 510640, China

T. Wang, *Meso-Neoproterozoic Geology and Petroleum Resources in China*,
Springer Geology, https://doi.org/10.1007/978-981-19-5666-9_9

similar to those of island arc basalts (IAB). In Yanbian area, both tholeiitic and calc-alkaline basalts are exposed, and their geochemical features are similar to those of the back-arc basin basalts (BABB). Overall, the Group 1 basaltic rocks surrounding the Yangtze Craton are likely formed in active continental margins. The Group 2 basaltic rocks formed during the early rift phase consist mainly of the tholeiitic and alkali basalts as well as dolerites around the Yangtze Craton margins, including the ca. 0.85 Ga Zhengzhushan bimodal volcanic rocks and the Shenwu dolerites in south-eastern margin, ca. 0.83–0.82 Ga Yiyang komatiitic basalts, the Guangfeng alkaline basalts and the Yingyangguan spilites in southern margin of the Yangtze Craton, and ca. 0.82–0.80 Ga Suxiong alkaline basalts, Bikou and Tiechuanshan tholeiitic basalts along the western to northwestern margin of the Yangtze Craton. In the Cathaysia Block, the Mamianshan bimodal volcanic rocks (alkaline basalts and rhyolites) in northwestern Fujian are also dated at ca. 0.82 Ga. The majority of the ca. 0.83–0.82 Ga bimodal volcanic rocks occurred at the bottom of the volcano-sedimentary sequences in a number of mid-Neoproterozoic rift basins. The geochemical compositions of the Group 2 basaltic rocks are either similar to those of typical oceanic island basalts (OIB), or transitional between OIB and IAB due to various degrees of contamination of the lithospheric mantle and/or continental crust materials. The Group 3 basaltic rocks formed in the major rift phase include the ca. 0.79 Ga Shangsu basalts and Daolinshan dolerites in southeastern margin of the Yangtze Craton, the ca. 0.77 Ga spilites in northern Guangxi and dolerites in western Hunan and the numerous 0.78–0.75 Ga mafic dykes occurring in the Kangdian rift in western margin of the Yangtze Craton. All these basaltic rocks belong to tholeiitic series and alkaline series, and their geochemical features are similar to those of OIB. The calculated mantle potential temperatures (T_p) of the Group 1 basaltic rocks and the ca. 0.85 Ga Shenwu dolerites (Group 2) range from 1355 to 1420 °C, which are consistent with those of the Neoproterozoic mid-oceanic-ridge basalt (MORB)-source mantle (ca. 1350–1450 °C). On the contrary, the ca. 0.82 Ga Yiyang komatiitic basalts were generated by melting of an anomalously hot mantle source with T_p up to ca. 1618 °C, ca. 260 °C higher than the MORB-source mantle, suggesting derivation from an abnormally hot mantle plume. The T_p for the other Group 2 and 3 basaltic rocks are ca. 25–140 °C higher than that of the MORB-sourced mantle, indicating various degrees of contribution from hot mantle plumes. The changes of rock types, geochemical compositions and the mantle potential temperatures are attributed to the regional tectonic transformation of the South China Block (SCB) from ca. 1.0–0.9 Ga Sibao Orogeny to ca. 0.85–0.75 Ga intracontinental rifting. Mantle plume/superplume activities play important roles in the mid-Neoproterozoic rift magmatism. Our interpretations of the geochemical results of basaltic rocks are consistent with other aspects of geological records. The Yangtze Cratob and Cathaysia block eventually amalgamated at ca. 0.9 Ga, forming a united South China continent that links Australia–East Antarctica and Laurentia in the Rodinia supercontinent. Termination of the Sibao Orogeny indicates that the Rodinia might have eventually assembled at ca. 0.9 Ga. The mid-Neoprterozoic (0.83–0.75 Ga) mantle plume/superplume activities resulted in the formation of widespread anorogenic magmatism and rift basins in South China.

Keywords South China · Neoproterozoic magmatism · Basaltic rocks · Orogeny · Rift baisn · Rodinia supercontinent

9.1 Introduction

There is a general consensus that the South China Block (SCB) was formed through amalgamation of the Yangtze Craton and Cathaysia Block, but the timing of amalgamation remains disputed. In the past decades, a number of studies demonstrated that the formation and evolution of the SCB is closely associated with assembly and break-up of Rodinia supercontinent. Therefore, the formation and evolution of the SCB is important for understanding the evolution of Rodinia supercontinent more than an issue of regional geology. In this chapter we integrate the research results for the pre-Neoproterozoic crystalline basements and magmatic rocks from the SCB including the isotopic ages, geochemical compositions and genesis of the basaltic rocks as well as the mantle compositions and potential temperatures to investigate the regional Neoproterozoic petroteconic associations and provide new constraints on the formation and evolution of the SCB and their relevance to the assembly and break-up of supercontinent Rodinia.

9.2 Pre-Neoproterozoic Crystalline Basement Rocks in South China

9.2.1 Yangtze Craton

The pre-Neoproterozoic crystalline basement rocks are sporadically exposed in the Yangtze Craton, including the Palaeoarchean-Palaeoproterozoic Kongling Complex in the Yangtze Gorge area, the Neoarchean Douling Complex in South Qinling, the Palaeoproterozoic Houhe Complex in northwestern Yangtze Craton, the Late Palaeoproterozoic to Mesoproterozoic metavolcanic-sedimentary rocks in southwestern Yangtze Craton, the Mesoproterozoic Tianli schists in southeastern Yangtze Craton, and minor amount of Late Mesoproterozoic bimodal volcanic rocks in southern and southwestern Yangtze Craton (Fig. 9.1).

The oldest crystalline basement rocks in the Yangtze Craton are the Kongling Complex near the Yangtze Gorge, consisting of Archean high-grade metamorphic TTG (tonalite, trondhjemite and granodiorite) gneisses, metasedimentary rocks and amphibolites (with minor granulitic mafic rocks) and Palaeoproterozoic granites. The majority of TTG gneisses in the Kongling Complex have been dated at 2.90–2.95 Ga (Qiu et al. 2000). The oldest rocks ever reported in the Yangtze Craton are trondhjemite gneisess which occur in minor amount and were dated at at 3.22–3.45 Ga (Jiao et al. 2009; Gao et al. 2011; Guo et al. 2014). All these TTG rocks were experienced high-grade metamorphism at ca. 2.0–1.9 Ga, which is consistent with

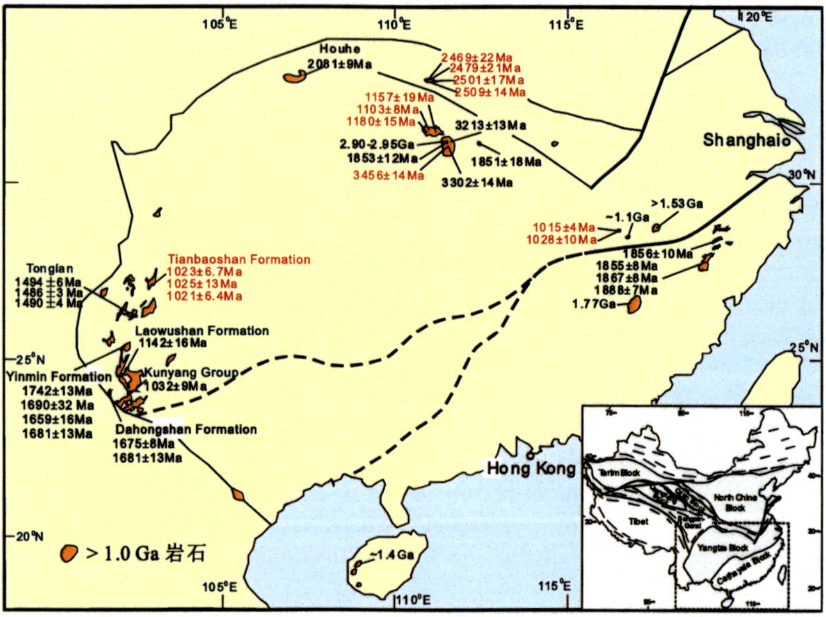

Fig. 9.1 Distributions of the pre-Neoproterozoic crystalline basement rocks in South China. *Data sources* Chen J et al. (1991), Jin and Sun (1997), Li X. H. (1997), Qiu et al. (2000), Liu et al. (2001), Li Z. X. et al. (2002, 2007, 2008b, 2010), Greentree (2006), Ye et al. (2007), Zhang C. H. et al. (2007), Greentree and Li (2008), Shu et al. (2008), Xiong et al. (2008), Chen Z. H. et al. (2009), Jiao et al. (2009), Li X. H. et al. (2009), Gao et al. (2011), Zhang L. J. et al. (2011), Wu et al. (2012), Yang et al. (2009), Yu et al. (2009), Peng et al. (2012), Zhu et al. (2016), Deng et al. (2017)

the timing of the granulite facies metamorphism at Huangtuling in Dabie Mountain (Sun and Zhou 2008; Wu et al. 2008, 2012). This regional high-grade metamorphism is interpreted as a Palaeoproterozic orogenic event that might be related to assembly of supercontinent Columbia/Nuna. The Quanyitang A-type granites intrusive into the Kongling Complex was dated at ca. 1.85 Ga (Xiong et al. 2008), which is coeval with the Huashanguan rapakivi granite in the northern margin of the Yangtze Craton (Zhang et al. 2011). These rocks are most likely the products of late- to post-orogenic magmatism formed during the waning stage of the Palaeoproterozoic orogeny in the Yangtze Craton.

The Douling Complex occurs in the northern Yangtze Craton with an exposed area of ca. 500 km^2. It is mainly composed of dioritic-granitic gneisses, amphibolites, schists, marbles, and quartzites that were intruded by large amount of Neoproterozoic granitoid rocks. U–Pb zircon age dating results suggest that dioritic-granitic gneisses were emplaced at ca. 2.5 Ga (Hu et al. 2013; Wu et al. 2014).

The Houhe Complex is located in the northwestern margin of the Yangtze Craton, consisting of tonalitic gneisses and minor amounts of amphibolites and marbles that experienced upper amphibolite facies metamorphism and migmatization. Zircon U–Pb dating results suggest that the tonalitic gneisses were emplaced at ca. 2.08 Ga, representing the timing for the formation of the Houhe Complex.

The pre-Neoproterozoic metamorphic volcano-sedimentary rocks are widely exposed in the southwestern Yangtze Craton, including the Late Palaeoproterozoic Dahongshan Group (and the equivalents such as the Hekou Group, the Dongchuan Group and the Lower Kunyang Group in different localities) and the Late Mesoproterozoic to Early Neoproterozoic Huili Group (and the equivalents such as the Upper Kunyang Group and the Yanbian Group). The Dahongshan Group and its equivalents are strongly deformed and experienced upper greenschist to lower amphibolite facies metamorphism. The volcanic rocks interlayered within the metasedimeary rocks are dated at ca. 1.74–1.66 Ga (Greentree and Li 2008; Zhao et al. 2010; Zhao and Zhou 2011), which are currently known as the oldest basement rocks in the southwestern Yangtze Craton. Metamorphic volcanic rocks from both the Upper Huili Group and the Upper Kunyang Group are dated synchronolously at ca. 1.03–1.00 Ga (Geng et al. 2007; Greentree et al. 2006; Zhang et al. 2007). In the Tong'an area, a large number of ca. 1.5 Ga gabbros and dikes were intruded into the shale and dolomite of the Heishan Formation of the Lower Huili Group (Fan et al. 2013), providing a minimum depositional age of ca. 1.5 Ga for the Lower Huili Group.

The Tianli schist is the only known Late Mesoproterozoic upper-greenschist facies metamorphic rock in the southern margin of the Yangtze Craton. As the basement of the Guangfeng rift basin, the Tianli schist was unconformably overlain by the ca. 825 Ma undeformed and unmetamorphed volcano-sedimentary sequences in the rift basin (Li W. X. et al. 2008b). The Tianli schist experienced two stages of deformation. In situ UV laser $^{40}Ar/^{39}Ar$ results from S1 and S2 muscovites suggest that the Tianli schists underwent metamorphism and deformation at ca. 1.04–1.02 Ga. Muscovite/biotite cooling ages of ca. 0.97–0.94 Ga are recorded by deformed and recrystallised muscovite and biotite, respectively, indicating tectonic reactivation before 0.9 Ga (Li Z. X. et al. 2007). The youngest U–Pb age of detrital zircons from the Tianli schist is 1.53 Ga. Therefore, deposition age for the protoliths of Tianli schist can be bracketed between 1.53 and 1.04 Ga (Li Z. X. et al. 2007).

In the southern Yangtze Craton, Li L. M. et al. (2013) recently reported that the Tieshanjie bimodal volcanic rocks in the Yiyang, Jiangxi Province were dated at ca. 1.16 Ga, coincident with the ca. 1.14 Ga bimodal volcanic rock of the Laowushan Group in Yunnan Province (Greentree et al. 2006; Zhang et al. 2007). The basalts within the bimodal volcanic rock are typically alkaline basalt formed in a rift basin. In the Shennongjia region in the northern Yangtze Craton, the calc-alkaline and alkaline volcanic rocks were dated at ca. 1.1 Ga, and interpreted as the products formed in an active continental margin related to collision between the Yangtze and the Australian cratons during the Grenville Orogeny (Qiu et al. 2011).

9.2.2 Cathaysia Block

The pre-Neoproterozoic crystalline basement rocks of the Cathaysia Block include the Badu Complex in the southwestern Zhejiang, the Chencai Complex in northern Zhejiang, and the Tianjinping amphibolite in northwestern Fujian as well as the Baoban Complex in the northwestern Hainan Island (Fig. 9.1).

The Badu Complex is composed mainly of metasedimentary rocks, amphibolite, migmatites and gneissic granite that intruded into the metasedimentary rocks. The detrital zircons from the metasedimentary rocks are mainly dated at ca. 2.5 Ga, with overgrowths during two phases of high-grade metamorphism at ca. 1.88 Ga and ca. 0.26–0.23 Ga (Xiang et al. 2008; Yu et al. 2012). The gneissic granites were dated at ca. 1.89–1.83 Ga (Liu et al. 2009; Yu et al. 2009; Li Z. X. et al.2010; Xia et al. 2012). These granites are the oldest crystalline rocks recognized in the Cathaysia Block, placing a minimum depositional age of ca. 1.89 Ga for the Badu metasedimentary rocks. Therefore, the Badu metasedimentary rocks should have been deposited between ca. 2.5 and ca. 1.9 Ga.

The crystallization ages of the amphibolite protoliths from the Tianjinping and Chencai are dated at ca. 1.78–1.77 Ga (Li Z. X. et al. 2010). The Tianjinping amphibolites have geochemical features similar to those of OIB and E-MORB, with initial $\varepsilon_{\mathrm{Nd}}(t)$ values are up to + 8.5 that is higher than the global depleted mantle value of + 5 at ca. 1.8 Ga. Thus, they were most likely derived from an extremely-depleted mantle source in an intracontinental rift setting (Li X. H. et al. 2000). Overall, the Cathaysia Block recorded the 1.89–1.88 Ga syn-orogenic S-type granites, 1.87–1.83 Ga late- to post-orogenic A-type granites and 1.78–1.77 Ga intracontinental rift-related basaltic magmatism.

The Baoban Complex in the northwestern Hainan Island is the oldest crystalline basement exposed in the southwestern Cathaysia Block, consisting of the amphibolite-facies metamorphic gneissic granites, sedimentary and volcanic rocks. SHRIMP zircon U–Pb results suggest that the gneissic granites and metamorphic volcanic rocks were emplaced at ca. 1.43 Ga (Li Z. X. et al. 2002, 2008b; Yao et al. 2017). The adjacent Shilu Group is also dated at ca. 1.43 Ga; they are coeval with the Baoban volcanic-sedimentary rocks rather than formed in Neoproterozoic as previously thought. The Baoban Complex experienced the Grenville-aged metamorphism during 1.3–1.0 Ga (Li Z. X. et al. 2002). Detrital zircon U–Pb age analysis results permist us to interpret the quartzite and quartz schist of Shihuiding Formation as locally-sourced Grenvillian foreland basin deposits at 1.2–1.0 Ga that unconformably overlies the ca. 1.43 Ga Shilu Group (Li Z. X. et al. 2008b; Yao et al. 2017).

9.3 Temporal and Spatial Distribution of Neoproterozoic Magmatic Rocks

In contrast to sporadic distribution of the pre-Neoproterozoic magmatic rocks, the Neoproterozoic intrusive rocks and volcanic-sedimentary rocks are widespread in the SCB (Fig. 9.2), with granitoid rocks covering a large area of the Yangtze Craton. These Neoproterozoic magmatic rocks provide important petrological records for the Neoproterozoic petrotectonic evolution of the SCB.

In this chapter, we compiled high quality isotopic age data published in the past decades, and use them to discuss the temporal and spatial distribution of the Neoproterozoic magmatic rocks in the Yangtze Craton and Cathaysia Block.

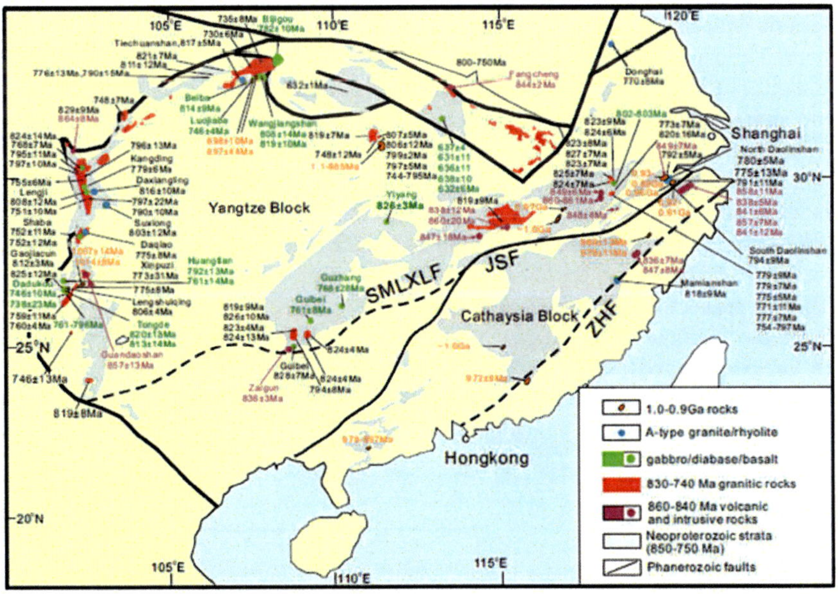

Fig. 9.2 Distribution of the Neoproterozoic intrusive and volcano-sedimentary rocks in South China. *Data source* Li X. H. (1999); Li X. H. et al. (2003a, 2003b, 2006, 2008a, 2009, 2010), Li Z. X. et al. (1999, 2003, 2010), Sinclair (2001), Shen et al. (2002), Zhou M. F. et al. (2002a, 2002b, 2006), Ling et al. (2003, 2006, 2007), Shu et al. (2006, 2011), Wu et al. (2006), Wang X. C. et al. (2007, 2008), Zhou et al. (2007), Bao et al. (2008), Huang et al. (2008, 2009), Li W. X. et al. (2008a, 2008b, 2010), Sun M. et al. (2008a); Sun W. H. et al (2008b), Wang X. L. et al. (2008, 2012), Zhao and Zhou (2008), Zhao X. F. et al. (2008), Zheng Y. F. et al. (2008b), Zhu et al. (2008, 2010), Xia et al. (2009), Zhang et al. (2009), Dong et al. (2010), Gao et al. (2010), Wang (2010), Dong et al. (2011, 2012), Xue et al. (2011), Zhang A. M. et al. (2012), Liu et al. (2013), Wang Y. J. et al. (2013)

9.3.1 Yangtze Craton

The Neoproterozoic magmatic rocks in the Yangtze Craton are dated between the Early Neoproterozoic (1.0–0.9 Ga) and late mid-Neoproterozoic (ca. 0.63 Ga), with a sugnificant magmatic flare-up being concentrated around 0.83–0.75 Ga (Fig. 9.3).

9.3.1.1 Early Neoproterozoic Magmatic Rocks

Early Neoproterozoic magmatic rocks crop out sporadically around the Yangtze Craton. In the southeastern Yangtze Craton, magmtic rocks are distributed along the northern side of the Jiangshao Fault in the Pingshui and Shuangxiwu areas of Zhejiang Province. They are dominantly calc-alkaline volcanic rocks (including basaltic andesite, andesite, dacite and rhyolite) and tonalite and granodiorite intrusions. The volcanic rocks experienced greenschist facies metamorphism and deformation. The Pingshui Formation volcanic rocks consist mainly of basaltic andesites and andesites dated at ca. 0.96 Ga (Chen et al. 2009; Li X. H. et al. 2009). These volcanic rocks were intruded by ca. 0.92–0.91 Ga tonalite and granodiorite (Ye et al. 2007). The Shuangxiwu Group volcanic rocks include, from bottom to top, the andesite and dacite of the Beiwu Formation, the pyroclastic rocks of the Yanshan Formation, and the dacite and rhyolite of the Zhangcun Formation. The volcanic rocks of the Beiwu and the Zhangcun Formations are dated at ca. 0.93 Ga and ca. 0.89 Ga, respectively (Li X. H. et al. 2009).

In the southwestern Yangtze Craton two earliest Neoproterozoic (Sibaoran or Greenville-aged) syn-tectonic granitoid intrusions are recognaized, i.e., the Huijinggou gneissic granite in the northeastern Panzhihua (Li Z. X. et al. 2002) and the deformed Yakou monzogranite near the Miyi County (Yang et al. 2009). They both are dated at ca. 1.0 Ga.

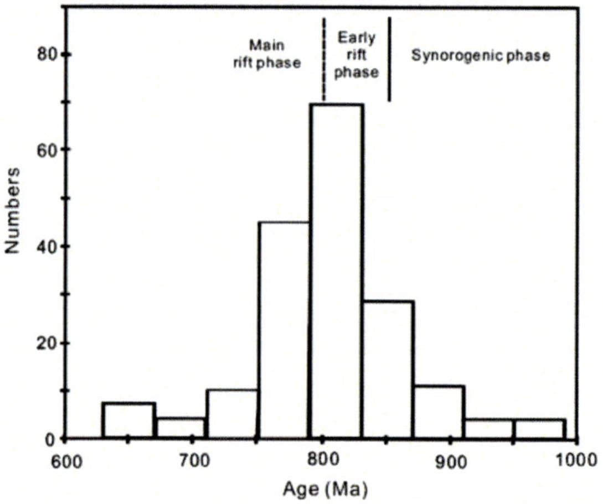

Fig. 9.3 Histogram of isotopic ages for the Neoproterozoic igneous rocks in South China

In the Hannan-Micangshan area of northern Yangtze Craton, the Early Neoproterozoic magmatic rocks are reported, including the ca. 0.9 Ga Liushudian gabbro (Dong et al. 2011; Zhou et al. 2018) and the ca. 0.9 Ga Guangwushan granite (Dong et al. 2012) as well as the similar-aged Xixiang Group meta-volcanic rocks (Ling et al. 2003). The Xixiang Group consists of ca. 0.95 Ga low-Ti tholeiites and high-Mg andesites in the lower part and ca. 0.89 Ga calc-alkaline basalt-andesite-dacite-rhyolite in the upper part (Ling et al. 2003), with the former being interpreted as a suite of fore-arc boninite association, while the latter magmatic association formed in an active continental margin.

9.3.1.2 Middle Neoproterozoic Magmatic Rocks

The Middle Neoproterozoic magmatic rocks are widespread in the Yangtze Craton. According to the isotopic age data, these magmatic rocks can be subdivided into four phases of 860–840 Ma, 830–800 Ma, 790–730 Ma and 700–630 Ma, respectively.

The first phase magmatic rocks in age of 860–840 Ma are relatively sparse in occurrence, and sporadically exposed along the Yangtze Craton margins. In the southeastern margin this phase of magmatism includes ca. 0.85 Ga Shenwu dolerites that intrude into Shuangxiwu Group (Li X. H. et al. 2008a, 2008b), ca. 0.85 Ga Ganbian alkaline complex in northeastern Jiangxi (Li X. H. et al. 2010), ca. 0.85 Ga Zhengzhushan bimodal volcanic rocks in Dexing (Li W. X. et al. 2010), ca. 0.86 Ga metamorphosed basalt-dacite-rhyolite suite in Wuyuan-Dexing (Liu et al. 2013), ca. 0.84 Ga spilite-keratophyre suite in Lushan (Dong et al. 2010), and ca. 0.85 Ga gabbro in northern Guangxi (Yao et al. 2014). In the Kangdian region of western Yangtze Craton, the ca. 0.86 Ga Gezong granite (Zhou M. F. et al. 2002b) and ca. 0.85 Ga Qiaotou gabbro (Shen et al. 2002) are reported. In addition, there are a small number of 0.86–0.84 Ga magmatic rocks occuring in the northern Yangtze Craton, including Tianpinghe pluton in South Hanzhong (Ling et al. 2006), several diorite and granodiorite intrusions in the Micangshan region (Dong et al. 2012), as well as the Fangcheng syenite in the eastern Qinling (Bao et al. 2008).

The second phase magmatic rocks dated at 830–800 Ma are widespread in the SCB, including large-scale granitoid batholiths, continental flood basalts (CFB) and extensive bimodal and/or felsic volcanic rocks. The granitoid rocks dated at ca. 830–820 Ma are mostly emplaced along the southern margin of Yangtze Craton, including the Xiuning granites in southern Anhui, the Jiuling granites in northern Jiangxi, the Bendong, Yuanbaoshan, Sanfang (Motianling) and Zhaigun granites in northern Guangxi, and the Ebian granite in Yunnan (Li X. H. 1999; Li X. H. et al. 2003b; Wu et al. 2006; Wang X. L. et al. 2006; Zheng Y. F. et al. 2008a). In the northern interior of Yangtze Craton, the Huangling granites dated at ca. 820 Ma are exposed near the Yangtze Gorge (Ma et al. 1989; Zhang et al. 2009). The ca. 830–820 Ma Xihe and Beiba granites cropped out in Micangshan area of northern Yangtze Craton (Dong et al. 2012). The same-aged basalts and mafic intrusions are also widespread in the Yangtze Craton, despite being volumetrically minor. This phase of basaltic rocks include the Xucun composite dikes in southern Anhui (Wang X. L. et al. 2012), the bimodal volcanic rocks in the Guangfeng Basin of northeastern Jiangxi (Li W. X. et al. 2008a), the Yiyang komatiitic basalts in Hunan (Wang X. C. et al. 2007; Wu et al. 2018), mafic dikes/sills in the northern Guangxi (Li Z. X. et al. 1999), the Fangjingshan basaltic lavas in Guizhou (Zhou J. C. et al. 2009), the Gaojiacun and Lengshuiqing maifc and ultra-mafic intrusions as well as the Tongde and the Lengji gabbros in the western Yangtze Craton (Sinclair 2001; Li X. H. et al. 2002; Zhu et al. 2007). Rare coeval intermediate volcanic rocks are exposed in Cangshuipu of Hunan Province (Wang J. et al. 2003; Zhang Y. Z. et al. 2013). In the northern Yangtze Craton, this phase of basaltic rocks includes the Wangjiangshan and Xijiaba gabbros in the Hannan region, (Zhou M. F. et al. 2002a; Dong et al. 2011), the extensive Bikou continental flood basalts (Wang X. C. et al. 2008), and the Tiechuanshan and Sunjiahe bimodal volcanic rocks in the Hannan region (Ling et al. 2003; Xia et al. 2009).

The third phase magmatic rocks in age of 790–730 Ma are also widespread throughout the SCB. This phase of granitoid intrusions includes the Daolinshan and Shiershan granites in the southeastern Yangtze Craton (Li X. H. et al. 2008a, 2008b; Wang et al. 2010; Li Z. X. et al 2003; Wang X. L. et al. 2012), numerous granitoid intrusions occurring along the NS-trending Kangdian rift (Zhou et al. 2002b; Li et al. 2003; Zhao and Zhou 2007a, 2007b; Huang et al. 2008; Zhao J. H. et al. 2008; Zhao X. F. et al. 2008; Huang et al. 2009), the granitoid rocks within the Hannan Complex in the northwestern Yangtze Craton (Ling et al. 2006) as well as the protoliths of the ultra-high pressure metamorphic orthogneisses in the Dabie-Sulu Belt (Zheng et al. 2009b). The coeval volcanic rocks as well as the mafic intrusions include the Shangshu and Puling bimodal volcanic rocks in northern Zhejiang and southern Anhui, respectively (Li X. H. et al. 2008a, 2008b; Wang X. L. et al. 2012), the dolerites in western Sichuan (Li Z. X. et al. 2003; Zhu et al. 2008, 2010; Lin et al. 2007), the mafic and ultra-mafic intrusions at Xijiaba, Youshui, Wangjianshan and Bijigou, which are associated with same-aged granites in Hannan region of northern Yangtze Craton (Zhou et al. 2002b; Dong et al. 2011; Zhao and Zhou 2008, 2009), mafic dikes in western Hunan (Zhou et al. 2007), the Xiaofeng composite dikes in the Huangling region (Li Z. X. et al. 2004), and the Wudangshan Group volcanic rocks in South Qinling (Ling et al. 2007).

The fourth phase of magmatic rocks in age of 700–630 Ma is mainly confined to the northern margin of the Yangtze Craton, including the Xijiaba and Xixiang granites (Dong et al. 2012), the volcanic rocks and mafic intrusions of Yaolinghe Group (Ling et al. 2007), the mafic and ultra-mafic dikes in the South Qinling (Xue et al. 2011), and the Zhou'an ultra-mafic rocks in southern Qinling (Wang M. X. et al. 2013).

9.3.2 *Cathaysia Block*

Neoproterozoic magmatic rocks are volumetrically minor in the Cathaysia Blcok possibly due to the widespread Phanerozoic sedimentary cover and magmatic overpriting. They were mainly formed in the Early Neoproterozoic (ca. 1.0–0.9 Ga) and Middle Neoproterozoic (0.85–0.72 Ga).

9.3.2.1 Early Neoproterozoic Magmatic Rocks

Early Neoproterozoic metamorphosed mafic and felsic rocks are sporadically exposed in the Cathaysia Block, including the ca. 1.0–0.98 Ga amphibolites in the Yunkai Mountain (Zhang A. M. et al. 2012; Wang Y. J. et al. 2013), ca. 0.98–0.97 Ga amphibolites and diabases in the Wuyi Mountain (Wang Y. J. et al. 2013), ca. 1.0 Ga Hezi gneissic granites in southern Jiangxi (Liu et al. 2001), and the ca. 0.97 Ga Jingnan rhyolites in the eastern Guangdong (Shu et al. 2008).

9.3.2.2 Mid-Neoproterozoic Magmatic Rocks

There are a number of ca. 0.86–0.84 Ga mafic intrusions cropped out in the eastern Cathaysia Block, including several small gabbro intrusions in northern Zhejiang (Shu et al. 2006, 2011; Li Z. X. et al. 2010) and two gneissic gabbros in Zhenghe of northwestern Fujian (Shu et al. 2011). In addition, mid-Neoproterozoic basalts and bimodal igneous rocks are also recognized in Chencai and Shaoxing of northern Zhejiang (Li Z. X. et al. 2010). The Chencai and Mayuan Complexes were previously considered as the Palaeoproterozoic metamorphic basement rocks, and the Mamianshan Group as the Mesoproterozoic sequence. However, recent zircon U–Pb analyses indicate the Chencai and Mayuan Complexes were formed during ca. 0.84–0.72 Ga (Wan et al. 2007; Li Z. X. et al. 2010), and subjected to amphibolite facies metamorphism in the Early Palaeozoic. The Mamianshan bimodal volcanic rocks were dated at ca. 0.82 Ga (Li W. X. et al. 2005). More recently, Yao et al. (2012) reported a zircon U–Pb age of ca. 0.83 Ga for a granidiorite pluton in Zhuji of northern Zhejiang.

A geochronological framework is presented in Table 9.1 based on high-quality isotopic age data for the Precambrian rocks from the Yangtze Craton and Cathaysia Block. As showed in the table, the two blocks archive different magmatic records before per-Neoproterozoic, but share similar magmatic activities and rock associations since ca. 0.85 Ga of mid-Neoproterozoic.

Table 9.1 Geochronological framework of Precambrian magmatism in the Yangtze Craton and Cathaysia block[a]

Eon	Cathaysia block	Yangtze Craton
Neoproterozoic		0.70–0.63 Ga: minor granites, volcanic rocks and mafic to ultra-mafic intrusions
	0.83–0.72 Ga: minor granites, bimodal volcanic rocks and orthogneiss	0.83–0.73 Ga: extensive felsic volcanic rocks and granites, association with komatiitic basalts and continental flood basalts, bimodal volcanic rocks, mafic intrusions; Nanhua rift basin, Kangdian rift basin; ca. 0.82 Ga "Fuchuan ophiolite"
	0.86–0.84 Ga: minor mafic intrusions, basalts and bimodal volcanic rocks	0.86–0.84 Ga: bimodal volcanic rocks, mafic dikes, gabbros, alkaline rocks and granites
		0.96–0.89 Ga: subduction-related calc-alkaline magmatism; ca. 1.0 Ga "NE Jiangxi ophiolite"
	1.0–0.97 Ga: minor amphibolites and metamorphosed felsic volcanic rocs (subduction-related igeous rocks?)	1.04–0.94 Ga: two phases of deformation and upper greenschist facies metamorphism recorded in the Tianli schist in the southeastern magrin; foreland basin in the southwestern margin
Mesoproterozoic		Ca. 1.1 Ga: alkaline and calc-alkaline intermediate to mafic volcanic rocks in northern margin; ca. 1.15 Ga: alkaline basalts and bimodal volcanic rocks in southern margin; ca. 1.1 Ga "Miaowan ophiolite"
	1.3–1.0 Ga: high grade metamorphic rocks in the Baoban complex	
	1.43 Ga: anorogenic granites and volcanic rocks in the Baoban complex	
		Ca. 1.5 Ga: gabbro-diabase dikes and intrusive stock (V-Ti magnetite)
Palaeoproterozoic	1.78–1.77 Ga: amphibolite (derived from OIB-type and MORB-type basalts, extreme depleted Nd isotopic featrue)	Ca. 1.7–1.6 Ga: volcanic rocks
	1.89–1.88 Ga: granulite facies metamorphism 1.89–1.83 Ga: granites	1.85 Ga: A-type granites and mafic dikes
		2.0–2.1 Ga: TTG and granulite facies metamorphism

(continued)

Table 9.1 (continued)

Eon	Cathaysia block	Yangtze Craton
Archaean		2.5 Ga dioritic-granitic gneisses; 2.7–2.6 Ga: A-type granites; 3.4–2.9 Ga: TTG and amphibolite facies metamorphism

[a] Apart from the ca. 1.1 Ga Miaowan ophiolite (Peng et al. 2012; Jiang et al. 2016; Deng et al. 2017), data sources in this table are cited and discussed in the main text

9.4 Neoproterozoic Petrotectonic Evolution in South China

Widespread Early to Midde Neoproterozoic magmatic rocks are closely associated in time and space with large-scale volcano-sedimentary basins developed in the South China. Granitoid and felsic volcanic rocks, together with their basaltic counterparts, culminated at 0.83–0.75 Ga of the mid-Neoproterozoic in the Yangtze Craton. The granitoid rocks include S-, I-, and A-types, along with minor amount of adakitic rocks. Interpretations of the petrogenesis and tectonic environments in which these granitoid rocks were emplaced are highly controversial. For instance, the extensive ca. 0.82 Ga granites are interpreted as syn-collisional (arc-continent and/or continentalcollision), late- to post-collisionbal (orogenic collapse), and anorogenic (mantle plume-related intraplate rift) magmatism by different researchers (e.g., Li et al. 2003a; Li X. H. 1999; Wang X. L. et al. 2004, 2006; Zhang et al. 2007, 2009; Zhang S. B. et al. 2012). Similarily, there are two competing models for the tectoinic settings of the 0.75–0.80 Ga A-type granites and adakitic rocks in the Kangdian region, i.e., ① subdutction-related setting, and ② intraplate rift setting (Zhou M. F. et al. 2006; Zhao and Zhou 2007b; Huang et al. 2008, 2009; Zhao X. F. et al. 2008). It is noted that the geochemical characteristics of granitoid rocks are not only reflective of their source compositions but also the melting and crystallization histories of the melts, thus could not be simplistically used to assess the tectonic setting. On the other hand, geochemistry of basaltic rocks (including basalts and basaltic dikes) reflects the compositions and thermal conditions of their mantle sources that are relevant to different petrotectonic associations. In this contribution we only use the geochemistry and isotope data for the Neoproterozoic basaltic rocks, in combination with other aspects of geological records, to investigate their genesis and regional petrotectonic evolution.

9.4.1 *Early Neoproterozoic (1.0–0.9 Ga) Basaltic Rocks*

9.4.1.1 Geochemical Characteristics

The Early Neoproterozoic basalts in the Yangtze Craton consist of the ca. 1.0 Ga Huili mafic dykes (Zhu et al. 2016) and the ca. 0.9 Ga Yanbian Group basalts (Li X.

H. et al. 2006) in the western magrin, the ca. 0.95–0.89 Ga Xixiang Group basalts (Ling et al. 2003) in the northwestern margin, and the ca. 0.96 Ga Pingshui Group spilites (Li X. H. et al. 2009) in the southeastern margin. These basaltic rocks are subjected to various degrees of deformation and metamorphism. In the Zr/TiO_2 versus Nb/Y diagram of Winchester and Floyd (1976), these basalts have low Nb/Y ratio (0.05–0.7), falling into the sub-alkaline field (Fig. 9.4a). The Yanbian basalts can be divided into two groups in terms of geochemical characteristics. The first group of rocks plot into the field of calc-alkaline basalts that are enriched in Th, U and LREE, but depleted in Nb, Ta, Ti, Zr, and Hf, similar to those of Pingshui and Xixiang basalts; the second group of rocks are tholeiitic basalts characterized by depletion in Th, U, Nb, Ta and LREE, similar to those of MORB. In the FeO^T/MgO versus TiO_2 diagram of Miyashiro (1974), the Pingshui spilites, the Xixiang basalts and the Yanbian calc-alkaline basalts are low in TiO_2 that decreases with the increasing FeO^T/MgO ratio (Fig. 9.4b). By contrast, the Yanbian tholeiitic basalts and the Huili mafic dykes have higher TiO_2 (>1.5%) that increases with the increasing FeO^T/MgO ratio, displaying a tholeiitic trend (Fig. 9.4b). The Xixiang and Pingshui basaltic rocks are characterized by enrichment in Th, U, and LREE, and depletion in Nb, Ta, Zr, Hf, and Ti, similar to the IAB (Ling et al. 2003; Li X. H. et al. 2009). In the diagram of Ti-Sm-V of Vermeesch (2006), the Pingshui spilites, the Xixiang basalts, and the Yanbian calc-alkaline basalts plot into the IAB field, whilst the Yanbian and Huili tholeiitic basaltic rocks fall into the MORB field (Fig. 9.5). Overall, the Early Neoproterozoic basalts in the Yangtze Craton are generally similar to the basalts formed in the active continental margins. The Huili mafic dykes are interpreted to have formed in an impact orogenic setting during the initial stage of collision between the Yangtze Craton and the Cathaysia Block prior to the Sibao Orogeny (Zhu et al. 2016), whereas the Yanbian tholeiitic basalts are likely produced by large degrees of partial melting of a depleted mantle source in a back-arc basin (Li X. H. et al. 2006).

The Early Neoproterozoic basaltic rocks in the Cathaysia Block include the amphibolites in the Yunkai Mountain and Wuyi Mountain areas. In the Zr/TiO_2 versus Nb/Y diagram, these amphibolites are low in Nb/Y (0.05–0.67), falling into the sub-alkaline field (Fig. 9.6a). In the FeO^T/MgO versus TiO_2 diagram, TiO_2 increases with increasing FeO^T/MgO, showing a tholeiitic trend. Wang Y. J. et al. (2013) divided these amphibolites into four groups based on their geochemical and Nd isotopic features (Fig. 9.6b). Group 1 is geochemically similar to fore-arc MORB-like basalt. Group 2 displays E-MORB geochemical signatures. Group 3 is geochemically similar to the Nb-enriched basalt in the Pickle Lake greenstone belt in northwestern Ontario. Group 4 is is characterized by geochemical features similar to arc volcanic rocks. In the diagram of Ti-Sm-V, all amphibolites from the first three groups fall into the MORB field; four of six samples of the Group 4 fall into the IAB field, and the remaining two samples into the MORB field (Fig. 9.7). Wang Y. J. et al. (2013) interpreted that these basaltic rocks (amphibolite protoliths) were formed in an arc-back-arc system. It should be noted, however, there seems a lack of Early Neoproterozic calc-alkaline basaltic rock in the Cathaysia Block, and most amphibolite samples are geochemically similar to MORB, which is different from the magmatic associations formed in active continental margin setting. Alternatively,

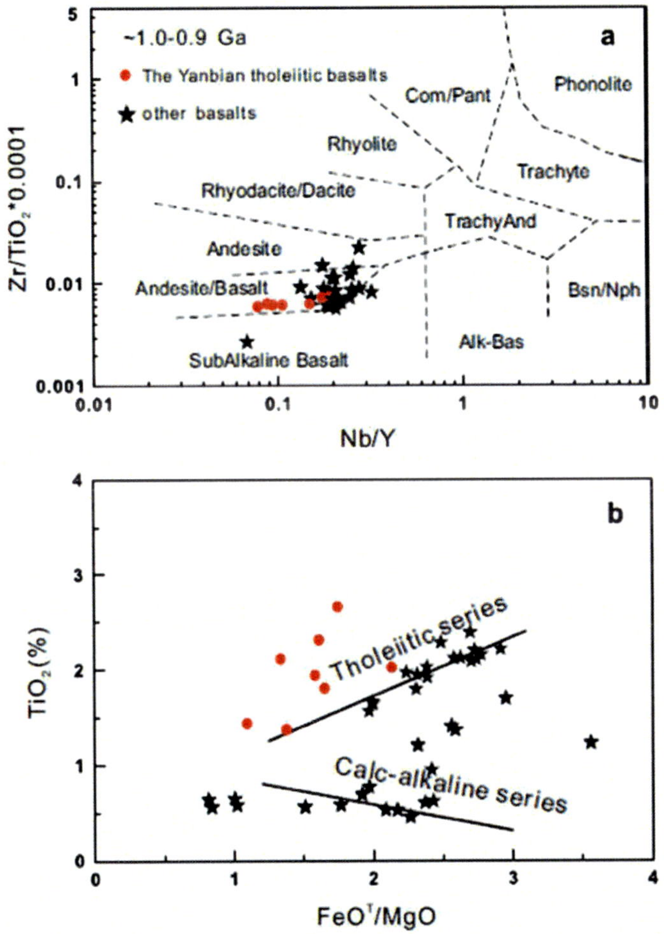

Fig. 9.4 The classification diagrams for early Neoproterozoic basalts in South China (modified after Li X. H. et al. 2008b). **a** Zr/TiO_2-Nb/Y; **b** FeO^T/MgO-TiO_2

these amphibolites might be the products formed in an extension setting, either in a back-arc or a rift basin.

9.4.1.2 Mantle Potential Temperature and Source Compositions

Chemical compositions of primary magmas have been widely used for estimating mantle melt conditions such as melting temperature, melting pressure, and mantle potential temperature (T_p) (Albarède 1992; Sugawara 2000; Herzberg and O'Hara 2002; Herzberg et al. 2007). Among them, T_p is the most important parameters to describe the mantle thermal condition.

Fig. 9.5 Ti-Sm-V discriminate diagram for early Neoproterozoic basalts of the Yangtze Craton (Li X. H. et al. 2008b). OIB—ocean island basalt; IAB—island arc basalt; MORB—mid-ocean ridge basalt

a. Zr/TiO_2-Nb/Y; b. FeO^T/MgO-TiO_2

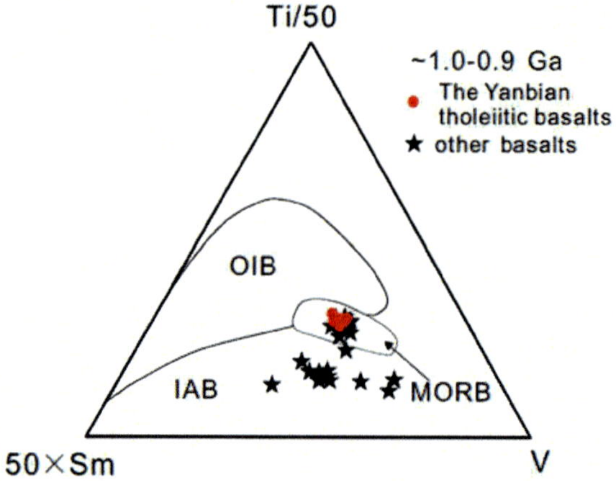

Li X. H. et al. (2008b) used the method of Herzberg et al. (2007) to calculate the compositions of primary magmas for the representative Early Neoproterozoic basaltic rocks (Table 9.2). The calculated compositions of primary magmas are high in SiO_2 (49–50%) and low in MgO (12.3–13.6%) and FeO (7.8–8.4%) for the Xixiang and Yanbian calc-alkaline basaltic rocks, but relatively low in SiO_2 (46%) and moderate in MgO (14.3%) and FeO (8.9%). MgO concentrations on primary basaltic melts are directly related to melting temperatures (e.g., Albarède 1992; Herzberg et al. 2007). According to the chemical compositions of primary magmas (Table 9.2), the melting temperature is calculated at ca. 1300 °C, ca. 1260 °C and 1420 °C for the Xixiang basalt, the Yanbian calc-alkaline basalts and tholeiitic basalts, respectively, using the equation of T (°C) = 2000 × MgO/(MgO + SiO_2) (wt%) + 969 (Albarède 1992). Mantle potential temperature (T_p) is an important parameter for characterizing the thermal conditions of the upper mantle. The average T_p is calculated at 1399 and 1355 °C for the Xixiang and Yanbian calc-alkaline basalts (Table 9.2), indicating that these basalts were most likely generated by melting of a low-temperature, hydrous mantle wedge source above the subduction zone. A slightly higher T_p of ca. 1470 °C is obtained for the Yanbian tholeiitic basalts, which is consistent with the Neoproterozoic ambient temperature of MORB-source asthenospheric mantle. These tholeiitic basalts have highly depleted Nd isotopes, with $\varepsilon_{Nd}(t)$ values ranging from + 5.7 to + 10.7. There are positive correlations between $\varepsilon_{Nd}(t)$ and MgO, and between Nb/Th and Nb/La, with the lowest Nb/Th ratio < 14 (Li X. H. et al. 2006), suggesting involvement of minor amount of sialic crustal components in the basaltic magmas. Combined with their high melting temperature (ca. 1420 °C) and T_p (ca. 1470 °C), the Yanbian tholeiitic basalts were likely attributed to melting of an asthenospheric mantle related to thinning of continental lithosphere in a back-arc basin.

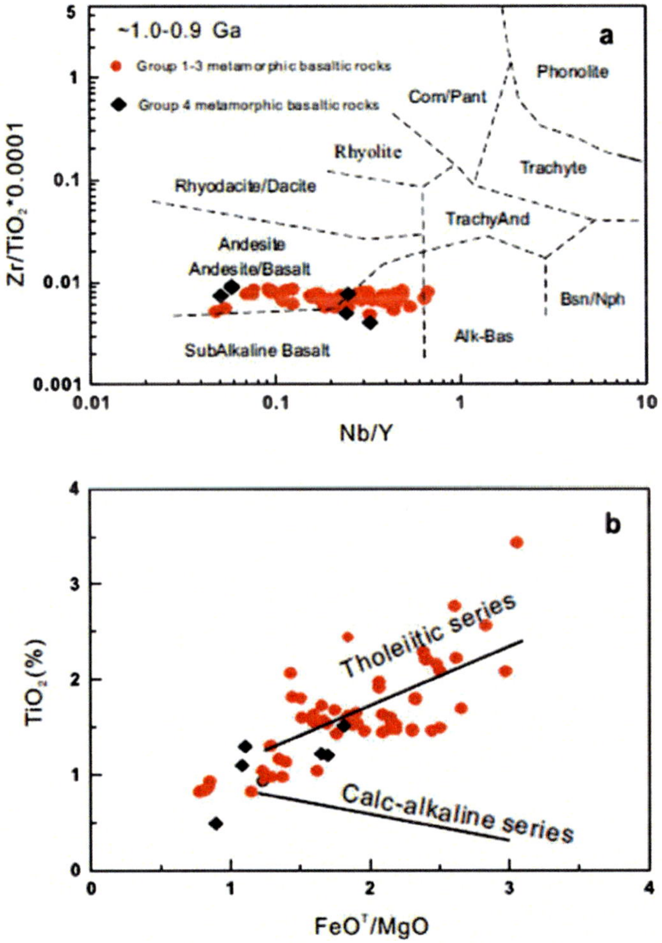

Fig. 9.6 Discrimination diagrams for the early Neoproterozoic basalts from the Cathaysia Block. **a** Zr/TiO_2 versus Nb/Y; **b** TiO_2 versus FeO^T/MgO. *Data source* Zhang A. M. et al. (2012), Wang Y. J. et al. (2013)

9.4.2 Mid-Neoproterozoic Basalts

9.4.2.1 Geochemical Characteristics

The mid-Neoproterozoic basaltic rocks occur sporadically in the Yangtze Craton. Most of them are voluminously minor, apart from the Bikou flood basalts that are exposed over an area of ca. 10,000 km^2, with the thickness estimates at different places vary from several hundred meters up to ca. 9 km (Wang X. C. et al. 2008). According to their temperal and spatial relationship with the Nanhua and Kangdian

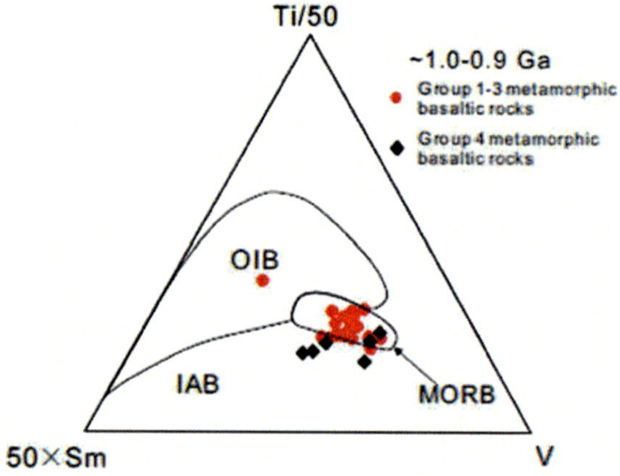

Fig. 9.7 Ti-Sm-V diagram for the early Neoproterozoic basalts from the Cathaysia Block. OIB—ocean island basalts; IAB—island arc basalts; MORB—mid-ocean ridge basalts. *Data source* Zhang A. M. et al. (2012), Wang Y. J. et al. (2013)

rift basins (Wang and Li 2003), these mid-Neoproterozoic basaltic rocks can be subdivided into two major phases: early rift phase (0.85–0.80 Ga) and major rift phase (0.79–0.75 Ga). The ca. 0.85 Ga Shenwu dolerites are included into the early rift phase because ① they are associated with the same-aged Gangbian alkaline complex that is most likely the products of anorogenic magmatism manifesting the initial rifting of South China after the Sibao Orogeny (Li X. H. et al. 2008a), and ② while the ca. 0.85 Ga Shenwu dolerites pre-date the mid-Neoproterozoic (ca. 0.83–0.82 Ga) unconformity, recent work demonstrates that this mid-Neoproterozoic unconformity is more likely attributed to a rapid regional crust uplifting, unroofing and basin subsidence related to the proposed ca. 825 Ma mantle plume (Yang et al. 2015) rather than the orogenesis as previously thought. Recently, several ca. 840 Ma volcanic rocks were reported in SE margin of the Yangtze Craton, including Sibao basalts, Zhangyuan basalts, Shexian basalts and Lushan basalts, Fuchuan basalts and Dengshan basalts (Zhang C. L. et al. 2013a; Yao et al. 2015; Li L. M. et al. 2016). Apart from the ca. 0.82 Ga Mamianshan basaltes from the Cathaysia Block and the ca. 0.81 Ga Suxiong alkaline basalts and the ca. 0.76 Ga Xiangxi diabases from the Yangtze Craton, the majority of the mid-Neoproterozoic basaltic rocks are low in Nb/Y (<0.7), plotting into the sub-alkaline basalt field on the Zr/TiO_2 versus Nb/Y classification diagram (Fig. 9.8a). All the mid-Neoproterozoic basaltic rocks are affinitive to tholeiitic basalts as shown by a positive trend between TiO_2 and FeO^T/MgO (Fig. 9.8b).

The early rift phase basaltic rocks have various trace element geochemical features, and can be classified as two groups (Li X. H. et al., 2008b). The first group includes the ca. 0.85 Ma Shenwu dolerite, the ca. 0.83 Ga Yiyang komatiitic basalts, the ca. 0.82 Ga Tiechuanshan basalts, and most ca. 0.82–0.81 Ga Bikou basalts. They possess trace element features transitional between OIB and IAB, characterized by enrichment in most incompatible elements such LILE and LREE and weak to moderate depletion in Nb and Ta. The second group includes the ca. 0.82 Ga Mamianshan alkaline basalts and the ca. 0.81 Ga Suxiong basalts and some Bikou

Table 9.2 Calculated chemical compositions, pressures, melting temperatures and mantle potential temperatures (T_p) for the primary melts of the Neoproterozoic basalts in South China (Li X. H. et al. 2008b; Wang X. C. et al. 2009)

Basalt type	Compositions (%)			Melting temperature (°C)		Mantle potential temperature (°C)	
	SiO_2	FeO	MgO	T	± 1	T_p	± 1
Early Neoproterozoic (1.0–0.9 Ga)							
Yanbian calc-alkaline basalts	50	7.8	12.6	1262	9	1355	45
Yanbian tholeiitic basalts	46	9.5	15	1417	10	1470	11
Xixiang basalts	49	8.4	13.8	1302	8	1399	44
Middle Neoproterozoic early rift phase (ca. 0.85–0.80 Ga)							
Shenwu diabase	49	7.9	12	1284	7	1353	19
Lower Bikou basalts	48	9.9	15	1369	9	1457	25
Upper Bikou basalts	47	10.8	17	1453	11	1535	26
Tiechuanshan basalts	49	7.9	13.8	1342	8	1425	20
Yiyang komatiitic basalts	47	10.3	20	1521	18	1618	46
Mamianshan basalts	47	9.8	15	1382	9	1457	23
Suxiong basalts	44	10.7	16	1445	13	1505	21
Major rift phase (0.79–0.75 Ga)							
Yanbian LREE-enriched mafic dikes	48	10.4	16	1386	11	1485	33
Daolinshan diabases	48	9.5	14	1353	8	1429	17
Shangshu basalts	47	10.2	16	1414	10	1494	22
Kangding LREE-enriched mafic dikes	48	10.7	17	1428	13	1529	32
Kangding LREE-depleted mafic dikes	47	11	18	1475	14	1565	33
Yanbian LREE-depleted mafic dikes	47	10	18	1470	14	1562	33
Xiangxi diabases	48	10.9	19	1446	18	1573	54
Guibei spilites	46	10.6	19	1507	16	1588	37

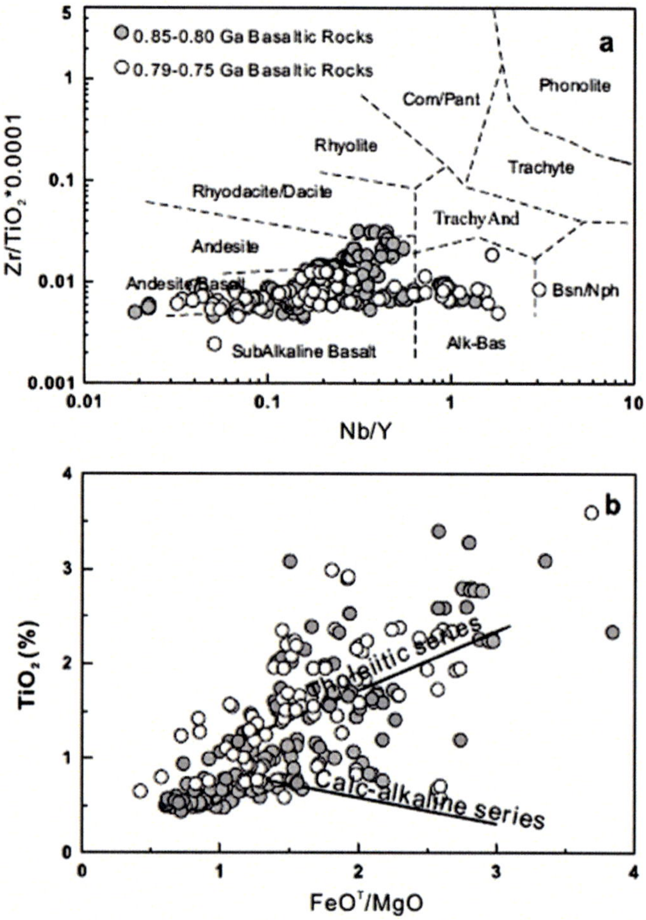

Fig. 9.8 The classification diagrams for the Middle Neoproterozoic basalts in South China (Li X. H. et al. 2008b, modified). **a** Zr/TiO_2-Nb/Y; **b** TiO_2-FeO^T/MgO

basalts. They are enriched in most incompatible trace elements without appreciable depletion in Nb and Ta, similar to OIB.

The major rift phase basaltic rocks can be classified into three groups in terms of their geochemical characteristics (Li X. H. et al. 2008b). Group 1 includes the ca. 0.79 Ga Shangshu basalts and Daolinshan diabases, the ca. 0.76 Ga Guibei spilites, and the ca. 0.79–0.76 Ga LREE-enriched dolerites from the Kangding and Yanbian regions. These basaltic rocks are characterized by enrichment in most incompatible elements such as LILE and LREE and weak to moderate depletion in Nb and Ta, similar to the first group basaltic rocks of the early rift phase. Group 2 is represented by the ca. 0.76 Ga Xiangxi dolerites that are featured by OIB-like trace element

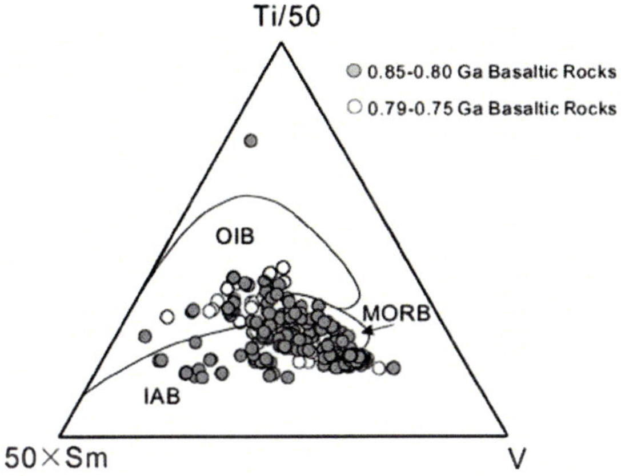

Fig. 9.9 Ti-Sm-V discrimination diagram (Vermeesch 2006) for the Middle Neoproterozoic basalts in South China (Li X. H. et al. 2008b, modified). OIB—ocean-island basalts; IAB—island arc basalts; MORB—mid-ocean range basalts

patterns. Group 3 includes some ca. 0.79–0.76 Ga dolerites that are depleted in LREE from the Kangding and Yanbian regions.

On the Ti-Sm-V tectonic discrimination diagram of Vermeesch (2006), the mid-Neoproterozoic basalts are plotted mainly into the OIB and MORB fields (Fig. 9.9), broadly similar to the intraplate basalts but different from IAB. These basalts are high in Ti/V ratio (>20) which is positively correlated with Nb/La ratio (not shown). Thus, enrichment in V is genetically related to assimilation of continental crustal materials which are low in Nb/La ratio.

9.4.2.2 Mantle Potential Temperature and Source Composition

Apart from a number of small mafic intrusions, the ca. 0.85 Ga basaltic eruptions and dikes are sparse, and represented by the ca. 0.85 Ga Shenwu dolerites (Li X. H. et al. 2008b). The calculated primary magma compositions of the Shenwu dolerites are low in MgO and FeO contents (12% and 7.9%, respectively), but high in SiO_2 (49%). Based on the primary magma compositions, the melting temperature and T_p are estimated at ca. 1300 °C and ca. 1353 °C, respectively.

The 0.83–0.80 Ga basaltic rocks are widespread in South China, including the Yiyang komatiitic basalts, the Bikou flood basalts and the Tiechuanshan basalts in the Yangtze Craton. Among these basalts, the primary magma compositions of the Yiyang komatiitic basalt have the highest MgO content (ca. 20%), attaining the highest melting temperature of 1521 °C and T_p of 1618 °C (Table 9.2). The primary magma compositions of the Suxiong and Upper Bikou Group basalts are also high in MgO (16–17%), giving the melting temperature of 1445–1453 °C and T_p 1505–1535 °C, respectively. In contrast, the calculatd primary magma compositions are relatively low in MgO (13.8–15%) for the basalts from Mamianshan, Tiechuanshan and Lower Bikou Group, corresponding to melting temperature of 1342–1382 °C, and T_p of 1425–1457 °C.

The ca. 0.79–0.75 Ga basaltic rocks are mainly exposed along the Yangtze Craton margins, including the bimodal volcanic rocks in Shangshu and Guangfeng areas and the dolerites in Daolingshan and Xiangxi areas in the southern marginas as well as abundant mafic dikes in the western margin. The primary magma compositions are generally low in MgO (14–16%) for the Dolinshan dolerites, the Shangshu basalts, and the Yanbian mafic dikes with enrichment in LREE, corresponding to melting temperatures of 1353–1414 °C, and T_p of 1429–1494 °C. On the contrary, primary magma compositions for a number of ca. 0.76 Ga basaltic rocks such as the Xiangxi dolerites, Guibei spilites, the Kangdian dolerites and the Yanbian doleirtes with depletion in LREE are characterized by high MgO (17–19%) content, corresponding to melting temperatures of 1428–1507 °C, and higher T_p of 1529–1588 °C (Table 9.2).

9.4.3 *Neoproterozoic Petrotectonic Association: From Sibao Orogen to Nanhua Rifting*

9.4.3.1 Ca. 1.0–0.9 Ga Syn-Orogenic Basaltic Magmatism

Apart from the Yanbian tholeiitic basalts that are likely fromed in a back-arc basin (Li X. H. et al. 2006), the Early Neoproterozoic basaltic rocks along the southern and northern margins of Yangtze Craton are predominantly calc-alkaline in compositions. Overall, these calc-alkaline rocks are geochemically similar to the magmatic rocks formed in active continental margins. Therefore, these Early Neoproterozoic basaltic rocks archieved the records of continental arc magmatism related to the double-side oceanic crust subduction beneath the northern and southern margins of Yangtze Craton during the Sibao Orogeny. The subcontinental lithospheric mantle (SCLM) of the Yangtze Craton was modified by subduction, resulting in hydration and enrichment in highly incompatible elementst. This interpretation is consistent with other geological records listed below.

(1) **Metamorphism recorded in the Tianli schists:** The Tianli schists is a unique metamorphic Complex that outcrops in the eastern segment of the Sibao Orogen at the southeastern margin of the Yangtze Craton. The Tianli schists consist predominantly of bluish-coloured quartz mica schists, with minor quartzite and marble. The schists are found in a high-angle unconformable contact with mid-Neoproterozoic rift volcanic-sedimentary successions dated at ca. 825 Ma (Li W. X. et al. 2008a). Sharp contrasts exist across this unconformity in both the degrees of metamorphism and styles and orientations of deformation structures. In the Tianli schists, the rocks were metamorphosed to middle-upper greenschist facies with the original layering almost entirely transposed by foliations. In-situ UV laser spot $^{40}Ar/^{39}Ar$ mica and SHRIMP U–Pb zircon dating results indicate that the depositional age of the protolith of the Tianli schists is younger than 1.53 Ga, as constrained by the youngest detrital zircon grains, but is older than

1.04 Ga as constrained by the oldest $^{40}Ar/^{39}Ar$ muscovite ages (Li Z. X. et al. 2007). The Tianli schists were metamorphosed at 1.04–1.01 Ga, and suffered tectonic reactivations before 0.90 Ga, close to the termination of the Sibao Orogeny.

(2) **Ca. 1.0 Ga NE Jiangxi ophiolite:** The NE Jiangxi ophiolite occurs along the southern margin of the Sibao Orogen, and is considered as the suture between Yangtze Craton and Cathaysia Block. The mafic rocks within the ophiolite were derived from a highly depleted mantle source at ca. 1.0 Ga (Chen et al. 1991; Zhang et al. 2015). The mafic and ultramafic ophiolitic rocks were intruded by ca. 0.97 Ga adakites (Li and Li 2003) and ca. 0.88 Ga "obduction-type" granites (Li W. X. et al. 2008b). The NE Jiangxi ophiolite is interpreted as products formed in a back-arc basin behind the ca. 1.0–0.9 Ga Shuangxiwu magma arc (Li X. H. et al. 2009). The back-arc basin is finally closed at ca. 0.88 Ga in response to the termination of the Sibao Orogeny.

(3) **EW-trend syn-orogenic deformation in the western Sibao Orogen:** Orogenic deformation in the eastern segment of the Sibao Orogen is mostly overprinted by younger orogenesis and magmatism during the Phanerozoic, but preserved in the western segment of the Sibao Orogen. There are a number of syn-orogenic gneissic granitoid rocks dated at ca. 1.0 Ga in southwestern Yangtze Craton (Li Z. X. et al. 2002; Yang et al. 2009). The basement rocks underlain the Kangdian rift basin in western Yangtze Craton are strongly deformed, and variably metamorphosed. On outcrop scale tight isoclinal folding is common; the axial planes of the folds are either subvertical or dipping steeply to the south, with fold hinges mostly subhorizontal or gently dipping. This suggests a general northward structural vergence. In the Yanbian area the easterly to northeasterly trending structures were cut by the ca. 0.86 Ga Guangdaoshan pluton, suggesting that a major fold and thrust event took place during the final stage of the Sibao orogenesis at ca. 0.90 Ga (Li X. H. et al. 2006). In contrast, the younger magmatic episode is represented by the non-deformed mafic–ultramafic intrusions and bimodal Suxiong Formation volcanic rocks dated at 0.82–0.80 Ga formed in an anorogenic setting.

(4) **Sedimenatry records and basin evolution:** There are distinct pre-1.0 Ga magmatic histories in the Yangtze Craton and Cathaysia Block (Table 9.1), with the Cathaysia Block being diagnostically featured by the occurrence of ca. 1.43–1.44 Ga granitoid intrusions and volcanic rocks (Li Z. X. et al. 2002, 2008b; Yao et al. 2017). Fragments of such diagnostic-aged magmatic rocks in one block might be transported to, and deposited on, the other once the two blocks were connected to each other. U–Pb age analysis shows a clear peak at ca. 1.43 Ga for detrital zircons from the ca. 1.0 Ga metasediments in the Kunyang and Huili Groups (Fig. 9.10; Li Z. X. et al. 2002; Li X. H. et al. 2014 and references therein). These ca. 1.43 Ga detrital zircons, well coincident in age with the same-aged granitoid rocks in Hainan of western Cathaysia, are interpreted as foreland basin deposits of the Sibao Orogen derived from the Cathaysia Block on the southwestern Yangtze side (Li Z. X. et al. 2002). Therefore, the western segments of Yangtze Craton and Cathaysia Block most likely collided

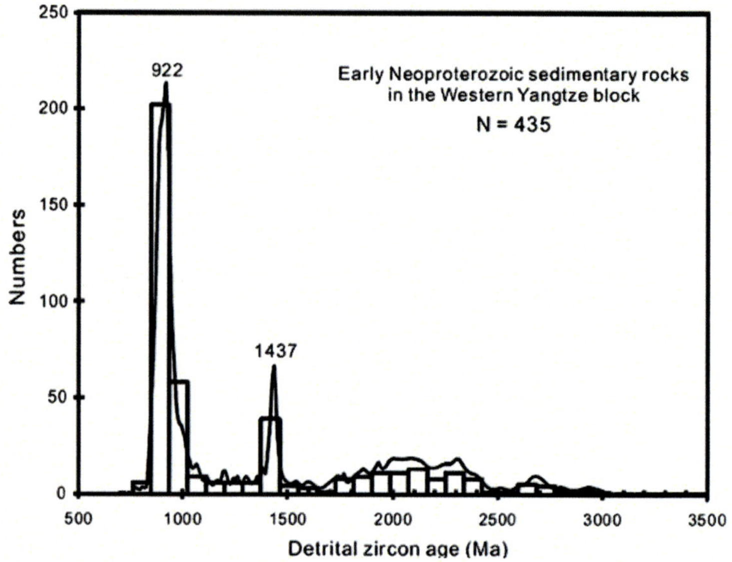

Fig. 9.10 Histogram of detrital zircon ages for the early Neoproterozoic sedimentary rocks in the western Yangtze Craton. *Data source* Li Z. X. et al. (2002), Greentree and Li (2008), Sun M. et al. (2008a), Sun W. H. et al. (2008b)

by 1.0 Ga, resulting in the formation of ca. 1.0 Ga syntectonic granitoids (Li Z. X. et al. 2002; Yang et al. 2009) and regional easterly to northeasterly-trending deformation (Li X. H. et al. 2006) in southwestern Yangtze Craton.

Middle Neoproterozoic sedimentary rocks are widespread in the southern Yangtze Craton. They are separated into the Sibao and the Danzhou Groups (and their equivalents such as the Lengjiaxi and Banxi Groups, respectively) by a ca. 0.82 Ga mid-Neoproterozoic angular unconformity. While there is a general consensus that the Danzhou Group and its lateral equivalents were deposited in a rift basin during the time interval between ca. 0.82 and ca. 0.716 Ga (Li Z. X. et al. 2003; Wang and Li 2003; Lan et al. 2014), it is highly controvercial for the nature of the baisn in which the sedimentary rocks from the Sibao Group (and equivalents) below the unconformity were deposited.

The SCB is generally considered to be formed through amalgamation of the Cathaysia and Yangtze Cratons during the Sibao (or Jiangnan) Orogeny, but the timing of the orogenesis and final assembly of the two blocks are an issue of hot debate. Integrated investigations of the petrogenesis of the Early Neoproterozoic basaltic rocks and other aspects of geological records including metamorphism, deformartion, sedimentary geochemistry, basin analysis, and ophiolite complex place constrains on the timing of 1.0–0.9 Ga for the Sibao orogenesis. The initial collosion between the two blocks likely occurred at southwestern Yangtze Craton, with development of foreland-basin deposits (Kunyang and Huili Groups) in the southwestern

Yangtze Craton derived from the Cathaysia Block and formation of the syn-orogenic gneissic granitoid rocks in a foreland fold-and-thrust-belt setting. Subduction of ceanic crust beneath the southeastern Yangtze Craton resulted in regional upper greenschist facies metamorphism (Tianli schists), the ca. 0.96–0.89 Ga Shuangxiwu arc magmatism, opening of back-arc basin represented by the ca. 1.0 Ga NE Jiangxi ophiolite. Slab-melting of the subducted back-arc basin leads to formation of the ca. 0.97 Ga adakites at Xiwan Formation "obduction-type granites" indicates closure of the back-arc basin and formation of the united SCB at of ca. 0.88 Ga.

Most 1.0–0.97 Ga amphibolites in the Cathaysia Block are geochemically similar to the basaltic rocks form in back-arc basins, suggesting that a Grenville-aged magmatic arc might have existed in the southern (Zhang A. M. et al. 2012) or northern (Wang Y. J. et al. 2013) margin of the Cathaysia Block. The ca. 0.98 Ga Jingnan meta-rhyolites with calc-alkaline composition are interlayered with meta-greywackes, both of which suffered strong deformation and metamorphsim. They are interpreted as the calc-alkaline volcanic rocks formed in an active cobtinental margin (Shu et al. 2008), similar to the ca. 1.0 Ga gneissic granites in southern Jiangxi (Liu et al. 2001). In addition, numerous Grenville-aged (1.2–0.96 Ga) detrital zircons were identified in sedimentary rocks of different ages in the Cathaysia Block, implying possible widespread exposures of the Grenville-aged magmatic rocks in the Cathaysia Block or surrounding regions (Wu et al. 2010). Because the Early Neoproterozoic rocks are sparse and suffered significant thermo-magmatic reworking during the Phanerzoic, the Early Neoproterozoic tectonic evolution of the Cathaysia Block awaits further investigations.

9.4.3.2 Ca. 0.9–0.85 Ga Transition from Sibao Orogen to Intraplate Rifting

WhC thoughts is that these magmatic rocks were products formed in a post-orogenic and/or anorogenic tectonic setting (Li X. H. et al. 2003a, 2008a; Bao et al. 2008; Li W. X. et al. 2010), the other considers that these rocks were formed at active continental margins (Zhou et al. 2002b; Shen et al. 2002; Sun et al. 2008a, 2008b; Dong et al. 2010). These magmatic rocks are dominated by bimodal volcanic rocks or alkaline complex with minor intermediate rocks. Such igneous rock associations are clearly different from those of magmatic rocks formed in active continental margins. It is noteworthy that the penetrative, easterly to northeasterly trending syn-orogenic deformation in southwestern Yangtze is consistent with the spatial distribution of the same-aged Sibao orogenic belt (Li X. H. et al. 2006), but perpendicular to the proposed SN-trending active continental margin (e.g., Zhou et al. 2002a, b; Zhao J. H. et al. 2008; Zhao and Cawood 2012). Such structural defoemation feature argues against an Early to Middle Neoproterozoic SN-trending active continental margin caused by easterward subduction beneath the Kangdian region. The ca. 0.85 Ga gabbros and metabasites in the Cathaysia Block are all characterized by geochemistry affinities with the intraplate basalts (Shu et al. 2006, 2011). Base on the magmatic association, geochemistry and concurrence of the same-aged intraplate basaltic magmatism in both the Yangtze Craton and the Cathaysia Block, the ca. 0.85 Ga magmatic rocks are most likely formed in an intraplate, anorogenic setting. If this interpretation is correct, the collision between the Cathaysia and Yangtze and formation of the united South China continent should be prior to ca. 0.85 Ga (Li X. H. et al. 2008a, 2009; Shu et al. 2011). Transition from the Sibao Orogen to the intraplate rifting likely occurs during ca. 0.9–0.85 Ga.

9.4.3.3 Ca. 0.83–0.75 Ga Intraplate Magmatism and Rift Basins

The mid-Neoproterozoic magmatic rocks dated at 0.83–0.75 Ga are widespread in the Yangtze Craton. As aforementioned the petrogenesis and tectonic implications of these rocks has been an issue of hot debates for the last two decades. Inaddition, the regional ca. 0.82 Ga mid-Neoproterozoic unconformity between the Sibao and Danzhou Group (and their equivalents) in southern Yangtze is another important issue for understanding the petrotectonic evolution in South China.

In view of magmatic associations, the magmatic rocks in age of 0.83–0.82 Ga are compositionally bimodal, with predominant felsic volcanic and intrusive rocks and subordinate mafic litholigoes, intermediate rocks are rare except for the limited Cangshuipu andesites in the Yiyang region of Hunan Province (Wang et al. 2003; Wang Y. J. et al. 2013). The basaltic rocks consisit mainly of high-temperature lavas including the Yiyang komatiitic basalts and the Bikou flood basalts together with some tholeiitic to alkaline basalts with rare occurence of calc-alkaline basalts. They are all characterized by geochemical features affinitive with the intraplate basalts.

The mid-Neoproterozoic sedimentary basins in South China are typical continental rift basins, including the north–south trending Kangdian basin in the western part, and the Yangtze-side of the northeasterly-trending Nanhua basin (Wang 2000; Wang and Li 2003). Stratigraphic and facies analyses demonstrate that the Neoproterozoic successions in these basins consist of four major sequence-sets, representing four phases of rifting (Wang and Li 2003). The first phase occurred at ca. 0.82 Ga after the 0.83–0.82 Ga bimodal magmatism. The second phase occurred at ca. 0.80 Ga. The third, a major rift phase, occurred at ca. 0.78–0.75 Ga. Syn-rift magamtism is widespread during the second and third phases. The final phase recorded the rift–drift transition during the Nantuo glaciation with little volcanism. It is noteworthy that eruption of the 0.83–0.82 Ga volcanic rocks is very close to the ca. 0.82 Ga mid-Neoproterozoic unconformity, following the regional crustal doming and unroofing (Li Z. H. et al. 1999; Wang and Li 2003).

The mid-Neoproterozoic unconformity is traditionally considered as an orogenic unconformity. Thus, a number of workers suggest that sedimentary rocks from the Sibao Group (and equivalents) beneath the unconformity were deposited in a back-arc basin (e.g., Gu et al. 2002; Wang X. L. et al. 2007; Wang W et al. 2012, 2013, 2016). However, some recent studies suggest that both the Sibao and Danzhou Groups (and their equivalents) were deposited in similar continental rift basins (e.g., Wang X. C. et al. 2008; Yang et al. 2015) based on the following arguments. ① The timing of the mid-Neoproterozoic unconformity is rigorously constrained at ca. 824–819 Ma, thus there is no significant age gap for the sedimentation across the regional unconformity; ② there is a striking similarity of the age distribution patterns, Hf–O isotopic features of detrital zircons from the Upper Siabo and the Lowest Danzhou Groups, indicating that they shared the same provenances and deposited within the similar tectonic settings; and ③ appearance of abundant low-$\delta^{18}O$ zircon grains from the sediments is coincident with regional intraplate anorogenic magmatism since ca. 850 Ma, suggesting that both the ca. 0.85–0.83 Ga Sibao Group and the ca. 0.82–0.716 Ga Danzhou Groups were most likely deposited in an intracontinental rift basin setting.

One of the salient features of the mid-Neoprtoterozoic magmatism is occurrence of extensive low-$\delta^{18}O$ igneous rocks along the northern and northwestern margins of the Yangtze Craton (Zheng et al. 2004, 2007, 2008a, 2008b; Liu and Zhang 2013). Although low-$\delta^{18}O$ granitoid rocks have not yet been identified in southern Yangtze, numerous low-$\delta^{18}O$ detrital zircons found from sedimentary rocks (Wang X. C. et al. 2011; Yang et al. 2015) suggest existence of low-$\delta^{18}O$ granitoid rocks in their sedimentary provenaces. Low-$\delta^{18}O$ granitoid rocks are characterized by $\delta^{18}O$ values below 5‰. The caldera and rift settings are considered as the most favorable places for water circulation to the depths and as heat sources to drive meteoric-hydrothermal systems, thus high temperature hydrothermal exchange of rocks with meteoric waters or seawater is the major cause of low-$\delta^{18}O$ water values in these rocks (e.g., Zheng et al. 2008a; Bindeman 2011). Large-scale occurrence of mid-Neoproterozoic low-$\delta^{18}O$ granitoid rocks along the Yangtze margins is well consistent with the development of rift basins associated with extensive intraplate basaltic rocks. However, a competing model for petrotectonic setting of these low-$\delta^{18}O$ granitoid rocks also exists (Wang W et al. 2017; Huang et al. 2019).

Because of significant overprinting by the Phanerozoic orogeny and magmattism, the mid-Neoproterozoic sedimentary basins are rarely preserved. The ca. 0.82 Ga Mamianshan bimodal volcanic rocks in northern Fujian Province are very similar in age, rock association and geochemical features to those of bimoal volcanic rocks in the Nanhua rift basin of Yangtze side (Li W. X. et al. 2005). Concurrence of the mid-Neoproterozoic bimodal magmatism in both Yangtze and Cathaysia indicates that along the northern margin of the Cathaysia Block there could have been a continental rift margin opposing the southeastern margin of the Yangtze Craton.

9.4.3.4 Neoproterozoic Petro-Tectonic Evolution in South China

Integrated investigation of petrogenesis of the Neoproterozoic basaltic rocks and other aspects of regional geological data, a geodynamic model was proposed to interpret of Neoproterozoic petro-magmatism evolution (Fig. 9.11; Li X. H. et al. 2008b; Wang et al. 2009).

During the Sibao Orogeny at ca. 1.0–0.9 Ga, double-side subduction of oceanic slabs occurred on the northern and southern margins of the Yangtze Craton, forming the Neoproterozoic magmatic arcs on the active continental margins. Subduction of the oceanic slabs modified the geochemical compositions of the subcontinental

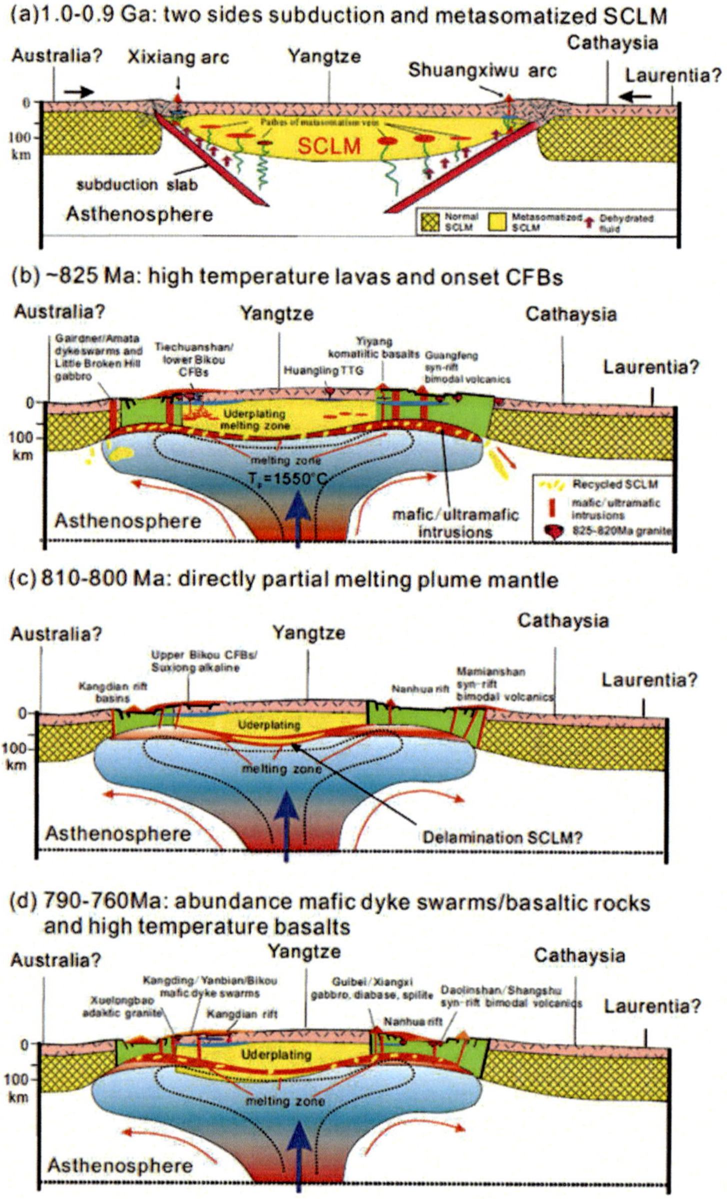

Fig. 9.11 A geodynamical modal for explaintion of early to middle Neoproterozoic geologic evolution (Li X. H. et al. 2008b; Wang et al. 2009, modified)

lithospheric mantle (SCLM), resulting in selective enrichment of strongly incompatible elements and formation of hydrous minerals. Long-term infiltration by asthenosphere-derived fluids/melts might also modify the geochemical compositions of the Yangtze's SCLM. These processes resulted in the formation of clinopyroxene, garnet and a small amount of rutile, apatite as well as possibly small amount of metamorphic garnet-pyroxenite in the interior of SCLM. Consequently, the modified SCLM has a lower solidus temperature.

The regional tectonic regime transformed from the Sibao Orogeny to intracontinental anorogenic rifting during 0.9–0.85 Ga. A rising mantle plume might arrive and impinge the lithosphere bottom in South China at ca. 0.85–0.83 Ga, causing heat transfer within the lithosphere interior and triggering partial melting of the hydrous SCLM. The mantle-derived magmas such as the ca. 0.85 Ga Shenwu dolerites emplaced along the weak zone of pre-existing orogenic belt. The ca. 0.825 Ga Yiyang komatiitic basalts are the direct petrological evidence for the mantle plume activity that causes large-scale crustal doming and unroofing. The rising mantle plume head also leads to thermal erosion of the overlying lithosphere, resulting in lithospheric thinning along the pre-existing suture zones. The heat from the mantle plume may trigger extensive melting of deeper crust to form voluminous granitoid rocks. Most of the ca. 0.83–0.82 Ga basaltic rokcs are characterized by poor FeO^T and depletion in high field-strength elements (HFSE). They are derived predominantly from the SCLM, similar in composition to the continental flood basalts in Karoo and Siberian (Lassiter and Depaolo 1997). The thickness of the South China lithosphere might have thined from ca. 100 km to $\leq$ 70 km accompanied by intensive continental rifting during 0.82–0.80 Ga. Voluminous 0.82–0.81 Ga basaltic magmas such as the Bikou flood basalts might underplate in the crust-mantle transition zone, resulting in crustal thickening.

The lithosphere thickness of South China further thinned during 0.79–0.75 Ga, and the mantle is anomalously hot with mantle temperature (T_p) of ca. 1520 °C. Extensive partial melting occurred between the bottom of lithosphere and the top of asthenosphere under extensional regime. High-temperature basaltic rocks such as the Tongde picritic dikes and the Guibei spilites were produced in response to another phase of mantle plume activity in the western and southern Yangtze Craton.

9.5 Palaeo-Position of South China in Rodinia Supercontinent

While it is generally accepted that Yangtze Craton and Cathaysia Block amalgamated into a united SCB which is part of Rodinia supercontinent in the Neoproterozoic, two major competing models exist regarding the timing of amalgamation of the two units and the palaeo-position of South China in the reconstruction of Rodinia supercontinent. One school of thoughts is the increasingly accepted scenario that the SCB was situated between Australia–East Antarctica and Laurentia (Li Z. X.

et al. 2002, 2008a) within the interior of Rodinia. In this model the Yangtze and Cathaysia firstly amalgamated in the west at ca. 1.0 Ga (Li Z. X. et al. 2002, 2007; Li X. H. et al. 2006), and then the two blocks finally amalgamated in the east at about ca. 0.89–0.88 Ga (Ye et al. 2007; Li W. X. et al. 2008b; Li et al. 2009). The ca. 0.83–0.75 Ga voluminous anorogenic magmatic rocks in South China and other continents are related to the mid-Neoproterozoic superplume activities which led to final fragmentation of Rodinia supercontinent (Li Z. X. et al. 2003; Wang X. C. et al. 2007, 2008). An alternative model prefers that the Yangtze and Cathaysia amalgamated through continental collision, or continental-arc-continental collision to form a united South China continent at 0.83–0.82 Ga (e.g., Li X. H. 1999; Wang X. L. et al. 2006, 2007, 2008; Zheng et al. 2007, 2008b; Zhao et al. 2011; Wang Y. J. et al. 2013), which was situated around the northern margin of Rodinia (Zhou et al. 2006; Yu et al. 2008; Cawood et al. 2013; Wang Y. J. et al. 2013). In addition, there is a "compromised" model that the mantle plume and the magmatic arc coexisted at ca. 0.82 Ga in southeastern Yangtze, leading to the simultaneous concurrence of plume- and arc-related magmatism (Zhang et al. 2013b).

Reconstruction of the palaeo-position of a specific continent and/or terrane within supercontinent Rodinia requires integration of multiple lines of evidence including understanding of pre-assembly basement provinces, orogenic histories, basin evolution, high-quality geochronological and palaeomagnetic data, paired rift margins, plume events, and plume-related volcanic rocks and mafic dykes, etc. (Li Z. X. et al. 2008a).

Placing the SCB on the northern margin of Rodinia would invoke a prolonged active continental margin along the northwestern margin of the Yangtze Craton during much of the Neoproterozoic time. Studies of the origin and genesis of basaltic rocks suggest a major tectonic transition from an active continental margin at ca. 0.95–0.89 Ga to an intracontinental rift at ca. 0.89–0.82 Ga (Ling et al. 2003; Zhou et al. 2018). The ca. 0.82–0.81 Ga Bikou basalts in northwestern Yangtze are most likely the remnants of a plume-related continental flood basalt province (Wang X. C. et al. 2008), coincident with the plume-derived ca. 0. 825 Ga Yiyang komatiitic basalts (Wang X. C. et al. 2007). In addition, many mid-Neoproterozoic igneous rocks from the Dabie-Sulu Belt around the northern to northeastern Yangtze Craton are characterized by exceptional low-δ^{18}O values, indicating intensive high-temperature water–rock interaction and generation of the low-δ^{18}O magmatism in rift tectonic zones (e.g., Zheng et al. 2009a, b; Liu and Zhang 2013). These independent lines of evidence are consistent with the existence of an intercontinental rift, rather than an active continental margin, for the northern margin of the Yangtze Craton during the mid-Neoproterozoic, and invalidate the proposed palaeo-position of the SCB on the fringe of Rodinia.

It can be seen from the geochronological frameworks of the Precambrian magmatism in the Yangtze Craton and the Cathaysia Block (Table 9.1) that the ca. 1.43 Ga anorogenic granitoid rocks in Hainan Island of southwestern Cathaysia Block is a unique record in South China (Li Z. X. et al. 2002, 2008a; Yao et al. 2017), making Cathaysia a possible western extension of ca. 1.5–1.3 Ga anorogenic transcontinental magmatic belt in southern Laurentia (e.g., Nyman et al. 1994) and likely the Late

Mesoproterozoic provenance links between Cathaysia and Laurentia (Li Z. X. et al. 2008b). No such ca. 1.4 Ga anorogenic granitoid rocks have been found in Yangtze. Minor ca. 1.4 Ga magmatic rocks have recently been identified in Central Australia (e.g., Kirkland et al. 2011) and possibly in East Antarctica (Goodge et al. 2008). These discoveries are in fact consistent with a proto-SWEAT-type Columbia/Nuna reconstruction with Cathaysia (but without Yangtze) siting between Australia-East Antarctica and Laurentia (Li Z. X. et al. 2008a; Zhang S. H. et al. 2012; Pisarevsky et al. 2013).

Metamorphism of the Tianli schists in southeastern Yangtze Craton dated at ca. 1.04–0.94 Ga is approximately synchronous with the 0.96–0.89 Ga Shuangxiwu magmatic arc rocks that suffered deformation and metamorphism (Ye et al. 2007; Li Z. X. et al. 2007; Li X. H. et al. 2009). The Shuangxiwu magmatic rocks are intruded by the ca. 0.85 Ga Shenwu dolerites that are unmetamorphosed and little deformed (Li X. H. et al. 2008a, 2008b). Concurrence of the same-aged, anorogenic Shenwu dolerites, the Zhenzhushan bimodal volcanic rocks (Li W. X. et al. 2010) and the Gangbian alkaline rocks (Li X. H. et al. 2010) in the southeastern Yangtze, as well as the intraplate basaltic intrusions in northern Cathaysia Block (Shu et al. 2006, 2011) suggests that the South China has been transformed from Sibao orogenic to the intracontinental anorogenic extension by ca. 0.85 Ga. The Sibao Group sedimentary rocks (and equivalents) dated at ca. 0.85–0.83 Ga (e.g., Gao et al. 2010; Yang et al. 2015) in southern Yangtze Craton are deposited in an anorogenic, extensional basin.

The SCB is placed between the Gondwana and Laurentia continents in terms of the summary of the IGCP-440 Project "Rodinia supercontinent assembly and break-up" (Li Z. X. et al. 2008a). The Yangtze and Cathaysia were finally amalgamated at ca. 0.9 Ga, forming the united SCB and connection of the Gondwana and Laurentia. Moreover, the Sibao Orogeny in South China is coincident with the Eastern Ghats Belt in eastern India and the Rayner Province Orogeny in eastern Antarctic, marking the final assembly of Rodinia supercontinent at ca. 0.9 Ga. The mid-Neoproterozoic mantle plume/superplume activities produced worldwide rifting and anorogenic magmatism at ca. 0.83–0.75 Ga (Fig. 9.12b). The SCB archives the best records of mid-Neoproterozoic plume-related intraplate magmatism, as it was most likey located at the plume/superplume center and then separated from the Rodinia supercontinent at ca. 0.75 Ga.

9.6 Summary

Based on integration of geochronology and geochemistry of Neoproterozoic basaltic rocks and other aspects of geological observations in South China Block, we suggest that the Group 1 basaltic rocks dated at 1.0–0.9 Ga surrounded the Yangtze Craton were formed in active continental margins as a result of double-side subduction of oceaic plates beneath the northern and southern Yangtze, whereas the same-aged basaltic rocks in the Cathaysia were fomed in an extension setting. Amalgamation between Cathaysia and Yangtze resulted in formation of the united South China Block

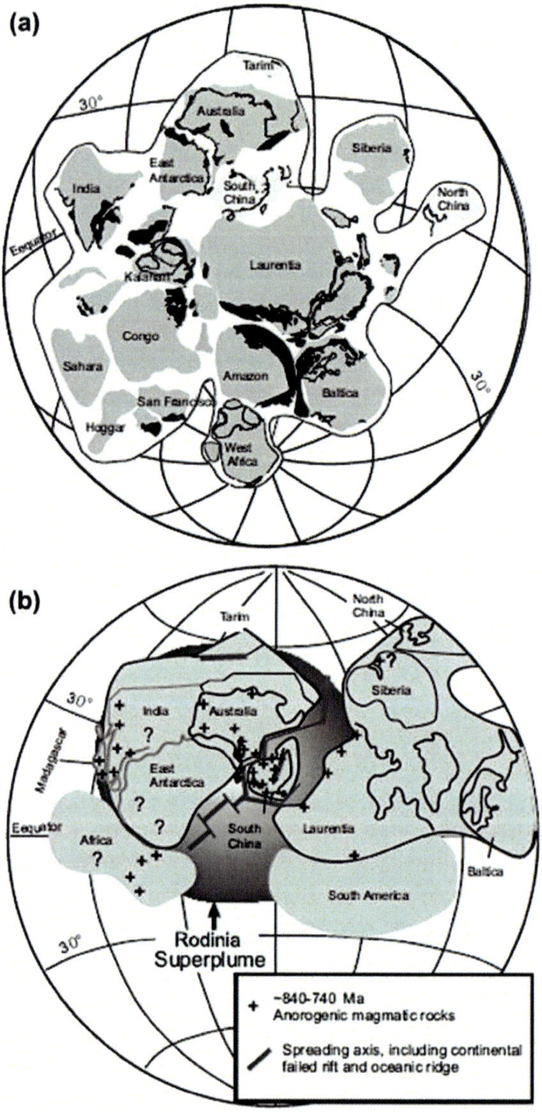

Fig. 9.12 **a** Reconstruction of supercontinent Rodinia at ca. 900 Ma; **b** a schematic diagram showing the proposed Neoproterozoic superplume beneath Rodinia at ca. 825–750 Ma (Li Z. X. et al. 2003, 2008a; modified)

during the final assembly of the Rodinia supercontinent in the Early Neoproterozoic at ca. 0.9 Ga. The mid-Neoproterozoic basaltic rocks consist of predominantly tholeiitic and subordinary alkaline basaltic rocks that initiated during the early rift phase at 0.85–0.80 Ga, and culminated during the major rift phase at 0.79–0.75 Ga. Mantle plume/superplume activities play important roles in generating the intensive and extensive mid-Neoproterozoic anorogenic magmatism in the SCB, and leading to final fragmentation of the SCB from Rodinia at ca. 0.75 Ga. While the palaeo-position of South China in Rodinia is ambiguous, we regard the Laurentia-Cathaysia-Yangtze-Australia-East Antarctica connection as the best solution based on a weight of all evidence.

References

Albarède F (1992) How deep do common basaltic magmas form and differentiate? J Geophys Res 97:10997–11009

Bao ZW, Wang Q, Bai GD, Zhao ZH, Song YW, Liu XM (2008) Geochronology and geochemistry of the fangcheng neoproterozoic alkali-syenites in east qinling orogen and its geodynamic implications. Chin Sci Bull 53:2050–2061

Bindeman IN (2011) When do we need pan-global freeze to explain ^{18}O-depleted zircons and rocks? Geology 39:799–800

Cawood PA, Wang Y, Xu Y, Zhao G (2013) Locating South China in Rodinia and Gondwana: a fragment of greater India lithosphere? Geology 41:903–906

Chen J, Foland KA, Xing F, Xu X, Zhou T (1991) Magmatism along the southeastern margin of the Yangtze block: Precambrian collision of the Yangtze and Cathaysia blocks of China. Geology 19:815–818

Chen ZH, Xing GF, Guo KY, Dong YG, Chen R, Zeng Y, Li LM, He ZY, Zhao L (2009) Petrogenesis of keratophyes in the Pingshui Gp., Zhejiang: constraints from zircon U-Pb ages and Hf isotopes. Chin Sci Bull 54:1570–1578

Deng H, Peng S, Polat A, Kusky T, Jiang X, Han Q, Wang L, Huang Y, Wang J, Zeng W, Hu Z (2017) Neoproterozoic IAT intrusion into Mesoproterozoic MOR Miaowan Ophiolite, Yangtze Craton: evidence for evolving tectonic settings. Precambr Res 289:75–94

Dong SW, Xue HM, Xiang XK, Ma LC (2010) Discovery and geological significance of the the Neoproterozoic spilite-kappiporphyry pillow lava in the Lushan area, northern Jiangxi. Geology China 37:1021–1033 (in Chinese with English abstract)

Dong YP, Liu XM, Santosh M, Zhang XN, Chen Q, Yang C, Yang Z (2011) Neoproterozoic subduction tectonics of the northwestern Yangtze Block in South China: constrains from zircon U-Pb geochronology and geochemistry of mafic intrusions in the hannan massif. Precambr Res 189:66–90

Dong YP, Liu XM, Santosh M, Chen Q, Zhang X, He D, Zhang GW (2012) Neoproterozoic accretionary tectonics along the northwestern margin of the Yangtze Block, China: constraints from zircon U-Pb geochronology and geochemistry. Precambr Res 196–197:247–274

Fan HP, Zhu WG, Li ZX, Zhong H, Bai ZJ, He DF, Chen CJ, Cao CY (2013) Ca. 1.5 Ga mafic magmatism in South China during the break-up of the supercontinent Nuna/Columbia: the Zhuqing Fe-Ti-V oxide ore-bearing mafic intrusions in western Yangtze Block. Lithos 168–169:85–98

Gao LZ, Dai CG, Liu YX, Wang M, Wang XH, Chen JS, Ding XZ, Zhang CH, Cao X, Liu JH (2010) SHRIMP U-Pb ages and stratigraphy significance of the Sibao Group tuffs in southwestern Guizhou, northern Guangxi. Geol Bull China 29:1259–1267 (in Chinese with English abstract)

Gao S, Yang J, Zhou L, Li M, Hu ZC, Guo JL, Yuan HL, Gong HJ, Xiao GQ, Wei JQ (2011) Age and growth of the Archean Kongling terrain. South China, with emphasis on 3.3 Ga granitoid gneisses. Am J Sci 311:153–182

Geng YS, Yang CH, Du LL, Wang XS, Ren LD, Zhou XW (2007) The age and formation environment of the tianbaoshan formation—zircon SHRIMP U-Pb age and geochemical evidence. Geol Rev 53:556–563 (in Chinese with English abstract)

Goodge JW, Vervoort JD, Fanning CM, Brecke DM, Farmer GL, Williams IS, Myrow PM, DePaolo DJ (2008) A positive test of East Antarctica-Laurentia juxtaposition within the Rodinia supercontinent. Science 321:235–240

Greentree MR, Li ZX (2008) The oldest known rocks in south–western China: SHRIMP U-Pb magmatic crystallisation age and detrital provenance analysis of the paleoproterozoic dahongshan group. J Asian Earth Sci 33:289–302
Greentree MR, Li ZX, Li XH, Wu H (2006) Late Mesoproterozoic to earliest neoproterozoic basin record of the Sibao orogenesis in western South China and relationship to the assembly of Rodinia. Precambr Res 151:79–100
Gu XX, Liu JM, Zheng MH, Tang JX, Qi L (2002) Provenance and tectonic settingof the Proterozoic turbidites in Hunan, South China: geochemical evidence. J Sediment Res 72:393–407
Guo JL, Gao S, Wu YB, Li M, Chen K, Hu ZC, Liang ZW, Liu YS, Zong KQ, Zhang W, Cehn HH (2014) 3.45 Ga granitic gneisses from the Yangtze Craton, South China: implications for Early Archean crustal growth. Precambr Res 242:82–95
Herzberg C, O'Hara MJ (2002) Plume-associated ultramafic magmas of phanerozoic age. J Petrol 43:1857–1883
Herzberg C, Asimow PD, Arndt N, Niu Y, Lesher CM, Fitton JG, Saunders AD (2007) Temperature in ambient mantle and plumes: constraints from basalts, picrites, and komatiites. Geochem Geophys Geosyst 8:Q02006
Hu J, Liu XC, Chen LY, Qu W, Li HK, Geng JZ (2013) A ~2.5 Ga magmatic event at the northern margin of the Yangtze craton: evidence from U/Pb dating and Hf isotopic analysis of zircons from the douling complex in the South Qinling Orogeny. Chin Sci Bull 58:3564–3579
Huang XL, Xu YG, Li XH, Li WX, Lan JB, Zhang HH, Liu YS, Wang YB, Li HY, Luo ZY, Yang QJ (2008) Petrogenesis and tectonic implications of Neoproterozoic, highly fractionated a-type granites from mianning, South China. Precambr Res 165:190–204
Huang XL, Xu YG, Lan JB, Yang QJ, Luo ZY (2009) Neoproterozoic adakitic rocks from Mopanshan in the western Yangtze Craton: partial melts of a thickened lower crust. Lithos 112:367–381
Huang DL, Wang XL, Xia XP, Wan YS, Zhang FF, Li JY, Du DH (2019) Neoproterozoic low-^{18}O zircons revisited: implications for Rodinia configuration. Geophys Res Lett 46:678–688
Jiang X, Peng S, Polat A, Kusky T, Wang L, Wu T, Lin M, Han Q (2016) Geochemistry and geochronology of mylonitic metasedimentary rocks associated with the Proterozoic Miaowan Ophiolite Complex, Yangtze Craton, China: implications for geodynamic events. Precambr Res 279:37–56
Jin WS, Sun DZ (1997) The deep crust structure and evolution of South China. Geological Publishing House, Beijing (in Chinese)
Jiao WF, Wu YB, Peng M, Wang J, Yang SH (2009) The U-Pb age and Hf isotopic composition of the oldest rocks in the Yangtze Block. Sci China Series d 39:972–978 (in Chinese)
Kirkland CL, Wingate MTD, Smithies RH (2011) 194764: monzogranite, Mount Scott. Geochronology Record, 965, Geological Survey of Western Australia, 4
Lan ZW, Li XH, Zhu MY, Chen ZQ, Zhang QR, Li QL, Lu DB, Liu Y, Tang GQ (2014) A rapid and synchronous initiation of the wide spread Cryogenian glaciations. Precambr Res 255:401–411
Lassiter JC, DePaolo DJ (1997) Plume/lithosphere interaction in the generation of continental and oceanic flood basalts: chemical and isotopic constraints. Geophys Monogr 100:335–355
Li LM, Lin SF, Xing GF, Davis DW, Davis WJ, Xiao WJ, Yin CQ (2013) Geochemistry and tectonic implications of Late Mesoproterozoic alkaline bimodal volcanic rocks from the Tieshajie Group in the southeastern Yangtze Block, South China. Precambr Res 230:179–192
Li LM, Lin SF, Xing GF, Davis DW, Jiang Y, Davis W, Zhang Y (2016) Ca. 830 Ma back-arc type volcanic rocks in the eastern part of the Jiangnan Orogen: implications for the Neoproterozoic tectonic evolution of South China Block. Precambr Res 275:209–224
Li WX, Li XH (2003) Adakitic granites within the NE Jiangxi Ophiolites, South China: geochemical and Nd isotopic evidence. Precambr Res 122:29–44
Li WX, Li XH, Li ZX (2005) Neoproterozoic bimodal magmatism in the Cathaysia block of South China and its tectonic significance. Precambr Res 136:51–66
Li WX, Li XH, Li ZX (2008a) Middle neoproterozoic syn-rifting volcanic rocks in Guangfeng, South China: petrogenesis and tectonic significance. Geol Mag 145:475–489

Li WX, Li XH, Li ZX, Lou FS (2008b) Obduction-type granites within the NE Jiangxi Ophiolite: implications for the final amalgamation between the Yangtze and Cathaysia blocks. Gondwana Res 13:288–301

Li WX, Li XH, Li ZX (2010) Ca. 850 Ma bimodal volcanic rocks in northeastern Jiangxi Province, South China: initial extension during the breakup of Rodinia. Am J Sci 310:951–980

Li XH (1997) Timing of the Cathaysia block formation: constraints from SHRIMP U-Pb zircon geochronology. Episodes 30:188–192

Li XH (1999) U-Pb zircon ages of granites from the southern margin of the Yangtze Block: timing of Neoproterozoic Jinning Orogeny in SE China and implications for Rodinia assembly. Precambr Res 97:43–57

Li XH, Sun M, Wei GJ, Liu Y, Lee CY, Malpas JG (2000) Geochemical and Sm-Nd isotopic study of amphibolites in the Cathaysia Block, SE China: evidence for extremely depleted mantle in the Paleoproterozoic. Precambr Res 102:251–262

Li XH, Li ZX, Zhou HW, Liu Y, Liang XR (2002) Zircon U-Pb chronology, elemental and Nd isotopic studies of Neoproterozoic basaltic magmatic rocks in western Sichuan: petrogenesis and geodynamic implications. Geosci Front 9:329–338

Li XH, Li ZX, Zhou HW, Liu Y, Liang XR, Li WX (2003a) SHRIMP U-Pb zircon age, geochemistry and Nd isotope of the Guandaoshan pluton in SW Sichuan: petrogenesis and tectonic significance. Sci China Ser D 46:73–83

Li XH, Li ZX, Ge W, Zhou H, Li W, Liu Y, Wingate MTD (2003b) Neoproterozoic granitoids in South China: crustal melting above a mantle plume at ca. 825 Ma. Precambr Res 122:45–83

Li XH, Li ZX, Sinclair JA, Li WX, Carter G (2006) Revisiting the "Yanbian Terrane": implications for Neoproterozoic tectonic evolution of the western Yangtze Block, South China. Precambr Res 151:14–30

Li XH, Li WX, Li ZX, Liu Y (2008a) 850–790 Ma bimodal volcanic and intrusive rocks in northern Zhejiang, South China: a major episode of continental rift magmatism during the breakup of Rodinia. Lithos 102:341–357

Li XH, Wang XC, Li WX, Li ZX (2008b) Petrogenesis and tectonic implications of oproterozoic basaltic rocks in South China: from orogeny to intracontinental rifting. Geochimica 37:382–298 (in Chinese with English abstract)

Li XH, Li WX, Li ZX, Lo CH, Wang J, Ye MF, Yang YH (2009) Amalgamation between the Yantze and Cathaysia blocks in South China: constraints from SHRIMP U-Pb zircon ages, geochemistry and Nd-Hf isotopes of the Shuangxiwu volcanic rocks. Precambr Res 174:117–128

Li XH, Li WX, Li ZX, Lo CH, Wang J, Ye MF, Yang YH (2010) Petrogenesis and tectonic significance of the ~850 Ma Gangbian alkaline complex in South China: evidence from in situ zircon U-Pb dating, Hf-O isotopes and whole-rock geochemistry. Lithos 114:1–15

Li XH, Li ZX, Li WX (2014) Detrital zircon U-Pb age and Hf isotope constrains on the generation and reworking of precambrian continental crust in the Cathaysia block, South China: a synthesis. Gondwana Res 25:1202–1215

Li ZX, Li XH, Kinny PD, Wang J (1999) The break up of Rodinia: did it start with a man the plume beneath South China. Earth Planet Sci Lett 173:171–181

Li ZX, Li XH, Zhou HW, Kinny PD (2002) Grenvillian continental collision in south China: new SHRIMP U-Pb zircon results and implications for the configuration of Rodinia. Geology 30:163–166

Li ZX, Li XH, Kinny PD, Wang J, Zhang S, Zhou H (2003) Geochronology of Neoproterozoic syn-rift magmatism in the Yangtze Craton, South China and correlations with other continents: evidence for a mantle superplume that broke up Rodinia. Precambr Res 122:85–109

Li ZX, Evans DAD, Zhang S (2004) A 90° spin on Rodinia: possible causal links between the Neoproterozoic supercontinent, superplume, true polar wander and low-latitude glaciation. Earth Planet Sci Lett 220:409–421

Li ZX, Wartho JA, Occhipinti S, Zhang CL, Li XH, Wang J, Bao C (2007) Early history of the eastern Sibao Orogen (South China) during the assembly of Rodinia: new mica $^{40}Ar/^{39}Ar$ dating and SHRIMP U-Pb detrital zircon provenance constraints. Precambr Res 159:79–94

Li ZX, Bogdanova SV, Collins AS, Davidson A, De Waele B, Ernst RE, Fitzsimons ICW, Fuck RA, Gladkochub DP, Jacobs J, Karlstrom KE, Lu S, Natapov LM, Pease V, Pisarevsky SA, Thrane K, Vernikovsky V (2008a) Assembly, configuration, and break-up history of Rodinia: a synthesis. Precambr Res 160:179–210

Li ZX, Li XH, Li WX, Ding SJ (2008b) Was Cathaysia part of Proterozoic Laurentia? New data from Hainan Island, south China. Terra Nova 20:154–164

Li ZX, Li XH, Wartho JA, Clark C, Li WX, Zhang CL, Bao CM (2010) Magmatic and metamorphic events during the Early Paleozoic Wuyi-Yunkai Orogeny, southeastern South China: new age constraints and pressure-temperature conditions. Geol Soc Am Bull 122:772–793

Lin GC, Li XH, Li WX (2007) SHRIMP U-Pb zircon age, geochemistry and Nd-Hf isotopes of the Neoproterozoic mafic dykes from western Sichuan: petrogenesis and tectonic implications. Sci China Ser D 50:1–16

Ling WL, Gao S, Zhang B, Li H, Liu Y, Cheng J (2003) Neoproterozoic tectonic evolution of the northwestern Yangtze craton South China: implications for amalgamation and break-up of the Rodinia Supercontinent. Precambr Res 122:111–140

Ling WL, Gao S, Cheng JP, Jiang LS, Yuan HL, Hu ZC (2006) Comparison of Neoproterozoic magmatic events in Yangtze block interior and margin and its tectonic significance—constraints on the U-Pb isotopic chronology of zircons from the Huangling and Hannan intrusion complexes. Acta Petrol Sin 22:387–396 (in Chinese with English abstract)

Ling WL, Ren BF, Duan RC, Liu XM, Mao XW, Peng LH, Liu ZX, Cheng JP, Yang HM (2007) Zircon U-Pb isotopic chronology and geological implications of the Wudangshan Group, Yaolinghe Group and basic intrusive rocks in the South Qinling. Chin Sci Bull 52:1445–1456 (in Chinese with English abstract)

Liu BX, Liu CG, Qiu YQ (2001) Pb-Pb isotopic ages and geological implications of the gneissic granite in Hezi, southern Jiangxi Province. Volcanol Miner Resour 22:264–268 (in Chinese with English abstract)

Liu JB, Zhang LM (2013) Neoproterozoic low to negative ^{18}O volcanic andintrusive rocks in the Qinling Mountains and their geological significance. Precambr Res 230:138–167

Liu R, Zhou H, Zhang L, Zhong Z, Zeng W, Xiang H, Jin S, Lu X, Li C (2009) Paleoproterozoic reworking of ancient crust in the Cathaysia Block, South China: evidence from zircon trace elements, U-Pb and Lu-Hf isotopes. Chin Sci Bull 54:1543–1554

Liu SW, Yang PT, Wang ZQ, Luo P, Wang YQ, Luo GH, Wang W, Guo BR (2013) Geochemistry and Zircon U-Pb age of Neoproterozoic low metamorphic volcanic rocks in Wuyuan-Dexing area, northeastern Jiangxi Province. Acta Petrol Sin 29:581–593 (in Chinese with English abstract)

Ma GG, Zhang ZC, Li HQ, Chen P, Huang ZX (1989) Study on Sinian isotope chronostratigraphy of Yangtze Platform. Yichang Inst Geol Miner Resour Published 14:83–123 (in Chinese with English abstract)

Miyashiro A (1974) Volcanic rock series in island arc and active continental margins. Am J Sci 274:321–355

Nyman MW, Karlstrom KE, Kirby E, Graubard CM (1994) Mesoproterozoic contractional orogeny in western North America: evidence from ca 1.4 Ga plutons. Geology 22:901–904

Peng SB, Kusky TM, Jiang XF, Wang L, Wang JP, Deng H (2012) Geology, geochemistry, and geochronology of the Miaowan Ophiolite, Yangtze Craton: implications for South China's amalgamation history with the Rodinian supercontinent. Gondwana Res 21:577–594

Pisarevsky SA, Elming SÅ, Pesonen LJ, Li ZX (2013) Mesoproterozoic paleogeography: supercontinent and beyond. Precambr Res 244:207–225

Qiu XF, Ling WL, Liu XM, Kusky T, Berkana W, Zhang YH, Gao YJ, Lu SS, Kuang H, Liu CX (2011) Recognition of Grenvillian volcanic suite in the Shennongjia region and its tectonic significance for the South China Craton. Precambr Res 191:101–119

Qiu YM, Gao S, McNaughton NJ, Groves DI, Ling WL (2000) First evidence of ~3.2 Ga continental crust in the Yangtze Craton of south China and its implications for Archean crustal evolution and Phanerozoic tectonics. Geology 28:11–14

Shen WZ, Gao JF, Xu SJ, Zhou GQ (2002) Geochemical characteristics and genesis of the Qiaotou basic complex, Luding County, western Yangtze Block. Geol J China Univ 8:380–389 (in Chinese with English abstract)

Shu LS, Faure M, Jiang SY, Yang Q, Wang YJ (2006) SHRIMP zircon U-Pb age, litho- and biostratigraphic analyses of the Huaiyu Domain in South China—evidence for a Neoproterozoic orogen, not Late Paleozoic-Early Mesozoic collision. Episodes 29:244–252

Shu LS, Deng P, Yu JH, Wang YB, Jiang SY (2008) The age and geochemical characteristics of rhyolite in the western Wuyi. Sci China Ser D 38:950–959 (in Chinese with English abstract)

Shu LS, Faure M, Yu JH, Jahn BM (2011) Geochronological and geochemical features of the Cathaysia Block (South China): new evidence for the Neoproterozoic breakup of Rodinia. Precambr Res 187:263–276

Sinclair JA (2001) A re-examination of the "Yanbian Ophiolite Suite": evidence for western extension of the Mesoproterozoic Sibao Orogen in South China. Geol Soc Australia Abstract 65:992100

Sugawara T (2000) Empirical relationships between temperature, pressure, and MgO content in olivine and pyroxene saturated liquid. J Geophys Res 105:8457–8472

Sun M, Chen NS, Zhao GC, Wilde SA, Ye K, Guo JH, Chen Y, Yuan C (2008a) U-Pb zircon and Sm-Nd isotopic study of the Huangtuling granulite, Dabie-Sulu belt, China: implication for the Paleoproterozoic tectonic history of the Yangtze Craton. Am J Sci 308:469–483

Sun WH, Zhou MF (2008) The 860-Ma, Cordilleran-type Guandaoshan dioritic pluton in the Yangtze Block, SW China: implications for the origin of Neoproterozoic magmatism. J Geol 116:238–253

Sun WH, Zhou MF, Yan DP, Li JW, Ma YX (2008b) Provenance and tectonic setting of the Neoproterozoic Yanbian Group, western Yangtze Block (SW China). Precambr Res 167:213–236

Vermeesch P (2006) Tectonic discrimination diagrams revisited. Geochem Geophys Geosyst 7:Q06017

Wan YS, Liu DY, Xu M, Zhuang J, Song B, Shi Y, Du L (2007) SHRIMP U-Pb zircon geochronology and geochemistry of metavolcanic and metasedimentary rocks in Northwestern Fujian, Cathaysia Block, China: tectonic implications and the need to redefine lithostratigraphic units. Gondwana Res 12:166–183

Wang J (2000) Evolution of Neoproterozoic Rift Basins in South China—the relationship with the break up of Rodinia. Geological Publishing House, Beijing (in Chinese)

Wang J, Li ZX (2003) History of Neoproterozoic rift basins in South China: implications for Rodinia break-up. Precambr Res 122:141–158

Wang J, Li XH, Duan TZ, Liu DY, Song B, Li ZX, Gao YH (2003) SHRIMP U-Pb dating of Cangshuipu volcanic zircon and new evidences of the base of Nanhua. Chin Sci Bull 48:1726–1731 (in Chinese)

Wang MX, Wang Y, Zhao JH (2013) Zircon U/Pb dating and Hf-O isotopes of the Zhouan ultramafic intrusion in the northern margin of the Yangtze Block, SW China: constraints on the nature of mantle source and timing of the supercontinent Rodinia breakup. Chin Sci Bull 58:777–787

Wang Q, Wyman DA, Li ZX, Bao ZW, Zhao ZH, Wang YX, Jian P, Yang YH, Chen LL (2010) Petrology, geochronology and geochemistry of ca. 780 Ma A-type granites in South China: petrogenesis and implications for crustal growth during the breakup of the supercontinent Rodinia. Precambr Res 178:185–208

Wang W, Zhou MF, Yan DP, Li JW (2012) Depositional age, provenance, and tectonic setting of the Neoproterozoic Sibao Group, southeastern Yangtze Block, South China. Precambr Res 192–195:107–124

Wang W, Zhou MF, Yan DP, Li L, John M (2013) Detrital zircon record of Neoproterozoic active-margin sedimentation in the eastern Jiangnan Orogen, South China. Precambr Res 235:1–19

Wang W, Zhou MF, Zhao JH, Pandit MK, Zheng JP, Liu ZR (2016) Neoproterozoic active continental margin in the southeastern Yangtze Block of South China: evidence from the ca. 830–810 Ma sedimentary strata. Sed Geol 342:254–267

Wang W, Cawood PA, Zhou MF, Pandit MK, Xia XP, Zhao JH (2017) Low-O rhyolites from the Malani Igneous Suite: a positive test for South China and NW India linkage in Rodinia. Geophys Res Lett 44:10298–10305

Wang XC, Li XH, Li WX, Li ZX (2007) Ca. 825 Ma komatiitic basalts in South China: first evidence for >1500°C mantle melts by a Rodinian mantle plume. Geology 35:1103–1106

Wang XC, Li XH, Li WX, Li ZX, Liu Y, Yang YH, Liang XR, Tu XL (2008) The Bikou basalts in northwestern Yangtze Block, South China: remains of 820–810 Ma continental flood basalts. Geol Soc Am Bull 120:1478–1492

Wang XC, Li XH, Li WX, Li ZX (2009) Variable involvements of mantle plumes in the genesis of mid-Neoproterozoic basaltic rocks in South China: a review. Gondwana Res 15:381–395

Wang XC, Li ZX, Li XH, Li QL, Tang GQ, Zhang QR, Liu Y (2011) Nonglacialorigin for low-18O Neoproterozoic magmas in the South China Block: evidence from new in-situ oxygen isotope analyses using SIMS. Geology 39:735–738

Wang XL, Zhou JC, Qiu JS, Gao JF (2004) Geochemistry of the Meso- to Neoproterozoic basic-acid rocks from Hunan Province South China: implications for the evolution of the western Jiangnan orogen. Precambr Res 135:79–103

Wang XL, Zhou JC, Qiu JS, Zhang W, Liu X, Zhang G (2006) LA-ICP-MS U-Pb zircon geochronology of the Neoproterozoic igneous rocks from Northern Guangxi, South China: implications for tectonic evolution. Precambr Res 145:111–130

Wang XL, Zhou JC, Griffin WL, Wang RC, Qiu JS, O'Reilly SY, Xu XS, Liu XM, Zhang GL (2007) Detrital zircon geochronology of Precambrian basement sequences in the Jiangnan Orogen: dating the assembly of the Yangtze and Cathaysia blocks. Precambr Res 159:117–131

Wang XL, Zhao GC, Zhou JC, Liu YS, Hu J (2008b) Geochronology and Hf isotopes of zircon from volcanic rocks of the Shuangqiaoshan group, South China: implications for the Neoproterozoic tectonic evolution of the eastern Jiangnan Orogen. Gondwana Res 14:355–367

Wang XL, Shu LS, Xing GF, Zhou JC, Tang M, Shu XJ, Qi L, Hu YH (2012) Post-orogenic extension in the eastern part of the Jiangnan Orogen: evidence from ca 800–760 Ma volcanic rocks. Precambr Res 222–223:404–423

Wang YJ, Zhang AM, Cawood PA, Fan WM, Xu J, Zhang G, Zhang Y (2013) Geochronological, geochemical and Nd-Hf-Os isotopic fingerprinting of an early Neoproterozoic arc–back-arc system in South China and its accretionary assembly along the margin of Rodinia. Precambr Res 231:343–371

Winchester JA, Floyd PA (1976) Geochemical magma type discrimination: application to altered and metamorphosed basic igneous rocks. Earth Planet Sci Lett 28:459–469

Wu L, Jia D, Li H, Deng F, Li Y (2010) Provenance of detrital zircons from the late Neoproterozoic to Ordovician sandstones of South China: implications for its continental affinity. Geol Mag 147:974–980

Wu RX, Zheng YF, Wu YB, Zhao ZF, Zhang SB, Liu XM, Wu FY (2006) Reworking of juvenile crust: element and isotope evidence from Neoproterozoic granodiorite in South China. Precambr Res 146:179–212

Wu T, Wang XC, Li WX, Wilde SA, Pang CJ, Li J (2018) The 825 Ma Yiyang high-MgO basalts of central South China: insights from Os-Hf-Nd data. Chem Geol 502:107–121

Wu YB, Zheng YF, Gao S, Jiao WF, Liu YS (2008) Zircon U-Pb age and trace element evidence for Paleoproterozoic granulite facies metamorphism and Archean crustal rocks in the Dabie Orogen. Lithos 101:308–322

Wu YB, Gao S, Zhang HF, Zheng JP, Liu XC, Wang H, Gong HJ, Zhou L, Yuan HL (2012) Geochemistry and zircon U-Pb geochronology of Paleoproterozoic arc related granitoid in the Northwestern Yangtze Block and its geological implications. Precambr Res 200–203:26–37

Wu YB, Zhou G, Gao S, Liu X, Qin Z, Wang H, Yang J, Yang S (2014) Petrogenesis of Neoarchean TTG rocks in the Yangtze Craton and its implication for the formation of Archean TTGs. Precambr Res 254:73–86

Xia LQ, Xia ZC, Ma ZP, Xu XY, Li XM (2009) Petrogenesis of volcanic rocks from Xixiang group in middle part of South Qinling Mountains. Northwest Geol 42:1–37 (in Chinese with English abstract)
Xia Y, Xu XS, Zhu KY (2012) Paleoproterozoic S- and A-type granites in southwestern Zhejiang: magmatism, metamorphism and implications for the crustal evolution of the Cathaysia basement. Precambr Res 216–219:177–207
Xiang H, Zhang L, Zhou HW, Zhong ZQ, Zeng W, Liu R, Jin S (2008) U-Pb zircon geochronology and Hf isotope study of metamorphosed basic-ultrabasic rocks from metamorphic basement in southwestern Zhejiang: the response of the Cathaysia block to Indosinian orogenic event. Sci China Ser D Earth Sci 51:788–800
Xiong Q, Zheng JP, Yu CM, Shu YP, Tang HY, Zhang ZH (2008) Yichang chair Chong a type granite zircon U-Pb age and Hf isotope and the Yangtze, Paleoproterozoic Tonghua carat effect. Chin Sci Bull 53:2782–2792 (in Chinese with English abstract)
Xue HM, Ma F, Song YQ (2011) The Neoproterozoic metamorphic rocks in the northern Yangtze Craton Suizhou-Jujube area, geochemistry and SHRIMP U-Pb zircon geochronology. Acta Petrol Sin 27:1116–1130 (in Chinese with English abstract)
Yang C, Li XH, Wang XC, Lan Z (2015) Mid-Neoproterozoic angular unconformity in the Yangtze Block revisited: insights from detrital zircon U-Pb age and Hf-O isotopes. Precambr Res 266:156–178
Yang CH, Geng YS, Du LL, Ren LD, Wang XS, Zhou XW, Yang ZS (2009) The identification of the Grenvillian granite on the western margin of the Yangtze Block and its geological implications. Geology China 26:647–657 (in Chinese with English abstract)
Yao JL, Shu LS, Santosh M, Li JY (2012) Precambrian crustal evolution of the South China Block and its relation to supercontinent history: constraints from U-Pb ages, Lu-Hf isotopes and REE geochemistry of zircons from sandstones and granodiorite. Precambr Res 208–211:19–48
Yao JL, Shu LS, Santosh M, Zhao GC (2014) Neoproterozoic arc-related mafic-ultramafic rocks and syn-collision granite from the western segment of the Jiangnan Orogen, South China: constraints on the Neoproterozoic assembly of the Yangtze and Cathaysia Blocks. Precambr Res 243:39–62
Yao JL, Shu LS, Santosh M, Li J (2015) Neoproterozoic arc-related andesite and orogeny-related unconformity in the eastern Jiangnan Orogenic Belt: constraints on the assembly of the Yangtze and Cathaysia blocks in South China. Precambr Res 262:84–100
Yao WH, Li ZX, Li WX, Li XH (2017) Proterozoic tectonics of Hainan Island in supercontinent cycles: new insights from geochronological and isotopic results. Precambr Res 290:86–100
Ye MF, Li XH, Li WX, Liu Y, Li ZX (2007) SHRIMP U-Pb zircon geochronological and geochemical evidence for Early Neoproterozoic Sibaoan magmatic arc along the southeastern margin of Yangtze Block. Gondwana Res 12:144–156
Yu JH, O'Reilly SY, Wang L, Griffin WL, Zhang M, Wang R, Jiang S, Shu L (2008) Where was South China in the Rodinia supercontinent? Evidence from U-Pb geochronology and Hf isotopes of detrital zircons. Precambr Res 164:1–15
Yu JH, Wang LJ, Griffin WL, O'Reilly SY, Zhang M, Li CZ, Shu LS (2009) A Paleoproterozoic orogeny recorded in a long-lived cratonic remnant (Wuyishan terrane), eastern Cathaysia Block, China. Precambr Res 174:347–363
Yu JH, O'Reilly SY, Zhou MF, Griffin WL, Wang LJ (2012) U-Pb geochronology and Hf-Nd isotopic geochemistry of the Badu Complex, Southeastern China: implications for the Precambrian crustal evolution and paleogeography of the Cathaysia Block. Precambr Res 222–223:424–449
Zhang AM, Wang YJ, Fan WM, Zhang YZ, Yang J (2012) Earliest Neoproterozoic (ca. 1.0 Ga) arc-back-arc basin nature along the northern Yunkai Domain of the Cathaysia Block: geochronological and geochemical evidence from the metabasite. Precambr Res 220–221:217–233
Zhang CH, Gao LZ, Wu ZJ, Shi XY, Yan QR, Li D (2007) SHRIMP U-Pb zircon age of tuff from the Kunyang Group in central Yunnan: evidence for Grenvillian Orogeny in South China. Chin Sci Bull 52:1517–1525

Zhang CL, Santosh M, Zou HB, Li HK, Huang WC (2013a) The Fuchuan Ophiolite in Jiangnan Orogen: geochemistry, zircon U-Pb geochronology, Hf isotope and implications for the Neoproterozoic assembly of South China. Lithos 179:263–274

Zhang CL, Li HK, Santosh M (2013b) Revisiting the tectonic evolution of South China: interaction between the Rodinia superplume and plate subduction? Terra Nova 25:212–220

Zhang CL, Zou HB, Zhu QB, Chen XY (2015) Late Mesoproterozoic to early Neoproterozoic ridge subduction along zsouthern margin of the Jiangnan Orogen: new evidence from the northeastern Jiangxi Ophiolite (NJO), South China. Precambr Res 268:1–15

Zhang LJ, Ma CQ, Wang LX, Se ZB, Wang SM (2011) Discovery of Paleoproterozoic rapakivi granite on the northern margin of the Yangtze block and its geological significance. Chin Sci Bull 2011:56 (in Chinese)

Zhang SB, Zheng YF, Zhao ZF, Wu YB, Yuan HL, Wu FY (2009) Origin of TTG-like rocks from anatexis of ancient lower crust: geochemical evidence from Neoproterozoic granitoids in South China. Lithos 113:347–368

Zhang SB, Wu RX, Zheng YF (2012) Neoproterozoic continental accretion in South China: geochemical evidence from the Fuchuan Ophiolite in the Jiangnan Orogen. Precambr Res 220–221:45–64

Zhang SH, Li ZX, Evans DAD, Wu HC, Li HY, Dong J (2012) Pre-Rodinia supercontinent Nuna shaping up: a global synthesis with new paleomagnetic results from North China. Earth Planet Sci Lett 353–354:145–155

Zhang YZ, Wang YJ (2016) Early Neoproterozoic (~840 Ma) arc magmatism: geochronological and geochemical constraints on the metabasites in the Central Jiangnan Orogen. Precambr Res 275:1–17

Zhang YZ, Wang YJ, Fan WM, Zhang AM, Ma LY (2013) Geochronological and geochemical constraints on the metasomatised source for the Neoproterozoic (similar to 825 Ma) high-Mg volcanic rocks from the Cangshuipu area (Hunan Province) along the Jiangnan domain and their tectonic implications. Precambr Res 220:139–157

Zhao G, Cawood PA (2012) Precambrian geology of China. Precambr Res 222–223:13–54

Zhao JH, Asimow PD (2014) Neoproterozoic boninite-series rocks in South China: a depleted mantle source modified by sediment-derived melt. Chem Geol 388:98–111

Zhao JH, Asimow PD (2018) Formation and evolution of a magmatic system in a rifting continental margin: Neoproterozoic arc- and MORB-like dike swarms in South China. J Petrol 59:1811–1844

Zhao JH, Zhou MF (2007a) Geochemistry of Neoproterozoic mafic intrusions in the Panzhihua district (Sichuan Province, SW China): implications for subduction related metasomatism in the upper mantle. Precambr Res 152:27–47

Zhao JH, Zhou MF (2007b) Neoproterozoic adakitic plutons and arc magmatism along the western margin of the Yangtze block, South China. J Geol 115:675–689

Zhao JH, Zhou MF (2008) Neoproterozoic adakitic plutons in the northern margin of the Yangtze Block, China: partial melting of a thickened lower crust and implications for secular crustal evolution. Lithos 104:231–248

Zhao JH, Zhou MF (2009) Secular evolution of the Neoproterozoic lithospheric mantle underneath the northern margin of the Yangtze Block, South China. Lithos 107:152–168

Zhao JH, Zhou MF (2013) Neoproterozoic high-Mg basalts formed by melting of ambient mantle in South China. Precambr Res 233:193–205

Zhao JH, Zhou MF, Yan DP, Yang YH, Sun M (2008) Zircon Lu-Hf isotopic constraints on Neoproterozoic subduction-related crustal growth along the western margin of the Yangtze Block, South China. Precambr Res 163:189–209

Zhao JH, Zhou MF, Yan DP, Zheng JP, Li JW (2011) Reappraisal of the ages of Neoproterozoic strata in South China: no connection with the Grenvillian Orogeny. Geology 39:299–302

Zhao XF, Zhou MF (2011) Fe-Cu deposits in the Kangdian region, SW China: a Proterozoic IOCG (iron-oxide-copper-gold) metallogenic province. Miner Deposit 46:731–747

Zhao XF, Zhou MF, Li JW, Wu FY (2008) Association of Neoproterozoic A- and I-type granites in South China: implications for generation of A-type granites in a subduction-related environment. Chem Geol 257:1–15

Zhao XF, Zhou MF, Li JW, Sum M, Gao JF, Sun WH, Yang JH (2010) Late Paleoproterozoic to Early Mesoproterozoic Dongchuan Group in Yunnan, SW China: implications for tectonic evolution of the Yangtze Block. Precambr Res 182:57–69

Zheng YF, Wu YB, Chen FK, Gong B, Li L, Zhao ZF (2004) Zircon U-Pb and oxygen isotope evidence for a large-scale ^{18}O depletion event in igneous rocks during the Neoproterozoic. Geochim Cosmochim Acta 68:4145–4165

Zheng YF, Zhang SB, Zhao ZF, Wu YB, Li XH, Li ZX, Wu FY (2007) Contrasting zircon Hf and O isotopes in the two episodes of Neoproterozoic granitoids in South China: implications for growth and reworking of continental crust. Lithos 96:127–150

Zheng YF, Gong B, Zhao ZF, Wu YB, Chen FK (2008a) Zircon U-Pb age and O isotope evidence for Neoproterozoic low-^{18}O magmatism during supercontinental rifting in South China: implications for the snowball earth event. Am J Sci 308:484–516

Zheng YF, Wu RX, Wu YB, Zhang SB, Yuan HL, Wu FY (2008b) Rift melting of juvenile arc-derived crust: geochemical evidence from Neoproterozoic volcanic and granitic rocks in the Jiangnan Orogen, South China. Precambr Res 163:351–383

Zheng YF, Gong B, Zhao ZF, Wu YB, Chen FK (2009a) Zircon U-Pb age and O isotope evidence for neoproterozoic low-^{18}O magmatism during supercontinental rifting in South China: implications for the snowball earth event. Am J Sci 308:484–516

Zheng YF, Chen RX, Zhao ZF (2009b) Chemical geodynamics of continental subduction-zone metamorphism: insights from studies of the Chinese Continental Scientific Drilling (CCSD) core samples. Tectonophysics 475:327–358

Zhou JB, Li XH, Ge WC, Li ZX (2007) Age and origin of middle Neoproterozoic mafic magmatism in southern Yangtze Block and relevance to the break-up of Rodinia. Gondwana Res 12:184–197

Zhou JC, Wang XL, Qiu JS (2009) Geochronology of Neoproterozoic mafic rocks and sandstones from northeastern Guizhou, South China: coeval arc magmatism and sedimentation. Precambr Res 170:27–42

Zhou JL, Li XH, Tang GQ, Gao BY, Bao ZA, Ling XX, Wu LG, Lu K, Zhu YS, Liao X (2018) Ca. 890 Ma magmatism in the northwest Yangtze block, South China: SIMS U-Pb dating, in-situ Hf-O isotopes, and tectonic implications. J Asian Earth Sci 151:101–111

Zhou MF, Yan DP, Kennedy AK, Li Y, Ding J (2002a) SHRIMP U-Pb zircon geochronological and geochemical evidence for Neoproterozoic arc-magmatism along the western margin of the Yangtze block, South China. Earth Planet Sci Lett 196:51–67

Zhou MF, Kennedy AK, Sun M, Malpas J, Lesher CM (2002b) Neoproterozoic arc-related mafic intrusions along the northern margin of South China: implications for the accretion of Rodinia. J Geol 110:611–618

Zhou MF, Yan DP, Wang CL, Qi L, Kennedy A (2006) Subduction-related origin of the 750 Ma Xuelongbao adakitic complex (Sichuan Province, China): implications for the tectonic setting of the giant Neoproterozoic magmatic event in South China. Earth Planet Sci Lett 248:286–300

Zhu WG, Zhong H, Li XH, Liu BG, Deng HL, Qin Y (2007) ^{40}Ar-^{39}Ar age, geochemistry and Sr-Nd-Pb isotopes of the Neoproterozoic Lengshuiqing Cu-Ni sulfide-bearing mafic-ultramafic complex, SW China. Precambr Res 155:98–124

Zhu WG Zhong H, Li XH, Deng HL, He DF, Wu KW, Bai ZJ (2008) SHRIMP zircon U-Pb geochronology, elemental, and Nd isotopic geochemistry of the Neoproterozoic mafic dykes in the Yanbian area, SW China. Precambr Res 164:66–85

Zhu WG, Li XH, Zhong H, Wang XC, He DF, Bai ZJ, Liu F (2010) The Tongde picritic dykes in the Western Yangtze Block: evidence for ca. 800 Ma mantle plume magmatism in South China during the breakup of Rodinia. J Geol 118:509–522

Zhu WG, Zhong H, Li ZX, Bai ZJ, Yang YJ (2016) SIMS zircon U-Pb ages, geochemistry and Nd-Hf isotopes of ca. 1.0 Ga mafic dykes and volcanic rocks in the Huili area, SW China: Origin and tectonic significance. Precambr Res 273:67–89

Chapter 10
Petrogenesis and Emplacement Age of the Gabbro-Diabase Sills Within the Mesoproterozoic Xiamaling Formation in Yanliao Faulted-Depression Zone, North China Craton

Li Su, Tieguan Wang, Xianhua Li, Shuguang Song, Shuwen Yang, and Hongyu Zhang

Abstract Basic intrusions are commonly found within the Xiamaling Formation in an E–W–trending belt of more than 400 km along the north of the Yanliao Faulted-Depression Zone (YFDZ), North China Craton (NCC). And in the Jibei Depression in particular, up to four lit-par-lit intruded gabbro-diabase sills (named as $\beta\mu1$–$\beta\mu4$ in ascending order) can be observed with a cumulative thickness of 117.5–312.3 m. By contrast, no basic intrusion has be seen within the Xiamaling Formation along the south of the YFDZ, indicating that the northern YFDZ experienced a period of intensive intrusion of mantle-derived magma. The distribution and scale of intrusive masses indicate that the center of basaltic magma intrusion should be in the Jibei depression and that the activity of basaltic magma was controlled by the discordogenic faults on the northern boundary of YFDZ. The SIMS 207Pb/206Pb dating results of baddeleyite revealed that the emplacement ages of $\beta\mu1$ and $\beta\mu3$ gabbro-diabase sills are 1327.5 $\pm$ 2.4 Ma and 1327.3 $\pm$ 2.3 Ma (equivalent to Mesoproterozoic) respectively. Petrological and geochemical features show that the $\beta\mu1$–$\beta\mu4$ gabbro-diabase sills are characteristically similar to those of typical within-plate basalt (WPB). These results suggest that the Xiamaling gabbro-diabase sills in Jibei Depression may be related to rifting of the supercontinent Columbia. Comprehensive analysis of the distribution and age of the bentonitic tuff (1366–1372 Ma) and diabase dike (dike groups; ca. 1345 Ma) in the same region suggests that the ca.

L. Su (✉) · S. Yang · H. Zhang
Institute of Earth Science and State Key Laboratory of Geological Process and Mineral Resources, China University of Geosciences, Beijing 100083, China

T. Wang
State Key Laboratory of Petroleum Resources and Prospect, University of Petroleum, Beijing, Beijing 102249, China

X. Li
State Key Laboratory of Lithospheric Evolution, Institute of Geology and Geophysics, Chinese Academy of Sciences, Beijing 100029, China

S. Song
School of Earth and Space Sciences, Peking University, Beijing 100871, China

T. Wang, *Meso-Neoproterozoic Geology and Petroleum Resources in China*, Springer Geology, https://doi.org/10.1007/978-981-19-5666-9_10

1327 Ma gabbro-diabase sills in the YFDZ likely manifaest the rift development and separation of the NCC from Supercontinent Columbia.

Keywords Gabbro-diabase sill · Xiamaling formation · Continental rift · North China Craton (NCC) · Mesoproterozoic

10.1 Regional Geological Setting

The North China Craton (NCC) is bounded by the Inner Mongolia-Hinggan Orogeny to the north, the Qinling–Dabie–Sulu Orogeny to the southwest, and the Tanlu Fault to the east, presents as a giant delta-shaped old continental block. The Lüliang Movement led to formation of several Meso-Neoproterozoic Faulted-Depression Zones in the North China Craton, including the Bayan Obo, Yanliao and Fanhe Faulted-Depression Zones, etc. in the north, the Xiong'er Faulted-Depression Zone in the southwest and the Jiao-Liao-Xu-Huai Faulted-Depression Zone in the east. Within these faulted-depression zones, extremely thick Late Palaeoproterozoic-Neoproterozoic sediments were deposited (Qiao and Gao 1999).

The Yanliao Faulted-Depression Zone (YFDZ) comprises the Liaoxi, Jibei, Jidong, Jingxi and Xuanlong Depressions as well as the Shanhaiguan and Mihuai Uplifts (Fig. 10.1). Within the depressions, the Meso-Neoproterozoic sedimentary covers can be divided into four stratigraphic units: Mesoproterozoic Changchengian (Pt21), Jixianian (Pt22), Xiamaling Formation (Pt23x), and Neoproterozoic Qingbaikouan (Pt31) with a total thickness of 4095–9260 m. Among these depressions, the stratigraphic sequences from the Changchengian to Qingbaikouan appears as a significant similarity either in lithology and in lithofacies, which indicates that whole the YFDZ should contain a unified structural-sedimentary palaeo-environment during the Meso-Neoproterozoic time, and there is no obvious separation by uplifts between depressions. The Meso-Neoproterozoic strata are thicker in the east (i.e. Jidong-Liaoxi-Jibei Depressions), but thinner in the west (i.e. Xuanlong-Jingxi Depressions) in the YFDZ, with the main depocenter located around the Jidong-Jibei Depressions where a total stratigraphic thickness up to 8143–9260 m is observed (Fig. 10.1).

In the YFDZ, the upper part of the Mesoproterozoic Xiamaling formation (Pt23x) consists dominantly of fine-grain clastic rocks containing greyish-black, yellowish-green shale, sandy shale, even oil-shale with sandstone and siltstone. Its basal interval is a layer of 2–4 m thick pure siliceous cement quartzose sandstone on the disconformable interface with the underlying limestone of Jixianian Tieling formation (Pt22t), nominated as "Qinyu Uplifting". Its top interval was more or less denuded and the overlaid by the sandstone of Neoproterozoic Qingbaikouan Luotuoling formation (Pt31l) with a disconformity in between, called "Yuxian Uplfting". Therefore, in different depressions of the YFDZ, the Xiamaling strata have variable residual thickness, ranging from 540.6 m in Xuanlong depression to 168 m in Jidong depression (Table 10.1).

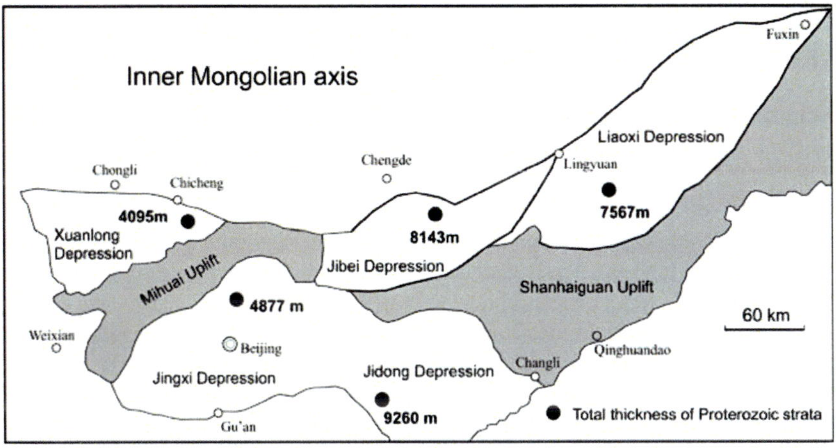

Fig. 10.1 A sketch showing the structural division and sedimentary thickness of the Yanliao faulted-depression zone on the northern margin of the North China Craton (Wang 1980, modified)

Table 10.1 Statigraphic thickness of Xiamaling formation in YFDZ

Stratigraphy	Stratigraphic thickness (m)				
	Liaoxi depression	Jibei depression	Xuanlong depression	Jidong depression	Jingxi depression
Xiamaling Formation (Pt23x)	303.4	369.5	540.6	168	249

The extensively outcropped 1.4–1.32 Ga tuff and gabbro-diabase sills in the Xiamaling formation (Gao et al. 2007, 2008a, b; Li et al. 2009a, b; Su et al. 2008; Zhang et al. 2009, 2012) suggest that the mantle-derived magmatism was at the summit in the YFDZ. Diabase intrusions of the Xiamaling formation can be seen in an EW-trending belt from Zhangjiakou to Lingyuan, extending more than 400 km along the Xuanlong-Jibei-Liaoxi depressions in the north of the YFDZ. As many as four layers of concordant intrusive gabbro-diabase sills are observed in the Xiamaling Formation of Jibei Depression, named βμ1–βμ4 from bottom upwards. Statistics of 11 stratigraphic outcrop sections of the Xiamaling Formation in Chengde, Pingquan, Kuancheng and Lingyuan Cities/Counties in the Jibei Depression indicate that only the Dishuiyan section in Chengde has 4 layers of gabbro-diabase sills, while other sections only have 3 layers or one layer. The thickness of indivedul sills varies from 13.3 m (βμ2 at Weizigou, Kuancheng) to 143.5 m (βμ1 at Shuangdong, Pingquan). Among these sills, βμ1 is the thickest one, typically more than 63.5 m thick. The cumulative thickness varies from 117.5 m (Zhenggou in Kuancheng) to 312.3 m (Dishuiyan in Chengde), acoungting for 42.7–62.2% of the total Xiamaling stratigraphic thickness (Table 10.2). Figure 10.2 shows a fence diagram illustrating the gabbro-diabase sills of 5 Xiamaling stratigraphic sections around Kuancheng and

Pingqian shows lateral variation of the occurrence mode of βμ1–βμ3 gabbro-diabase sills.

As compared with the sills in Jibei Depression, it is noteworthy that not only the Xiamaling diabase intrusions in the Xuanlong Depression resulted in more diabase sills/veins, but also the thickness of indivedul sill became thiner so as to be veins or thin-bedded sills, e.g. these at Zhaojiashan section, in Huailai, Hebei province shows totally 9 layers of the Xiamaling diabase veins/sills, and the indivedual thickness ranges from 0.85 to 31.5 m with a cumulative thickness of only 91 m. The thickness of wall-rock alteration zone is also limited either on the roof- or at the foot-walls, which indicates a weak thermo-optical alteration effect to the Xiamaling black shale.

Moreover, no basic intrusion has been found within the Xiamaling Formation in the Jingxi-Jidong Depressions in the south of YFDZ. Therefore, the center of Late Mesoproterozoic basic magmatism in YFDZ would be most probably located around the Jibei Depression in terms of distribution of the basic igneous rock; and the magmatism is obviously controlled by the discordogenic faults on the northern margin of the YFDZ.

Taking the Longtangou outcrop section in Jibei Depression as an example, the thermo-optical alteration on roof- and at foot-walls of βμ1–βμ3 sills is remarkable, among which, the thickest βμ1 shows the most notable alteration effect with a 16.1 m thick foot-wall alteration zone where shale was altered into slate or hornfels, sandy and lime bandings were turned into quartzite and marble bands, respectively. While the roof-wall alteration zone is up to 60 m thick, but the alteration effect becomes weaker and heterogeneous, shale was only transformed into slate or carboniferous shale. As a result of intrusion of βμ1–βμ3 sills, organic matter of the Xiamaling black shale is in an over-mature phase and the hydrocarbon potential has been totally lost (Fig. 10.3).

A geological shallow borehole, Jiqian-2 Well, was deployed at Beizhangzi Village, Kuancheng County on the northern margin of the Guozhangzi monocline belt (Fig. 10.4a). The drilling commenced from the central sandstone interval of Xiamaling Formation, penetrating 115 m thick cryptocrystalline βμ1 dia base sill. Within the βμl sill, there is local facies change from cryptocrystalline gabbro-diabase to medium-crystal gabbro at the well depth of 66.1 m and 111.3 m respectively (Fig. 10.5).

10.2 Petrology of Gabbro-Diabase Sills

10.2.1 Petrological Characteristics of Wall-Rock Alteration Zones

Petrological observation on the core of Jiqian-2 Hole and surface outcrops discovered that there are wall-rock alteration zones found on the roof- and at the foot-walls of gabbro-diabase sills in the Jibei Depression, which are characterized mainly by

Table 10.2 Statistics of number and thickness of gabbro-diabase sills within the Xiamaling Formation in Jibei Depression[a]

Lithology	Chengde/m	Kuancheng (m)						Pingqian (m)			Lingyuan (m)
	Dishuiyan	Laoyemiao	Shanchakou	Erdaogou	Zhenggou	Weizigou	Yaodinggou	Shangzhuang	Xiaojinzhangzi	Shuangdong	Longtangou
Wall-rock	1.3	44.4	250.3	136.1	122	70.5	67.7	222.5	221.6	absence	61.2
βμ4 sill	67.1										
Wall-rock	74.8										
βμ3 sill	47.2	54.4				58.7	83			66.7	65.9
Wall-rock	47.0	28.3				66.7	43.76			79.1	38.9
Bμ2 sill	113.9	33.3				13.3	66			32.5	19.0
Wall-rock	22.1	100.1				42.8	44.74			111.6	150.2
βμ1 sill	84.1	116.3	81.1	109.4	63.5	97.0	77.0	69.5	78.9	143.5	79.6
Wall-rock	45.6	43.5	33.8	87.6	89.9	112.5	126.21	33.9	45.9	44.9	37.5
Total sill thickness /m	312.3	204.0	231.2	142.2	117.5	169.0	226.0	161.6	165.4	242.7	164.5

Note [a]Wang T G, Gao Z Z, Liu H B, et al. 1979. Principal characteristics of petroleum geology in the eastern segment of Yansan region (Internal Report). The 3rd Regional Geological Explorational Team, Jingzhou: Jianghan Institute of Petroleum (in Chinese)

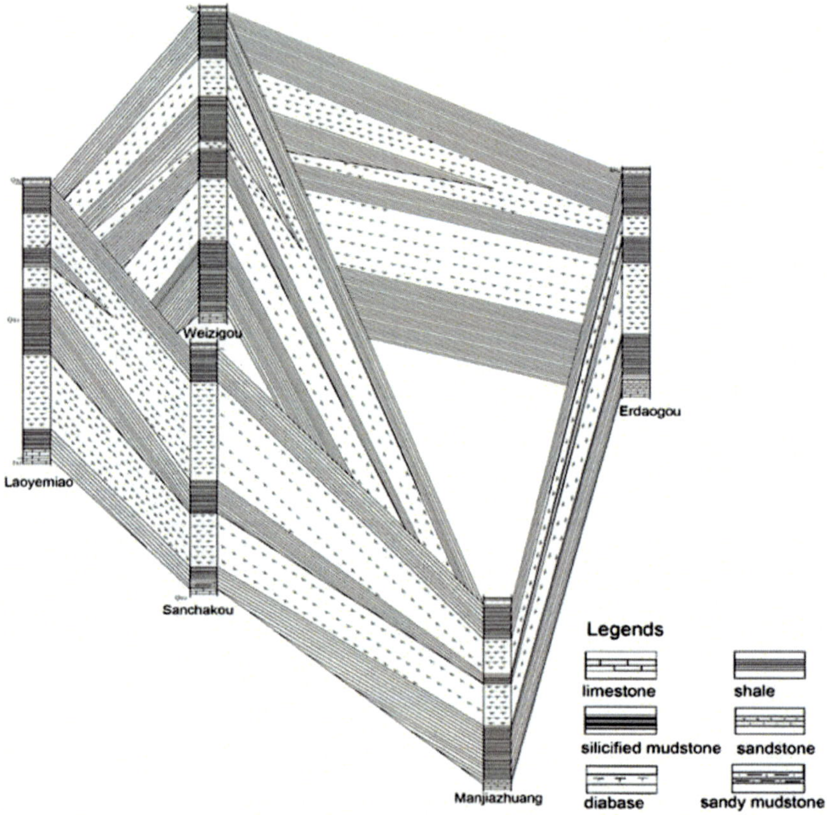

Fig. 10.2 A fence diagram of the gabbro-diabase sills in the Xiamaling Formation at the Huapi anticline, Kuancheng, Hebei Province[a]
[a]Wang T G, Gao Z Z, Liu H B, et al. (1979). Principal characteristics of petroleum geology in the eastern segment of Yansan region (Internal Report). The 3rd Regional Geological Explorational Team, Jingzhou: Jianghan Institute of Petroleum (in Chinese)

reworking of thermo-optical alteration. The alteration zones are 16–60 m thick and the altered rank is decreasing with increasing distance from the sill (Fig. 10.3). Lit-par-lit sills were conformably intruded into wall-rocks (Plate. 10.1a, b). Litherlogically, the wall-rock contains mainly slate, siliceous mudstone, feldspathic quartzose siltstone, calcareous mudstone and thin-bedded limestone, which are resulted from the thermo-optical alteration along with intrusion of magmatic melts, bringing about thin-bedded siliceous slate, calcareous slate (Plate. 10.1d, e) as well as sericite, chlorite, and aggregates of calcareous and ferruginous grains, and appearing as oval mottles within recrystalline minerals, such as mottled slate (Plate. 10.1f, g), and porphyroblastic texture due to quartz synneusis seen in part of siliceous slate (Plate. 10.1h).

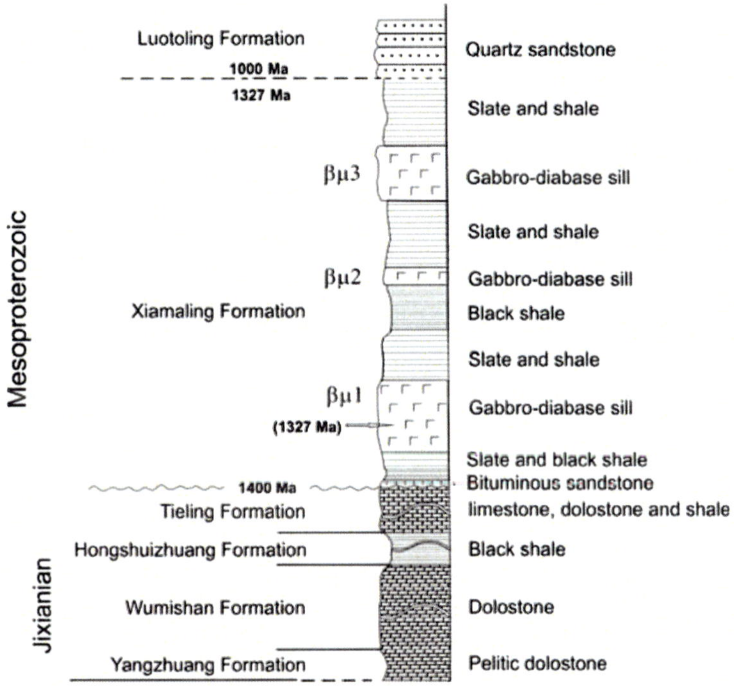

Fig. 10.3 The Xiamaling gabbro-diabase sills at Longtangou outcrop section in Lingyuan County, Hebei Province (Wang and Simoneit, 1995)

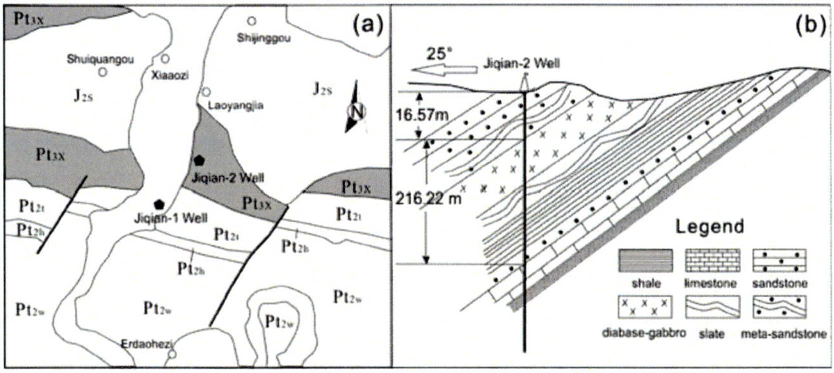

Fig. 10.4 Local geological map showing well site (**a**) and geological cross section (**b**) of Jiqian-2 Well (Wang Tieguan, unpublished date)

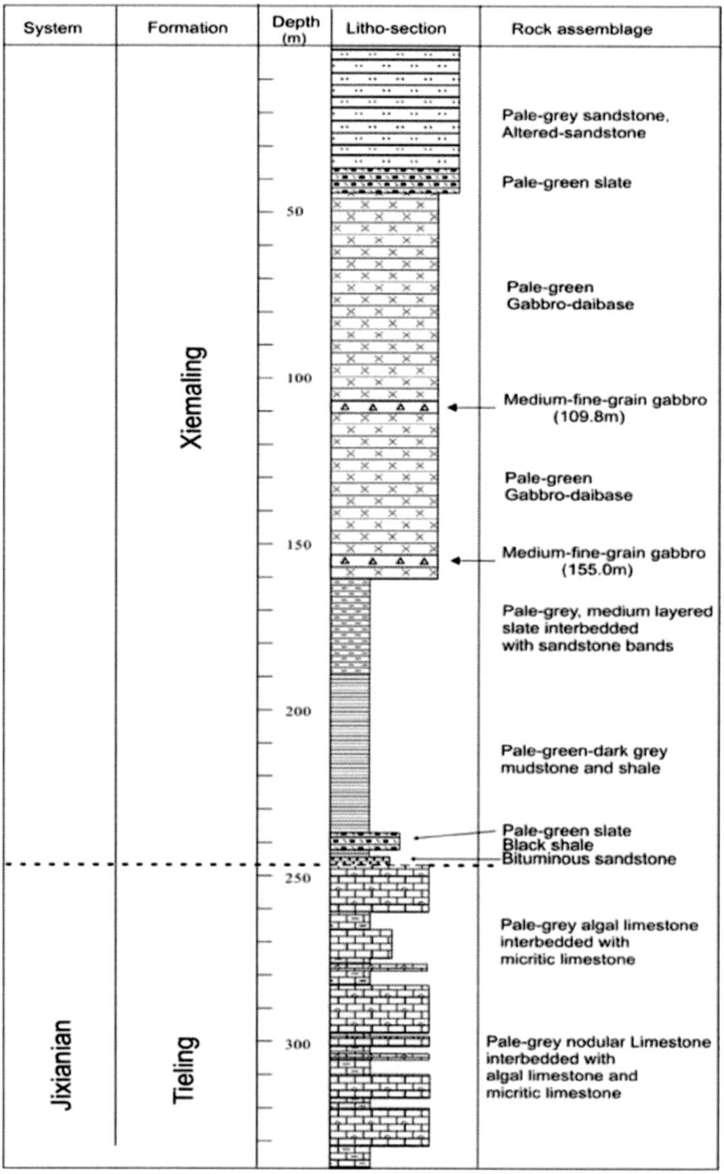

Fig. 10.5 Stratigraphic columnar section of Jiqian-2 Well (1:500) (Wang Tieguan, unpublished date)

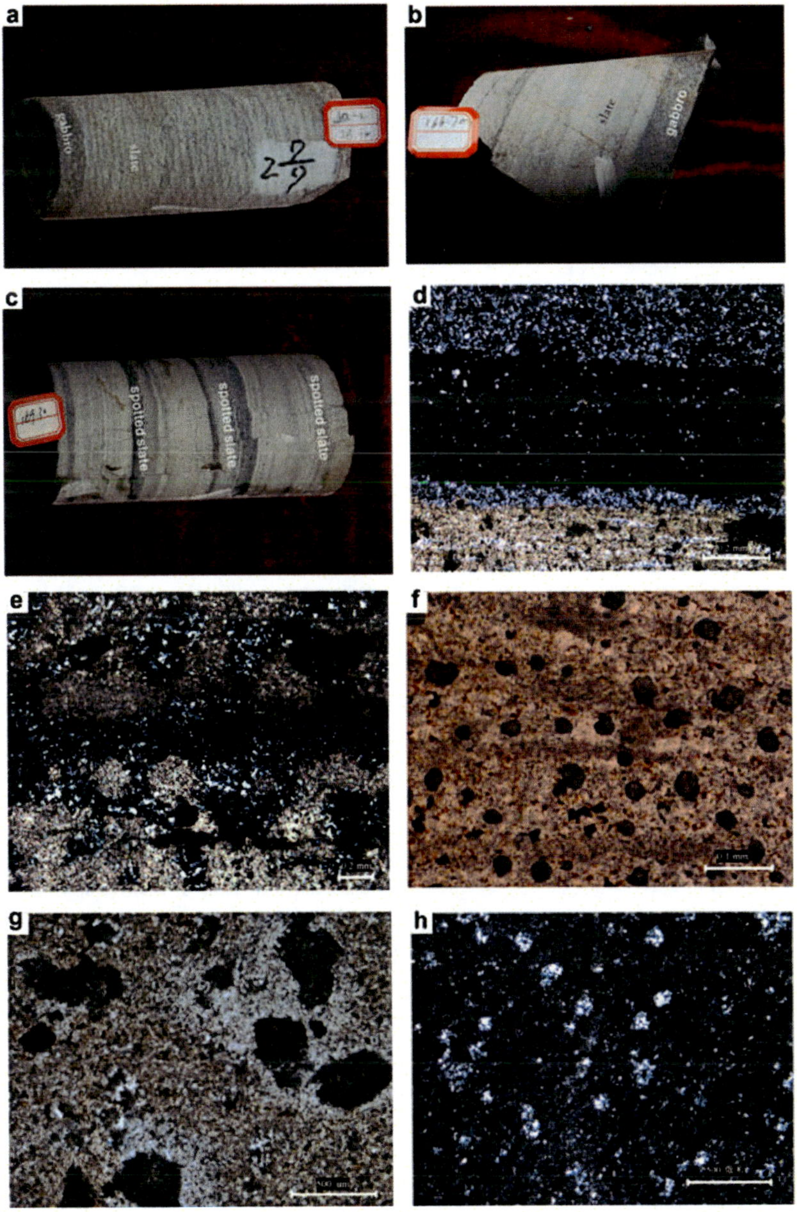

Plate 10.1 Microphotographs of contact of gabbro-diabase sill with Xiamaling wall-rocks and rocks in the outer contact zone. **a** Contact of gabbro-diabase with Xiamaling roof-wall, **b** contact of gabbro-diabase with Xiamaling foot-wall, **c** mottled slate at the outer contact of foot-wall, **d** calcareous-siliceous slate at roof-wall, **e** siliceous-calcareous slate at foot-wal, **f** calcareous-siliceous mottled slate at foot-wall, **g** siliceous mottled slate at roof-wall, **h** quartz agglomerates of siliceous slate at roof-wall. *Note* all samples collected from Jiqian-2 Well

The contact relationship between sill and wall-rock shows that the intrusion of gabbro-diabase was mainly characterised by the occurrence of parallel stratification and the occurrence of penetrating stratification is seldem found. Except for the penetrating occurrence of sill with its wall-rock, no clear brittle fracture of the Xiamaling shale has been caused, implying that the intrusion of basalt magma was almost contemporaneous with the burial diagenesis of the Xiamaling Formation. The inner contact zone of the basic intrusive body with wall-rock could contain fine-grained gabbro, gabbro-diabase, such as feldspathic quartzose sandstone occasionally obsevered as xenoliths (Plate. 10.2e).

10.2.2 Petrology of Gabbrid

Gabbro-diabase sills of the Xiamaling Formation show uniform lithological and mineral compositions, gabbrid with varying grain sizes constitutes the main sill body (Plate 10.2a, c, d). Medium- to fine-crystalline gabbro-diabase (Plate 10.2g), cryptocrystalline diabase (Plate 10.2h) and gabbro-proterobase are found in the inner contact zone. The major diagenetic minerals include clinopyroxene and basic plagioclase with a small amount of enstenite and hornblende in some rocks, and 1–5% magnetite is observed in a few samples (Plate 10.2f). The gabbroite usually have gabbro-diabasic texture (Plate 10.2a, c, d), appearing as medium- to fine-crystalline and coarse-crystalline textures. The graded bedding formed by grain-size variation is well developed, showing rhythmic bedding texture (Plate 10.2d), and the graded bedding is identical with that of the strata of the Xiamaling Formation (Plate 10.1 a, b).

Numerous zircon and baddeleyite crystals are sorted from the gabbro, gabbro-diabse and altered diabase core samples of βμ1 sill in the Jixian-2 Well as well as the gabrro and gabbro-diabase rock samples of βμ3 sill at the outcrops in Shangdong (at Pingquan) and Huapi anticlines (at Kuancheng; Table 10.3). In particular, totally 650 and 310 baddeleyite crystals of βμ1 sill in Jixian-2 Well and of βμ3 sill in Shangdong outcrop have been respectively picked up for isotopic dating.

10.3 Geochemical Characteristics

10.3.1 Analytic Methods

Whole-rock major and trace elements were determined at the Laboratory of Element Geochemistry in the School of Scientific Research, China University of Geosciences, Beijing. Alkaline fusion was used for chemical pretreatment of powdered rock samples; major elements werte analyzed using a Leeman Prodigy ICP-OES (inductively coupled plasma-optical emission spectroscopy). The USGS standard rock

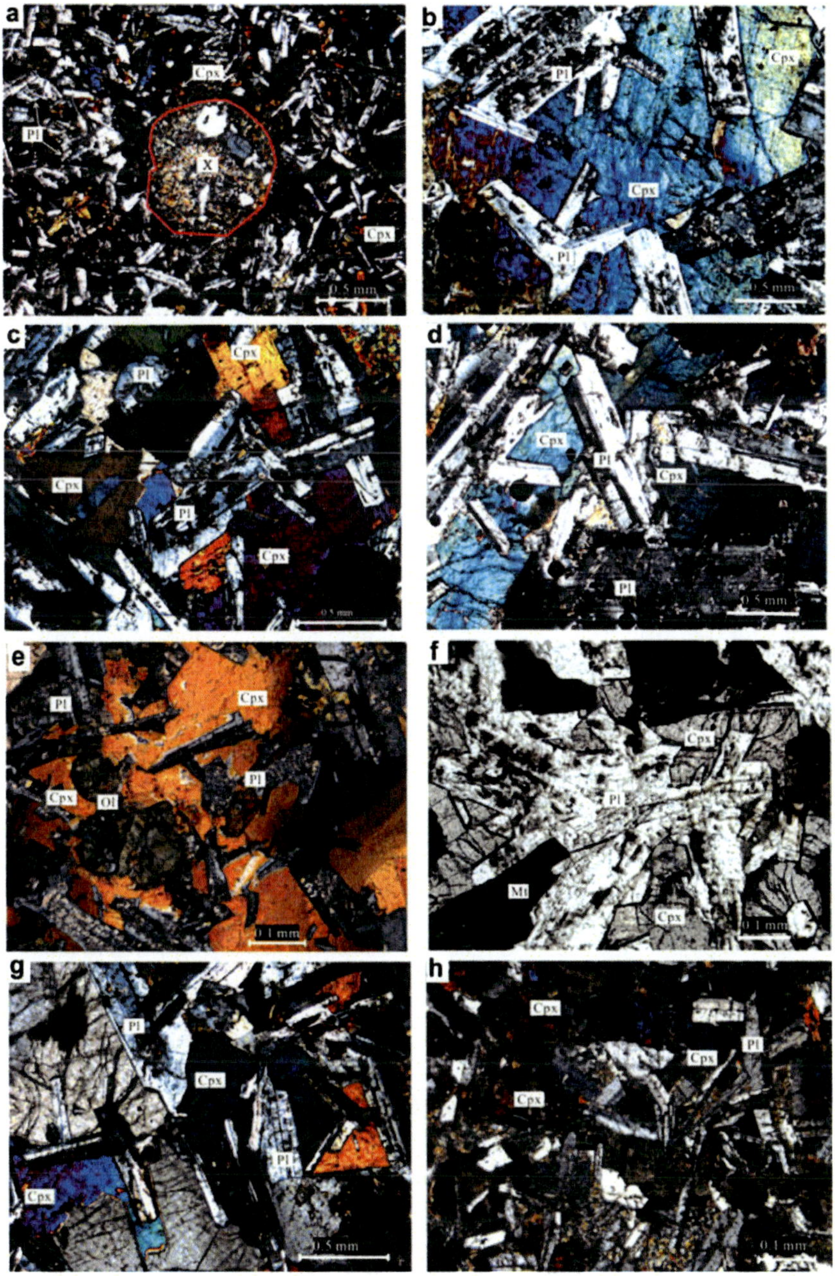

Plate. 10.2 Microphotographs of gabbro-diabase and gabbro, showing crystalline grains increasing from the contact zone to the interior. Cpx. clinopyroxene; Pl. plagioclase; Mt. magnetite; X. xenolith

Table 10.3 Statistics of zircon and baddeleyite crystals sorted from the Xiamaling gabbro-diabase sills in Jibei Depression

Sample No.	Sampling site and well depth	Sill	Litherlogy	Rock sample weight (kg)	Mineral crystal number (grain)
JQ2-3	Jiqian-2 Well, 47.61 m	βμl	Aletered sandstone	ca. 4.0	Zircon: 20
JQ2-6	Jiqian-2 Well, 109.8 m		Medium-crystalline gabbro	ca. 5.0	Zircon: 40; Baddeyite: 310
JQ2-4	Jiqian-2 Well, 155.0 m		Medium-crystalline gabbro-diabase	ca. 5.0	Zircon: 16
PQ-SD-1	Pingquan, Shuangdong (outcrop)	Bμ3	Medium-fine crystalline gabbro	ca. 11.0	Zircon: 1000; Baddeleyite: 650
KC-HP4-3	Kuangcheng, Huapi (outcrop)	–	Medium-fine crystalline gabbro-diabase	ca. 6.0	Zircob: 100

sample, AGV-2, and the standard rock samples of the Chinese National Research Center for GeoAnalysis, GSR-1 and GSR-3, were used for quality control. Analytical precision is < 1.5% for TiO2, ca. 2.0% for P2O5 and < 1.0% for other major elements. Rock powders were digested by using two-acid (HNO3 + HF) pressure autoclave method; trace elements were determined using the Agilent 7,500a ICP-MS. The content calibration curve was constructed using analyses of standard solutions STD-1, STD-2 and STD-4 distributed by the American Bureau for Standard. Data quality was controlled using a set of rock standards including AGV-2 and W-2 from USGS and GSR-2 from the Chinese National Research Center for GeoAnalysis. The analytical precision is < 15% for Ta, Tm and Gd, < 10% for Cu, Sc, Nb, Er, Th, U and Pb, and < 5% for all other elements reported.

Rock samples of the gabbro-diabase sills of the Xiamaling formation in the Jibei depression were collected from medium-grained gabbro in βμ1 sill at different depths of Jiqian-2 Well at Kuancheng, Hebei Province (Figs. 10.4 and 10.5) and from altered gabbro-diabase of βμ3 sill at the Shuangdong Anticline in Pingquan. The major and trace element data are listed in Table 10.3. By comparison, gabbro-diabase samples from sill βμ1 of Jiqian-2 Well, except those near the contact, show significant increase in TiO2* and Fe2O3.

10.3.2 Major Elements

It is shown in Table 10.4 that the gabbro-diabase sills in the Xiamaling Formation have limited variation in SiO2[a] and Al2O3 contents. The content of SiO2 ranges from 48.01% to 51.17%, falling in the basalt field in the TAS classification diagram

(Fig. 10.6a). The Mg^b values vary greatly from 0.16 to 0.65, indicating enrichment in Fe2O3T and depletion in MgO. All samples have TiO2 content > 1.5%, with some exceeding 3.0%. By comparison, gabbro samples from sill βμ1 of Jiqian 2 Well, except those near the contact, show significant increase in $TiO2^a$ and Fe2O3 from lower units upwards. The result suggests that the parental magma was enriched in TiO2 and that interstitial minerals magnetite and titanomagnetite started to crystallize with increasing oxygen fugacity at the late stage of crystallization. In the SiO2-FeOT/MgO discrimination diagram (Fig. 10.6b), all samples fall in the tholeiitic field.

10.3.3 Trace Elements

Trace element analysis (Table 10.4) shows that REE (rare-earth element) abundance is relatively high.

In the Zr/Y-Zr discrimination diagram (Fig. 10.7a), all samples fall in the WPB field; in the Ti/1000-V dΣREE ranging from 86.3 to 226.4 μg/g. Gabbro samples of sill βμ1 in Jiqian 2 Well with well-defined crystallization sequence consistently exhibit light rare-earth elnment (LREE)-enriched patterns (Fig. 10.8a), with [La/Yb]N of 3.8–4.4. Weak negative Eu anomalies (δEu value of 0.84–0.98) are observed for samples from the upper part of the intrusion, with the strongest anomaly being documented by magnetite-bearing gabbro in the upper part. The ΣREE increases from the lower units upwards. Samples containing magnetite cumulates, JQ2-8B, 10B and 11B in particular, have very high ΣREE (up to 190 μg/g) as compared with others. They also have relatively low δEu value (<0.9) and unusually high Th content of 0.967–1.217 μg/g, likely a result of crustal contamination during late stage of magmatic evolution, or alternatively obvious change of oxygen fugacity in magmatic chambers. The gabbro samples from sill βμ3in the Pingquan area also have LREE-enriched pattern similar to those of sill βμ1, but Eu depletion is significant (δEu: 0.63). In the trace element spidergram (Fig. 10.8b) all samples are characterized by enrichment in LILE and Pb, and depletion in Sr. Abundance and pattern of REE and trace elements show that basaltic magmas of this stage are clearly different from N-MORB (normal-mid-ocean ridge basalt) or E-MORB (enriched-mid-ocean ridge basalt), but similar to OIB. The high Pb content and weak Nb–Ta depletion may result from crustal contamination, whereas low Sr and Eu contents possibly reflect early-stage fractional crystallization of plagioclase. Similatity in trace elements between sill βμ1 in the Jiqian-2 Well and βμ3 in Pingquan suggests that these coeval mafic magmas may be comagmatic.

In the Zr/Y-Zr discrimination diagram (Fig. 10.7a), all samples fall in the WPB field; in the Ti/1000-V diagram, most samples fall in the field of CFB (continental flood basalt), while a few fall in the OIB (ocean-island basalt) field. Thus, the gabbro-diabase sills were most likely the products of intraplate mafic magmatism related to continental rifting during the breakup of Supercontinent Columbia.

Table 10.4 Whole-rock geochemical compositions of gabbro-diabase sills in the Jibei depression

Samples	JQ2-3	JQ2-6B	JQ2-7B	JQ2-8B	JQ2-10B	JQ2-11B	JQ2-13B	JQ2-14B	JQ2-15B
Sampling location	No2 DH 47.61 m	No2 DH 52.48 m	No2 DH 55.43 m	No2 DH 59.70 m	No2 DH 62.33 m	No2 DH 73.26 m	No2 DH 81.40 m	No2 DH 87.75 m	No2 DH 100.26 m
SiO2	48	49.26	49.14	49.43	49.31	50.15	48.64	49.96	47.76
TiO2	2.44	2.42	2.9	3.2	3.52	3.07	1.63	2.23	1.82
Al2O3	13.46	12.81	12.45	11.24	11.19	10.66	12.29	12.64	13.89
Fe2O3T	16.32	16.51	19.06	19.09	18.97	18.16	14.96	13.19	13.29
MnO	0.21	0.21	0.25	0.24	0.26	0.25	0.19	0.2	0.19
MgO	6.15	4.86	4.38	3.49	3.3	2.86	6.12	5.69	6.48
CaO	8.26	9.05	8.47	7.59	7.43	6.78	8.81	9.18	10.34
Na2O	1.97	2.1	2.07	1.89	1.96	2.11	2.25	2.25	2.37
K2O	0.13	0.85	0.82	1.05	1.43	1.35	1.15	1.55	0.93
P2O5	0.22	0.21	0.27	0.31	0.31	0.35	0.15	0.16	0.17
LOI	2.02	1.44	1.11	1.75	1.53	1.72	3.18	1.94	1.93
SiO2*	49.00	49.98	49.69	50.32	50.08	51.05	50.25	50.96	48.71
TiO2*	2.49	2.46	2.93	3.26	3.58	3.13	1.68	2.27	1.86
Mg	47	41	35	30	29	27	49	50	53
Li Sc Ti	18.33 40.32 13,854.0	9.726 42.58 15,759	7.956 41.56 17,606	10.66 40.50 20,628	8.888 39.14 20,718	11.01 38.74 19,926	12.46 43.15 10,162	10.73 45.08 13,396	11.60 43.44 10,728
Sc	40.32	42.58	41.56	40.50	39.14	38.74	43.15	45.08	43.44
Ti	13,854.0	15,759	17,606	20,628	20,718	19,926	10,162	13,396	10,728
V	445.8	453.0	490.8	462.2	354.8	275.2	387.7	462.8	387.0

(continued)

Table 10.4 (continued)

Samples	JQ2-3	JQ2-6B	JQ2-7B	JQ2-8B	JQ2-10B	JQ2-11B	JQ2-13B	JQ2-14B	JQ2-15B
Sampling location	No2 DH 47.61 m	No2 DH 52.48 m	No2 DH 55.43 m	No2 DH 59.70 m	No2 DH 62.33 m	No2 DH 73.26 m	No2 DH 81.40 m	No2 DH 87.75 m	No2 DH 100.26 m
Cr	110.3	108.8	27.10	5.086	2.686	0.813	100.6	142.9	349.6
Co	47.90	46.66	46.20	44.56	41.54	41.06	40.11	46.50	48.48
Ni	65.68	45.56	31.14	19.23	11.39	7.464	48.68	57.10	88.36
Cu	105.4	127.6	146.9	173.9	133.1	114.2	77.66	95.96	81.32
Zn	159.0	127.1	141.1	157.8	159.3	164.1	87.68	104.1	100.9
Ga	22.54	23.54	24.16	24.84	24.66	27.12	20.77	20.96	21.12
Rb	9.602	27.52	25.58	51.00	59.10	57.66	50.45	57.06	32.74
Sr	178.4	248.2	206.6	197.3	194.1	209.0	218.5	245.0	300.0
Y	33.24	37.76	42.62	50.06	52.32	61.88	26.29	30.24	26.60
Zr	162.5	182.2	202.7	255.8	268.9	318.3	112.7	132.9	114.5
Nb	16.06	19.52	22.34	27.59	30.74	33.81	11.86	14.97	12.56
Cs	1.443	2.006	3.002	3.520	3.412	3.890	3.257	0.642	1.370
Ba	101.3	470.2	354.8	372.2	402.4	511.2	300.0	246.0	181.9
La	18.07	20.34	23.02	27.74	30.00	35.74	13.13	15.30	13.35
Ce	40.26	44.90	51.08	61.48	65.42	78.70	29.00	34.02	29.74
Pr	5.348	6.002	6.856	8.240	8.680	10.41	3.904	4.560	4.008
Nd	23.90	26.30	30.04	35.88	37.78	45.04	17.16	20.08	17.72
Sm	6.054	6.602	7.552	8.970	9.386	11.09	4.435	5.138	4.570

(continued)

Table 10.4 (continued)

Samples	JQ2-3	JQ2-6B	JQ2-7B	JQ2-8B	JQ2-10B	JQ2-11B	JQ2-13B	JQ2-14B	JQ2-15B
Sampling location	No2 DH 47.61 m	No2 DH 52.48 m	No2 DH 55.43 m	No2 DH 59.70 m	No2 DH 62.33 m	No2 DH 73.26 m	No2 DH 81.40 m	No2 DH 87.75 m	No2 DH 100.26 m
Eu	1.916	2.126	2.294	2.746	2.852	3.220	1.418	1.656	1.524
Gd	6.472	7.286	8.296	9.814	10.32	12.16	4.978	5.746	5.122
Tb	1.010	1.132	1.284	1.530	1.593	1.874	0.778	0.902	0.798
Dy	6.334	7.132	8.004	9.514	9.914	11.65	4.921	5.652	5.022
Ho	1.251	1.449	1.618	1.923	2.002	2.356	0.992	1.153	1.016
Er	3.534	4.038	4.504	5.380	5.588	6.576	2.783	3.198	2.826
Tm	0.489	0.555	0.619	0.748	0.766	0.906	0.382	0.434	0.386
Yb	3.162	3.562	3.978	4.774	4.966	5.780	2.460	2.812	2.482
Lu	0.463	0.521	0.581	0.698	0.734	0.845	0.357	0.407	0.362
Hf	3.937	4.617	5.195	6.489	6.889	8.140	3.002	3.551	3.072
Ta	0.983	1.246	1.423	1.792	1.977	2.187	0.773	0.949	0.802
Pb	2.534	5.616	5.232	7.080	7.224	6.686	1.682	4.464	3.364
Th	2.892	3.418	3.818	4.808	5.036	6.170	2.217	2.562	2.340
U	0.565	0.694	0.769	0.967	0.999	1.217	0.432	0.510	0.444
∑REE	118.3	131.9	149.7	179.4	190.0	226.4	86.71	101.1	88.92
[La/Yb]N	4.1	4.1	4.2	4.2	4.3	4.4	3.8	3.9	3.9
δEu	0.93	0.93	0.88	0.89	0.88	0.84	0.92	0.93	0.96

(continued)

Table 10.4 (continued)

Sample	JQ2-16B	JQ2-6	JQ2-19B	JQ2-20B	JQ2-25B	JQ2-26B	JQ2-4	KC-HP4-3	PQ-SD-2
Sampling location	No2 DH 107.55 m	No2 DH 109.80 m	No2 DH 125.10 m	No2 DH 128.90 m	No2 DH 151.95 m	No2 DH 152.65 m	No2 DH 155.03 m	Kuancheng (Outcrop)	Pinquan (Outcrop)
SiO2	47.64	47.66	48.65	47.04	48.66	47.53	47.82	49.44	46.76
TiO2	1.78	1.66	1.79	1.83	2.09	2.14	2.28	2.3	2.66
Al2O3	14.81	14.28	14.45	14.1	13.38	13.42	13.06	13.67	14.36
Fe2O3T	14.76	14.03	14.6	15.72	15.94	15.96	16.18	15.79	13.68
MnO	0.19	0.18	0.19	0.18	0.19	0.2	0.23	0.2	0.1
MgO	6.62	6.86	6.21	6.59	6.46	6.18	5.66	4.94	8.85
CaO	9.17	9.03	8.7	8.58	8.1	8.22	8.3	7.38	2.31
Na2O	2.28	2.35	2.08	2.45	1.98	1.95	2.03	1.95	3.71
K2O	0.98	0.73	0.88	0.66	1.32	1.32	0.89	0.91	0.34
P2O5	0.17	0.2	0.17	0.18	0.18	0.18	0.23	0.23	0.24
LOI	1.76	2.12	1.59	2.01	2.04	2.12	1.75	3.39	6.33
SiO2*	48.49	48.70	49.44	48.01	49.67	48.57	48.69	51.17	49.94
TiO2*	1.81	1.70	1.82	1.87	2.13	2.19	2.32	2.38	2.84
Mg	51	53	50	49	49	47	45	42	60
Li	12.65	17.36	13.67	13.61	15.76	16.48	16.56	20.38	76.84
Sc	32.92	33.78	32.52	32.56	34.64	37.36	37.28	36.52	43.32
Ti	10,193	9,818.0	11,581	11,954	13,140	13,864	13,406.0	13,352.0	1,5064.0

(continued)

Table 10.4 (continued)

Sample	JQ2-16B	JQ2-6	JQ2-19B	JQ2-20B	JQ2-25B	JQ2-26B	JQ2-4	KC-HP4-3	PQ-SD-2
Sampling location	No2 DH 107.55 m	No2 DH 109.80 m	No2 DH 125.10 m	No2 DH 128.90 m	No2 DH 151.95 m	No2 DH 152.65 m	No2 DH 155.03 m	Kuancheng (Outcrop)	Pinquan (Outcrop)
V	322.6	336.8	346.4	348.2	385.6	406.2	426.0	416.6	479.4
Cr	189.5	187.2	104.1	96.10	95.66	100.3	97.54	85.74	120.5
Co	52.06	58.86	54.74	56.58	53.58	55.00	56.80	51.88	56.98
Ni	101.9	120.3	114.3	121.2	108.6	111.6	109.5	84.84	100.4
Cu	82.58	97.08	93.72	96.38	104.4	109.6	129.9	119.7	136.6
Zn	101.0	139.2	109.8	106.8	97.40	117.1	125.1	129.2	257.6
Ga	21.72	21.42	22.74	21.56	21.82	21.44	22.50	22.06	24.74
Rb	33.74	29.14	27.06	29.20	42.80	40.48	31.88	35.80	19.91
Sr	276.0	229.8	262.4	271.6	244.6	252.4	195.1	210.2	58.22
Y	25.44	24.60	27.98	28.62	33.32	33.14	33.30	33.30	27.22
Zr	113.9	120.2	127.7	132.3	166.8	151.2	155.4	156.5	167.9
Nb	12.53	10.37	14.15	14.74	16.95	17.10	16.12	16.21	17.74
Cs	0.943	4.820	1.320	3.022	1.571	1.580	1.717	1.862	2.942
Ba	171.2	148.9	174.7	208.6	393.0	412.8	340.2	443.4	133.2
La	13.85	13.22	14.53	15.30	18.30	17.82	17.50	17.96	17.15
Ce	30.52	29.38	32.44	33.82	40.06	39.18	39.30	39.50	38.36
Pr	4.070	3.914	4.370	4.522	5.300	5.212	5.196	5.264	5.098
Nd	17.70	17.17	19.10	19.74	22.94	22.72	23.38	23.38	22.12

(continued)

Table 10.4 (continued)

Sample	JQ2-16B	JQ2-6	JQ2-19B	JQ2-20B	JQ2-25B	JQ2-26B	JQ2-4	KC-HP4-3	PQ-SD-2
Sampling location	No2 DH 107.55 m	No2 DH 109.80 m	No2 DH 125.10 m	No2 DH 128.90 m	No2 DH 151.95 m	No2 DH 152.65 m	No2 DH 155.03 m	Kuancheng (Outcrop)	Pinquan (Outcrop)
Sm	4.440	4.378	4.838	4.994	5.746	5.710	5.868	5.880	5.388
Eu	1.463	1.458	1.603	1.607	1.798	1.824	1.875	1.868	1.135
Gd	4.860	4.688	5.306	5.480	6.268	6.244	6.322	6.384	5.578
Tb	0.761	0.734	0.828	0.847	0.975	0.969	0.996	0.991	0.892
Dy	4.740	4.700	5.158	5.288	6.108	6.040	6.362	6.234	5.588
Ho	0.960	0.929	1.042	1.063	1.240	1.217	1.271	1.266	1.119
Er	2.694	2.650	2.904	2.972	3.458	3.428	3.552	3.530	3.166
Tm	0.372	0.359	0.397	0.410	0.484	0.470	0.491	0.487	0.449
Yb	2.366	2.374	2.548	2.600	3.092	2.988	3.154	3.104	2.950
Lu	0.346	0.348	0.374	0.381	0.447	0.438	0.465	0.466	0.421
Hf	3.034	2.896	3.300	3.422	4.305	3.868	3.868	3.823	4.142
Ta	0.804	0.715	0.857	0.928	1.098	1.052	1.064	1.135	1.171
Pb	3.396	3.300	4.806	4.156	3.744	3.932	4.184	4.188	59.30
Th	2.258	2.092	2.392	2.528	3.332	2.866	2.876	2.980	3.128
U	0.480	0.404	0.482	0.508	0.660	0.589	0.568	0.417	0.613
∑REE	89.14	86.30	95.44	99.03	116.2	114.3	115.7	116.3	109.4
[La/Yb]N	4.2	4.0	4.1	4.2	4.2	4.3	4.0	4.2	4.2
δEu	0.96	0.98	0.96	0.93	0.91	0.93	0.94	0.93	0.63

Notes: Trace element concentration, g/g; Major element oxide, wt %. *Recalculation after deducting the LOI

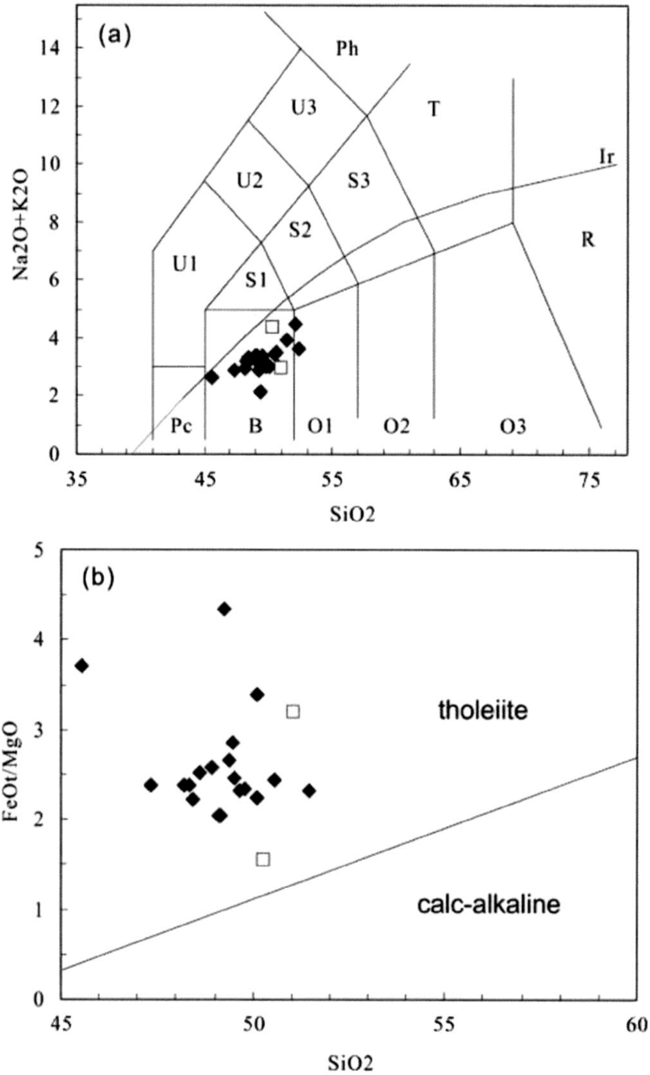

Fig. 10.6 Geochemical classification for gabbro-diabase sills in the Xiamaling Formation. **a** TAS diagram (after 12), **b** SiO2-FeOT/MgO diagram (after 13). TAS. total alkali versus silica, i.e. SiO2-(Na2O + K2O); Pc. picrite; B. basalt; O1. andesitic basalt; O2. andesite; R. rhyolite; S1. trachybasalt; S2. basaltic trachyandesite; S3. trachyandesite; T. trachyte; U1. basanite; U2. phonotephrite; U3. pollenite; Ph. phonolite; Ir. boundary of alkaline rock and subalkaline rock.

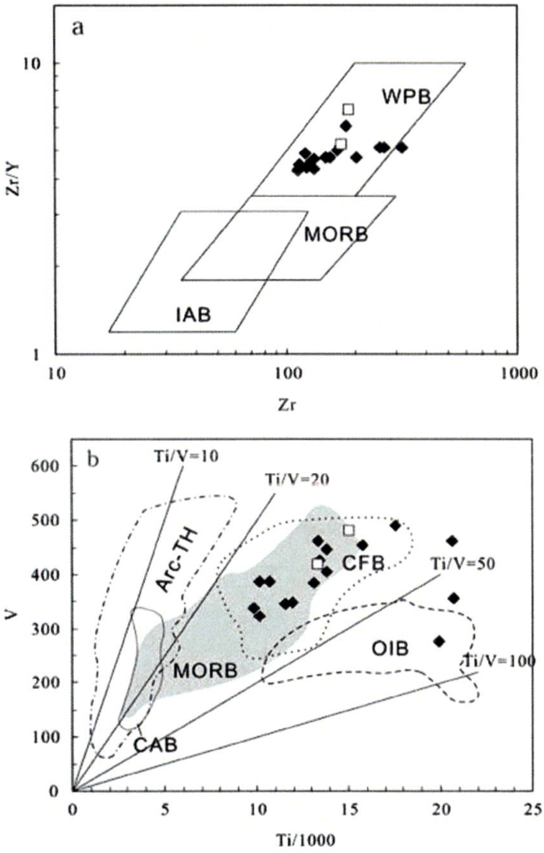

Fig. 10.7 **a** Zr-Zr/Y diagram (Pearce and Norry 1979) and **b** Ti/1000-V diagram (Shervais 1986). ◆: Drilling sample; □: surface sample; WPB. within-plate basalt; MORB. mid ocean ridge basalt; IAB. island arc basalt; Arc-TH. arc-tholeiite; CFB. continental flood basalt; OIB. oceanic island basal

10.4 Emplacement Age of Magmatism: Baddeleyite Geochronology

Two gabbro-diabase sill samples were collected for thermo-ionization mass spectrometer (SIMS) Pb/Pb dating of baddeleyite. One is a medium-grained gabbro sample JQ2-6 collected from βμ1 sill in Jiqian-2 Well at a depth 109.8 m; another is an altered gabbro-diabase sample PQ-SD-2 collected from an outcrop section of βμ3 sill at Shuangdong anticline in Pingquan county, Hebei Province (Table 10.3).

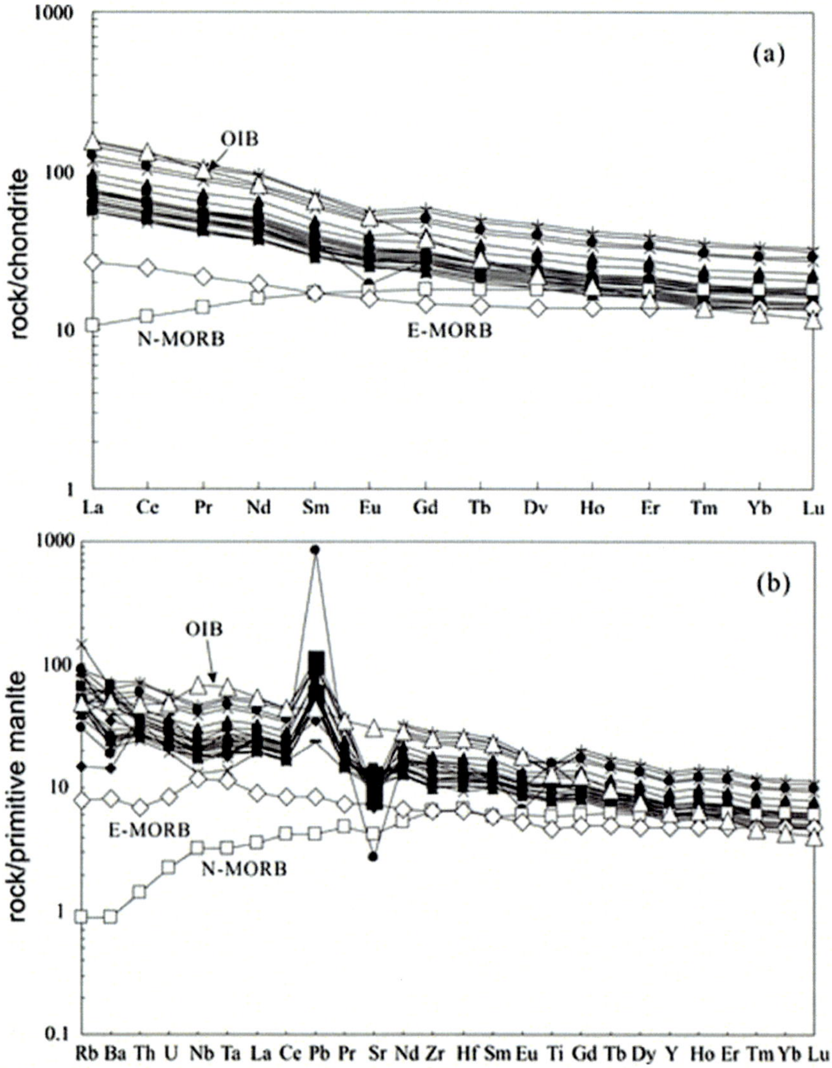

Fig. 10.8 **a** REE pattern and **b** trace element spidergram of trace elements for the gabbro-diabase sills within the Mesoproterozoic Xiamaling Formation. Trace element data for OIB, E-MORB, N-MORB are after Niu 2004; Boudinier and Goodard 2003. The chondrite and primitive mantle values are from Sun and McDonough 1989. ■: drilling sample; ●: surface sample; × : magnetite-bearing sample

10.4.1 Analytic Methods

Heavy mineral separation was performed using conventional gravity and magnetic separation methods at the Chemistry Laboratory of the Institute of Regional Geological Survey, Hebei Academy of Geological Survey. Baddeleyite was further concentrated by hand picking under a binocular microscope. The mineral grains, together with baddeleyite standard, were mounted on double sided tape. A standard one-inch diameter epoxy mount was made, then polished to expose the interior of crystals. Baddeleyite grains were documented with transmitted and reflected light micrographs, followed by cathodoluminescence (CL) images that were made using a high-resolution CL analyzer fitted to a Quanta 200F field emission ESEM-Gatan Mono CL3-CL spectrograph at the Laboratory of Electron Microscopy in the School of Physics, Peking University. The method is similar to that described in Chen et al. (2006). The Ranisow RM-100 Raman spectrometer was used to analyze Raman spectrum characteristics of baddeleyite at School of Earth and Space Sciences, Peking University. High-resolution Cameca IMS-1280 SIMS at the Institute of Geology and Geophysics, Chinese Academy of Sciences was employed for $^{207}Pb/^{206}Pb$ dating on baddeleyite. The method was the same as that described in Li et al. (2009a, b).

10.4.2 Crystal Characteristics of Baddeleyite

The chemical composition of baddeleyite is zirconium dioxide (ZrO2). Different from zircon, baddeleyite crystallizes mainly from Si-undersaturated magmass. It serves as the most reliable mineral for chronological research of basic igneous rocks such as gabbro.

A total of 310 and 650 baddeleyite grains were separated from a medium-crystalline gabbro sample JQ2-6 of $\beta\mu 1$ sill from Jiqian-2 Hole and a diabase sample PG-SD-1 from an outcropped $\beta\mu 3$ sill at Shuangdong anticline in Pingquan, Hebei Province (Table 10.3). Baddeleyite crystals from above two samples are small in size, and present mostly as tetragonal prism-shape with an aspect ratio of > 2. Baddeleyites of JQS-6 are dark-brown under transmitted light and their CL images do not show clear crystalline oscillatory zoning, but are mostly dark-gray in color (Plate 10.3a, c), whereas those of PQ-SD-1 are brownish-green and their CL images do not show oscillatory zonings, either (Plate 10.3b, d).

10.4.3 Raman Spectroscopy of Baddeleyite

It is shown in Fig. 10.9 that baddeleyite samples from the two samples appear primarily as the same Raman spectrum as that of the baddeleyite standard. By comparison, baddeleyites of JQ2-6 have Raman spectral peak and intensity more similar to

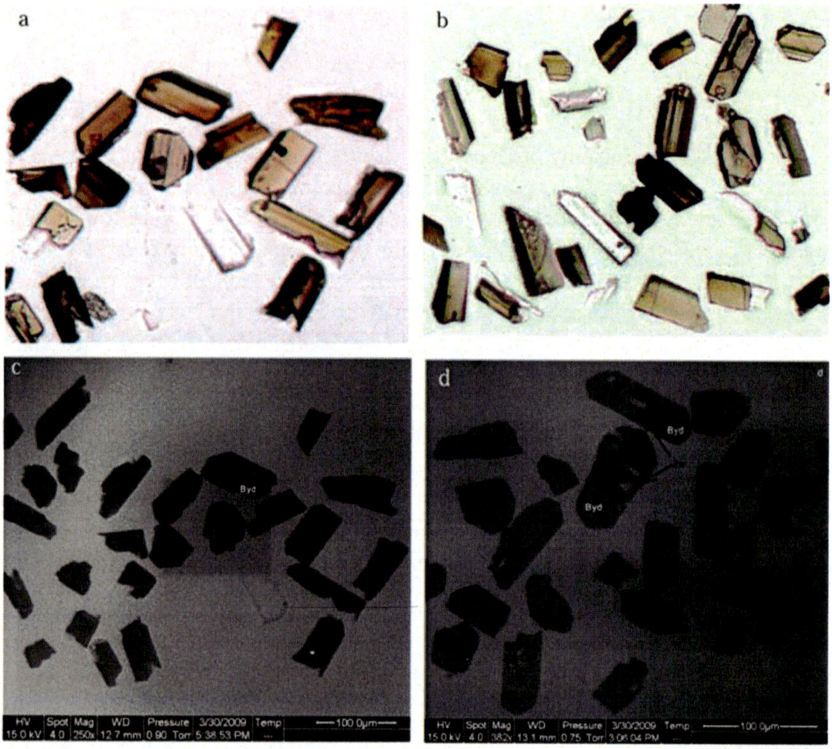

Plate 10.3 TEM and CL images of baddeleyite. **a** Micrograph of JQ2-6 baddeleyite, **b** micrograph of PQ-SD-1 baddeleyite, **c** CL image of JQ2-6 baddeleyite, **d** CL image of PQ-SD-1 baddeleyite. *Note* Byd. baddeleyite; Zir. zircon

pure baddeleyite (Fig. 10.9b). While the main composition of baddeleyite from PQ-SD-1 is ZrO2, it contains relatively more zircon micellae (Fig. 10.9c). Fine zircon inclusions in baddeleyite are also visiable in CL images of baddeleyite (Plate 10.3c).

10.4.4 $^{207}Pb/^{206}Pb$ Dating of Baddeleyite

The SIMS 207Pb/206Pb measurements were conducted on 19 and 17 baddeleyite grains, respectively for βμ1 and βμ3 sills that were concordantly intruded into the Xiamaling Formation in the Jibei Depression. Analytical results are listed in Table 10.5 and illustrated in Fig. 10.10.

The $^{207}Pb/^{206}Pb$ age of the medium-cryatalline gabbro of βμ1 sill in Jiqian-2 Well is 1327.5 ± 2.4 Ma (Fig. 10.10a) and that of the medium-fine crystalline gabbro-diabase of the outcropped βμ3 sill in Shuangdong anticline is 1327.3 ± 2.3 Ma (Fig. 10.10b). Thus above two sills show as the identical emplacement ages

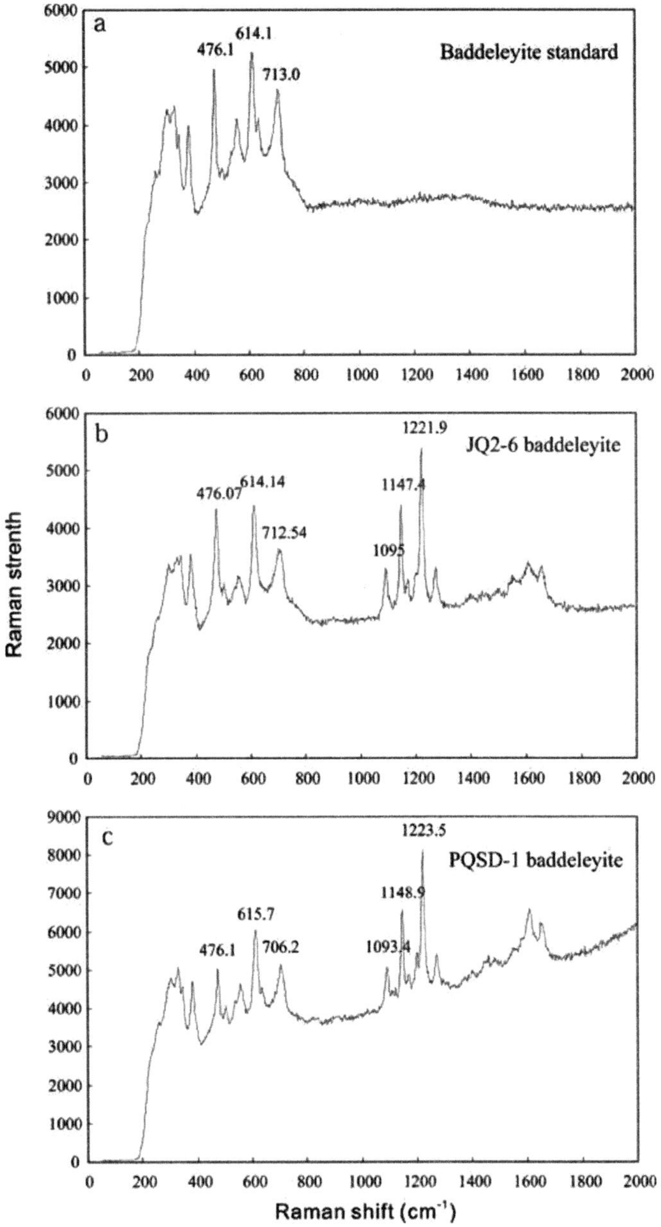

Fig. 10.9 Raman spectrum of baddeleyite. **a** Standard sample of baddeleyite, **b** baddeleyite from medium-crystalline gabbro in βμ1, **c** baddeleyite from altered gabbro-diabase in βμ3

Table 10.5 U–Pb analyses of baddeleyite by SIMS

Spot	$^{204}Pb/^{206}Pbm$	± 1σ (%)	$^{207}Pb/^{206}Pbm$	± 1σ (%)	$^{207}Pb/^{206}Pb$	± 1σ (%)	$t_{207/206}$(Ma)	± 1σ
JQ2-6@1	2.6 × 104	6	0.086,21	0.13	0.085,69	0.13	1331	4
JQ2-6@2	5.0 × 104	7	0.086,02	0.10	0.085,75	0.10	1333	4
JQ2-6@3	1.2 × 104	12	0.087,00	0.19	0.085,91	0.24	1336	5
JQ2-6@4	1.1 × 104	6	0.086,66	0.14	0.085,37	0.17	1324	4
JQ2-6@5	9.0 × 104	8	0.085,48	0.09	0.085,33	0.09	1323	4
JQ2-6@6	4.5 × 104	8	0.085,86	0.14	0.085,56	0.14	1328	4
JQ2-6@7	7.9 × 104	7	0.085,86	0.08	0.085,69	0.08	1331	4
JQ2-6@8	2.0 × 105	27	0.085,12	0.25	0.085,05	0.26	1317	5
JQ2-6@9	1.6 × 105	15	0.085,78	0.13	0.085,70	0.14	1331	4
JQ2-6@10	8.6 × 104	12	0.085,50	0.15	0.085,34	0.15	1323	4
JQ2-6@11	3.5 × 104	10	0.086,11	0.19	0.085,72	0.20	1332	4
JQ2-6@12	6.2 × 105	28	0.085,56	0.12	0.085,54	0.12	1328	4
JQ2-6@13	1.2 × 105	12	0.085,74	0.15	0.085,63	0.15	1330	4
JQ2-6@14	1.8 × 105	15	0.085,42	0.12	0.085,35	0.12	1323	4
JQ2-6@15	1.3 × 105	19	0.085,97	0.21	0.085,86	0.21	1335	4
JQ2-6@16	1.9 × 105	17	0.085,40	0.14	0.085,33	0.14	1323	4
JQ2-6@17	7.7 × 105	48	0.085,42	0.19	0.085,40	0.19	1325	4
JQ2-6@18	2.1 × 105	17	0.085,23	0.13	0.085,16	0.13	1319	4
JQ2-6@19	4.4 × 105	24	0.085,61	0.14	0.085,58	0.14	1329	4
PQ-SD-1@1	3.0 × 105	19	0.085,56	0.12	0.085,51	0.12	1327	4
PQ-SD-1@2	6.5 × 104	14	0.086,08	0.20	0.085,87	0.21	1335	4

(continued)

Table 10.5 (continued)

Spot	$^{204}Pb/^{206}Pbm$	± 1σ (%)	$^{207}Pb/^{206}Pbm$	± 1σ (%)	$^{207}Pb/^{206}Pb$	± 1σ (%)	$t_{207/206}$(Ma)	± 1σ
PQ-SD-1@3	6.5 × 104	16	0.085,61	0.22	0.085,40	0.22	1325	4
PQ-SD-1@4	1.1 × 105	17	0.085,81	0.17	0.085,69	0.17	1331	4
PQ-SD-1@5	1.2 × 104	7	0.087,03	0.20	0.085,88	0.22	1336	4
PQ-SD-1@6	2.5 × 104	11	0.085,72	0.27	0.085,18	0.28	1320	5
PQ-SD-1@7	5.6 × 104	9	0.085,65	0.14	0.085,41	0.14	1325	4
PQ-SD-1@8	4.8 × 104	10	0.085,74	0.14	0.085,46	0.15	1326	4
PQ-SD-1@9	7.3 × 105	27	0.085,66	0.12	0.085,64	0.12	1330	4
PQ-SD-1@10	2.3 × 105	22	0.085,67	0.19	0.085,61	0.19	1329	4
PQ-SD-1@11	2.8 × 104	5	0.086,11	0.10	0.085,63	0.11	1330	4
PQ-SD-1@12	2.4 × 105	22	0.085,28	0.17	0.085,22	0.17	1321	4
PQ-SD-1@13	4.7 × 105	24	0.085,47	0.13	0.085,44	0.13	1326	4
PQ-SD-1@14	3.0 × 105	25	0.085,68	0.15	0.085,64	0.15	1330	4
PQ-SD-1@15	2.1 × 105	27	0.085,28	0.22	0.085,21	0.22	1320	4
PQ-SD-1@16	4.1 × 105	24	0.085,47	0.17	0.085,43	0.17	1325	4
PQ-SD-1@17	4.5 × 105	23	0.085,50	0.13	0.085,47	0.13	1326	4

Notes: $^{204}Pb/^{206}Pbm$ and $^{207}Pb/^{206}Pbm$ are the measured values; $^{207}Pb/^{206}Pb$ is the calculated value after common-lead correction; analytical procedures are similar to those reported by Li et al. 2009a, b

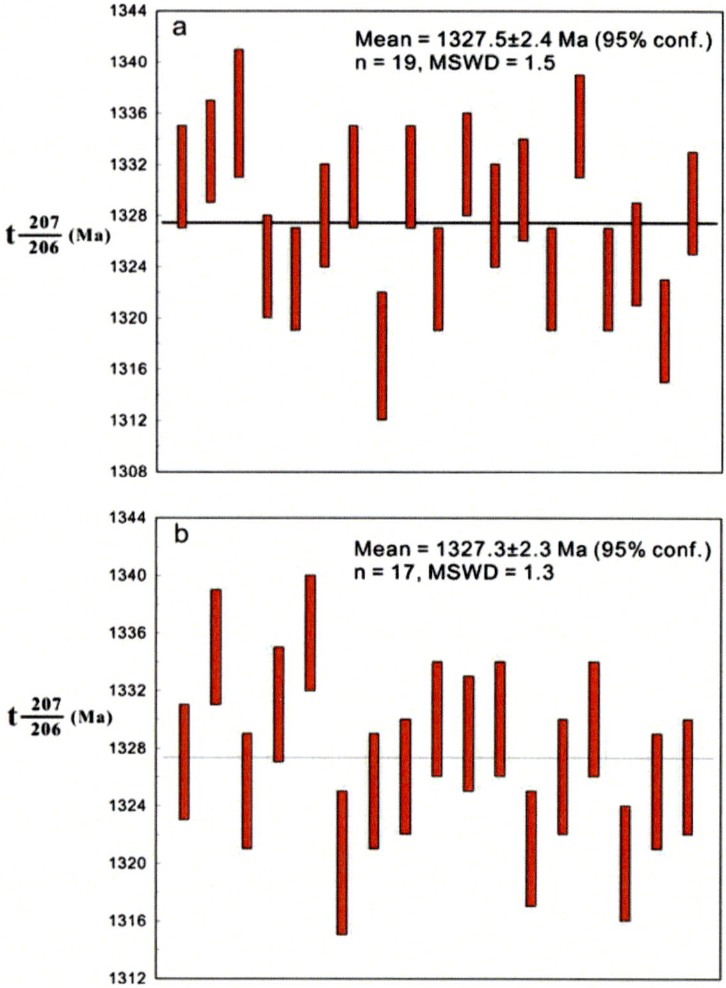

Fig. 10.10 $^{207}Pb/^{206}Pb$ concordant age spectrum of baddeleyite **a** baddeleyite from JQ2-6 in βμ1 sill, **b** baddeleyite from PQ-SD-1 in βμ3 sill. *Note* Mean weighted mean age; MSWD mean standard weighted deviation

within analytical errors, suggesting that βμ1–βμ3 sills of the Xiamaling Formation are emplaced synchronously. these results suggest that a more extensive episode of basaltic intrusion with CFB characteristics took place at ca. 1327 Ma, following the volcanic extrusion (tuff) and subvolcanic intrusion (i.e., diabase dyking) respectively reported at 1368 Ma by Gao et al. (2007, 2008a, b) and 1345 Ma by Zhang et al. (2009) during the initial rifting of the YFDZ.

10.5 Petrogenesis of 1400–1300 Ma Basic Igneous Rocks in the YFDZ

The 1380–1360 Ma volcanic rocks and sills are extensively distributed also in Africa and the Canadian shield. For instance, igneous rocks dated at 1380–1370 Ma by Ernst et al. (2013) have been found around Angola in the Congo Craton, comprising Cunene basic-ultrabasic complex and concurrent A-type granite, i.e. alkaline, anhydrate, anorogenic granite (e.g. Mayer et al. 2004; Ernst and Bleeker 2010). In the Canadian shield, both the 1386–1380 Ma Hart River basaltic dikes and the 1379 Ma Salmon River gabbro sills have compositions similar to those formed in rift environments (Ernst and Bleeker 2010). Recent studies revealed that magmatism dated at ca. 1400–1300 Ma in the northern margin of the North China Craton was widespread, including the 1,366–1,372 Ma bentonitic tuff in the Jibei Depression of the central YFDZ (Gao et al. 2007, 2008a, b, 2009), the 1320 Ma diabase "dikes" (Li et al. 2009a, b), the 1345 Ma diabase (Zhang et al. 2009, 2012) and the 1327 Ma diabase sills (in the present study), the 1354 ± 59 Ma carbonatites in the Bayan Obo and Zha'er Taishan Faulted-Depression Zones (Yang et al. 2011), the 1313–1231 Ma diabase (Yang et al. 2011; Zhang et al. 2012), and the 1331–1324 Ma granite (Zhang et al. 2012), etc. It has been proposed that these globally 1400–1300 Ma mantle-derived magmas were likely related to the break-up of the supercontinent Columbia (Zhang et al. 2017).

The 1400–1300 Ma magmatism in the YFDZ occurred mainly as intermittent volcanic eruption in the early stage, resulting in the formation of multilayered tuff, such as bentonitic tuff layers at 1485 Ma (Li et al. 2011) and 1437 Ma (Su et al. 2010) in the Tieling and Wumishan Formations as well as bentonitic tuff at 1366–1372 Ma (Gao et al. 2007, 2008a, b, 2009) in the Xiamaling Formation. At ca. 1350 Ma, the North China Craton was separated from the supercontinent Columbia to form several intracontinental or epicontinental rift zones. Continuous extension led to intrusion of mantle-derived basaltic magma in continental rifts and formation of large-scale concordantly intruded gabbro and diabase sills near the surface.

The rifting event in the Meso-Neoproterozoic led to at least 5 sedimentation hiatuses and unconformity in the sedimentary sequence in the NCC. The genertic mechanism of these unconformable planes is a response of the NCC to the break-up of Columbia. The magmatism and sedimentation hiatuses of this time period were associated with mineral deposit formation, such as the Bayan Obo rare-earth mineral deposit. Recently, bitumen-bearing dolostone beds have been found in the Tieling-Xiamaling Formation in the YFDZ and in the southwestern margin of the Ordos Basin (Li et al. 2011). Further study is needed to understand their relationship with rifting and intrusion of mantle-derived basaltic magma in the NCC.

Geological and geochemical research of 1327 Ma gabbro-diabase and gabbro sills in the northern margin of the NCC, especially those in the Jibei Depression of the central YFDZ has also revealed: ① a latest depositional age of the Xiamaling Formation would be at 1327 Ma because the wall-rock alteration zones of the Xiamaling shale was induced by basaltic intrusions (Figs. 10.3 and 10.5); ② the potential of hydrocarbon generation of Xiamaling black shale source bed would be lost due to the strong thermo-optical alteration caused by basic βμ1–βμ4 sills in the Jibei Depression.

References

Boudinier JL, Goodard M (2003) Orogenic, ophiolitic and abssal peridotites. In: Carlson RW (ed). Treatise on Geochemistry, vol 2. Mantle and Core. Elsevier Science Ltd, Amsterdam, pp 103–170
Chen L, Xu J, Su L (2006) Application of cathodoluminecence to zircon in FEG-ESEM. Prog Nat Sci 16(9):919–924
Ernst RE, Bleeker W (2010) Large igneous provinces (LIPs), giant dyke swarms, and mantle plumes: significance for breakup events within Canada and adjacent regions from 2.5 Ga to present. Can J Earth Sci 47:695–739
Ernst RE, Pereira EMA, Hamilton SA, Pisarevsky J, Rodriques CCG, Tassinari W, Teixeira V, Van-Dunem V (2013) Mesoproterozoic intraplate magmatic "barcode" record of the Angola portion of the Congo craton: newly dated magmatic events at 1500 and 1110 Ma and implications for Nuna (Columbia) supercontinent reconstructions. Precambr Res 230:103–118
Gao LZ, Zhang CH, Shi XY, Zhou HR, Wang ZQ (2007) Zircon SHRIMP U-Pb dating of the tuff bed in the Xiamaling formation of the Qingbaikouan system in North China. Geol Bull China 26(3):249–255 (in Chinese with English abstract)
Gao LZ, Zhang CH, Shi XY, Song B, Wang ZQ, Liu YM (2008a) Mesoproterozoic age for Xiamaling formation in North China Plate indicated by zircon SHRIMP dating. Sci Bull 53(21):2617–2623 (in Chinese)
Gao LZ, Zhang CH, Yin CY, Shi XY, Wang ZQ, Liu YM, Liu PJ, Tang F, Song B (2008b) SHRIMP zircon ages: basis for refining the chronostratigraphic classification of the Mesozoi- and Neoproterozoic strata in North China Old Land. Acta Geosci Sin 29(3):366–376 (in Chinese with English abstract)
Gao LZ, Zhang CH, Liu PJ, Ding XZ, Wang ZQ, Zang Y (2009) Recognition of Meso- and Neoproterozoic stratigraphic framework in North and South China. Acta Geosci Sin 30(4):433–446 (in Chinese with English abstract)
Le Maitre RW, Bateman E, Dudek P, Keller A et al (1989) A classification of igneous rocks and glossary of terms: recommendations of the international union of geological sciences subcommission on the systematics of igneous rocks. John Wiley and Sons, Miley-Blackwell
Li HK, Lu SN, Li HM, Sun LX, Xiang ZQ, Gen JZ, Zhou HY (2009) Zircon and beddeleyite U-Pb precision dating of basic rock sills intruding Xiamaling formation. North China Geol Bull China 28(10):1396–1404 (in Chinese with English abstract)
Li RX, Liang JW, Weng K (2011) Palaeo-reservoir bitumen of Middle Protozoic Jixian System in the southwest margin of the Ordos Basin. China. Petrol Explor Dev 38(2):168–172 (in Chinese with English abstract)
Li XH, Liu Y, Li QL, Guo CH, Chamberlain KR (2009) Precise determination of Phanerozoic zircon Pb/Pb age by multi-collector SIMS without external standardization. Geochem Geophys Geosyst 10:Q04010
Mayer A, Hofmann AW, Sinigoi S, Morais E (2004) Mesoproterozoic Sm-Nd and U-Pb ages for the Kunene anorthosite complex of SW Angola. Precambr Res 133:187–206
Miyashiro A (1974) Volcanic rock series in island arc and active continental margin. Am J Sci 247:321–355
Niu Y (2004) Bulk-rock major and trace element complsitions of abyssal-peridotites: implications for mantle melting, melt extraction and past melting process beneath mid-ocean ridge. J Petrol 45(12):2423–2458
Pearce JA, Norry MJ (1979) Petrogenetic implications of Ti, Zr, Y and Nb variations in volcamc rocks. Contrib Miner Petrol 69(1):33–47

Qiao XF, Gao LZ (1999) Earthquake events in Neoproterozoic and Early Palaeozoic and its relationship with supercontinental Rodinia in North China. Sci Bull 44(16):1753–1758 (in Chinese)

Shervais JW (1986) Ti-V plots and the petrogenesis of modern and ophiolitic lavas. Earth Planet Sci Lett 59:101–118

Su WB, Zhang SH, Huff WD, Li HK, Ettensohn FR, Chen XY, Yang HM, Han YG, Song B, Santosh M (2008) SHRIMP U-Pb ages of K-bentonite beds in the Xiamaling Formation: implications for revised subdivision of the Meso- to Neoproterozoic history of the North China Craton. Gondwana Res 14:543–553

Su WB, Li HK, Huff WD, Etttensohn FR, Zhang SH, Zhao HY, Wan YS (2010) SHRIMP U-Pb dating for a K-bentonite bed in the Tieling Formation. North China Chinese Sci Bull 55(29):3312–3323

Sun SS, McDonough WF (1989) Chemical and isotope systematics of oceanic basalts: implication for mantle composition and processes, In: Saunders AD, Norry MJ (eds), Magmatism in the ocean Basins. Geological Society, vol 42. Special Publications, London, pp 313–345

Wang TG (1980) On the indigenous nature of Sinian Suberathem oil seep and its significance in the Yanshan region. Pet Explor Dev 2:34–52 (in Chinese)

Wang TG, Simoneit BRT (1995) Tricyclic terpanes in Precambrian bituminous sandstone, eastern Yanshan region, North China. Chem Geol 120:155–170

Yang KF, Fan HR, Santosh M, Hu FF, Wang KY (2011) Mesoproterozoic carbonatitic magmatism in the Bayan Obo deposit, Inner Mongolia, North China: constraints for the mechanism of super accumulation of rare earth elements. Ore Geol Rev 40:122–131

Zhang SH, Zhao Y, Yang ZY, He ZF, Wu H (2009) The 1.35 Ga diabase sills from the northern North China Craton: implications for breakup of the Columbia (Nuna) Supercontinent. Earth Planet Sci Lett 288:588–600

Zhang SH, Zhao Y, Santosh M (2012) Mid-Mesoproterozoic bimodal magmatic rocks in the northern North China Craton: implications for magmatism related to breakup of the Supercontinent Columbia. Precambr Res 222–223:339–367

Zhang SH, Zhao Y, Li XH, Ernstd RE, Yang ZY (2017) The 1.33–1.30 Ga Yanliao large igneous province in the North China Craton: Implications for reconstruction of the Nuna (Columbia) Supercontinent, and specifically with the North Australian Craton. Earth Planet Sci Lett 465:112–125

Chapter 11
The State-of-Art of Global and Chinese Meso-Neoproterozoic Petroleum Resources

Tieguan Wang, Daofu Song, Chengyu Yang, and Ronghui Fang

Abstract Since 1960s, the significant research progress on the Proterozoic early lives and life diversity as well as on the Meso-Neoproterozoic shales and carbonate source rocks have established a material foundation for the studies of indigenous Meso-Neoproterozoic sedimentary organic matter and petroliferous nature, and providing the prerequisites for the prospectivity of indigenous petroleum resources. So far at least dozens oil and gas fields, some of considerables size, have been discovered in the Meso-Neoproterozoic strata, oil and/or gas of which were sourced from the Infracambrian source beds. Based on uncompleted global statistics, there are four countries, i.e., Lena-Tounguska Petroleum Province (LTPP) in Siberia Craton (Russia), Oman Basins in Arabian Craton (Oman), Baghewala Oilfield in Indian Craton (India), and Anyue Gasfield in western Yangtze Craton (China), in the world, containing proven geological reserves and/or commercial production of indigenous Infracambrian petroleum; nine regions/countries having been confirmed indigenous Meso-Neoproterozoic oil flow, oil-seep and/or asphalt, but so far no commercial production has been obtained yet; five regions/countries being revealed to possess hydrocarbon generation potential in the Infracambrian strata. China is one of the countries where the Meso-Neoproterozoic sequences are most completely developed and preserved, and Chinese Meso-Neoproterozoic geology was earlier studied in the word. However, both geological research and exploration of Meso-Neoproterozoic petroleum resources are faced with a challenging reality of more age-old strata, more complicated geological tectonics and more extensive scientific innovative possibility. Therefore, it is a considerable urgent question to study and evaluate the prospectivity of Meso-Neoproterozoic oil and gas resources in China. In this chapter, the distribution, exploration and development of Meso-Neoproterozoic oil and gas respectively in Russian LTPP, Sultanate of Oman, Pakistan and India basins, East European Craton, African Taoudenni Basin, Australian Centralian Basins, American Midcontinent Rift System as well as Chinese basins have been compiled, and their prospectivity approached.

T. Wang (✉) · D. Song · C. Yang · R. Fang
State Key Laboratory of Petroleum Resources and Prospecting, China University of Petroleum-Beijing, Beijing 102249, China

T. Wang, *Meso-Neoproterozoic Geology and Petroleum Resources in China*, Springer Geology, https://doi.org/10.1007/978-981-19-5666-9_11

Keywords Meso-neoproterozoic · Lena-tounguska petroleum province (LTPP) · Anyue gasfield · Oman basins · Yanliao faulted-depression zone (YLFDZ)

11.1 Introduction

As for fossil fuels, oil and gas are generally preserved within various porous or fractured reservoir rocks with different geological age. However, it is common knowledge that petroleum is generated only within sedimentary strata rich-in organic matter, i.e., so called source beds/rocks. Global indigenous oil and gas fields are merely found in sedimentary basins as yet. In geological time, palaeo-organisms would be the sole material source for sedimentary organic matter. The Meso-Neoproterozoic sequences (1800–541 Ma in age) are the age-old sediments in the Earth. As recently as 1950s, it was generally accepted that sedimentary rock of Precambrian age located within basins, geosynclines, platforms and cratons could not contain hydrocarbon deposits. The absence of pre-Palaeozoic Era life characteristics in the Precambrian stratigraphic column was the most often presented reasons why Precambrian terrain should be ignored by the petroleum geologists (Dickes 1986a).

In recent 60 years, however, the researches of Proterozoic early life and life diversity in the Earth have made much headway (Diches 1986b; Chen et al. 1996; Hou et al. 1999; Chen 2004; Sun 2006; Du et al. 2009; Shu et al. 2016). In the meanwhile, petroleum geological and geochemical studies revealed that not only the age-old Meso-Neoproterozoic dark-colored shale and carbonate could contain abundant organic matter so as to become excellent source rocks, but also the source rock maturities would indicate various thermo-evolutional stages from marginal mature though mature, high-mature to over-mature phases, so that up to present some Precambrian sedimentary organic matter could be still within the category of "liquid window" for hydrocarbon generation. In some regions, therefore, numerous indigenous oil and gas seeps can be found, which could be very significant for the research of Meso-Neoproterozoic petroliferous nature and petroleum resources potential.

As a matter of fact, globally at least there are several dozen oil, gas-condensate and/or gas fields found within the Precambrian sequences in Russia, China, Oman and India since 1960s (Beijing Petroleum Exploration and Development Institute and North China Petroleum Bureau 1992; Fedorov 1997; IHS Energy Group. 2005. International Petroleum Exploration and Development Database. Colorado; Craig et al. 2009; Bhat et al. 2012). Based on the global statistics of petroleum resources published since 1990s, however, the geological reserve portion of indigenous Early Proterozoic petroleum only amount to 1–2% (Hunt 1991; Klemme and Ulmishek 1991). So far still no oil and gas reserve portions of Meso-Neoproterozoic sequences can be found in the published statistics in China (e.g., Fig. 11.1).

Menchikoff (1949) and Pruvost (1951) initially proposed an informally used geological term "Infracambrian". As currently used, the "Infracambrian" is loosely confined to formations including Precambrian-earliest Cambrian sedimentary

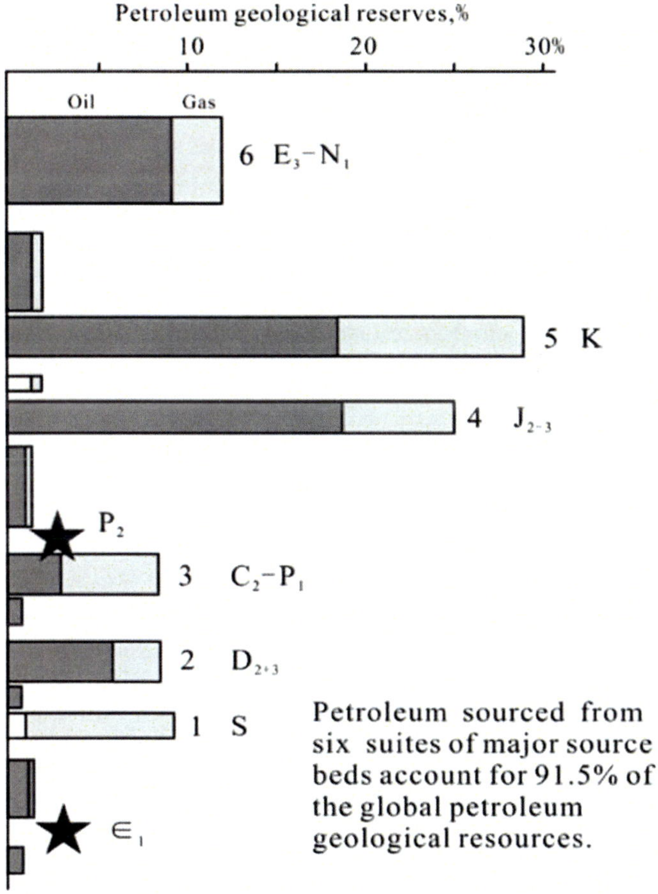

Fig. 11.1 Statistics of global petroleum geological reserves generated from the source beds of different geological time. [Liang (2008). Research progress of marine facies hydrocarbon generation and entrapment in south China (unpublished lecture notes)]

sequences that underlay known Cambrian strata and unconformably overlay crystalline basement (Smith 2009). In 2006, the Global Infracambrian Petroleum Systems Conference has been held at the Geological Society of London, the goal of which was to review current knowledge about Neoproterozoic–Early Cambrian petroleum systems worldwide (Craig et al. 2009). As a result of the conference, the Society Special Publication 326 entitled *Global Neoproterozoic Petroleum Systems: the Emerging Potential in North Africa* was published by the London Geological Society in 2009.

The Meso-Neoproterozoic sequences have well developed and more completely preserved in China, and the Meso-Neoproterozoic stratigraphy was earlier investigated by Chinese geologists in the world (Lee and Chao 1924; Kao et al. 1934).

However, quite a large area of the Meso-Neoproterozoic stratigraphic distribution is under the complicated tectonic and highly thermo-evolutional conditions in China, resulting in more difficulties for indigenous oil and gas preservation and exploration. Therefore, the researches of Meso-Neoproterozoic petroleum resources in China are facing the favorable advantages of completely developed strata and profoundly accumulation of scientific results as well as the challenging reality of more age-old strata and more complicated geological tectonics and more scientific innovative possibility. Therefore, it is a considerable urgent question that how to study and evaluate the prospectivity of Meso-Neoproterozoic petroleum resources in China.

In this chapter, the authors try to compile the distribution of currently known Meso-Neoproterozoic petroleum resources, investigate their petroleum geological conditions, and approach the prospectivity of indigenous Meso-Neoproterozoic oil/gas.

11.2 Global Distribution of Meso-Neoproterozoic Petroleum Resources

Based on incomplete global statistics, there are four regions/countries, i.e., ① Lena-Tounguska Petroleum Province (LTPP) in Siberia Craton (Russia), ② Anyue Gasfield in western Yangtze Craton (China), ③ Oman Basins in eastern Arabian Craton (Oman), and ④ Baghewala Oilfield in Indian Craton (India), in the world, containing proven geological reserves and/or commercial production of indigenous Meso-Neoproterozoic petroleum (Fig. 11.2).

Nine regions/countries, i.e., ① Moscow and Kama-Belsk Basins in East European Craton (Russia), ② Yanliao Faulted-Depression Zone (YFDZ, oil-seeps) in North China Craton (China), ③ Longmenshan Fronthill Belt (large asphalt veins) in Yangtze Craton (China), ④ Vindhyan uperbasin in Indian Craton (India), ⑤ Nonesuch (oil-seeps) in North American Craton (USA), ⑥ McArthur Basin and Centralian Superbasin in Australian Cratons (Australia), ⑦ Taoudenni Basin in West African Craton (Mauritania, Mali and Algeria), ⑧ Tindouf Basin (Morocco and Algeria), and ⑨ Sirte/Cyrenaica Basin in West African Craton (Libya), having confirmed indigenous Meso-Neoproterozoic oil flow, oil-seep, and/or asphalt, but no commercial production obtained yet.

Five regions/countries. i.e., ① Potwar and Mijalar Basins in Indian Craton (Pakistan), ② Saudi Arabia in Arabian Craton, ③ Sào Francisco Basin (Brazil) and ④ Argentina, Bolivia and Paraguay in Amazon Craton, ⑤ Al Kufra Basin in West African Craton (Libya), being revealed to possess the hydrocarbon generation potential of Meso-Neoproterozoic sequences (Craig et al. 2009; Ghori et al. 2009; Lottaroli et al. 2009; Wang and Han 2011).

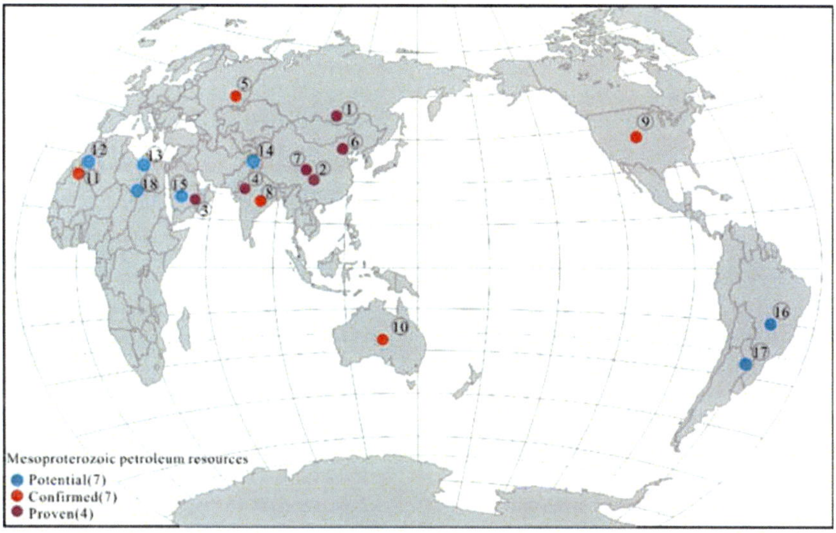

Fig. 11.2 Global distribution of proven, confirmed and potential Meso-Neoproterozoic petroleum resources (Craig et al. 2009; Smith 2009, modified). Proven resources (①–④): ① Lena-Tounguska Petroleum Province (LTPP, Russia); ② Anyue gasfield (China); ③ Oman Basins (Oman); ④ Baghewala Oilfield (India). Confirmed resources (⑤–(13)): ⑤ Moscow and Kama-Belsk Basins (Russia); ⑥ Yanliao oil-seeps (China); ⑦ Longmenshan asphalt veins (China); ⑧ Vindhyan Superbasin (India); ⑨ Nonesuch oil-seeps (USA); ⑩ McArthur Basin and Centralian Superbasin (Australia); (11) Taoudenni Basin (Mali, Mauritania and Algeria); (12)Tindouf Basin (Morocco and Algeria); (13) Sirte/Cyrenaica (Libya). Potential resources ((14)–(18)): (14) Potwar and Mijalar Basins (Pakistan); (15) Saudi Arabia; (16) Sào Francisco Basin (Brazil); (17) Argentina, Bolivia and Paraguay; (18) Al Kufra Basin (Libya)

11.2.1 Lena-Tunguska Petroleum Province in Siberian Craton

11.2.1.1 Regional Geological Setting

With areal extent of 4.5×10^6 km^2, the Siberia Craton is located in northeast Russia between the Yenisey and Lena Rivers. As a most importent hydrocarbon-producing area in the craton, the Lena-Tunguska region is commonly called the Lena-Tunguska Petroleum Province (LTPP) in Russia, and covers an area of about 2.8×10^6 km^2, which is bounded on the east and northeast by the Archean-Early Proterozoic Aldan and Anabar Shields, and also encircled elsewhere by a series of Precambrian foldbelts, i.e., the west, southwest and south sides respectively by Yenisey, East Yenisey and Baikal-Patom Foldbelts (Fig. 11.3; Kuznetsov 1997).

In the LTPP, the principal source and reservoir beds are within Meso-Neoproterozoic to Early Cambrian sequences. Meso-Neoproterozoic sedimentary strata of great volume extend over a large territory of Siberia Craton, reaching 5 km

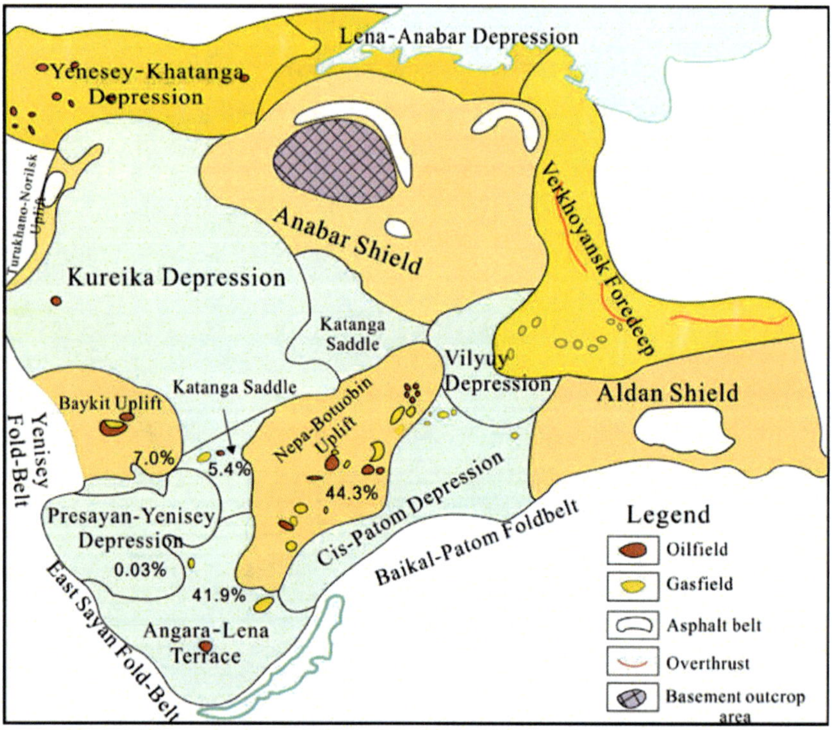

Fig. 11.3 Tectonic unit division and oil/gas field distribution of Siberia Craton in Russia (Tong and Xu 2004; Kuznetsov 1997 modified). The percentages indicate the portions of Meso-Neoproterozoic petroleum reserve in each tectonic unit (basic data based on IHS Energy Group 2005, International Petroleum Exploration and Development Database, Colorado)

or more in thickness. They have been formed lasting app. 1650–1600 to 540–530 Ma (Frolov et al. 2015). The oil- and gas-bearing sedimentary sequences are Riphean, Vendian and Early Cambrian in age. A thick Cambrian salt succession would provide regional super-seal that facilitated the preservation of hydrocarbons (Ghori et al. 2009).

11.2.1.2 Effective Source Beds

Hydrocarbon accumulations and their Riphean source rocks within the pericratonic Riphean palaeo-basins along the margin of Siberia Craton (i.e., the Cis-Patom Trough, Turukhansk and Udzha Basins in Fig. 11.4) were most likely destroyed by Baikal orogeny event with intensive erosion. For instance, even if the Riphean source rocks are characterized by up to 4% TOC content in Turukhansk Uplift, the generation potential (S_2) and hydrogen index (HI) are still very low (<0.1 mg HC/g Rock and 50 mg HC/g TOC respectively), implying that organic matter of those rocks is completely exhausted (Forov et al. 2015).

By way of exception, the organic-rich shale of Upper Riphean Irmeken Formation in the intracratonic Riphean palaeo-basins (i.e., Irkineeva-Vanavara and Kureika-Anabar Basins in Fig. 11.4) are unlikely to be affected by the significant pre-Vendian erosion event. So far the Iremeken shale (sampled from the interval 2182–2186 m of Yurubchen-104 Well on Baykit Uplift, Fig. 11.4) are still chracterized by the

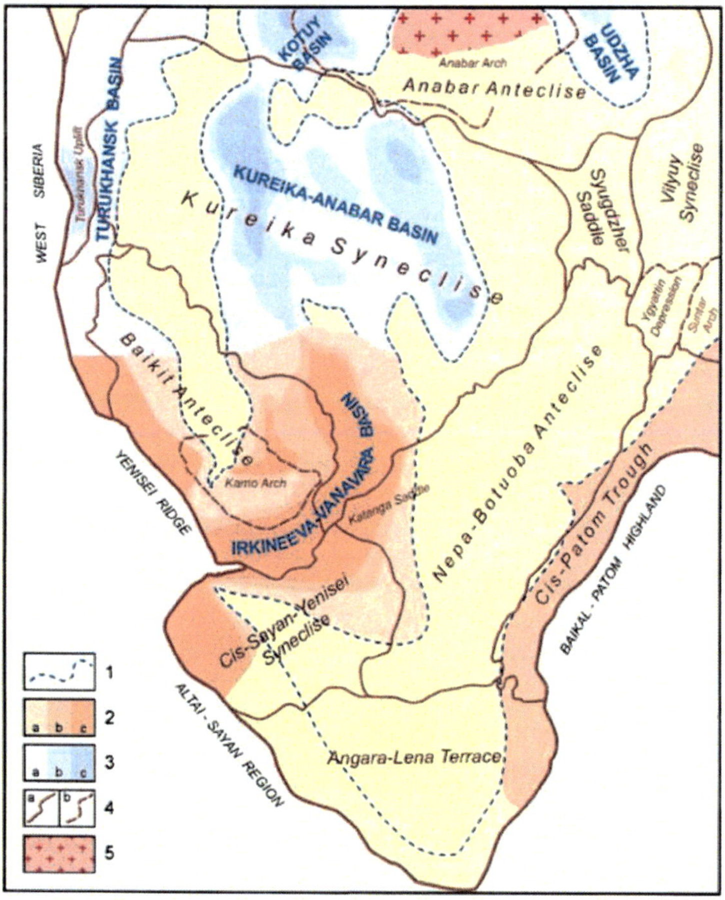

Fig. 11.4 Distribution of the Riphean palaeo-basins in the LTPP (Frolov et al. 2015). 1. Riphean palaeo-basin boundary; 2. Riphean palaeo-basin in the southern and central Siberia Craton and the inferred stratigraphic ages exposed under pre-Vrndian unconformaty surface: **a** Lower-Middle Riphean, **b** Middle-Upper Riphean, **c** Upper Riphean; 3. Riphean palaeo-basin in the northern part of the Craton and their assumed sedimentary infill thichness: **a** < 2 km, **b** 2–3 km, **c** > 3 km; 4. boundaries of recent major structural elements of the Siberian Craton: **a** 1st order structures, **b** 2nd order structures; 5. Modern outcrops of basement recks

following parameters: TOC 12.6%, chloroforn extractable bitumen content 0.56%, hydrocarbon index 463 mg HC/g TOC and T_{max} 445 °C (Larichev et al. 2004), which show a fair to high organic abundance and moderate maturity for hydrocarbon generation. Therefore, the Iremeken shale can act as a source bed for the oil and gas reservoirs/fields (e.g. the reservoirs of Riphean Kamov Group) in Baykit Uplift and Katanka Saddle, espetially for the Yurubchen-Tokhomo Zone (Fig. 11.3; Filipstsov et al. 1999; Ulmishek 2001; Melanie 2010; Frolov et al. 2015).

On the whole, Vendian and Lower Cambrian deposits would be the main source beds for present day oil fields in the central and northern parts of the Nepa-Botuobin Uplift and Turukhano-Norisk Uplift (cf. Fig. 11.3 and 11.4; Frolov et al. 2015).

11.2.1.3 Petroleum Resources

According to the basic data of IHS[①] (IHS Energy Group 2005. International Petroleum Exploration and Development Database. Colorado), totally there are 64 oil- and/or gas-fields composed of 168 individual reservoirs and found within the Meso-Neoproterozoic to Lower Cambrian strata in the LTPP. The proven plus probable petroleum reserves are 2.027×10^{12} m^3 for gas, 0.76×10^8 t for gas-condensate, 5.52×10^8 t for oil, and the total geological reserves would be up to 22.36×10^8 t in oil equivalent.

As assessed in twice by United States Geological Survey (USGS), however, the preserved volumes of both oil and gas in the Siberia Craton range from 2.8 bbbl (ca. 3.8×10^8 t) of oil and 48.9 Tcf (ca. 1.38×10^{12} m^3) of gas to 11.3 bbbl (ca. 15.5×10^8 t) of oil and 175 Tcf (ca. 5.0×10^{12} m^3) of gas (Ulmishek 2001). The main reason for the difference in the assessments is the assumption that much of undiscovered resources will be accounted for by reserve growth in two discovered giant fields, the Yurubchen-Takhoma Oilfield and the Kovyktin Gasfield (Ghori et al. 2009). Furthermore, Efimov et al. (2012) estimated that the proven recoverable reserves in Meso-Neoproterozoic silicaclastic and carbonate reaervoirs are about 6×10^8 t of oil and more than 2.7×10^{12} m^3 of gas in Siberia Craton (Ghori et al. 2009).

Based on statistics of the proven plus probable petroleum reserves in different tectonic units by IHS[①] (IHS Energy Group. 2005. International Petroleum Exploration and Development Database. Colorado), 86.3% of the total petroleum reserves are concentrated on two tectonic units, i.e., Nepa-Botuobin Uplift (individually accounting for 44.4%) and Angara-Lena Terrace (41.9%), while 7.0% of the total reserves are distributed at Baykit Uplift, 5.5% found at Khatenga Saddle and only 0.03% at Cis-Sayan Depression[①] (Fig. 11.3; IHS Energy Group. 2005. International Petroleum Exploration and Development Database. Colorado; Wang and Han 2011).

The stratigraphic horizon distribution of proven plus probable petroleum reserves are 5.31% in Cambrian, 1.26% in Cambrian plus Vendian, 86.77% in Vendian, 0.01% in Vendian plus Riphean, 6.58% in Riphean, and 0.07% still uncertain respectively[①] (IHS Energy Group. 2005. International Petroleum Exploration and Development

Table 11.1 Reserves of the largest oil and gas fields in the Siberia Craton[①]

Type	Field	Oil/10^6 t		Gas/10^9 m^3	
		Output from beginning of development	Reserve (A + B + C_1)[a]	Output from beginning of development	Reserve (A + B + C_1)[a]
Oil–Gas	Middle Botuobin	0.9	100	5.7	164
Gas	Chayandin	–	50	0.1	1200
Gas-Oil	Yurubcheno-Tokhm	0.7	168	–	140
Oil	Kuyumbin	0.3	110	–	20
Oil	Upper-Chona	16.1	150	0.5	16
Oil–Gas	Talakane	19.2	167	0.5	43
Gas	Kovyktin	–	115	0.5	1900

[a] A. proven develop and production reserve (PDP); B. proven undeveloped reserve (PUD); C_1. probable reserve

① Galimov (2014). Petroleum geology research of Precambrian in Russia (Eastern Siberia, Eastern European Platforms). An unpublished lecture notes in Beijing (in Russian)

Database. Colorado). Obviously the Vendian reservoirs contain the largest portion of proven plus probable oil and gas reserves. The major Vendian oil- and gasfields (incl. such large ones as Chayandin, Middle Botuobin and Upper Chona Fields) are located on the Nepa-Botuobin Uplift and some are situated on the Angara-Lena Terrace (incl. the giant Kovyktin Gasfield) and the Cis-Patom Depression. While the Riphean carbonate reservoirs only account for a minor reserve portion. To date, Riphean fields have been discovered only in one place on Baykit Uplif, i.e., the Yurrubchen-Tokhom Oil–Gas Field (Fig. 11.4), the oil and gas reserves in above mentioned major fields are listed in Table 11.1.

11.2.1.4 Oil and Gas Accumulation

Taking the large Middle Botuobin Oil–Gas Field as an example, it is located at the north of Nepa-Botuobin Uplift (Fig. 11.5), and tectonically referred to the structure type of complicated long-axis-anticline, containing a massive gas-condensate reservoir with oil-ring. The areas of gas reservoir and oil-ring account for 800 km^2 and 600 km^2 respectively, while the anticlinic area would be up to 1570 km^2 (Fig. 11.6). The field had been discovered at July 1970, somehow, it was developed after one decade. At least 108 wells have been drilled (Tong and Xu 2004) and its A + B + C_1 reserves are 164×10^9 m^3 of gas and 100×10^6 t of oil[①] (Table 11.1. Galimov (2014). Petroleum geology research of Precambrian in Russia (Eastern Siberia, Eastern European Platforms). An unpublished lecture notes in Beijing (in Russian).), both oil and gas were supplied by the source bed of Vendian shale. The reservoir driving types are attributed to water and gas drives as well as partly dissolved-gas drive. The gas output of individual well could be up to 24 t/d of oil and 24 × 10^4–26 × 10^4 m^3/d of gas (Tong and Xu 2004).

As the second case, Yurubchen-Tokhm Oil–Gas Field situated on Baykit Uplift is also a large gas-condensate field with oil-ring and attributed to a karst-fissure type of reservoir confined by stratigraphic unconformable contact and faulting plane (Figs. 11.5 and 11.7). The first oil was tested in 1977. Since the high production of oil and gas flows was obtained from the Ю-2 exploratory well in 1982, it has been proven to be light oil (42°–45° API) with low sulphur content (0.2–0.3%; Ghori et al. 2009), at least 101 wells have been drilled, and it is ascertained that oil and gas are accumulated within the Vendian reservoir, while gas-condensate in the Upper Riphean reservoir (Tong and Xu 2004). Its proven oil- and gas-containing area is up to 3100 km^2 with A + B + C_1 reserves 1.68 × 10^8 t of oil and 140 × 10^9 m^3 of gas① (Table 11.1. Galimov (2014). Petroleum geology research of Precambrian in Russia (Eastern Siberia, Eastern European Platforms). An unpublished lecture notes

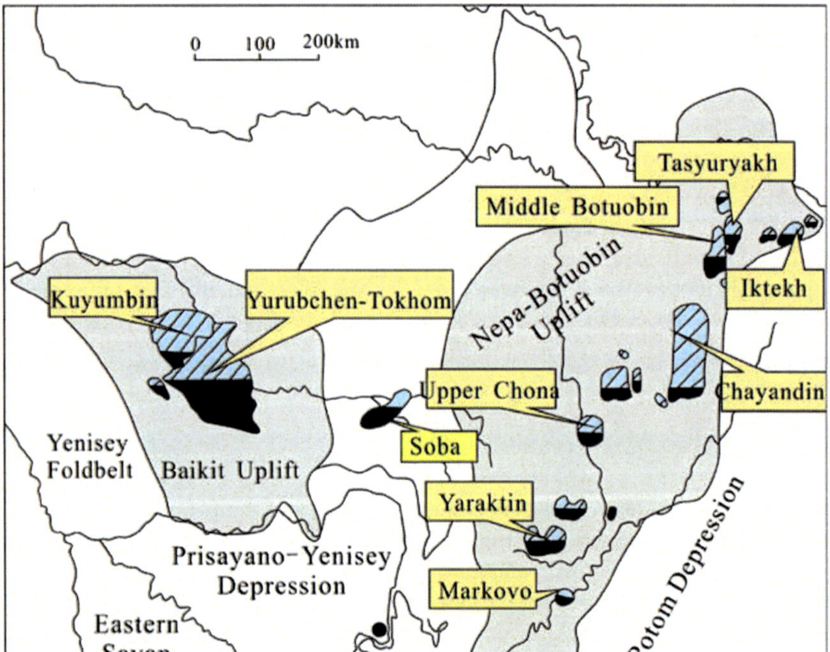

Fig. 11.5 Distribution of major oil- and gasfield at LTPP in Siberia Craton, Russia [Galimov (2014). Petroleum geology research of Precambrian in Russia (Eastern Siberia, Eastern European Platforms). An unpublished lecture notes in Beijing (in Russian)]. (Fedorov 1997, modified)

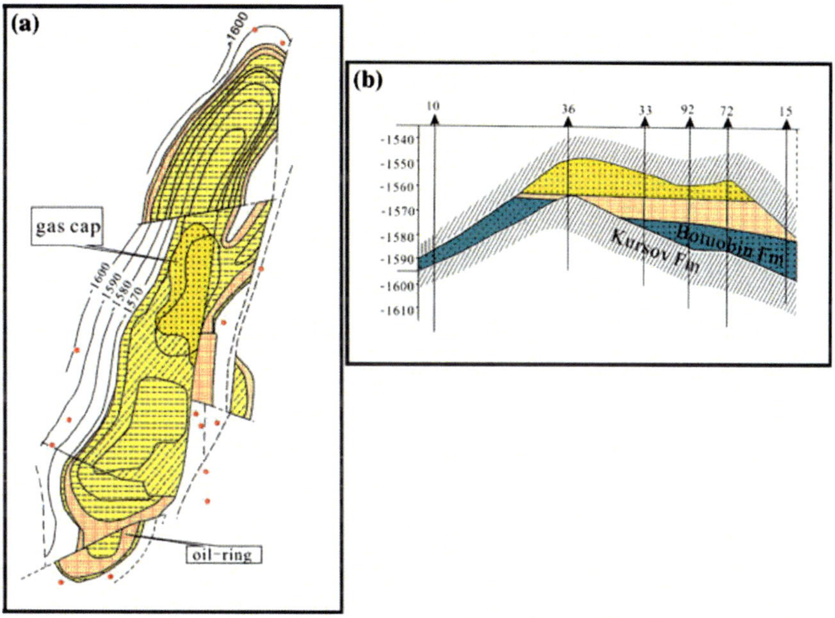

Fig. 11.6 Structural contour map (**a**) and oil and gas reservoir section (**b**) of Middle Botuobin Oil–Gas Field (Tong and Xu 2004)

in Beijing (in Russian).), the reservoir was sourced by Riphean muddy carbonate source bed② (Tong and Xu 2004; IHS Energy Group. 2005. International Petroleum Exploration and Development Database. Colorado; Frolov et al. 2015).

As for the oil and gas accumulation conditions, Kuznetsov (1997) proposed that Riphean source and reservoir beds are unconformably overlain by Vendian succession, which are distributed in the Baykit Uplift and Katanga Saddle. In this region, all the fields (such as Yurubchen-Tokhm Oil–Gas Field) produce oil and gas from topmost Riphean dolostone reservoirs and regional seals are provided by carbonate, mudstone and salt of Vendian and Early Cambrian age. Riphean thick shale beds may serve as intraformation seals. Frolov et al. (2015) proposed that the Riphean reached the maximum burial around mostly in Palaeozoic time in the intracratonic palaeobasins. While a modelling calculation shows that most of the hydrocarbons migrated prior to the Devonian and that subsequent migration was significant (Kuznetsov 1997).

11.2.1.5 Timing of Oil Entrapment

Based on oil geochemistry and basin modeling, Everett et al. indicate that the Eastern Siberian source beds as being Late Riphean to Early Vendian in age, the critical moment or timing of oil entrapment was controlled primarily by deposition of a

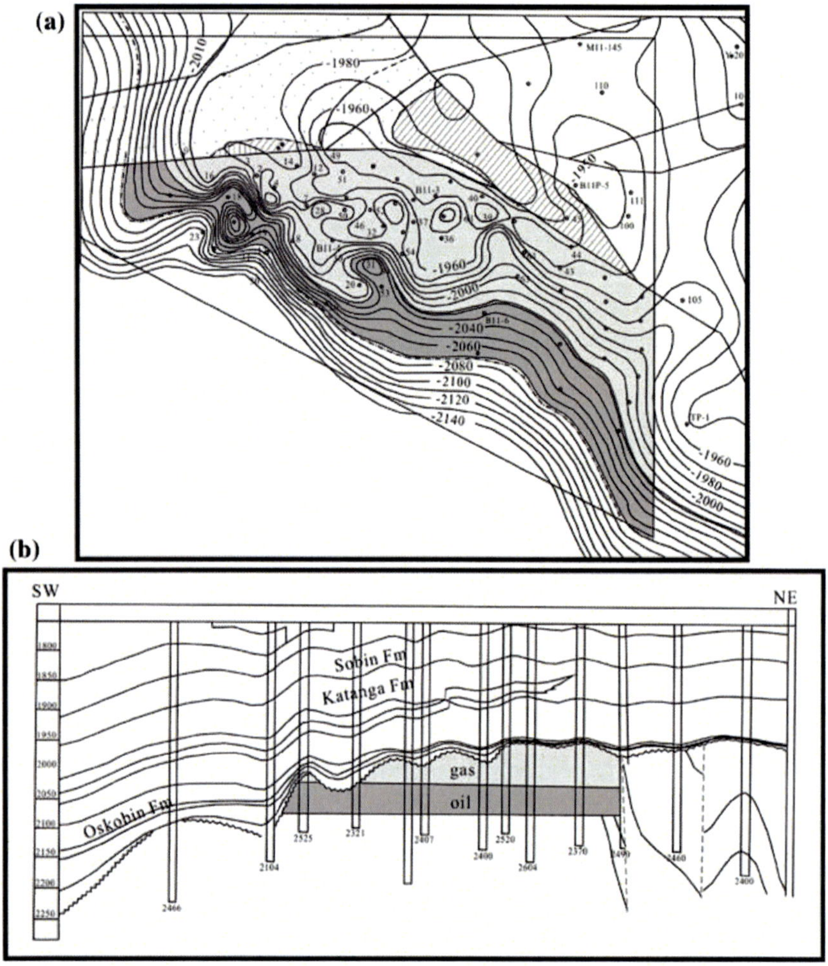

Fig. 11.7 Structural contour map (a) and oil and gas reservoir section (b) of Yurubchen-Tokhm Oil–Gas Field (Tong and Xu 2004)

salt seal in Early Cambrian, hydrocarbon accumulation was effectively ended in the Late Ordovician–Early Silurian, fifty percent of the Nepa-Botuobin gas samples may be classified as wet gases, implying source bed thermal maturity levels less than a vitrinite reflectance value of 2.0% with the remaining samples as thermogenic dry gases, most likely resulting from the relatively rapid over-maturation of the source rock during regional metamorphism and uplift.

11.2.2 Moscow and Kama-Belsk Basins in East European Craton (Russia)

East European Craton is also called Russia Craton, where oil and gas fields have been found in the Palaeozoic Devonian, Carboniferous and Permian strata. To present, however, no commercial hydrocarbon accumulations have been found within the Proterozoic sequences, while, in many cases, oil-seeps and low oil production in the Neoproterozoic Vendian sequence has been found in some exploratory wells. A number of large oil prospective regions could present in the East European Craton, e.g., Moscow and Kama-Belsk Basins① (Fig. 11.8; Fedorov 1997. Galimov (2014). Petroleum geology research of Precambrian in Russia (Eastern Siberia, Eastern European Platforms). An unpublished lecture notes in Beijing (in Russian).).

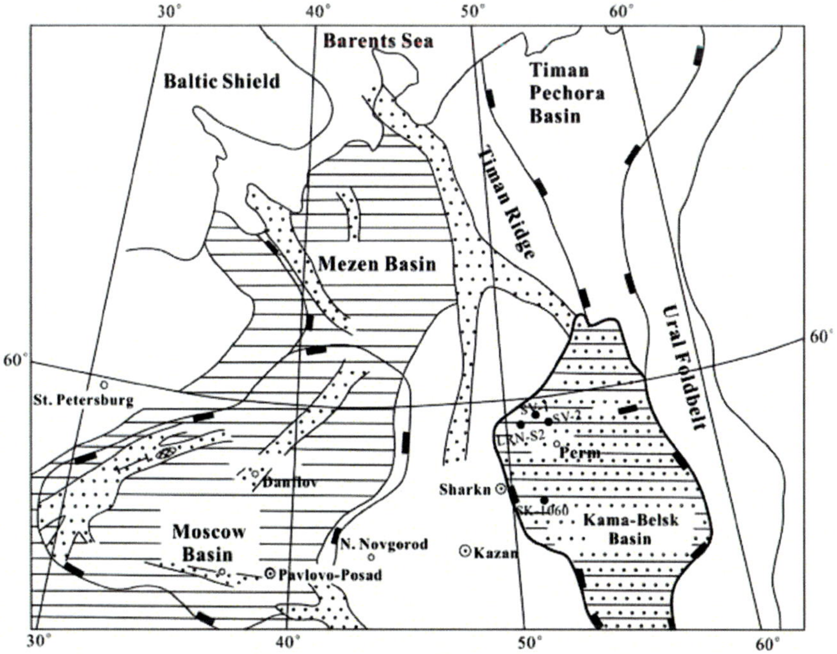

Fig. 11.8 Distribution of Proterozoic sedimentary basins on the East European Craton and related exploratory wells. LRN. Larionov; SK. Sharkan; SV. Sivinsk; ●.well site; ○.city. (Galimov (2014). Petroleum geology research of Precambrian in Russia (Eastern Siberia, Eastern European Platforms). An unpublished lecture notes in Beijing (in Russian)). (Fedorov 1997, compiled)

11.2.2.1 Moscow Basin

As a typical intracratonic depression, Moscow Basin is located at the central part of the East European Craton (Fig. 11.8). Stratigraphic column of the basin is up to 5 km thick, in which the Proterozoic and Palaeozoic intervals are of approximately equal thickness, while the overlying Mesozoic is relatively thin. Within the Proterozoic succession, the Riphean sequence is the thickest one, whereas the Vendian strata the most widespread① (Galimov (2014). Petroleum geology research of Precambrian in Russia (Eastern Siberia, Eastern European Platforms). An unpublished lecture notes in Beijing (in Russian).).

Deep Seismic Sounding (DSS) data indicate that a mantle diapir (i.e., an uplift of Moho discontinuity) may be present beneath the basin, resulting in a system of EN-SW striking grabens with Riphean sediments. The graben system is identified as a rift origin by DSS investigation (Fedorov 1997).

Light oil flows have been recorded from the sandstones of Upper Vendian Redkino Series in Wells 1, 4 and 9 in the Danilov area (Fig. 11.8). Oil flows of 50 L/d, with simultaneous 1000 m^3/d of gas containing 92% hydrocarbons, has been recorded in Well 9. The oil density is 0.79–0.83 g/cm^3 with gasoline fraction 26% to 35–42% and sulphur content 0.04–0.39%, while a dominance of paraffin (63–82%) over aromatics (3.7–3.9%) is distinctive. These light oils are referred to paraffinic-naphthenic hydrocarbons, its low-density fraction may not be removed, or to which a low-density fraction could be added by recent relatively young migration, possibly Meso-Cenozoic? (Fedorov 1997).

11.2.2.2 Kama-Belsk Basin

The Kama-Belsk Basin is a pericratonic depression located at the eastern margin of the East European Craton (Fig. 11.8). The basin contains the greatest thickness of Riphean sequence (up to 10 km), and is overlaid by Vendian sequence, according to drilling and geophysical data① (Galimov (2014). Petroleum geology research of Precambrian in Russia (Eastern Siberia, Eastern European Platforms). An unpublished lecture notes in Beijing (in Russian).).

In addition, heavy oil flows with output of 1 m^3/d–7 t/d have been recorded from the Vendian reservoirs respectively in Sivinsk-1, Sivinsk-2, Larionov-52 and Sharkan-1060 Wells (Fig. 11.8). Most of the oils belong to naphthenic-base crude. The oil in Sharkan Field is typical: its oil density is as high as 0.97 g/cm^3, resin plus asphaltene contents up to 30% and gasoline fraction 26% to 35–42%, which would be similar to biodegraded oil (Beijing Petroleum Exploration and Development Institute and North China Petroleum Bureau 1992; Fedorov 1997), whereas the oils are also "light" in carbon isotopic ratios showing a $\delta^{13}C$ value $-31‰$, which differs from that of Palaeozoic oils and sedimentary organic matter in the East European Craton, but is still similar to Riphean and Vendian oils and sedimentary organic matter on the Siberia Craton (Vysotsky et al. 1993).

11.2.3 Sultanate of Oman Basins in Eastern Arabian Craton

11.2.3.1 Regional Geological Setting

With a territory area of about 30×10^4 km^2, Sultanate of Oman is situated on the eastern margin of Arabian Peninsula or Arabian Craton (cf. the inset in Fig. 11.9). Oil exploration in Oman began in 1925, however the first discovery was not made until 1962. During early 1998, the average daily oil production in Oman was up to 900,000 bbl/d (ca. 12.3×10^4 t/d; Knott 1998).

A series of sedimentary basins are developed within the interior of Oman, and constitute a NNE-trending system of restricted basins, incl. South Oman Salt Basin (SOSB), Fuhud and Al Ghabah Salt Basins (Fig. 11.9; Peters et al. 2003; Amthor et al. 2005), which may have stretched from the Indian Craton across the eastern extremity of the Arabian Craton to the Hormuz Salt Basin of North Iran and possibly beyond (Amthor et al. 2005). Seismic data indicate that the western margin of the SOSB is delineated by structurally complex transpressional deformation fronts (Immerz et al. 2000). On the contrary, its eastern margin is characterized by onlap and thinning of basin strata onto a structural high close to modern-day coast of Oman (Fig. 11.9; Amthor et al. 2005).

11.2.3.2 Stratigraphic Division

Table 11.2a shows a generalized stratigraphic column of the Neoproterozoic–Cambrian in Oman (Amthor et al. 2005; Grosjean et al. 2009). The salt-bearing Late Neoproterozoic to Early Cambrian succession is represented by the Huqf Supergroup in Arabian Craton. The supergroup overlays unconformably on an old crystalline basement (Bowring et al. 2007) and is divided into three groups (from base to top): Abu Mahara, Nafun and Ara Groups. The Abu Mahara and Nafun Groups are subdivided into six lithostratigraphic units (not in stratigraphic order): siliciclastics (Ghadir Manqil, Masirah Bay and Shuram Formations) and carbonates (Khufai and Buah Formations).

The Ara Group consists of a thick cyclic sequence of carbonates, evaporites and associated siliciclastics, which can be subdivided into five formations., i.e., Birba, "U"-shale/Athel/Al Noor and Dhahaban Formations from base up (Table 11.2a). These formations are correlated to at least seven 3rd-order evaporite-carbonate sequences (A0 to A6 Cycles) established at the basin-center of the SOSB (Table 11.2a; Amthor et al. 2005), among which, Birba Formation comprises carbonate and evaporites of A0–A3 Cycles; the "U" Formation (in A4 Cycle) consists of organic-rich shales (ca. 80 m thick), carbonates, evaporates and ash interbeds, the ash beds could be used for providing U–Pb chronostratigraphic data; the Athel Formation (in A4 Cycle) comprises three members: ① Al Shomou Silicilyte Member, a thick (300–400 m) organic-rich chert; ② Thuleilat Shale Member; and ③ Athel Carbonate

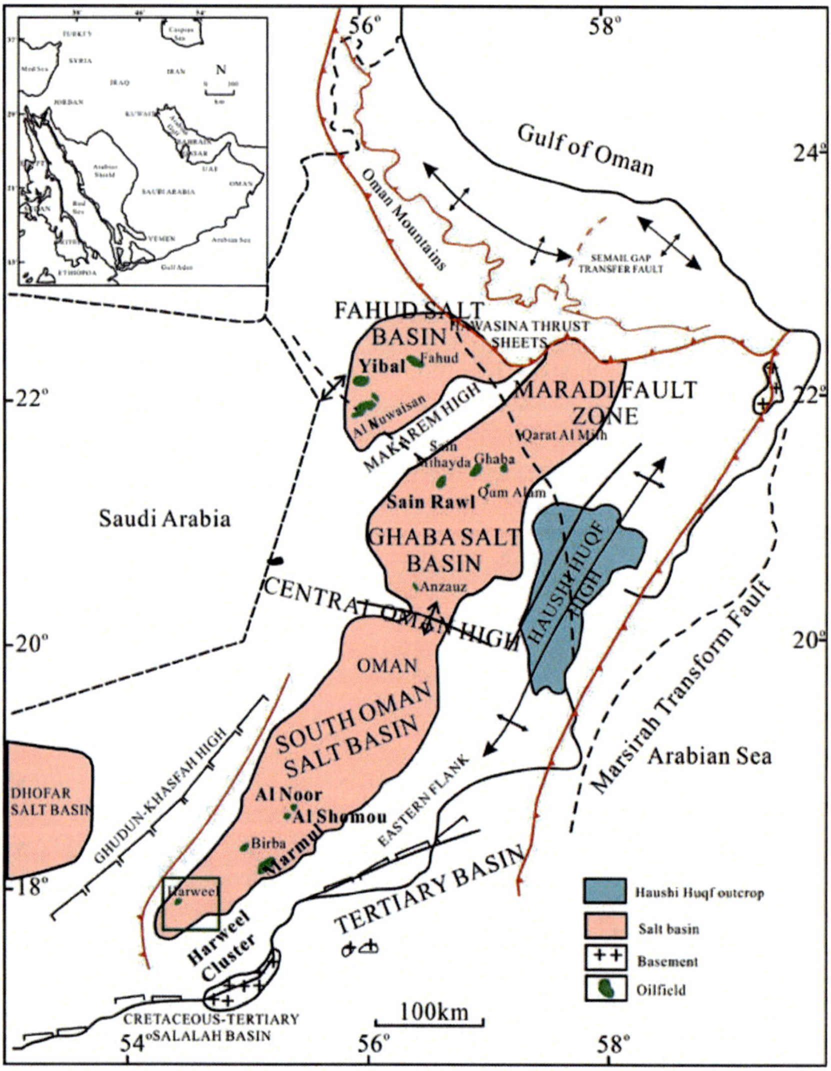

Fig. 11.9 Schematic tectonic map showing the distribution of the Neoproterozoic–early Cambrian salt basins and major oilfields in Oman (Peters et al. 2003; Amthor et al. 2005 compiled)

Member (Fig. 11.9b). The Al Shomou Silicilyte Member is stratigraphically "sandwiched" by two regional extensive organic-rich shale mems, the underlying "U"-Shale Formation and the overlying Thuleilat Shale Member (Table 11.2b; Amthor et al. 2005).

Table 11.2 Generalized stratigraphic column for neoproterozoic–Cambrian sequences in Oman

a

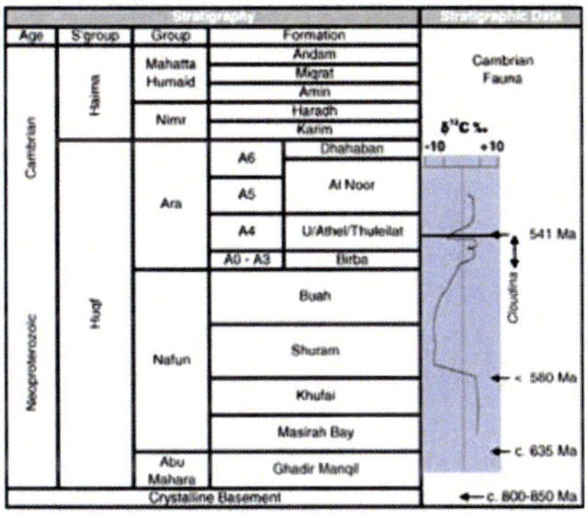

b

Group	Formation	Member	Sequence	Lithology
Huqf	Al Noor	Al Noor Salt	A6	Salt
			A5	Silicilyte
				Salt
	Athel	Thuleilat	A4	Thuleilat Shale \ Athel Carbonate
		Al Shomou Silicilyte		Silicilyte
	"U"	"U"-Shale		Shale
		Salt		Salt

Note a. the composition of A4–A6 sequences in Ara Group (Grosjean et al. 2009); b. Amthor et al. 2005, simplified

11.2.3.3 Source Beds and Oil Types

Most of the South Oman oils have been proven to be associated with the source bed of the Huqf Supergroup by the correlation between the organic-rich rock units and reservoir fluids (Grosjean et al. 2009). Geochemically, two major oil types, the "Huqf" oil and "Q" oil, were recognized (Grantham et al. 1987), Both oil types show conspicuous mid-chain monomethyl-alkanes, trivially named "X compounds", similar to the organic matter within the Huqf source rocks, being considered a distinct character of Precambrian–Cambrian sedimentary organic matter. Accordingly, a straightforward correlation has been established between these oils and the Huqf Supergroup (Grantham 1986; Grantham et al. 1987; Terken et al. 2001).

11.2.3.4 Silicilyte and Carbonate "Stringer" Reservoirs

Huqf oils are found respectively in the Ara carbonate "stringer" and the intra-salt Athel silicilyte reservoirs (Figs. 11.10 and 11.11) as well as in the Cambrian Haima and Permo-Carboniferous Haushi reservoirs, particularly along the eastern flank of the SOSB. In contract, "Q" oils occur mostly in central Oman, but can also be found in the northernmost part of the SOSB, where encountered in the same well, "Q" oils occur in reservoir that overlie those of "Huqf" oils (Grosjean et al. 2009).

The most significant hydrocarbon reservoirs in Huqf Supergroup are the carbonate "stringer" and "silicilyte" reservoirs found at depth up to 4–5 km, both of which are stratigraphically intimately related and sealed by Ara salt in the SOSB (Figs. 11.10 and 11.11).

The "silicilyte" is a new petrological term specially for the microcrystalline silicon matrix, or called "microcrystalline chert", consisting of 80–90% microcrystalline

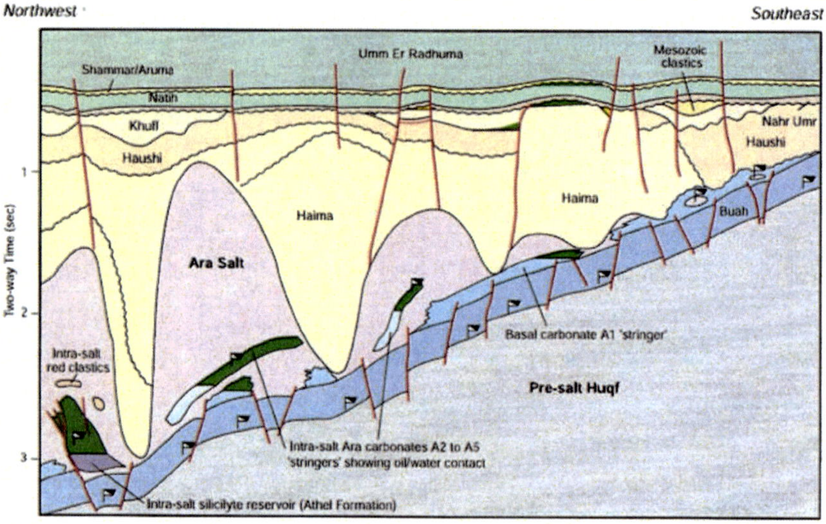

Fig. 11.10 Simplified geological cross-section through the SOSB to illustrate the geological occurrence of Ara "stringer" and "silicilyte" oil reservoirs enclosed by large, irregular salt. Bodies oil is indicated in green (Peters et al. 2003)

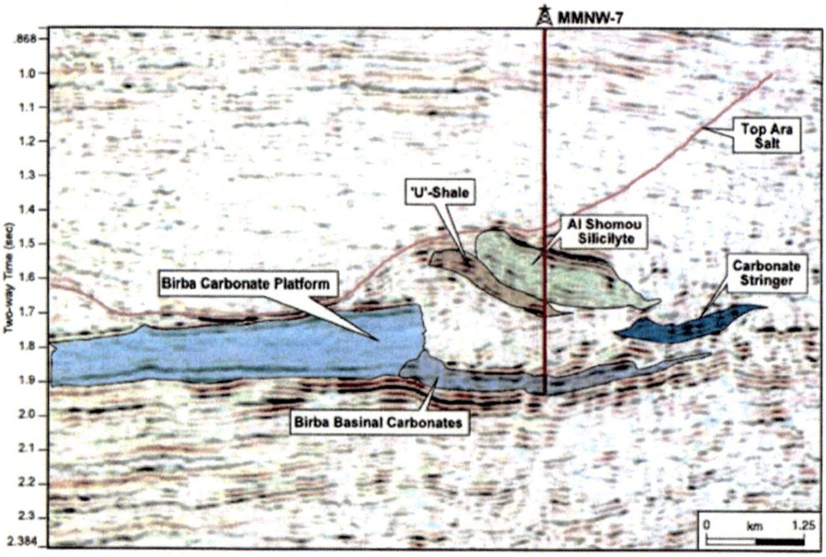

Fig. 11.11 Seismic section showing the geological occurrence of the carbonate stringer and silicilyte reservoir (Amthor et al. 2005)

quartz (crystal-size 2–3 μm), and commonly constitutes sheet-like aggregates, i.e., an up to 400 m thick and several kilometers wide slabs rich-in organic matter (TOC 7% in average). The Al Shomou silicilyte reservoir is typically characterized by light (48° API) sour oil, and finely laminated porous reservoir rocks with high-microporosity (up to 30%), low-permeability (only 0.02 mD), high overpressures (19.8 kPa/m in average gradient) and high present-day oil saturation (>80%; Amthor et al. 1998).

Whereas the Al Shomou Silicilyte is a unique source and reservoir rocks, and "sandwiched" by "U"-shale and Thuleilat shale (Fig. 11.11), all of which are rich-in organic matter, the 'silicilyte' reservoir should be self-charged, e.g., those in Al Noor and Al Shomou Fields (Fig. 11.9; Amthor et al. 2005), where unstimulated well initial flows of 250–600 bbl/d (ca. 34–82 t/d) could be increased to over 2500 bbl/d (ca. 342 t/d) by applying massive hydraulic fracturing technology (Wong et al. 1998). In August of 2000, the first oil was produced from the Al Noor Field (Fig. 11.9; Amthor et al. 2005).

Furthermore, the intra-salt Ara carbonate "stringer" reservoirs in the southern part of the SOSB (Figs. 11.10 and 11.11) are frequently dolomitized with an average porosity of 8–12% and 80–120 m-thick. Since they commonly contain intervals rich-in organic matter, "stringers" may form a self-charged oil reservoir encased in salt (Fig. 11.10), and such reservoirs can yield more than 6000 bbl/d (ca. 842 t/d) of oil in production tests, particularly if the reservoir is highly overpressured (Peters et al. 2003).

11.2.3.5 Petroleum Resources and Case Studies

Based on an earlier assessment of USGS, total petroleum resources of Ghaba and Fahud Salt Basins in north-central Oman were estimated as 11.3 bbbl of oil equivalent (ca. 15.5 × 10^8 t), The largest ones are Saih Rawl Field in the Ghaba Salt Basin and Yibal Field in Fahud Salt Basin (Fig. 11.9; Pollastro 1999).

During 1997–2002, there has been a string of exploration successes in discovering some 3.5 × 10^8 m^3 (ca. 2 × 10^{12} bbbl) of oil only at the Harweel Cluster region in SSOB (Fig. 11.9). This oil is deep and high pressure, comes with moderate gas-oil-ratio (GOR, 185 m^3/m^3), and the fluids contain 15% CO_2 and 5% H_2S. The Harweel Cluster consists of nine 'stringer' reservoirs found within intra-salt carbonate of Ara Group A2 and A3 Cycles in seven distinct oil fields. The fields are Dafag, Ghafeer, Harweel Deep, Rabab, Sakhiya, Sarmad and Zalzala. Only Ghafeer and Sakhiya have reservoirs in both of the Cycles. The reservoirs generally contain low permeability (1–10 mD), and a wide range of fluid properties from retrograde gas-condensate to black oil (O'Dell and Lamers 2003).

It is reported that the Marmul Oilfield, located in the south of SOSB (Fig. 11.9), contains an estimated 2 bbbl (ca. 2.74 × 10^8 t) of heavy oil (21.5° in average API gravity). Its glacial and periglacial reservoirs range from Cambrian to Permo-Carboniferous in age, with the oil being sourced by the source bed, which is ubiquitous in the middle potion of the Huqf Supergroup, and the oils display differing degrees of biodegradation (Katz and Everett 2016). To add up the petroleum reserves in Marmul Oilfield, Ghaba and Fahud Salt Basins as well as Harweel Cluster region, it would be more than 21.2×10^8 t of oil equivalent in the aggregate.

11.2.3.6 Timing of Petroleum Generation and Reservoir Charging

As to oil generation, tacking the middle Huqf source bed at the Marmul Oilfield in South Oman Salt Basin (SOSB) as an example, Katz and Everett (2016) reviewed that based on its burial and petroleum generation histories, the oil generation was initialed at 520 Ma (i.e., Early Cambrian), occurring largely during the Ordovician and Silurian, and was complete by 350 Ma (i.e., Early Carboniferous; Terken et al. 2001). Apatite fission track data suggest a significant uplifting during the Devonian (Visser 1991), and so almost the entire Silurian, Devonian and Carboniferous are missing (Aley and Nash 1984). The equivalent vitrinite reflectance eqR_o values suggest that the maximum burial temperature was achieved during the Middle Cambrian–Early Silurian (Terken et al. 2001). Regionally, oil generation may have preceded trap development, thus requiring oil storage and remigration to develop the accumulation. Oil migration into the structure appears to have taken place during the Tertiary (Heward 1989).

Moreover, Pollastro (1999) considered that burial history reconstruction suggests an early minor stage of oil generation in Middle-Lower Huqf source bed during the Early Silurian and peak oil generation during Late Permian/Early Triassic (ca. 250 Ma) at the Yibal Field in Fahud Salt Basin, North Oman (Visser 1991). While Modelling suggests that cracking gas expelled from Huqf source bed charged structures across the Fahud Basin during a period ranging from 80 Ma to present day in the Fahud Basin (Amthor et al. 1998).

As a typical case for a successful Infracambrian-sourced play with Early Palaeozoic hydrocarbon generation, the structural stability of Oman, and the presence of a salt super-seal would facilitate the preservation of these early generated oils.

11.2.4 India and Pakistan Basins in Indian Craton

The Meso-Neoproterozoic sedimentary basins in India and Pakistan share similar tectonic settings and stratigraphic depositional environments, containing three major petroliferous regions, i.e., Central Indus, Vindhyan and Cuddapah Basins (Fig. 11.12).

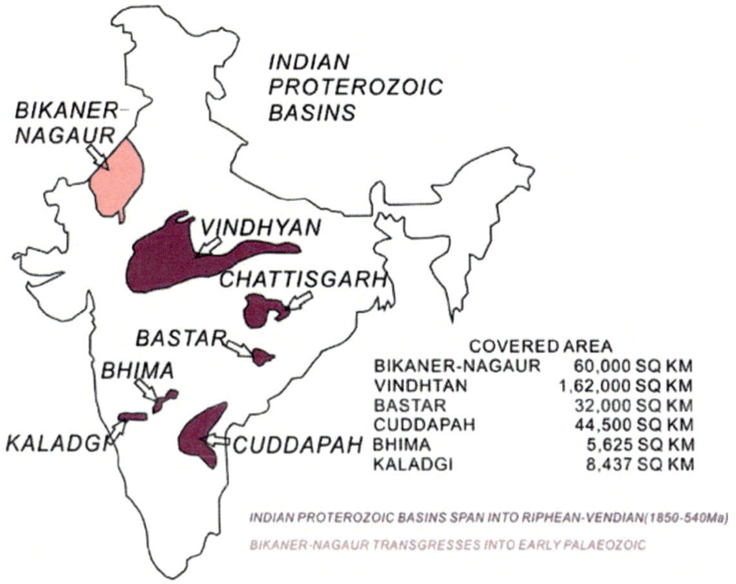

Fig. 11.12 Distribution of the major Meso-neoproterozoic sedimentary basins in the Indian Craton (Bhat et al. 2012)

11.2.4.1 Central Indus Basin: Punjab Platform and Bikaner-Nagaur Basin

(1) **Regional geological setting**: The Central Indus Basin (CIB) is situated geographically at the border area of Pakistan and India, and geologically on the northwestern segment of the Indian Craton. The limits of CIB are defined by the NE-SW trending Aravallis Range to the east and the Solaiman Range to the west (Fig. 11.13a). This large sedimentary basin covers approximately 29×10^4 km^2. There are different geological nomenclatures used in CIB, it is tectonically named the Punjab Platform on the Pakistan side within the centro-western CIB, and also the Bikaner-Nagaur Basin on the India side within eastern CIB, respectively (Fig. 11.13). On the Pakistan side, the CIB could be divided into three tectonic units from east to west: ① the Punjab Platform-Bikaner Basin, ② the Sulaiman Foredeep/Depression, and ③ the Sulaiman Range/Fold Belt (Qadri 1995; Fig. 11.13)

The Punjab Platform is a broad monocline dipping gently to the west towards the Sulaiman Depression (Fig. 11.13; Asim et al. 2015) covering an area of more than 10×10^4 km^2, and also extends eastward into India where it is called as Bikaner-Nagaur Basin (Fig. 11.13; Aadil and Sohail 2011; Bhat et al. 2012). Stratigraphically, evidence of drilled wells confirms the presence of the Neoproterozoic-Cambrian successions along and across the eastern border between Pakistan and India, and its thickness of ca. 1500 m is reported from Marvi-1 Well drilled at the southeastern part

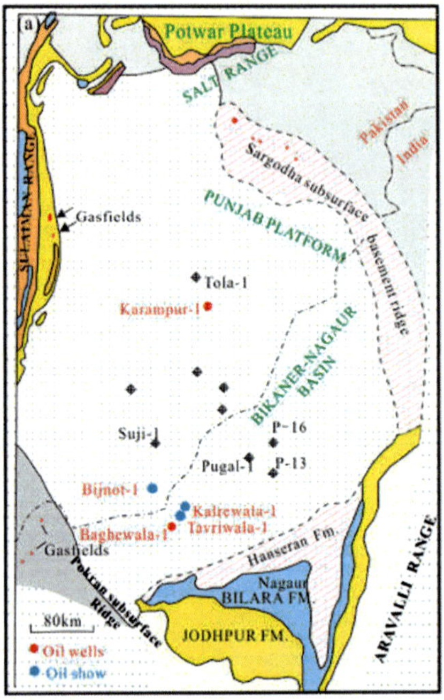

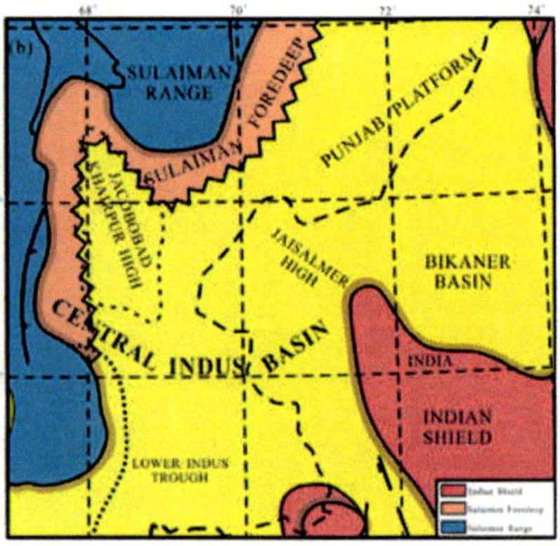

Fig. 11.13 The location and oil-exploratory well sites of the Bikaner-Nagaur Basin, India and the Punjab Platform, Pakistan (**a**) and the tectonic units of Central Indus Basin (**b**). **a** Ghori et al. 2009; **b** Aadil and Sohail 2011, modified

of Pakistan (Bhat et al. 2012). The location, oil-exploratory well sites and generalized stratigraphic correlation of the CIB are shown in Fig. 11.13 and Table 11.3 (Ghori et al. 2009; Aadil and Sohail 2011).

Table 11.3 Neoproterozoic-lower Cambrian stratigraphy of CIB (Ghori et al. 2009, modified)

Age	Punjab platform group/Formation	Bikaner-Nagau basin group/Formation
Cambrian	Kussak Khewra	Upper carbonate
Mesoprotereozoic	Salt range formation (Karampur-I Well)	Hanseran formation, evaporite
		Bilara Formation, laminated-dolostone (Baghewala-I Well)
		Jodhpur formation, sandstone (Baghewala-I Well)

Note Oil-producing wells are indicated within brackets

(2) **Petroleum geology and resources**: Oil exploration in the Punjab Platform started in 1950s. Neoproterozoic and Cambrian reservoirs have been unsuccessfully explored although only 12 wells have been drilled for Neoproterozoic and Cambrian (Hasany et al. 2012). Despite these failures, heavy asphaltic oil was first produced from fractured dolostone within the Neoproterozoic Salt Range Formation at the Karampur-1 Well of Punjab Platform in 1959, but no commercial hydrocarbon output obtained yet (Fig. 11.13a; Table 11.3; Ghori et al. 2009). However, the major heavy oil discovery was made at a shallower depth (1103–1117 m) in the Baghewala-1 Well, Bikaner-Nagaur Basin in 1991, only some 7 bbl (ca. 0.96 t) of sulfur-rich heavy oil (17.6°API, ca. 0.953 g/cm^3) was originally produced from Neoproterozoic Jodhpur sandstone and Bilara carbonate reservoirs (Fig. 11.13a; Peters et al. 1995; Ghori et al. 2009; Hasany et al. 2012; Asim et al. 2015). The oil is non-biodegraded as evident from the abundant presence of n-paraffin and acyclic isoprenoids (Peters et al. 1995; Asim et al. 2015), it should be the marginal mature oil. However, good oil shows were also reported from the Bijnot-1 Well on the Punjab Platform as well as the Kalrewala-1 and Tavriwala-1 Wells in the Bikaner-Nagaur Basin (Fig. 11.13a; Ghori et al. 2009). While the isopach maps show that the Cambrian and Precambrian depocenter were in southeast part of the Punjab Platform, and sediments would be buried at deeper level in west towards Sulaiman Depression in CIB (Aadil and Sohail 2011).

The Baghewala-1 oil is considered to be sourced from sulphur-rich organic matter in the laminated carbonates within the Neoproterozoic Bilara Formation which contain moderate to high TOC and HI typical of oil-prone organic matter, up to 400 mg HC/g TOC, and show T_{max} values of ≥436 °C, indicating marginally mature source bed within the early oil window based on the source biomarkers (Peters et al. 1995; Ram 2012; Asim et al. 2015). In addition, the geochemical characteristics of Baghewala-1, Karampur-1 and Kalrewala-1 oils are very similar to that of oils and source rocks from the Neoproterozoic-Lower Cambrian Huqf Supergroup in Oman (Peters et al. 1995; Sheikh et al. 2003; Ghori et al. 2009).

The reported geological reserves for the Baghewala Oilfield are 628 Mbbl (ca. 8540 × 10^4 t) of oil, and it contains four separate reservoirs. The oldest reservoir consists of sandstone within the Ediacaran Jodhpur Formation with a porosity of 16–24% and oil saturation ranging from 65 to 80%. The Early Cambrian upper Carbonate Formation dolostone forms the youngest reservoir with porosities ranging from 7 to 15% (Peters et al. 1995; Sheikh et al. 2003; Ghori et al. 2009).

11.2.4.2 Vindhyan Superbasin

(1) **Regional geological setting**: The arcuate-shaped Vindhyan Superbasin is located at the central segment of Indian Craton (Fig. 11.12). It contains a more than 5000 m-thick sequence of sandstones, shales and limestones with the Late Mesoproterozoic–Early Cambrian ages, and occupies an area of ca. 16.2 ×

10^4 km^2, among which ca. 8000 km^2 extends into the Ganga Valley Basin in the north and northeast beneath the Cenozoic sediments of the Himalayan Foredeep. To date, however, no commercial hydrocarbon has been discovered in the Vindhyan Superbasin yet (Ojha 2012).

The Ganga Valley Basin comprises the northern continuation of the Vindhyan Superbasin, as a Precambrian–Lower Cambrian interior sag basin overlain by the Cenozoic Himalayan Foredeep Basin (Fig. 11.14; Table 11.4). There totally are five subsags identified in the Ganga Valley Basin based upon gravity anomalies. As a larger one, the Sarda Subsag aligned east–west, with a sediment thickness of more than 7000 m. A schematic WE geological profile across the Sarda Subsag is shown in Fig. 11.15.

(2) **Petroleum geology and exploration:** Geochemical studies indicate that TOC contents in Ganga Valley Basin is fair to good for hydrocarbon generation. The Lower Vindhyan Supergroup Madhubani Formation represents alternating shale, sandstone and limestone of subtidal lagoon facies, within this, the interval 4670–5144 m of the Bilaspur Well in the Sarda Subsag gives TOC in the range 1.62–2.24% for oil source rock and also potential source rocks in the basin are essentially confined to the upper Vindhyan Subgroup Ujhani Formation.

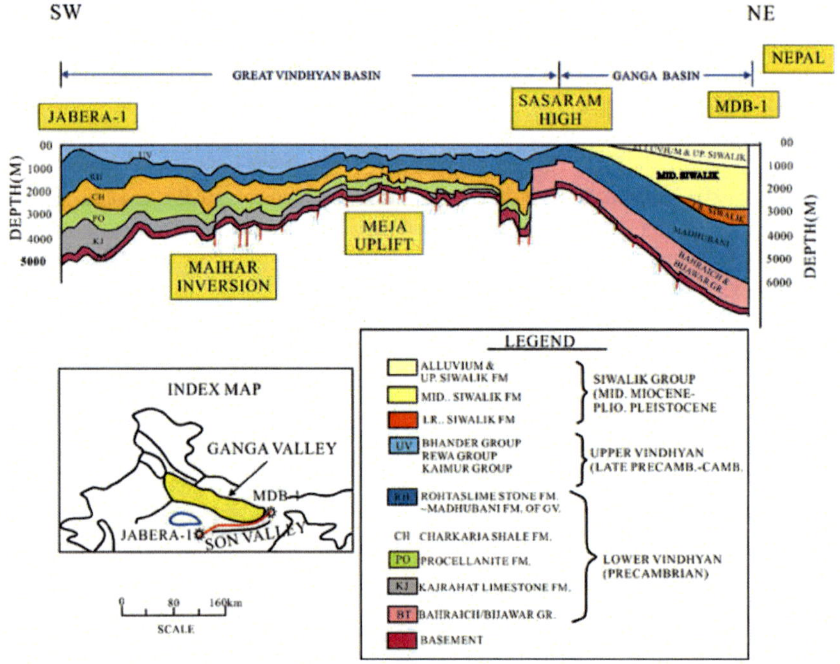

Fig. 11.14 A SW-NE trending schematic geological profile along the eastern Vindhyan Superbasin to the Ganga (Valley) basin (Ojha 2012)

Table 11.4 Generalized stratigraphic column of the Ganga Valley in Vindhyan Superbasin (Ojha 2012)

Period/Age	Group/Supergroup	Formation	Lithology in brief	Sedimentation cycle	Environment
Cenozoic Siwalik Sedimentation lower miocene onwards					
410–16 (394) Ma Hiatus					
Middle Silurian to lower devonian	Extension of upper vindhyan subgroup	Karnapur formation	Dominantly arenacious with fissile shale		Semi-arid, carbonate, shelf
Lower Ordovician to Lower Silurian		Tilhar formation	Limestones in association with calcarious shale	Puranpur-Gandak	
480–454 (26) Ma Hiatus					
Lower Ordovician to neoproterozoic	Upper vindhyan subgroup	Ujhani formation	Subgreywacke to arenite, fissile shales, occasionally intercalated with thin argillaceous-dolomitic limestones		Shallow, marine
1130–800 (330) Ma Hiatus					
Mesoproterozoic	Lower vindhyan subgroup (Semri Group)	Madhubani formation, Bahraich formation	Dolostone, limestone, Shale, siltstone and quartzite phyllites/schists	Bahraich-Sahaspur	Marginal, marine
Archaean	Basement		Granitic/gneiss rocks	Craton	

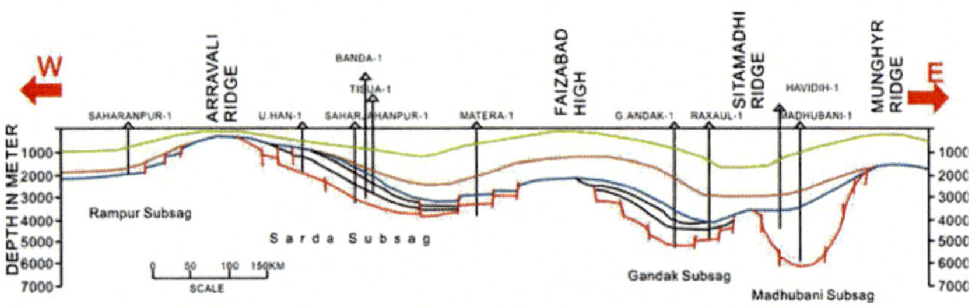

Fig. 11.15 A schematic WE geological profile across the Ganga (Valley) Basin showing the distribution of four subsags within the basin (Ojha 2012)

While the Madhubani Subsag also developed more than 7000 m thick sedimentary succession. As the deepest one, Madhubani-1 Well has been drilled up to a depth of 5957 m in this subsag (Fig. 11.15), and minor gas show was reported from the Madhubani Formation in this well with a C_1 count greater than 97% probably as deeply buried dry gas. Many zones have also displayed fluoresce, indicating the presence of hydrocarbons during drilling operations (Ojha 2012).

The Son Valley Basin constitutes the most-eastern segment of the Vindhyan Superbasin (see the index map in the Fig. 11.14). It shares the similar stratigraphic and petroleum geological conditions with the Vindhyan Superbasin. In addition, non-commercial hydrocarbon was discovered in the first exploratory well, Jabera-1 Well, within the Son Valley Basin. It was drilled to a depth of 3597.7 m, and showed the presence of gas at many levels during drilling (δC_1 value 39.0‰) and also flowed a non-commercial quantity of 2000–3000 m^3/d of gas from the upper part of the Jardepahar Formation in the Semri Group (or called Lower Vindhyan Subgroup; Ram 2012), but it is also reported that the flowed gas at 4000 m^3/d from the Lower Vindhyan Subgroup Charkaria Formation of Jabera-1 Well during DSTs (drill stem test). Its chemical composition is as follows: C_1 65.2–75.4%, C_2 5.3–13.6%, C_{3+} 0.8–17.8%, CO_2 0–0.3%, N_2 0–27.7%, He 0–0.21% as wet gas. However, these substantiated the hydrocarbon generation potential of the Vindhyan Superbasin (Ojha 2012).

11.2.5 Taoudenni Basin in West African Craton

11.2.5.1 Regional Geological Setting

As the largest broad intracratonic basin in the West and North Africa, with an area of ca. 2×10^6 km^2, the Taoudenni Basin developed over part of the West African Craton in southern and eastern Mauritania, and extends into northwestern Mali and southwestern Algeria (Fig. 11.16; Rahmani et al. 2009; Albert-Villanueva et al. 2016). The basin is bounded in the east by the Proterozoic Trans-Suharan Suture Zone and in the west by the Hercynian Mauritanides Fold Belt. To the north is the Reguibat Shield and to the south the Ivory Coast Shield, both of which are composed of Archean metamorphic and granitic basement (Fig. 11.16; Bronner et al. 1980; Albert-Villanueva, et al. 2016).

The Taoudenni Bain preserves up to 1300 m of gently dipping (<1°) unmetamorphosed and virtually undeformed Proterozoic strata (Kah et al. 2009). The stratigraphic fill of the Taoudenn Basin can be divided into four supergroups which are bounded by four regional unconformities (Albert-Villanueva et al. 2016).

The Supergroup 1 of Meso-Neoproterozoic ages consists of continental to marine successions, which can be subdivided into five units, from bottom to top:

(1) The Mesoproterozoic Char Group onlaps the Archean basement, and comprises fluvial, aeolian and shallow-marine coarse- to fine-grained siliciclastics together with minor carbonate deposits, reaching a maximum thickness of 400 m at outcrop (Kah et al. 2009).

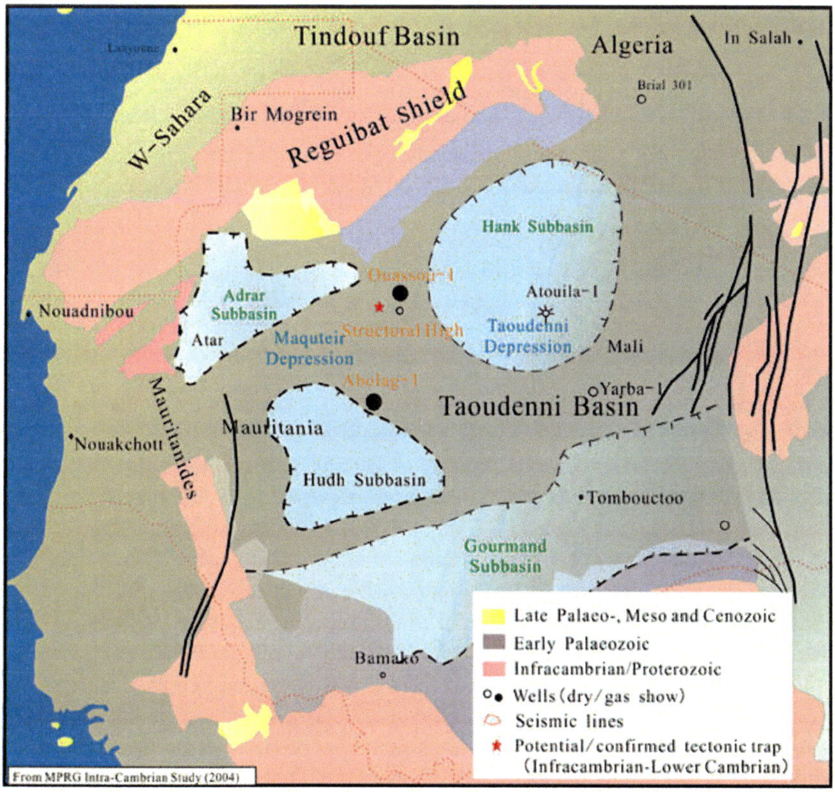

Fig. 11.16 Location of outcrops, wells and the extent of the Taoudenni Basin across Mauritania, Mali and Algeria (Lottaroli et al. 2009, compiled)

(2) The Late Mesoproterozoic-Early Neoproterozoic Atar Group consists of dolomitic carbonate, fine-grained siliciclastics and minor evaporitic strata, representing craton-wide flooding and shallow marine deposition (Kah et al. 2009), which has been reported in the Abolag-1 and Yarba-1 Wells with thicknesses of about 350 m and up to 600 m respectively (Gaters 2005). Rb–Sr analyses of glauconite and illite indicated that the Atar Group was deposited in Tonian-Cryogenian time (890 ± 35–775 ± 52 Ma, i.e. Early Neoproterozoic; Clauer 1981), but recent Re-Os geochemistry suggests that the group is some 200 Ma older, i.e., the Late Mesoproterozoic age (Rooney et al. 2010).
(3) Late Neoproterozoic Assabet el Haasiane Group is composed of prograding fluvio-deltaic sediments, including the Assabet Formation (thick fine-grained sandstones and siltstones), the Jebliat Formation (glacial deposits) and the Teniagouri Formation (thick Shales) in ascending order (Jean-Pierre et al. 2014). The fine-grained siliciclastic strata rich in Ediacara-type fauna (620–590 Ma, i.e., Late Ediacaran; Albert-Villanueva et al. 2016).

The Supergroup 2 of the Late Cambrian–Ordovician ages is composed of marine black sandy shales with epidote, sericite and siliceous shales, and siltstone interbedded with sandstones. The Supergroup 3 of Early Silurian age is constituted by quartzose sandstone and siltstone, while the Supergroup 4 of Devonian–Carboniferous age by marine and continental sandstone and shale successions (Albert-Villanueva et al. 2016). Furthermore, the Neoproterozoic–Carboniferous succession are covered by a thin sequence of Mesozoic–Cenozoic rocks in the center of this basin.

11.2.5.2 Petroleum Geology and Exploration

Organic-rich shales and limestones in the Taoudenni Basin have been reported from the Late Mesoproterozoic Atar Group in Mauritania nomenclature; equivalent to the Hank and Dar Cheikh Groups in Algrian nomenclature, and to the EI Mreiti Group in Khatt-Mauritanian Hank (Lahondére et al. 2005). These sediments are believed to have deposited between 890 and 620 Ma in northwest Africa. Infracambrian type I hydrocarbon source rocks, with TOC of 17–20% and hydrogen index (HI) of 800 mg HC/g TOC indicating a good oil generation potential, were penetrated in a water well drilled at the northwest part of the Taoudenni Basin (Gaters 2005).

Intense hydrocarbon exploration took place in this basin during 1972–1985. Only two deep exploratory wells have been drilled at the Taoudenni Basin, i.e., Abolag-1 (Texaco) and Quasa-1 (AGIP) Wells, in 1974 (Figs. 11.16 and 11.17). Gas show was recorded in the Abolag-1 Well at the interval of 2300–3000 m, and about 480 Mcf/d (ca. 13.6 × 10^6 m^3/d) of gas produced from the fractured limestones in the upmost Middle Infracambrian at ca. 3000 m depth during an "open hole" test, indicating the general viability of the Infracambrian play in the Taoudenni Basin. But the Quasa-1 Well was failed to reach the Infracambrian objective, and only recorded minor uneconomic gas shows. It is likely that Quasa-1 Well did not test a valid closure (Figs. 11.16 and 11.17; Ghori et al. 2009).

In recent years, three wells were further drilled at the northwestern part of Taoudenni Basin, i.e., R Well (Albert-Villanueva et al. 2016), "A" (western well) and "B" Wells (eastern well), but no economical discovery was made so far (Fig. 11.17; Jean-Pierre et al. 2014). The R Well drilled by Repsol on the northwest margin of the

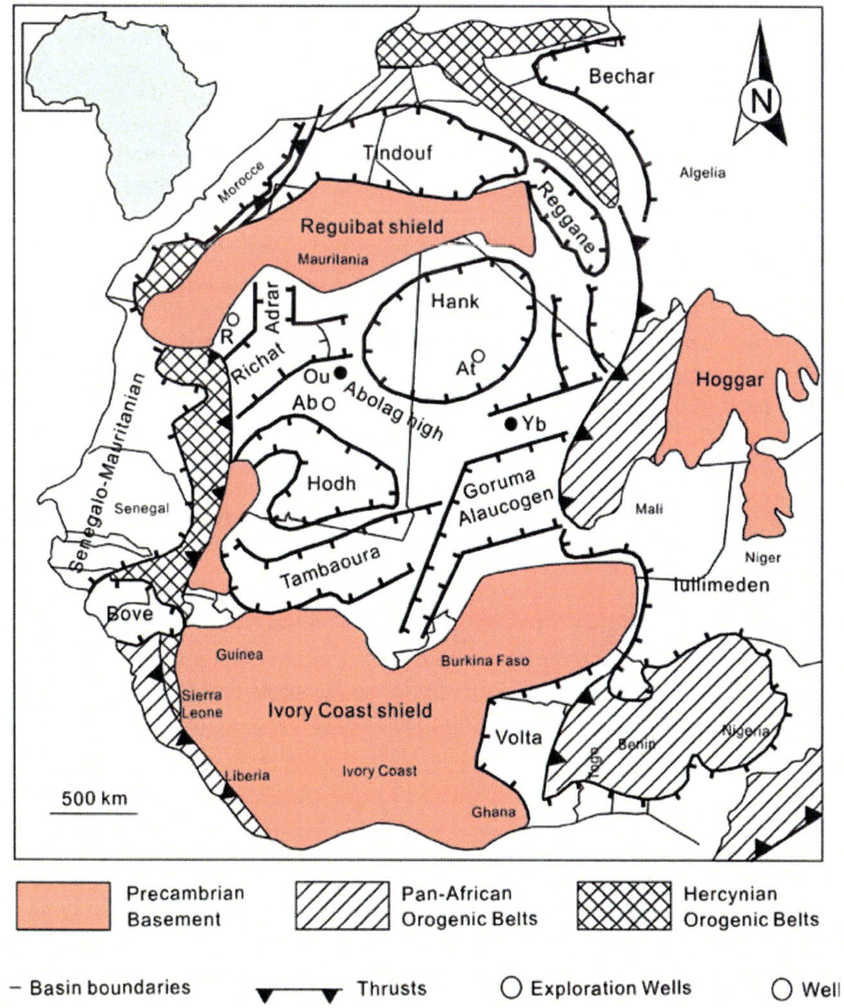

Fig. 11.17 Simplified structural map and oil exploratory well distribution of the Taoudenni Basin (Albert-Villanueva et al. 2016, modified). R. R Well; Ab. Abolag-1 Well; Qu. Quassou-1 Well; At. Atouila-1 Well; Yb. Yarba-1 Well

basin in 2000, which recorded solid bitumen and pyrobitumen within dolomitized stromatolitic carbonates in the Meso-Neoproterozoic Atar Group.

The presence of abundant pyrobitumen shards with a "jigsaw" disposition suggests that pyrobitumen may have been affected by hydraulic fracturing and hydrothermal process which affected the reservoir rocks possibly during Mesozoic times, causing hydraulic fracturing and the local thermal cracking of bitumen to pyrobitumen together with gas formation (Albert-Villanueva et al. 2016). Both "A" and "B" Wells yielded gas shows along the Assabet Formation sandstones and the Atar Group carbonates, but no significant hydrocarbon accumulation was found (Jean-Pierre et al. 2014).

Palynological and geochemical analyses indicate an age for the gas-bearing sequence ranging from Tonian to Early Cryogenian, with low organic abundance and high maturity (Kolonic et al. 2004. Infracambrian hydrocarbon potential of the Taoudenni Basin (Mauritania-Algeria-Mali). Maghreb Petroleum Research Group (MPRG). London-Bremen, 51 (unpublished)). As for the maturity of the Infracambrian source bed, the maximum values of equivalent vitrinite reflectance eqR_o are over 3.6% at the base, and over 2.6% on the top of the middle Infracambrian, while more than 2.0% on the top of the upper Infracambrian, indicating that an over-mature phase for the source bed (Gang 2009). Assuming present-day heatflow and a simple burial model, the Infracambrian should be within the gas-generating window in the deeper parts of the Taoudenni Basin (Gaters 2005).

11.2.6 McArthur Basin and Centralian Superbasin in Australian Craton

The Proterozoic strata are well developed within the boundaries of Australia, among which the Palaeo-Mesoproterozoic is distributed in the northern McArthur Basin and the southern Adelaide Fold Belt, while the Neoproterozoic principally in the Centralian Superbasin (or called "Centralian System"; Fig. 11.18).

11.2.6.1 McArthur Basin

(1) **Regional geological setting:** The McArthur Basin covers an area of ca. 20 × 10^4 km^2 and contains mainly Mesoproterozoic unmetamorphosed flat lying to folded sedimentary rocks that form a cratonic cover sequence near the eastern edge of the north Australian Craton (Fig. 11.18).

The Mesoproterozoic sequence has been subdivided into four lithostratigraphic sequences, i.e., Tawallsh, McArthur, Nathan and Roper Groups in ascending order, each is separated by regional unconformities (Jakson et al. 1988). The Tawallsh Group consist of over-mature arkose or quartz-rich sandstone about 4500 m thick. The McArthur and Nathan Groups have a total combined thickness of ca. 5500 m,

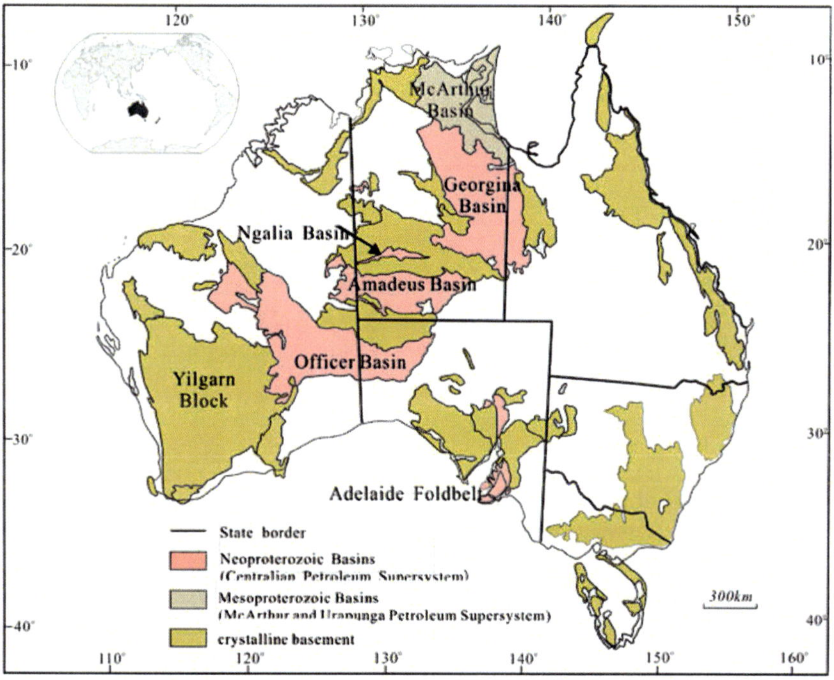

Fig. 11.18 Location of the proterozoic sedimentary basins in Australia (Ghori et al. 2009)

and consist mainly of evaporitic and stromatolitic cherty dolostones interbedded with dolomitic siltstone and shale that were deposited in a variety of marginal marine, lacustrine and fluvial environments, in which only the Barney Creek Formation in the middle of the McArthur Group has been dated. Three dates have been published: a U–Pb date of 1690 ± 30 Ma on Zircon crystals (i.e., Late Palaeoproterozoic age); two Rb–Sr dates of 1589 ± 28 Ma and 1537 ± 52 Ma on illite (i.e., Early Mesoproterozoic age). The uppermost Roper Group consist of 10–300 m thick quartz arenite, siltstone and commonly 100–400 m thick shale deposited in a stable marine setting, its oldest Rb–Sr age measured on glauconite sampled near the base of the group is 1390 ± 20 Ma; while a minimum age of 1280 Ma is indicated by K–Ar dating of dolostone sills that intrude the upper part of the group (ca. middle Mesoproterozoic age; Jackson et al. 1988).

(2) **Petroleum geology**: The most organic-rich source rocks in the Palaeo-Mesoproterozoic McArthur Basin are reported from the lacustrine Barney Creek Formation in the McArthur Group and from the marine Velkerri Formation in the Roper Group (Jackson et al. 1986; Womer 1986; Rawlings 1999a, b). Source rocks with comparable thickness and potential to Phanerozoic source rocks are

present in these sequences with TOC up to 7% containing types I and II kerogens, with thermal maturities ranging from marginal mature to over-mature (Crick et al. 1988).

Weeping oil and gas blowouts occurred in several shallow wells drilled in the McArthur Basin for lead–zinc exploration in the mid-1970s. Two different oil types have been observed: a heavily biodegraded oil containing associated galena, sphalerite and barite, which was probably generated during the phase of Pb–Zn mineralization; and a "golden honey color", very volatile oil generated during the later tectonic events (Wilkins 2007; Ghori et al. 2009). The Roper Group of McArthur Basin was one of the oldest sequence currently explored for HCs in Australia due to the presence of extensive oil and gas shows reported in stratigraphic and petroleum exploration wells drilled during the 1980s (Jackson et al. 1988; Ghori et al. 2009).

11.2.6.2 Centralian Superbasin

(1) **Regional geological setting**: The Centralian Superbasin includes the mid-Late Neoproterozoic fill (840–545 Ma) of the Amadeus, Geogina, Ngalia and Officer Basins, which developed as a single depositional system but separated into different structural units (Fig. 11.18; Walter and Gorter (1993). Centralian Superbasin, Australia. Petroconsultants Australia, Sydney (unpublished); Ghori et al. 2009). The superbasin contains four Neoproterozoic supersequences (Table 11.5):

① The Supersequence 1 (SS1; Early Cryogenian) commenced with a thick sand sheet, overlain by dolostones, limestones, evaporates and fine siliciclastics, which reaches a total thickness of more than 3000 m in the Officer Basin. This sequence is considered to be the most prospective for oil and gas exploration.

② The base of Supersequence 2 (SS2; mid-Late Cryogenian) is defined by the Sturtian glaciation deposits, which are overlain by widespread igneous sills and by shales with interbedded carbonates and sandstones. This supersequence is mainly developed within the Amadeus Basin.

③ The base of Supersequence 3 (SS3; Late Cryogenian) is defined by the Marinoan glaciation deposits which are continent-wide (Priss and Forbes 1981), and may be coeval with extensive tillites found over much of the globe as the products of global glaciation.

④ The basal part of Supersequence 4 (SS4: Edicaran) consists predominantly of sandstone, containing an "Ediacara fauna" and the upper part is Cambrian in age (Ghori et al. 2009).

Taking the Officer Basin as an example, this basin is elongated with a NW–SE trend, and contains over 8000 m of Neoproterozoic strata, overlain by Lower Palaeozoic rocks. Above three bounded sedimentary successions, i.e., the Supersequences 1, 3 and 4, exist unconformity throughout most of the Officer Basin, and these

Table 11.5 Generalized time and seismic stratigraphy, tectonic events, source, reservoir and seal rocks for the Officer Basin, western Australia (Ghori et al. 2009)

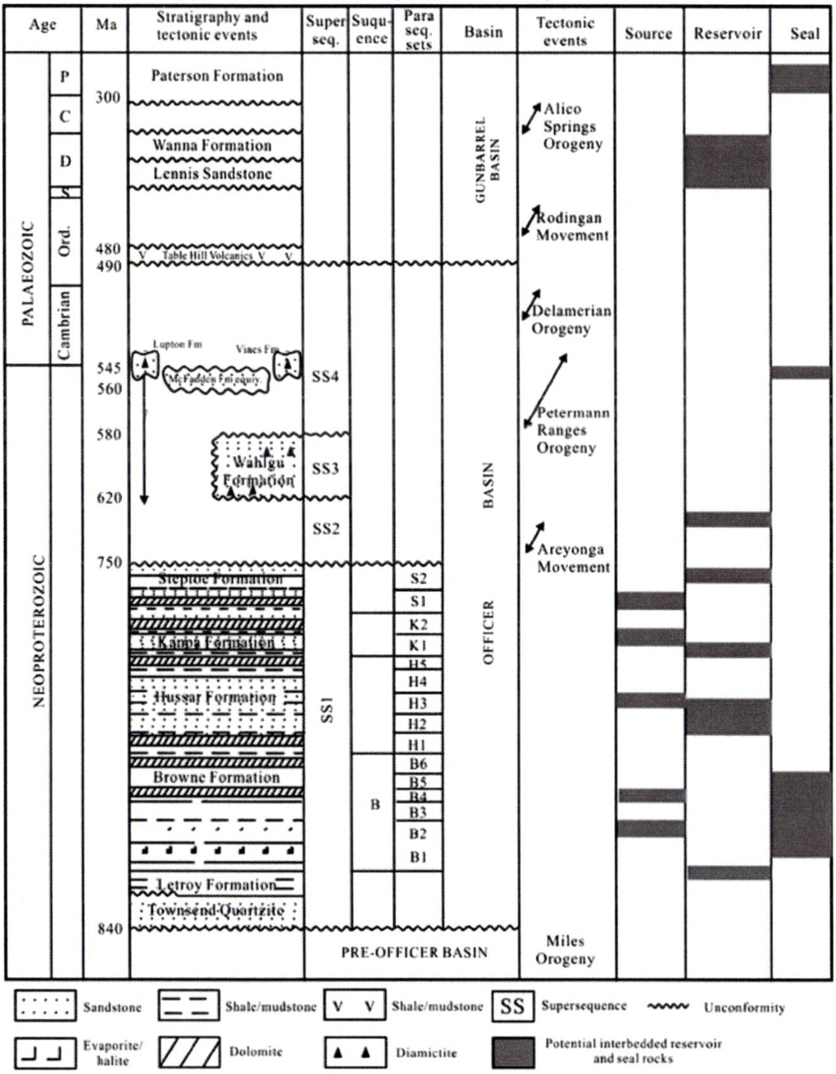

can be correlated with key tectonic episodes (Table 11.5). The Areyonga Movement appears to be responsible for the larger structures in the Officer Basin, and separates SS1 from SS3. Structural and stratigraphic variations within the overlying SS3 and SS4 are respectively attributed to later deformation related to the Petermann Ranges Orogeny and Delamerian Orogeny (Table 11.5; Wade et al. 2005). The overlying units within SS1 consist of conformable and laterally correlative genetic parasequence units bounded by flooding surfaces. These units comprise the Browne (B1–6), Husssr (H1–5), Kanpa (K1–2) and Steptoe Formations (S1–2; Table 11.5). In most seismic lines, SS1 is characterized by continuous parallel reflectors that are traceable across most of the basin (Ghori et al. 2009).

(2) **Petroleum geology and exploration**: During three stages of petroleum exploration in the late 1960s, early 1980s and late 1990s, 16 wells and 19 seismic surveys have been undertaken (Fig. 11.19). The available geochemical data indicate the presence of thin source beds with fair-excellent HC generating potential in the Neoproterozoic successions in some exploratory wells and mineral drill holes. The organic-rich source beds occur in the B2 and B4 of the Browne Formation, the H3 of the Hussar Formation, the K1 of the Kanpa Formation and the S1 of the Steptoe Formation. Pyrolysis gas chromatography and extract analyses indicate that most of the organic matter in the source beds is oil- and gas-generating type II kerogen with TOC contents 0.93–2.05% (max. up to 21.5%), HI 131–498 mg HC/g TOC (min. 68–77 mg HC/g TOC) and T_{max} 413–471 °C. Basin modelling studies suggest that most of the HC generation potential of the Browne Formation was exhausted during the Neoproterozoic. However, the Hussar, Kanpa and Steptoe Formations were not so deeply buried, and the HC generation from these units extends into the Phanerozoic (Ghori et al. 2009).

Chronostratigraphy, tectonic events and location of source, reservoir and seal rocks are summarized in Table 11.5. Minor oil shows and numerous bitumen occurrences have been reported in many of the petroleum exploratory wells drilled in the basin (Table 11.6; Ghori et al. 2009). Despite early optimism for petroleum exploration, the Officer Basin has not been fulfilled yet (Jackson et al. 1988).

Since 1963, over 50 petroleum exploratory wells have been drilled in the Amadeus Basin, resulting in the significant discovery of the Dingo Gasfield in the Neoproterozoic-Lower Cambrian strata in 1981. Amadeus Basin is an EW-trending elongate downwarp about 17×10^4 km^2 in the central Australia (Figs. 11.18 and 11.20a). The basin margins are well defined to the north and south by igneous and metamorphic rocks of the Proterozoic Arunta and Musgrave Blocks. To the east and to the west, the basin margin is obscured by a cover of younger rocks.

The basin's major stratigraphic units are shown in Table 11.7. The stratigraphy reflects a basal Neoproterozoic succession of shelf, lagoonal, continental and shallow-marine sediments, including carbonate and evaporate, overlain by Palaeozoic sediments. The Neoproterozoic fine-grained clastic carbonate and evaporate rocks contain mainly gas-prone matter. A small amount of oil-prone kerogen derived from the Pertatataka and Areyonga Formations and Gillen Member of the Better Springs Formation suggests some oil-source potential for the Neoproterozoic succession.

Dingo Gasfield is located at approximately 75 km south of Alice Springs NT (Fig. 11.20b). Tectonically it appears as an anticline/dome structure with areal closure 68.9 km^2 and vertical closure 160 m (Fig. 11.20c). Stratigraphically, there are two gas-producing units discovered in the Dingo-1 Well, i.e., the Arumbera sandstone

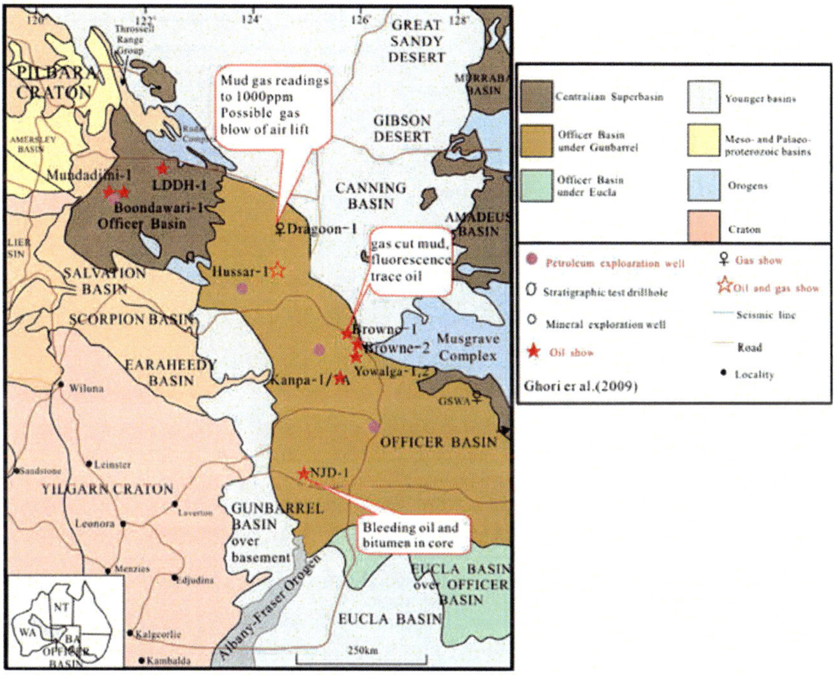

Fig. 11.19 Structural units, seismic survey and petroleum shows in the Officer Basin, western Australia (Ghori et al. 2009)

Table 11.6 Neoproterozoic hydrocarbon shows in Officer Basin, western Australia (Ghori et al. 2009)

Well	Quality of show	Formation
Boondawari-1	40% oil fluorescence, trace oil in core	Spearhole formation
Browne-1	Gas cut mud, cut fluorescence, trace oil in core	Browne formation
Dragoon-1	Mud gas to 10%, methane equivalent, including hydrocarbons up to pentane	Browne formation
Hussar-1	Mud gas readings to 1000 ppm; possible gas blow on air lift; trip gas to 4.6% total gas; 72% oil saturation from log analysis	Kanpa formation Hussar formation
Kanpa-1A	Dull yellow-orange sample fluorescence, light yellow-white cut fluorescence, brown oil stains in sandstone and dolostone cuttings	Kanpa formation
LDDH-1	Bitumen in core	Tarcunyah formation
Mundadjini-1	10% oil fluorescence in core	Spearhole formation
NJD-1	Bleeding oil and bitumen in core	Neoproterozoic
OD-23	Bleeding oil and bitumen in core	
Vines-1	Total gas peaks 25 times background	

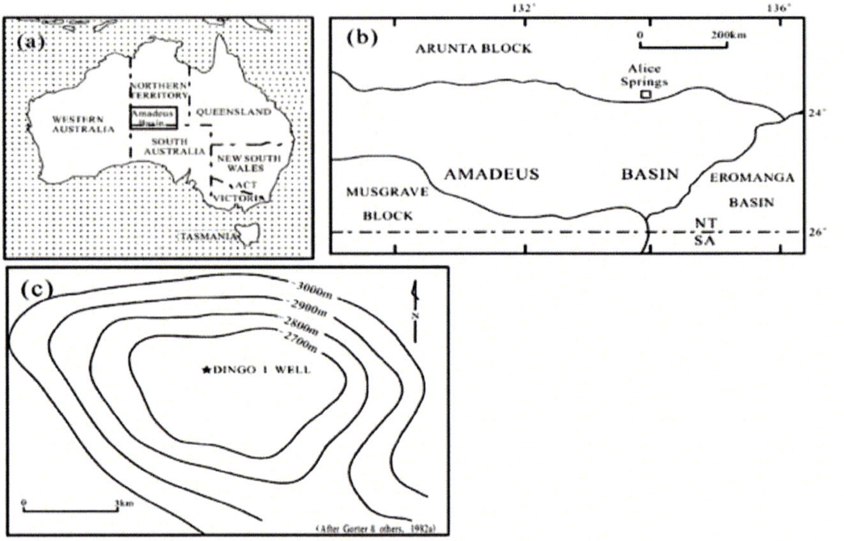

Fig. 11.20 Location of the Amadeus Basin (**a**) and Dingo Gasfield (**b**) as well as the structure contour map of Dingo Gasfield (**c**) (Ozimic et al. 1986)

Table 11.7 Stratigraphy of the Amadeus Basin (Wells et al. 1970)

Erathem/System/Series	Group/Formation/Member	Max. Thickness/m
Lower carboniferous–middle-upper devonian	Pertnjara group	3658
Lower-middle devonian–upper Silurian	Mereenie sandstone	975
Upper ordovician	Carmichael sandstone	91
Lower-middle ordovician	Larapinta group	2500
Cambrian	Pertaoorta group	>2102
Neoproterozoic	Julie formation Pertataka formation	610–1829
	Areyonga formation	396
	Bitter springs formation "Johnny Ck Beos"	914
	Heavitree Quartzite	457
Arunta complex		

at the base of the Lower Cambrian Pertaoorta Group, and the Julie Formation at the top of Neoproterozoic sequence. However, Dingo is an economic and undeveloped gasfield (Ozimic et al. 1986).

11.2.7 *Midcontinent Rift System in the North American Craton (USA)*

11.2.7.1 Regional Geological Setting

Although Precambrian rocks are distributed throughout the United States and commonly are present in the deeper parts of sedimentary basins, their hydrocarbon source rocks and resources are poorly known (Palacs 1997). However, the source rock potential and even some oil-seeps of the Meso-Neoproterozoic strata have been reported respectively in the Midcontinent Rift System in the North American Craton.

As a major structure unit in the North American Craton, the Midcontinent Rift System is delineated by strong gravimetric and magnetic anomalies, and infilled with up to 15 km of rift-related mafic volcanic rocks and up to 10 km of overlying sediments (Behrendt et al. 1988). Rocks of the rift system are exposed in the Lake Superior region of Michigan, northern Wisconsin and Minnesota, and extend in the subsurface through Iowa, Nebraska into northeastern Kansas (Dickas 1986c; Fig. 11.21). The 1500 km-long Midcontinent Rift System is a failed rift, characterized by a series of asymmetric basins filled with clastic rocks in places as thick as 9754 m (Anderson 1989; Palacs 1997). The rocks belong to the Meoproterozoic Keweenawan Supergroup, comprising a lower Oronto and an upper Bayfield Groups. The Oronto Group consists of the Copper Harbor Conglomerate, Nonesuch and Freda Formations in ascending order (Daniels 1982; Elmore et al. 1989). The Copper Harbor Conglomerate Formation is a red sandstone and conglomeratic unit up to 2 km thick, with an U–Pb age of 1087.2 ± 1.6 Ma (i.e., Late Mesoproterozoic age; Daniels 1982; Palacs 1997; Table 11.8), the 40–300 m thick green to gray siltstones and shales of the Nonesuch Formation have a whole-rock age of 1044 ± 45 Ma (Mauk and Hieshina 1992), while the Freda Formation is composed of up to 4 km thick red sandstones (Table 11.8; Mauk and Hieshina 1992).

11.2.7.2 Petroleum Geology

The potential for petroleum resources in the Midcontinent Rift System has long been recognized because of oil seeps and pyrobitumen within the Nonesuch Formation at the White Pine copper mine in the Lake Superior segment (Fig. 11.21; Dickas 1986c), where field and petrographic studies have established six types of petroleum occurrences:

① Liquid oil inclusions in veins filling or adjacent to thrust and tear faults;
② Droplet-shaped inclusions of solid pyrobitumen (altered petroleum) in veins, spatially associated with thrust and tear faults;
③ Subparallel trains of liquid oil inclusions in microfractures genetically related to microthrusts;
④ Solid pyrobitumen cement in sandstone;
⑤ Pore-filling petroleum in the "lower sandstone";

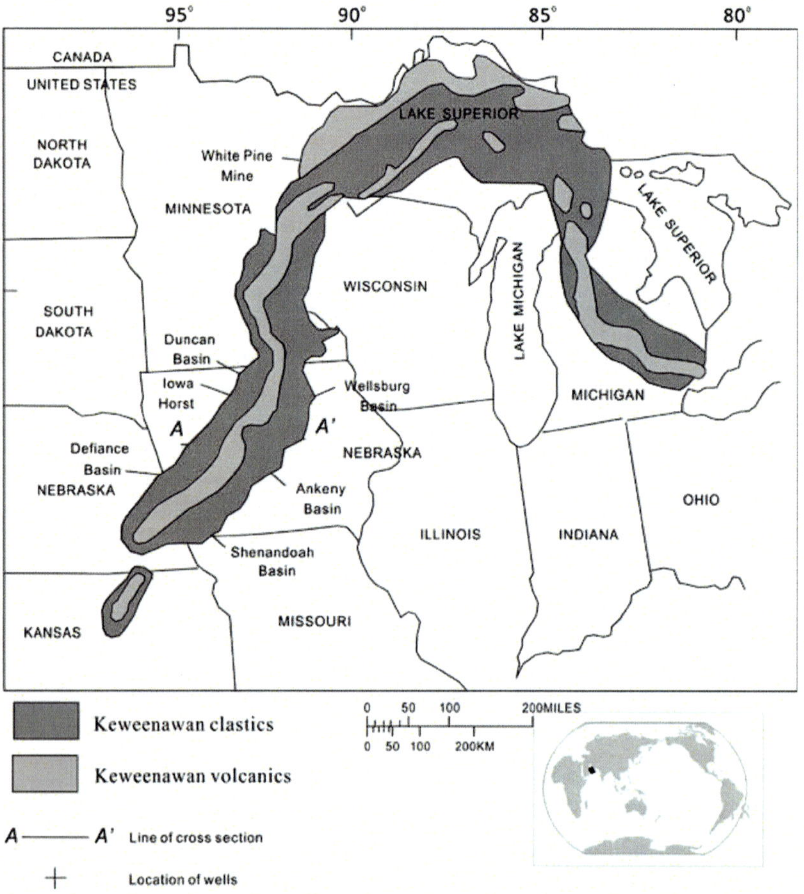

Fig. 11.21 Map showing the general location of major rock types of the midcontinent rift system. The four locations: ① Lake Superior segment outcrop belt; ② Minnesota segment, Lonsdale 65-1 Well, Rice County; ③ Iowa segment, Amoco Eischeid No. 1 Well, Carroll County; ④ Kansas segment, Texaco 1 Poersche Well, Washingron County and the Producers 1–4 Finn Well, Marshall County, which is 21 miles northeast of the 1 Poersche Well (Palacas 1997)

⑥ Active oil seeps occurred at shallow depth and in thin intervals of silty shales of the Nonesuch Formation throughout the mine, irrespective of local tectonic fabric. Oil seeps appear as 1–5 m-wide oozes of liquid oil emanating from faults, joints and roof bolt holes within the mine (Mauk and Hieshima 1992).

All seeps emerge from the back roof of the mine, suggesting that source or reservoir beds lie above the ore horizon. Oil also occurs locally in pores of sandstone, spatially associated with pyrobitumen cements, but no any bedding-plane parallel seep sources were observed (Palacs 1997). Types-①, ② and ③ inclusions, and Type-④ pyrobitumen cements in sandstone, are spatially associated with compressional

Table 11.8 Stratigraphic position of the Oronto Group Nonesuch Formation in Midcontinent Rift and the Chuar Group in Grand Canyon

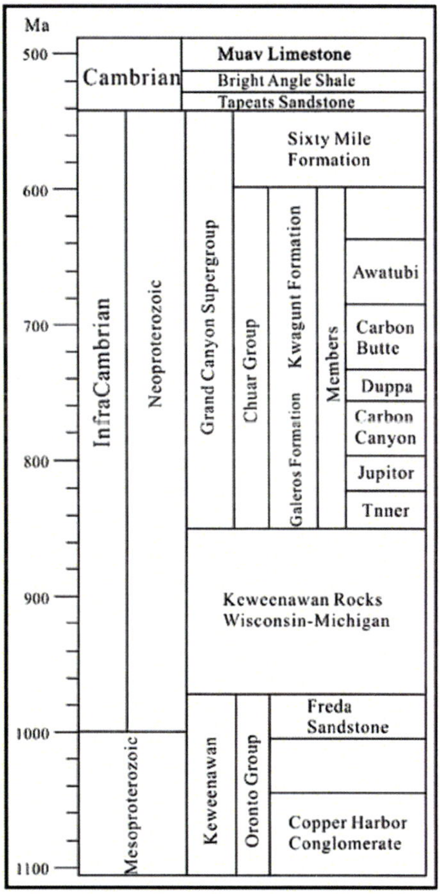

faults, suggesting that oil migrated into the White Pine mine along fault-related conduits synchronously with thrusting and mineralization (Mauk and Hieshima 1992).

In the Nonesuch Formation, the middle unit of the Oronto Group, ranging from ca. 76–213 m in thickness and averaging 183 m, consists of interbedded dark-gray to -green sandstone, siltstone and silty shale (Palacs 1997). The TOC for Nunesuch rocks is generally less than 0.3%. However, the TOC for finely laminated silty or calcareous shales in thin intervals ranges 0.25–2.8% and averages 0.6%, while the dichloromathane extractable bitumen contents averaged ca. 20 ppm with maximum of ca. 300 ppm (Pratt et al. 1991), while Rock–Eval T_{max} values of kerogen isolates were 435–440 °C, indicating maturity equivalent to the initial phase of oil generation for Type II kerogen (Pratt et al. 1991).

Palacs (1997) proposed that if thicker section of these finely laminated hydrocarbon- generating shales are downdip from the outcrop belt, and if those shales were subjected to high level of thermal maturation in the geologic past, the hydrocarbon source potential for the Lake Superior and the adjoining area in northern Wisconsin should be fair to good.

11.3 Distribution of Meso-Neoproterozoic Petroleum Resources in China

The Chinese Meso-Neoproterozoic sedimentary basins are distributed in three regions, i.e., the Yangtze Craton (YC) in South China, North China Craton (NCC) and Tarim Block (TB) in NW China (Fig. 11.22). So far two known Chinese Infracambrian gasfields, i.e., Weiyuan and Anyue Gasfields, are geographically situated within the centro-southern part of current Sichuan Basin, and tectonically referred to the Chuanzhong Uplift (used to be called Leshan-Longnvsi Uplift in geological literature) in western Yangtze Craton (cf. Chap. 14), while many asphaltic veins, with different size, sourced from the black shale of Neoproterozoic Sinian Doushantuo Formation, have been found in Longmenshan Nappe Zone on the northwest margin of western Yangtze Craton (Fig. 11.22; cf. Chap. 15). In addition, numerous Meso-Neoproterozoic indigenous liquid oil-seeps, asphalt, oil sandstone and bituminous sandstone have been discovered in the Jibei Depression within the Yanliao Faultd-Depression Zone (YFDZ), North China Craton (Fig. 11.22; cf. Chap. 12). However, so far there only are some potential Neoproterozoic-Lower Cambrian source beds found in Tarim Block (Fig. 11.22; cf. Chap. 6).

11.3.1 Chuanzhong Uplift and Anyue Gasfield in Western Yangtze Craton

11.3.1.1 Geological setting

In western Yangtze Craton, the Infracambrian sequences consist of Sinian Doushantuo and Dengying Formations as well as Lower Cambrian Maidiping, Qiongzhusi (or Jiulaodong), Canglangpu and Longwangmiao Formations in ascending order (Table 11.9), among which the black shales of Doushantuo and Qiongzhusi Formations are two major potential effective source beds. However, three stratigraphic/exploratory wells (i.e., Nvji, Wei-117 and Ziyang-1 Wells), have been drilled at the structural axial region of Chuanzhong Uplift, where there is an obvious trend of the Doushantuo stratigraphic thinning and/or lithofacies change from black shale of deeper water reducing sedimentary environment to purplish-red shale, dolostone and grey sandstone of shallow water oxidizing environments

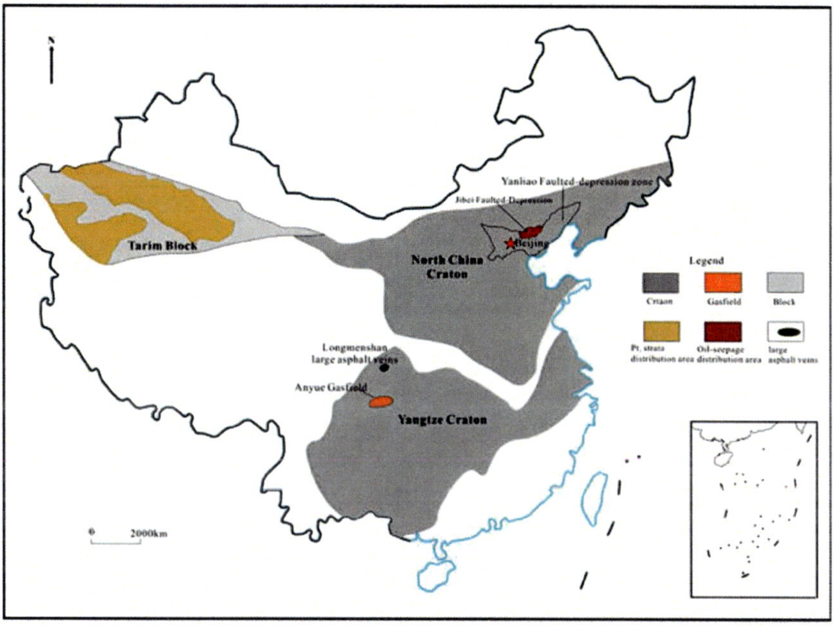

Fig. 11.22 Distribution of the Meso-neoproterozoic sedimentary basins in China

towards Chuanzhong Uplift by comparison with the surrounding depressions, which implies the emergence of Chuanzhong palaeo-uplift since early Sinian, and so Doushantuo shale source bed would be only distributed around the uplift. Moreover, the tops of Deng-2 and Deng-4 Members, Maidiping Formation as well as Longwangmiao Formation all appear as a disconformable contact with their overlying strata respectively (Table 11.9; cf. Chap. 14).

The seismic reflector structural contour map of Dengying top surface shows the current structural form of Chuanzhong Uplift, which occurs as a NEE-trending structural axis and shows the structural features of a large uplift with two secondary salients plus one faulted-sag. The western large domal salient occupies an area of ca. 1200 km^2 with the current structural highest point at Weiyuan Anticline, while the eastern dumbbell-like salient contains two high points respectively at Moxi and Gaoshiti Anticlines with a total trap area up to 3500 m^2, in between there is a structural saddle called Deyang-Ziyang Faulted-Sag (Fig. 11.23) which is constrained by a series of multi-order growth faults during the early sedimentary period of Qiongzhusi Formation. These growth faults can be clearly seen on cross seismic sections (Fig. 11.24), but no display on the seismic reflector contour map with contour interval of 100 m (Fig. 11.23), indicating their limited fault throws. However, these multi-order normal faults could obviously increase the stratigraphic thickness of Qiongzhusi Formation inside the faulted-sag so as to be the Qiongzhusi depocenter with a maximum stratigraphic thickness up to 540 m (at Gaoshi-17 Well; Fig. 11.24) as the main

Table 11.9 Sinian–lower Cambrian stratigraphy in Chuanzhong uplift

Stratigraphy					Lithology
System/Series	Formation/Member		Symbol	Thick/m	
Middle Cambrian	Gaotai		$\in_2 g$	0–200	Grayish-yellow shale and dolomitic sandstone
Lower Cambrian	Longwangmiao		$\in_1 l$	0–300	Gray grain dolostone, muddy dolostone and limestone
	Canglangpu		$\in_1 ch$	0–300	Grayish-yellow, yellowish-green sandy shale, shale and sandstone
	Qiongzhusi		$\in_1 q$	170–560	Gray and black muddy siltstone, carbonaceous shale
	Maidiping		$\in_1 m$	0–200	Dark-gray and black diamict interval
Sinian	Dengying	Deng-4	$Z_2dn_4^2$	110–200	Clotted dolostone with laminated dolostone, dolarenite and muddy dolostone interbeds argillaceous
			$Z_2dn_4^1$	100–170	Dolarenite, muddy dolostone and algal dolostone
		Deng-3	Z_2dn_3	50–100	Dark-colored shale, blueish-gray mudstone with dolostone and tuff interbeds
		Deng-2	Z_2dn_2	440–520	Upper section: micritic dolostone; Lower section: dolostone with grape-rim texture rim
		Deng-1	Z_2dn_1	20–70	Muddy micritic to very fine-grained dolostone, algal-laminated dolostone and partial gypseous salt

(continued)

Table 11.9 (continued)

Stratigraphy				Lithology
System/Series	Formation/Member	Symbol	Thick/m	
	Doushantuo	Z_1ds	10–50	Dark-gray and black shale, lime-shale with dolostone interbeds

hydrocarbon source kitchen within the Deyang-Ziyang Faulted-Sag, Chuanzhong Uplift.

The Chuanzhong Uplift is one of the palaeo-uplifts in western Yangtze Craton (cf. Chap. 13). As a Sinian-Early Palaeozoic palaeo-uplift, its stratigraphic denuded area of Silurian would exceed 6×10^4 km^2. It can also be seen from its pre-Permian palaeo-geological map that Chuanzhong Uplift appears as a palaeo-structural configuration of large south-steep and north-slow semi-anticline (or structural nose), with NEE-trending extension and eastwards pitching, keeping on Sinian–Cambrian as its core region, while Ordovician–Silurian as the pericline structure, showing the existence of a palaeo-uplift of early Sinian to Silurian time (Fig. 11.25).

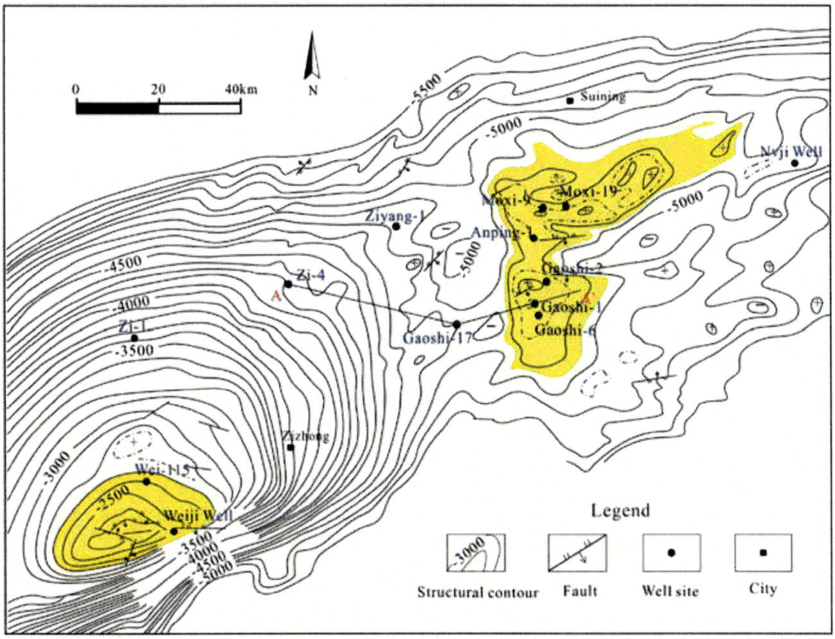

Fig. 11.23 Seismic reflector structural contour map of Sinian top surface on Chuanzhong Uplift in western Yangtze Craton with the contour interval 100 m

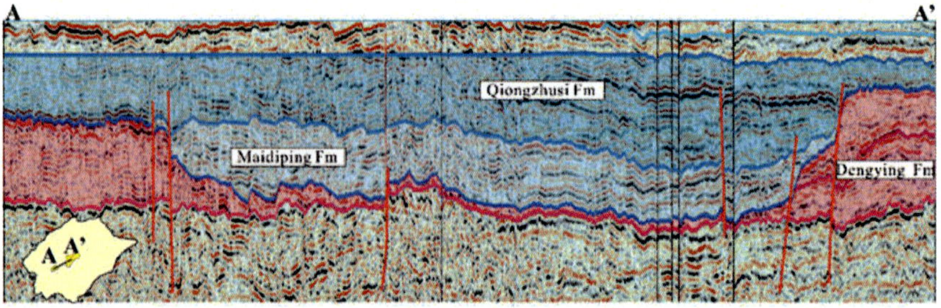

Fig. 11.24 Seismic cross section of Deyang-Ziyang Faulted-Sag. [Zou et al. (2014). Oil and gas characteristics of old carbonate and discovery of large Anyue Gasfield. China-Russia Academic Exchange Report Set on Old Carbonate Oil and Gas Geology. Hebei, Langfang: PetroChina Research Institute of Petroleum Exploration and Development (in Chinese and Russian)]

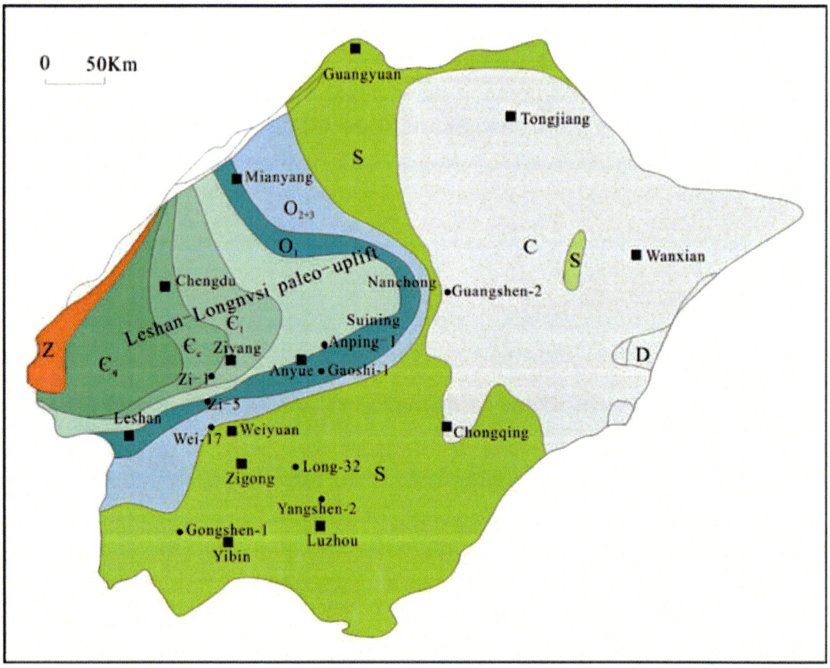

Fig. 11.25 Pre-Permian palaeo-geological map of Chuanzhong Uplift (Wei et al. 2013). The map range is only limited to the current Sichuan Basin

So far two known Chinese Infracambrian gasfields, Weiyuan and Anyue Gasfields, are situated on the Chuanzhong Uplift in the western Yangtze Craton. Weiyuan Gasfield is located at the western salient of the uplift (cf. Chap. 13), while the Anyue Gasfield at the eastern salient near the uplift axis, which is composed of two adjacent

Moxi and Gaoshiti Anticlines with Moxi at the north and Gaoshiti at the south, the latter is slightly higher than the former (Fig. 11.23; cf. Chap. 14).

11.3.1.2 Anyue Gasfield

(1) **Geological setting and petroleum geology:** The commercial gas flows have been produced from carbonate reservoirs at both the tops of Deng-2 and Deng-4 Members as well as the Longwangmiao Formation in Moxi and Gaoshiti Anticlines respectively (Table 11.9). The Deng-2 and -4 Members had developed dolostone building-up of algal mound and shoal facies, and then were reformed by late karstification, resulting in well-developed pore space of dolostone reservoir. As the first discovery well, a high production of gas flow of 102×10^4 m^3/d was obtained from the Deng-2 gas reservoir at Gaoshi-1 Well in 2011. While the Longwangmiao Formation contains the grain dolostone reservoir of shoal facies with multi-layers overlapped sheet-like occurrence, making these reservoir beds to be well-connected and highly productive, its gas-containing range would beyond the current structural trap area, resulting in the structural-lithological type of gas reservoirs at Moxi Anticline① (Wen et al. (2014). Studies on oil and Gas Entrapment and Play Evaluation in the Longnvsi Palaeo-uplift, Sichuan Basin (Internal Report). Chengdu: Institute of Petroleum Exploration and Development, PetroChina Southwest Oil and Gasfield Company (in Chinese)). A highly productive gas flows of 190.68 $\times 10^4$ m^3/d was gained from both upper and lower intervals of Longwangmiao Formation at Moxi-8 Well in 2012. Consequently, the Anyue Gasfield was further confirmed and named.

As major producing intervals, the Longwangmiao gas reservoir beds in Moxi Anticline are tectonically a south-steep and north-gentle brachyanticline with a reservoir high point at Moxi-9 Well (asl. −4226.3 m) and the reservoir bottom limit of −4458.3 m at Moxi-16 Well with a gas reservoir altitude 232 m (Fig. 11.26). While the gas-bearing altitude in individual wells averages 53.5 m. It has a proven gas reserve of 4403.8×10^8 m^3, and the natural gas is attributed to over-mature dry gas, with dry index 0.99–1.00, resulted from liquid oil cracking① (Wen et al. (2014). Studies on oil and Gas Entrapment and Play Evaluation in the Longnvsi Palaeo-uplift, Sichuan Basin (Internal Report). Chengdu: Institute of Petroleum Exploration and Development, PetroChina Southwest Oil and Gasfield Company (in Chinese)).

(2) **Reservoir pyrobitumen and fossil-oil-reservoir**: Reservoir pyrobitumen is wide-spread in the Chuanzhong Uplift (Liu et al. 2009; Du et al. 2016; Yang 2018). Du et al. (2016) reported that the pyrobitumen-containing cores of gas reservoir beds, with equivalent vitrinite reflectance eqR_o values 2.5–3.5%, had been recorded in 22 gas wells in Anyue Gasfield including Moxi and Gaoshiti Anticlines, among which 8 wells for the Dengying gas reservoirs (Fig. 11.27a) and 14 wells for the Longwangmiao gas reservoir (Fig. 11.27b). While the total numbers of pyrobitumen-sampling wells are increasing up to 73 wells in Moxi

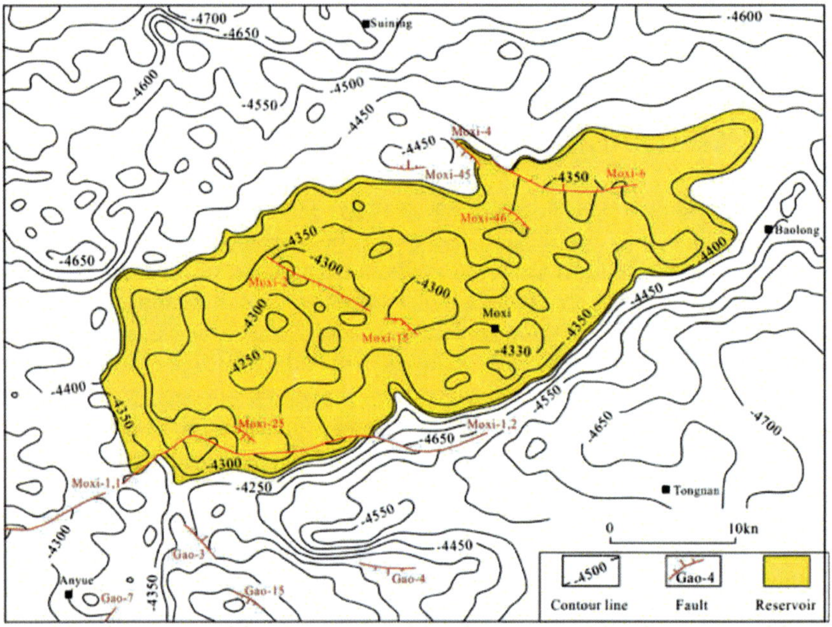

Fig. 11.26 Seismic reflector contour map of lower Cambrian Longwangmiao top surface indicating the distributional range of gas reservoir (shown in yellow color) in Moxi Anticline. MX. Moxi. [Wen et al. (2014). Studies on oil and Gas Entrapment and Play Evaluation in the Longnvsi Palaeo-uplift, Sichuan Basin (Internal Report). Chengdu: Institute of Petroleum Exploration and Development, PetroChina Southwest Oil and Gasfield Company (in Chinese)]

and Gaoshiti Anticlines (Fig. 11.27c; Yang 2018). The pyrobitumen is associated with over-mature cracking gas as a pair of paragenetic products, resulted from the disproportionation of liquid oil during the thermo-evolutional process in the palaeo-oil-reservoir. In other words, the original liquid oil was cracked into dry gas and meanwhile polymerized into solid reservoir pyrobitumen respectively. As a whole, the in-situ reservoir pyrobitumen could be attributed to the occurrence of fossil-oil-reservoir.

(2) **Palaeo-oil filling pathway and source kitchen**: Based on the analysis of chloroform extract in reservoir pyrobitumen, DMDBTs (i.e., dimethyldibenzothiophenes) can be used as molecular tracing indicators of oil-filling in oil reservoir (Wang et al. 2004). The palaeo-oil-filling pathway has been traced for the fossil-oil-reservoirs of both Deng-4 Member and Longwangmiao Formation in Anyue Gasfield (Fig. 11.28a, b). As the effective molecular tracing parameters, the isopleth map of 4,6-/(1,4 + 1,6-)DMDBT ratio shows the main oil-filling pathways from west to east in the fossil-oil-reservoirs of Deng-4 Member and Longwangmiao Formation, taking Moxi-12 Well as well as Gaoshi-7 and Gaoshi-10 Wells as oil-filling points respectively at the centro-western

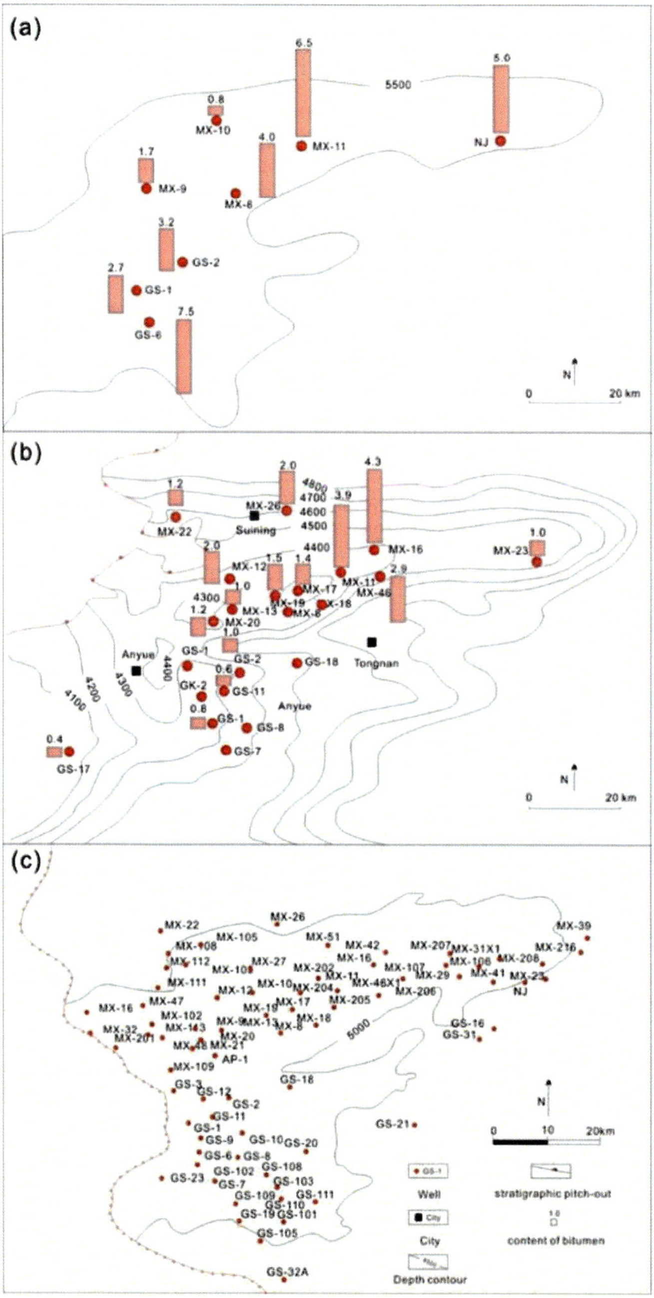

Fig. 11.27 Distribution of reservoir pyrobitumen (**a**, **b**) and pyrobitumen-sampling well sites (**c**) within Dengying and Longwangmiao reservoir beds in Moxi and Gaoshiti Anticlines. **a** Dengying gas reservoir, 8 wells; **b** Longwangmiao gas reservoir, 14 wells (Du et al. 2016, modified); **c** Pyrobitumen-sampling will sites in Longwangmiao plus Dengying Formations, totally 73 wells. MX. Moxi; GS. Gaoshiti; NJ. Nvji Well

segment of Moxi and Gaoshiti Anticlines, and tracing the major Qiongzhusi source kitchen back to the Deyang-Ziyang Faulted-Sag (Fig. 11.28a, b). In addition, two secondary oil-filling pathways are also traced from north to south within both the fossil-oil-reservoirs at the eastern segment of Moxi and Gaoshiti Anticlines, respectively with Moxi-11 and Moxi-39 Wells as well as Moxi-26 and Moxi-39 Wells as the oil-filling points for the fossil-oil-reservoirs of Deng-4 Member (Fig. 11.28a) and of Longwangmiao Formation respectively (Fig. 11.28b), The secondary oil-filling pathways would make a significant contribution to the oil entrapment for palaeo-oil-reservoir, especially in the east of Moxi Anticline. Moreover, the secondary filling pathways have also revealed a potential oil source kitchen and predicted a favorable exploration direction on the north of Moxi Anticline (Fig. 11.28).

Based on geological interpretation of seismic data, Guo① (Guo (2016). Carbonate reservoir bed, formational rules and explorational evaluation of large gasfield in south China. National Science and Technology Major Project (internal report). Sinopec Exploration Branch (in Chinese)) had reported some potential depression(s) on the north and/or northwest of Chuanzhong Uplift, where the Sinian Doushantuo and/or Lower Cambrian Qiongzhusi shale source beds could be available (Fig. 11.29a). Moreover, one exploratory well, Mashen-1 Well, at the northernmost margin of a potential northern depression revealed the Qiongzhusi black shale at well interval of 7090–8058 m contains TOC 1.89–8.95% (avg. 4.99%), and the black shale with TOC > 2.0% is up to 128 m-thick (Fig. 11.29b), thus a potential source kitchen could be preliminarily revealed, but it still needs further study due to the lack of enough drilling data.

However, a high producing gas flow of 51.62×10^4 m^3/d was obtained from the sandstone gas reservoir of Lower Cambrian Canglangpu Formation (cf. Table 11.9) in another new exploratory well, Jiaotan-1 Well, at 3 well intervals between 6972 m and 7026 m in 2020, which just lies to ca. 126 km north of Gaoshi-1 Well at Moxi-Gaoshiti Anticlines and on the way from the potential source kitchen to the Anyue Gasfield for hydrocarbon migrating and filling.

(4) **Palaeo-oil entrapment age:** Using fission track dating, combined with equivalent vitrinite reflectance eqR_o data, a terrestrial heatflow curve has been measured by Qiu et al.① (Qiu et al. (2015). Temperature and pressure fields for the formation of typical large gas fields. Technical summary report on major national science and technology projects. Beijing: China University of Petroleum-Beijing (in Chinese)) at Nvji Well in the eastern ends of the Chuanzhong Uplift (Fig. 11.30) where the regional thermo-evolutional history could be reconstructed (Liu et al. 2018), which can be divided into three thermo-evolutional periods, including the slow and steady rising, peak and decay periods (Fig. 11.30; Table 11.10).

Based on the previous long-term sstudies, a correlativity of terrestrial heatflow and hydrocarbon-generating threshold depth for the Mesoproterozoic strata in the Jibei Depression, North China Craton has been established, i.e., the heatflow values of

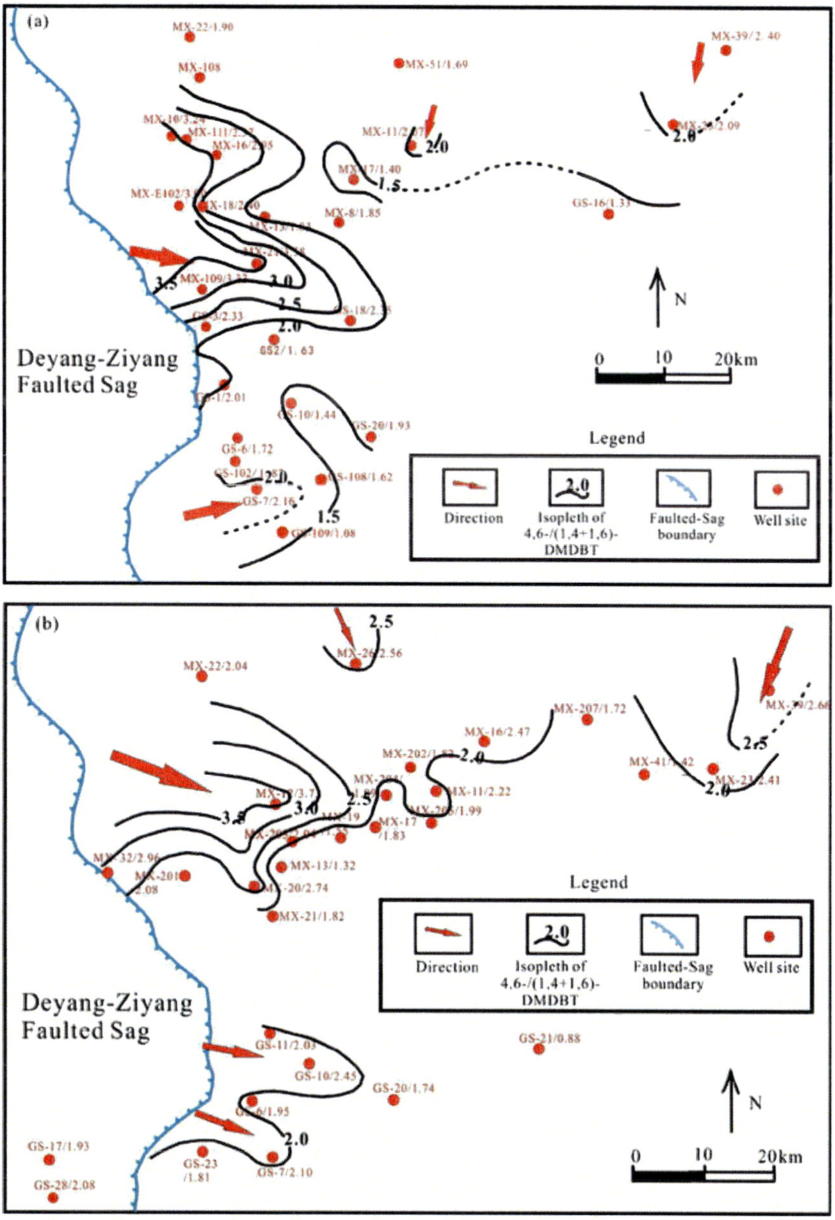

Fig. 11.28 Isograms of 4,6-/(1,4 + 1,6)DMDBT molecular parameter tracing the oil-filling pathways in fossil-oil-reservoirs. a. The fossil-oil-reservoir of Sinian Deng-4 Member, totally 28 wells; b. the fossil-oil-reservoir of Longwangmiao Formation, totally 29 wells. MX. Moxi; GS. Gaoshiti; DMDBT. Dimethyldibenzothiophene

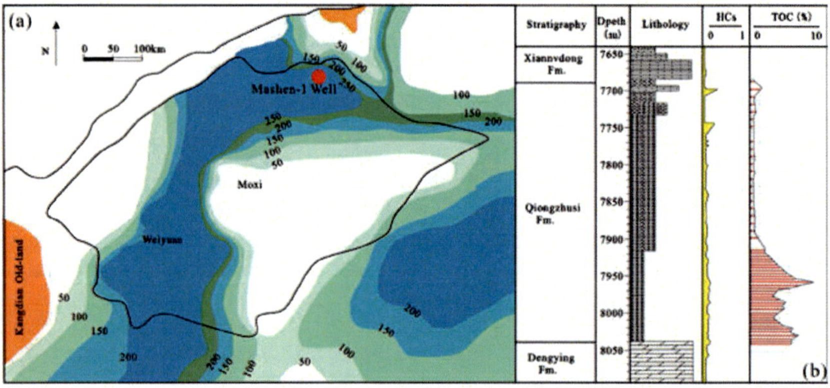

Fig. 11.29 The distribution of potential source kitchen on the north of the Chuanzhong Uplift (**a**) and a geochemical profile of Mashen-1 Well (**b**) HCs. total hydrocarbons. [Guo (2016). Carbonate reservoir bed, formational rules and explorational evaluation of large gasfield in south China. National Science and Technology Major Project (internal report). Sinopec Exploration Branch (in Chinese)]

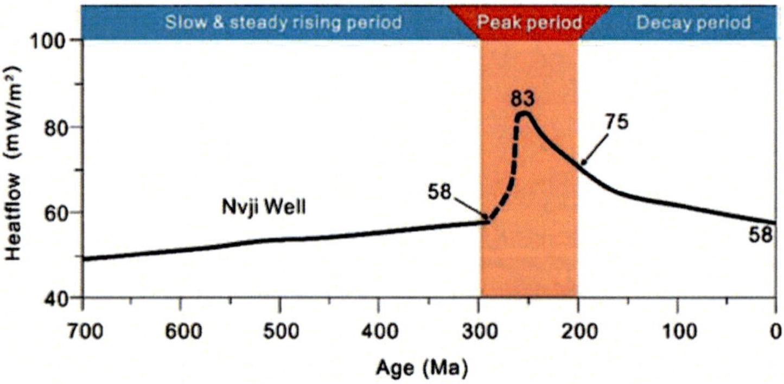

Fig. 11.30 The measured terrestrial heatflow curves at Nvji Well by means of fission track dating (Qiu et al. (2015). Temperature and pressure fields for the formation of typical large gas fields. Technical summary report on major national science and technology projects. Beijing: China University of Petroleum-Beijing (in Chinese), modified)

30–54 mW/m^2 possess corresponding threshold depth ≥3500 m (Table 11.11; Wang et al. 2016). While a distributional range of 60.7–79.5 mW/m^2 for the measured Eogene terrestrial heatflow values has corresponding threshold depth range from 2300 to 3000 m respectively in many oilfields such as Dagang, Shengli, Huabei and Jidong Oilfields in the north of NCC (Table 11.11; Zhu and Chen 2002; Hao et al. 2006; Gang et al. 2012; Cai 2012).

In the basis of geological analogy on the correlativity of terrestrial heatflow and threshold depth, according to the terrestrial heatflow curve of Nvji Well, the threshold depths of different geological times can be determined in the Qiongzhusi source

Table 11.10 The thermo-evolutional periods of terrestrial heatflow at Nvji Well (data cited from Qiu et al. 2015[a]

Thermo-evolutional periods	Slow and steady rising period	Peak period	Decay period
Terrestrial heatflow/mW/m^2	40–58	58–83–75	75–58
Geological age/Ma	700–300	300–200	200–0
Geological time	Nanhuan–Early Permian	Early Permian–Triassic	Jurassic–present

[a] Qiu et al. (2015). Temperature and pressure fields for the formation of typical large gas fields. Technical summary report on major national science and technology projects. Beijing: China University of Petroleum-Beijing (in Chinese)

Table 11.11 The geological analogy of terrestrial heatflow value and threshold depth

Region		Terrestrial heatflow value/(mW/m^2)	Threshold depth/m	References
Jibei depression (Mesoproterozoic)		30–54	$\geq$ 3500	Wang et al. (2016)
Chuanzhong Uplift	Nanhuan–early permian	40–58	ca. 3500	Geological analogy
	Early permian–triassic	58–83–75	in Btwn. 2500 and 2800	
Adjacent oilfields in the great North China plains (Eogene)		60.7–79.5	2300–3000	Zhu and Chen (2002) etc

kitchen, Chuanzhong Uplift (Table 11.11). The terrestrial heatflow values of Nvji Well are 40–58 mW/m^2 for the slow and steady rising period during Nanhuan-Early Permian (700–300 Ma; Fig. 11.30; Table 11.10), which are equivalent to these of the Mesoproterozoic in Jibei Depression, and accordingly, the threshold depth of Chuanzhong Uplift could be assign to ca. 3500 m (Table 11.11). While the terrestrial heatflow values of 58–83–75 mW/m^2 for the heatflow peak period during Early Permian–Triassic in Chuanzhong Uplift (Fig. 11.30; Table 11.11) are comparable with those of 60.7–79.2 mW/m^2 at the Eogene oilfields in NCC, which contain the threshold depth between 2300 m and 3000 m, thus the threshold depth could be approximately assigned in between 2500 m and 2800 m in Nvji Well for the peak period of Chuanzhong Uplift (Table 11.11).

Using BasinMod I software, the timing of palaeo-oil generation, migration and entrapment events can be estimated by means of single-well numerical modelling on the stratigraphic burial history and associated hydrocarbon threshold depth of source bed. Thus, the Gaoshi-17 Well within the Qiongzhusi source kitchen has been preferentially selected for reconstruction of the burial history of Qiongzhusi source bed in Deyang-Ziyang Faulted-Sag, and for presumption of oil entrapment age of fossil-oil-reservoirs in Moxi-Gaoshiti Anticlines. The modelling result shows

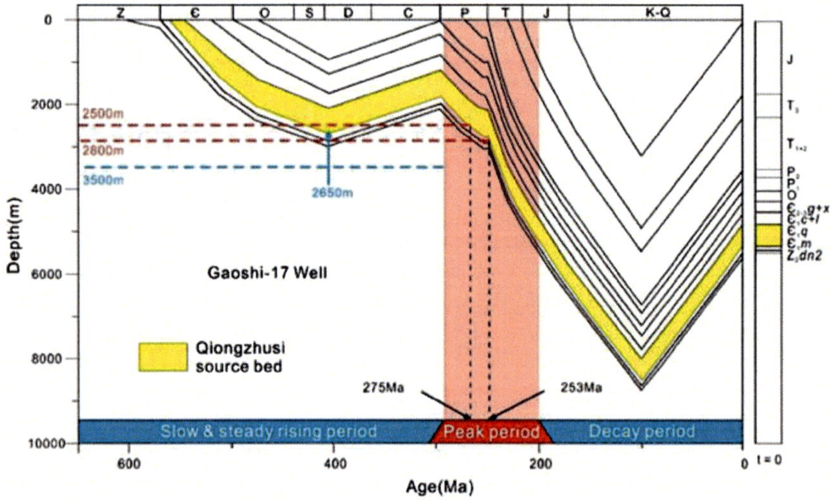

Fig. 11.31 Reconstructed burial history of Gaoshi-17 Well in Deyang-Ziyang Faulted-Sag and presumed oil entrapment age of Qiongzhusi fossil-oil-reservoir in Moxi-Gaoshiti Anticlines (Yang 2018; Yang et al. 2020 modified)

that the Qiongzhusi source bed had reached a maximum burial depth only 2650 m at the end of Silurian (Fig. 11.31; Yang 2018; Yang et al. 2020), which is less than the threshold depth ca. 3500 m, during the slow and steady rising period of terrestrial heatflow (Table 11.11), when the Qiongzhusi source bed should still be immature one, no oil/gas could be generated and sourced to the fossil-oil-reservoir at that period (i.e., Nanhuan to Early Permian; Table 11.11).

Due to the thermal effect of Emeishan mantle plume or large Igneous Province (Zhu et al. 2016) during the heatflow peak period, however, the heatflow values of Qiongzhusi source bed quickly raised to 58–83–75 mW/m^2 in Early Permian to Triassic with corresponding threshold depth in between 2500 m and 2800 m (Table 11.11), consequently the initial oil entrapment age could be assigned in between 275 and 253 Ma (in btwn. late Early and Late Permian) based on the threshold depth and reconstructed burial history of Qiongzhusi source bed in Gaoshi-17 Well (Fig. 11.31).

(5) **Emplacement age of cracking gas reservoir:** Based on microscopic observation, the Dengying and Longwangmiao gas reservoirs contain numerous fluid inclusions, including gas (methane) and its paragenetic gas–water two-phase inclusions, in authigenic minerals of quartz and late saddle dolomite (Fig. 11.32b, c). Moreover, pure methane inclusion can be distinguished from other gas inclusions by Raman spectrum (Fig. 11.32a).

A prerequisite for fluid entrapment/emplacement timing is that the fluid was trapped as a single-phase, accordingly, the dominant homogenization temperature of gas–water two-phase inclusion could be considered as a minimum trapping temperature (Emerry and Robinson 1993), and viewed approximately as the lower

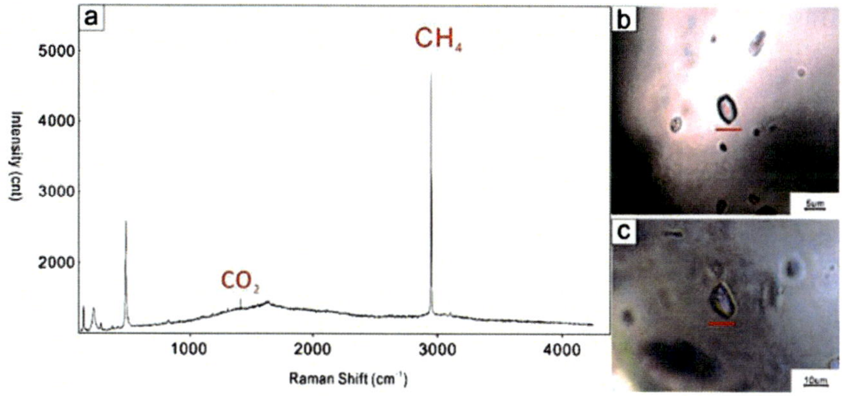

Fig. 11.32 Microphotograph and Raman spectrum of fluid inclusions in the dolostone of Deng-4 gas reservoir at Gaoshi-6 Well (Yang et al. 2018). **a** Raman spectroscopes of methane inclusions, well depth 5049 m; **b** Methane inclusion under plane-polarized transmitted light microscope, well depth 5049 m; **c** Gas–water two-phase inclusions under plane-polarized transmitted light microscope, well depth 5048.97 m

limit of entrapment/emplacement temperature for the oil/gas. Taking the dolostone gas reservoir of Deng-4 Member at Gaoshi-6 well, Gaoshiti Anticline as a case study to determine the emplacement age of cracking gas reservoir by means of microthermometry.

The microthermometric result shows that the homogenization temperature histogram of paragenetic gas–water two-phase inclusions appears as a unimodal distribution, with a dominant frequency at 170–180 °C (avg. 180 °C), for the Deng-4 gas reservoir at Gaoshi-6 Well (Fig. 11.33). As for the phase analysis of cracking gas reservoir, however, the dominant homogenization temperature of two-phase inclusions should be just a critical temperature of phase transformation from the two-phase towards the single phase, and also the minimum trapping temperature of fluid inclusions, which need to be carefully corrected into a real in-situ fluid trapping temperature by pressure correction of inclusion (Emeryy and Robinson 1993).

On a heating–cooling microscope stage, Hanor (1980), Emerry and Robinson (1993), Aplin et al. (2000), Liu et al. (2003) and Ni et al. have respectively approached the pressure correction methodology of homogenization temperature to acquire the real in-situ trapping temperature of fluid inclusions, especially for gas reservoir. In this way, Yang (2018) obtained the corrected homogenization temperature of fluid inclusion in the Deng-4 cracking gas-reservoir at Gaoshi-6 Well (well depth 5049 m; Yang et al. 2018). The research procedure is as follows:

First, the measured homogenization temperature and salinity (or freezing point) of paragenetic methane and two-phase inclusions are respectively measured.

Second, two sets of isochores respectively for methane and two-phase inclusions are established on the pressure–temperature plot (Fig. 11.34), and the intersection for both sets of isochores would indicate the really trapping temperature and pressure

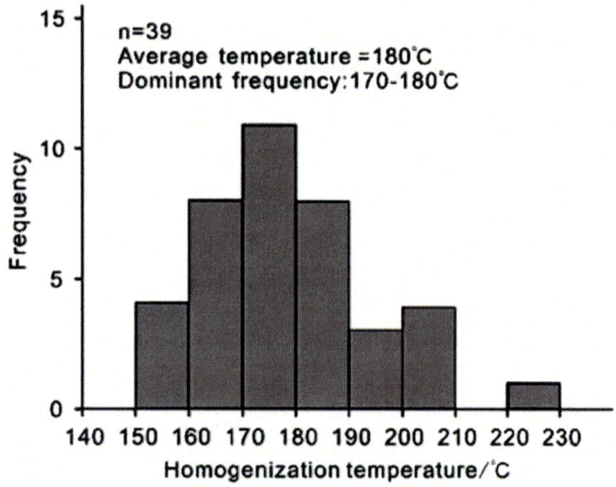

Fig. 11.33 Homogenization temperature histogram of gas–water two-phase inclusion in the Deng-4 dolostone gas reservoir, Gaoshi-6 Well (Yang 2018; Yang et al. 2018)

of paragenetic fluid inclusions. In this case, however, there are two intersections for both sets of isochores on the plot, i.e., the lower one at 185–227 °C/484–700 Bar and the higher one at 249–319 °C/1619–2300 Bar (Fig. 11.34).

Third, by comparison to the measured homogenization temperatures with dominant frequency at 170–180 °C (avg. at 180 °C; Fig. 11.33), the lower intersection is comparable and reasonable to indicates the corrected fluid trapping temperature/pressure for the Deng-4 gas-reservoir in Anyue Gasfield (Yang 2018; Yang et al. 2018). While the higher intersection is far beyond the measured range of homogenization temperature in fluid inclusions (Fig. 11.33), which should be noting to the

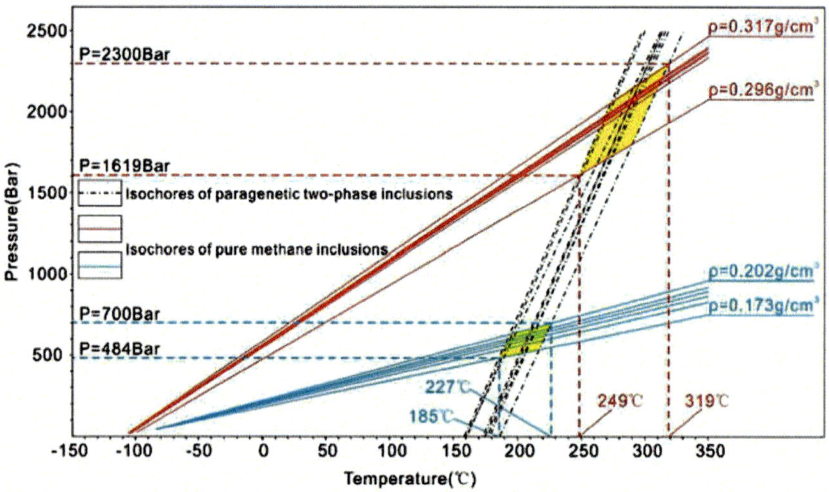

Fig. 11.34 Temperature–pressure plot of Anyue Gasfield data. The intersections of two sets isochores for pure methane and gas–water two-phase inclusions are indicated

origin of cracking gas as well as to the fossil-oil-reservoir. Since the host minerals (quartz and late saddle dolomite) of fluid inclusions belong to hydrothermal authigenic minerals and the presence of Emeishan mantle plume/large Igneous Province, nevertheless, it would be most probably that the higher temperature/pressure methane inclusions are associated with a mesothermal origin.

In order to transfer the corrected gas trapping temperature into the gas trapping age or cracking gas emplacement age for the Deng-4 gas reservoir, the Gaoshi-6 well in Gaoshiti Anticline is selected to reconstruct its stratigraphic burial-thermal histories of Dengying and Longwangmiao Formations by means of single-well numerical modelling (Fig. 11.35). The burial-thermal histories are similar either at Gaoshi-6 or at Gaoshi-17 Wells (Fig. 11.31), both wells are 20 km apart on the axis of Chuanzhong Uplift. The Deng-4 gas reservoir has experienced two subsidence-uplift tectonic cycles, the palaeo-geotherm of Deng-4 palaeo-oil reservoir had been lower than 140°C, which was not going to cause palaeo-oil cracking and thermos-alteration during the first tectonic cycle, while the burial depth was beyond 8500 m with the palaeo-geotherm higher than 220 °C and overstepping the temperature scope of oil cracking during the subsidence of the second tectonic cycle (Fig. 11.35). Consequently, the corrected trapping temperature 185–227°C for the fluid inclusions in Deng-4 gas reservoir can be intuitively transferred into the trapping age of gas inclusions, i.e., 175–144 Ma (late Middle Jurassic to early Early Cretaceous), which can be considered to be the emplacement age of cracking gas reservoir.

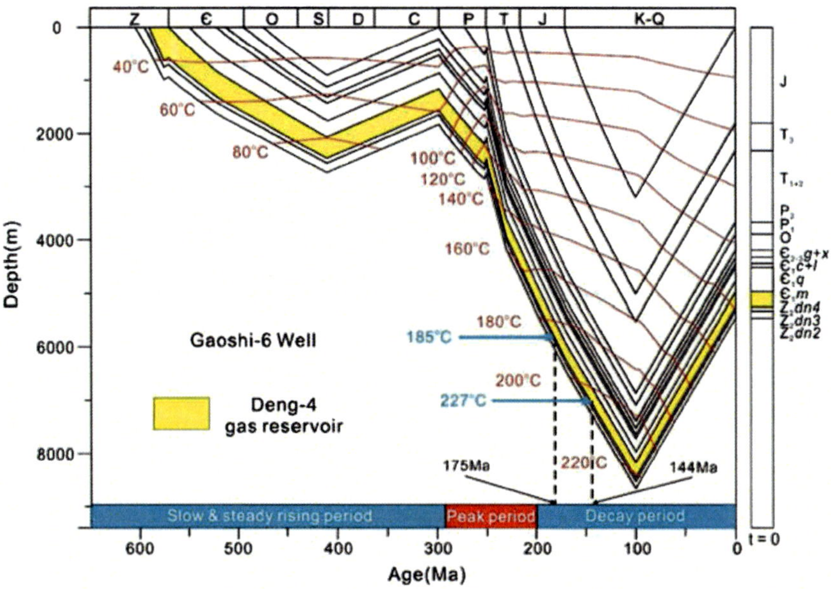

Fig. 11.35 Reconstructed stratigraphically burial and thermal histories of Gaoshi-6 Well based on single-well numerical modelling using Basin Mod I software (Yang et al. 2018)

11.3.2 Yanliao Faulted-Depression Zone (YFDZ) in North China Craton

11.3.2.1 Regional Geological Setting

The YFDZ is situated on the north margin of the North China Craton (NCC). It shows EW-trending and stretches across Hebei, Beijing, Tianjin and Liaoning Provinces/Cities, occupying totally an area of ca. 10.6×10^4 km^2. Tectonically, its central parts are mainly Shanhaiguan and Mihuai Uplifts, composed of the crystalline basement of Archean metamorphic rocks and multiphase granites, while two Meso-Neoproterozoic depressional belts are respectively developed on its north and south sides, which could be divided into Liaoxi-Jibei-Xuanlong Depressions on the north side as well as Jidong-Jingxi Depressions on the south side from east to west (Fig. 11.36), where Mesoproterozoic Changchengian (Pt_2^1), Jixianian (Pt_2^2), Xiamaling Formation (Pt_2^3x) and Neoproterozoic Qingbaikouan (Pt_3^1) sequences are deposited, and overlaid by Palaeozoic and Mesozoic strata (Table 11.12).

On the whole, the total Meso-Neoproterozoic sequences are relatively thick at the eastern segment and thin at the western one of the YFDZ, i.e., the maximum thicknesses up to 9260 m in Jidong Depression, 8043 m in Jibei Depression and 7567 m in Liaoxi Depression, as well as the minimum only 4877 m in Jingxi Depression and 4095 m in Xuanlong Depression (Fig. 11.36 and Table 11.12), among which the main sediments are attributed to Mesoproterozoic stromatolitic carbonates and partial clastic rocks. Obviously, the depocenter of Mesoproterozoic sequences would

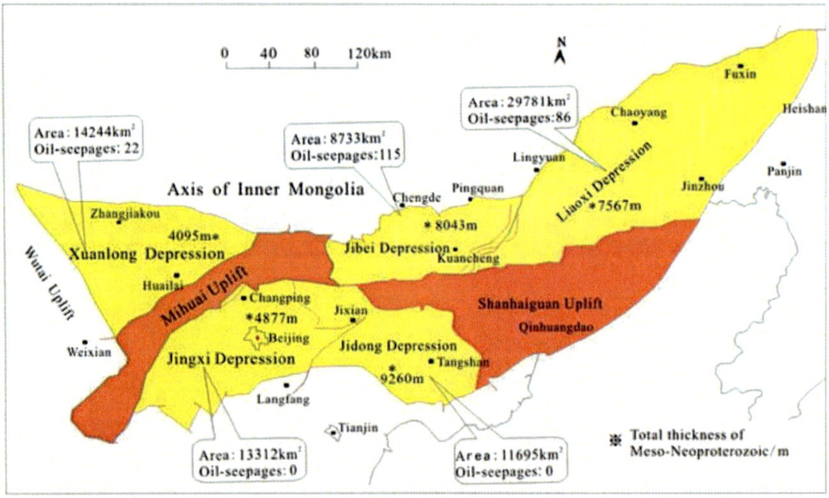

Fig. 11.36 Meso-Neoproterozoic tectonic division and stratigraphic thickness-oil seep distribution in YFDZ (Wang 1980, modified). Symbol * indicates the Meso-Neoproterozoic stratigraphic thickness in each depression

Table 11.12 Meso-neoproterozoic stratigraphic sequences and stratum thicknesses in YFDZ

Depression		Southern belt		Northern belt		
		Jiangxi	Xuanlong	Jibei	Liaoxi	Jidong
Stratigraphiy	Qingbaikouan (760–1000 Ma)	193.2	71.9	111.8	168.2	230[a]
	Xiamaling formation (1320–1400 Ma)	249.0	540.6[a]	369.5	303.4	168
	Jianxianian (1400–1600 Ma)	3448.1	2967.3	4519.0	4563.8	6175[a]
	Tieling formation	209.7	213.9	211.1	328.8	325
	Hongshuizhuang formation	100.9	41.6	101.7	92.1	131
	Wumishan formation	2168.4	1874.6	2947.2	2936.4	3416
	Yangzhuang formation	78.3	36.0	322.4	255.8	707
	Gaoyuzhuang formation	890.8	801.2	936.6	950.7	1596
	Changchengian (1600–1670 Ma)	387.6	515.0	3042.6[a]	2586.5	2687
Total stratigraphic Thickness/m		4280	4095	8403	7621.9	9260

Note [a] indicates depocenter of different strata

be around Jidong-Jibei Depressions especially at Jibei for Changchengian, Jidong for Jixianian and Qingbaikouan, while the depocenter of Xiamaling Formation is at Xuanlong (Table 11.12). Their stromatolite and macrofossil assemblages, lithology-lithofacies and stratigraphic division show as a highly correlativity in whole the YFDZ, which implies that the palaeo-oceanic waters are well connected and so the palaeo-oceanic sedimentary environments unified in all the depressions during the Mesoproterozoic time. The current Shanhaiguan and Mihuai Uplifts, which have separated both northern and southern depression belts (Fig. 11.36), are predominantly referred to the late uplifting tectonic units in the YFDZ, and never completely separated the Meso-Neoproterozoic palaeo-oceanic waters.

11.3.2.2 Petroleum Geology

Numerous oil-seeps, asphalt and bituminous sandstone are widely distributed within the three depressions of northern depression belt. Taking Jibei Depression as an example, so far totally 115 sites of oil-seeps, asphalt and bituminous sandstone have been found, among which 98 sites are attributed to Mesoproterozoic strata, accounting for 85.2% of the total sites, and mainly appearing as liquid oil phase (Plate 11.1b; Table 11.13).

Table 11.13 Oil-seep phases and their productive layers in Jibei Depression, YFDZ

No	Oil-seep occurrence: Era	System	Formation	Oil-seep phase	Oil-seep number ∑115 sites		Percentage /%		
1	Mesozoic	Cretaceous	Xiguayuan, K_1x	Oil, asphalt	2		1.7		
2	Lower Palaeozoic	Ordovician	Majiagou, O_2m	Asphalt	1	3	0.9		2.6
3			Yel, O_1y	Oil, asphalt	2		1.7		
4		Cambrian	Changshan, $\in_3c$	Asphalt	1	13	0.9		10.5
5			Mantou, $\in_1m$	Oil, asphalt	8		7.0		
6			Fujunshan, $\in_1f$	Asphalt, oil	3		2.6		
7	Mesoproterzoic	Xiamaling, Pt_2^3x		Asphalt, oil	20		17.4		85.2
8		Jixianian	Tieling, Pt_2^2t	Oil, asphalt	60	77	52.2	66.9	
9			Hongshuizhuang, Pt_2^2h	Oil	2		1.7		
10			Wumishan, Pt_2^2w	Oil, asphalt	15		13.0		
11			Gaoyuzhuang, Pt_2^2g	Asphalt	1		0.9	0.9	

However, Mesoproterozoic oil-seeps are never found in the Jingxi and Jidong Depressions of southern depression belt during (Wang 1980; Wang and Han 2011; Fig. 11.36).[1,2,3]

Occurrence of oil-seep and bituminous sandstone in Jibei Depression, YFDZ. **a** Bituminous sandstone in the basal sandstone of Xiamaling Formation (photograph on outcrop); **b** liquid oil-seep in the dolostone of Wumishan Formation (photograph at mine drift); **c, d** bituminous sandstone in the basal sandstone of Xiamaling Formation (microphtograph)

As major source beds in Jibei Depression, both Gaoyuzhuang black micritic dolostone and Hongshuizhuang black shale contain high organic abundance with

[1] Wang et al. (1978). The bright prospects of Sinian primary oil and gas reservoirs. Oral Presentation on the Scientific Conference of Petroleum Geology and Seismic Exploration in East China. Beijing: China Petroleum Chemistry Industry Ministry (in Chinese).

[2] Wang et al. (1979). Principal characteristics of petroleum geology in the eastern segment of Yansan region (Internal Report). The 3rd Regional Geological Survey Team, Jingzhou: Jianghan Institute of Petroleum (in Chinese).

[3] Wang et al. (2009). Petroleum prospectivity and regional play predication of the lower stratigraphic assemblage in North China Platform (Internal Report). Beijing: China University of Petroleum-Beijing.

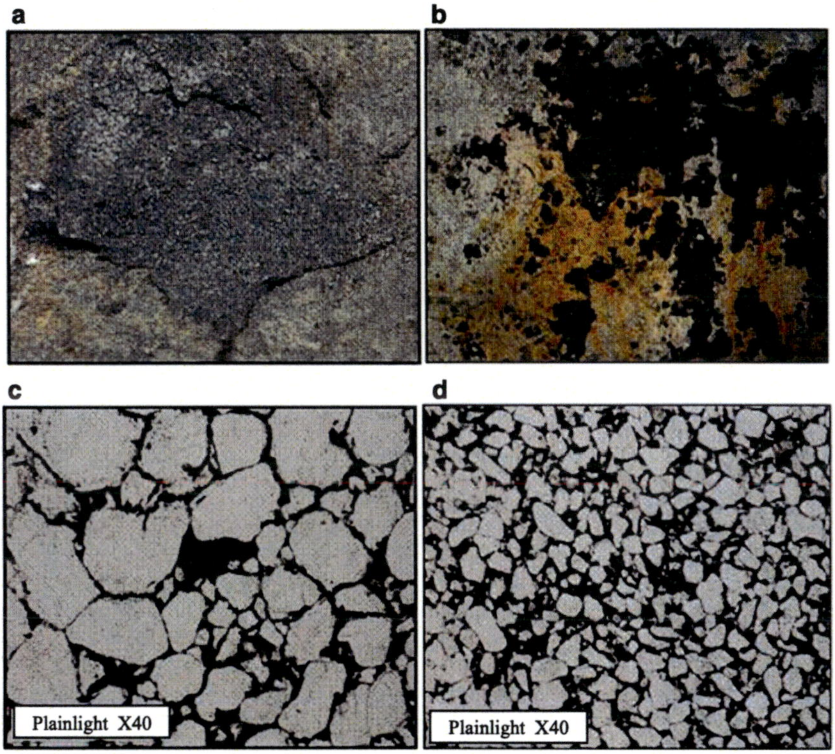

Plate 11.1 The *m/z* 123 mass chromatographs for oil-source correlation of the Mesoproterozoic bituminous sandstone and asphalts in Jibei Depression. **a–c**. Degradation products of catalytical hydrocracking for source rock kerogens; **d–f**. Aliphatic fractions of oil-seeps and bituminous sandstone; C_{18}–C_{23}. 13α(*n*-alkyl)-tricyclic terpanes

TOC 1.16% and 4.65% in average (max. 4.29% and 7.21%) as well as chloroform extractable bitumen 63 ppm and 265 ppm (max. 152 ppm and 4510 ppm) respectively. The measured equivalent vitrinite reflectance eqR_o values 1.38–1.75% (avg. 1.59%) and 0.9–1.42% (avg. 1.19%) respectively, which are referred to high-mature to over-mature phase for Gaoyuzhuang and mature to high-mature phase for Hongshuizhuang source beds. If taking TOC 0.5% as the lower limit of effective source rock, the cumulative thicknesses of source beds would be 164 and 60 m for Gaoyuzhuang and Hongshuizhuang Formations respectively (Table 11.14).

In Xiamaling Formation, the basal medium- to thin-bedded sandstone layers or lens are composed of quartz and chert, and well-cemented by silica and/or asphalt, appearing as white tight and hard quartzose sandstone or black bituminous sandstone, respectively (Plate 11.1a, c, d), even uncemented loose bituminous sands could be found at both south and north flanks of the central syncline belt (called Dangba Syncline Belt) in Jibei Depression. The 3.8 m-thick bituminous sandstone is outcropped at Longtangou (in Lingyuan, western Liaoning), while the outcrop

Table 11.14 Geochemical parameters of effective source beds in Jibei Depression, YFDZ

Parameters	Gaoyuzhuang formation	Hongshuizhuang formation
TOC/%	0.50–4.29/1.16 (69)[a]	0.50–7.21/4.65 (36)
Chloroform bitumen content/ppm	26–152/63 (11)	34–4510/2650 (10)
Hydrogen index/(mg HC/g rock)	11–45/21 (61)	97–311/233 (36)
Oil-generating potential $S_1 + S_2$/(mg HC/g rock)	0.09–2.39/0.32 (61)	0.52–18.2/12.2 (36)
Equivalent vitrinite reflectance eqR_o/%	1.38–1.75/1.59	0.90–1.42/1.19
Cumulative source rock thickness/m	164	60

[a] Minimum–maximum/average (sample number)

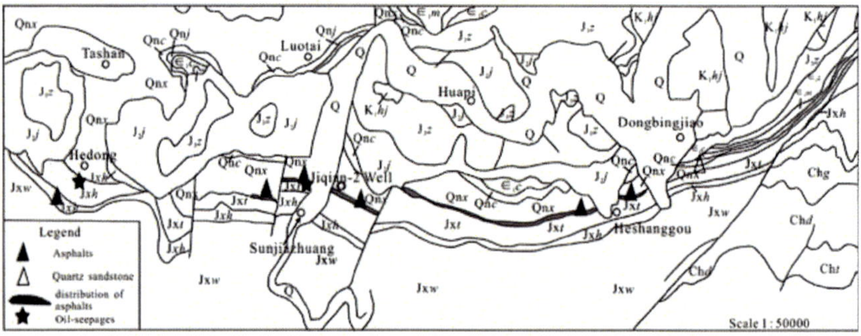

Fig. 11.37 Geological map showing the distribution of outcrop sites for Xiamaling bituminous sandstone at Lujiazhuang (Kuancheng) in Jibei depression

of bituminous sandstone can be traced for 10 km at Lujiazhuang on surface (in Kuancheng, northern Hebei; Fig. 11.37). The discovery of extensive bituminous sandstone would symbolize the actual existance of the oldest Mesoproterozoic fossil-oil-reservoirs (1400 Ma; Wang and Simoneit 1995; cf. Chap. 12).

11.3.2.3 Asphalt/Bituminous Sandstone-Source Rock Correlation

A novel biomarker series of C_{18}–C_{23} 13α(*n*-alkyl)-tricyclic terpanes had been initially detected and identified from the aliphatic fractions of Mesoproterozoic bituminous sandstone, oil-seep and asphalt in Jibei Depression (Fig. 11.38d–f; Wang 1991). So far the biomarker series is only found in the Meso-Neoproterozoic sedimentary organic matters, either in YFDZ, NCC (Wang and Simoneit 1995; Zhang et al. 2007; Wang et al. 2011) and in the Sinian-sourced asphaltic veins at Longmenshan, western Yangtze Craton (Huang and Wang 2008; cf. Chap. 15), which may be act as a specific biomarker series for the organic matter of Precambrian age.

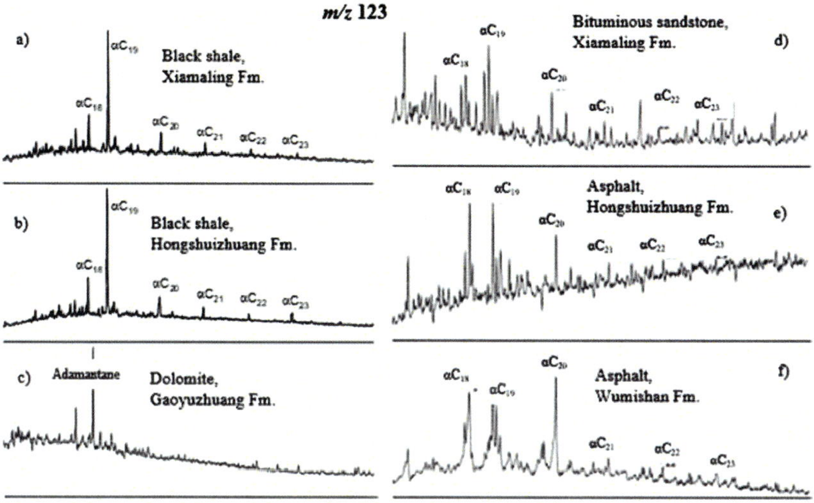

Fig. 11.38 The *m/z* 123 mass chromatographs for oil-source correlation of the Mesoproterozoic bituminous sandstone and asphalts in Jibei Depression. **a–c** Aliphatic fractions of source rocks; **d–f** Aliphatic fractions of oil-seeps and bituminous sandstone. C_{18}–C_{23}. 13α(*n*-alkyl)-tricyclic terpanes

Based on a systematically geochemical investigation of source beds in the Jibei Depression, 13α(*n*-alkyl)-tricyclic terpanes had also been detected from the aliphatic fractions of Hongshuizhuang and Xiamaling black shales (Fig. 11.38a, b), which are comparable to above oil-seeps, bituminous sandstone and asphalt (Fig. 11.39d–f), but never found in the Gaoyuzhuang black micrite dolostone (Fig. 11.38c).

By way of degradation technique of catalytical hydrocracking for kerogens isolated and purified from above source rocks, 13α(*n*-alkyl)-tricyclic terpanes were only detected in the kerogen-degraded products of Hongshuizhuang black shale (Fig. 11.39b), and well correlated with the bituminous sandstone, oil-seeps and asphalt in Jibei Depression (Fig. 11.39d–f), indicating the Hongshuizhuang black shale as the sole or major source bed for the bituminous sandstone, oil-seeps and asphalt in Jibei Depression (Fig. 11.38a).

11.3.2.4 Significance of the Xiamaling Basal Bituminous Sands

Obviously, asphalt or solid bitumen is not fluid and can't directly be filled into the intergranular opening of Xiamaling basal sandstone (Plate 11.1a, c, d), and thus, the Xiamaling basal bituminous sandstone should actually be the original oil sandstone, and then be altered into bituminous sandstone by the late thermo-alteration (cf. Chaps. 10 and 12). Therefore, the bituminous sandstone itself would act as the symbol of fossil-oil-reservoir, particularly at Longtangou (in Lingyuan, western

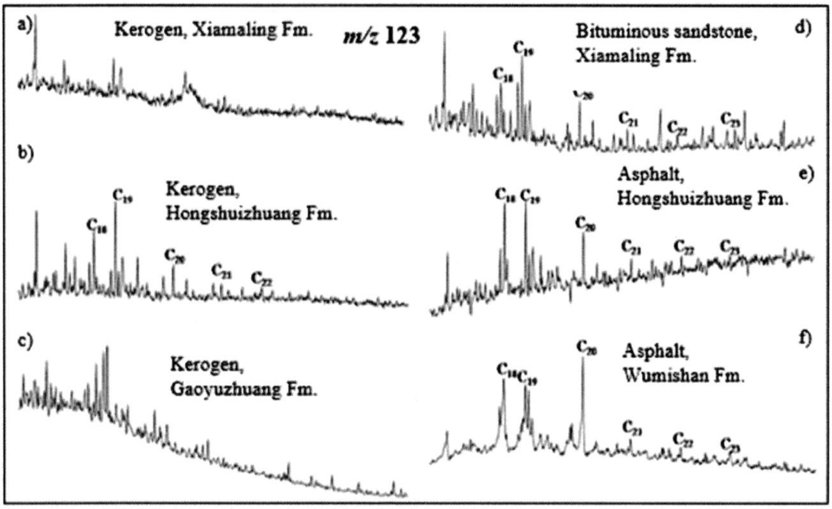

Fig. 11.39 The *m/z* 123 mass chromatographs for oil-source correlation of the Mesoproterozoic bituminous sandstone and asphalts in Jibei Depression. **a–c**. Degradation products of catalytical hydrocracking for source rock kerogens; **d–f**. Aliphatic fractions of oil-seeps and bituminous sandstone; C_{18}–C_{23}. 13α(*n*-alkyl)-tricyclic terpanes

Liaoning), where flesh uncemented bituminous sands are also found by means of artificial trenching operation on the outcrop of Xiamaling basal sandstone, which implies that when liquid oil was filling and entrapping into the Xiamaling basal sand-body, the sand-body had been just commenced to be cemented and its cementation was still unfinished yet during the early diagenesis. Owing to the known bottom limit age of Xiamaling Formation is 1400 Ma, i.e., equivalent to the basal boundary age of Mesoproterozoic, hence, it could also date the oil entrapment age of Xiamaling basal fossil-oil-reservoir back to 1400 Ma.

11.4 Conclusions

(1) As for global Meso-Neoproterozoic indigenous petroleum resources, there are four regions/countries, (Lena-Tounguska Petroleum Province in Russia, Anyue Gasfield in China, Oman Basins and Baghewala Oilfield in India) containing proven geological reserves and/or commercial production; nine regions/countries having confirmed the indigenous oil, oil-seep and/or asphalts, but still no commercial production found yet; five regions/countries being revealed to possess hydrocarbon generating potential in the Infracambrian strata in the word. All the indigenous petroleum was sourced from Meso-Neoproterozoic source beds, in some cases, it could be derived from Early Cambrian source bed, and thus it is also called Infracambrian petroleum.

(2) Mesoproterozoic to Early Cambrian indigenous petroleum occurrence includes oil, gas-condensate, gas and even asphalt (bitumen)/pyroasphalt (pyrobitumen). In view of organic matter thermo-evolution stages, the indigenous petroleum appears as different thermo-maturities from marginal mature (e.g., Baghewala oil), through mature (e.g., Oman oil and wet gas) and high-mature (e.g., Lena-Tounguska oil, gas-condensate and gas) to over-mature (e.g., Anyue cracking dry gas and pyrobitumen) phases.
(3) Almost all the known Infracambrian oil and gas reservoirs could be attributed to late entrapment/emplacement causes. In many cases, the entrapment/emplacement ages of Meso-Neoproterozoic oil and gas reservoirs are not earlier than Early Palaeozoic time, even the oil/gas field could be formed during Mesozoic time, which would facilitate the preservation of Infracambrian oil and/or gas resources.
(4) The Cambrian wide-spread salt beds in Lena-Tounguska Petroleum Province (Russia) and large, irregular salt bodies (salt dome) in Oman are regional superseal favorable to Infracambrian oil and gas entrapment and preservation.
(5) So far some distinct biomarkers such as conspicuous mid-chain monomethylalkanes (so called "X compounds") and 13α(*n*-alkyl)-tricyclic terpanes are found in Precambrian oil and sedimentary organic matter, which may be significant to distinguish the indigenous Meso-Neoproterozoic oil and asphalt for oil-source rock correlation.
(6) As new types of Infracambrian oil reservoir found in Oman, both intra-salt "silicilyte" and carbonate "stringer" reservoirs are self-sourced, self-reservoired and enclosed by large salt bodies. "Silicilyte" is a new petrological term specially for the microcrystalline silicon matrix, or called microcrystalline chert.

References

Aadil N, Sohail GM (2011) Stratigraphic correlation and isopach of Punjab Platform in Middle Indus Basin, Pakistan. GEOIndia, pp 1–6

Albert-Villanueva E, Permanyer A, Tritlla J, Levresse G, Salas R (2016) Solid hydrocarbons in Proterozoic dolostones, Taoudenni Basin, Mauritania. J Petrol Geol

Aley AA, Nash DF (1984) A summary of the geology and oil habitat of the eastern flank hydrocarbon province of south Oman. In: Proceedings of seminar on the source and habitat of petroleum in Arab countries. Kuwait, pp 521–541

Amthor JE, Ramseyer K, Faulkner T, Lucas P (2005) Stratigraphy and sedimentology of a chert reservoir at the Precambrian-Cambrian boundary: the AI shomou silicilyte, South Oman salt basin. GeoArabia 10:89–122

Amthor JE, Smith W, Nederlof NI, Frewin N L, Lake S (1998) Prolific oil production from a source rock—the Athel silicilyte source rock play in south Oman. Am Assoc of Petrol Geol Annu Convention A22

Anderson RR (1989) Gravity and magnetic modeling of central segment of Mid-continent Rift in Iowa—new insights into its stratigraphy, structure, and geological history. Am Asso Petrol Geol Bull 73(8):1043

Aplin AC, Larter SR, Bigge MA, Macleod G, Swarbrick RE, Grunberger D (2000) PVTX history of the North Sea's Judy oilfield. J Geochem Explor 69:641–644
Asim S, Zhu P, Qureshi SN, Naseer MT (2015) A case study of Precambrian to Eocene sediments' hydrocarbon potential assessment in Central Indus Basin of Pakistan. Areb J Geosci 8:10339–10357
Beijing Petroleum Exploration, Development Institute and North China Petroleum Bureau (1992) Reservoir beds and phase forecast of deep oil and gas pool-taking Jizhong depression and South Caspian Basin as examples. Petroleum Industry Press, Beijing, pp 273–357 (in Chinese)
Behrendt JC, Green AG, Cannon WF, Hutchinson DR, Lee MW, Milkereit B, Ahena WF, Spencer C (1988) Crustal structure of the Mid continent rift system: results from GLIMPCE deep seismic reflection profile. Geology 16:81–85
Bhat GM, Craig J, Hafiz M, Hakhoo N, Thurow JW, Thusu B, Cozzi A (2012) Geology and hydrocarbon potential of neoproterozoic-Cambrian Basins in Asia: an introduction. In: Bhat GM et al (eds) Geology and hydrocarbon potential of neoproterozoic-Cambrian Basins in Asia. Geological Society Special Publication 366, The Geological Society, London
Bowring SA, Grotzinger JP, Condon DJ, Ramezani J, Newall M (2007) Geochronologic constraints on the chronostratigraphic framework of the neoproterozoic Huqf supergroup, sultanate of Oman. Am J Sci 307:1097–1145
Bronner G, Roussel J, Trompette R (1980) Genesis and geodynamic evolution of the Taoudenni Cratonic Basin (Upper Precambrian and Paleozoic), Western Africa. Dyn Plat Inter Geodyn Ser 1:73–80
Cai XY (2012) Hydrocarbon generation-expulsion mechanisms and efficiencies of lacustrine source rocks: a case study from the Dongying sag, Bohai Bay Basin. Oil Gas Geol 33(3):329–334 (in Chinese with English abstract)
Chen JY (2004) Dawn of animal would. Jiangsu Science Press, Nanjing (in Chinese)
Chen JY, Zhou GQ, Zhu MY, Ye KY (1996) The Chengjiang Biota: a unique window of the Cambrian explosion. State Science Nature Museum, Taiwan (in Chinese)
Clauer N (1981) Rb-Sr and K-Ar dating of Precambrian clays and glauconies. Precambr Res 15:53–71
Craig J, Thurow J, Thusu B, Whitham A, Abutarruma Y (2009) Global neoproterozoic petroleum systems: the emerging potential in North Africa. In: Craig J et al (eds) Global Neoproterozoic Petroleum systems: the emerging potential in North Africa. Geological Society Special Publication 326, The Geological Society, London, pp 1–25
Crick IH, Boreham CJ, Cook AC (1988) Petroleum geology and geochemistry of Middle Proterozoic McAuthur Basin, Northern Australia II: assessment of source rock potential. AAPG Bull 72:1495–1514
Daniels PA Jr (1982) Upper Precambrian sedimentary rocks: oronto group, Michigan-Wisconsin. In: Wold RJ, Hinze WJ (eds) Geology and tectonics of the lake superior basin, vol. 156. Geological Society of America Memoir, pp 107–133
Dickes AB (1986a) Precambrian as a hydrocarbon exploration target. Geosci Wis 11:5–7
Dickes AB (1986b) Worldwide distribution of Precambrian hydrocarbon deposits. Geosci Wis 11:8–11
Dickes AB (1986c) Comparative Precambrian stratigraphy and structure along the Mid-continent Rift. Am Assoc Geol Bull 70(3):225–238
Du RL, Tian LF, Hu HB, Sun LM, Chen J (2009) Chinese Precambriam Palaeontology research: neoproterozoic Qinbaikouan Longfengshan Biota. Science Press, Beijing (in Chinese)
Du JH, Wang ZC, Zou CN, Xu CC, Shen P, Zhang BM, Jiang H, Huang SP (2016) Discovery of intra-cratonic rift in the Upper Yangtze and its control effect on the formation of Anyue giant gas field. Acta Patrolei Sin 37(1):1–16 (in Chinese with English abstract)
Efimov AS, Cert AA, Mel'nikov PN, Starosel'tcev VS, Vymyatin AA, Akimov VG, Cherepanova II, Brazhnikov MV (2012) About current state and trends of hydrocarbon Resource potential, geological exploration and licensing in East Siberia and Sakha Republic (Yakutia). Geol Neft Gasa, 5:57–74 (in Russian)

Elmore RD, Milavec GJ, Imbus SW, Engel MH (1989) The Precambrian Nonesuch Formation of the North American Mid-continent rift, sedimentology and organic geochemical aspects of lacustrine deposition. Precambr Res 43:181–213
Emerry D, Robinson A (1993) Inorganic geochemistry: application to petroleum geology. Blackwell Scientific Publication, Oxford, pp 41–66
Fedorov DI (1997) The stratigraphy and hydrocarbon potential of the Riphean-Vendian (Middle-Late Proterozoic) succession on the Russian Platform. J Petr Geol 20(2):205–222
Filipstove YA, Petrishina YV, Bogorodskaya LI, Kontrorovich AA, Krinin VA (1999) Evaluation of maturity and oil- and gas-generation properties of the organic matter in Riphean and Vendian rocks of the Baykit and Katanga petroleum regions. Geol 1 Geofiz 40:1362–1374
Frolov SV, Akhmanov GG, Bakay EA, Lubnina NV, Korobova NI, Karnyushina EE, Kozlova EV (2015) Meso-neoproterozoic petroleum systems of the Eastern Siberian sedimentary basins. Precambr Res 259:95–113
Gang WZ (2009) Hydrocarbon generation conditions and exploration potential of the Taoudenni Basin, Mauritania. Pet Sci 6:29–37 (in Chinese with English abstract)
Gang WZ, Wu Y, Gao G, Ma Q, Pang XQ (2012) Geochemical features and geologic significances of source rocks in Nanpu Sag, Bohai Bay Basin. Pet Geol Exp 34(1):57–61 (in Chinese with English abstract)
Gaters G (2005) Hydrocarbon project in mauritania and Mali, West Africa. Technical experts Report in farmout brochure by Baraka Petroleum. South Perth, Australia
Ghori KAR, Craig J, Thusu B, Lüning S, Geiger M (2009) Global Infracambrian petroleum system: a review. In: Craig J et al (eds) Global Neoproterozoic petroleum systems: The emerging potential in North Africa. Geological Society Special Publication 326, The Geological Society, London, pp 110–136
Grantham PJ (1986) The occurrence of unusual C_{27} and C_{29} sterane predominances in two types of Oman crude oil. Org Geochem 9(1):1–10
Grantham PJ, Lijmbach GWM, Postthuma J, Hughes Clarke MW, Willink RJ (1987) Origin of crude oils in Oman. J Pet Geol 11:61–80
Grosjean E, Love GD, Stalvies C, Fike DA, Summons RE (2009) Origin of petroleum in the Neoproterozoic-Cambrian South Oman salt basin. Org Geochem 40:87–110
Hanor J (1980) Dissolved methane in sedimentary brines; potential effect on the PVT properties of fluid inclusions. Econ Geol 75(4):603–609
Hao F, Zou HY, Fang Y, Hu JW (2006) Kinetics of organic matter maturation and hydrocarbon generation in overpressure environment. Acta Petrolei Sinica 27(5):9–18 (in Chinese with English abstract)
Hasany ST, Aftab M, Siddiqui RA (2012) Refound Exploration Opportunities and Cambrian Sediments of Punjab Platform. Pakistan Petroleum Limited, Pakistan. Karachi
Heward AP (1989) Early Ordovician alluvial fan deposits of the Marmul oil field south Oman. J Geol Soc London 146:557
Hou XG, Bergstrom J, Wang HF, Feng XH, Chen AL (1999) The Chengjiang Fauna: exceptionally well-preserved animals from 530 million years ago. Yunnan Science and Technology Press, Kunming (in Chinese)
Huang DF, Wang LS (2008) Geochemical characteristics of bituminous dike in Kuangshanliang area and its significance. Petrolei Sinica 29(1):23–28 (in Chinese with English abstract)
Hunt J (1991) Generation of gas and oil from coal and other terrestrial organic matter. Org Geochem 17(6):673–680
Immerz P, Oterdoom WH, Yonbery El (2000) The Huqf/Haima hydrocarbon system of Oman and the terminal phase of the Pan-African Orogeny: evaporite deposition in a compressive setting. In: The 4th middle east geoscience conference, GEO2000. GeoArabia, Abstract, pp 387–433
Jackson JM, Powell TG, Summons RE, Sweet IP (1986) Hydrocarbon shows and petroleum source rocks in sediments as old as 1.7×10^9 years. Nature 322:727–727

Jackson JM, Sweet IP, Powell TG (1988) Studies on petroleum geology and geochemistry of the of Middle Proterozoic McArthur Basin, northern Australia I: petroleum potential. Aust Petrol Explor Assoc J 28:283–302

Jean-Pierre G, Herbert E, Amir K (2014) Petroleum system, migration and charge history in the Neo- and Meso-Proterozoic series of the Taoudenni Basin, Adrar: insights from fluid inclusions. In: International petroleum technology conference, IPTC-18011-MS, pp 1–5

Kah CL, Bartley JK, Stagner AF (2009) Reinterpreting a Proterozoic enigma: conophyton-Jacutophyton stromatolites of the Mesoproterozoic Atar Group, Mauritania. Spec Publ Int Assoc Sedimentol 41:277–296

Kao CS, Hsiung YH, Kao P (1934) Preliminary notes on Sinian stratigraphy of North China. Bull Geol Soc China 13(2):243–288

Katz BJ, Everett MA (2016) An overview of pre-Devonian petroleum systems—unique characteristics and elevated risks. Mar Pet Geol 73:492–516

Klemme HD, Ulmishek GF (1991) Effective petroleum source rocks of the world: stratigraphic distribution and controlling factors. AAPG Bull 75(12):1809–1851

Knott DJ (1998) Omam prepares for oil expansion and gas production for LNG export. Oil Gas J 96:29–34

Kuznetsov VG (1997) Riphean hydrocarbon reservoir of the Yurubchen-Tokhom zone, Lena-Tunguska Province, NE Russia. J Pet Geol 20(4):459–474

Lahondére D, Roger J et al (2005) Notice explicative des cartes geologiques á 1/20,000 et 1/500,000 de l' extréme sud de la Mauritanie. DMG, Ministére des mine et de l'Industrie, Nouakchott, 610

Larichev AI, Melenevskii VN, Shvedenkov GY, Sukhoruchko VI (2004) Aquapyrolysis of organic matter from the Riphean carbon-rich argillite of the Yurubchen-Takhom oil and gas accumulation zone. Dokl Earth Sci 398(7):961–963

Lee LS, Chao YT (1924) Geology of the Gorge district of the Yangtze (from Ichang to Tzeckuei) with special reference to the development of the Gorges. Bull Geol Soc Chin 3(3–4):351–391

Liu DH, Xiao XM, Mi JK, Li XQ, Shen JK, Song ZG, Peng PA (2003) Determination of trapping pressure and temperature of petroleum inclusions using PVT simulation software—a case study of Lower Ordovician carbonates from the Lunnan Low Uplift, Tarim Basin. Mar Pet Geol 20(1):29–43

Liu SG, Ma YS, Cai XY, Guo S, Wang GZ, Yong ZQ, Sun W, Yuan HF, Pan CL (2009) Characteristic and accumulation process of the natural gas from Sinian to Lower Palaeozoic in Sichuan Basin, China. J Chengdu Univ Technol (sci Technol Ed) 36(4):345–354 (in Chinese with English abstract)

Liu W, Qiu NS, Xu QC, Liu Y (2018) Precambrian temperature and pressure system of Gaoshiti-Moxi block in the central palaeo-uplift of Sichuan Basin, southwest China. Precambr Res 313:91–108

Lottaroli F, Craig J, Thusu B (2009) Neoproterozoic–early Cambrian (Infracambrian) hydrocarbon prospectivity of North Africa: a synthesis. In: Craig J et al (eds) Global neoproterozoic petroleum systems: The emerging potential in North Africa. Geological Society Special Publication 326, The Geological Society, London, pp 137–156

Mauk JL, Hieshima GB (1992) Organic matter and copper mineralization at White Pine, Michigan, USA. Chem Geol 99:189–211

Melanie AE (2010) Characterizing the Precambrian petroleum systems of Eastern Siberia: evidence from oil geochemistry and basin moleling. SPE 136334:1–11

Menchikoff N (1949) Quelques traits de I'hisoire geologique du Sahara occidentail. Annales Hebert Et Haug 7:303–325

O'Dell M, Lamers E (2003) Subsurface uncertainty management in the Harweel Cluster, south Oman. SPE 84189. Int Soc Petrol Eng

Ojha PS (2012) Precambrian sedimentary basins of India: an appraisal of their petroleum otential. In: Bhat GM, et al (eds) Geology and hydrocarbon potential of neoproterozoic–Cambrian basins in Asia. Geological Society Special Publication, vol. 366. Geological Society, London, pp 19–58

Ozimic S, Passmore VL, Pain L, Lavering IH (1986) Australian petroleum accumulation report 1: amadeus basin, central Australia. Australian Government Publication Service, Canberra
Palacs JG (1997) Source-rock potential of Precambrian rocks in selected Basin of United States. US Geol Surv Bull 02146-J:125–134
Peters KE, Watters CC, Gupta Das U, McCaaffrey MA, Lee CY (1995) Recognition of an Infracambrian source rock based on biomarkers in the Baghewala Oil field, India. AAPG 79:1481–1494
Peters JM, Filbrandt JB, Grotzinger JP, Newall MJ, Shuster MW, Al-Syabi HA (2003) Surface-piercing salt domes of interior North Oman, and their significance foe the Ara carbonate "stringer" hydrocarbon play. Geo Arab 8:231–270
Pollastro RM (1999) Ghaba Salt Basin province and Fuhud Salt Basin province, Oman—geological overview and total petroleum systems. US Geol Surv Bull 2167:1–41
Pratt LM, Summons RE, Hieshima GB (1991) Sterane and triterpane biomarkers in the Precambrian Nonesuch Formation, North American Midcontinent Rift. Geochimca Et Cosmochimica Acta 55:911–916
Priss WV, Forbes BG (1981) Stratigrahpy, correlation and sedimentary history Adelaidean (Latest Proterozoic) basin in Australia. Precambr Res 15:255–304
Pruvost P (1951) L'Infracambrien. Bull de la Soc Belge Geol Palaeontologie Hydrol 60:43–65
Qadri IB (1995) Petroleum geology of Pakistan. Pakistan Petroleum Limited, Karachi
Rahmani A, Goucem A, Boukhallat S, Saadallah N (2009) Infracambrian petroleum play elements of the NE Taoudenni Basin (Algeria). In: Craig J et al (eds) Global neoproterozoic petroleum systems, the emerging potential in North Africa. Geological Society Special Publication 326, Geological Society, London, pp 221–229
Ram J (2012) Neoproterozoic successions in Peninsular India and their hydrocarbon prospectivity. In: Bhat GM et al (eds) Geology and hydrocarbon potential of neoproterozoic-Cambrian Basins in Asia. Geological Society Special Publications 366. Geological Society, London, pp 75–90
Rawlings DJ (1999a) Stratigraphic resolution of a multiphase intracratonic basin system: the McArthur Basin, northern Australia. Aust J Earth Sci 46:1–17
Rawlings DJ (1999b) Stratigraphic resolution of a multiphase intracraton basin system: the McArthur Basin, northern Australia. Aust J Earth Sci 46:703–723
Rooney AD, Selby D, Houzay JP, Renne PR (2010) Re-Os geochemistry of a Mesoproterozoic sedimentary succession, Taoudenni Basin, Mauritania: implications for basin-wide correlations and Re-Os organic-rich sediments systematic. Earth Planet Sci Lett 289:486–496
Sheikh RA, Jamil MA, McCann J, Saqi MI (2003) Distribution of Infracambrian reservoirs on Punjab Platform in central Indus Basin of Pakistan. ATC 2003 conference and oil show. Society of Petroleum Engineers (SPE) and Pakistan Association of Petroleum Geoscientists (PAPG), Islamabad, pp 1–17
Shu DG et al (2016) Ancestors from Cambrian explosion. Northwest University Press, Xian (in Chinese)
Smith AG (2009) Neoproterozoic timescales and stratigraphy. In: Craig J et al (eds) Global neoproterozoic petroleum systems: the emerging potential in North Africa. Geological Society Special Publication, vol. 326, The Geological Society, London, pp 27–54
Sun SF (2006) Meso- to Neoproterozoic Micropaleobotany in Jixian, China. Geology Press, Beijing (in Chinese)
Terken JMJ, Frewin NL, Indrelid SL (2001) Petroleum systems of Oman: charge timing and risks. AAPG Bull 85:1817–1845
Tong XG, Xu SB (2004) Global petroleum exploration and development atlas: Fascicle of former soviet union countries. Petroleum Industry Press, Beijing, pp 138–163 (in Chinese)
Ulmishek GF (2001) Petroleum geology and resources of the Baykit High Province, Eastern Siberia, Russia. US Geological Survey Bulletin 2201-F, U S Geological Survey, Washington
Visser W (1991) Burial and thermal history of Proterozoic source rocks in Oman. Precambrian Researches 54:15–36

Vysotsky IV, Korehagina YuI, Sokolov BA (1993) Genetic aspects of assessment of the petroleum potential of the Moscow syneclise. Geol Nefit i Gaza 12:26–29 (in Russia)

Wade BP, Hand M, Barovich KM (2005) Nd isotopic and geochemical constraints on provenance of sedimentary rocks in the eastern Officer Basin, Australia; implications for the duration of the intracratonic Precambrian Orogeny. J Geol Soc Lond 162:513–530

Wang TG (1980) Primary properties of Sinian Suberathem oil-seep and its petroleum geological significance in Yanshan region. Pet Explor Dev 7(2):34–52 (in Chinses)

Wang TG (1991) A novel tricyclic terpane biomarker series in the Upper Proterozoic bituminous sandstone, eastern Yanshan region. Sci China (series B) 34(4):479–489

Wang TG, Han KY (2011) On Meso-Neoproterozoic primary petroleum resources. Acta Petrolei Sin 32(1):1–7 (in Chinese with English abstract)

Wang TG, Simoneit BRT (1995) Tricyclic terpanes in Precambrian Bituminous sandstone from the eastern Yanshan region, North China. Chem Geol 120:155–170

Wang TG, He FQ, Li MJ, Hou Y, Guo SQ (2004) Alkyl-dibenzothiophenes: molecular tracers for filling pathway in oil reservoirs. Chin Sci Bull 49(22):2399–2404

Wang TG, Zhong NN, Wang CJ, Zhu YX, Liu Y, Song DF (2016) Source bed and oil entrapment-alteration histories of fossil-oil-reservoirs in the Xiamaling Formation basal sandstone, Jibei depression. Pet Sci Bull 1(1):24–36 (in Chinese with English abstract)

Wang CJ, Wang M, Xu J, Li YL, Yu Y, Bai J, Dong T, Zhang XY, Xiong XF, Gai HF (2011) 13α(*n*-alkyl)-tricyclic terpanes: a series of biomarkers for the unique microbial mat ecosystem in the Middle Mesoproterozoic (1.45–1.30 Ga) North China Sea. Mineral Mag 75:2, 114

Wei GQ, Shen P, Yang W, Zhang J, Jiao GH, Xie WR, Xie ZY (2013) Formation conditions and exploration prospects of Sinian large gas fields, Sichuan Basin. Pet Explor Dev 40(2):129–138 (in Chinese with English abstract)

Wells AJ, Forman DJ, Ranferd LC, Cooks PJ (1970) Geology of the Amadeus Basin. Bur Miner Resour Aust Bull

Wilkins N (2007) Proterozoic evangelist tries to convent the unbelievers. Oil and gas gazette, Dec 2006-Jan 2007, 2–3

Womer MB (1986) Hydrocarbon occurrence and diagenetic history within proterozoic sediments, McArthur river area, Northern territory, Australia. Aust Petrol Explor Assoc J 26:363–374

Wong SW, Ford S, Turner B (1998) Massive fracture stimulation in deep, high-pressure Athel formation. Soc Petrol Eng Pap 50614:407–412

Yang CY (2018) Petroleum entrapment and evolution history of Leshan-Longnvsi uplift, SW China. China University of Petroleum-Beijing, Beijing (in Chinese with English abstract)

Yang CY, Wen L, Wang TG, Wang B, Luo B, Li MJ, Tian XW, Ni ZY (2020) Timing of hydrocarbon accumulation for palaeo-oil reservoir in Anyue gasfield in Chuanzhong uplift. Oil Gas Geol 41(3):48–58 (in Chinese with English abstract)

Yang CY, Ni ZY, Wang TG, Chen ZH, Hong HT, Wen L, Luo B, Wang WZ (2018) A new genetic mechanism of natural gas accumulation. Sci Rep 8(1):8, 336

Zhang SC, Zhang BM, Bian LZ, Jin ZJ, Wang DR, Chen JF (2007) The Xiamaling oil shale generated through Rhodophyta over 800 Ma ago. Sci China Ser D Earth Sci 50(4):527–535

Zhu MZ, Chen JY (2002) Hydrocarbon-generating threshold of the source rocks in Palaeogene of Linnan subsag. Petrol Geol Recovery Eff 9(2):35–37 (in Chinese with English abstract)

Zhu CQ, Hu SB, Qiu NS, Rao S, Yuan YS (2016) The thermal history of the Sichuan Basin, SW China: evidence from the deep boreholes. Sci China Earth Sci 59:70–82

Zou CN, Du JH, Xu CC, Wong ZC, Zhang BM, Wei GQ, Wang TS, Yao GS, Deng SW, Liu JJ, Zhou H, Xu AT, Yang Z, Jiang H, Gu ZD (2014) Formation, distribution, resource potential and discovery of the Sinian-Cambrian giant gas field, Sichuan basin, SW China. Petrol Explor Dev 41(3):278–293 (in Chinese with English abstract)

Chapter 12
Source Beds and Oil Charging to Alteration Histories of Fossil-Oil-Reservoirs in the Basal Sandstone of Xiamaling Formation, Jibei Depression

Tieguan Wang, Ningning Zhong, Chunjiang Wang, Yixiu Zhu, Yan Liu, and Daofu Song

Abstract As for the Mesoproterozoic sequence (1800–1320 Ma) in the Jibei Depression of Yanliao-Faulted-Depression-Zone (YFDZ), there are 2 sets of oil source beds, i.e., black micritic-dolostone in the Gaoyuzhuang Formation and black shale in the Hongshuizhuang Formation, as well as 3 sites of basal bituminous sandstone as fossil-oil-reservoirs in the Xiamaling Formation respectively at Longtangou, Shuangdong and Lujiazhuang. The bituminous sandstone appears as two phases of solid bitumen components (SBCs): the early SBC with 1.68–2.52% in bitumen reflectance R_b and the late one with 0.81–1.01%. Owing to the intrusion of gabbro-diabase sills into Xiamaling Formation, original oil-reservoir had been altered into bituminous sandstone, i.e., so-called fossil-oil-reservoir, whereas viscous oil-seeps are still found within wall-rock alteration zone which provide an evidence for late oil charging after the magma cooling and consolidation. The discovery of bituminous sands indicates an early oil charging and entrapment process at early diagenetic stage of the basal sandstone in Xiamaling Formation (ca. 1400 Ma), while its alteration age could be determined according to the emplacement age of gabbro-diabase sills in Xiamaling Formation (1327 Ma). Based on the analyses of stratigraphic sequence and thickness, it is defined that early oil charging and entrapment in the Xiamaling basal sandstone can only be sourced from the Gaoyuzhuang source bed, and it's hydrocarbon-generating threshold depth would be around 3600 m. Based on this threshold depth, the oil entrapment age sourced by the Hongshuizhuang source bed should be at the Mesozoic period. Oil-source rock correlation results show that all the liquid oil-seeps found in the Wumishan and Tieling Formations and the soluble bitumen fraction of late SBC in the Xiamaling basal bituminous sandstone would be sourced by Hongshuizhuang source bed in Jibei Depression.

T. Wang (✉) · N. Zhong · C. Wang · Y. Zhu · Y. Liu · D. Song
State Key Laboratory of Petroleum Resources and Prospecting, Chinan University of Petroleum, Beijing 102249, China

Y. Liu
Institute of Mud Logging Technology and Engineering, Yangtze University, Hubei, Jingzhou 434023, China

T. Wang, *Meso-Neoproterozoic Geology and Petroleum Resources in China*, Springer Geology, https://doi.org/10.1007/978-981-19-5666-9_12

Keywords Mesoproterozoic · Xiamaling Formation · Basal bituminous sandstone · Fossil-oil-reservoir · Oil entrapment age · Hydrocarbon-generating threshold depth

12.1 Introduction

Geographically, the Jibei Depression is situated at the Xinglong, Chengde, Kuancheng and Pingquan Counties/Cities of Hebei Province and the Lingyuan City of Liaoning Province, North China, belonging to Yanshan Mountains, and geologically, it is a Meso-Neoproterozoic sedimentary depression in the north of Yanliao Faulted-Depression Zone (YFDZ) with an area of ca. 8700 km^2. It is separated from the Inner Mongolian Axis to the north by the Chengde-Pingquan-Lingyuan Fault, and also encircled elsewhere by a series of Precambrian uplifts and depressions along the east, south and west sides respectively by Liaoxi Depression, Shanhaiguan and Mihuai Upliftss. On the whole, the tectonic framework of YFDZ consists of "five depressions and two uplifts", i.e., the Xuanlong, Jibei, Liaoxi, Jingxi and Jidong Depressions as well as Shanhaiguan and Mihuai Uplifts (cf. Fig. 11.36).

An unusually thick Mesoproterozoic sequence was developed in the five depressions of YFDZ, including Changchengian ($Pt_2{}^1$; 1800–1600 Ma), Jixianian ($Pt_2{}^2$; 1600–1400 Ma) and Xiamaling Formation ($Pt_2{}^3x$; 1400–1320 Ma), with a depocenter at Jidong-Jibei Depressions and a sedimentary hiatus of the $Pt_2{}^4$ strata (1320–1000 Ma) on the stratigraphic column. While the Neoproterozoic sequence only has Qingbaikouan ($Pt_3{}^1$; 1000–800 Ma) with limited sedimentary thickness and without Nanhuan ($Pt_3{}^2$; 800–630 Ma) and Sinian ($Pt_3{}^3$; 630–542 Ma) sediments (Table 12.1).

In the Jibei Depression, the Mesoproterozoic sequence is up to 7931.1 m-thick, while the Neoproterozoic only 111.8 m in thickness. With a total thickness of 8042.9 m, The Meso-Neoproterozoic strata are composed of predominate carbonate and minor clastic rocks overlaid by Palaeozoic and Mesozoic strata (Table 12.1).

The Xiamaling Formation used to be taken as the basal stratigraphic unit of Neoproterozoic $Pt_3{}^1$ Qingbaikouan sediments. Based on the new evidences of zircon U–Pb and baddeleyite $^{207}Pb/^{206}Pb$ isotopic dating, however, it has recently been attributed to the top unit of Mesoproterozoic $Pt_2{}^3$ sequence (Table 12.1; Gao et al. 2008; Li et al. 2009; cf. Chap. 10). With a stratigraphic thickness of 540 m, the Xiamaling Formation is mainly composed of shales and also its depocenter is at the Xuanlong Depression. The Xiamaling Formation can be subdivided into four members, i.e., Xia-1 to Xia-4 Members in ascending order. Black shale of Xia-3 Member constitutes the main source bed of Xiamaling Formation. However, the Xiamaling Formation was mostly denuded by the later crustal movement of "Yuxian Uplifting" so that the Xia-3 source bed is completely lost, only the remnant basal sandy shale of Xia-1 Member was surviving in Jibei and Jidong Depressions. So far totally 115 sites of oil-seeps, including oil, asphalt and bituminous sandstone, have been found in the Jibei Depression, among which 78 sites are distributed within the Mesoproterozoic strata, accounting for 85.2% of the total sites (Wang 1980; Wang

Table 12.1 Meso-neoproterozoic stratigraphic chart in the Jibei Depression

Era	System/series		Formation	Thickness (m)		Lithology	Age (Ma)
Palaeozoic	Low-Cambrian	ϵ_1	Fujunshan	–		Greyish-white limestone containing trilobite	
Neoproterozoic	Sinian	Pt_3^3	257 Ma stratigraphic hiatus (780–542 Ma)				←542
	Nanhuan	Pt_3^2					←635
	Qingbaikouan	Pt_3^1	Jing'eryu	39.2	111.8	Egg-green, greyish-white dolomitic limestones	←780
			Luotuoling	72.6		Feldspathic quartzose sandstone, glauconitic sandstone and variegated shale	
Mesoproterozoic	Pt_2^4		327 Ma stratigraphic hiatus (1327–1000 Ma)				←1000
	Pt_2^3		Xiamaling	369.5	369.5	Black/green shales and variegated sandstone	←1320
	Jixianian	Pt_2^2	Tieling	211.1	4519.0	Manganiferous dolostone, variegated shale and stromatolitic limestone	←1400
			Hongshuizhuang	101.7		Black shale and light-color muddy dolostone	
			Wumishan	2947.2		Chert-stripped and stromatolitic dolostones	
			Yangzhuang	322.4		Purplish-red and greyish-white muddy-sandy dolostones	

(continued)

Table 12.1 (continued)

Era	System/series		Formation	Thickness (m)		Lithology	Age (Ma)
			Gaoyuzhuang	936.6		Chert-stripped and manganiferous dolostones, dolomitic limestone	↙1600
	Changchengian	Pt_2^1	Dahongyu	442.6	3042.6	Sandstone and potassium-rich basic volcanic rock	
			Tuanshanzi	259.2		Iron-rich dolostone	
			Chuanlinggou	293.0		Green and black shales	
			Changzougou	2047.8		Conglomerate and quartzose sandstone	↙1800

and Han 2011). From the view point of oil-seep occurrence, most of the oil-seep and asphalt sites were discovered in the carbonates of both Tieling (60 sites, 52.2% of the total) and Wumishan Formations (20 sites, 13%), while the bituminous sandstone only occurs at the basal sandstone of Xiamaling Formation (20 sites, 17.4%).

As a case study of fossil-oil-reservoir, the basal bituminous sandstone of Xiamaling Formation at the Longtangou (Lingyuan, Liaoning Province) has been reported in succession (Wang 1980; Wang et al. 1988; Wang and Simoneit 1995; Wang and Han 2011; Liu et al. 2011). Subsequently, the investigations on the Xiamaling fossil-oil-reservoir are extended to Lujiazhuang (Kuancheng) and Shuangdong (Pingquan, Hebei Province).

At the present study, the authors try to analyze the source beds, oil charging and alteration histories of the Xiamaling basal sandstone fossil-oil-reservoir as well as its petroleum geological significances.

12.2 Petrologic Characteristics of the Basal Sandstone in Xiamaling Formation

Due to the effect of "Qinyu Uplifting", the Xiamaling Formation shows a disconformable contact with the underlying Tieling Formation in the Jibei Depression. The thickness of basal sandstone deposited on the disconformable contact is from 0.5 m (Lujiazhuang in Kuancheng) to 7.58 m (Hezhuangzi in Lingyuan). The lithology of basal sandstone is vertically or laterally variable, often appearing as greyish-white and white coarse- to fine-grained siliceous quartzose sandstones or siltstone, locally containing granule, which consist dominantly of mono-crystal quartz (accounting for 90% or more) and siliceous debris (i.e., chert and polycrystal quartz, accounting for 1–10%, with little feldspar and mica), and occur as primary high sphericity and roundness, indicating a high maturity of mineral composition and a depositional environment of foreshore (Plate 12.1a).

The basal sandstone is principally cemented by silica (i.e., the quartz overgrowth in Plate 12.1a) with little muddy and carbonatic cements. The muddy cement was derived from feldspar alteration products or intergranular matrix. The cement content is variable from 1 to 12% (avg. 2–3%) with very uneven cementation degree (Plate 12.1b–d).

The black, greyish-black and blackish-brown solid bitumen is filled into the intergranular pores and fractures of Xiamaling basal sandstones, so that its fresh outcrop or fracture appears as black color, while the weathered surface shows yellowish-green and blackish-green colors. The solid bitumen in intergranular pores could act as the secondary cement for the Xiamaling basal sandstone (Plate 12.1b, d), and the bituminous fillings in fractures occur as vein- and banding- or ribbon-shapes (Plate 12.1c). The bitumen content of basal sandstone obviously varies from 1 to 25% (avg. 13.6%) estimated under microscope, which can result in the occurrence of bituminous sandstones. The basal sandstone with bitumen content $\geq$ 10% usually shows porous-cementation, while those with bitumen content < 10% appear as contact- and mosaic-cementations.

On the whole, there is a complementary relationship between the bitumen and silica contents in the Xiamaling basal sandstone. The basal sandstones cemented mainly by silica are very tight and rigid without bitumen or only with trace bitumen (Plate 12.1a), while those by bitumen with minor silica are loose and fragile, even a coexisting uncemented bituminous sands (Plate 12.1d) and siliceous cemented bituminous sandstone (Plate 12.1b, c) has been found at the Longtangou (Lingyuan) and Lujiazhuang (Kuancheng) respectively.

12.3 Occurrence of Fossil-Oil-Reservoir in the Xiamaling Basal Sandstone

The secondary tectonic units of Jibei Depression are comprised by Xiejiaying Syncline, Pingquan Anticline, Dangba Syncline and Guozhangzi Monocline Belts, extended in a NE–SW trending, and appearing as an alternating tectonic framework of positive and negative units from NW to SE. The Meso-Neoproterozoic sequences are exposed on the surface in the positive Pingquan Anticline and Guozhangzi Monocline Belts, while they are overlaid by Palaeozoic and Mesozoic sequences in the negative Xiejiaying and Dangba Syncline Belts, which is favorable for the preservation of Meso-Neoproterozoic oil and/or gas reservoirs. This tectonic framework was formed during the Mesozoic time.

The known Xiamaling basal bituminous sandstones are distributed at the north and south limbs of the Dengba Syncline Belt in the Jibei Depression, resulting in three fossil-oil-reservoirs respectively at Longtangou, Lujiazhuang and Shuangdong (Fig. 12.1).

12.3.1 Longtangou Fossil-Oil-Reservoir

The terrain of Longtangou is a valley area at the Lingyuan City in Liaoning Province, ca. 700 m apart from the boundary between Hebei and Liaoning Provinces. Geologically, it is on the north border of Guozhangzi Monocline Belt (Figs. 12.1 and 12.2).

The Xiamaling black shale contains medium- to thin-bedded bandings or lens of quartzose sandstone, while bituminous sandstone occurs within its basal sandstone. Laterally, the basal sandstone is steadily distributed, even there are six outcrop sites of basal bituminous sandstone of Xiamaling Formation found within the 5 km range in Longtangou, which can be referred to a litho-stratigraphic type of fossil-oil-reservoir (Figs. 12.1 and 12.2). The basal sandstone is 3.8 m (Longtangou) to 7.58 m

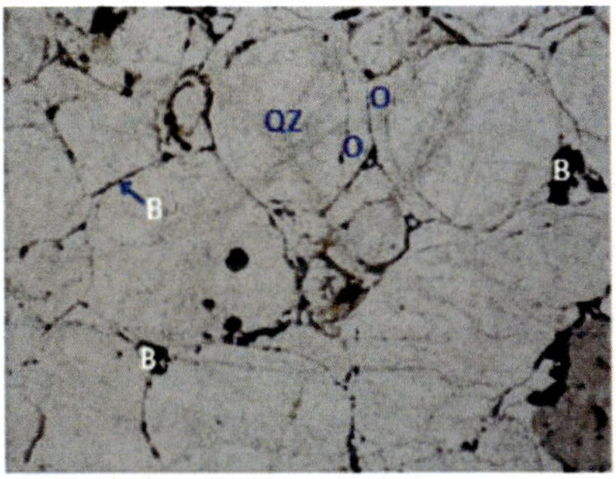

a. Silliceous quartzose sandstone, cement content: 12%, bitumen content: 4%; ×50

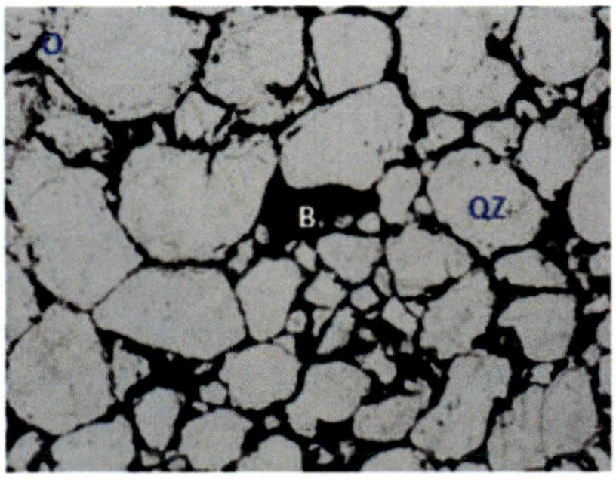

b. Bituminous sandstone, cement content: 5%, bitumen content: 20%–25%; ×40

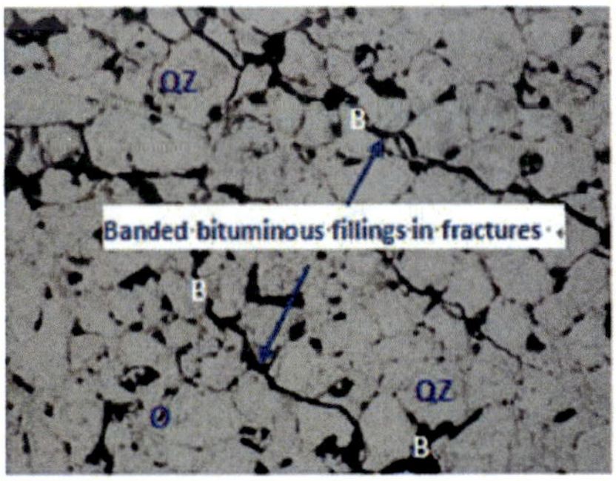

c. Banding bitumen fillings, cement content: 10%, bitumen content: 10%; ×40.

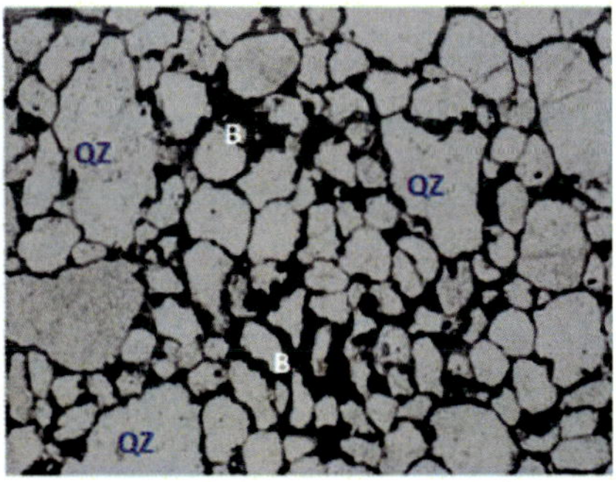

d. Bituminous sands, bitumen content: 20%–30%, cement content: trace; ×40

Plate 12.1 Microphotographs of basal sandstone **a**, bituminous sandstone **b**, **c** and bituminous sands (**d**, artificial cemented) under plane polarized microscope. *QZ* quartzose sands; *O* quartz overgrowth (siliceous cement); *B* solid bitumen

(Hezhangzi) in thickness, and its bitumen and silica contents show as a complementary relationship, resulting in an obvious disparity in bituminous enrichment level, e.g., the bituminous sands in the Hezhangzi appear as greyish-black or black color, loose and fragile, uncemented or only with trace siliceous cements (Plate 12.1d), while the basal sandstone in the Mashilingou and Songzhangzi shows palely-red color, tight and rigid without bitumen or only with few bituminous fillings observed under microscope (Plate 12.1a).

12.3.2 *Shuangdong Fossil-Oil-Reservoir*

The Shuangdong fossil-oil-reservoir is situated in the Pingquan City, Hebei Province. Its tectonic location is referred to the Shuangdong Brachyanticline at the northeast end of Pingquan Anticline Belt (Fig. 12.1). Its surface structure is composed of

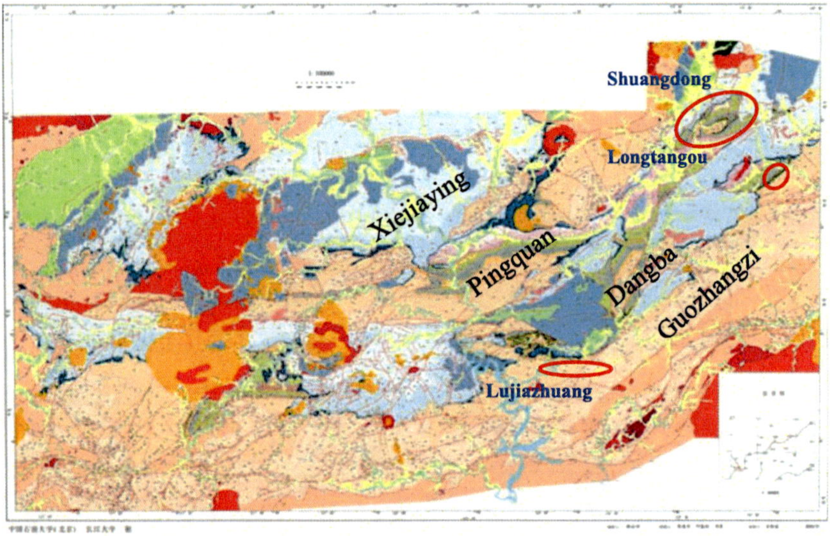

Fig. 12.1 Distribution of tectonlc units and fossil-oil-reservoirs in Jibei depression

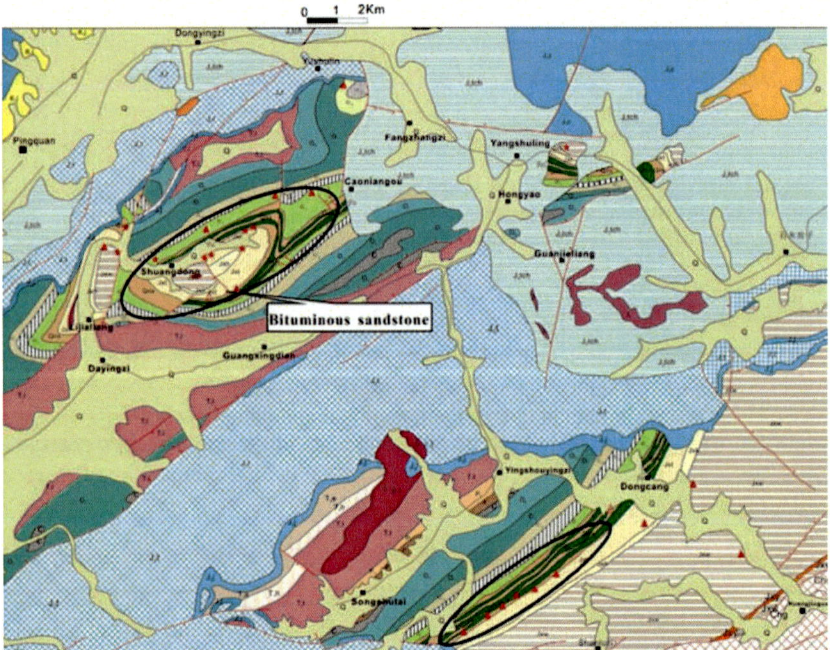

Fig. 12.2 Distribution of Longtangou and Shuangdong fossil-oil-reservoirs in Jibei depression[1]. ★: Liquid oil-seepage outcrop sites; ▲: Solid bitumen outcrop sites

the Wumishan Formation to Middle Ordovician strata. Oil-seep is wide-spread on its outcrop, especially oozed "live" oil-seep or immoveable viscous oil and associated solid asphalt found from carbonate fractures of the Jixianian Wumishan and Tieling Formations as well as the Lower Cambrian Fujunshan Formation on artificial outcrops (such as fluorite adits and opencast stone pits). This kind of oil-seep has totally been discovered up to 30 outcropping sites composing an anticline type of fossil-oil-reservoir (Fig. 12.2).

The loose Xiamaling basal bituminous sandstone is also found at the south limb of main anticline high, which has little cement with siliceous cement content of 1.5–2.5%, muddy content 1–2.5% and bitumen content 5–8%.

12.3.3 Lujiazhuang Fossil-Oil-Reservoir

The Lujiazhuang fossil-oil-reservoir is located in Kuancheng County, Hebei Province. Just like the Longtangou fossil-oil-reservoir, tectonically it is also situated at the north border of Guozhangzi Monocline Belt (Fig. 12.1). Along with regional stratigraphic strike, the thickness of Xiamaling basal sandstone varies from 0.5 m to 7.67 m, and 8 outcrop sites of bituminous sandstone can be successively traced within 10 km range (Figs. 12.1 and 12.3), it can be referred to a litho-stratigraphic type of fossil-oi-reservoir. As the central part of fossil-oil-reservoir, bituminous sandstone is 4.32 m thick, accounting for 97.9% of the total sandstone thickness, and consists of thick- to medium-bedded bebbly-, coarse- to medium-grained sandstones characterized by loose lithology and good permeability at Lujiazhuang, while bituminous sandstone is 7.67 m thick, accounting for 100% of the total sandstone thickness, and it is dominant by medium- to fine-grained sandstones at Jiqian-2 Well (Fig. 12.3). From the central part to both east and west sides, the basal sandstones become gradually more compact, and the sandstone color changes from greyish-dark, through grey to greyish-white or red.

At the south limb of the Dangba Syncline Belt, i.e., the north border of the Guozhangzi Monocline Belt, the occurrence type and bitumen properties of Lujiazhuang fossil-oi-reservoir are identical with the Longtangou fossil-oil-reservoirs. Thus, it can prove a larger distributional range for the potential fossil-oil-reservoir.

12.4 Origin of Bituminous Sandstone and Alteration of Fossil-Oil-Reservoir

12.4.1 Oil Charging Age of Fossil-Oil-Reservoir

Obviously, the basal bituminous sandstone can act as an indicator for fossil-oil-reservoirs, and its distribution spread to the NW and SE limbs of the Dangba Syncline

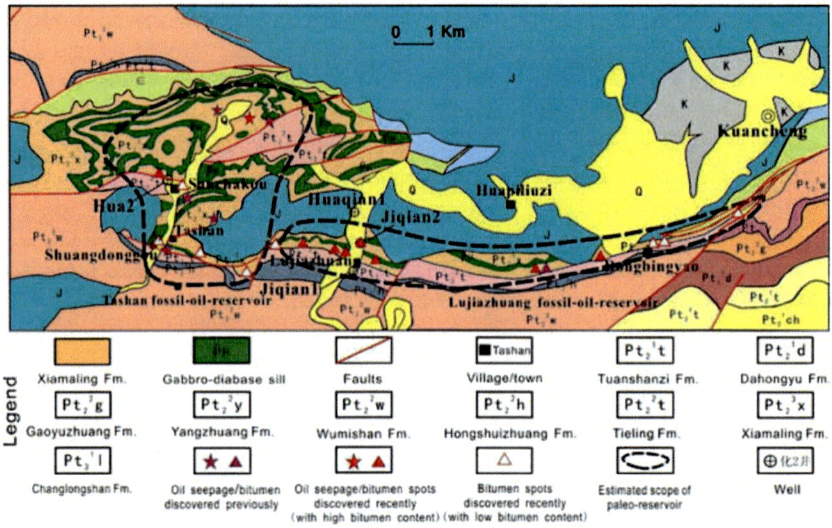

Fig. 12.3 Distribution of Lujiazhuang fossil-oil-reservoir in Jibei depression (Liu et al. 2011)

Belt (Fig. 12.1). Laterally its outcrop can be continuously traced as long as 5 km (Longtangou, Fig. 12.2) to 10 km (Lujiazhuang, Fig. 12.3), while vertically the thickness of bituminous sandstone is up to 3.80 m (Longtangou field section, Fig. 12.4) to 7.67 m (Jiqian-2 Well in Fig. 12.3).

It is noteworthy that the paragenetic occurrence of uncemented fresh bituminous sands (Plate 12.1d) and tightly cemented bituminous sandstone (Plate 12.1b) has been observed at both Longtangou and Lujiazhuang, which indicate that the liquid oil has partially been charged into the uncemented basal sand bed at the early diagenesis stage, its cementation process was still unfinished yet, and then the liquid oil could displaced primary pore water within the sand bed, stopped the precipitation and cementation of neogenic minerals. Thus, the cementation process of sand bed would be abruptly terminated, resulting in the coexisting of oil sands and oil-bearing sandstone, which would be the precursors of fossil-oil-reservoir in Xiamaling basal sandstone, Therefore, the bituminous sands can also act as the indicator to dating the oil initial charging and entrapment ages, which should be near the basal boundary age of the Xiamaling Formation i.e., ca. 1400 Ma (Table 12.1).

[1] Zhong NN, Zhang ZH, Huang ZL et al (2010) Meso-neoproterozoic thermo-evolution, hydrocarbon generation and charging histories (internal report). China University of Petroleum-Beijing, Beijing (in Chinese), modified.

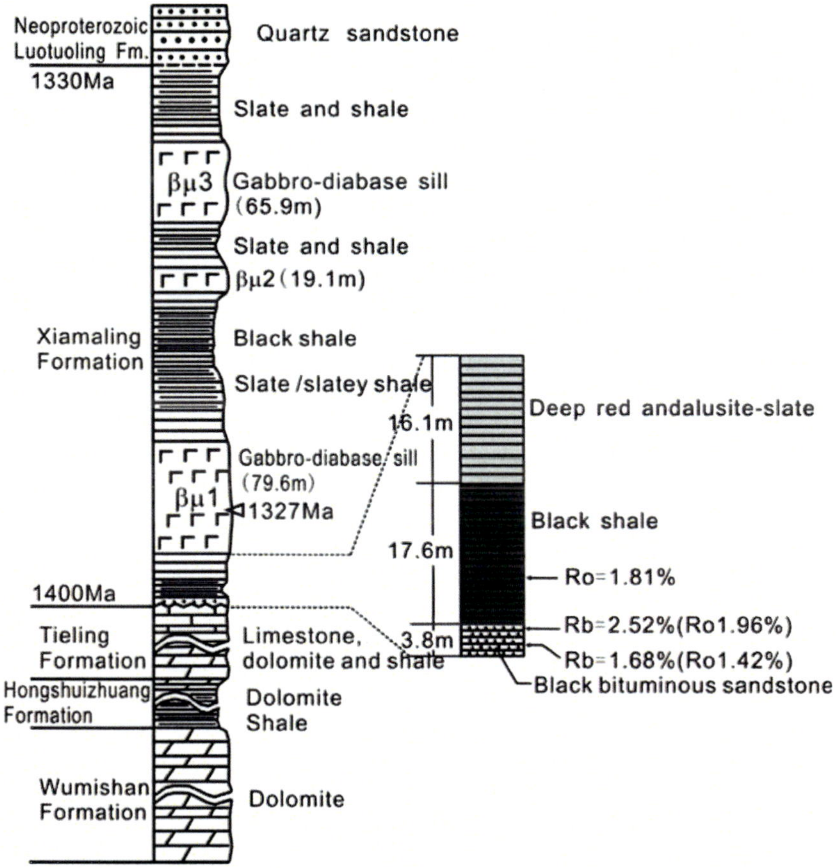

Fig. 12.4 Occurrence of the Xiamaling gabbro-diabase sills and basal bituminous sandstone at Longtangou outcrop section, Lingyuan[2] (Wang and Simoneit 1995, modified)

12.4.2 Emplacement Age and Wall-Rock Alteration Zones of Gabbro-Diabase Sills

12.4.2.1 Occurrence of Wall-Rock Alteration Zones

There are 1–4 gabbro-diabase sills interbedded within the Xiamaling shale in the Jibei Depression, which are named βμ1–βμ4 sills in ascending order. According to the

[2] Wang TG, Zhong NN, Zhu SX et al (2009) Petroleum prospectivity and regional play predication of the lower stratigraphic assemblage in North China platform (internal report). China University of Petroleum-Beijing, Beijing (in Chinese), modified.

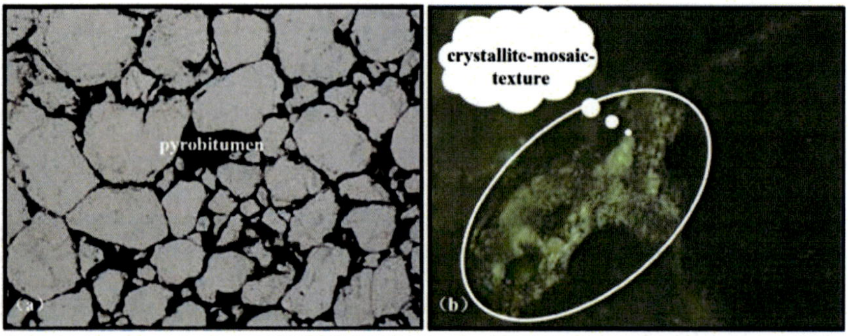

Plate 12.2 Bituminous sandstone and its crystallite-mosaic-texture (CMT) of the Xiamaling basal sandstone in Xiamaling Formation[3]. **a** Bituminous sandstone with bitumen 15% under plane-polarized microscope, ×40; **b** Reservoir pyrobitumen under cross-polarized microscope with oil immersion objective, ×500

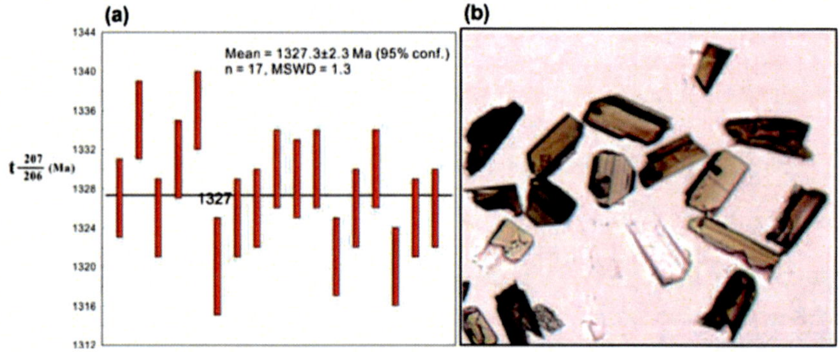

Plate 12.3 Baddeleyite in the Xiamaling gabbro-diabase sills, Jibei depression (**b**) and its U–Pb isotopic age by model IMS-1280 Cameca high resolution secondary ion mass spectrometry (**a**)[4] (cf. Chap. 10)

statistics of 11 field stratigraphic sections plus Jiqian-2 Well drilling profile, the individual sill is ca. 13.3–143.5 m-thick with βμ1 as the thickest one (more than 63.5 m-thick) and the cumulative thickness of sills up to 117.5–312.3 m, accounting for half stratigraphic thickness in the Xiamaling Formation[5] (ca. 43–62%; cf. Chap. 10).

---

[3] Zhong NN, Zhang ZH, Huang ZL et al (2010) Meso-neoproterozoic thermo-evolution, hydrocarbon generation and charging histories (internal report). China University of Petroleum-Beijing, Beijing (in Chinese).

[4] Wang TG, Zhong NN, Zhu SX et al (2009) Petroleum prosectivity and regional play predication of the lower stratigraphic assemblage in North China platform (internal report). China University of Petroleum-Beijing, Beijing (in Chinese).

[5] Wang TG, Zhong NN, Zhu SX et al (2009) Petroleum prospectivity and regional play predication of the lower stratigraphic assemblage in North China platform (internal report). China University of Petroleum-Beijing, Beijing (in Chinese).

Taking the Longtangou section as an example, there only are $\beta\mu 1$–$\beta\mu 3$ sills in the Xiamaling Formation, each sill contains obvious wall-rock alteration zones respectively on the roof- and at the foot-walls (Fig. 12.4). $\beta\mu 1$ has the thickest alteration zones. Its roof-wall is up to 60 m-thick with weak and uneven alteration rank and a coexisted occurrence of slate and carbonaceous shale, while its foot-wall only 16.1 m thick with strong alteration rank, in which, the original shale has been altered into crimson andalusite-slate. Its underlying strata are 17.6 m thick dark-grey shale and 3.8 m thick basal bituminous sandstone (Fig. 12.4). Obviously, the thermo-optical alteration of $\beta\mu 1$ can transcends bottom wall-rock alteration zone plus dark-grey shale, resulting in the alteration from the basal oil-sands to bituminous sandstone (Fig. 12.4), even no liquid oil-seep has been found at the underlying Tieling limestone, only solid bitumen fillings in the limestone sutures could be seen under microscope.

12.4.2.2 Significance of Bitumen Reflectance R_b

The Xiamaling basal bituminous sandstone was sampled from three outcrop sections of the Longtangou fossil-oil-reservoir, and the bitumen reflectance R_b was measured for maturity evaluation (Table 12.2). As shown in Table 12.2, the solid bitumen in the bituminous sandstone samples of Tuzishan section can be distinguished into two R_b distributional ranges, indicating two remarkably different thermo-maturity phases of solid bitumen components (SBCs), i.e., One early SBC with much higher R_b values 1.68–2.52% for over-matured phase, another late SBC with lower R_b values 0.81–1.01% for mature phase which is corresponding to the range of oil-generating window. In Longtangou and Hezhangzi sections, however, the SBC of basal bituminous sandstone only shows R_b values of 1.98–2.39%, which is well corresponding to that of the early SBC, but no late SBC by comparison to the Tuzishan basal bituminous sandstone (Table 12.2). All the five sandstone samples listed in Table 12.2 contain the early SBC but only two of them (i.e., L1 and L2) have the late SBC. Moreover, there totally are up to 116 measured points for the early SBC, while only 52 points for the late SBC, the early ones would be 2.2 times more than the late ones. Thus, it is assumed that the fossil-oil-reservoir is dominant by the early SBC, while the late SBC would be contaminated by and superimposed on the early SBC.

As for the early SBC, the R_b values at the upper position of basal bituminous sandstone (1.8 m beneath the top of basal sandstone) are up to 2.52%, while the values at the lower position (2.9 m beneath the top) only accounts for 1.68% in the Tuzishan section, in between the stratigraphic interval is just 1.1 m-thick, the R_b distribution actually appears as an inverse thermal anomaly, i.e., "the high R_b value at the upper, while the low R_b value at the lower" (Table 12.2). In this case, there must exist a "local heater" at the overlying stratum. In fact, there really is a 79.6 m-thick $\beta\mu 1$ sill overlaid on the Xiamaling basal sandstone in Longtangou with the stratigraphic interval of 33.1 m-thick, which could serve as a "local heater" resulting in an inverse and stronger thermo-optical alteration at the upper sandstone than that at the lower one (Fig. 12.4).

Table 12.2 Distinction of two phases of solid bitumen components (SBCs) within the Xiamaling basal bituminous sandstone in Longtangou, Lingyuan

Sample No.	Lithology	Section	Sampling position	SBC phases	Measured bitumen reflectance, R_b (%)[a]	Mosaic crystallite texture
L2	Black bituminous quartz sandstone (basal sandstone)	Tuzishan	1.8 m beneath the top of basal sandstone	Early	2.52 (16)	Numerous
				Late	0.81 (35)	None
L1			2.9 m beneath the top of basal sandstone	Early	1.68 (17)	A little
				Late	1.01 (17)	None
LY-LT-17		Longtangou	0.5 m beneath the top of basal sandstone	Early	2.39 (30)	Numerous
LY-LT-16			2.0 m beneath the top basal sandstone	Early	2.22 (35)	Numerous
LY-2-1		Hezhangzi	0.2 m beneath the top of basal sandstone	Early	2.38 (11)	Numerous
				Late	1.98 (7)	None

[a]Numbers in the parentheses represent the measured points of solid bitumen under microscope

However, the R_b values of late SBC in the Tuzishan section vary from 1.01 to 0.81% in ascending order, which is a normal R_b variation, e.g., "the high R_b value at the lower and the low R_b value at the upper", without the effect of βμ1 "local heater". Therefore, the formation of late SBC in Tuzishan should be after the magma cooling and the βμ1 sill consolidation (Table 12.2).

12.4.2.3 Oil Coke and Its Significance

Under cross polarized microscope, different quantity of mosaic-crystallite-texture (MCT) can be observed in the early SBC (Plate 12.2), in which, the SBC of high R_b values (up to 2.22–2.52%) contain numerous MCT. Whereas the one of lower R_b values (respectively 1.68% and 1.98%) only a little, even no MCT, i.e., all the late SBC appears as complete extinction, no MCT could be found (Table 12.2).

The continued thermo-optical alteration of magma results in the cracking and gasification of aliphatic fraction and the polymerization of polar fractions (aromatics, N, S, O-compounds and asphaltene), which can be referred to the disproporanation reaction. While both cracking gas (methane) reservoir and reservoir bitumen/pyrobitumen would be the terminal products of disproporanation reaction. Since hydrogen atom has been drastically lost and carbon atom gradually plymerized into MCT, consequently the reservoir bitumen/pyrobitumen would become fossil-oil-reservoir, even natural oil coke, which can be observed under cross polarizing microscope (Plate 12.2). Natural oil coke would be the high-rank product of pyrobitumen in the over-mature phase, which could be a significant proof for the thermo-optical alteration of fossil-oil-reservoir.

In general, the early oil charging and entrapment process in the Xiamaling fossil-oil-reservoir had certainly occurred before the emplacement of gabbro-diabase sills, the sill intrusion resulted in the alteration of fossil-oil-reservoir and the emerging of MCT in the early SBC. While the late SBC without MCT should be attributed to the product of late oil filling into and impregnatiing on the fossil-oil-reservoir.

The fossil-oil-reservoir of Xiamaling bituminous sandstone was mainly contributed by the early charged oils, i.e., quantitatively the early SBC predominate over the late one, but the early SBF is almost exhausted by thermo-alteration. Although the contribution of late SBC to the fossil-oil-reservoir would be very limited, the late SBF content is in far excess of the early SBF.

12.4.2.4 Emplacement Age of Gabbro-Diabase Sills

The fresh meso-crystalline gabbro in βμ1 and βμ3 sills have been found and sampled respectively from the Jiqian-2 Well (βμ1 drilling core, Fig. 12.3) and the Shuangdong field section (βμ3 outcrop rock, Fig. 12.2), from which totally 960 grains of baddeleyites crystals are identified and picked up (Plate 12.3b).

Using Model IMS-1280 Cameca high resolution secondary ion mass spectroscopy (SIMS) in the Institute of Geology and Geophysics, Chinese Academy of Sciences, the U–Pb dating of 36 baddeleyitic crystals was performed. A corresponding magmatic emplacement age 1327 Ma of the βμ1 and βμ3 gabbro-diabase sills was acquired[6] (Plate 12.3a; cf. Chap. 10), which indicate that the multiple gabbro-diabase sills in Xiamaling Formation are also attributed to the identical origin of coeval magmatic intrusion, this magmatic emplacement age should also be the wall-rock alteration age of Xiamaling basal bituminous sandstone.

To sum up, the starting age of early oil charging and entrapment in the Xiamaling basal sandstone should be at 1400 Ma, while the magmatic emplacement resulted in wall-rock alteration, the alteration age is dated at 1327 Ma. Hence, the actual existing and sustained time of original oil-reservoir in the Xiamaling basal sandstone would be only 72.5 Ma (i.e., from 1400 to 1327 Ma). Based on this alteration age, it could be determined that the fossil-oil-reservoir of Xiamaling basal sandstone in Jibei Depression is probably the oldest fossil-oil-reservoir in the world.

[6] Wang TG, Zhong NN, Zhu SX et al (2009) Petroleum prospectivity and regional play predication of the lower stratigraphic assemblage in North China platform (internal report). China University of Petroleum-Beijing, Beijing (in Chinese).

12.4.2.5 Two Phases of Oil Charging Before and After Magmatic Emplacement

Gabbro-diabase sills of the Xiamaling Formation are wide-spread in Jibei Depression, resulting in the high temperature thermo-optical alteration of Xiamaling shale, and impelling the original oil sands/sandstone to be altered into bituminous sands/sandstone at wall-rock alteration zone, the black viscous oil-seep, with occurrences of penetrating bedding (Fig. 12.5a) and parallel bedding (Fig. 12.5b), is also found. Obviously, it was only after magma cooling and sill consolidation that a late oil charging and impregnating can result in the viscous oil-seep. Therefore, these viscous oil would provide an important evidence of late oil charging, it is the late oil charging that can serve as the material source for the later SOC and its SBF.

12.5 Hydrocarbon Source of the Fossil-Oil-Reservoir

12.5.1 Effective Source Beds in the Jibei Depression

There are three sets of Mesoproterozoic source beds respectively within Gaoyuzhuang, Hongshuizhuang and Xiamaling Formations in the YFDZ (Table 12.1). As for the Jibei Depression, only the Hongshuizhuang black shale and Gaoyuzhuang micrite-dolostone could be the effective source beds.

Taking total organic content (TOC) of 0.5% as the lower limit of effective source rock, the content of corresponding soluble bitumen fraction (SBF, i.e., chloroform-extractable bitumen) is up to 340–4510 ppm (avg. 265 ppm) for the Hongshuizhuang black shale, and appearing as high organic abundance features with an equivalent vitrinite reflectance eqR_o value from 0.90 to 1.42% (avg. 1.19%), which are basically within the phase of "oil-generating window". Therefore, the black shale should be attributed to an effective mature to high mature source bed.

While the black or greyish-black micrite-dolostone is up to 164 m-thick, and also has a high TOC value (0.5–4.29%, avg. 1.16%) and a low SBF (26–152 ppm, avg. 63 ppm) in the Gaoyuzhuang Formation. Its equivalent vitrinite reflectance eqR_o value up to 1.38–1.75% (avg. 1.59%), occurring as the feature of high organic abundance and over-mature source bed.

Due to the influence of regional "Yuxian Uplifting", however, most of the Xiamaling Formation strata have been denuded, even the best shale source rocks of the Xia-3 Member have not survived the denudation as well, only the interval of Xia-1 Member has remained. In addition, 2–4 gabbro-diabase sills are wide-spread and interbedded within the Xiamaling Formation (Figs. 12.3, 12.4 and 12.5), the cumulative thickness of sills could account for 43–63% of the total Xiamaling thickness. The muddy roof and bottom wall-rocks have been altered into slate and/or hornfel by high temperature thermo-optical alteration, and the wall-rock alteration zones is ca. 16–60 m-thick. Therefore, the hydrocarbon potential of remained Xiamaling black

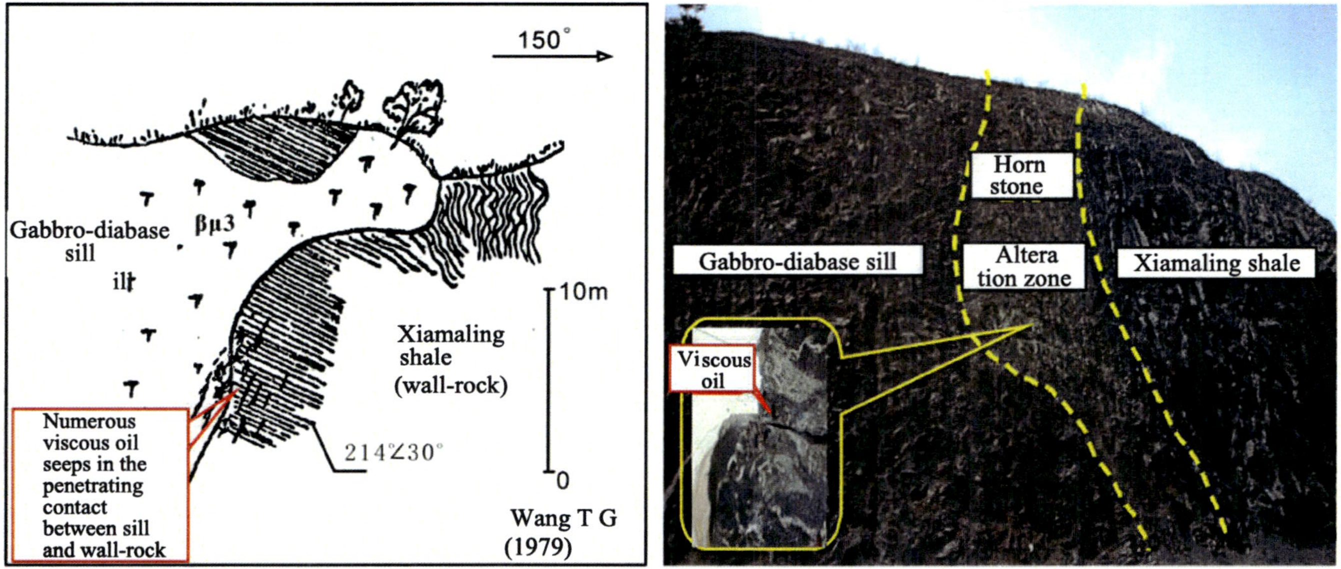

Fig. 12.5 Occurrences of penetrating and parallel beddings of viscous oil-seep within the Xiamaling wall-rock alteration zones. **a** Penetrating bedding, at Dadonggou, Kuancheng (outcrop sketch; Wang TG, Gao ZZ, Liu HB et al (1979) Principal characteristics of petroleum geology in the eastern segment of Yansan region (internal report). In: The 3rd regional geological explorational team. Jianghan Institute of Petroleum, Jingzhou (in Chinese)); **b** Parallel bedding, at Sanchakou, Xinglong (photograph; Zhong NN, Zhang ZH, Huang ZL et al (2010) Meso-neoproterozoic thermo-evolution, hydrocarbon generation and charging histories (internal report). China University of Petroleum-Beijing, Beijing (in Chinese))

shale has already been lost and could not serve for effective source bed in the Jibei Depression.

12.5.2 Hydrocarbon Source Correlation

12.5.2.1 Hydrocarbon-Generating Threshold Depth

Since there only are two Mesoproterozoic effective source beds, i.e., Gaoyuzhuang and Hongshuizhuang Formations (Tables 12.1 and 12.3), the key moment for oil to charge into the Xiamaling basal sandstone should be at 1400 Ma, when the buried depth of Hongshuizhuang black shale was just 312.8 m at most, which would be far away from the hydrocarbon—generating threshold depth, the Hongshuizhuang source bed was still within immature phase and impossible to supply oil for the Xiamaling basal sandstone. Therefore, the Hongshuizhuang black shale could not be the effective source bed for early original oil-reservoir in the basal sandstone.

At the key moment of 1400 Ma, however, the stratigraphic interval is up to 3582 m-thick between the Gaoyuzhuang and Xiamaling Formations (Tables 12.1 and 12.3), the Gaoyuzhuang source bed should already reach its hydrocarbon threshold depth. Hence, based on the above threshold depth analysis, the Gaoyuzhuang black micrite-dolostone should contain the conditions to supply oil for the Xiamaling basal sandstone and serve as an effective source bed for its original oil-reservoir.

Using the approximate value 3600 m, in the stead of the Gaoyuzhuang buried depth 3582 m, as the hydrocarbon threshold depth for the Hongshuizhuang source beds in Jibei Depression, the age could be calculated according to its overlain stratigraphic thickness for Hongshuizhuang source bed to be buried up to 3600 m deep, or rather to reach its hydrocarbon threshold depth.

In view of the Pt_2^4 stratigraphic hiatus (time scope 1327–1400 Ma) during the late Mesoproterozoic, the related regional uplifting period would be continued as long as 327 Ma in the YFDZ (Table 12.1). As for the Pt_3^2 Nanhuan and Pt_3^3 Sinian hiatus (800–543 Ma) in late Neoproterozoic, the corresponding uplifting period will be 257 Ma (Table 12.1). Once more, the Middle Ordovician to Early Carboniferous hiatus (465–310 Ma), its uplifting period is 155 Ma (Table 12.1). The YFDZ could have a totally regional uplifting period up to 739 Ma from the late Mesoproterozoic to the Middle Palaeozoic, and short of the corresponding sediments. In this case, therefore, the age to reach its threshold depth 3600 m should be delayed until the Mesozoic Era for the Hongshuizhuang source bed. This is most probably the main reason why the indigenous liquid oil could be well preserved in the very old Mesoproterozoic sequence (Table 12.3).

Table 12.3 Buried depth of two sets of the Mesoproterozoic source beds at the early entrapment age 1400 Ma for the Xiamaling fossil-oil-reservoir, Jibei Depression

Stratan units			Tieling Formation	Hongshuizhuang Formation	Wumishan Formation	Yangzhuang Formation	Gaoyuzhuang Formation
Thickness (m)			211.1	102.7	2947.2	322.4	939.6
Buried depth (m)	Hongshuizhuang Formation	Top	211.1	–	–		
		Bottom	313.8		–		
	Gaoyuzhuang Formation	Top	3582.4				–
		Bottom	4523.0				

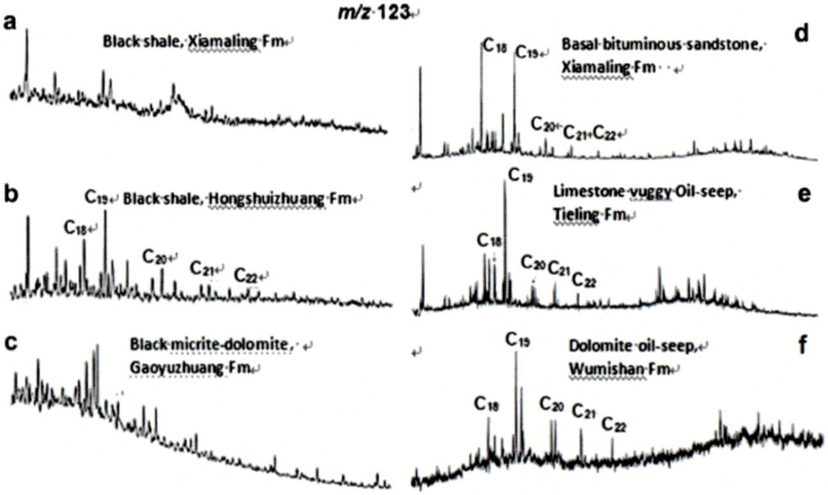

Fig. 12.6 The *m/z* 123 mass chromatographs for the Mesoproterozoic oil-seep-source rock correlation[7]. **a–c** Kerogen-degraded product by artificially catalytical hydrocracking for source rock; **d–f** aliphatic fraction of oil-seepes and bituminous sandstone. C_{18}–C_{23} the 13α(*n*-alkyl)-tricyclic terpanes

12.5.2.2 Hydrocarbon-Source Correlations of Oil-Seep and Bitumen

Wang (1991a; b) first reported a novel biomarker series of C_{18}–C_{23} 13α(*n*-alkyl)-tricyclic terpanes in the SBF (i.e., chloroform-extract) of the Xiamaling basal bituminous sandstone, Jibei Depression (Fig. 12.6d). The biomarker series was subsequently detected from the Hongshuizhuang black shale and the oil-seep/asphalt of Wumishan and Tieling Formations in Jibei and Xuanlong Depressions (cf. Fig. 12.6e, f; Zhang et al. 2007; Wang et al. 2011) as well as from the asphalt veins in the Longmenshan region, northwest Yangtze Craton (cf. Chap. 15). Since the Longmenshan asphalt veins are derived from the black shale of Sinian Doushantuo Formation (cf. Chap. 15), so far all these reports of 13α(*n*-alkyl)-tricyclic terpanes seem only to be limited within the Meso-Neoproterozoic sourced hydrocarbons. Therefore, it is inferred that the 13α(*n*-alkyl)-tricyclic terpanes could serve as the distinctive biomarkers for Meso-Neoproterozoic hydrocarbons.

Kerogen had been separated and purified from above-mentioned three potential source beds, i.e., the Gaoyuzhuang black micrite-dolostone, and the Hongshuizhuang and Xiamaling black shales. By means of artificially catalytical hydrocracking pretreatments and GC–MS analysis, the kerogen-degraded products of above source rocks were obtained, and the biomarker series of 13α(*n*-alkyl)-tricyclic terpanes was only detected from the kerogen-degraded product of the Hongshuizhuang black

[7] Zhong NN, Zhang ZH, Huang ZL et al (2010) Meso-neoproterozoic thermo-evolution, hydrocarbon generation and charging histories (internal report). China University of Petroleum-Beijing, Beijing (in Chinese).

shale (Fig. 12.6b), however, none of which was found in the kerogen products of Gaoyuzhuang and Xiamaling source rocks (Fig. 12.6a, c).

Moreover, the 13α(*n*-alkyl)-tricyclic terpanes are also detected from the Wumishan and Tieling oil-seeps and the SBF of Xiamaling basal bituminous sandstone (Fig. 12.6d–f), which could be well correlated with the Hongshuizhuang black shale, but irrelative to the Gaoyuzhuang and Xiamaling source beds (Fig. 12.6). In this case, the series of 13α(*n*-alkyl)-tricyclic terpene series can be the specific biomarker for the Hongshuizhuang source bed in Jibei Depression. Therefore, it is verified that the Mesoproterozoic liquid oil-seep and the SBF of Xiamaling bituminous sandstone must have been derived from the Hongshuizhuang black shale source bed.

12.5.2.3 Two Phases of Solid Bitumen Components (SBC)

As discussed above, the early charged oil in the Xiamaling basal fossil-oil-reservoir was supplied from the Gaoyuzhaung source bed. At the key moment of 1327 Ma, the magmatic intrusion into the Xiamaling basal sandstone resulted in the thermo-alteration of fossil-oil-reservoir, oil sands or oil-containing sandstone was altered into over-mature bituminous sands/sandstone with predominant early SBC and trace early soluble bitumen fraction (SBF).

Until the Mesozoic Era, deeply buried Hongshuizhuang black shale could reach and overstep its hydrocarbon threshold depth (ca. 3600 m), impelling the late matured oil generation, migration and entrapment, and resulting in the potential oil reservoirs and surface liquid oil-seep in the Wumishan and Tieling Formations as well as the late SBC and associated SBF in the Xiamaling basal bituminous sandstone.

Although the late matured SBC is quantitatively limited, but its associated SBF is quantitatively enough to mask the biomarker information of trace early SBF. Therefore, the SBF of Xiamaling bituminous sandstone shows as oil source features of the Hongshuizhuang black shale based on the above oil-source correlation (Fig. 12.5b, d).

12.5.2.4 Lithosphere Depth and Hydrocarbon Generation Threshold

Xu (1996) has accomplished a study of the lithosphere dynamics in Chinese sedimentary basins based on 5000 detected points of the magnetiotelluric sounding, and plotted an isobath map of the conductive layer within the upper mantle in Chinese mainland (Fig. 12.7). Xu (1996, 2003a, b) interpreted that the upper mantle conductive layer (UMCL) is caused by fusion of the hydrous mantle, and hydrous magmatic melt would increase the mantle conductivity, especially under high temperature and pressure conditions. Moreover, there is a good consistent relationship between the variations of terrestrial heat flow and UMCL depth, the deeper the UMCL, the lower the geotemprature. Therefore, the lithosphere thermo-behavior can be inferred by means of the buried depth of UMCL.

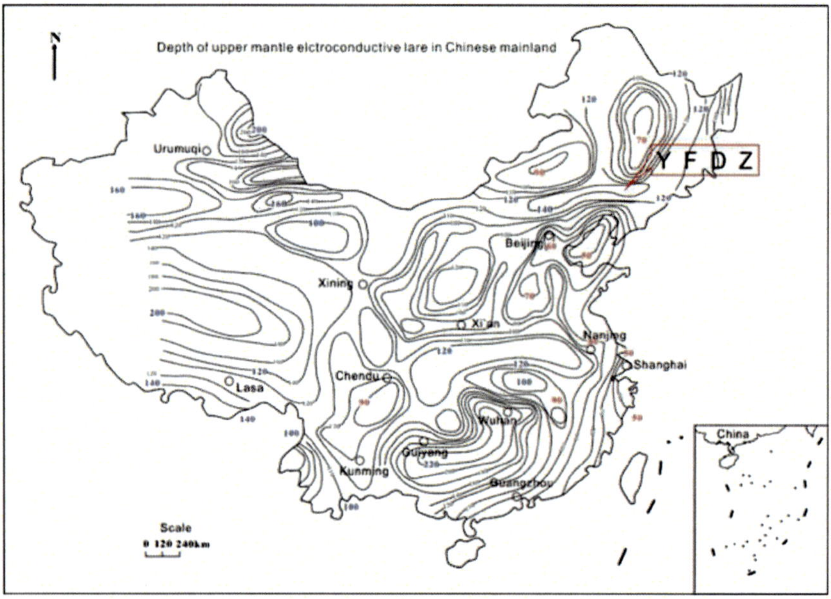

Fig. 12.7 The isobath map of upper mantle conductive layer (UMCL) in China (Xu 1996, modified)

In East China, the buried depths of UMCL for Mesozoic-Cenozoic sedimentary basins are respectively 70–100 km (e.g., northern Songliao Basin, Northeast China) or 50–90 km (e.g., Bohai Bey Basin, North China). As the "hot lithosphere" and "hot basin", the average geothermal gradients are 13–20 °C/km (Songliao) or 15–28 °C/km (Bohai Bay) so that the common hydrocarbon threshold depth would be ca. 1500 m (Songliao) or ca. 2500 m (Bohai Bay), which are not exceed 3000 m at the most.

As an exception, however, the largest buried depth of UMCL has been detected more than 140 km-deep in the northern YFDZ, or rather, at the Xuanlong-Jibei-Liaoxi Depressions. Consequently, these depressions contain the thickest lithosphere and the lowest palaeo- and present-geothermal gradients, their overlaid crust cover has made this area the sole "cool lithosphere" and "cool basin" in the East China (Fig. 12.7; Xu 1996). Since these depressions have the largest buried depth of UMCL and the largest hydrocarbon threshold depth so that the threshold depth of Gaoyuzhuang Formation can be near 3600 m, it should be enough to make the Gaoyuzhuang black micrite-dolostone to be the early effective source bed for the Xiamaling basal fossil-oil-reservoir at the oil entrapment age of 1400 Ma. Furthermore, this threshold depth can be the foundation to stipulate the hydrocarbon threshold depth and oil entrapment age for the Hongshuizhuang black shale source bed as well.

12.6 Conclusions

(1) There are 2 Mesoproterozoic effective source beds, i.e., the over-mature black micrite-dolostone of Gaoyuzhuang Formation with a cumulative thickness of 164 m and the mature black shale of Hongshuizhuang Formation with a cumulative thickness of 60 m, in the Jibei Depression.

(2) Three sites of known fossil-oil-reservoirs have been discovered in the basal bituminous sandstone of Mesoproterozoic Xiamaling Formation respectively at Longtangou (Lingyuan City, Liaoning), Shuangdong (Pingquan City, Hebei) and Lujiazhuang (Kuancheng County, Hebei) in the Jibei Depression. The bituminous sandstone has 2 phases of solid bitumen components (SBCs) with different origins, i.e., the predominant early SBC with bitumen reflectance R_b of 1.68–2.52%, in which the natural cokes with mosaic-crystallite-texture (MCT) can be observed under cross polarizing microscope, and the minor late SBC with R_b value of 0.81–1.01% can be seen only in partial bituminous sandstones and no natural cokes and MCT found, in which the paragenetic and coexisted late soluble bitumen fraction (SBF) appears as a relatively higher content, and conceal the biomarker information of the early SBF.

(3) The Wumishan and Tieling liquid oil-seeps and the late soluble bitumen fraction (SBF) of the Xiamaling basal bituminous sandstone were derived from the Hongshuizhuang source bed based on the hydrocarbon-source correlation, while the solid bitumen component (SBC) of the Xiamaling basal bituminous sandstone was mainly sourced by the Gaoyuzhuang source bed according to the analysis of stratigraphic thickness and hydrocarbon threshold depth.

(4) In Jibei Depression, there are multiple regional gabbro-diabase sills and associated wall-rock alteration zones interbedded within the Xiamaling Formation. In the Xiamaling basal sandstones, the early SBC, oil cokes and viscous oil-seeps of wall-rock alteration zone indicate two episodes of oil migration and charging histories during the Mesoproterozoic Era (1400–1327 Ma).

(5) In the fossil-oil-reservoirs of Xiamaling basal sandstone, the discovery of bituminous sands reveals that the early oil charging and entrapment process would be started at the early diagenetic period of the basal sandstone (ca. 1400 Ma). The emplacement of gabbro-diabase resulted in the early charged oils was altered into the early SBC and even the oil coke, the oil sands/sandstone became bituminous sands/sandstone at 1327 Ma.

(6) Based on the analysis of stratigraphic sequence and thickness, it is assumed that the early charged oil in Xiamaling basal sandstone could be only sourced by the Gaoyuzhuang source bed, its hydrocarbon threshold depth would be ca. 3600 m. The megnetotelluric sounding result reveals that the buried depth of lithosphere is more than 140 km in the Xuanlong-Jibei-Liaoxi Depressions, this region should have a “cool lithosphere” and “cool basin” structure. According to its threshold depth of 3600 m, the oil entrapment age of Hongshuizhuang source bed has be stipulated at the Mesozoic Era.

Acknowledgements The authors thank Academician Ma Zongjin and Senior Researcher Xu Changfang in the Institute of Geology, China Seismological Bureau for providing the UMCL results and helpful discussion. We also wish to thank Prof. Su Li in the China University of Geosciences for her assistance on petrological identification and geochemical study, Academician Li Xianhua in the Institute of Geology and Geophysics, Chinese Academy of Sciences, for his assistance on isotopic dating, Profs. Li Suyuan and Yue Changtao in China University of Petroleum-Beijing for providing the laboratory and experimental guidance on the artificially catalytical hydrocracking of kerogen.

References

Gao LZ, Zhang CH, Shi XY, Song B, Wang ZQ, Liu YM (2008) Mesoproterozoic age for Xiamaling formation in North China plate indicated by SHRIMP dating. Chin Sci Bull 53(17):2665–2671

Li HK, Lu SN, Li HM, Sun LX, Xiang ZQ, Gen JZ, Zhou HY (2009) Zircon and beddeleyite U–Pb precision dating of basic rock sills intruding Xiamaling formation, North China. Geol Bull China 28(10):1396–1404 (in Chinese with English abstract)

Liu Y, Zhong NN, Tian YJ, Qi W, Mu GY (2011) The oldest (1327 Ma) oil accumulation in China: meso-proterozoic Xiamaling fromation bituminous sandstone reservoir. Pet Explor Dev 38(4):503–512

Wang TG (1980) Primary properties of Sinian Suberathem oil-seep and its petroleum geological significance in Yanshan region. Pet Explor Dev 7(2):34–52 (in Chinses)

Wang TG (1991a) A novel tricyclic terpane biomarker series in the Upper Proterozoic bituminous sandstone, eastern Yanshan region. Sci China B 34(4):479–489

Wang TG (1991b) Geochemical characteristics of Longtangou bituminous sandstone, eastern Yanshan region, North China—approach to a Precambrian reservoir bitumen. J SE Asian Earth Sci 120:155–170

Wang TG, Han KY (2011) On meso-neoproterozoic primary petroleum resources. Acta Petrolei Sinica 31(1):1–7 (in Chinese with English abstract)

Wang TG, Simoneit BRT (1995) Tricyclic terpane biomarker series in the Upper Proterozoic bituminous sandstone, eastern Yanshan region, North China. Chem Geol 120:155–170

Wang TG, Huang GH, Xu ZY (1988) Approach to the Proterozoic Xiamaling formation basal sandstone fossil-oil-reservoir. Oil Gas Geol 9(3):479–489 (in Chinese with English abstract)

Wang CJ, Wang M, Xu J, Li YL, Yu Y, Bai J, Dong T, Zhang XY, Xiong XF, Gai HF (2011) 13α(*n*-alkyl)-tricyclic terpanes: a series of biomarkers for the unique microbial mat in the middle Mesoproterozoic (1.45–1.30 Ma) North China Sea. Mineral Mag 75:2–114

Xu CF (1996) The electrical structure of earth crust and mantle and the earthquake distribution (1). Acta Seismol Sinica 18(2):254–261 (in Chinese)

Xu CF (2003a) China continental conductive layers and its relationship with basin evolution. Earth Sci J China Univ Geosci 8(1):16–20 (in Chinese with English abstract)

Xu CF (2003b) The study of lithosphere tectonics and basin formation of Chinese mainland and migration of oil and gas. Earth Sci Front 10(3):115–127 (in Chinese with English abstract)

Zhang SC, Zhang BM, Bian LZ, Jin ZJ, Wang DR, Chen JF (2007) The Xiamaling oil shale generated through Rhodophyta over 800 Ma age. Sci China Ser D Earth Sci 50(4):527–535

Chapter 13
Sinian Gas Prospectivity in the Western Yangtze Craton, Southwest China

Keyou Han, Wei Sun, and Dan Li

Abstract Based on the research results of petroleum geological exploration during the past half century, the Sinian natural gas prospectivity and exploration potential in the western Yangtze Craton (YC) are discussed by means of the analyses of litho-facies and palaeotectonics. The black shale of early Sianian Doushantuo Formation would be a qualified hydrocarbon source bed and the algal dolostone and shoal facies dolorudite of Late Sinian Dengying Formation are high-quality reservoir bed, which can constitute the oldest realistic source-reservoir assemblage in the western Yangtze Craton. Regionally, Dengying Formation can be divided into two subformations: the upper one includes Deng-3 and Deng-4 Members, Deng-3 Member contains bluish-gray shale or clastic rocks, its bottom is disconformably overlaid on the Deng-2 Member; while the lower one consists of Deng-1 and Deng-2 Members. Among which, the Deng-2 and Deng-4 Members are optimal reservoir beds. During the sedimentary period of Lower Cambrian Qiongzhusi (or called Jiulaodong) Formation, black shale was deposited, which could act as a perfect source bed and direct seal bed for the Sinian gas reservoirs. Since the wide-spread Sinian-Lower Cambrian sequence contains superior source beds, well preserved oil/gas reservoirs and successively developed palaeo-uplifts, the geological conditions for large-scale gasfield formation should be available in the western Yangtze Craton. Before or during Triassic, the Chuanzhong palaeo-uplift would be significant for oil entrapment and accumulation of original oil-reservoirs. Before Cenozoic, the gas-reservoir with abnormally high pressure was formed due to the influences of persistently deep-burying, geotemperature rising and the oil-cracking. The oil and gas were redistributed during the compressional uplifting course. Finally, the present gas reservoirs are formed. The key points for the future petroleum exploration of Dengying Formation should be focused on the Chuanzhong, Huayingshan and Tianjingshan palaeo-uplifts.

K. Han (✉) · D. Li
Institute of Petroleum Exploration and Development, PetroChina Southwest Oil and Gasfield Company, CNPC, Chengdu 610051, China

W. Sun
Chengdu University of Technology, Chengdu 610059, China

T. Wang, *Meso-Neoproterozoic Geology and Petroleum Resources in China*, Springer Geology, https://doi.org/10.1007/978-981-19-5666-9_13

Keywords Western Yangtze Craton · Chuanzhong Uplift · Palaeo-uplift · Sinian · Doushantuo Formation · Dengying Formation

13.1 Introduction

Since the discovery of Weiyuan Gasfield in 1964, petroleum geologists never stop to probe the Sinian petroleum prospectivity and exploratory potential in the western Yangtze Craton (YC). Besides the exploration and development of Weiyuan Gasfield, a series of petroleum geological survey, seismic exploration, deep exploratory well drilling and relevant investigation were deployed in the vast western Yangtze Craton. Consequently, certain knowledge on the Neoproterozoic gas-bearing geological conditions in western YC have been obtained, especially for the optimistic evaluation on the gas prospectivity in Chuanzhong Uplift (used to call it "Leshanlongnvsi Uplift"). Recently, gas exploration has made breakthrough on the uplift, and a high yield Anyue Gasfield, with large gas-bearing area, be found during 2011–2012. However, in view of the facts that this large gasfield located on the palaeo-uplifts of Jinning Period (720 Ma), lacking of Nanhuan deposition, and the lower Sinian Doushantuo Formation was not well developed in Chuanzhong Uplift, only few drilled exploratory wells have penetrated the upper Sinian Dengying Formation, some deep seismic data are not satisfactory, so far the lower Sinian and its underlying strata are still poorly understood in the western YC. In order to obtaining a general idea, the authors try to discuss and forecast the Sinian gas prospectivity in western YC from a broader point of view in combination with the exploration results and geological data.

13.2 Outline and Progress of Sinian Gas Exploration

13.2.1 Exploratory Progress

For decades, the Sinian petroleum exploration were mainly deployed at Sichuan Province and Chongqing City in the western YC, but some exploratory wells have been drilled in west and central Hubei, south Shanxi, central Guizhou and other provinces in southwest China. Sun (2008) and Du et al. (2015) have summarized above exploratory results as follows.

(1) From 1964 on, totally 108 gas wells have been drilled and obtaining the geological reserves 408.61×10^8 $m^3$of natural gas in Weiyuan Gasfield.
(2) Totally 12 exploratory wells were drilled respectively at 6 anticlines/blocks on the Chuanzhong Uplift. Among which, A stratigraphic deep-well, Nvji Well, at Longnvsi Anticline in the east of the uplift and an exploratory well, Heshen-1

Well at Hebaochang Anticline on the south of the uplift have obtained commercial gas flows, the Zi-1 Well at a high position of Deyang-Ziyang Faulted-Sag was acquired a low-output gas–water reservoir, and the Anping-1 Well at Anpingdian Anticline has gas show, whereas two exploratory wells (Gaoshi-1 Well and Moxi-8 Well) discovered a large high yield Anyue Gasfield, which has confirmed the Chuanzhong Uplift to be a favourable gas-productive region.

(3) Unfortunately, numerous exploratory wells had been deployed at some geological structures, such as Ziliujing, Tiangongtang, Laolongba, Panlongchang, Hanwangchang and Dawoding surrounding Chuanzhong Uplift, all of which only produced water. Thus it is clear that the gas preservational conditions are not perfect in the surrounding region.

(4) At the periphery of western YC, all the exploratory wells produced fresh water, including 3 wells at the anticlines, i.e., Zengjiahe at Guangyuan and Dalianghui at Nanjiang in northern Sichuan, Nanshanling at Ningqiang in southern Shaanxi on the north fringe of current Sichuan Basin; and 2 wells at the Changning Anticline in southern Sichuan on the south fringe as well as 4 wells at the anticlines, i.e., Dinshan and Lintanchang in southeast Chongqing, Yupize in eastern Sichuan, Lichuan in western Hubei on the southeast fringe, resulting in all the gas exploration failure due to the poor preservational conditions so that no gas residue.

To sum up, as concerns natural gas, the gas preservation conditions of above-mentioned local structures are not good enough, only the Chuanzhong Uplift contained the best exploratory result in the western YC, because it was a successional palaeo-uplift for 635 Ma since the early Sinian on, and experienced crustal oscillation in Early Palaeozoic, denudation in Late Palaeozoic, uplifting in Mesozoic and finally stratigraphic folding in Cenozoic. The specific geological evolutional history leads to its favorable conditions for oil and gas enrichment, and the well preservation conditions of thick Middle–Lower Triassic gypsum-salt seal beds overlying the Sinian strata, resulting in the present features of giant petroliferous uplift and the excellent gas exploratory effect (Fig. 13.1).

13.2.2 Major Gas Exploratory Results on Chuanzhong Uplift

13.2.2.1 Weiyuan Gasfield

Tectonically, it is the largest anticline in the southwest of the Chuanzhong Uplift, with a structural trap area of 800 km^2 and closure 800 m on the top surface of Sinian Dengying Formation. Its first exploratory well, Wei-1 Well, was drilled according to gas-seep information as far back as 1938, and the well was finished without any result until 1942. In 1956, a stratigraphic well, named Weiji Well, was drilled at a new site near the surface anticlinal high point, only 18 m apart from the original site of Wei-1 Well (Figs. 13.1 and 13.2), but finally the drilling had to be ended at the

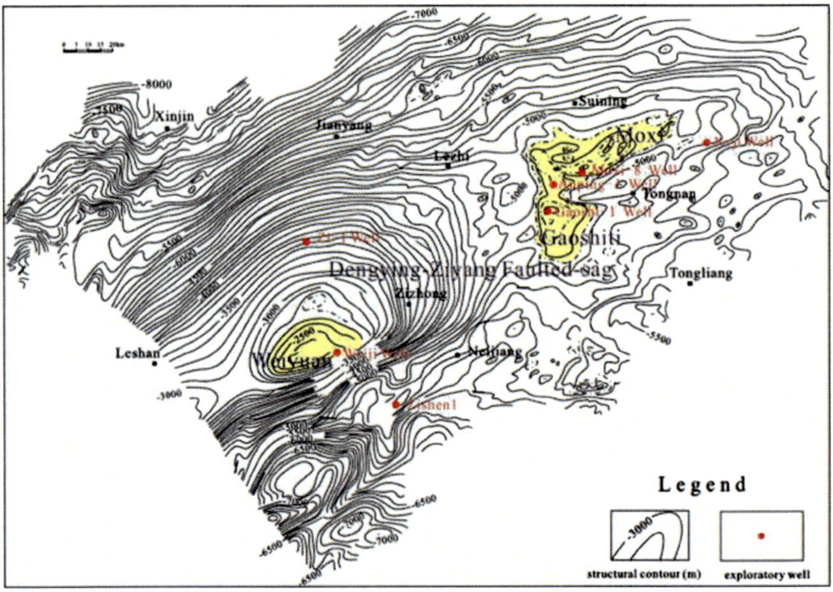

Fig. 13.1 Seismic reflector structural contour map of Sinian top surface in Chuanzhong Uplift, western YC

Lower Cambrian interval due to the drilling rig limitation. Moreover, the Weiji Well was deepened again in 1964, and then mud loss happened in the dolostone interval of Sinian Dengying Formation, and drill stem test (DST) caused gas blowout with gas output of 14.5×10^4 m^3/d. Consequently, a bottom-water gas reservoir was discovered in Deng-2 Member. In the same year, Wei-2 Well was drilled 8.5 km apart from the Wei-1 Well at the same high point on subsurface seismic structural map, and gained a high output gas flow of 74.5×10^4 m^3/d in the well interval of 2835.5–3005 m. Which has confirmed the Weiyuan Anticline as a medium-sized Sinian Gasfield. After further exploration, totally 108 exploratory wells have been drilled on Weiyuan Anticline, among which 88 gas-producing wells are successfully obtained, and the gas production of 72 gas wells are higher than 10^4 m^3/d (cf. Fig. 13.2 for partial well sites), the total proven gas-containing area would be up to 216 km^2 with gas column height of 244 m and gas fillness level of 25%. It's originally proven gas geological reserve is 408.61×10^8 m^3.

As the largest gasfield with single-reservoir and the first old Neoproterozoic gasfield in China, Weiyuan Gasfield has been developed since 1968, its gas producing capacity reached the production peak with gas output 11.6×10^8 m^3/a in 1976, and the cumulative gas production was up to 145.94×10^8 m^3 with gas recovery coefficient 38% until 2001. Owing to reasons of producible resources exhausting and economic efficiency, however, Weiyuan Gasfield was entirely shut down with remanent recoverable reserve 1.88×10^8 m^3 in 2001. So far only few gas wells are still working for the production of rare helium gas.

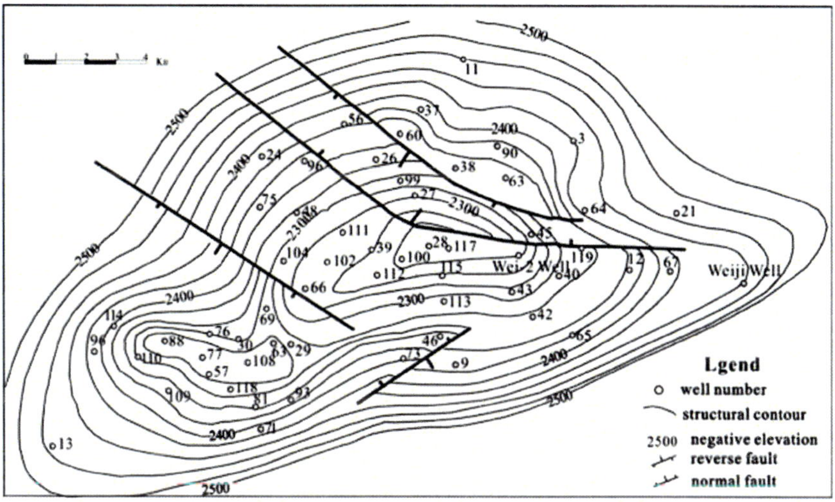

Fig. 13.2 The top surface contour map of Deng-2 Member in Weiyuan Gasfield and the distribution of gas well sites (based on the geological data of PetroChina Southwest Oil and Gasfield Company)

13.2.2.2 Ziyang Gas–Water Reservoir

Tectonically, it is a local monocline mostly on the current north peripheral slope of Weiyuan Anticline (Fig. 13.1), but it is verified by seismic data during 1992–1993 that this area was a palaeo-anticlinal trap on the top of palaeo-uplift during Early Palaeozoic, its elevation is 1200 m lower than that of Weiyuan Anticline. In 1993, the Zi-1 Well had started drilling (Fig. 13.2), and producing a gas flow of 13.13×10^4 m^3 by DST from Deng-2 Member at well depth of 4080.5 m. Hereafter, the well completing test (WCT) of Zi-2 Well produced gas of 5.33×10^4 m^3/d and water of 86 m^3/d at well depth of 4534.57 m. Therefore, the Ziyang low-production gas–water reservoir was found. Until 1996, totally 7 exploratory wells were drilled, controlling a gas-bearing area of 50 km^2, with gas geological P_2 reserve ca. 102.1×10^8 m^3 and P_3 reserve ca. 423×10^8 m^3.

13.2.2.3 Longnvsi Anticline

It lies on the east of Gaoshiti and Moxi Anticlines, all the anticlines are attributed to the secondary local structures on Chuanzhong Uplift. The Longnvsi Anticline is just in the easternmost segment of the uplift (Fig. 13.1). In 1972, a stratigraphic deep-well, Nvji Well, was drilled at Longnvsi Anticline (Fig. 13.1). The Nvji Well penetrated Sinian up to Nanhuan rhyodacite with a finishing drilling depth of 6011 m as the deepest exploratory well in China in 1977. This deep stratigraphic well has provided following significant geological achievements.

(1) The strata penetrated by the Nvji Well are attributable to 7 systems and 13 formations from Neoproterozoic to Mesozoic, and affording systematic petrophysic and stratigraphic parameters as well as oil, gas and water information, e.g., electrical properties, acoustic wave and velocity, formation pressure and geotherm, which has provided various parameter basis for further geological studies and petroleum exploration in western YC.

    Particularly, the systematically equivalent vitrinite reflectance eqR_o data could characterize the thermal evolutional level of sedimentary organic matter and clarify the significance of commercial gas output of 1.85×10^4 m^3/d produced from the Sinian Dengying Formation in the Nvji Well under overmature conditions of organic matter, i.e., eqR_o values up to 3.65%, which would suggest a thermo-evolutional lower limiting value for effective natural gas exploration.

(2) Nvji Well totally found 14 gas-show horizons, in which, the gas-shows of Neoproterozoic to Palaeozoic account for 64% of the total gas-shows, and there are two gas-bearing horizons of Sinian Dengying Formation, two of Cambrian and one of Ordovician. The WCT of Nvji Well had found three potential commercial gas-reservoirs, i.e., Sinian Deng-4 Member producing gas flow of 1.85×10^4 m^3/d (at the well interval of 5206–5248 m), Lower Ordovician 3.09 $\times 10^4$ m^3/d (at 4523.8–4534.8 m) and Permian Qixia Formation upper dolostone 4.68×10^4 m^3/d (at 4400.5–4408 m). Especially, Ordovician, Cambrian and Sinian Dengying Formation would constitute a lower reservoir—seal bed assemblage for the natural gas preservation and exploration, and thus showing the gas productivity in western YC.

(3) Nvji Well could confirm the existence of Chuanzhong palaeo-uplift during the Nanhuan to Early Palaeozoic. At the well depth 5934 m, Nvji Well had penetrated 9 m-thick Sinian Doushantuo Formation and revealed the underlying rhyodacite of Nanhuan Suxiong Formation (isotopic geological age of 701.5 Ma; Luo 1986) without Nanhuan glacial sediments, in between there is a stratigraphic hiatus of 150 Ma. In addition, Permian is directly overlaid on the remanent 29 m thick Lower Ordovician at the well depth of 4518 m, and resulting in the hiatus of Middle–Upper Ordovician, Silurian, Devonian and Carboniferous sequences. Consequently. As the oldest palaeo-uplift in western YC, the Chuanzhong Uplift was continuously uplifting since early Sinian Doushantuo sedimentary period on. This kind of palaeo-uplift ought to be the Neoproterozoic and Palaeozoic oil and/or gas migrational target.

13.2.2.4 Anyue Gasfield

Structurally, this gasfield consists of Gaoshiti and Moxi Anticlines, which are main secondary local structures and linked each other by a small and narrow structural saddle on the Chuanzhong Uplift. Anyue Gasfield is bounded by Longnvsi Anticline on the east and separated from Weiyuan Anticline by a long and narrow faulted-sag (named Deyang-Ziyang Faulted-Sag) on the west, but its elevation is

2000 m lower than that of Weiyuan Anticline on the top surface of Sinian Deng-3 Member (Fig. 13.1). Due to a minor denuded thickness by the Episode II of Tongwan Movement at the end of Sinian, the Dengying Formation has been more completely preserved, so that the stratigraphic thickness of Deng-3 plus Deng-4 Members would be 200 m thicker in Gaoshiti and Moxi than that in Weiyuan. By comparison to Weiyuan and Ziyang, there would be an extra gas reservoir bed of Deng-4 Member in Gaoshiti and Moxi Anticlines.

In the last decade, natural gas exploration has made considerable headway in Chuanzhong Uplift. A commercial gas flow of 102×10^4 m^3/d were produced from the Deng-2 Member carbonate reservoir of Sinian Dengying Formation in Gaoshi-1 Well at the Gaoshiti Anticline in 2011. While the WCT of the Moxi-8 Well at Moxi Anticline has obtained high-output gas flow in 2012, the total gas production amounted to 190.68×10^4 m^3/d from two dissolved-pore dolarenite reservoirs of Lower Cambrian Longwangmiao Formation.

By the end of 2013, there were 23 commercial gas-producing wells at Moxi and Gaoshiti Anticlines. Consequentially, there totally are three commercial gas-producing reservoirs respectively in Longwangmiao Formation, Deng-4 and Deng-2 Members proved in the Anyue Gasfield (including Moxi and Gaoshiti Anticlines), with the gas P_1 geological reserve 8488×10^8 m^3 and P_3 reserve 1.2×10^{12} m^3 and the gas production 130×10^8 m^3/a.

13.3 Potential Source Beds in Western Yangtze Craton

13.3.1 Lithology and Lithofacies

During the early Sinian Doushantuo sedimentary period, transgression had caused sea level rising and marine sedimentary range extension, so that the original Nanhuan old-land was submerged or became relic islands, and 100–500 m-thick muddy sediments of the Doushantuo Formation were extensively deposited in the sea waters, resulting in four depocenters, i.e., ① Pingwu (at northwest Sichuan) to Kongding (western Sichuan), ② Dazhou to Wanzhou (eastern Sichuan), ③ western Hubei to western Hunan and ④ Shanghai to Hangzhou (Zejiang), respectively (Fig. 13.3). The Doushantuo stratigraphic thickness would be up to 300–500 m in each depocenter, in which carbonaceous shale, black silicalite, dark-gray dolostone, phosphorite and manganolite are developed, and containing a lot of bacterial and algal fossils, among which maticellular macroalgae, sponge and metazoan are well developed, such as these in Weng'an, Miaohe and Lantian biotas, which fully testifies the Doushantuo sedimentary period as a quick biological multiplying and evolution epoch, it may be called as the time of eukaryotic-radiation. Under strong reducing conditions on seabed, the palaeo-oceanographical sediments would preserve a large amount of organic matter. Based on the studies of iron- and manganese-contents of sediments as well as silicalite, phosphorite and manganolite, it is shown that the hydrothermal

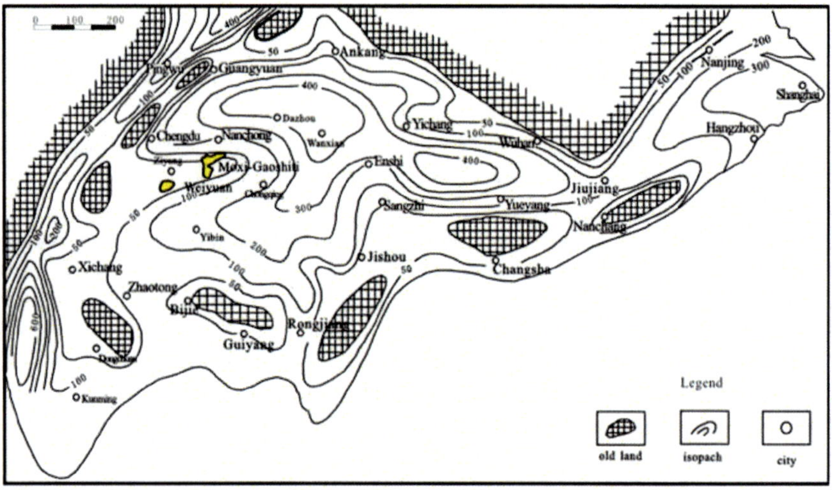

Fig. 13.3 Stratigraphic isopach map of Sinian Doushantuo Formation in YC. The map was compiled based on 42 date points (including outcrop profiles and seismic data)

leakage plus biochemical actions had resulted in numerous phosphorite in the Doushantuo Formation, phosphatic sea water could nourish and promote marine life growth and luxuriance so that the Doushantuo Formation is not only an important mineralizing sequence, but also an optimal source bed for petroleum generation (Hu 1997; Yu et al. 2005; Wang and Han 2008).

Based on the characteristics of lithology, rock textures and palaeontology, the lithofacies palaeogeographic map shows a certain consistency among the variations of sedimentary area, stratigraphic thickness and lithofacies of the Doushantuo Formation. The thick strata are always deposited within the abysmal area, while the thin strata in the neritic area (Figs. 13.3 and 13.4). Black shale would be well developed with abundant organic carbon in the abysmal to bathyal area, leading to a widely regional distribution of source bed and constituting the favorable region for petroleum generation. While the neritic area would be poor in petroleum-generating conditions, but better for phosphorite enrichment. Most phosphorites appear as spherulitic texture, this kind of ore bed with spherulitic texture could be only enriched under certain hydrodynamic conditions. Therefore, the Doushantuo sedimentary period should be an important marine sedimentation epoch of mineralization and petroleum generation.

The lithology of Lower Cambrian Qiongzhusi Formation mainly shows a gradually upward transformation from black and carbonaceous shales to dark-colour silty mudstone and siltstone in western YC. This set of strsta is stably distributed with thickness about 300–500 m around Mianyang-Changning, while it is just 120–170 m thick in other area.

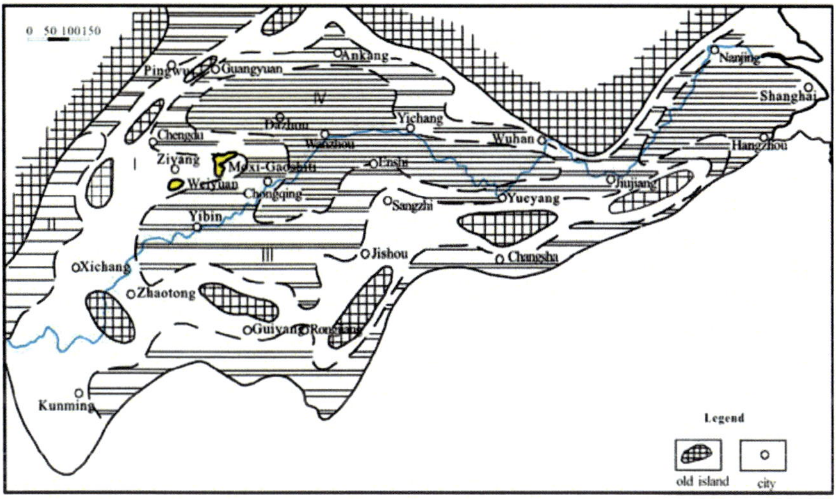

Fig. 13.4 Lithofacies-palaeogeography map of Sinian Doushantuo Formation in YC. I—Terrigenous and littoral sand-mudstone facies belt: quartzose and bunter sandstones with stratum thickness > 100 m, there sometimes are argillaceous dolostone. II—Neritic dolostone and red sand-mudstone facies belt: more clastic rocks, oxidizing environment, containing phosphorus at somewhere. III—Neritic black phosphorite facies belt: black dolostone, carbargillite and silicalite, phosphoric abundance zone. IV—Abysmal-bathyal facies belt: black carbargillite, silicalite and dark-grey micritic-limestone containing phosphonium and manganese

13.3.2 Hydrocarbon-Generating Potential

It is well known that organic matter of Sinian Doushantuo Formation was principally derived from bacterial and algal source input in western YC, and thus the Doushantuo black shale shows very high TOC value in the following six stratigraphic outcrop sections. e.g., TOC value of 3.5% at Weng'an (Guizhou), 3.2% at Chengkou (Chongqing), 2.8% at Mianxian (Shanxi), 2.4% at Shimen (Hunan), 2.1% Mianzhu (Sichuan), 2% at Jingshan (Hubei) profiles (Table 13.1).

It can be seen from Table 13.1 that Type I (sapropel-type) kerogen is referred to a high-quality hydrocarbon-generating source material, the hydrocarbon-generating rate of thermal simulation can be over 42% (Wang et al. 1997). The thickness of Doushantuo source bed ranges from 20 to 140 m, generally 30–60 m (avg. 55.8 m). According to the statistics of 18 stratigraphic sections, Doushantuo source bed accounts to about 25% of its stratigraphic thickness, and its maximum hydrocarbon-generating intensity could be up to 478 × 10^4 t/km^2 in average.

The Qingzhusi Formation is rech-in Types I (sapropel-type) and II (mixture-type) kerogens, generally with TOC 1–2% (avg. 1.71%) in western YC (Table 13.1), especially around the Mianyang to Changning area, its hydrocarbon potential could reach 140 × 10^8 m^3/km^2, while it also can be 30 × 10^8–50 × 10^8 m^3/km^2 in other areas.

Table 13.1 The composition of source beds of Sinian Doushantuo and Lower Cambrian Qingzhusi Formations in western YC

Strata	TOC (%)	Kerogen composition			Soluble organic matter composition (%)				
		Sapropel (%)	Bituminite (%)	Kerogen type	Ali	Aro	N, S, O	Asp	HCs
Qingzhusi Formation	1.71	53.5	46.5	I, II	40.7	12.6	31.0	17.2	53.3
Doushantuo Formation	1.97	71	29	I	42.7	15.1	28.1	13.4	57.9

Note Ali aliphatics; *Aro* aromatics; *Asp* asphaltene; *N, S, O* Nan-hydrocarbons; *HCs* total hydrocarbons

13.4 Characteristics and Distribution of Dengying Reservoir Bed 1

13.4.1 Vertical Distribution of Reservoir Beds

Sinian Dengying Formation is a set of 500–1000 m-thick grey to greyish-white algal dolostone of plateau facies and dolorudite of shoal facies interbedded with thin silicalite in western YC and adjacent regions, which has been divided into four members as follows (in ascending order).

(1) **Deng-1** Member: Used to be called "lower algae-lean interval", ca. 40–80 m-thick grey to dark-grey medium-bedded and lamellated crypto- to very-fine-crystalline dolostone. Its bottom appears as continuous sedimentation with the underlying Doushantou Formation or unconformity on Nanhuan granite in somewhere else.
(2) **Deng-2** Member: Its lower interval is also called "algae-rich interval", ca. 80–360 m-thick greyish-white algae-rich medium-bedded and massive very fine- to fine-crystalline dolostones with botryoidal structure. While its upper interval is called "upper algae-lean interval", ca., 0–160 m-thick grey to greyish-white fine- to medium-crystalline dolorudites, occasionally containing oolite- and saccharoidal-dolostones, solution openings, karst caves and dry asphalt, which constitutes the first reservoir bed and also the major gas pay in Weiyuan Gasfield and Ziyang gas–water reservoir. The top of Deng-2 Member appears as a disconformity with its overlying Deng-3 Member, caused by the Episode I of Tongwan Movement.
(3) **Deng-3** Member: It consists of ca. 0–60 m thick, dark muddy and blueish-grey shale, interbedded dolostone and tuff. In Chuanzhong Uplift, its bottom has 40 m-thick blueish-grey shale or black shale and interbedded dolostone and sandstone, containing birdeye structure.
(4) **Deng-4** Member: It has ca. 0–350 m thick, dolarenite, algal dolostone containing siliceous bandings and stromatolite. Its upper interval contains grey to greyish-white fine- to medium-crystalline algal fragment dolostone and dolarenite, containing solution openings, karst caves and dry asphalt, occasionally with birdeye and oolitic textures, which constitutes the second reservoir bed in Chuanzhong Uplift. The top of Deng-4 Member shows a disconformity contact with the overlying Lower Cambrian resulted from the Episode II of Tongwan Movement.

Based on the investigations of field survey and drilled wells, the Dengying reservoir beds are developed at the top surface of both Deng-2 and Deng-4 Members. The reservoir bed of Deng-2 Member was developed at the "upper algae-lean interval", while the reservoir bed of Deng-4 Member at the algal fragment dolostone and dolarenite interval. In view of the sedimentary cycle, both the reservoir beds are within the algal fragment dolarenite interval at the upper part of sedimentary cycle, and both the top surfaces are disconformably erosional surface, which has suffered the reworking of epigenetic karstification and favorable to the formation of solution pores and caves. Thus, the reservoir spaces would combine the primary intergranular plus intercrystal pores and the secondary solution pores and caves, making a mixed type of reservoir space in the Dengying reservoir beds (Wei et al. 2010; Cao et al. 2011).

To sum up, the development of Dengying reservoir beds has be dually constrained by lithology-lithofacies and karstification, especially the denudation of disconformable surface resulted from Tongwan Movement at the end of Sinian period. The karstification of Deng-4 Member reservoir bed is particularly significant.

Figure 13.5 shows the differential denudation effects on the disconformable surface of the Tongwan Movement at various regions. The denudation effect was not limited within the Deng-4 Member, even the Dengying Formation had been totally denuded out in somewhere else, such as at Ya'an and Baoxing in the southwest corner of western YC, so that the Middle Devonian was directly overlaid on Mesoproterozoic complex. To a certain extent, the development level of Dengying reservoir bed was closely associated with the denudation extent. The Deng-4 Member was almost totally denuded out, only Deng-2 Member reservoir bed remained at Weiyuan Gasfield and Ziyang gas–water reservoir, while the denudation extent of Dengying Formation was relatively low, both reservoir beds of Deng-2 and Deng-4 Members can still be preserved in Gaoshiti and Moxi Anticlines of Anyue Gasfield.

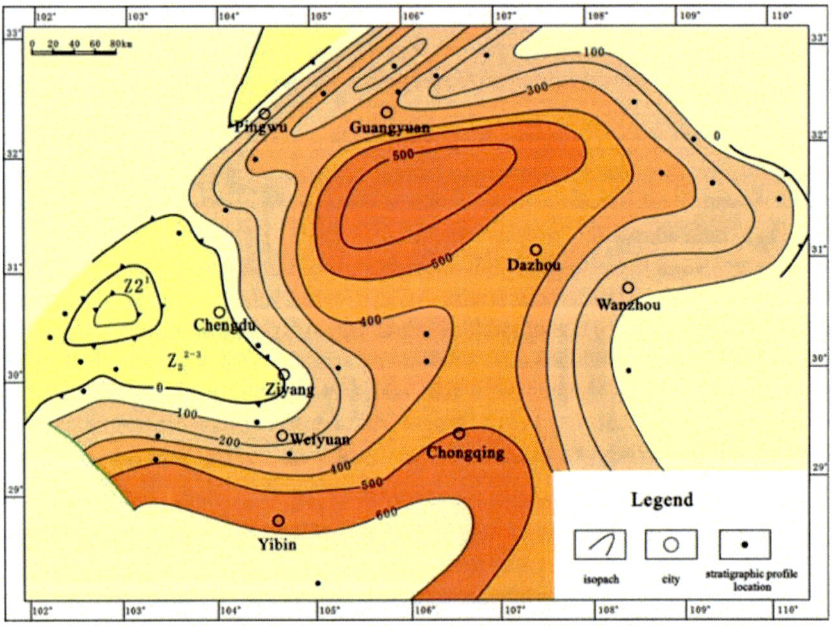

Fig. 13.5 The remanent stratum isopach map of Deng-4 Member in western YC showing the Tongwan Movement denudation effect on the Dengying Formation

13.4.2 Reservoir Beds and Lithofacies

Dengying sedimentary period was the most extensive transgression epoch during Sinian, when the sea water covered all the YC. In view of remanent stratigraphic isopach map of Deng-4 Member after the Tongwan Movement in western YC (Fig. 13.5), the craton still inherited the Nanhuan regional tectonic fractures, which generally appears as a NW–SE trending framework of "two uplifts plus two depressions" (Fig. 13.6):

(1) **Pingwu Depression** along the Longmenshan Backhill Belt on the northwest fringe of Western YC with thicker Sinian stratigraphic thickness 700–1200 m.
(2) **Xichang-Chengdu-Guangyuan-Hanzhong Uplift** on the northwest fringe of western YC with thinner Sinian stratigraphic thickness only 0–600 m.
(3) **Kunming-Yibin-Dazhou Depression** in the western YC with thicker Sinian stratigraphic thickness 500–1200 m.
(4) **Jishou-Nanchang-Jiujiang Uplift** in the middle YC with the thinnest Sinian stratigraphic thickness, e.g., the Sinian Liuchapo Formation is just 80–200 m (Fig. 13.6; Tang et al. 2009).

Dengying Formation is a wide-spread carbonate sequence of plateau facies which mainly composed of dolostone, its lithofacies variation shows rock color, stratification features, silicalite and algae developments and clast (including sandy clast and algal fragment) contents.

The lithofacies-palaeogeographic map of Sinian Dengying Formation in YC shows that the western YC appears as a large dolostone plateau (Fig. 13.7). As

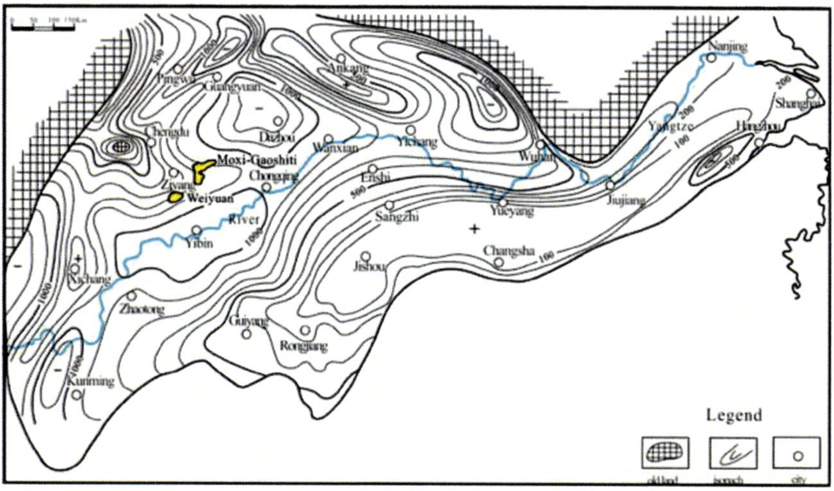

Fig. 13.6 Stratigraphic isopach map of Sinian Dengying Formation in YC. The map was compiled based on 84 date points (including outcrop profiles and drilled well data)

compared with the stratigraphic isopach map of Dengying Formation, it can be seen that along with the Kunming–Yibin–Dazhou line, there was a NW-trending plateau depocenter with stratigraphic thickness more than 1000 m (facies belt I in Fig. 13.6). There is a 312 m-thick salt bed within the lower Dengying Formation in the Ning-2 well and a gypsum bed in the Dafang Well (Guizhou), both indicating the early lagoon deposits. While in surrounding areas (facies belt II), there is a plateau marginal beach facies with neritic waters and strong hydrodynamic condition, where clast carbonates were well developed and resulting in a reservoir bed developing zone. Since the waters gradually deepening eastwards, the carbonate plateau facies in western YC would be transited into interbedded dark-grey dolostone and silicalite of neritic slop facies and hydrothermal silicalite of abysmal basin facies in middle-eastern YC (Fig. 13.7; Cao et al. 2011; Chen et al. 2002; Yu et al. 2003).

In view of the characteristics of lithofacies, only the semi-restricted plateau facies (facies belt II in Fig. 13.6) contains the conditions for oil accumulation, and matches the erosional surface, so that a complex type of reservoir bed, with primary intergranular and intercrystalline pores as well as secondary solution pores and caves, would be available. This kind of Dengying reservoir bed features is very remarkable either in surface outcrops or on well core profiles, e.g., the 80 m-thick Deng-4 Member reservoir bed contains porosity of 2.18–10.69% at the Fandian outcrop section (in

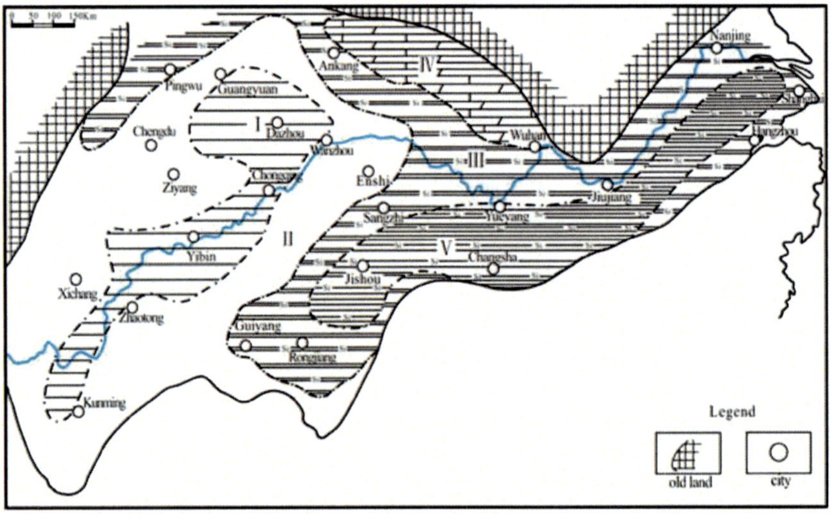

Fig. 13.7 Lithofacies-palaeogeography map of Sinian Dengying Formation in YC. I—Carbonate plateau facies belt: including basal lagoon gypsum-salt, middle algal dolostone and top massive micritic-dolostone with total thickness more than 1000 m. II—Semi-restricted plateau facies belt: bottom algae-rich dolostone with botryoidal structure, top tidal flat and beach facies containing algal fragments, pore and cave developed in fine-crystalline saccharoidal dolostone. III—Neritic slop facies belt: bottom grey micritic-dolostone, top silicalitic interbedings andbands, algae-rich. IV—Abysmal basin facies belt: black silicalite with thickness 80–150 m, containing basal dolostone in somewhere

E'meishan), in which an interval of 23.75 m was carefully described bearing 7.55 m-thick cavern layer and accounting for 30.8% of the interval thickness. The cavern appears as four geological occurrences, i.e., pinhole, banding, cancellate and patch, which are unevenly distributed showing dispersed caves and crowded pores, most of which are referred to the reservoir bed type of intergranular and intercrystalline pores.

The 167 m-thick Deng-4 Member reservoir bed in the well interval 3019–3186 m at Wei-28 Well (in Weiyuan Gasfield) has recorded 2 m-thick pinhole dolostone and 4 m-thick vuggy dolostone intervals, drill tool dropping for 4 m, once well kike at the depth 3186 m, and tested gas output 41.28×10^4 m^3/d plus water output 98.2 m^3/d.

While the 360 m-thick Deng-4 Member reservoir bed at the Qiangjiadong outcrop section (at Ningqiang, southern Shaanxi) possess vuggy-cavern dolostone and inter-crystalline sccharoidal dolostone with various vuggy and cavern types, e.g., pinhole, cancellate, brecciated and lace-like structures, etc. Both pores and caves were filled with asphalt. Besides, a 310 m-thick asphalt-bearing interval was found at both outcrop and Qiang-1 Well on the Nanshanling Anticline at the same region.

In addition, a 13 m-thick cancellate cavern-bearing core and asphalt-bearing inter-crystalline pores, with porosity 3.4–8.68%, in the well interval 5036–5091 m was found in the Deng-4 Member dolostone at Anping-1 Well on the Anpingdian Anti-cline (near Moxi). While on the south of the Anping-1 Well, the Gaoshiti-1 Well has gas output of 2.85×10^4 m^3/d from the same horizon.

All the above cases fully illustrate that the Dengying Formation contains a complex reservoir bed type of primary pores and secondary vuggy and cavern, which possess wide-spread regional features (Table 13.2).

Dengying dolostone reservoir beds were distributed in most of the YC (Table 13.2), among which its porosity in Weiyuan Gasfield would be the lowest one (e.g., only 4.85% in Wei-113 Well, Table 13.2), while Zi-2 Well at Ziyang gas–water reservoir and Gaoshi-1 Well at Gaoshiti Anticline show the highest porosities (11.05–8.36%; Table 13.2). Dengying dolostone reservoir bed and Doushantuo black shale source bed could constitutes a perfect regional source-reservoir assemblage, providing a geological foundation for large gasfield formation.

13.5 Oil and Gas Entrapment

Based on geological surveys and petroleum explorations, it is confirmed that Sinian Dengying Formation had be involved in twice important periods of oil and gas entrap-memt and evolution, i.e., original oil-reservoir formation as well as cracking-gas reservoir and fossil-oil-reservoirs formation (Wang et al. 2001; Sun et al. 2007).

Table 13.2 Sinian Dengying dolostone reservoir bed parameters in YC

No.	Region	Profile location	Reservoir thickness (m)	Visual porosity (%)	Bitumen content (%)	[a]Total visual porosity (%)
1	Ningqiang, Shanxi	Qiang-1 Well	310	4.5	4.5	9
2	Ningqiang, Shanxi	Kengjia Cave	360	6.0	6.0	12
3	Chengkou, Chongqing	Mukui River	200	3.2	3.6	6.8
4	Ziyang, Sichuan	Zi-2 Well	80	5.76	5.29	11.05
5	Weiyuan, Sichuan	Wei-113 Well	120	3.92	0.93	4.85
6	Anyue, Sichuan	Gaoshi-1 Well	260	4.2	4.5	8.7
7	Jinsha, Guizhou	Yankong	68	3.36	> 5	8.36
8	Cili, Hunan	Nanshanping	29	12	5	17
9	Yuhang, Zhejiang	Taishan	70	8–24	2–7	15

[a]The remanent visual porosity after bitumen and secondary mineral filling

13.5.1 Fossil-Oil-Reservoirs Formation

Under the regional geological conditions of perfect source-reservoir assemblage, Early Palaeozoic Caledonian Movement had resulted in a series of large palaeo-uplifts (Fig. 13.8) as the targets of oil migration during the geological periods, which would be favorable for original oil-reservoir formation.

Owing to the stratigraphic burying and thermo-evolution, the liquid oil in original oil-reservoir would be altered into gas/light-medium oil and asphalt/reservoir bitumen by disproportionation, resulting in gas/light-medium oil reservoir and fossil-oil-reservoir respectively.

Therefore, numerous fossil-oil-reservoirs were found within Sinian Dengying Formation and Lower Palaeozoic Cambrian to Silurian strata on the palaeo-uplifts. So far 24 fossil-oil-reservoirs have been reported, among which 16 in Sinian Dengying reservoir or in its overlying strata associated with Doushantuo source bed, and 8 in Lower Palaeozoic (Fig. 13.8; Ying and Huang 1989; Zhang and Zhang 1993; Zhao et al. 2001; Zhao et al. 2008; Dai et al. 2009).

The geological reserve of original crude oil in these fossil-oil-reservoirs has a massive scale, e.g., 137 known asphaltic veins found principally in Lower Cambrian and partially in Ordovician–Silurian strata constituting two original oil-reservoirs of fracture type respectively at the Kuangshanliang (No. 4 in Fig. 13.8) and Tianba (No.

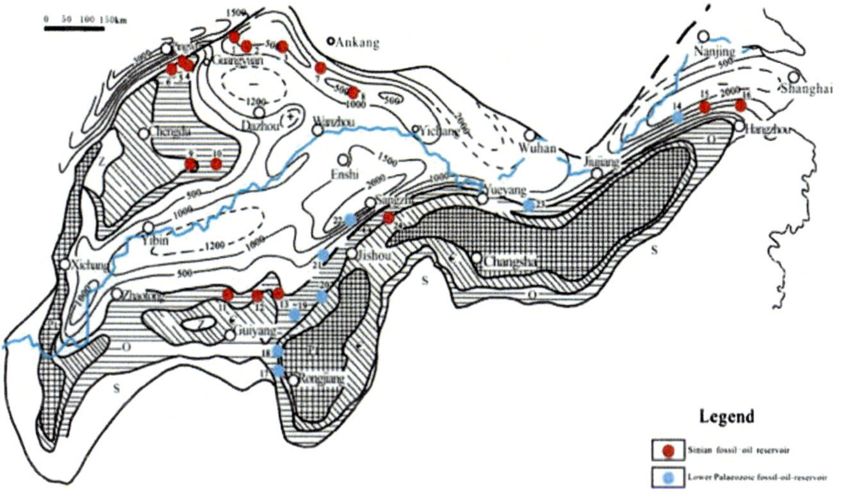

Fig. 13.8 The remanent Silurian stratum isopach map in YC showing Early Palaeozoic fossil-oil-reservoirs distribution. Fossil-oil-reservoirs as follows: 1—Kenghaodong, 2—Ningqiang-1 Well (southern Shaanxi); 3—Micangshan at Nanjing (northern Sichuan); 4—Kuangshanliang (KSL) at Guangyuan, 5—Tianba (TB) at Qingchuan, 6—Houba at Jiangyou (northwest Sichuan); 7—Zhenba (southern Shaanxi); 8—Chengkou (northern Zhongqing); 9—Ziyang; 10—Moxi-Gaoshiti (central Sichuan); 11—Yankong at Jinsha, 12—Yangshui at Kaiyang, 13—Baidoushan at Weng'an (Guizhou); 14—Taiping (southern Anhui); 15—Kangshan at Anji; 16—Taishan ar Yuhang (Zhejiang); 17—Danzhai; 18—Majiang; 19—Kaili; 20—Tongren (Guizhou); 21—Xiushan (Zhongqing); 22—Wangcun at Yongshun (western Hunan); 23—Tongshan (Hubei); 24—Cili (western Hunan)

5) Anticlines (KSL and TB in Fig. 13.9), which would be sourced by the Doushantuo black shale source bed (Wang et al. 2005).

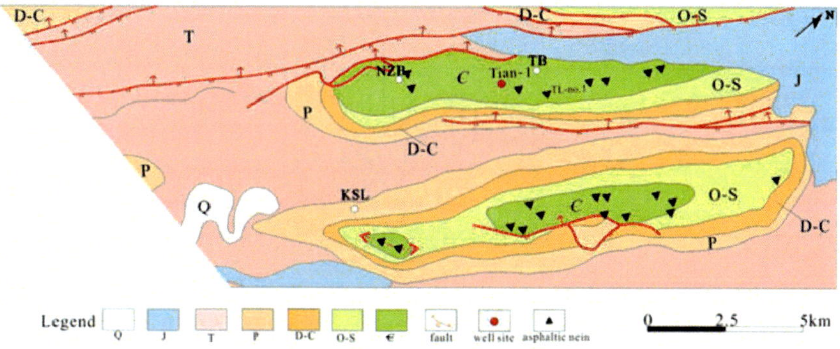

Fig. 13.9 Geological map and the distribution of asphaltic veins in the Kuangshanliang (KSL) and Tianba (TB) Anticlines at Qingchuan, northwest Sichuan (Dai et al. 2009). *KSL* Kuangshanliang; *TB* Tiamba; *NZB* Nianziba. Stratum symbol: *J* Jurassic; *T* Triassic; *P* Permian; *D–C* Devonian–Carnoniferous; *O–S* Ordovician–Silurian; ∈. Cambrian

Original liquid oil had been charged into various faults and fractures resulting in the original fractured oil-reservoirs, since then the liquid oil was further thermo-optical alterated into light-medium oil and numerous solid asphaltic veins in-situ at Kuangshanliang and Tianba Anticlines respectively. These asphaltic veins are 0.9–1.2% in equivalent vitrinite reflectance eqR_o and analogous to maltha of medium mature phase. Besides the asphaltic veins, therefore, the disproportionation products of thermo-optical alteration should be light-medium oil instead of gas.

Above-memtioned numerous asphaltic veins constitute the fossil-oil-reservoirs in both the anticlines. Among which, 32 asphaltic veins are more than 1 m-thick, the largest one would be the TL-No. 1 Vein with 8 m in thickness, 160 m in height and ca. 970 m in length, the initially estimated asphaltic reserve was ca. 146 × 10^8 t in Tianba (Figs. 13.9 and 13.10; cf. Chap. 15). About 100 m apart from the TL-No. 1 asphaltic vein at Tianba Anticline, a geological shallow well, Tian-1 Well, was drilled in 1966. A paragenic relationship of medium oil and asphaltic vein was found in the Tian-1 Well (for the vein occurrence and the well sites cf. Figs. 13.9 and 13.10), when 30 L oil with density of 0.882 g/cm^3 was produced from the well interval of 333–335 m, and a 15.3 m-thick asphaltic vein recorded within the interval of 149–164 m.

As the second case, the Micangshan fossil-oil-reservoirs at Nanjiang, northern Sichuan (No. 3 in Fig. 13.8) is tectonically situated on the Micangshan Uplift, which occur as a giant NW trending anticline with structural area of 4500 km^2. Metamorphic complex rocks of Pre-Sinian Houdiya Group outcrop at the anticlinal core area, which is surrounded by the Sinian to Lower Palaeozoic strata on surface outcrop

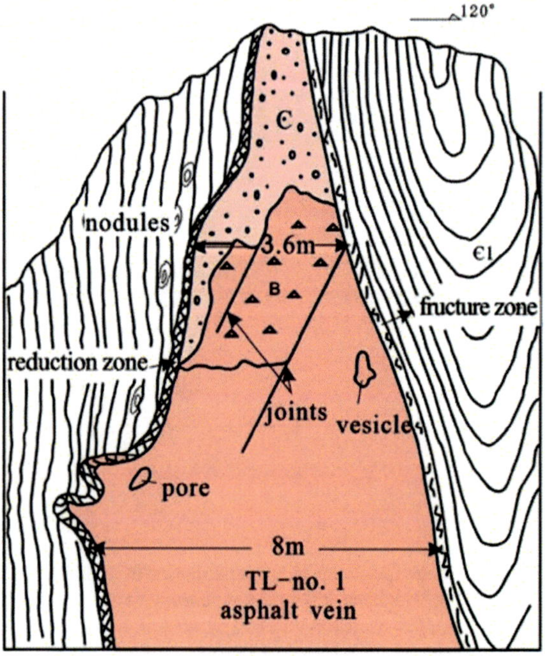

Fig. 13.10 A sketch of TL-No. 1 vein on the Tianba Anticline at Qingchuan, northwest Sichuan

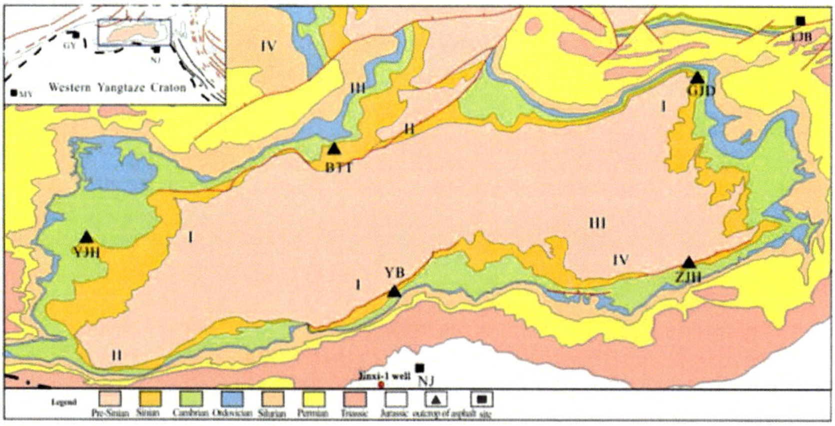

Fig. 13.11 Geological map of Micangshan Uplift and fossil-oil-reservoir localities (Dai et al. 2009). *GJD* Guangjiadian; *LJB* Luojiaba; *YB* Yangba; *YJH* Yanjinghe; *ZJH* Zhujiahe in the geological; *GY* Guangyuan; *NJ* Nanjiang; *MY* Mianyang at the inset

(Fig. 13.11; Dai et al. 2009). The Sinian Dengying dolostone is a principal asphalt-rich reservoir bed, and the top of Deng-4 Member had suffered obvious weathering and denudation, resulting in the development of vuggy-cavernous dolostone. Original liquid oil was filled into solution vugs, karst caves and bedding fissures in the Dengying vuggy-cavernous dolostone, and then liquid oil was thermo-optical alterated into solid asphalt as reservoir pyrobitumen (Dai et al. 2009).

Based on an investigation of 8 stratigraphic sections, asphalt is principally distributed within the Deng-4 Member ca. 100 m apart from its top surface, especially at the high point of secondary anticlines, such as at the Yangba (YB), Yanjinghe (YJH) and Guangjiadian (GJD in Fig. 13.11).

Where the effective asphalt-bearing interval is more than 30 m-thick with asphaltic visual porosity more than 5% in average, and its equivalent vitrinite reflectance eqR_0 values are as high as 2.48–2.84%, which should be referred to pyrobitumen of over-mature phase duo to its deeply burial geological history (Dai et al. 2009).

The third case is Anyue Gasfield, where the large-scale cracking-gas reservoirs and associated fossil-oil-reservoirs in Deng-2 and Deng-4 Members as well as in the Lower Cambrian Longwangmiao Formation are typical oil disproportionation products in over-mature phase. These gas reservoirs produce dry gas, and fossil-oil-reservoirs composed of pyrobitumen with a series of "mesophase" mosaic-microcrystalline-textures, i.e., grain, domain and fibrous mosaic-microcrystalline-textures. "Mesophase" reflects a highly organic thermo-evolutional level between amorphous asphalt and crystal shungite to graohite (Yang 2018; cf. Chaps. 11 and 14).

All the fossil-oil-reservoirs shown in Fig. 13.8 were constrained by the palaeo-uplifts since Early Palaeozoic on, which would be the most important conditions for the formation of large oil reservoirs in Neoproterozoic to Lower Cambrian.

In comparison with Lower Palaeozoic, Sinian contains more fossil-oil-reservoirs (Fig. 13.8), which imply that the distribution of Sinian reservoir beds are more wide-spread and more steady.

Sinian Dengying reservoir beds in western YC was sandwiched by Sinian Doushantuo and Lower Cambrian Qiongzhusi source beds. The oil generation and entrapment age in Chuanzhong Uplift would be during the late Early to Late Permian period (275–253 Ma), while oil cracking into dry gas and asphalt during the late Middle Jurassic to early Early Cretaceous period (175–144 Ma; Yang 2018; cf. Chaps. 11 and 14).

13.5.2 Secondary Cracking Gas Reservoirs

The petroleum exploration had confirmed that the most of marine gas reservoirs in western YC could be attributed to secondary cracking gas reservoirs formed by original oil cracking. Along with the burial deepening and geotemperature rising of original oil-reservoir, the original liquid oil would be gradually cracked into gas when it reach the over-mature phase, and the relevant vitrinite reflectance eqR_o would be much higher than 1.2%. As reservoir pyrobitumen, the affluent solid asphalt in Sinian Dengying and Lower Cambrian Longwangmiao Formations reveal that lots of oil have been cracked into gas, and constrained by palaeo-uplifts. All the known craking-gas reservoirs, such as these in Weiyuan and Anyue Gasfields, are distributed on the high position of palaeo-uplifts. However, numerous exploratory wells didn't produce oil and gas, only water and nitrogen gas were available on the local anticlines either outside the palaeo-uplift or within the depressions.

In order to date the gas reservoir emplacement age and to know the features of craking-gas reservoirs, tectonic evolution, gas properties, hydrocarbon inclusions of reservoir beds and terrestrial heat flow have been more deeply investigated by previous researchers, the major results are summarized as follows.

13.5.2.1 Chuanzhong Uplift Evolution

It is a long-term successive palaeo-uplift initially formed as early as ca. 780 Ma before, since than it experienced a uplifting period until the end of Nanhuan (ca. 635 Ma), and an extensive transgression period happened in Sinian Dengying sedimentary period when it was totally submerged by sea-water. After that, the palaeo-uplift was raised and denuded again during Tongwan Movement (ca. 542 Ma). Owing to its long geohistory and multiple tectonic movements, its structural high point and axial line on the top surface of Sinian palaeo-uplift were constantly changeable, its trap area and closure also increased. According to the vitrinite reflectance R_o isogram of Upper Triassic coal seam in western YC, and the correlation curve between R_o value/buried depth and palaeotemperature of Nvji Well, both sedimentary isopach map of Mesozoic terrestrial strata and the palaeo-structural map of the Sinian

Table 13.3 Tectonic evolutional parameters of Chuanzhongi palaeo-uplift

Tectogenesis	Age	Uplifting range (m)	Trap area (km^2)	Depth of high point (m)	Trap location
Caledonia	Pre-Permian	650	480	500	Dayi, Hongya-Ya'an
Hercynian	Pre-Late Triassic	750	480	1200	Dayi-Ya'an, Ziyang-Moxi
Early Indosinian	Pre-Late assic	800	160	2500–3000	Ziyang, Anyue-Longnvsi
Late Indosinian	Pre-Jurassic	850	18,800	3000–3500	Ziyang-Longnvsi
Yanshan	Pre-Eocene	1200	19,600	6500	Leshan-Longnvsi

top surface during Jurassic to Cretaceous in western YC have been reconstructed, and also the structural parameters of Chuanzhong Uplift during geological times obtained (Table 13.3). Moreover, besides the Chuanzhong palaeo-uplift, there are two additional palaeo-uplifts during Jurassic to Cretaceous, i.e., the Huayingshan palaeo-uplift in E Sichuan and Tianjingshan palaeo-uplifts in northwest Sichuan (Fig. 13.12).

It can be seen from Table 13.3 that the trap area and closure of Chuanzhong Uplifts are up to 19,600 km^2 and 1200 m respectively as the maximums at pre-Eocene, and the trap was located at Leshan-Longnvsi with a burial depth of Dengying Formation more than 6500 m and a converted palaeo-geotemperature up to 215 °C, which indicates an evolutional phase of oil-cracking into gas, and also the optimal period for craking-gas reservoir emplacement. Based on its large trap area before Eocene, it is inferred that the cracking gas potential of Chuanzhong Uplift contains considerable scope.

13.5.2.2 Fluid Inclusions in Dengying Reservoir Beds

By way of the investigation on fluid inclusions of reservoir beds in gas wells of Weiyuan and Ziyang regions, three generations of authigenic minerals associated respectively with oil, gaseous hydrocarbon- or bitumen-inclusions have been found as follows (Table 13.4).

Waples (2000) reports that the temperature of in-reservoir oil destruction and cracking gas formation would be 150–200 °C. Table 13.4 shows that the authigenic mineral of generation II contains mainly the oil inclusions with homogenization temperature (HT) 160–190 °C and the generation III contains only inclusions of gaseous hydrocarbon phase plus asphalt with HT 200–210 °C, which would be in the hydrocarbon evolutional phase of oil-cracking into gas. Moreover, by comparing with the temperature evolutional curve of Dengying Formation in the Chuanzhong Uplift, the geotemperatures of Dengying Formation were just within 160–200 °C during Triassic (Liu et al. 2018). Therefore, the gas was emplaced and accumulated

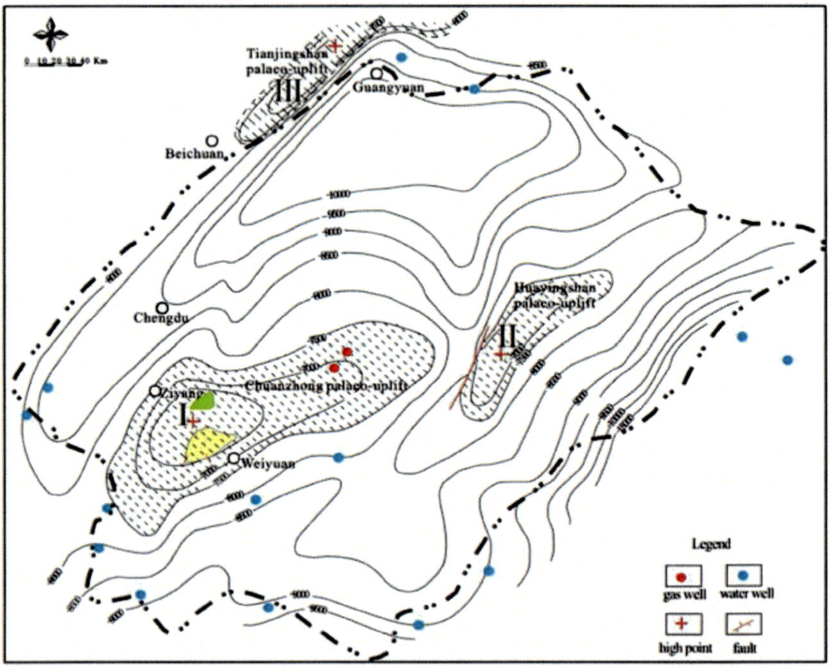

Fig. 13.12 Palaeo-structural map of the Sinian top surface during Juressic to Cretaceous showing three palaeo-uplift distribution in western YC

into cracking-gas reservoir at the latest in the Mesozoic with a average HT 205 °C and a calculated burial depth 6500–6900 m as consistent with both the results of gas exploration and tectonic research. Therefore, the Mesozoic palaeo-uplifts should be the optimal period and space for cracking-gas entrapment (Wei et al. 2010; Sun et al. 2007).

13.6 Conclusion

Yangtze Craton (YC) is the most stable tectonic unit in southern China, in which the western YC is an important oil- and gas-producing region occupying the largest area of 20 × 10^4 km^2. By means of the explorational results in western YC, the conclusions have been made as follows.

(1) During the Sinian transgression period, the Doushantuo Formation was deposited as the first set of hydrocarbon source bed, and followed by a set of massive algal dolostones in the western YC, and then the craton was uplifting and eroded during Tongwan Movement. Early Cambrian transgression resulted in the sedimentation of Qiongzhusi source bed, which constituts a high-quality

Table 13.4 Comprehensive data of hydrocarbon inclusions in the Dengying dolostone reservoir beds in Weiyuan-Ziyang regions (Wang et al. 1997)

Generation	Characters of mineral	Inclusion phase	Fluore-scence	Gas to liquid ratio	Main gas composition		CH_4/CO_2	Homogenization temperature (HT) (°C)
					CH_4 (%)	CO_2 (%)		
I	Micrite fiberous dolostone, vertical to wall	Mainly oil plus liquid–gas	Yellow, purple	10–25	10.1–22.0	50.7–75.9	0.2–0.38	(135) 120–150
II	Coarse grained dolostone, parallel to wall	Liquid–gas phase HC inclusion	Brownish yellow	15–60	27.6–40.4	44.7–59.4	0.61–0.68	(175) 160–190
III	Coarse grained dolostone, quartz, vein and massive	Gaseous HC, with asphalt	Colourless	> 60	55.3–74.0	18.8–36.1	2.14–2.53	(205) 200–210

source-reservoir assemblage with the overlying Dengying Formation, and lay the foundation for numerous fossil-oil-reservoir and large gasfields in Dengying and Longwangmiao Formations.

(2) As the Chuanzhong palaeo-uplift in the Caledonian period became a main target of oil and gas migration, original liquid oil generated from the Lower Cambrian Qiongzhusi and probably Sinian Doushantuo source beds has migrated and entrapped, through the disconformable contacts of Dengying Formation, so as to result in the original oil reservoir. With the successive deep buried action, the original liquid oil was cracked into dry gas reservoirs and reservoir pyrobitumen/fossil-oil-reservoir.

References

Cao JW, Liang B, Chen HF, Zhang QY (2011) Development characteristics and controlling factors of Sinian Dengying formation in west of Xuefeng Mountain. Sci Technol Inf 30:122 (in Chinese with English abstract)

Chen DF, Chen GQ, Chen XP (2002) Palaeozoic sea level changes and the formation of large-super large hydrothermal sedimentary deposits in Southern China. Sci China (ser D) 32(Supp 1):120–126 (in Chinese)

Dai HS, Liu SG, Sun W, Han KY, Luo ZL, Xie ZL, Huang YZ (2009) Study on characteristics of Sinian-Silurian bitumen outcrops in the Longmenshan-Micangshan area, Southwest China. J Chengdu Univ Technol (sci Technol Ed) 36(6):687–696 (in Chinese with English abstract)

Du JH, Wang ZC, Zou CN, Xu CC, Wei GQ, Zhang BM, Yang W, Zhou JG, Wang TS, Deng SH (2015) Geologic theory and exploration practice of ancient large carbonate gas field. Petroleum Industry Press, Beijing (in Chinese)

Hu NF (1997) Features of oil-source rocks of Sinian Doushantuo formation, Guizhou. Guizhou Geol 14(3):244–252 (in Chinese with English abstract)

Liu M, Qiu NS, Xu QC, Liu YF (2018) Precambrian temperature and pressure system of Gaoshiti-Moxi block in the central paleo-uplift of Sichuan Basin, Southwest China. Precambr Res 313:91–108

Luo ZL (1986) Is there a palaeocontinental nuclear in central Sichuan? J Chengdu Coll Geol 13(3):65–73 (in Chinese with English abstract)

Sun W (2008) The research on the formation process and mechanism of gas pools in proterozoic to lower palaeozoic Erathem, Sichuan Basin. Chengdu University of Technology, Chengdu (in Chinese with English abstract)

Sun W, Liu SG, Ma YS, Cai XY, Xu GS, Wang GZ, Yong HF, Pan CL (2007) Determination and quantitative simulation of gas pool formation process of Sinian cracked gas in Weiyuan-Ziyang area, Sichuan Basin. Acta Geol Sin 81(8):1153–1159 (in Chinese with English abstract)

Tang ZY, Duan QF, Zhou XW, Li K (2009) Preliminary approach on the stratabound lead-zinc deposits and lithofacies palaeogeographic framework of the Dengyingxia age, Sinian (Ediacaran), in Western Hunan-Western Hubei Area. Geol Rev 55(5):712–721 (in Chinese with English abstract)

Wang TG, Han KY (2008) On meso-neoproterozoic primary petroleum resources. Acta Petrolei Sin 32(1):1–7 (in Chinese with English abstract)

Wang LS, Gou XM, Liu GY, Wang L, Wang WM, Wang MY (1997) The organic geochemistry and origin of natural gases in Sichuan Basin. Acta Sedimentol Sin 15(2):49–53 (in Chinese with English abstract)

Wang YG, Chen SJ, Xu SQ (2001) Natural gas accumulation conditions and exploration techniques of upper proterozoic in the Sichuan Basin. Petroleum Industry Press, Beijing (in Chinese)
Wang LS, Han KY, Xie BH (2005) Reservoiring conditions of the oil and gas fields in the north section of Longmenshan mountain nappe structural belts. Nat Gas Ind 25(Supplement A):1–5 (in Chinese with English abstract)
Waples DW (2000) The kinetics of in-reservoir oil destruction and gas formation: constrains from experimental and empirical data, and from thermodynamics. Org Geochem 31(6):1137–1152
Wei GQ, Jiao GH, Yang W, Xie ZY, Li DJ, Xie WR, Liu MC, Zeng FY (2010) Hydrocarbon pooling conditions and exploration potential of Sinian-lower palaeozoic gas reservoirs in the Sichuan Basin. Nat Gas Ind 30(12):5–9 (in Chinese with English abstract)
Yang CY (2018) History of oil and gas entrapment and evolution in Leshan-Longnvsi Uplift, SW China. China University of Petroleum-Beijing, Beijing (in Chinese with English abstract)
Ying WH, Huang LH (1989) Preservation condition of the lower palaeozoic natural gas in Shangzhi-Shimen Synclinore, Northwest Hunan. Oil Gas Ind 10(2):170–181 (in Chinese with English abstract)
Yu MY, He MH, Wang Y, Zhao YL (2005) Sedimentary successions and environment of the Sinian Doushantuo formation in Jiangkou, Guizhou Province. Geol Sci Technol Inf 24(3):38–42 (in Chinese)
Yu XQ, Shu LS, Deng P, Zhi LG (2003) The middle-shallow water sedimentary environment of the late Sinian of South Anhui-based on the olistotrome and the siliceous storm sediments. Acta Sedimentol Sin 21(3):398–403
Zhang L, Zhang HX (1993) A discussion on oil and gas bearing conditions of Sinian and lower palaeozoic on Dabashan mountain frontal border. Nat Gas Ind 3(1):41–98 (in Chinese with English abstract)
Zhao ZH, Zhang GQ, Xue XL (2008) Fossil oil pools and residual oil and gas pools in the lower assemblage of Qianzhong Uplift. Nat Gas Ind 28(8):39–42 (in Chinese with English abstract)
Zhao ZJ, Feng JL, Chen XS, Zhou JG (2001) Discovery of Dengyin formation fossil pool in Cili, Hunan and its significance. Oil Gas Geol 22(2):114–118 (in Chinese with English abstract)

Chapter 14
Sinian–Lower Cambrian Anyue Gasfield in Western Yangtze Craton

Jinhu Du, Guoqi Wei, Caineng Zou, Wei Yang, Zengye Xie, Zhihong Wang, Wuren Xie, and Saijun Wu

Abstract Anyue Gasfield is located in the western Yangtze Craton (YC), i.e., the current Sichuan Basin. It is a marine carbonate gasfield with the oldest stratigraphic age, the highest thermo-evolutional level and the largest individual gas reserve in China. By the end of 2016, its P1 gas geological reserve is up to 8488×10^8 m^3 and 3P reserve 1.2×10^{12} m^3 respectively in Sinian Dengying and Lower Cambrian Longwangmiao Formations. The emplacement of Sinian to Lower Cambrian gas reservoirs have been constrained by the giant successional Chuanzhong palaeo-uplift, especially at Moxi-Gaoshiti Anticlines, which have been always situated at a high position on the uplift and providing favorable geological conditions for the early entrapment of fossil-oil reservoirs and the late enrichment of cracking-gas reservoirs as well as for the formation of Anyue Gasfield. The Deyang-Ziyang Faulted-Sag is just between the Weiyuan Anticline and the Moxi-Gaoshiti Anticlines, where the fine-grained sediments of Lower Cambrian Qiongzhusi (or called Jiulaodong) Formation was filled, and controlling the occurrence of Qiongzhusi black shale/mudstone as the main hydrocarbon source kitchen, while the development of Dengying algal dolostone and Longwangmiao grain-dolostone at both sides of the faulted-sag as the high-quality reservoir beds. There are two sets of main hydrocarbon source beds, i.e., Lower Cambrian Qiongzhusi black-greyish black shale/mudstone and Doushantou black shale in the western YC. However, the Qiongzhusi shale/mudstone would be the effective source bed for both Dengying and Longwangmiao cracking-gas reservoirs, while the Doushantou black shale would be a potential one in Chuanzhong Uplift. In addition, three sets of reservoir beds are developed in Anyue Gasfield, i.e., two algal dolostone reservoir beds of Deng-2 and Deng-4 Members respectively with reservoir bed thickness 5.1–69.1 m and 60–110 m, and one grain-dolostone reservoir bed of Longwangmiao Formation with reservoir bed thickness 10.8–61.1 m. Three types of gas reservoirs have been found in Anyue Gasfield, including the structural-lithologic type of Longwangmiao gas reservoir, the structural-stratigraphic type of Deng-4 gas reservoir as well as the structural type of Deng-2 gas reservoir, all of which

J. Du (✉)
PetroChina Exploration & Production Company, CNPC, Beijing 100007, China

G. Wei · C. Zou · W. Yang · Z. Xie · Z. Wang · W. Xie · S. Wu
Research Institute of Petroleum Exploration and Development, CNPC, Beijing 100083, China

T. Wang, *Meso-Neoproterozoic Geology and Petroleum Resources in China*, Springer Geology, https://doi.org/10.1007/978-981-19-5666-9_14

can be attributed to medium–low sulfur-contented and medium CO_2-contented dry gas reservoir. Moreover, the Longwangmiao gas reservoir is also a deeply buried reservoir with high temperature and high pressure, while the Deng-2 and Deng-4 gas reservoirs are ultra-deeply buried reservoirs with high temperature and normal pressure.

Keywords Moxi-Gaoshiti Anticlines · Dengying Formation · Qiongzhusi Formation · Longwangmiao Formation · Chuanzhong Uplift · Deyang-Ziyang Faulted-Sag

14.1 Introduction

Since natural gas is the predominant hydrocarbon resources, the western Yangtze Craton (YC) is geologically a huge petroliferous superimposed basin and geographically corresponding to the current Sichuan Basin, occupying an area of ca. 18×10^4 km^2. It has involved two tectonic evolutional stages, i.e., the Sinian–Middle Triassic cratonic depression stage and the Late Triassic–Cenozoic foreland basin stage. Therefore, a set of completely stratigraphic sequences are developed in the western YC. The Sinian to Middle Triassic strata are marine sequence dominated by 6000–7000 m-thick marine carbonate rocks. While the Upper Triassic to Neogene deposits are mainly 2000–5000 m-thick non-marine clastic sequences (Editorial Committee of Petroleum Geology of China 1989). Among which, the 2000–3000 m-thick Sinian–Cambrian sequences are wide-spread and contain favourably geological conditions for petroleum enrichment in the western YC (Yang et al. 2014; Zhou et al. 2015; Li et al. 2016; Xu et al. 2016; Xie et al. 2017).

The western YC has a long and tortuous historical course of petroleum exploratory more than 70 years, its initial petroleum exploration could be traced back to 1940s. Until July 2011, however, one high yield commercial gas flow of 1.02×10^6 m^3 had been first obtained in Gaoshi-1 Well from the Sinian Dengying Formation, which preluded a large-scale natural gas exploration in the Sinian–Cambrian sequences. Since then, another high yield commercial gas flow of 1.90×10^6 m^3 was gained in Moxi-8 Well from the Lower Cambrian Longwangmiao Formation in September, 2012. Both discovery wells are located respectively at the Moxi and Gaoshiti Anticlines on the Chuanzhong Uplift, where a new large gasfield has been named as Anyue Gasfield.

By the end of 2016, the Anyue Gasfield has the P1 gas geological reserves up to 8488×10^8 m^3 and 3P gas reserves about 1.2×10^{12} m^3, including the gas reservoirs in Dengying and Longwangmiao Formations as a recent significant achievement for gas exploration within old marine carbonate sequences in China (Du et al. 2014).

14.2 Geological Background

14.2.1 Reginal Stratigraphy

Reginal stratigraphic division of Sinian–Lower Cambrian in the western YC is listed in Table 14.1. Sinian stratigraphic column includes Doushantuo and Dengying Formations. The main lithologies of Lower Sinian Doushantou Formation are black shale, siltstone, micritic- and muddy-dolostones with high organic abundance. It has been confirmed that Doushantuo black shale would be the effective source bed for numerous asphaltic veins in the Longmenshan Fronthill Zone at the northwest margin of western YC (cf. Chap. 15). Up to the present, however, no Doushantuo black shale has be revealed by any drilled wells on Chuanzhong Uplift yet, but this source bed may be most probably distributed in the surrounding depression(s) of Chuanzhong Uplift as a potential source bed, especially on the north or northwest of the uplift based on the seismic information.

The upper Sinian Dengying Formation mainly consists of algal, stromatolitic, grain and micritic dolostones. The Dengying Formation is commonly more than 600 m thick on the Chuanzhong Uplift, but gradually thickened towards the surrounding depression(s). It can be subdivided into four lithological members from Deng-1 to Deng-4 Members in ascending order (Table 14.1), among which, both Deng-2 and Deng-4 Members constitute two algal dolostone reservoir beds with a disconformable interface respectively on their tops resulted from the Episodes I and II of the Tongwan Movement. In addition, the black and blueish-grey shales of Deng-3 Member may act as a potential local source bed due to its limited thickness and distributional range.

Table 14.1 Regional stratigraphy in Chuanzhong Uplif and its surrounding region (Yang 2018, modified)

Stratigraphy				Thickness (m)	Lithology	Age (Ma)	Tectonic movement	Note
System	Fm./Mem.		Sym.					
Cambrian	Xixiangchi		$Є_{2-3}x$	0-700	Grey dolostone and Grain dolostone			
	Gaotai		$Є_{2}g$	0-200	Greyish-yellow shale and dolo-sandstone			
	Longwangmiao		$Є_{1}l$	70-200	Grey muddy-dolostone and Grain dolostone	513**		Reservoir bed
	Canglangpu		$Є_{1}c$	65-300	Greyish-green fine sandstone			
	Qiongzhusi		$Є_{1}q$	90-540	Greyish-Black shale & carbonate shale	520**	Tongwanian III	Source bed
	Maidiping		$Є_{1}m$	0-380	Diamict with Black shale	525* / 541*	Tongwanian II	Source bed
Sinian	Dengying	Deng-4	$Z_{2}dn^{4}$	0-350	Grey algal dolostone			Reservoir bed
		Deng-3	$Z_{2}dn^{3}$	0-50	Black bluish-gray shale		Tongwanian I	Source bed
		Deng-2	$Z_{2}dn^{2}$	20-950	Grey algal dolostone			Reservoir bed
		Deng-1	$Z_{2}dn^{1}$	20-500	Grey dolostone			
	Doushantuo		Z_{1}	10-420	Blackish-grey shale, siltstone and dolostone	635*	Chengjiang -Jinningian	Source bed
Nanhuan	Nantuo		$Nh_{2}n$		Red siliceous clastic rock and green tillite			
Pre-Nanhuan basement				>6000	Crystalline basement			

Note: *-mesured **-inferred

Lower Cambrian includes four formations, i.e., Maidiping, Qiongzhusi, Changlangpu and Longwangmiao Formations (Table 14.1). Maidiping Formation is 0–380 m-thick rhythmic diamict of carbonate, phosphatite, silicalite and clastic rocks, which would be attributed to senke-filling deposits, and only found on the lower position of the Tongwan Movement disconformable contact. Therefore, it could not be a significant source bed for the Chuanzhong Uplift even though its shale may contain higher TOC.

The Qiongzhusi black and greyish-black shales and carbonaceous shale rich-in trilobite fossils constitute a wide-spread main hydrocarbon source bed, while the Longwanmiao grain-, micrite- and muddy-dolostones and dolomitic limestone would be an important reservoir bed in Chuanzhong Uplift.

14.2.2 Chuanzhong Uplift

A successional giant Chuanzhong Uplift (used to call it "Leshan-Longnvsi Uplift") has constrained on Sinian–Cambrian gas entrapment and enrichment in western YC (Wei et al. 2010; Du et al. 2014; Wang et al. 2014; Liu et al. 2016).

The Doushantuo stratigraphic pinching and facies-change from deep-water shale to shallow-water shale and sandstone towards the Chuanzhong Uplift are revealed by Nvji and Ziyang-1 Wells, and implying the existence of a palaeo-uplift embryonic-form during early Sinian. The palaeo-uplift was keeping on an uplifting state before Early Cambrian Longwangmiao sedimentary period, showing two palaeo-salients respectively at the west and east of Chuanzhong palaeo-uplift (Fig. 14.1), and the palaeo-uplift became more remarkable before Permian (Fig. 14.2). Consequently, the uplifting process was continued until the sedimentary period of non-marine Upper Triassic Xujiahe Formation. Owing to the effect of tectonic dynamism, further uplifting of Sinian top surface made Weiyuan Anticline to be the highest point on the uplift in the Late Palaeozoic, and maintaining Moxi-Gaoshiti Anticlines still to be the structural high points up to present (Fig. 14.3). The long-term uplifting palaeo-tectonic and geomorphological framework has constrained on the development of hydrocarbon source and reservoir beds. The successional uplifting would provide the advantageous place for late cracking-gas enrichment and large-scale Anyue Gasfield formation, i.e., the early entrapment of original oil-reservoirs and the late emplacement of cracking-gas reservoirs, especially for the Gaoshiti and Moxi Anticlines on eastern high points of the uplift.

The current seismic reflector structural contour map of Sinian top surface in western YC appears as an uplifting core area located at Weiyuan Anticline and Moxi-Gaoshiti Anticlines, with total area of 4×10^4 km^2, which resulted from a successional tectonic development. In between, a structural saddle, so called Deyang-Ziyang Faulted-Sag, has separated above core area into two individual salients, i.e., Weiyuan Anticline at the west and Moxi-Gaoshiti Anticlines at the east (Figs. 14.2 and 14.3). Both the anticlinal traps of Sinian top surface respectively occupies an area

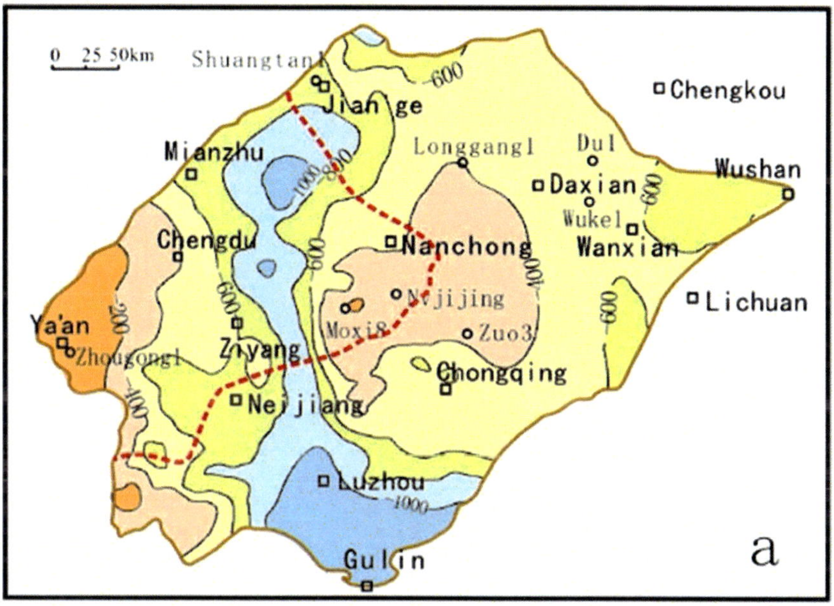

Fig. 14.1 Palaeo-structural contour map of Sinian top surface on Chuanzhong Uplift before the sedimentary period of Longwangmiao Formation (ϵ_1l) showing the distribution of Deyang-Ziyang Faulted-Sag (in light-blue colour), Moxi-Gaoshiti Anticlines (in light-orange colour) and Weiyuan Anticline (in dark-orange colour). The map range is only limited to current Sichuan Basin

of 1800 km^2 in Weiyuan, and of 3500 km^2 in Moxi-Gaoshiti. Thus, Anyue Gasfield are emplaced.

In addition, there is a large slope zone, with total area of 8×10^4 km^2, surrounding the giant Chuanzhong Uplift ever since Sinian. The slope generally had a simple monoclinal setting without any structural trap until Cenozoic, while numerous local structural traps of ca. 4000 km^2 in total closed area are emerged on the slope due to the Himalayan Movement, and respectively distributed in the east and the south of Chuanzhong palaeo-uplift.

14.2.3 Deyang-Ziyang Faulted-Sag

As an important secondary tectonic unit, the Deyang-Ziyang Faulted-Sag is situated in between the Weiyuan Anticline and Moxi-Gaoshiti Anticlines on the Chuanzhong Uplift (Figs. 14.1 and 14.4), and filled principally with the sediments of Lower Cambrian Maidiping Formation (ca. 541–525 Ma in age) and Qiongzhusi-Canglangpu Formations (ca. 520–513 Ma, Table 14.1; Fig. 14.5). Sedimentologically, the lower-middle intervals of Qiongzhusi Formation are a segment of the

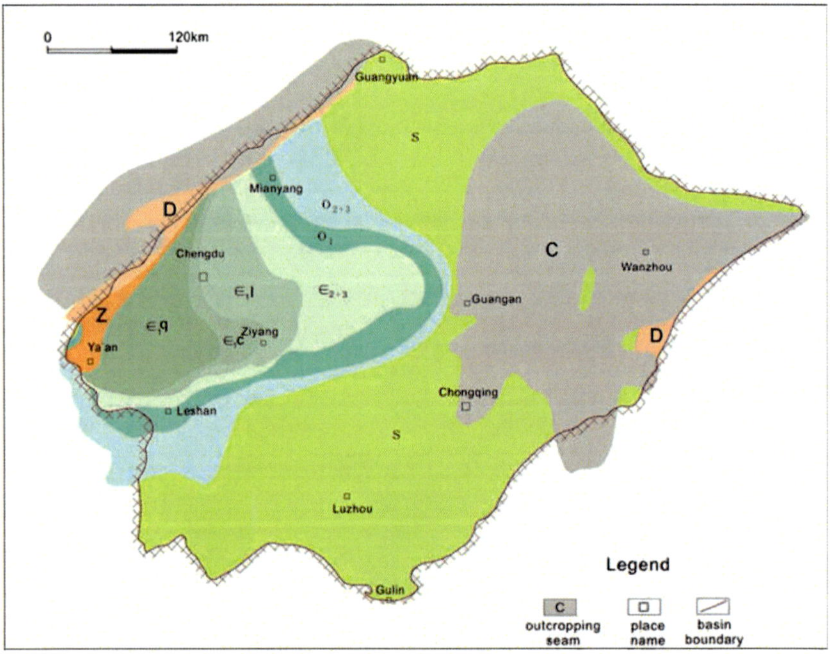

Fig. 14.2 Pre-Permian palaeo-geological map of Chuanzhong Uplift. Stratigraphic symbol: *Z* sinian; ∈ Cambrian; *O* Ordovician; *S* Silurian; *D* Devonian; *C* Carboniferous. The map range is only limited to current Sichuan Basin

Mianyang-Gulin fine-grained sedimentary system, which shows as near NNW-trending extension from Gulin to Mianyang (Fig. 14.4). Its middle part is just across the Chuanzhong Uplift between Longchang and Ziyang with a minimum lateral width of the sedimentary system ca. 50–55 km, while the lateral width of its northern part would be up to 250 km and more (Fig. 14.4). Moreover, there are multiple synsedimentary longitudinal faults developed within the fine-grained sedimentary system (Fig. 14.4), resulting in the obviously thickening of Qiongzhusi-Canglangpu Formations towards the central part of Deyang-Ziyang Faulted-Sag, especially for Qiongzhusi Formation with a maximum thickness up to 540 m (Ziyang-1 Well) and more (Gaoshi-17 Well; Fig. 14.5). Consequently, the Deyang-Ziyang Faulted-Sag would act as the Qiongzhusi depocenter and hydrocarbon source kitchen.

It can be seen from the well-tie section of stratigraphic correlation through Deyang-Ziyang Faulted-Sag (Fig. 14.5) that as basement rocks of the faulted-sag, the Sinian strata are systematically thinning towards the center of Deyang-Ziyang Faulted-Sag, even pinch-out. Moreover, the Deng-4 Member is obviously denuded by Episode II of Tongwan Movement within the faulted-sag, while the Maidiping Formation is just deposited at the low position on the disconformable surface of denudation as senke-filling diamict (Fig. 14.5). Both of which would indicate the existence of an ancient tectonic high point for the Chuanzhong palaeo-uplift at the position of

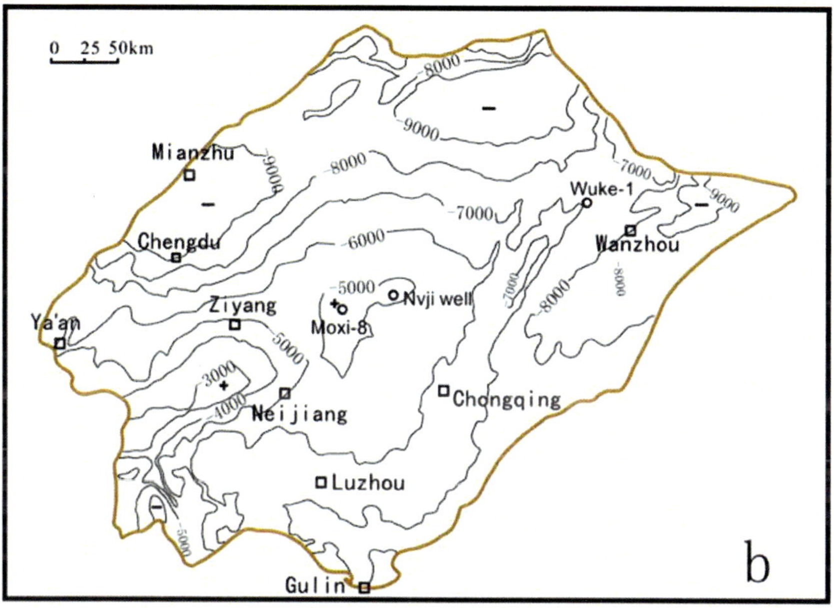

Fig. 14.3 The seismic reflector structural contour map of Sinian top surface on Chuanzhong Uplift with the contour interval 1000 m. The map range is only limited to current Sichuan Basin

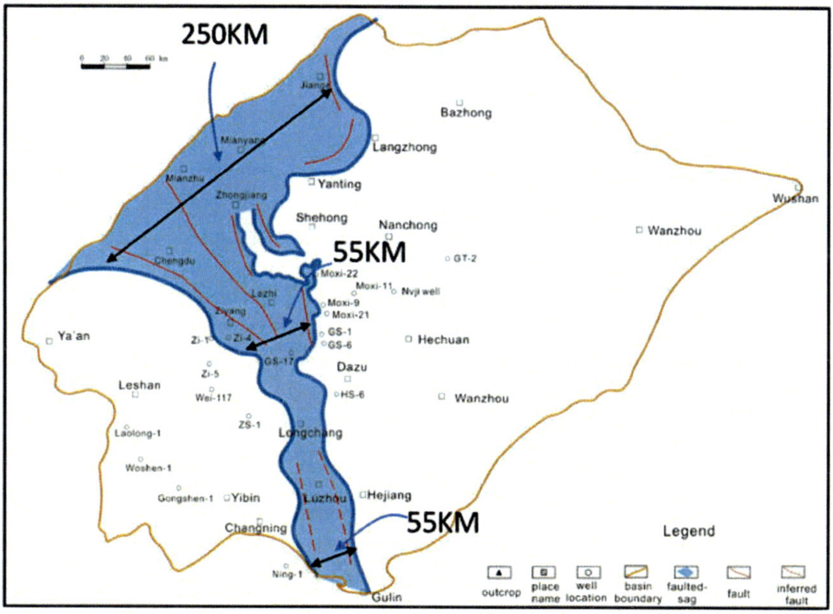

Fig. 14.4 Mianyang-Gulin fine-grained sedimentary system of Lower Cambrian Qiongzhusi Formation. The map range is only limited to current Sichuan Basin

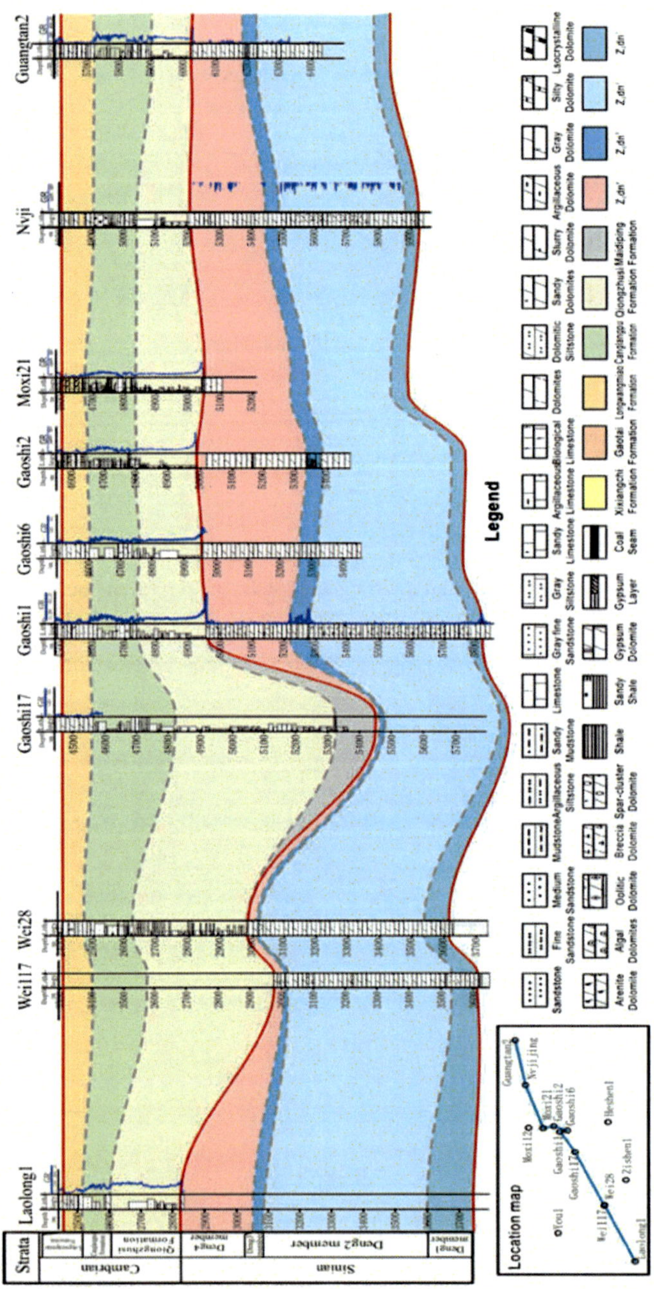

Fig. 14.5 The well-tie section of stratigraphic correlation through Deyang-Ziyang Faulted-Sag (after Wei et al. 2015a)

Deyang-Ziyang Faulted-Sag during Sinian, and thus the Deyang-Ziyang Faulted-Sag is just superposed over the Sinian palaeo-structural high point of Chuanzhong Uplift (Fig. 14.5).

The Deyang-Ziyang Faulted-Sag constrained on the geological conditions of Sinian-Cambrian hydrocarbon emplacement and preservation. The thick-bedded black shale of Lower Cambrian Qiongzhusi Formation are well developed within the faulted-sag as main source kitchen, and both dolostone building-up of Sinian Dengying Formation and grain-dolostone of Lower Cambrian Longmangmiao Formation are at the margin of palaeo-platform on both sides of the faulted-sag as major reservoir beds.

14.3 Hydrocarbon Source Beds

Previous studies commonly accepted that the Sinian natural gas reservoir in western YC is sourced mainly by Lower Cambrian source bed and entrappted and enriched at the Sinian top weathered crust (Bao 1988; Chen 1992; Dai et al. 2003). Based on the investigation of Sinian-Cambrian source rock on downhole cores and peripheral outcrops, it was proposed that there are four sets of potential source beds respectively in Lower Cambrian Qiongzhusi (ϵ_1q) and Maidiping (ϵ_1m) Formations, upper Sinian Deng-3 Member (Z_2dn^3) and lower Sinian Doushantuo Formation (Z_1ds) in the Chuanzhong Uplift (Table 14.1).

14.3.1 Lower Cambrian Qiongzhusi Formation (ϵ_1q)

The Qiongzhusi Formation (ϵ_1q) black and greyish-black shale/mudstone, carbonaceous shale and dolomitic shale, generally with stratigraphic thickness 100–540 m, would be the most effective source bed in western YC, particularly at the Deyang-Ziyang Faulted-Sag as a main source kitchen, where the Qiongzhusi Formation would be up to 540 m thick (in Ziyang-1 Well) and its effective source rocks with TOC $\geq$ 0.5% are up to 300 m-thick, TOC values of its 409 source rock samples average as high as 1.95%. Its macerals are predominantly composed of amorphous sapropelinite (accounting for 95% and more in total macerals), which belonging to Type-I kerogen (sapropel-type) and lower aquatic organism source input. Organic matter appears as floccule under scanning electron microscope (SEM). The isotopic composition of kerogen is slightly lighter with carbon isotopic δ^{13}C values from -36.4‰ to -30‰ (avg. -32.8‰), also indicative of typical sapropel-type of source rock. The corresponding equivalent vitrinite reflectance eqR_o values are between 1.84 and 2.42% (avg. 2.12%), and referred to over-mature phase (Table 14.2; e.g., Gaoshi-17 Well in Fig. 14.6), all of which made Qiongzhusi black shale to be an effective over-mature source kitchen at the Deyang-Ziyang Faulted-Sag.

Table 14.2 Geochemical parameters of potential hydrocarbon source beds in Chuanzhong Uplift

Formation/member	Lithology	Thickness (m)	TOC (%)	$\delta^{13}C_{kerogen}$ (‰)	eqR_o (%)	OM type	Maturity
Qiongzhusi	Shale	100–540	0.50–8.49/1.95[a] (409)	−36.4–−30/−32.8 (60)[a]	1.84–2.42/2.12[a]	Type-I (sapropel-type)	Over-mature
Maidiping		ca. 130	0.52–4.00/1.68	−36.4–−32.0/−34.3	2.23–2.42		
Deng-3		0–37	0.50–4.73/0.87 (62)	−33.4–−28.5/−32.0	3.16–3.21		
Doushantuo		9–30	0.56–14.2/2.91 (95)	−31.0–−30.7/−30.9 (23)	2.08–3.82		

[a]Minimun–maximun/average (analyzed sample number)

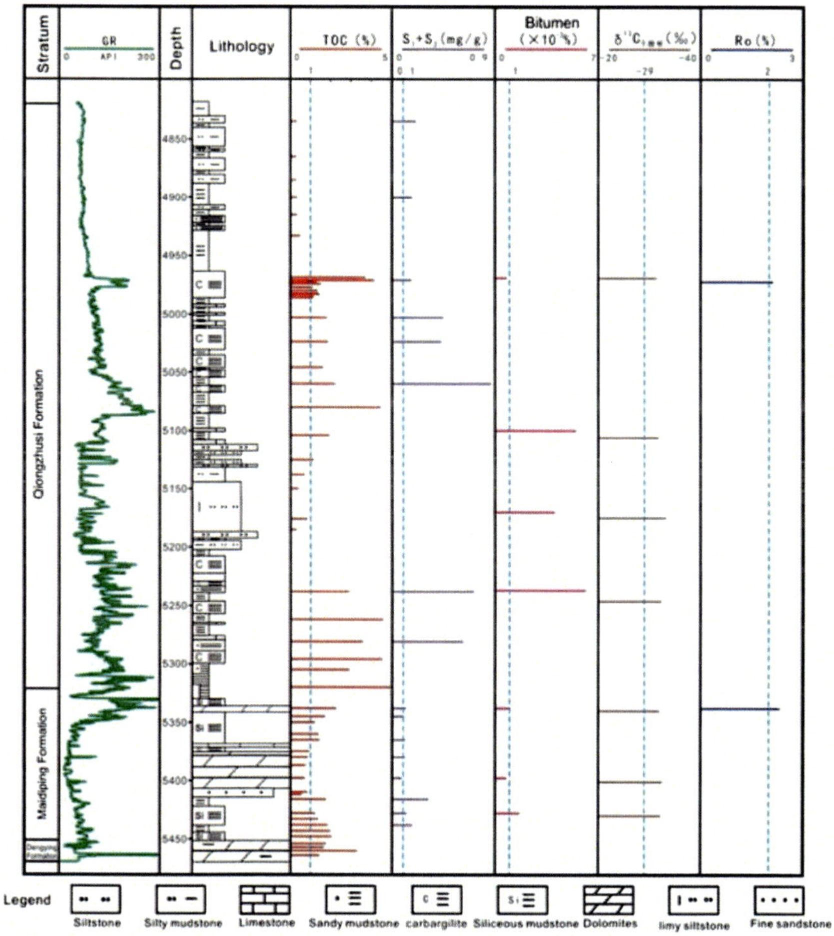

Fig. 14.6 Geochemical profile of Maidiping and Qiongzhusi Formations in Gaoshi-17 Well

14.3.2 Lower Cambrian Maidiping Formation ($\in 1m$)

The potential source bed of Maidiping Formation ($\in_1$m) consists of rhythmic diamict of carbonate, phosphatite, silicalite and clastic rocks with scattered carbonaceous mudstone. The mudstone contains higher organic abundance with TOC values 0.52–4.00% (avg. 1.68%), its kerogen isotopic δ^{13}C values are around −36.4%—−32.0‰ (avg. −34.3%), which is attributed to the Type-I (sapropel-type) source rock. The eqR_o values are 2.23–2.42%, indicating the over-mature phase (Table 14.2; e.g., Gaoshi-17 Well in Fig. 14.6).

In Fact, however, Maidiping Formation belongs to the senke-filling sediments, resulting in its distribution only in a limited area, and its mudstone of high organic abundance is scattered within the rhythmic diamict so that its source rocks are not concentrated. Therefore, the Maidiping source bed is referred to an insignificant source bed in Chuanzhong Uplift.

14.3.3 Deng-3 Member (Z_1dn^3) in Dengying Formation

The lithology of source bed in Deng-3 Member (Z_1dn^3) mainly includes black shale intercalated with scattered thin-bedded grey dolomitic mudstone in Chuanzhong Uplift. The Deng-3 Member source bed contains relatively high organic abundance, the TOC values of 62 source rock samples are around 0.50–4.73% (avg. 0.87%), kerogen isotopes range from −33.4 to −28.5‰ (avg. −32.0‰) also showing the Type-I (sapropel-type) source rock and the eqR_o values are around 3.16–3.21% indicating the over-mature phase (Table 14.2).

As the whole, however, the stratigraphic thickness of Deng-3 Member ranges from 10 to 50 m. Relatively speaking, so far its thicker source bed is only found in few exploratory wells such as a 35.5 m thick black mudstone source bed in Gaoke-1 Well. While the Deng-3 Member in the periphery region of Chuanzhong Uplift is also thinning with bluish-grey mudstone as major lithology so that the Deng-3 source bed is only of local significance.

14.3.4 Lower Sinian Doushantuo Formation (Z_1ds)

Doushantuo Formation (Z_1ds) is a set of dolostone and shale alternative stratum, its stratigraphic thickness is variable from 9 to 220 m on Chuanzhong Uplift and surrounding region. Doushantuo shale is mainly developed around periphery depressions of the Chuanzhong Uplift. However, it was scarcely encountered by drilled wells on Chuanzhong Uplift, and the disclosed stratigraphic thickness of Doushantuo Formation is very thin, generally 10–30 m, even only 9 m-thick in Nvji Well, and its lithology and lithofacies are strikingly variable from deep-water black shale to shallow-water purple and greyish-green shale, very-fine crystal dolostone, even grey sandstone, and thus no source rock is available in Wei-117, Ziyang-1, Wei-28 and Nvji Wells within the Deyang-Ziyang Faulted-Sag.

The Doushantuo black shale contains high organic abundance with TOC values around 0.50–14.2% (avg. 2.91%) and carbon isotopic $\delta^{13}C$ values ranging from −31.2 to −30.7‰ (avg. −30.9‰), showing the Type-I (sapropel-type) source rock. Its corresponding eqR_o values are between 2.08 and 3.82% within the over-mature phase (Table 14.2).

Although the Doushantuo black shale source bed has high organic abundance and eqR_o value indicates the over-mature phases (cf. Chap. 13), so far the Doushantuo black shale is not disclosed by any drilled wells on the Chuanzhong Uplift yet. However, its source kitchen may be still available surrounding the Chuanzhong

Uplift, especially on the north or northwest of the uplift based on seismic data (cf. Fig. 11.29 in Chap. 11). Therefore, Doushantuo black shale would still be a potential effective source bed for Chuanzhong Uplift.

14.4 Reservoir Beds

14.4.1 Lower Cambrian Longwangmiao Formation ($\in_1 l$)

14.4.1.1 Lithology and Reservoir Space

The lithological types of reservoir rocks in the Longwangmiao Formation ($\in_1$l) are mainly composed of grain-dolostones, i.e., dolarenite and crystal dolostone, including fine-medium-coarse crystal dolostone, arenitic very-fine-crystal dolostone, micritic and very-fine-crystal dolarenite (Plate 14.1).

Its reservoir space includes karst caves, intergranular and intercrystalline solution vugs/pores, and fractures (Plate 14.1). Solution vugs/pores are the most important reservoir space with a long axis of 0.2–12.0 mm (mainly 4–8 mm). Intergranular pores are mainly developed in dolarenite and crystal dolostone, generally with pore diameter of 0.02–0.08 mm, appearing mostly as irregular polygon-shape, often with fibrous and bladed dolomitic cement, while the intercrystalline pores are only found within crystal dolostone, their pore diameter is closely related to the crystal-grain size, generally with 0.003–0.004 mm, occurring as triangle- or irregular polygon-shape often with half-filled bitumen. Fractures can be seen on the cores, including structural joints, diagenetic fractures and sutures. Structure joints are generally more straight mostly with high angle, while solution fractures generally resulted from the corrosion of fresh water or underground water, slot walls are not straight sometimes with embayed or beaded-shape.

14.4.1.2 Physical Properties of Reservoir Bed

As concerns the physical porosities, the measured reservoir porosity values are just between 2.0 and 18.5% (avg. 4.3%) and the matrix permeability ranges from 0.001 to 2 mD (avg. 1.59 mD) for the core plugs.

However, the statistic histograms of whole core samples show that the measured reservoir porosity values are distributed within the range of 0 to > 10%, and its dominant frequency is in the range of 4–6%. The porosity values of 2.0–4.0% account for 27.8% in the total measured samples, 4.0–6.0% for 41.7% and > 6.0% for 20.5%, with an average of 4.81% (Fig. 14.7a).

While the measured reservoir permeability values are within the range of 0.010, 1–78.5 mD, its dominant frequency is in the range of 0.01–10 mD with the overall permeability average of 3.91 mD for total measured whole core samples (Fig. 14.7b)

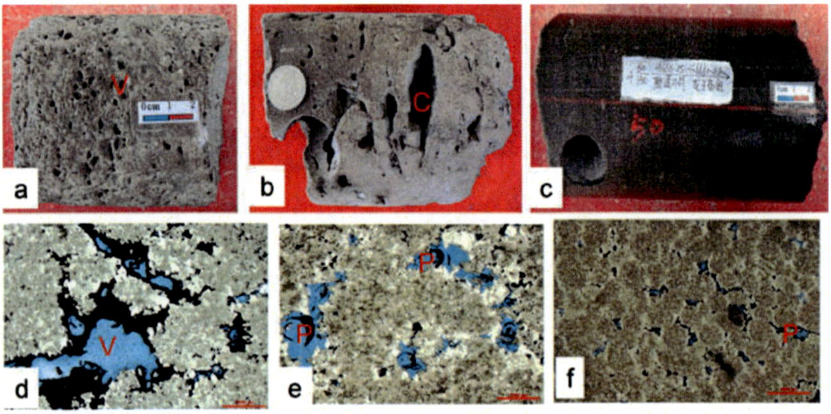

Plate 14.1 Typical lithological types of Longwangmiao Formation in Chuanzhong Uplift. **a** Fine-crystal dolostone, small vugs (V) and pinhole, Moxi-13 Well, 4607.68 m; **b** medium-coarsely crystal dolostone, solution caves (C), Moxi-204 Well, 4667.27 m; **c** oolitic dolostone, Moxi-21 Well, 4660.25 m; **d** fine-medium crystal residual dolarenite, intergranular solution vugs (V, with cast), Moxi-17 Well, 4623.24 m, single polar, ×20; **e** fine-crystal dolarenite, intergranular solution Vugs (V, with cast), Gaoshi-10 Well, 4624.2 m, single polar, ×20; **f** dolostone, intercrystalline solution pores (P, with cast), Moxi-202 Well, 4660.3 m, single polar, ×20. **a–c** Photos; **d–f** microphotos; cast in light blue colour

and with the average permeability values of 0.534–17.73 mD for individual wells, all of which would be obviously better than that of core plugs. According to the physical property evaluation, however, the Longwangmiao Formation would be referred to a set of low-porosity and low-permeability carbonate reservoir beds.

14.4.1.3 Occurrence and Distribution

There vertically are four sets of superposed and continuously gran shoals in the Longwangmiao Formation (Fig. 14.8), which have been reformed by three episodes of karstification respectively during syndiagenetic, epidiagenetic and burial stages, resulting in a set of high-quality reservoir beds with a large distributional area.

The Longwangmiao reservoir bed mainly consists of pore type of reservoir rocks, locally with fracture-pore/cave type of reservoir rocks. The cumulative thickness of reservoir rocks ranges from 10.8 to 61.1 m (avg. 39.5 m) in individual wells. The distributional area with the largest thickness of reservoir bed is located around Moxi Anticline, followed by the vicinity of Gaoshiti Anticline. The Longwangmiao reservoir bed is distributed along a NE–SW trend, which is consistent with the orientation of the gain-shoals and the distributional area of Types I (good) and II (better) reservoir beds would be up to 52,700 km^2. However, there is an underdeveloped zone of reservoir beds just in between Moxi and Gaoshiti Anticlines.

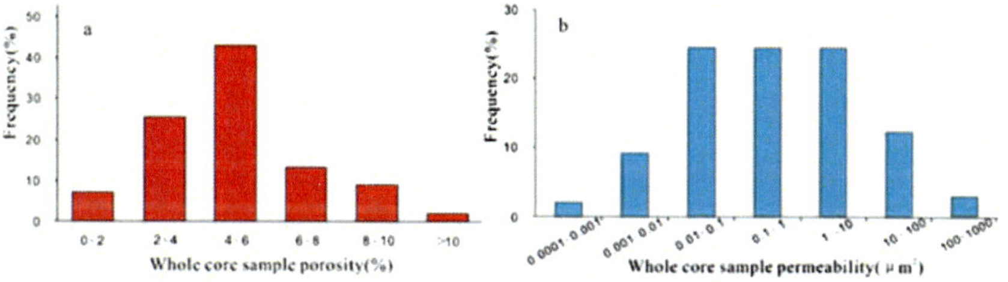

Fig. 14.7 Statistic histogram of porosity **a** and permeability **b** of whole core samples for the reservoirs bed of Longwangmiao Formation in Chuanzhong Uplift

14.4.2 Upper Sinian Dengying Formation (Z_2dn)

14.4.2.1 Lithology and Reservoir Space

The Upper Sinian Dengying Formation consists of primary algal dolostone with platform marginal and restricted platform facies as well as slope-shelf facies, which are constrained by the synsedimentary Chuanzhong palaeo-uplift.

The lithology of Dengying Formation mainly includes five types, i.e., micritic to very-fine-crystal dolostones, algal dolarenite, laminited dolostone, thrombolitic dolostone and snowflake dolostone. The micritic to very-fine-crystal dolostones are referred to microbial dolostone formed by cyanobacteria-participating dolomitic deposition (Plate 14.2c). The clasts in the algal dolarenite mainly consist of micritic and microcrystal dolostones or microbial microcrystal dolostone, and showing irregular round- and angular-shapes (Plate 14.2a). The laminited dolostone is rich-in microbial mat ichnites with flat lamellar- or wavy-bedding (Plate 14.2b, e, f), and mainly situated at shallow tidal flats and the top of lime-mud-mound. The thrombolitic dolostone is a sort of microbiota totally different from stromatolites, its macroscopic thrombolite result from medium-sized thrombolitic texture, and mostly composed of indistinguishable calcified globular-cyanobacterial and discontinuous microbiota. If its organic content is low, it would contain internal sphaerolitic texture and small dissolution pores, while the organic content is high, the internal fenestrae are developed (Plate 14.2f). The snowflake dolostone belongs to one kind of laminites, when the algal mat is unevenly growing or appears as size-variable patches, the snowflake spots would occur on rock fractures, which are actually the contemporaneous or succeeding dolomite filled within the body-cavity after the algal mat decomposition.

The reservoir spaces of Dengying Formation are basically composed of residual pinholes, solution vugs, karst caves and Yanshanian or Himalayan fractures, among which medium-small size vuggy-cavern-layer would be the most important reservoir spaces (Plate 14.3) so that it can be characterized as the fracture-vug-cave type of reservoir bed.

However, the lithology and reservoir spaces still have a certain diversity between the reservoir beds respectively in Deng-2 and Deng-4 Members. The major lithologies of Deng-4 reservoir bed are syanobacterial stromatolitic, laminited and thrombolitic dolostones, their reservoir spaces are residual intergranular vugs as well as residual karst fractures, pores and caves. While the syanobacterial mound or shoal dolostone and grain dolostone constitute the major lithology of Deng-2 reservoir bed and the residual karst vugs and fractures of bedding distribution would be the main reservoir space.

14.4.2.2 Physical Properties of Reservoir Bed

The reservoir bed of Dengying Formation is characterized by low porosity and low permeability locally with high porosity intervals. Overall, its porosity averages 3.10%, while its horizontal permeability as high as 6.24 mD and vertical permeability only 0.81 mD for whole core samples (Fig. 14.9).

As to the Deng-4 reservoir bed, the porosity ranges 0.97–8.02% (avg. 4.28%) for 28 whole core samples, and the permeability 0.001–6.32 mD (avg. 1.49 mD) for 14 whole core samples in Gaoshi-1 Well. While the porosity is 0.95–7.89% (avg. 2.02%) for 24 whole core samples, and the permeability 0.001–9.32 mD (avg. 0.63 mD) for 18 whole core samples in Moxi-8 Well.

Concerning the Deng-2 reservoir bed, its porosity is 4.9–8.8% (avg. 6.2%) for 6 whole core samples in Gaoke-1 Well, and 2.7–5.51% (avg. 3.8%) for 6 whole core samples in Pan-1 Well.

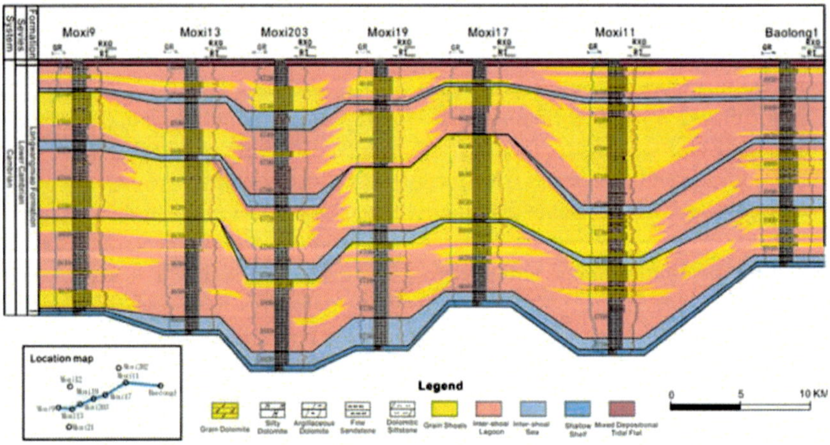

Fig. 14.8 The well-tie geological section showing the vertical occurrence of Longwangmiao carbonate reservoir beds in Moxi Anticline, Anyue Gasfield

14.4.2.3 Occurrence and Distribution

The development of Sinian Dengying Formation would be constrained by the Tongwan palaeo-geomorphological pattern. Since the Dengying Formation had experienced the overall uplifting course of YC through the Episodes I, II and III of Tongwan Movement respectively at the end of Deng-2 and Deng-4 Members in Maidiping sedimentary periods (Table 14.1; Wang et al. 2014; Wu et al. 2016), the Deng-2 and Deng-4 Members suffered the weathering, leaching, eroding and reforming respectively, resulting in two sets of widely distributed Deng-2 and Deng-4 karst reservoir beds, with the cumulative thickness of 36–148 m (avg. 70 m) in the Moxi and Gaoshiti Anticlines as the Anyue Gasfield, but the Weiyuan Anticline/Gasfield only contains Deng-2 karst reservoir bed without Deng-4 reservoir bed (Fig. 14.10).

The superior reservoir beds of Dengying Formation are mainly distributed along both sides of the platform margin and the distributional ranges of types I and II reservoir beds are 148,000 km^2 for the Deng-2 Member and 125,000 km^2 for the Deng-4 Member, respectively.

Plate 14.2 Main lithological types of Sinian reservoir rocks in Chuanzhong Uplift. **a** Algal dolarenite with round algal fragments, Z_2dn^2, Gaoshi-1 Well, 5298 m, single-polar, ×100; **b** spume laminited dolostone, Z_2dn^2, Gaoshi-1 Well, 5369 m, single-polar, ×100; **c** very-fine-crystal dolostone, Gaoshi-1 Well, Z_2dn^4, 4972.8 m, single-polar, ×40; **d** spume laminited dolostone, Z_2dn^2, Wei-117 Well, 3044.8 m, single-polar, ×40; **e** spume laminited dolostone, Z_2dn^2, Gaoke-1 Well, 5209 m, single-polar, ×40; **f** cotted dolostone, Z_2dn^4, Moxi-8 Well, 5106.93 m, single-polar, ×40. **a–f** Microphotos

14.5 Gas and Gas Reservoirs

14.5.1 Gas Composition

The gas composition includes generally hydrocarbons and non-hydrocarbons. The hydrocarbon components of Dengying and Longwangmiao gas reservoirs in Chuanzhong Uplift predominantly consist of methane (79.1–97.2%, Table 14.3) with a trace amount of ethane (0.03–0.90%) and without C_3^+ hydrocarbons in most case. Its dry index (C_1/C_{1-5}) is up to 1.0, which are indicative of typical cracking dry gas in over-mature phases.

The non-hydrocarbon components of Dengying-Longwangmiao gas reservoirs are composed of CO_2, N_2, H_2S, rare gases He and Ar in Chuanzhong Uplift, among which CO_2 (1.4–14.7%) and N_2 (0.4–15.5%) contents are relatively the major components, while H_2S (0.8–7.2%) is minor one, and He (0.01–0.4%) plus Ar (0–0.10) are just in trace (Table 14.3), among which, Ar is only detectable at Weiyuan Gasfield.

As the minor composition, there still are some differences in the non-hydrocarbons of Dengying and Longwangmiao gas reservoirs between the Anyue and Weiyuan Gasfields (Table 14.3). The CO_2 content (1.4–14.7%) is a dominant component, N_2 (0.4–2.5%) and H_2S (0.8–2.8%) are minor, and He content (0.01–0.0.06%) is in trace, and no Ar is detectable in Moxi-Gaoshiti Anticlines (Anyue Gasfield). Whereas the N_2 content (2.7–15.5%) is relatively dominant, H_2S (1.1–7.2%) and CO_2 (2.1–6.8%) are minor, and rare gas He (0.1–0.4%) could reach production grade (Table 14.3) and even Ar content (0–0.15%) could be detected in Weiyuan Anticline. These differences may result from the diversities in source rocks and/or gas evolution process.

14.5.2 Gas Reservoirs

Stratigraphically, Anyue Gasfield is composed of the upper Sinian Deng-2, Deng-4 and Lower Cambrian Longwangmiao gas reservoirs on the Chuanzhong Uplift. Tectonically, its Deng-2 gas reservoir is distributed respectively at Weiyuan, Moxi and Gaoshiti Anticlines, the Deng-4 gas reservoir only at Moxi and Gaoshiti Anticlines (Fig. 14.10), and the geological reserve of Longwangmiao gas reservoir is proven just at Moxi Anticline and predicted at Gaoshiti and Longnvsi Anticlines on the south and east of Moxi Anticline respectively (Fig. 14.11), The Longwangmiao reservoir beds is generally 10–60 m-thick, among which, the thickest reservoir bed is developed at the Moxi Anticline.

Table 14.3 Natural gas composition in Chuanzhong Uplift

Reservoir (Well)		Gas composition (%)						
		CH_4	C_2H_6	Dry index (C_1/C_{1-5})	CO_2	N_2	H_2S	He
Anyue Gasfield (Moxi-Gaoshiti Anticlines)								
ϵ_1l	Longwangmiao	96.3–97.2	0.13–0.14	1.00	1.4–3.4	0.6–2.4	1.4–9.6 g/m^3	0.01
Z_2dn	Dengying	82.7–94.6	0.03–0.05	1.00	4.1–14.7	0.4–2.5	0.8–2.8	0.01–0.06
Weiyuan Gasfield/Anticline								
Z_2dn	Dengying	79.1–97.2	0.04–0.90	1.00	2.1–6.8	2.7–15.5	1.1–7.2	0.1–0.4

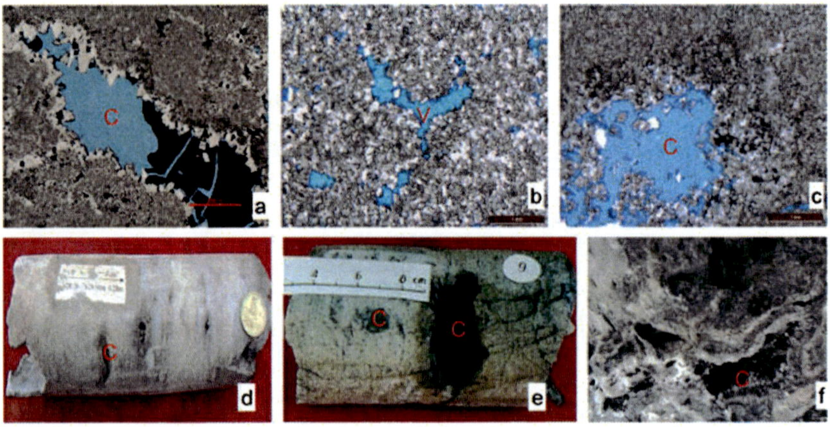

Plate 14.3 The characteristics of reservoir space in Dengying Formation, Chuanzhong Uplift. **a** Very-fine-crystal dolostone, solution cave (C, cast) with half-filled with reservoir bitumen, Z_2dn^4, Gaoshi-1 Well, 4984.9 m, single-polar, ×40; **b** medium-coarse-crystal dolostone, intergranular Vugs (V, cast), Z_2dn^4, Gaoshi-7 Well, 5282 m, single-polar, ×20; **c** very-fine-crystal dolostone, intergranular solution caves (C, cast), Z_2dn^2, outcrop section at Ronjing, single-polar, ×20; **d** dolostone, solution caves (C), Z_2dn^4, Pan-1 Well, 5628.6 m; **e** very-fine-crystal dolostone, karst caves (C), Z_2dn^2, Zi-1 Well, 5628 m; **f** dolostone, solution caves (C) filled with bitumen, Z_2dn^2, outcrop at Songlin, Guizhou. **a–c** Microphotos; **d–f** photos; cast in light blue colour

14.5.3 Longwangmiao Gas Reservoir

It is a karst dolomitic gas reservoir developed on the anticlinic background with an average reservoir thickness of 36 m and a gas-column height up to 232 m (Figs. 14.12 and 14.13).

The gas reservoir contains an area up to 800 km^2, which is beyond the range of lowest anticlinic closed contour (i.e., so-called "spill point") on the east side of the anticline, and has a lithologic sealing zone on the west side due to the worsening reservoir quality (Fig. 14.12). While its gas–water-contacts (GWC) occur at different depth in the northern marginal region, e.g., the GWC at the depth of −4385 m in Moxi-47 Well, at −4459 m in Moxi-27 Well and at −4593 m in Moxi-51 Well

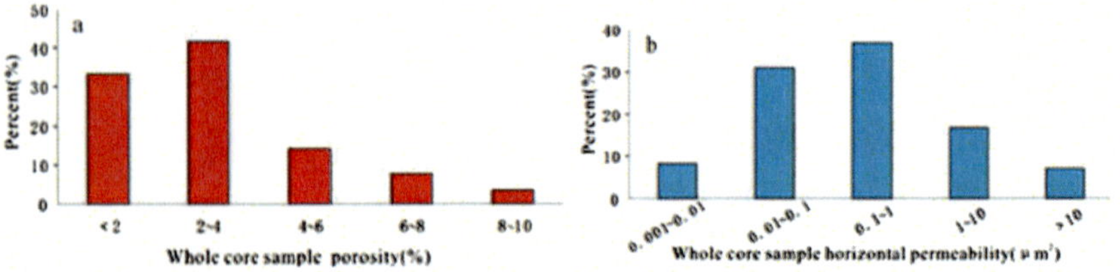

Fig. 14.9 Statistic histogram of porosity **a** and permeability **b** of whole core samples for the Dengying Formation

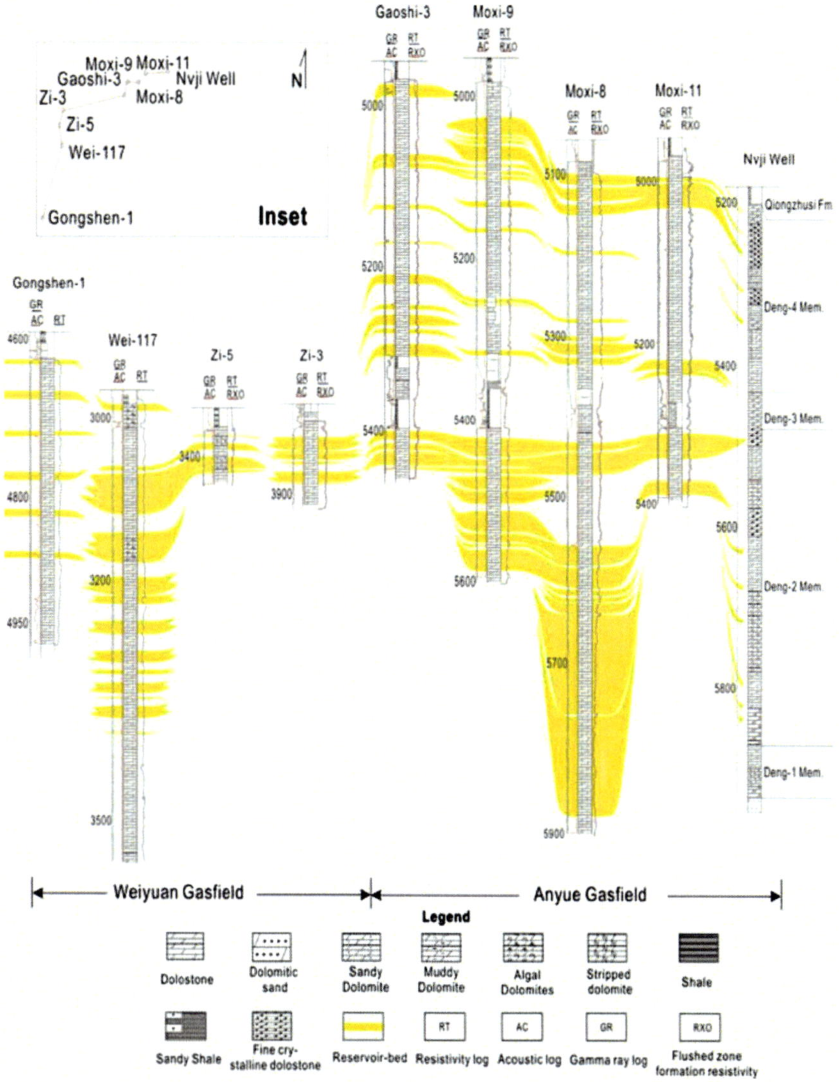

Fig. 14.10 The well-tie geological section showing the vertical distribution of Deng-2 and Deng-4 karst carbonate reservoir beds in Anyue and Weiyuan Gasfields

because of the lithologic constraint of grain-shoal reservoir bed (Figs. 14.11 and 14.13). Therefore, the Moxi Longwangmiao gas reservoir would be attributed to a structural-lithologic type of gas reservoir.

The Longwangmiao gas reservoir is a deeply buried system, with high formation temperature and high formation pressure, in the Moxi Anticline, its burial depth exceeds 4600–4700 m with the average formation temperature up to 140.3–150.4 °C and formation pressure 75.7 MPa at the middle interval of the gas reservoir with a high pressure coefficient of 1.65.

The P1 gas geological reserve of Moxi Longwangmiao gas reservoir is 4403.83 × 10^8 m^3, and thus the gas reservoir is attributed to a large-scale individual carbonatic gas reservoir with the largest geological reserve in China.

14.5.3.1 Deng-4 Gas Reservoir

The Deng-4 reservoir beds are only preserved in the Anyue Gasfield (Fig. 14.10), and the Deng-4 gas reservoir is referred to a lithologic-stratigraphic type of gas reservoir developed on the background of Moxi-Gaoshiti-Longnvsi Anticlines on Chuanzhong Uplift. Structurally, the three Anticlines share a common closed line along the structural contour of −5010 m with a unified structural trap area of 3474 km^2 (Fig. 14.14). While the Deng-4 gas reservoir contains a GWC along the contour of −5230 m, and occupying a gas-bearing area up to 7500 km^2, which is remarkably larger than the unified structural trap area (Fig. 14.14). A bottom water layer has been encountered within the lower interval of Deng-4 Member only in Moxi-22 Well at the northern low site of Moxi Anticline, while no bottom water was found by drilling at the southern part of the anticline (Fig. 10.15; Wei et al. 2015b; Du et al. 2016).

Moreover, the Deng-4 gas reservoir appears as two asymmetric gas-column heights, i.e., 370 m at Gaoshi-19 Well on its south side and 590 m at Moxi-22 Well on its north side (Figs. 14.14 and 14.15). A gas test production of more than 1 × 10^6 m^3/d has been obtained from the upper interval in the Moxi-22 Well.

During the sedimentary period of Deng-4 Member, the rimmed sedimentary platform was well developed on the Chuanzhong palaeo-uplift, its western marginal facies tract has an area of 1500 km^2, where the karst vuggy-cavern-layers at the top erosional surface of Deng-4 Member could be vertically extended down to 300 m beneath the top surface, resulting in a concentratively distributed range of superior karst dolostone reservoir beds with the accumulative reservoir thickness as high as 60–110 m, becoming the high yield sweet spot zone of the Deng-4 gas reservoir at the marginal facies tract (Figs. 14.14 and 14.15).

In addition, there still is a favorable gas-bearing area of 6000 km^2 on the east of the marginal facies tract, where the Deng-4 Member shows as thin-bedded reservoir beds generally with the cumulative reservoir thickness of less than 40 m as compared to the western platform marginal facies tract or gas sweet spot zone (Figs. 14.14 and 14.15).

The high-quality hydrocarbon source kitchen of Lower Cambrian Qiongzhusi black shale is mainly developed within the Deyang-Ziyang Faulted-Sag, which adjoins the western marginal facies tract of Dengying rimmed platform (Fig. 14.14), both of which could constitute a laterally sourced configurational relationship between source and reservoir, and also acts as a function of lateral sealing for the Deng-4 gas reservoir entrapment.

The burial depth of Deng-4 gas reservoir reaches 5000–5100 m. At the middle interval of Deng-4 gas reservoir, its formation temperature is up to 149.6–161.0 °C

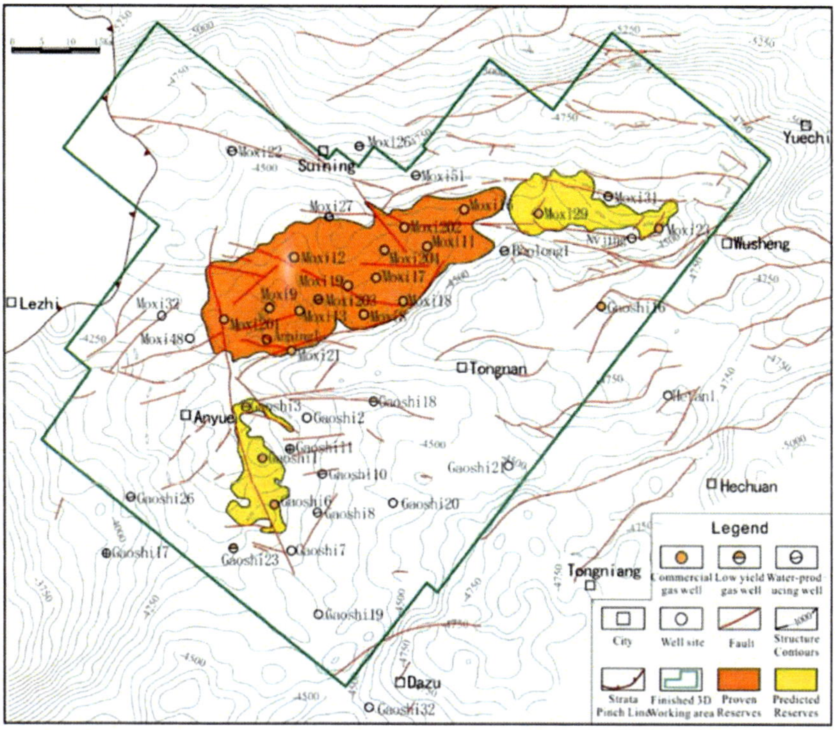

Fig. 14.11 Distribution of proven and predicted Longwangmiao gas reservoirs in Anyue Gasfield

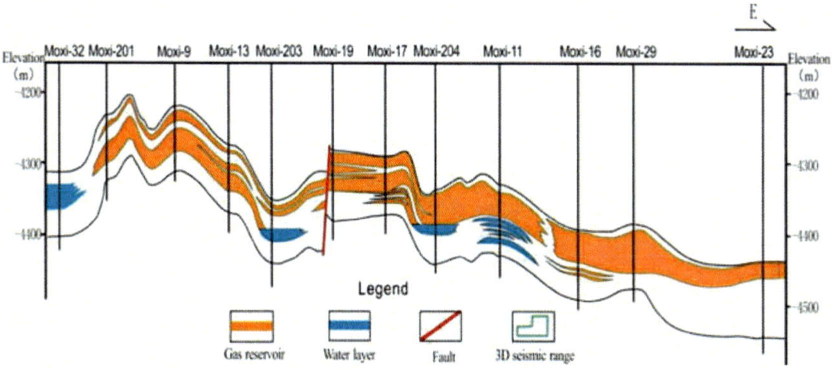

Fig. 14.12 A near WE tend geologic cross section shoeing the Longwangmiao structural-lithologic gas reservoir in Moxi Anticline

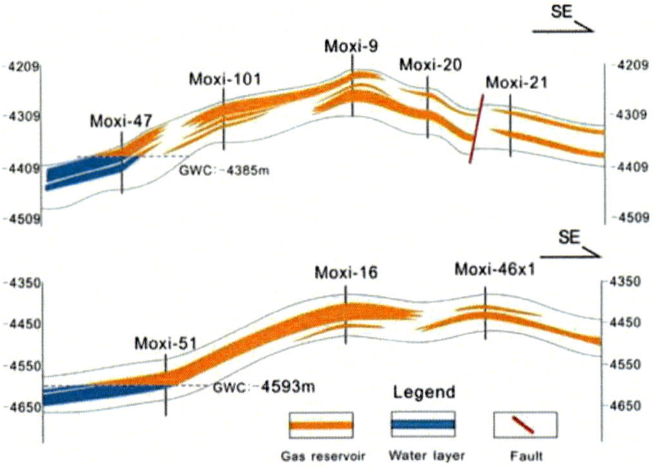

Fig. 14.13 The NW–SE tendency geologic cross section shoeing the Longwangmiao structural-lithologic gas reservoir in Moxi Anticline

and formation pressure 56.57–56.63 MPa with a normal pressure coefficient of 1.06–1.13. Therefore, it is an ultra-deep buried gas reservoir with high formation temperature and normal formation pressure.

14.5.3.2 Deng-2 Gas Reservoir

The Deng-2 reservoir beds are widely distributed in both Weiyuan and Anyue Gasfields (Fig. 14.10). In Anyue Gasfield, however, the Deng-2 gas reservoir is discontiguous and separated by a structural saddle with two-step faults between the Moxi and Gaoshiti Anticlines, which could be attributed to the structural type of gas reservoirs with bottom water layer (Figs. 14.16 and 14.17).

On the structural contour map of the top surface of Deng-2 Member, there is no unified structural closed line in the NEE-trending Moxi and SN-trending Gaoshiti Anticlines, resulting in two individual trap structures respectively with the lowest closed lines along the −5170 m contour at the Moxi and along the −5150 m at the Gaoshiti. Their trap area and closure are 586 km^2 and 140 m for the Moxi, and 540 km^2 and 160 m for the Gaoshiti. Both Moxi and Gaoshiti gas reservoirs are obviously constrained by their anticlinic structures and have respective unified GWCs, e.g., along the −5160 m contour at Moxi-8 Well and along the −5150 m contour at Gaoshi-1 Well (Fig. 14.17; Du et al. 2016).

So far there totally are 13 commercial gas wells drilled at Gaoshiti and Moxi Anticlines, controlling the total gas-bearing area of 970 km^2. Totally, 5.1–69.1 m-thick (avg. 34.3 m) effective Deng-2 reservoir bed is revealed by drilling and its lithology and reservoir space are basically similar to these of Deng-4 reservoir bed, but it is laterally more stable with better continuity as compared to the Deng-4

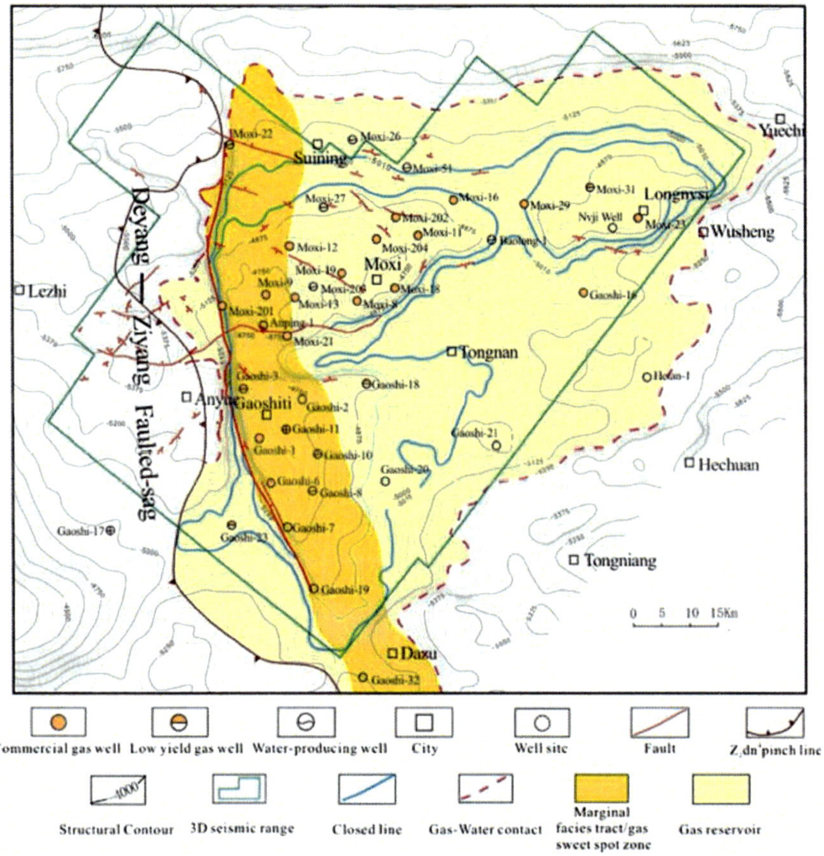

Fig. 14.14 The distribution of marginal facies tract (sweet spot zone) and favorable gas-bearing area in Deng-4 gas reservoir, Anyue Gasfield

gas reservoir. The upper interval of Deng-2 Member produces gas and the lower interval generally contains water as the bottom water layer. The gas-bearing range is respectively constrained by current Gaoshiti and Moxi anticline structures with a similar GWC at the depth of −5150 m at Gaoshi-1 Well and −5160 m at Moxi-8 Well (Fig. 14.16).

Just like the Deng-4 Member, the Deng-2 gas reservoir is also an ultra-deep buried system with high formation temperature and normal formation pressure. Its burial depth reaches 5300–5400 m. At its middle interval, the formation temperature is 155.8–159.9 °C and formation pressure 56.58–59.08 MPa with a normal pressure coefficient of 1.06–1.10.

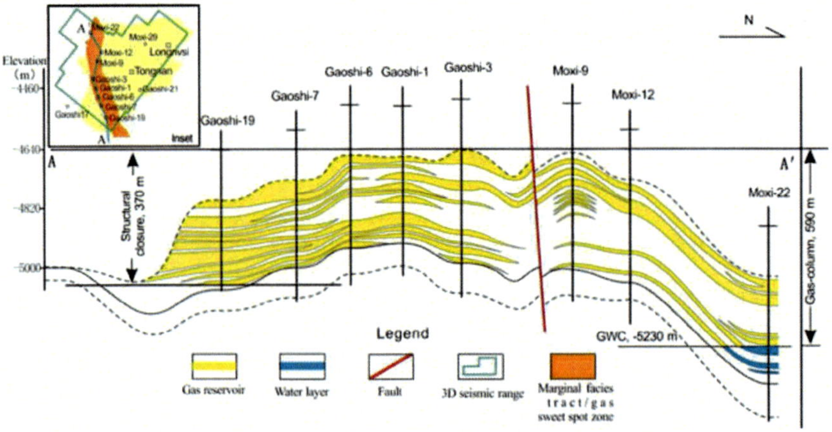

Fig. 14.15 A geological section of Z_2dn^4 Deng-4 gas reservoir in Moxi-Gaoshiti Anticlines

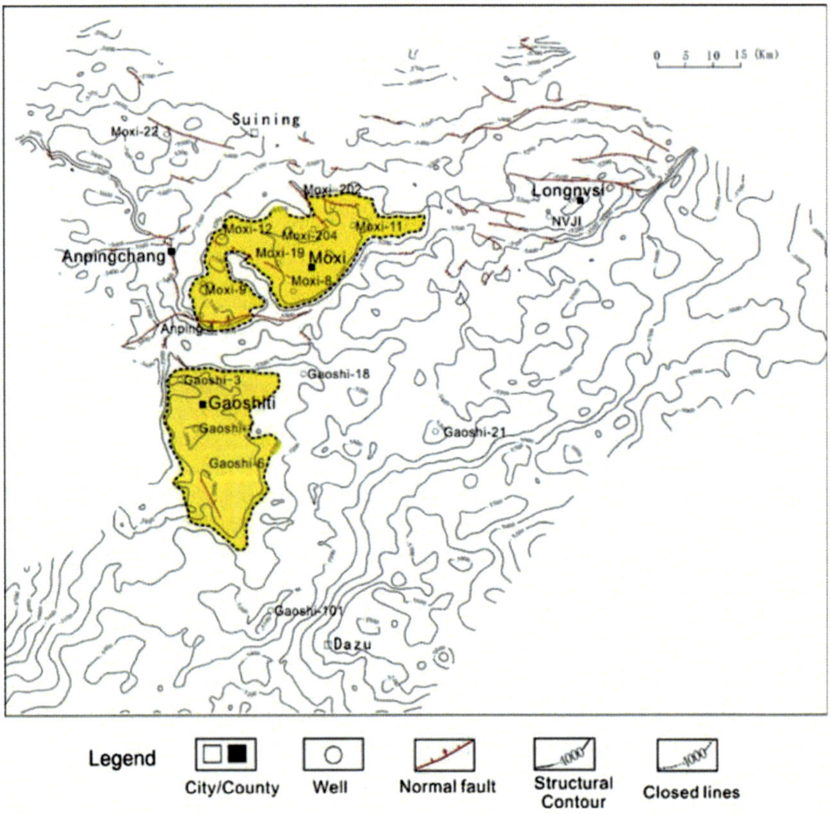

Fig. 14.16 Structural contour map of the top surface and the distribution of gas reservoirs for Deng-2 Member in Moxi-Gaoshiti-Longnvsi Anticlines

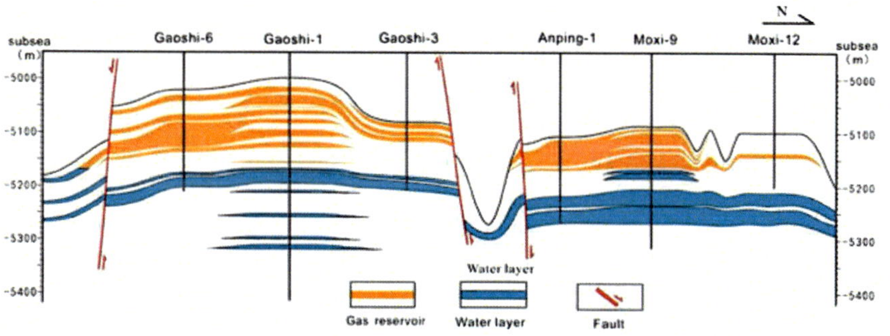

Fig. 14.17 A geological section of Deng-2 gas reservoirs in Gaoshiti and Moxi Anticlines

14.6 Conclusion

(1) Anyue Gasfield is located at the Moxi-Gaoshiti Anticlines in the east salient of the Chuanzhong Uplift, western YC, southwest China. It is the karst carbonate gasfield with the oldest stratigraphic age, the highest thermo-evolution level and the largest individual gas reserve in China. By the end of 2016, its P1 gas geological reserve is up to 8488×10^8 m^3 and 3P gas reserves about 1.2×10^{12} m^3.

(2) There are four sets of potential source beds respectively in Lower Cambrian Qiongzhusi and Maidiping Formations, Upper Sinian Deng-3 Member and Lower Sinian Doushantuo Formation on the Chuanzhong Uplift, among which the black shale of Qiongzhusi Formation would be the main source bed and the Deyang-Ziyang Faulted-Sag acts as the major source kitchen.

(3) Stratigraphically there are three gas-bearing sequences in Anyue Gasfield, i.e., the Lower Cambrian Longwangmiao Formation and the Deng-2 and Deng-4 Members of Upper Sinian Dengying Formation. The lithological types of reservoir rocks are mainly composed of dolarenite and crystal dolostone in the Longwangmiao Formation and micritic to very-fine-crystal dolostones, algal dolarenite, laminited, thrombolitrtic and snowflake dolostones in the Dengying Formation.

(4) The Longwangmiao Formation develops a unified structural-lithologic type of gas reservoir in the Moxi-Gaoshiti-Longnvsi Anticlines. The Deng-4 Member contains lithologic-stratigraphic type of gas reservoir developed on the background of Moxi-Gaoshiti Anticlines, and the Deng-2 Member have two structural types of gas reservoirs with bottom water in Weiyuan and Moxi-Gaoshiti Anticlines.

(5) All the gas reservoirs produce dry gas with medium to low sulfur content and medium CO_2. The Longwangmiao gas reservoir is attributed deep buried high-temperature and high-pressure gas reservoir, while the Deng-2 and Deng-4 gas reservoirs to ultra-deep buried high temperature and normal pressure ones.

References

Bao C (1988) Natural gas geology. Science Press, Beijing (in Chinese)

Chen WZ (1992) Discussing again on the source of Sinian gas pool in Weiyuan Gasfield in Sichuan Basin. Nat Gas Ind 12(6):28–32 (in Chinese with English abstract)

Dai JX, Chen JF, Zhong NN, Pang XQ, Qing SF et al (2003) Large gasfield and their gas sources in China. Petroleum Industry Press, Beijing, pp 16–30 (in Chinese)

Du JH, Zou CN, Xu CC, He HQ, Shen P, Yang YM, Li YL, Wei GQ, Wang ZC, Yang Y (2014) Theoretical and technical innovations in strategic discovery of a giant gas field in Cambrian Longwangmiao Formation of central Sichuan palaeo-uplift, Sichuan Basin. Pet Explor Dev 41(3):268–277 (in Chinese with English abstract)

Du JH, Wang ZC, Zhou CN, Sheng P, Zhang BM, Jiang H, Huang SP (2016) Discovery of intercraton rift in the Upper Yangtze and its control effect on the formation of Anyue giant gas field. Acta Petrolei Sinica 37(1):1–16 (in Chinese with English abstract)

Editorial Committee of Petroleum Geology of China (1989) Petroleum geology of China, vol 10: Sichuan oil and gas province. Petroleum Industry Press, Beijing (in Chinese)

Li WZ, Zhou JG, Zhang JY, Hao Y, Zeng YY, Ni C, Wang F, Tang S (2016) Main controlling factors and favorable zone distribution of Xixiangchi Formation reservoirs in the Sichuan Basin. Nat Gas Ind 36(1):52–60 (in Chinese with English abstract)

Liu SG, Sun W, Zhong Y, Tian YH, Wu J, Wang GZ, Song JM, Deng B, Ren B, Li ZW (2016) Discussion on the theories of the hydrocarbon formation and distribution of the deep-seated marine carbonates in the Sichuan superimposed basin, China. China Pet Explor 21(1):15–27 (in Chinese with English abstract)

Wang ZC, Jiang H, Wang TS, Lu WH, Gu ZD, Xu AN, Yang Y, Xu ZH (2014) Palaeo-morphology formed during Tongwan tectonization in Sichuan Basin and its significance for hydrocarbon accumulation. Pet Explor Dev 41(3):305–312 (in Chinese with English abstract)

Wei GQ, Jiao GH, Yang W, Xie ZY, Li DJ, Xie WR, Liu MC, Zeng FY (2010) Hydrocarbon pooling conditions and exploration potential of Sinian-Lower Palaeozoic gas reservoirs in the Sichuan Basin. Nat Gas Ind 30(12):5–9 (in Chinese with English abstract)

Wei GQ, Yang W, Du JH, Xu CC, Zou CN, Xie WR, Zeng FY, Wu SJ (2015a) Geological characteristics of the Sinian-Early Cambrian intracratonic rift Sichuan Basin. Nat Gas Ind 35(1):24–35 (in Chinese with English abstract)

Wei GQ, Du JH, Xu CC, Zhou CN, Yang W, Sheng P, Xie ZY, Zhang J (2015b) Characteristics and accumulation modes of larger gas reservoirs in Sinian-Cambrian of Gaoshiti-Moxi region, Sichuan Basin. Acta Petrolei Sinica 36(1):1–12 (in Chinese with English abstract)

Wu SJ, Wei GQ, Yang W, Xie WR, Zeng FY (2016) Tongwan movement and its geologic significances in Sichuan Basin. Nat Gas Geosci 27(1):60–69 (in Chinese with English abstract)

Xie ZY, Wei GQ, Zhang J, Yang W, Zhang L, Wang ZH, Zhao J (2017) Characteristics of source rocks of the Datangpo Formation, Nanhua system, at the south–eastern margin of Sichuan Basin and their significance to oil and gas exploration. Nat Gas Ind 37(6):1–11 (in Chinese with English abstract)

Xu AN, Hu SY, Wang ZC, Bao DM, Li M, Lu WH, Zhai XF (2016) Sedimentary mode and reservoir distribution of the Cambrian carbonate and evaporite paragenesis system in the Sichuan Basin. Nat Gas Ind 36(6):11–20 (in Chinese with English abstract)

Yang CY (2018) Petroleum entrapment and evolution history of on Leshan-Longnvsi Uplift, southwest China. China University of Petroleum-Beijing, Beijing (in Chinese with English abstract)

Yang W, Wei GQ, Zhao RR, Liu MC, Jin H, Zhao ZA, Shen JH (2014) Characteristics and distribution of karst reservoirs in the Sinian Dengying Formation, Sichuan Basin. Nat Gas Ind 34(3):55–60 (in Chinese with English abstract)

Zhou JG, Xu CC, Yao GS, Yang G, Zhang JY, Hao Y, Wang F, Pan LY, Gu MF, Li WZ (2015) Genesis and evolution of Lower Cambrian Longwangmiao Formation reservoirs, Sichuan Basin, southwest China. Pet Explor Dev 42(2):175–184 (in Chinese with English abstract)

Chapter 15
Occurrence and Genetic Mechanism of Large Asphaltic Veins at the Longmenshan Fronthill Belt, Western Yangtze Craton, South China

Keyou Han, Guangli Wang, Tieguan Wang, and Lansheng Wang

Abstract The Longmenshan Fronthill Belt is the most front anticlinal tectonic belt in the Longmenshan Nappe Zone, it is geohistorically in a low thermo-evolutional region due to a long-term crust uplifting state, where there are numerous asphaltic veins with different scales, the remarkable large-scale ones are directly sourced from the original dolostone oil-reservoir in the Late Sinian Dengying Formation, while the hydrocarbon composition of asphaltic veins is well correlated with that of the early Sinian Doushantuo Formation black shale, both of which appear as unusually predominance of relative abundance of C_{21}–C_{22} pregnane and homopregnane, C_{29} 30-norhopane, 13α(*n*-alkyl)-tricyclic terpanes and C_{24} tetracyclic terpane biomarkers. Therefore, Doushantuo black shale should be the source bed for these asphaltic veins. The geological occurrence of Erchangliang large asphaltic veins reveal three geological conditions for its genetic mechanism: ① there is an excess of liquid oil as the original material for the asphaltic veins, ② a fault and fracture system resulted from the nappe structure would provide the passageway and reservoir space, ③ an episodic hydraufracturing system driven by abnormal high-pressure in a very short duration. Which is proved by the tectonic stress analysis of Kuangshanliang Anticline and by the investigation of Changjianggou-Kuangshanliang-Xianshuigou Nappe-Overthrust, and there is a genetic relationship between the asphaltic veins and the dual-layer thin-skinned structure of Longmenshan Nappe Zone as the results of the Middle-Late Triassic Indosinian Movement. As a special oil accumulation zone, undoubtedly the nappe zone of Indosinian epoch suffered a larger destruction on the shallow-layer local structures as well as on its inside oil and gas reservoirs, but the deep-layer local structures and associated oil- and gas-reservoirs might still be well preserved, providing geological and geochemical foundation for the exploration of Neoproterozoic indigenous oil and gas reservoirs in the Longmenshan Fronthill Belt.

K. Han (✉) · L. Wang
Institute of Petroleum Exploration and Development, PetroChina Southwest Oil and Gasfield Company, CNPC, Chengdu 610051, China

G. Wang · T. Wang
State Key Laboratory of Petroleum Resources and Prospecting, China University of Petroleum-Beijing, Beijing 102249, China

T. Wang, *Meso-Neoproterozoic Geology and Petroleum Resources in China*, Springer Geology, https://doi.org/10.1007/978-981-19-5666-9_15

Keywords Longmenshan Nappe zone · The Longmenshan Fronthill Belt · Dual-layer thin-skinned structure · Kuangshanliang Anticline · Large asphaltic veins · Doushantuo source bed

15.1 Introduction

As a NE-trending long and narrow region with 380 km in length and 40–70 km in width, covering an area of ca. 2.3 × 10^4 km^2, the Longmenshan Nappe Zone is situated on the northwest margin of Yangtze Craton (YC), and asttributed to the collisional origin between Songpan-Ganzi Fold Belt and YC. Whereas its northern segment is extended from Anxian and Beichuan at sourth to Guangyuan and Qingchuan at north with ca. 180 km-long and ca. 40–75 km-wide and an area of ca. 1.2 × 10^4 km^2 (Fig. 15.1), which contains three major faulted belts, i.e., Qingchuan, Beichuan and Jiangyou-Guanyuan Faulted Belts from west to east in turn. Accordingly, the Longmenshan Nappe Zone can be subdivided into three secondary tectonic units (Fig. 15.1).

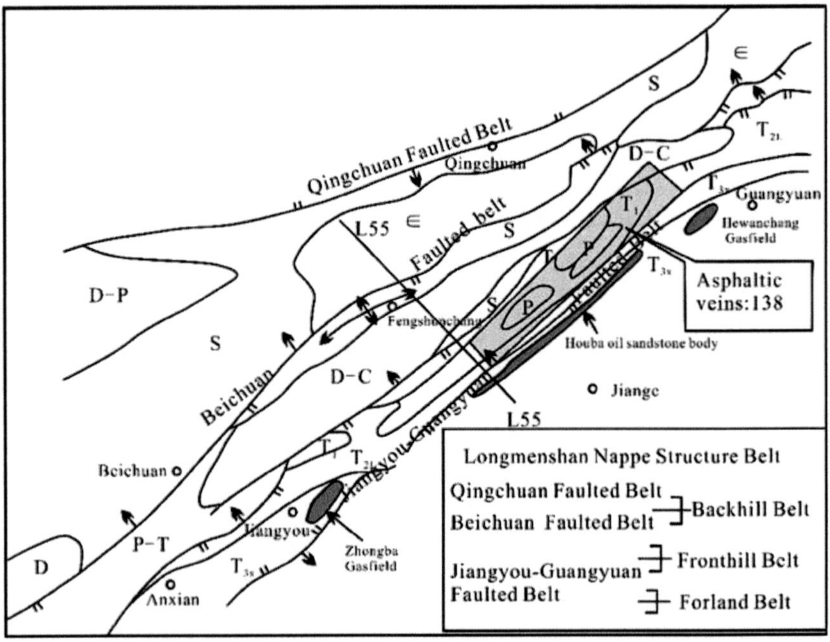

Fig. 15.1 Indosinian epoch palaeogeological map in the northern segment of Longmenshan Nappe Zone showing the distribution of major faulted belts and the secondary tectonic units. L55–L55′ indicates the plane position of seismic section in Fig. 15.2

1. **The Longmenshan Backhill Belt** between the Qingchuan and Beichuan Faulted Belts is composed of Cambrian, Silurian, Devonian, Carboniferous and Permian strata. Its Palaeozoic sequence is totally more than 7000 m-thick, and lithologically predominananted by clastic rocks. Owing to dynamic metamorphism, argillite has been metamorphosed into phyllite so that it would be referred to a light matamorphic belt of sedimentary strata.
2. **The Longmenshan Fronthill Belt** between the Beichuan and Jiangyou-Guangyuan Faulted Belts is a long and narrow nappe structure unit, which consists of Silurian, Devonian, Carboniferous, Permian and Triassic strata, and covers an area of 3500 km^2 with 170 km-long and 15–20 km-wide. Geohistorically multiple crust upliftings had resulted in multiple hiatus of Palaeozoic sediments so that the cumulative sedimentary thickness from Lower Cambrian to Triassic is only 800–1200 m. Consequently, the thermo-evolutional level of Sinian to Lower Palaeozoic sedimentary organic matter would not be so high, which could be favourable to the preservation of oil and gas resources, as well as asphalt veins.
3. **The Longmenshan Foreland Belt** is a narrow region on the east of Jiangyou-Guangyuan Faulted Belt, where Hewanchang Gasfield and Houba oil-sandstone-body were discovered. Within the Middle Jurrasic Shaximiao Formation, the oil-sandstone-body appears as NE-trending extension with 33 km-long, 4–5 km-wide and 27–43.9 m-thick, and its oil sandstone is 13–17% in porosity and 11.2–30.8% in oil saturation. The calculated residual oil resources of oil-sandstone-body would be up to 858×10^4–2340×10^4 t. It is considered that the Houba oil-sandstone-body may be congeneric with the asphaltic veins in the Longmenshan Fronthill Belt, and the original liquid oil was sourced by the black shale of Sinian Doushantuo Formation (Wang et al. 2005; Dai et al. 2007). Therefore, it would be very hopeful for the Longmenshan Foreland Belt to be an oil enrichment area of "Neoproterozoic-sourced and younger stratum-reservoired" oil and gas resources.

The northern segment of Longmenshan Fronthill Belt is well-known for its numerous oil-seep and asphalt veins. According to the incomplete statistics, there are 77 sites of oil-seep, 33 sites of gas-seep, and 166 asphaltic veins found in this segment, among which, the major producing area of asphaltic veins are in between the Shangsi (Jian'ge County) and Zhuyuan (Qingchuan County). The Longmenshan Fronthill Belt has developed the anticlinal type of local structures, e.g., the Zhongba, Haitangpu and Daoliuhe Anticlines at its southern segment as well as the Tianjingshan, Kuangshanliang and Tianba Anticlines at its northeastern segment, which would be the earliest oil exploratory area in the YC, the first exploratory well had been drilled at the Haitangpu Anticline in 1944. The large Lower Palaeozoic asphaltic veins were found at the Kuangshanliang and Tianba Anticlines in 1966, meanwhile Tian-1 Well was drilled at the Tianba Anticline, and black oil of 30 L produced from the well interval of 333–335.5 m. Moreover, the narrow and long region of the Longmenshan Fronthill Belt would be a petroliferous area of lower geotemperature between the hot Sichuan Basin on the east and the light metamophized Longmenshan Backhill Belt on the west.

Based on the geological investigation of the asphaltic veins in the Longmenshan Fronthill Belt, this chapter will approach the occurrence and genetic mechanism of large asphaltic veins, and try to provide a geological and geochemical foundation for the petroleum exploration in future.

15.2 Specific Geological Settings

15.2.1 Geological Evolution History of Palaeotectonics

Since Neoproterozoic Nanhuan on, the northern segment of Longmenshan Nappe Zone has been the crust uplifting region on the northwest margin of YC. The total sedimentary thickness of Palaeozoic sequences could be up to 7000 m and more in the Longmenshan Backhill Belt, while only 800–1200 m-thick in the Longmenshan Fronthill Belt which consists of Lower Cambrian Guojiaba Formation, Middle Ordivician Baota Formation and Middle Silurian strata without Middle-Upper Cambrian, Lower and Upper Ordivician, and Lower and Upper Silurian strata. Which is favourable for oil and gas to be well preserved and free of metamorphism in the Longmenshan Fronthill Belt during the geological time. Particularly at the end of Middle Triassic, the Indosinian Movement caused the crust uplifting more remarhable in the Longmenshan Fronthill Belt with a ciosure up to 400 m, which made the Longmenshan Fronthill Belt to be a peripheral palaeo-uplift belt between the YC and the Longmenshan Backhill Belt.

Until the the Late Triassic, the Longmenshan region started to be folded resulting in a rudiments of the nappe structure belt. In the basis of initial palaeo-uplift belt, The local superficical structures, such as Tianjingshan, Kuangshanliang and Tianba Anticlines, were emerging, where oil and gas could be entrapped. Hereafter, along with the stratigraphic denudation, planation and unloading, the Lower Palaeozoic strata were gradually exposed to the suface resulting in oil reservoir destruction, oil and gas volatilization and asphaltic vein formation in somewhere, and then covered by the Jurassic sediments.

15.2.2 Unique Dual-Layers Thin-Skinned Nappe Structure

The Longmenshan Nappe Zone includes the overthrust belts and associated fold structures, which constitutes a thin-skinned nappe structural belt, caused by the napping of Indosinian palaeo-structure from west to east. After the Mesozoic sedimentation, the Longmenshan Nappe Zone has undergone Himalayan folding again until Palaeogene, resulting in the current Longmenshan nappe structural system. It can be seen from the L55 seismic cross-section in Fig. 15.2 that whole Longmenshan Nappe Zone contains one set of Palaeozoic fault system, and a detachment surface within Cambrian-Silurian strata as a top-steep and bottom-gently low-angle plough-style overthrust plane. The detachment surface of major thrust is buried 2–4 km-deep. It has separated the nappe zone into deep and shallow layers. The shallow layer structure is very complicated, while the deep layer structure quite simple, which would constitute a typical thin-skinned nappe structure (Wang et al. 2005). The L55 seismic section shows that the shallow layer appears as an nappe mat composed of four imbricated and inverted folds, i.e., Baicaogou Synticline, Jiaoziding Anticline, Yangtianwo Syncline and Tianjingshan Anticline from east to west in turn (Fig. 15.2). The Longmenshan Fronthill Belt is just the Tianjingshan nappe mat at the end of the nappe belt in Fig. 15.2, which suffered complex compression deformation, and also dynamic metamorphism at the west side (Song 1989). It is confirmed by equilibrium profile investigation that the shallow layer appears as a nappe-traction distance up to 24 km, fold shortening distance 18 km, total shortening distance 42 km and compression rate 43.3%. While the deep layer could be attributed to the normal fold structures without overturning appearance and probably no dynamic metamorphic phenomenon, which are totally different from the shallow one. On L55 section, the deep layer consists of Fengshunchang Buried Anticline, Yangtianwo Syncline and Tianjingshan Buried Anticline from west to east in turn (Fig. 15.2), and its nappe-traction distance only 10 m, fold shortening distance 12 m, total shortening distance 22 m and compression rate 28.5%.

Owing to the traction of overthrust, the shallow layer Tianjingshan, Kuangshanliang and Tianba Anticlines are cross-cut off in the Longmenshan Fronthill Belt, resulting in the destruction of shallow original oil reservoirs and the emerging of asphaltic veins with oil-seep along faults and fractures (Fig. 15.3). It is verified by field survey and drilled well that the shallow layer structures are complicated and destructed, and oil- and gas-seeps are wide-spread, while the deep buried structures would be slightly gentle, well preserved and so favourable for oil and gas preservation as compared with the shallow one.

Taking the Fengshunchang buried anticline as an example of the deep local structure (Figs. 15.2 and 15.4). It is ascertained by seismic exploration that the top surface of Sinian appears as a comb-shaped high-and-steep anticline with long-axis 40 km,

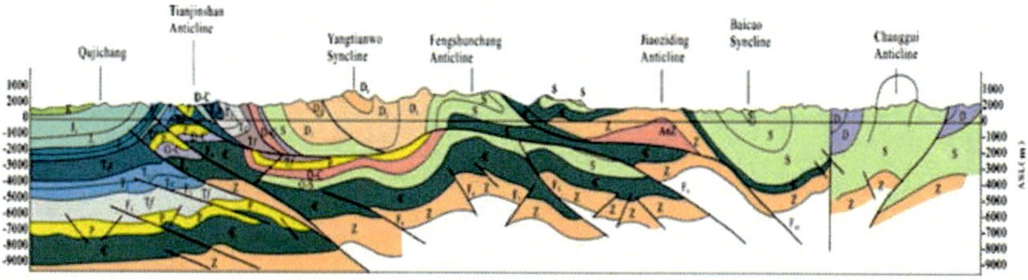

Fig. 15.2 Geological structure cross-section of thrust belt at the north segment of Longmenshan Nappe Zone (Song 1989, modified)

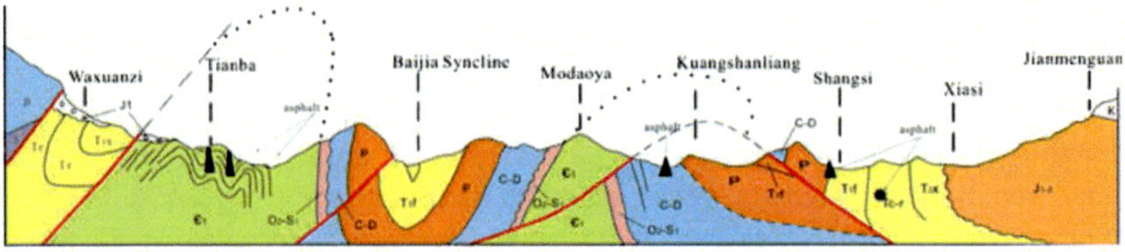

Fig. 15.3 Kuangshanliang-Tianba structural cross-section shows its relationship with the asphaltic veins. T_1f Feixianguan Formation; P_2 Upper Permian oil-seeps are found in coal pits

brachy-axis 19 km, NE–SW axial direction, obvious southwestward pitching end and unclear northeastward pitching end. The buried depth of top Dengying Formation is up to 4300 m. This anticline is well preserved on the whole, there only is a axial reverse-fault with southeastwards napping and fault displacement ca. 600–1200 m, at the anticlinal SE-margin.

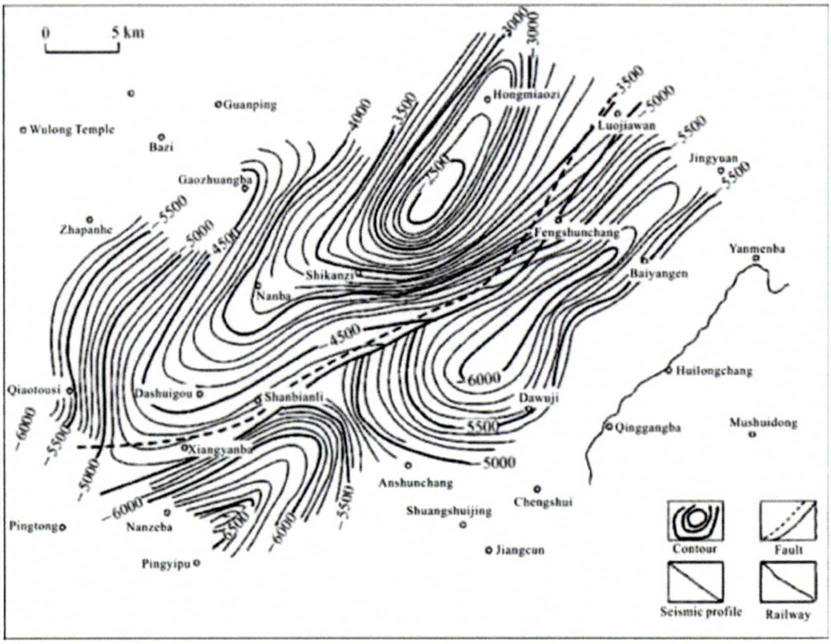

Fig. 15.4 Sinian top surface structural map of the Fengshunchang buried anticline

15.2.3 Low Thermo-Evolutional Zone

It is well known that the western YC is overall attributed to a high thermo-evolutional region of organic matter, which is rich in natural gas resources due to a great quantity of cracking gas drived from original Palaeozoic crude oil. However, the Longmenshan Nappe Zone contains a very thick crust and a low-velocity zone, which could hinder the upwards conducted terrestrial heat flow from the deep heater and make the Longmenshan Fronthill Belt to be a low thermo-evolutional zone. Meanwhile, the cumulative thickness of Lower Cambrian and overlying Palaeozoic and Triassic strata is only ca. 1000 m due to the lack of Middle-Late Cambrian, Early and Late Ordovician, Early and Late Silurian and Late Carboniferous sedimentation in the geohistorical time. The Sinian and Lower Cambrian strata are never deeply buried in the Longmenshan Fronthill Belt thereby. The above two factors have made the Longmenshan Fronthill Belt occuring as the geological features of a low thermo-evolutional zone, and thus its measured equivalent vitrinite reflectance eqR_o values range 0.99–1.5% for Lower Palaeozoic, 0.75–1.3% for Upper Palaeozoic and 0.42–0.65% for Mesozoic, which are most probably the lowest thermo-evolutional level of organic matter in the western YC, and very favourable for the preservation of Sinian and Palaeozoic petroleum.

15.3 Occurrence and Characteristics of the Asphaltic Veins

15.3.1 Distribution and Occurrence

15.3.1.1 Distribution

Based on the field survey in 1966, totally 138 Palaeozoic asphaltic veins were found in the Longmenshan Fronthill Belt, which occur at the core area of three anticlines, i.e., Tianjingshan, Kuangshanliang and Tianba Anticlines respectively, with NE-axial direction (Figs. 15.3 and 15.5).

(1) **Tianjingshan Anticline**: It has an intact surface structure with long-axis 20 km and brachy-axis 2.5 km. The oldest stratum exposed on outcrop would be the 225 m-thick sandy shale of Lower Cambrian Guojiaba Formation. It is confirmed by seismic data and drilled well that its deep layer structure has been cross-cut off. There only is a longitudinal fracture type of maltha/viscous oil vein found at its Hongyazui structural high point.
(2) **Kuangshanliang Anticline**: Its long-axis is 15 km and brachy-axis 3.8 km. The 485 m-thick stratum of Guojiaba Formation exposed at the anticlinal core. The anticline was dissected by a tongue-shaped Changjianggou-Kuangshanliang-Xianshuigou napping overthrust (Fig. 15.6) so that the anticlinal surface high

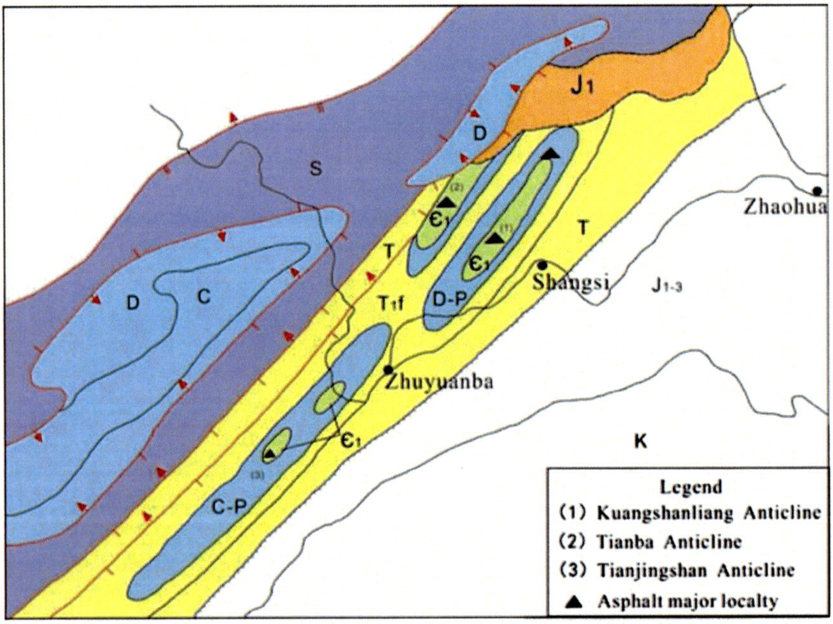

Fig. 15.5 The structural location of Kuangshanliang, Tianba and Tianjingshan anticlines with centrally distributed asphaltic veins

pointis clear on structure. So far there are totally 100 asphaltic veins found within the anticline, but all the scale is limited.

(3) **Tianba Anticline**: It is a duplex anticline of many secondary small folds, with long-axis 12.7 km and brachy-axis 3 km. Its southwestward pitching end is located on the south of Nianziba, while the northeast end and northwest flank are covered by an unconformity surface, the Lower Cambrian is overlaid by Lower Jurassic, and upright/inverted strata are present at the southeast flank. There are 37 asphaltic veins at the axial area, where large-scale asphaltic veins are concentrated on Erchangliang, such as the Guojiaba TL-No. 1 and TL-No. 2 large asphaltic veins.

According to the statistics of geologic age and producing formation, totally 138 asphaltic veins have been found in the Palaeozoic strata, among which, 122 veins in Cambrian and 16 veins in Ordovician–Silurian. Accordingly, it is inferred that Cambrian stratum may be closer to hydrocarbon source than Ordovician–Silurian ones.

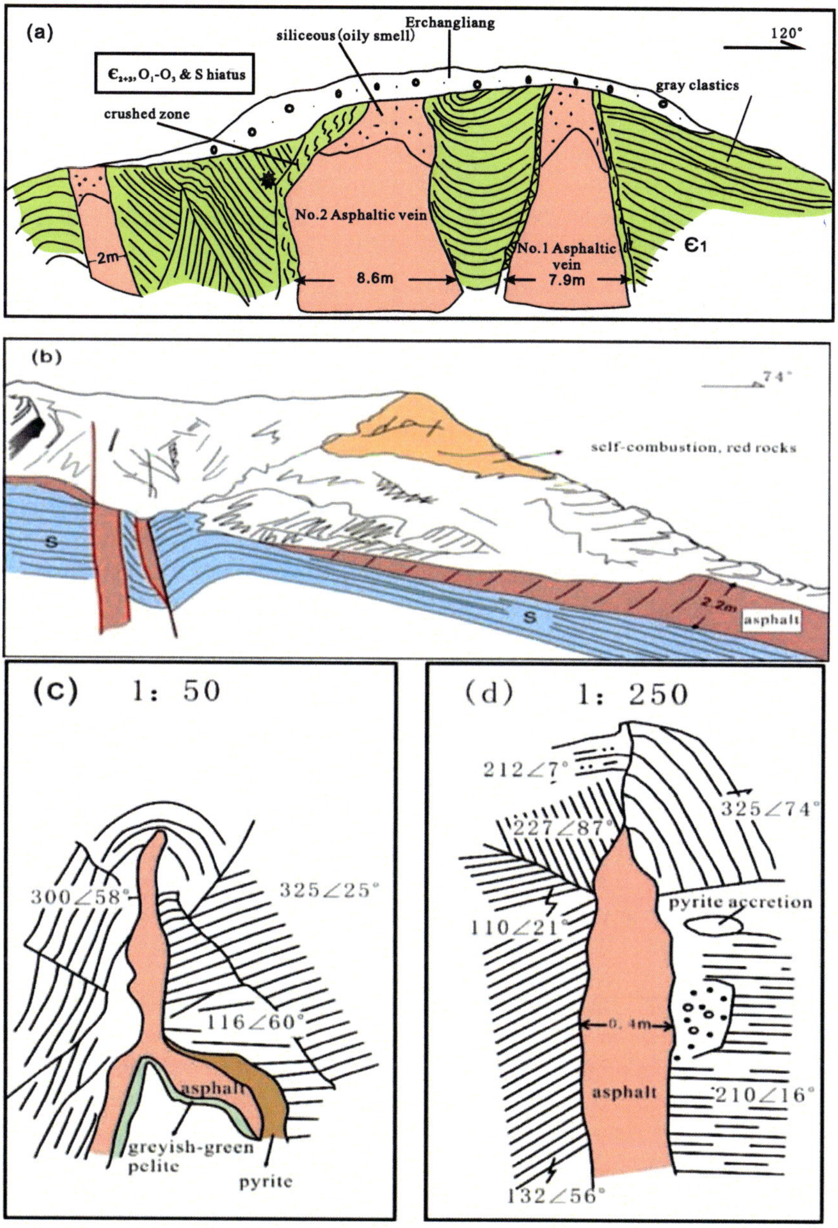

Fig. 15.6 Field sketches of three type of asphaltic veins in Longmenshan Front Belt (Han et al. 2016). **a** The fault type of veins at Erchangliang, Tianba anticline and the TL-No. 1–TL-No. 3 veins are shown from right to left in turn; **b** the interlayer type of veins at Macun; **c**, **d** the fracture type of veins at Tianba

15.3.1.2 Occurrence Types

The geological occurrence of asphaltic veins can be classified into three types, i.e., fault, interlayer and fracture types (Fig. 15.6a–c), among which, 54% for fault type, 31% for fracture type and 15% for interlayer type, respectively. The fault-fracture associated veins account for 85% of the total, implying that faults and fractures should be the major passageway and reservoir space of the numerous asphaltic veins in the Longmenshan Fronthill Belt.

(1) **The fault type of asphaltic veins** (Fig. 15.6a): The most typical fault type of asphaltic veins should be the Changjianggou vein at Jian'ge County. It is situated in a EW-trending deep-cut river valley covered by recent sediments. At south hillside of the valley, the Changjianggou fault plane occurs as NNE-tendency, its upper plate of Middle Devonian-Lower Permian strata was directly overthrusted on the lower plate of Lower Triassic Feixianguan Formation, which made the lower plate mudstone to be altered into phyllite by dynamic metamorphism, and also the Feixianguan phyllite was compressed into small drag-folds (Fig. 15.7a). While the Kuangshanliang-Xianshuigou Fault at the north hillside appears as a SEE-tendency fault plane, its upper plate of Middle Devonian-Lower Permian strata overthrusted on the lower plate Feixianguan Formation (Fig. 15.7b). Changjianggou Fault was connected with Kuangshanliang-Xianshuigou Fault from the south hillside through the valley to the north hillside, resulting in an integrated overthrust (Fig. 15.7c), and it is rich in asphalt-bearing mixture or thin asphaltic fillings, even liquid oil-seep, along its fault plane (Fig. 15.7d). Although these asphaltic occurrences are in small size, they are still very significant for the genetic mechanism investigation and the dating of asphaltic veins. The asphaltic fillings should be the contemporaneous product with the Changjianggou-Kuangshanliang-Xianshuigou overthrust.

    In addition, the TL-No. 1 and TL-No. 2 large asphaltic veins in Guojiaba Formation occur closely at Erchangliang in Tianba Anticline, and both can be attributed to the typical fault type (Fig. 15.6a; Plates 15.1a, b). According to the field survey in 1966, both veins are 7.9 m and 8.6 m in thickness respectively, with TL-No. 2 vein is the thickest one in the Longmenshan Fronthill Belt. Moreover, an additional TL-No. 3 vein is a smaller one existing on their north side (Fig. 15.6a). In the same year, a geological shallow drill hole, Tian-1 Well, was drilled ca. 100 m apart from Erchangliang outcrop (Plate 15.1c). An 15.3 m-thick asphaltic layer of in apparent thickness was recorded at the well interval of 149.0–164.3 m. Since then, after long-term informal mining, both the large asphaltic veins have changed into ca. 1–4 m thick on surface until 2007 (Plates 15.1a–c).

    As for the fault type of asphaltic veins, the older the stratum, the larger the vein scale is, and also the more rock debris the asphaltic vein contains, which implies that on the way of liquid oil infiltration upwards along fault or fracture, the oil would become lighter, and the component of associated asphaltic veins more pure, reflecting the effect of material differentiation during liquid oil migration.

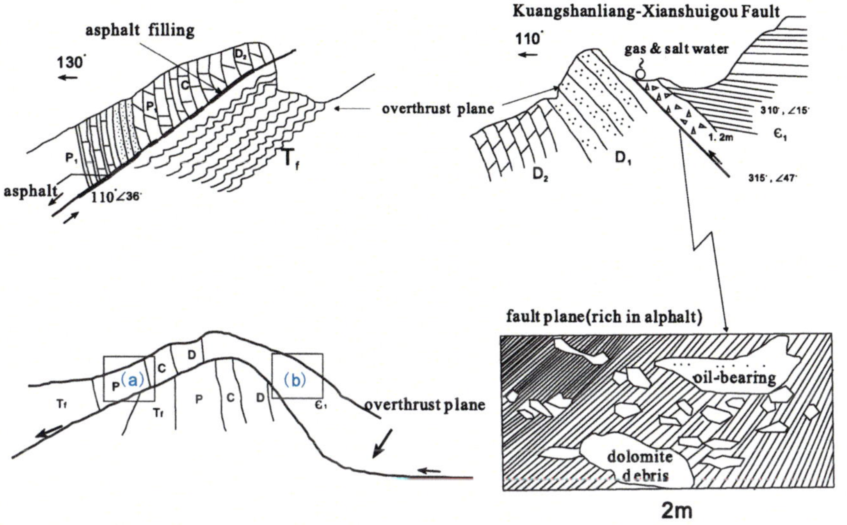

Fig. 15.7 Field sketch of the Changjianggou-Kuangshanliang-Xianshuigou napping overthrust (**c**) where Changjianggou Fault (**a**) was connected with Kuangshanliang-Xianshuigou fault (**b**) and the sketch of faulting plane occurrence (**d**) (Han et al. 2016). There is ca. 5–10 cm-thick asphaltic bandings along the overthrust plane and the mudstone wall-rock was altered into phyllite with T_1f small drag-folds (**a**), T_1f–Lower Triassic Jialingjiang Formation (**a**)

(2) **The interlayer type of asphaltic veins** (Fig. 15.6b): This type is only found within the interbedded fractures, and often appears as asphaltic veins of 0.1–0.5 m-thick middle size, these veins occur mostly in the Ordovician–Silurian strata around Macun at the northern pitching end of Kuangshanliang Anticline, while the producing stratum of KL-No. 1 large asphaltic vein occurs within the younger Middle Silurian strata as a 2.2 m-thick more pure bright-black asphalt with specular gloss.

(3) **The fracture type of asphaltic veins** (Fig. 15.6c): It generally appears as numerous black and pure asphaltic veins, accounting for ca. 60% of the total veins, but all these veins are in small size, usually with thickness of 0.5 m and more, and the most veins occur within the Lower Cambrian strata.

Besides the above-described three types of asphaltic veins, there is an additional asphaltic occurrence, i.e., the synsedimentary isolated asphaltic bandings (or so-called "asphaltic pie"; Plate 15.2), it occurs only within a layer of ca. 4 m-thick grey lime-mudstone and muddy siltstone with wavy bedding at 180 m beneath the top of Guojiaba Formation (Plate 15.2). This kind of asphaltic bandings have been found in Tianjingshan, Kuangshanliang and Tianba Anticlines, it would indicate the early existence of viscous oil/maltha before the Guojiaba sedimentary period, and the local leaking and/or destruction of original oil-reservoir during or before early Cambrian age, resulting in the transportation of viscous oil/maltha into sedimentary waters so as to be deposited along bedding plane as the syngenetic asphaltic bandings.

Plate 15.1 Occurrence of Erchangliang large asphaltic veins at Tianba Anticline (Photographs shot in 2007). **a** Distant view, the TL-No. 1 and TL-No. 2 veins are shown from right to left; **b** close view of the TL-No. 1 vein; **c** abandoned asphalt ore and slagheap, the well site of Tian-1 Well is ca.100 m apart

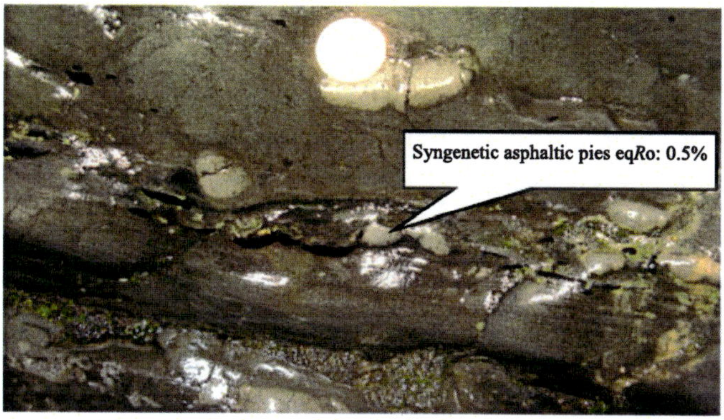

Plate 15.2 Syngenetic asphaltic bandings of Guojiaba formation in the Longmenshan Fronthill Belt

15.3.2 Reserve Scale of Asphalt Veins

A geological investigation shows that the occurrence and reserve scale of asphaltic veins are related to their producing stratum. The asphaltic veins contain different scale from several centimeters to 8 m in thickness, among which, 40 asphaltic veins of large and middle scale, with thickness 0.5–1 m and more, account for 29% of the total known veins, the large ones are attributed to the fault type of asphaltic veins produced in older strata. Recently some private operators keep working on the asphaltic mining for building and industry materials.

The No. 127 geological team of former Sichuan Petroleum Administration had conducted a geological survey of the 138 asphaltic veins in 1966. According to the initial budget estimate, the asphaltic geological reserve would be 47×10^4 m^3. In the same year, Tian-1 Well revealed a 15.3 m-thick asphaltic layer during drilling in Guojiaba Formation, which could be correlated with and traced to the TL-No. 1 and TL-No. 2 veins, in between there are 100 m apart. If calculated by a distance of 100 m, the asphaltic geological reserve at Erchangliang might be up to 127×10^4 m^3.

A geological exploratory team of former Sichuan Bureau of Land and Resources delegated authority to a private operator for exploring asphaltic resources around Majiagou about 4 km away from TL-No. 1 and TL-No. 2 asphaltic veins in 2015, the proven subsurface asphaltic reserve might be 180×10^4 m^3. However, the Majiagou exploratory area only accounts for ca. one fifth of the potentially distributional area of asphaltic veins. Accordingly, the geological reserve of asphaltic veins would be quite sizable.

15.4 Enlightenments of Asphaltic Veins

Field survey combined with laboratory analytical data of the asphaltic veins would be helpful to obtaining below enlightenments for the Longmenshan Fronthill Belt:

(1) Owing to different occurrences and producing strata, the asphaltic veins would contain 3 kind of colors, i.e., dark-black, black and bright-black asphalts, due to compositional diversities in the Longmenshan Fronthill Belt (Table 15.1).

Taking Tianba and Kuangshanliang asphaltic veins as examples, it can be seen from Table 15.1 that among the three kinds of asphaltic veins, the TL-No. 1 vein at Tianba was produced from the oldest stratum (i.e., lower-middle Lower Cambrian), and the KL-No. 1 vein at Kuangshanliang from the youngest one (i.e., Middle Silurian), while the KL-No. 81 vein also at Kuangshanliang just in between (i.e., upper Lower Cambrian), which reflect a rule that the younger vein the more pure in asphaltic component. Which would imply that the oldest producing stratum would be closer to its hydrocarbon source.

(2) As results of field observation, the TL-No. 1 vein appears as dark-black without gloss, and it contains joints, vesicles, dolomitic and siliceous debris (up to 14 cm as the large size) as well as the sandy shale debris of Lower Cambrian Guojiaba Formation. There are brownish-grey oil-immersed chart and chalcedony debris with very strong oily smell in the asphaltic vein.

Based on laboratory analytical data, the TL-No. 1 vein generally consists of solid bitumen (accounting for 67%) and chert (17%) plus minor mineral debris (including chalcedony, argillite, dolomite and gypsum). While total content of dolomite, chert and chalcedony plus gypsum can be up to 30% and more in the vein. Since above minerals are not the mineral components in the wall-rocks of Guojiaba Formation for the TL-No. 1 vein, but these minerals could be found in the deep strata of the Sinian Dengying Formation. As the allogenic "xenoliths", therefore, the existence of above mineral debris would indicate that the asphaltic material should be directly supplied from the underlying carbonate strata of Sinian Dengying Formation.

(3) As for the bulk composition of bitumen (i.e., chloroform extract) of TL-No. 1 asphaltic vein, the saturate, aromatic, resin and asphaltene contents are 32.4%, 23.1%, 43.8% and 23.1%, respectively, and the H/C ratio 0.12, saturates/aromatics 1.4 and risen/asphaltene 6.6. Which should attribute the asphalt to the maltha (soft bitumen) in the stead of gilsonite or pyrobitumen, showing the features of medium mature asphalt (Table 15.2).

Moreover, the measured equivalent vitrinite reflectance eqR_o value is 0.99% for the asphaltic vein, which would indicate the thermos-evolutional level in the scope of "liquid window" for hydrocarbon generation, and also corresponding to the above bulk composition.

Therefore, the asphalt and its source bed in Longmenshan Fronthill Belt should still contain oil-generating potential, from which liquid oil could be generated by means of pyrolysis and/or distillation.

Table 15.1 Comparison of asphaltic colors and compositional diversities (%)

Asphalt colour	Burning loss	SO_2	MgO_2	Gypsum	Dolomite	Acid insoluble substance	Oxide	Case of asphaltic vein	Producing horizon
Dark-black	53.4	2.47	6.26	4.2	5.76	36.3	2.5	TL-No. 1	lower-middle lower Cambrian
Black	81.4	5.66	0.67	9.62	3.48	6.02	1.0	KL-No. 81	Upper lower Cambrian
Bright-black	99.0	0	0	0	0	0.8	0	KL-No. 1	Middle Silurian

Note TL Tianba asphalt; *KL* Kuangshanliang asphalt

Table 15.2 Analytical data of asphaltic veins and associated oils in Longmenshan Fronthill Belt

Analytic item	Oil component and physical property	Analyzed sample				
		Distilled medium oil of Majiagou asphalt (400 °C)		Medium oil produced from Tian-1 Well	TL-No. 1 asphaltic vein	
					Bitumen (asphaltic extract)	
Elemental composition	C (%)	–		–	75.3	H/C 0.12
	H (%)	–		–	9.03	
Buck composition of crude oil or bitumen	Aliphatic HCs (%)	–		–	32.4	Ali/Aro 1.4
	Aromatic HCs (%)	–		–	23.1	
	Resin (%)	9.09	Res/Asp 13.8	–	43.8	Res/Asp 6.6
	Asphaltene (%)	0.66		–	6.68	
	Paraffin (wax) content (%)	0.57		–	–	
	Sulphur content (%)	2.90		–	–	
	Saltness (%)	1.84		–	–	
Physical property of crude oil	Kinetic viscosity (mm^2/s)	3.03 (40 °C)		–	–	
	Dynamic viscosity (MPa•s)	2.72 (40 °C)		12.8 (50°C)	–	
	Density (g/cm^3)	0.8981		0.882	–	
	Freezing point (°C)	–		28.4	–	
Simulated distillation	Gasoline fraction (%)	25		10	–	
	Kerosene fraction (%)	70		29	–	
	Diesel fraction (%)			19	–	
Pyrolysates	Pyrolysis oil (kg/t)	–		–	–	
	CO_2 (m^3/t)	–		–	–	
	Natural pitch coke (kg/t)	–		–	–	
Analytical time		2015		1960s		

Note Ali aliphatic HCs; *Aro* aromatic HCs; *Res* resin; *Asp* asphaltene; *HCs* hydrocarbons

In fact, by way of a thermo-simulation at 480 °C, the simulation products of TL-No. 1 asphaltic vein include cracking oil of 157 kg/cm^3, CO_2 of 132–209 m^3/t and natural pitch coke of 689 kg/t (Table 15.2), which would confirm the oil-generating potential of the asphaltic vein.

(4) Liquid oil of 30 L has been produced also from the Guajiaba Formation at Tian-1 Well within the well interval of 333–335.5 m in 1966. The oil physical property appears as 0.882 g/cm^3 in density, 12.8 MPa•s in dynamic viscosity (50 °C) and 28.4 °C in freezing point. While its simulated distillation yields gasoline of 10%, kerosene of 29% and diesel of 19%. Accordingly, it can be referred to the medium oil with medium density and viscosity, and high melting point (Table 15.2).

In addition, some dark-brown oil was obtained by a simplified distillation (at 400 °C) of a sample of Majiagou asphaltic vein in 2015, the bulk composition of this oil contains resin of 9.09%, asphaltene of 0.66% (with resin/asphaltene ratio 13.8) and paraffin content of 0.57%. The physical properties of dark-brown oil show 0.8981 g/cm^3 in density, 3.03 mm^2/s and 2.72 MPa•s in kinetic and dynamic viscosities (40 °C) respectively. By way of simulated distillation, the distillates yield gasoline of 25% and kerosene plus diesel of 70%. The distilled oil can also be referred to medium oil with some physio-chemical properties roughly corresponding to the Tian-1 Well medium oil (Table 15.2), which may imply that the Tian-1 Well medium oil should be congenetic with the asphaltic veins in the Longmenshan Fronthill Belt.

15.5 Approach to Material Source of the Asphaltic Veins

It is reported that the Sinian Doushantuo Formation would be the source bed for the asphaltic veins in the Longmenshan Fronthill Belt (Wang et al. 2005; Huang and Wang 2008), but another reporter proposed the contribution of Cambrian mudstone (Dai et al. 2007). However, in this chapter, the authors would like to explore the hydrocarbon material source for the asphaltic veins based on the geological and geochemical results.

15.5.1 Original Dengying Oil-Reservoir as Direct Oil Source

As above, the asphaltic veins are wide-spread in the Longmenshan Fronthill Belt, and the following evidences indicate that the hydrocarbon material would be directly supplied by the original oil-reservoir in the dolostone of Sinian Dengying Formation.

(1) By way of the comparison between field sketch (Fig. 15.6a, drawn by Han et al. 2016) and photographs (Plate 15.1a, b, shot in 2007) of the Erchangliang asphaltic veins at Tianba Anticline, It can be recognized that after more than 40 years civil opencast mining from shallow to deep, great changes have taken

place for the asphaltic veins, not only a positive landform become a negative one, but also the quantities, occurrence and scales of asphaltic veins have been altered. There originally were three veins, only two remained, and their thicknesses were reduced from 7–8 m (Fig. 15.6a) to 1–4 m (Plate 15.1a), respectively.

(2) The close view of TL-No. 1 vein photograph (Plate 15.1b) clearly shows that there are two sets of very smooth joint planes, with face-to-face tendency, at both sides of the remained ca. 4 m-thick TL-No. 1 asphaltic vein, which constitute a pair of high-angle X-conjugate joints on profile, displaying the peculiarity of shearing-compression fractures, within which as a filling, the TL-No. 1 asphaltic vein should be extruded into the X-conjugate joint system by high-pressure from the deep along the shearing-compression joint planes.

(3) Obviously, the formation of so thick asphaltic veins in the shearing-compression joint system must have a modern hydrofracturing-like driving system and an excess of medium. The medium would be liquid oil in the stead of solid asphalt. It was driven by an abnormal high-pressure hydrofracturing system that could extrude an excess of liquid oil into the shearing-compression fractures in a short duration, and strutting the fractures to a certain scale (e.g., 4–8.6 m-wide), occupying and maintaining the interstitial space.

(4) As for routine source rock expelling hydrocarbon and oil secondary migration process, it could neither have the driving condition of an abnormal high-pressure, nor contain the supply condition of an excess of oil. Only if under a strong episodic tectonic stress, could an excess of oil from a original oil-reservoir be driven into the joint system to constitute large asphaltic veins in a short time.

15.5.2 Doushantuo Black Shale as Oil Source Bed

15.5.2.1 Specific Molecular Information

(1) **Special *n*-alkane distributional pattern**: In the Longmenshan asphaltic veins, the carbon number distribution of *n*-alkane series ranges from nC_{12} to nC_{29}, taking nC_{16} as the major peak, and showing nC_{15}–nC_{17} three strong peaks on the GC trace (Fig. 15.8a; Huang and Wang 2008), this kind of distributional pattern is rarely reported. As the authors' recall, it has only been found from three cases in literatures, the first case is in the Palaeoproterozoic Eulamina carbonaceous shales (2700 Ma) of the McArthur Basin, North Australia Craton (Mckirdy 1974; Fig. 15.8c), the second one in the Mesoproterozoic Xiamaling basal bituminous sandstone (1327 Ma) of the Yanliao Faulted-Depression Zone (YFDZ), North China Craton (Wang 1990; Fig. 15.8b), and the third one just in the Longmenshan asphaltic veins (Huang and Wang 2008; Fig. 15.8a). Usually nC_{21}^{-} alkanes are considered as the derivatives of lower aquatic organisms (such as bacteria, cyanobacteria and algae), and the *n*-alkane series may have individual strong peak within the nC_{14}–nC_{22} range on GC trace and the nC_{17} alkane as the major peak often appears in many Precambrian hydrocarbons,

which is believed to be the cyanobacterial contribution. Above mentioned the nC_{15}–nC_{17} specific distributional pattern found in Proterozoic *n*-alkane series may reflect the characteristics of Precambrian hydrocarbon source input, which may be an evidence for the Precambrian origin of Longmenshan asphaltic veins.

(2) **Novel tricyclic terpane series**: It is noticeable that two series of different tricyclic terpanes have been detected from aliphatic fraction in the Kuangshanliang asphaltic veins, i.e., conventional C_{19}–C_{26} tricyclic terpane series (Fig. 15.9a) and a new C_{19}–C_{20} 13α(*n*-alkyl)-tricyclic terpane series (Fig. 15.9b; Huang and Wang 2009). As novel biomarkers, the C_{18}–C_{23} 13α(*n*-alkyl)-tricyclic terpane series were firstly identified from the Mesoproterozoic Xiamaling basal bitumen sandstone, North China (Wang 1990, 1991a, b). Since then 13α(*n*-alkyl)-tricyclic terpanes have been reported only for the Meso-Neoproterozoic oil seeps, source rocks, asphalts and the catalytic hydropyrolytic product of kerogen in the YFDZ (Wang and Simoneit 1995; Zhang et al. 2007; Wang et al. 2011; cf. Chap. 11) and also in the Longmenshan Fronthill Belt (Huang and Wang 2008; Fig. 15.9b). Consequently, the 13α(*n*-alkyl)-tricyclic terpanes may be diagnostic for the hydrocarbons of Precambrian age, and the occurrence of the 13α(*n*-alkyl)-tricyclic terpanes may indicate the asphaltic veins to be the Precambrian origin in Longmenshan Fronthill Belt.

15.5.2.2 Asphalt-Source Rock Correlation

Both Erchangliang large asphaltic veins and Doushantuo black shale were sampled during the field investigation in the Longmenshan Fronthill Belt in 2007. By means of GC–MS analysis, the asphalt veins would be remarkably correlative with the black shale based on the following specific biomarker characteristics:

(1) Sterane series contain an obvious predominance of C_{21} pregnane and C_{22} homopregnane over C_{27}–C_{29} regular steranes on *m*/*z* 217 mass chromatogram (Figs. 15.10a and 15.11a); C_{14}–C_{26} Conventional tricyclic terpane series predominate over C_{27}–C_{35} (no C_{28}) hopane series with the predominant peak at C_{23} tricyclic terpane on *m*/*z* 191 mass chromatogram (Figs. 15.10b and 15.11b).
(2) As to the hopane series, C_{29} hopane could surpass the C_{30} hopane in abundance on *m*/*z* 191 mass chromatogram (Figs. 15.10b and 15.11b).
(3) C_{24} tetracyclic is higher than C_{26} tricyclic terpane in abundance on *m*/*z* 191 mass Chromatogram (Figs. 15.10b and 15.11b).

The above unusual characteristics can seldom appear simultaneously in a pair of oil (or asphalt) and rock samples, showing a very close relationship between Doushantuo black shale and Erchangliang larger asphaltic vein, indicating the Sinian Doushantuo black shale should be the source bed for the asphaltic veins in Longmenshan Fronthill Belt.

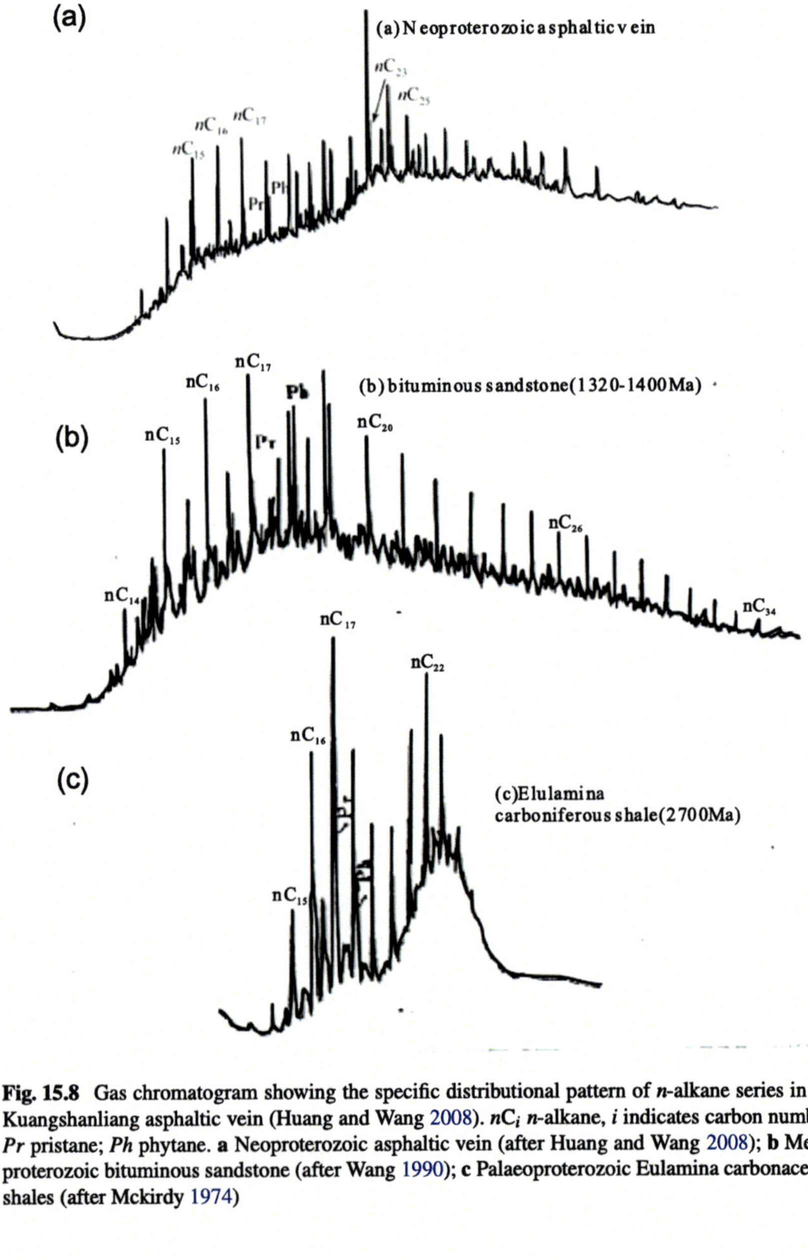

Fig. 15.8 Gas chromatogram showing the specific distributional pattern of *n*-alkane series in the Kuangshanliang asphaltic vein (Huang and Wang 2008). nC_i *n*-alkane, *i* indicates carbon number; *Pr* pristane; *Ph* phytane. **a** Neoproterozoic asphaltic vein (after Huang and Wang 2008); **b** Mesoproterozoic bituminous sandstone (after Wang 1990); **c** Palaeoproterozoic Eulamina carbonaceous shales (after Mckirdy 1974)

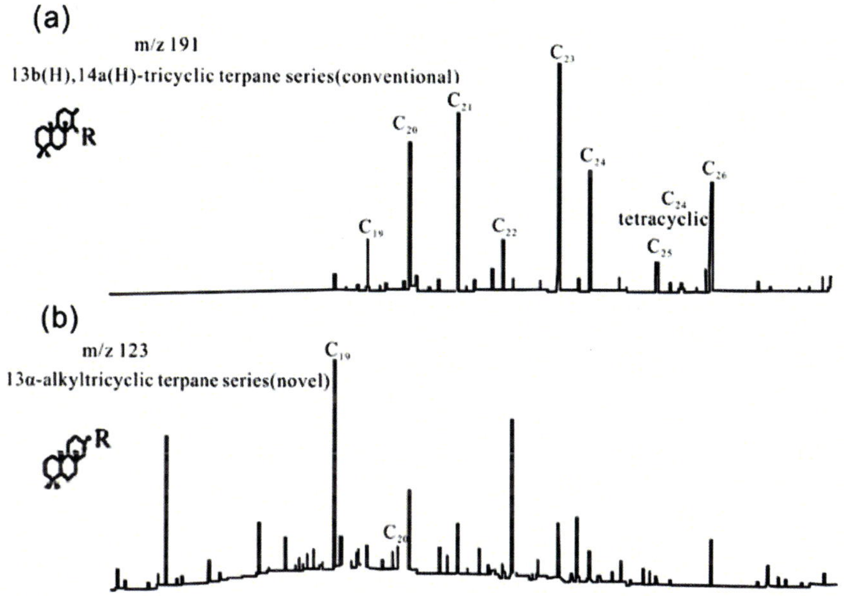

Fig. 15.9 Two tricyclic terpane series detected in the Kuangshanliang asphaltic veins (Huang and Wang 2008). **a** Conventional C_{19}–C_{26} tricyclic terpane series based on *m*/*z* 191 mass chromatogram; **b** new C_{19}–C_{20} 13α(*n*-alkyl)-tricyclic terpane series based on *m*/*z* 123 mass chromatogram

15.5.2.3 Hydrocarbon Potential of the Doushantuo Source Bed

The Doushantuo Formation, with total stratigraphic thickness 0–1200 m but generally 130–300 m, can be divided into three sections: ① the lower section is motley clastic rocks and black shale; ② the middle section grey limestone; and ③ the upper section black shales with Lime-aggregates, plant fossils and phosphorous layers in Longmenshan Fronthill Belt. While the upper black shale section contains Type-I kerogen (sapropelic type) with sapropelinite of 64–89% (avg. 81%) and bituminite of 13–36% (avg. 20%) and high organic abundance with TOC 0.2–7% (avg. 2.98%), but higher maturity with equivalent vitrinite reflectance eqR_o values 1.9–4.24% (avg. 3.16%), which should be referred to an over-mature phases of source bed.

The potential Doushantuo source bed has an average thickness ca. 67 m, its distributional range would cover an area of 1.2×10^4–1.5×10^4 km^2 with a depocenter just at the northern segment of Longmenshan Fronthill Belt.

15.6 Genetic Mechanism of the Large Asphaltic Veins

To sum up, the genetic mechanism of the large asphaltic veins at Longmenshan Fronthill Belt should include the following constraint factors:

(1) As an indicator for early existence of viscous oil or original oil-reservoir, the syngenetic asphaltic bandings (or called asphaltic "pies") along bedding plane could verifier the local leaking and/or destruction of an original oil-reservoir during or before the Early Cambrian age, so that making early liquid oil/maltha to be transported into sedimentary waters and constituting the syngenetic asphaltic occurrence. Obviously, the emergence of asphaltic bandings would be earlier than the Erchangliang larger asphaltic veins, and at least it can confirm the early existence of viscous oil or original oil-reservoir at the underlaying stratum of Guojiaba Formation.

(2) The occurrence of Erchangliang asphaltic veins shows that there must be three geological conditions for the formation of larger asphaltic veins, which were extruded into the X-conjugate joint system along the shearing-compression fracture planes:

① An original oil-reservoir supplies the excess of oil as the precursory material of asphaltic veins;

② A nappe structure derived fault and fracture system provides the passageway and reservoir space for the excess of oil charging;

③ A hydrofracturing system was driven by episodic abnormal high-pressure in a short duration.

Only if the above three conditions are available, the excess of oil enriched in the original oil-reservoir can be extruded into the shearing-compression fractures, and strut the fractures to a certain scale, occupying and maintaining the interstitial space, then the large asphaltic veins could be formed.

(3) The large asphaltic veins contain allogenic "xenoliths" of dolostone, silicalites and gypsum debris (Figs. 15.6d and 15.7a), which are not the components of Guojiaba wall-rocks, but can be found in the deep strata of the Dengying Formation, indicating that the precursory material of asphaltic vein can be directly supplied from the underlying original Dengying carbonate-oil-reservoir.

(4) The upper black shale section of Doushantuo Formation belongs to sapropelic type of source rock with TOC 0.2–7% (avg. 2.98%) and higher maturity of R_o values 1.9–4.24% (avg. 3.16%), which should be referred to a source bed of over-mature phase. The biomarker composition of asphaltic veins is well correlated with that of the Dengying black shale, indicating the asphaltic veins should be derived from the Dengying source bed.

(5) According to Huang and Wang (2008), there are 65 asphaltic veins plotted on the geological map of Kuangshanliang Anticline (Fig. 15.12a). It is a NE-trending anticline, on which the asphaltic veins can be classified into 3 sets: one set is filled into the tension fractures (i.e., cross fractures) along the direction of anticlinal brachy-axis, and other two sets occur within the shearing-compression X-conjugate fractures, both of which oblique to the anticlinal long-axis (Fig. 15.12), but no vein appears within the compression fractures (i.e., longitudinal fractures) along the direction of long-axis. The distribution of asphaltic veins on the geological map (Fig. 15.12a) is well consistent with

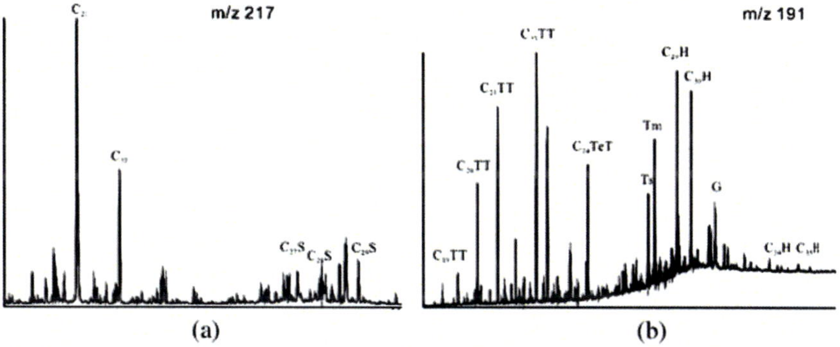

Fig. 15.10 The *m/z* 217 and *m/z* 191 mass chromatograms showing the distribution of steranes (**a**) and terpanes (**b**) of the asphaltic veins at Erchangliang, Tianba. C_{21}P pregnane; C_{22}P homopregnane C_{27}–C_{29}S regular sterane series; C_{19}–C_{26}T.T conventional tricyclic terpane series; C_{24}Te.T tetracyclic terpane; T_m. C_{27} trisnorhopane; C_{29}–C_{35}H hopane series; *G* gamacerene. C_i, *i* indicates carbon number

the field occurrence of asphaltic veins (Plate 15.2a, b). As a nappe structure, the Lower Palaeozoic Cambrian-Ordovician–Silurian are overthrusted on the Upper Palaeozoic Devonian-Carboniferous-Permian (Fig. 15.6), its upper plate constructs the axial region of the Kuangshanliang Anticline, where the Cambrian strata are outcropped on surface. The tectonic stress analysis shows that the Kuangshanliang Anticline had been compressed by pressure stress, a driving stress came from the NW-direction of the anticline, while passive stress from the SE direction, resulting in the anticline was overthrusted from NW to SE direction (Fig. 15.12b), meanwhile the cross, longitudinal and X-conjugate fractures were taking shape on the anticline. The formation of Kuangshanliang Anticline should be at the Late Triassic epoch, resulting from the Indosinian Movement, and its associated asphaltic veins are the contemporaneous with the anticline.

(6) There are some 5–10 cm-thick asphaltic bandings as small syngenetic fillings found within the bending and undulant fault plane of Changjianggou-Kuangshanliang-Xianshuigou overthrust in the Longmenshan Fronthill Belt (Fig. 15.6) as well as the asphaltic mixture and liquid oil within the faulted belt (Fig. 15.6d), which provide the evidences for the genetic mechanism of the asphaltic veins in Longmenshan Fronthill Belt, i.e., the tectonic stress and the fault-fracture system of Nappe structure resulted in the high pressure driving force and passageway for the original liquid oil to be extruded into the Guojiaba strata constituting the asphaltic veins. Therefore, both nappe structural system and asphaltic veins are cogenetic products. Therefore, these asphaltic veins should also be the products of the Late Triassic Indosinian Movement (Figs. 15.13 and 15.14).

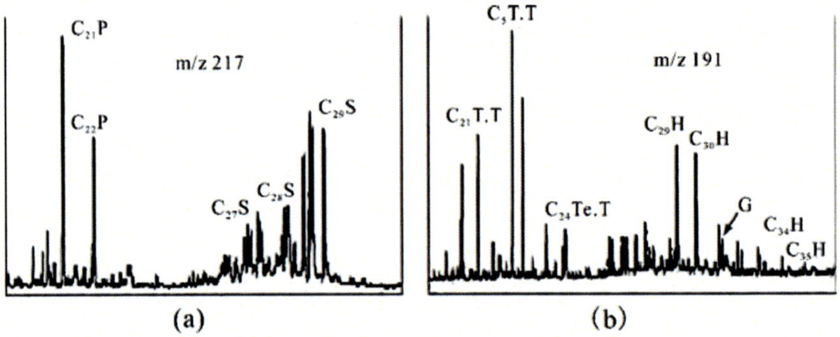

Fig. 15.11 The *m*/*z* 217 and *m*/*z* 191 mass chromatograms showing the distribution of steranes (**a**) and terpanes (**b**) of the Doushantuo black shale sampled at Qingchuan. C_{21}P pregnane; C_{22}P homopregnane C_{27}–C_{29}S regular sterane series; C_{19}–C_{26}T.T conventional tricyclic terpane series; C_{24}Te.T tetracyclic terpane; C_{29}–C_{35}H hopane series; G gamacerene. C_i, *i* indicates carbon number

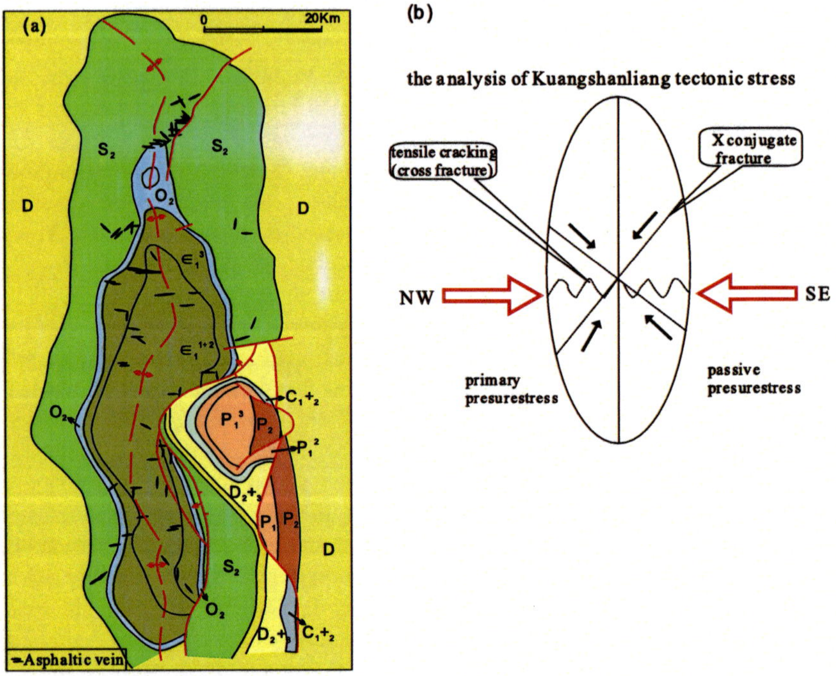

Fig. 15.12 Geological map of the Kuangshanliang Anticline (Huang and Wang 2008) showing the distribution of 65 asphaltic veins (**a**) and the analysis of Kuangshanliang tectonic stress (**b**)

References

Dai HM, Liu WL, Yang YM, Li YG, Duan Y (2007) The origin of oil-soaked sandstones in the piedmont zone of Longmenshan, the Sichuan Basin. Pet Explor Dev 29(604–608):1–5 (in Chinese with English abstract)

Han KY, Wang GL, Wang TG, Wang LS (2016) Petroleum geological characteristics of asphalt veins in the Front Mountain of Longmenshan, Northwest Sichuan. In: Sun S, Wang TG (eds) Meso-neoproterozoic geology and petroleum resources in East China. Science Press, Beijing, pp 542–559 (in Chinese)

Huang DF, Wang LS (2008) Geochemical characteristics of bituminous dike in Kuangshanliang area of the northwestern Sichuan Basin and its significance. Acta Petrolei Sinica 29:23–28 (in Chinese with English abstract)

Mckirdy DM (1974) Organic geochemistry in Precambrian research. Precambr Res 1(2):75–137

Song WH (1989) On nappe structure at northern sector of Longmen Mountains and its oil and gas prospecting. Nat Gas Ind 9:2–9 (in Chinese with English abstract)

Wang TG (1990) Approach to biomarker geochemistry. China University of Geoscience Press, Wuhan, pp 137–145 (in Chinese)

Wang TG (1991a) A novel tricyclic terpane biomarker series in the upper proterozoic bituminous sandstone, eastern Yanshan region. Sci China (ser B) 34(4):479–489

Wang TG (1991b) Geochemical characteristics of Longtangou bituminous sandstone, eastern Yanshan region, North China—approach to a Precambrian reservoir bitumen. J SE Asian Earth Sci 5(1–4):373–379

Wang TG, Simoneit BRT (1995) Tricyclic terpanes in Precambrian bituminous sandstone from the eastern Yanshan region in China. Chem Geol 120:155–170

Wang LS, Han K, Xie BH, Zhang J, Du M, Wan MX, Li D (2005) Reservoir conditions of the oil and gas fields in the north section of Longmen Mountain nappe structural belts. Natual Gas Ind (Supp A):1–5 (in Chinese with English abstract)

Wang CJ, Wang M, Xu J, Li YL, Yu Y, Bai J, Dong T, Zhang XY, Xiong XF, Gai H (2011) 13α(*n*-alkyl)-tricyclic terpanes: a series of biomarkers for the unique microbial mat ecosystem in the middle Mesoproterozoic (1.45–1.30 Ga) North China Sea. Mineralgical Mag 75:2114

Zhang SC, Zhang BM, Bian LZ, Jin ZJ, Wang DR, Chen JF (2007) The Xiamaling oil shale generated through rhodophyta over 800 Ma age. Sci China Ser D Earth Sci 50(4):527–535

Index

T. Wang, *Meso-Neoproterozoic Geology and Petroleum Resources in China*, Springer Geology, https://doi.org/10.1007/978-981-19-5666-9